Magnetism and Magnetic Materials

Covering basic physical concepts, experimental methods, and applications, this book is an indispensable text on the fascinating science of magnetism, and an invaluable source of practical reference data.

Accessible, authoritative, and assuming undergraduate familiarity with vectors, electromagnetism and quantum mechanics, this textbook is well suited to graduate courses. Emphasis is placed on practical calculations and numerical magnitudes – from nanoscale to astronomical scale – focussing on modern applications, including permanent magnet structures and spin electronic devices.

Each self-contained chapter begins with a summary, and ends with exercises and further reading. The book is thoroughly illustrated with over 600 figures to help convey concepts and clearly explain ideas. Easily digestible tables and data sheets provide a wealth of useful information on magnetic properties. The 38 principal magnetic materials, and many more related compounds, are treated in detail.

J. M. D. Coey leads the Magnetism and Spin Electronics group at Trinity College, Dublin, where he is Erasmus Smith's Professor of Natural and Experimental Philosophy. An authority on magnetism and its applications, he has been awarded the Gold Medal of the Royal Irish Academy and the Charles Chree Medal of the Institute of Physics for his work on magnetic materials.

Magnetism and Magnetic Materials

J. M. D. COEY
Trinity College, Dublin

CAMBRIDGE
UNIVERSITY PRESS

CAMBRIDGE
UNIVERSITY PRESS

University Printing House, Cambridge CB2 8BS, United Kingdom

One Liberty Plaza, 20th Floor, New York, NY 10006, USA

477 Williamstown Road, Port Melbourne, VIC 3207, Australia

314-321, 3rd Floor, Plot 3, Splendor Forum, Jasola District Centre, New Delhi - 110025, India

79 Anson Road, #06-04/06, Singapore 079906

Cambridge University Press is part of the University of Cambridge.

It furthers the University's mission by disseminating knowledge in the pursuit of education, learning and research at the highest international levels of excellence.

www.cambridge.org
Information on this title: www.cambridge.org/9781108717519

First published 2010
7th printing 2017
First paperback edition 2018

A catalogue record for this publication is available from the British Library

Library of Congress Cataloging in Publication data

ISBN 978-0-521-81614-4 Hardback
ISBN 978-1-108-71751-9 Paperback

Contents

List of tables of numerical data

Preface

This book offers a broad introduction to magnetism and its applications, designed for graduate students and advanced undergraduates as well as practising scientists and engineers. The approach is descriptive and quantitative, treating concepts, phenomena, materials and devices in a way that emphasises numerical magnitudes, and provides a wealth of useful data.

Magnetism is a venerable subject, which underwent four revolutionary changes in the course of the twentieth century – understanding of the physics, extension to high frequencies, the avalanche of consumer applications and, most recently, the emergence of spin electronics. The reader probably owns one or two hundred magnets, or some billions if you have a computer where each bit on the hard disc counts as an individually addressable magnet. Sixty years ago, the number would have been at best two or three. Magnetics, in partnership with semiconductors, has created the information revolution, which in turn has given birth to new ways to research the subject – numerical simulation of physical theory, automatic data acquisition and web-based literature searches.

The text is structured in five parts. First, there is a short overview of the field. Then come eight chapters devoted to concepts and principles. Two parts follow which treat experimental methods and materials, respectively. Finally there are four chapters on applications. An elementary knowledge of electromagnetism and quantum mechanics is needed for the second part. Each chapter ends with a short bibliography of secondary literature, and some exercises. SI units are used throughout, to avoid confusion and promote magnetic numeracy. A detailed conversion table for cgs units, which are still in widespread use, is provided inside the back cover. There is some attempt to place the study of magnetism in a global context; our activity is not only intellectual and practical, it is also social and economic.

The text has grown out of courses given to undergraduates, postgraduates and engineers over the past 15 years in Dublin, San Diego, Tallahassee, Strasbourg and Seagate, as well as from the activities of our own research group at Trinity College, Dublin. I am very grateful to many students, past and present, who contributed to the venture, as well as to numerous colleagues who took the trouble to read a chapter and let me have their criticism and advice, and correct at least some of the mistakes. I should mention particularly Sara McMurry, Plamen Stamenov and Munuswamy Venkatesan, as well as Grainne Costigan, Graham Green, Ma Qinli and Chen Junyang, who all worked on the figures, and Emer Brady who helped me get the whole text into shape.

Outlines of the solutions to the odd-numbered exercises are available at the Cambridge website www.cambridge.org/9780521816144. Comments, corrections and suggestions for improvements of the text are very welcome; please post them at www.tcd.ie/physics/magnetism/coeybook.php.

Finally, I am grateful to Wong May, thinking of everything we missed doing together when I was too busy with this.

J. M. D. Coey
Dublin, November 2009

Acknowledgements

The following figures are reproduced with permission from the publishers:

American Association for the Advancement of Science: 14.18, p.525 (margin), p.537 (margin),14.27; American Institute of Physics: 5.25, 5.31, 6.18, 8.5, 8.33, 10.12, 11.8; American Physical Society: 4.9, 5.35, 5.40, 6.27a, 6.27b, 8.3, 8.8, 8.9, 8.15, 8.17, 8.18, 8.21, 8.22, 8.26, 8.29, 9.5, p.360 (margin), 11.15, 14.16; American Geophysical Union p.572 (margin); United States Geological Survey Geomagnetism Program: 15.18, p.572 (margin); American Society for Metals: 5.35; Cambridge University Press: 4.15, 4.17, 7.8, 7.18, 9.12, 10.16, p.573 (margin); Elsevier: 6.23, 8.2, 8.4, 11.22, 14.22, 14.23, 14.26, 15.22; Institute of Electrical and Electronics Engineers: 5.32, 8.31, 8.34, 8.35, 9.6, 11.6, 11.7; MacMillan Publishers: 14.17, 15.4c; Oxford University Press: 5.26; National Academy of Sciences:15.1; Springer Verlag: 4.18, 14.13, 14.21, 15.8, 15.21; Taylor and Francis: 1.6, 2.8b, 10.2; Institution of Engineering and Technology: 11.20; University of Chicago Press: 1.1a; John Wiley: 5.21, 6.4, 6.15, 8.11a,b, 9.9, 12.10

Fermi surfaces are reproduced with kind permission of the University of Florida, Department of Physics, http://www.phys.ufl.edu/fermisurface.

Thanks are due to Wiebke Drenckhan and Orphee Cugat for permission to reproduce the cartoons on pages 161 and 531.

Figure 15.3 is reproduced by courtesy of Johannes Kluehspiess. Figure 15.5 is reproduced by courtesy of L. Nelemans, High Field Magnet Laboratory, Nijmegen. Figure 15.5 is reproduced by permission of Y. I.Wang, Figure 15.17 is repoduced by courtesy of N. Sadato; Figure 15.23 is reproduced by courtesy of P. Rochette.

1 Introduction

After a short historical summary, the central concepts of magnetic order and hysteresis are presented. Magnet applications are summarized, and magnetism is situated in relation to physics, materials science and industrial technology.

1.1 A brief history of magnetism

The history of magnetism is coeval with the history of science. The magnet's ability to attract ferrous objects by remote control, acting at a distance, has captivated countless curious spirits over two millenia (not least the young Albert Einstein). To demonstrate a force field that can be manipulated at will, you need only two chunks of permanent magnet or one chunk of permanent magnet and a piece of temporary magnet such as iron. Feeble permanent magnets are quite widespread in nature in the form of lodestones – rocks rich in magnetite, the iron oxide Fe_3O_4 – which were magnetized by huge electric currents in lightning strikes. Priests and people in Sumer, ancient Greece, China and pre-Colomban America were familiar with the natural magic of these magnets.

A lodestone carved in the shape of a Chinese spoon was the centrepiece of an early magnetic device, the 'South pointer'. Used for geomancy in China at the beginning of our era (Fig. 1.1), the spoon turns on the base to align its handle with the Earth's magnetic field. Evidence of the South pointer's application can be seen in the grid-like street plans of certain Chinese towns, where the axes of quarters built at different times are misaligned because of the secular variation of the direction of the horizontal component of the Earth's magnetic field.

A propitious discovery, attributed to Zheng Gongliang in 1064, was that iron could acquire a thermoremanent magnetization when quenched from red heat. Steel needles thus magnetized in the Earth's field were the first artificial permanent magnets. They aligned themselves with the field when floated or suitably suspended. A short step led to the invention of the navigational compass, which was described by Shen Kua around 1088. Reinvented in Europe a century later, the compass enabled the great voyages of discovery, including the European discovery of America by Christopher Columbus in 1492 and the earlier Chinese discovery of Africa by the eunuch admiral Cheng Ho in 1433.

Shen Kua, 沈括 1031–1095.

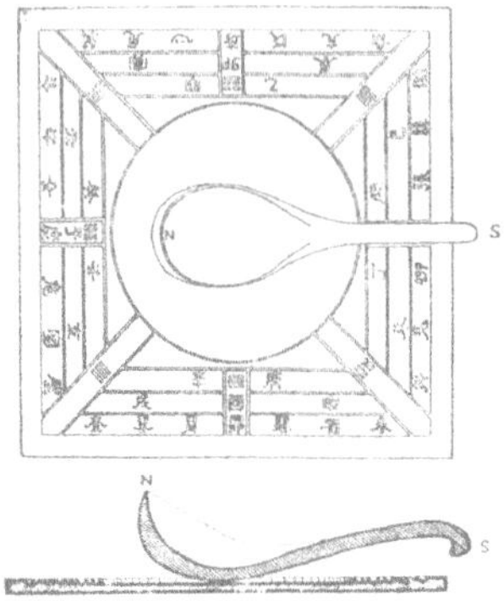

Figure 1.1

Some early magnetic devices: the 'South pointer' used for orientation in China around the beginning of the present era, and a Portuguese mariner's compass from the fifteenth century.

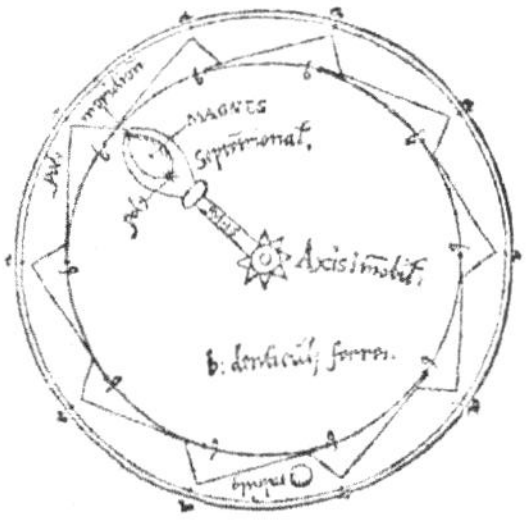

A perpetuum mobile, proposed by Petrus Peregrinus in 1269.

William Gilbert, 1544–1603.

When we come to the middle ages, virtues and superstitions had accreted to the lodestone like iron filings. Some were associated with its name.[1] People dreamt of perpetual motion and magnetic levitation. The first European text on magnetism by Petrus Peregrinus describes a *perpetuum mobile*. Perpetual motion was not to be, except perhaps in the never-ending dance of electrons in atomic orbitals with quantized angular momentum, but purely passive magnetic levitation was eventually achieved at the end of the twentieth century. Much egregious fantasy was debunked by William Gilbert in his 1600 monograph *De Magnete*, which was arguably the first modern scientific text. Examination of the direction of the dipole field at the surface of a lodestone sphere or 'terella', and relating it to the observation of dip which by then had been measured at many points on the Earth's surface, led Gilbert to identify the source of the magnetic force which aligned the compass needle as the Earth itself, rather than the stars as previously assumed. He inferred that the Earth itself was a great magnet.[2]

The curious Greek notion that the magnet possessed a soul – it was animated because it moved – was to persist in Europe well into the seventeenth century, when it was finally laid to rest by Descartes. But other superstitions regarding the benign or malign influences of magnetic North and South poles remain alive and well, as a few minutes spent browsing the Internet will reveal.

Magnetic research in the seventeenth and eighteenth centuries was mostly the domain of the military, particularly the British Navy. An important civilian advance, promoted by the Swiss polymath Daniel Bernoulli, was the invention in 1743 of the horseshoe magnet. This was to become magnetism's most enduring archetype. The horseshoe is an ingenious solution to the problem of making a reasonably compact magnet which will not destroy itself in its own demagnetizing field. It has remained the icon of magnetism up to the present

[1] In English, the word 'magnet' is derived through Latin from the Greek for Magnesian stone (Ὁ μαγνης λῖθος), after sources of lodestones in Asia Minor. In Sanscrit 'चुम्बक' and Romance languages – French 'l'aimant', Spanish 'imán', Portuguese 'imã' – the connotation is the attraction of opposite poles, like that of man and woman.

[2] '*Magnus magnes ipse est globus terrestris*'.

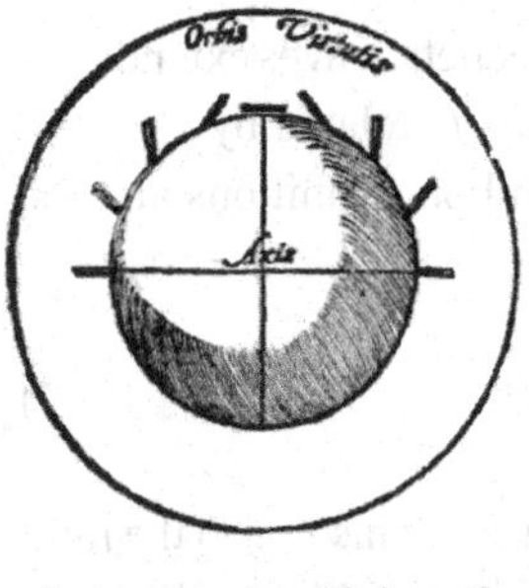

A lodestone 'terella' used by Gilbert to demonstrate how the magnetic field of the Earth resembles that of a magnet.

Réné Descartes, 1596–1650.

An eighteenth century horseshoe magnet.

day. Usually red, and marked with 'North' and 'South' poles, horseshoe magnets still feature in primary school science books all over the world, despite the fact that these horseshoes have been quite obsolete for the past 50 years.

The obvious resemblances between magnetism and electricity, where like or unlike charges repel or attract, led to a search for a deeper connection between the two cousins. Luigi Galvani's 'animal electricity', stemming from his celebrated experiments on frogs and corpses, had a physical basis – nerves work by electricity. It inspired Anton Messmer to postulate a doctrine of 'animal magnetism' which was enthusiastically embraced in Parisian salons for some years before Louis XVI was moved to appoint a Royal Commission to investigate. Chaired by Benjamin Franklin, the Commission thoroughly discredited the phenomenon, on the basis of a series of blind tests. Their report, published in 1784, was a landmark of scientific rationality.

It was in Denmark in 1820 that Hans-Christian Oersted eventually discovered the true connection between electricity and magnetism by accident. He demonstrated that a current-carrying wire produced a *circumferential* field capable of deflecting a compass needle. Within weeks, André-Marie Ampère and Dominique-François Arago in Paris wound wire into a coil and showed that the current-carrying coil was equivalent to a magnet. The electromagnetic revolution was launched.

The remarkable sequence of events that ensued changed the world for ever. Michael Faraday's intuition that the electric and magnetic forces could be conceived in terms of all-pervading fields was critical. He discovered electromagnetic induction (1821) and demonstrated the principle of the electric motor with a steel magnet, a current-carrying wire and a dish of mercury. The discovery of a connection between magnetism and light followed with the magneto-optic Faraday effect (1845).

All this experimental work inspired James Clerk Maxwell's formulation[3] of a unified theory of electricity, magnetism and light in 1864, which is summarized in the four famous equations that bear his name:

$$\nabla \cdot \boldsymbol{B} = 0, \tag{1.1a}$$

$$\epsilon_0 \nabla \cdot \boldsymbol{E} = \rho, \tag{1.1b}$$

$$(1/\mu_0)\nabla \times \boldsymbol{B} = \boldsymbol{j} + \epsilon_0 \partial \boldsymbol{E}/\partial t, \tag{1.1c}$$

$$\nabla \times \boldsymbol{E} = -\partial \boldsymbol{B}/\partial t. \tag{1.1d}$$

These equations relate the electric and magnetic fields, $\boldsymbol{E}$ and $\boldsymbol{B}$ at a point in *free space* to the distributions of electric charge and current densities, ρ and $\boldsymbol{j}$ in surrounding space. A spectacular consequence of Maxwell's equations is the existence of a solution representing coupled oscillatory electric and magnetic

[3] 'From a long view of the history of mankind there can be little doubt that the most significant event of the nineteenth century will be judged as Maxwell's discovery of the laws of electrodynamics' (R. Feynman *The Feynman Lectures in Physics*. Vol. II, Menlo Park: Addison-Wesley (1964)).

André Marie Ampère, 1775–1836.

fields propagating at the speed of light. These electromagnetic waves extend over the entire spectrum, with wavelength Λ and frequency f, related by $c = \Lambda f$. The electric and magnetic constants ϵ_0 and μ_0 depend on definitions and the system of units, but they are related by

$$\sqrt{\epsilon_0\mu_0} = \frac{1}{c}, \tag{1.2}$$

where c is the speed of light in vacuum, 2.998×10^8 m s^{-1}. This is also the ratio of the average values of E and B in the electromagnetic wave. Maxwell's equations are asymmetric in the fields $\boldsymbol{E}$ and $\boldsymbol{B}$ because no magnetic counterpart of electric charge has ever been identified in nature. Gilbert's idea of North and South magnetic poles, somehow analogous to Coulomb's positive and negative electric charges, has no physical reality, although poles remain a conceptual convenience and they simplify certain calculations. Ampère's approach, regarding electric currents as the source of magnetic fields, has a sounder physical basis. Either approach can be used to *describe* ferromagnetic material such as magnetite or iron, whose magnetism is equally well represented by distributions of magnetic poles or electric currents. Nevertheless, the real building blocks of electricity and magnetism are *electric charges* and *magnetic dipoles*; the dipoles are equivalent to electric current loops. Dielectric and magnetic materials are handled by introducing two auxiliary fields $\boldsymbol{D}$ and $\boldsymbol{H}$, as discussed in Chapter 2.

Hans-Christian Oersted, 1777–1851.

An additional equation, due to Lorentz, gives the force on a particle with charge q moving with velocity $\boldsymbol{v}$, which is subject to electric and magnetic fields:

$$\boldsymbol{f} = q(\boldsymbol{E} + \boldsymbol{v} \times \boldsymbol{B}). \tag{1.3}$$

Units of $\boldsymbol{E}$ are volts per metre (or newtons per coulomb), and the units of $\boldsymbol{B}$ are newtons per ampere per metre (or tesla).

A technical landmark in the early nineteenth century was William Sturgeon's invention of the iron-cored electromagnet in 1824. The horseshoe-shaped core was temporarily magnetized by the magnetic field produced by current flowing in the windings. Electromagnets proved more effective than the weak permanent magnets then available for excitation of electric motors and generators. By the time the electron was discovered in 1897,[4] the electrification of the planet was already well advanced. Urban electrical distribution networks dispelled the tyranny of night with electric light and the stench of public streets was eliminated as horses were displaced by electric trams. Telegraph cables spanned the Earth, transmitting messages close to the speed of light for the equivalent of €20 a word.

Michael Faraday, 1791–1867.

[4] The decisive step for the discovery of the electron was taken in England by Joseph John Thompson, who measured the ratio of its charge to mass. The name, derived from ἤλεκτρον the Greek word for amber, had been coined earlier (1891 in Dublin) by George Johnston Stoney.

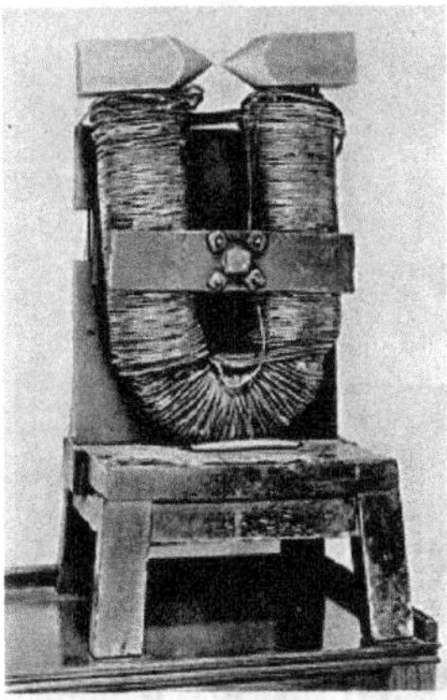

A nineteenth century electromagnet.

James Clerk Maxwell, 1831–1879.

Despite the dazzling technical and intellectual triumphs of the electromagnetic revolution, the problem of explaining how a solid could possibly be ferromagnetic was unsolved. The magnetization of iron, $M = 1.76 \times 10^6$ amperes per metre, implies a perpetually circulating Ampèrian surface current density of the same magnitude. Currents of hundreds of thousands of amperes coursing around the surface of a magnetized iron bar appeared to be a wildly implausible proposition. Just as preposterous was Pierre Weiss's molecular field theory, dating from 1907, which successfully explained the phase transition at the Curie point where iron reversibly loses its ferromagnetism. The theory postulated an internal magnetic field parallel to, but some three orders of magnitude greater than, the magnetization. Although Maxwell's equation (1.1a) proclaims that the magnetic field $\boldsymbol{B}$ should be continuous, no field remotely approaching that magnitude has ever been detected outside a magnetized iron specimen. Ferromagnetism therefore challenged the foundations of classical physics, and a satisfactory explanation only emerged after quantum mechanics and relativity, the twin pillars on which modern physics rests, were erected in the early years of the twentieth century.

Strangely, the Ampèrian currents turned out to be associated with quantized angular momentum, and especially with the intrinsic spin of the electron, discovered by George Uhlenbeck and Samuel Goudsmit in 1925. The spin is quantized in such a way that it can have just two possible orientations in a magnetic field, 'up' and 'down'. Spin is the source of the electron's intrinsic magnetic moment, which is known as the Bohr magneton: $\mu_B = 9.274 \times 10^{-24}$ A m^2. The magnetic properties of solids arise essentially from the magnetic moments of their atomic electrons. The interactions responsible for ferromagnetism represented by the Weiss molecular field were shown by Werner Heisenberg in 1929 to be *electrostatic* in nature, originating from the quantum mechanics of the Pauli principle. Heisenberg formulated a Hamiltonian to represent the interaction of two neighbouring atoms whose total electronic spins, in units of Planck's constant $\hbar = 1.055 \times 10^{-34}$ J s, are $\boldsymbol{S}_i$ and $\boldsymbol{S}_j$, namely

$$\mathcal{H} = -2\mathcal{J}\boldsymbol{S}_i \cdot \boldsymbol{S}_j, \tag{1.4}$$

where $\mathcal{J}$ is the exchange constant; $\mathcal{J}/k_B$ is typically in the range 1–100 K. Here k_B is Boltzmann's constant, 1.3807×10^{-23} J K^{-1}. Atomic magnetic moments are associated with the electronic spins. The quantum revolution underpinning modern atomic and solid state physics and chemistry was essentially complete at the time of the sixth Solvay Congress in 1930 (Fig. 1.2). Filling in the details has proved to be astonishingly rich and endlessly useful.[5] For instance, when the exchange interaction $\mathcal{J}$ is negative (antiferromagnetic) rather than

[5] Already in 1930 there was the conviction that all the basic problems of the physics of solids had been solved in principle; Paul Dirac said 'The underlying physical phenomena necessary for a mathematical explanation of a large part of physics and all of chemistry are now understood in principle, the only difficulty being that the exact application of these laws leads to equations much too complicated to be soluble' (P. Dirac, *Proc. Roy. Soc.* **A123**, 714 (1929)).

Photo Benjamin Couprie

A. PICCARD W. GERLACH C. DARWIN P.A. DIRAC
E. HENRIOT MANNEBACK H.A. KRAMERS J.H. VAN VLECK W. HEISENBERG
E. HERZEN J. VERSCHAFFELT A. COTTON J. ERRERA O. STERN H. BAUER P. KAPITZA L. BRILLOUIN P. DEBYE W. PAULI J. DORFMAN E. FERMI
Th. DE DONDER P. ZEEMAN P. WEISS A. SOMMERFELD Mme CURIE P. LANGEVIN A. EINSTEIN O. RICHARDSON B. CABRERA N. BOHR W.J. DE HAAS

Figure 1.2

Participants at the 1930 Solvay Congress, which was devoted to magnetism.

positive (ferromagnetic) there is a tendency for the spins at sites i and j to align antiparallel rather than parallel. Louis Néel pointed out in 1936 and 1948 that this leads to antiferromagnetism or ferrimagnetism, depending on the topology of the crystal lattice. Magnetite, the archetypal natural magnetic material, is a ferrimagnet.

Louis Néel, 1904–2000.

One lesson from a study of the history of magnetism is that fundamental understanding of the science may not be a prerequisite for technological progress. Yet fundamental understanding helps. The progression from the poorly differentiated set of hard and soft magnetic steels that existed at the start of the twentieth century to the wealth of different materials available today, with all sorts of useful properties described in this book, owes more to metallurgy and systematic crystal chemistry than it does to quantum physics. Only since the rare-earth elements began to be alloyed with cobalt and iron in new permanent magnets from the late 1960s onwards has quantum mechanics contributed significantly to magnetic materials development. Much progress in science is made empirically, with no recourse to basic theory. One area, however, where quantum mechanics has been of central importance for magnetism is in its interaction with electromagnetic radiation in the radiofrequency, microwave and optical ranges. The discovery of magnetic resonance methods in the 1940s

Table 1.1. The seven ages of magnetism

Period	Dates	Icon	Drivers	Materials
Ancient period	−2000–1500	Compass	State, geomancers	Iron, lodestone
Early modern age	1500–1820	Horseshoe magnet	Navy	Iron, lodestone
Electromagnetic age	1820–1900	Electromagnet	Industry/infrastructure	Electrical steel
Age of understanding	1900–1935	Pauli matrices	Academic	(Alnico)
High-frequency age	1935–1960	Magnetic resonance	Military	Ferrites
Age of applications	1960–1995	Electric screwdriver	Consumer market	Sm-Co, Nd-Fe-B
Age of spin electronics	1995–	Read head	Consumer market	Multilayers

Samuel Goudsmit, 1902–1978.

Georg Uhlenbeck, 1900–1988.

and 1950s and the introduction of powerful spectroscopic and diffraction techniques led to new insights into the magnetic and electronic structure of solids. Technology for generating and manipulating microwaves had been developed in Great Britain for the Second World War.

Recent decades have witnessed an immense expansion of magnetic applications. The science developed over a century, mostly in Europe, was ripe for exploitation throughout the industrialized world. Advances in permanent magnetism, magnetic recording and high-frequency materials underpin much of the progress that has been made with computers, telecommunications equipment and consumer goods that benefit most people on Earth. Permanent magnets have come back to replace electromagnets in a billion tiny motors manufactured every year. Magnetic recording sustains the information revolution and the Internet. There have been seminal advances in earth science, medical imaging and the theory of phase transitions that can be laid at the door of magnetism. This long and promising history of magnetism can be envisaged as seven ages, which are summarized in Table 1.1. The third millenium sees us at the threshold of the seventh age, that of spin electronics. Conventional electronics has ignored the spin on the electron. We are just now beginning to learn how to manipulate spin currents and to make good use of them.

1.2 Magnetism and hysteresis

The most striking manifestation of magnetism in solids is the spontaneous magnetization of ferromagnetic materials such as iron or magnetite. Spontaneous magnetism is usually associated with hysteresis,[6] a phenomenon studied by James Ewing, and named by him in 1881.[7]

[6] 'Hysteresis' was coined from the greek ὑστερειν, to lag behind.

[7] Ewing, a Scot, was appointed as a foreign Professor of Engineering at the University of Tokyo by the Meiji government in 1878. He is regarded as the founder of magnetic research in Japan.

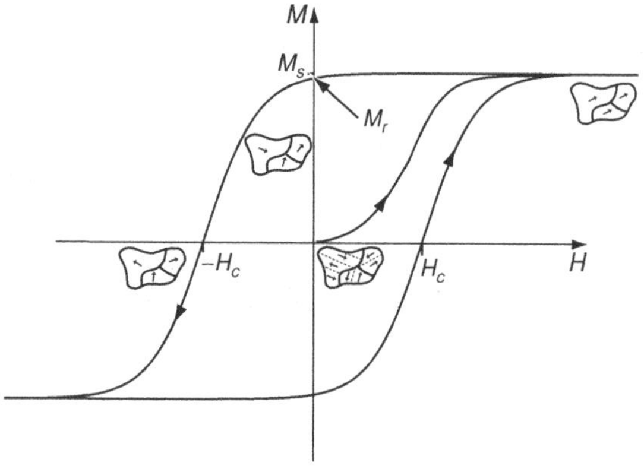

Figure 1.3

The hysteresis loop of a ferromagnet. Initially in an unmagnetized, virgin state. Magnetization appears as an imposed magnetic field H, modifies and eventually eliminates the microstructure of ferromagnetic domains magnetized in different directions, to reveal the spontaneous magnetization M_s. The remanence M_r which remains when the applied field is restored to zero, and the coercivity H_c, which is the reverse field needed to reduce the magnetization to zero, are marked on the loop.

1.2.1 The ferromagnetic hysteresis loop

The essential practical characteristic of any ferromagnetic material is the irreversible nonlinear response of magnetization $\boldsymbol{M}$ to an imposed magnetic field $\boldsymbol{H}$. This response is epitomized by the hysteresis loop. The material responds to $\boldsymbol{H}$, rather than $\boldsymbol{B}$, for reasons discussed in the next chapter where we distinguish the applied and internal fields. In free space, $\boldsymbol{B}$ and $\boldsymbol{H}$ are simply proportional. Magnetization, the magnetic dipole moment per unit volume of material, and the H-field are both measured in amperes per metre (A m^{-1}). Since this is a rather small unit – the Earth's magnetic field is about 50 A m^{-1} – the multiples kA m^{-1} and MA m^{-1} are often employed. The applied field must be comparable in magnitude to the magnetization in order to trace a hysteresis loop. The values of spontaneous magnetization M_s of the ferromagnetic elements Fe, Co and Ni at 296 K are 1720, 1370 and 485 kA m^{-1}, respectively. That of magnetite, Fe_3O_4, is 480 kA m^{-1}. A large electromagnet may produce a field of 1000 kA m^{-1} (1 MA m^{-1}), using coils carrying currents of order 100 A.

Hard magnetic materials[8] have broad, square $M(H)$ loops. They are suitable for permanent magnets because, once magnetized by applying a field $H \geq M_s$ sufficient to saturate the magnetization, they remain in a magnetized state when the field is removed. Soft magnetic materials have very narrow loops. They are temporary magnets, readily losing their magnetization as soon as the field is removed. The applied field serves to unveil the spontaneous ferromagnetic order that already exists on the scale of microscopic domains. These domain structures are illustrated schematically on the hysteresis loop of Fig. 1.3 for the unmagnetized state at the origin, the saturated state where $M = M_s$, the remanent state in zero field where $M = M_r$ and the state at $H = H_c$, the coercive field where M changes sign. M_r and H_c are known as the remanence and the coercivity. Magnetic domains were proposed by James Ewing and the principles of domain theory were established by Lev Landau and Evgenii Lifschitz in 1935.

James Ewing, 1855–1935.

[8] The terms *hard* and *soft* for magnets originated from the mechanical properties of the corresponding magnetic steels.

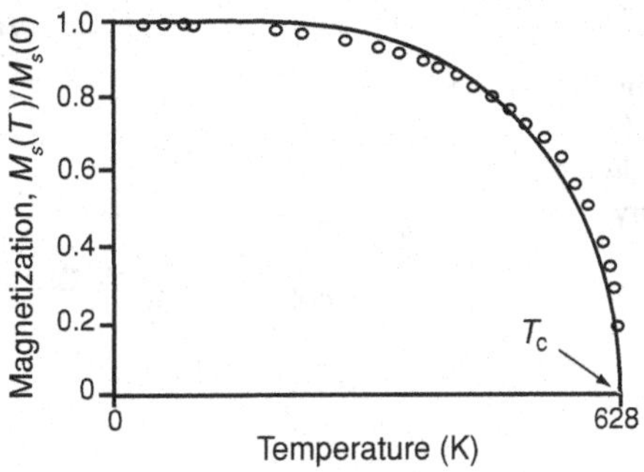

Figure 1.4

Temperature dependence of the spontaneous magnetization of nickel. The Curie point at 628 K is marked.

The hysteresis loop is central to technical magnetism; physicists endeavour to explain it, materials scientists aim to improve it and engineers work to exploit it. The loop combines information on an intrinsic magnetic property, the spontaneous magnetization M_s which exists within a domain of a ferromagnet, and two extrinsic properties, the remanence M_r and coercivity H_c, which depend on a host of extraneous factors including the sample shape, surface roughness, microscopic defects and thermal history, as well as the rate at which the field is swept in order to trace the loop.

1.2.2 The Curie temperature

The spontaneous magnetization due to alignment of the atomic magnetic moments depends on temperature, and it falls precipitously to zero at the Curie temperature T_C. The magnetic ordering is a continuous thermodynamic phase transition with a λ-shaped anomaly in specific heat, associated with disordering of the atomic dipole moments. Above T_C, $M_s(T)$ is zero; below T_C, $M_s(T)$ is reversible. This behaviour is illustrated for nickel in Fig. 1.4.

The Curie temperatures of the three ferromagnetic metals, iron, cobalt and nickel, are 1044 K, 1388 K and 628 K, respectively. No material is known to have a higher Curie temperature than cobalt. Magnetite has a Curie temperature of 856 K.

Pierre Curie, 1859–1906.

1.2.3 Coercivity

The progress in the twentieth century which has spawned such a range of magnetic applications can be summarized in three words – *mastery of coercivity*. No new ferromagnetic material has been discovered with a magnetization greater than that of 'permendur', $Fe_{65}Co_{35}$, for which $M_s = 1950$ kA m^{-1}, but coercivity which barely spanned two orders of magnitude in 1900, from the softest soft iron to the hardest magnet steel, now ranges over eight orders of

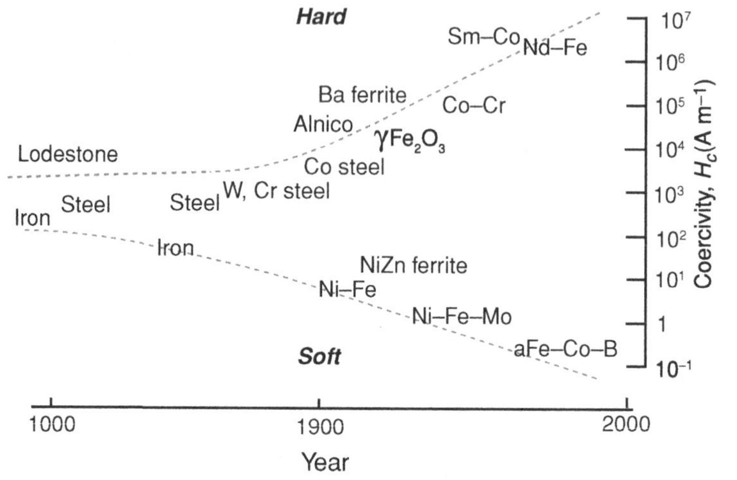

Figure 1.5

Progress in expanding the range of coercivity of magnetic materials during the twentieth century.

magnitude, from less than 0.1 A m^{-1} to more than 10 MA m^{-1}, as shown in Fig. 1.5.

1.2.4 Anisotropy

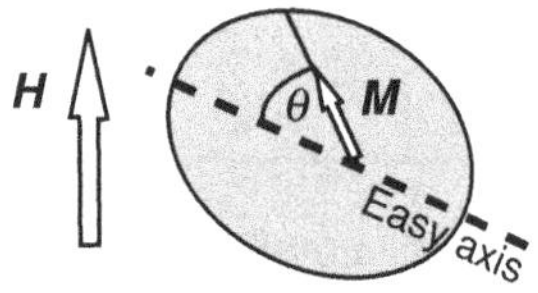

Magnetization is not necessarily parallel to applied field, unless H is applied in an easy direction.

The natural direction of magnetization in a microscopic ferromagnetic domain is usually constrained to lie along one or more easy axes. Since magnetism is associated with circulating electron currents, time reversal symmetry requires that a state with a certain magnetization distribution $\boldsymbol{M}(r)$ should have the same energy as the state with reversed magnetization along the same axis, $-\boldsymbol{M}(r)$. This tendency is represented by the anisotropy energy E_a, of which the leading term is

$$E_a = K_u \sin^2 \theta, \tag{1.5}$$

where θ is the angle between the direction of $\boldsymbol{M}$ and the easy axis. Here E_a and K_u, the anistropy constant, are measured in J m^{-3}. Typical values range from less than 1 kJ m^{-3} to more than 10 MJ m^{-3}. Anistropy sets on upper bound on the coercivity available in hard magnets. We show in Chapter 7 that

$$H_c < 2K_u/\mu_0 M_s, \tag{1.6}$$

where the magnetic constant μ_0 is $4\pi \times 10^{-7}$ J A^{-2} m^{-1}. Anisotropy also leads to unwanted coercivity in soft magnets. It may be noted from the units that μ_0 always multiplies H^2 or MH in expressions for magnetic energy per unit volume.

Atomic densities in solids are around $n = 10^{29}$ m^{-3}, so if anisotropy energy per atom is expressed in terms of an equivalent temperature using $E_a/n = k_B T$, it is in the range 1 mK–10 K. The energy is usually small in relation to the Curie temperature, but it is nevertheless decisive in determining the hysteresis.

1.2.5 Susceptibility

At temperatures above T_C, where the ferromagnetic order collapses, and the material becomes paramagnetic, the atomic moments of a few Bohr magnetons experience random thermal fluctuations. Although M_s is zero, an applied field can induce some alignment of the atomic moments, leading to a small magnetization M which varies linearly with H, except in very large fields or very close to the Curie point. The susceptibility, defined as

$$\chi = M/H, \tag{1.7}$$

is a dimensionless quantity, which diverges as $T \to T_C$ from above. Above T_C it often follows a Curie–Weiss law

$$\chi = C/(T - T_C), \tag{1.8}$$

where C is known as the Curie constant. Its value is of order 1 K.

The magnetic response to an applied field of materials which do not order magnetically may be either paramagnetic or diamagnetic.[9] In isotropic paramagnets, the induced magnetization $\boldsymbol{M}$ is in the same direction as $\boldsymbol{H}$, whereas in diamagnets it is in the opposite direction. Superconductors exhibit diamagnetic hysteresis loops below their superconducting transition temperature T_{sc}, and their susceptibility can approach the limiting value of -1.

The susceptibility of many paramagnets follows a Curie law,

$$\chi = C/T, \tag{1.9}$$

but for some metallic paramagnets and almost all diamagnets χ is independent of temperature. The sign of the room-temperature susceptibility is indicated on the magnetic periodic table (Table A, see endpapers) and the molar susceptibility χ_{mol} is plotted for the elements in Fig. 3.5. There it is appropriate to look at the molar susceptibility because some of the elements are gasses at room temperature. A cubic metre of a solid contains roughly 10^5 moles, so χ_{mol} is approximately five orders of magnitude less than χ. From Table A, it can be seen that the transition metals are paramagnetic, whereas main group elements are mostly diamagnetic.

1.2.6 Other types of magnetic order

The spontaneous magnetization of a ferromagnet is the result of alignment of the magnetic moments of individual atoms. But parallel alignment is not the only – or even the most common – type of magnetic order. In an antiferromagnet, the atomic moments form two equivalent but oppositely oriented

[9] Faraday was the first to classify solids as diamagnetic, paramagnetic or ferromagnetic, according to their response to a magnetic field.

magnetic sublattices. Although $M_s = 0$, the material nonetheless exhibits a phase transition with a λ-shaped specific heat anomaly where the moments begin to order. The antiferromagnetic transition occurs at the Néel temperature T_N. Occasionally it is possible to switch an antiferromagnet into a ferromagnet if a sufficiently large field is applied. This discontinuous change of magnetic order is known as a metamagnetic transition.

If the sublattices are inequivalent, with sublattice magnetizations M_A and M_B where $\boldsymbol{M}_A \neq -\boldsymbol{M}_B$, there is a net spontaneous magnetization. The material is a ferrimagnet. Most of the useful magnetic oxides, including magnetite, are ferrimagnetic.

The alignment of the atomic moments in the ordered state need not be collinear. Multiple noncollinear sublattices are found in manganese and some of its alloys. Other materials such as MnSi or Mn_3Au have helical or spiral magnetic structures that are incommensurate with the underlying crystal lattice. In some disordered and amorphous materials the atomic moments freeze in more or less random directions. Such random, noncollinear magnets are known collectively as spin glasses. The original spin glassses were magnetically dilute crystalline alloys, but several different varieties of random spin freezing are encountered in noncrystalline (amorphous) solids.

Finally, we remark on the behaviour of ferromagnetic fine particles whose volume V is so small that the product $K_u V$ is less than or comparable to the thermal energy $k_B T$; in that case the total sum moment, $\mathfrak{m}$, of all the coupled atoms fluctuates randomly like that of a large paramagnetic atom or macrospin. The susceptibility follows a Curie law with a huge value of C. The name superparamagnetism was coined by Néel for this phenomenon, which is important for ferrofluids (magnetic liquids which are really colloidal suspensions of ferrimagnetic fine particles) and in rock magnetism.

Figure 1.6 portrays the magnetic family tree, summarizing the behaviour of the magnetization or susceptibility for the different types of magnetic order in crystalline and amorphous solids.

1.2.7 The magnetic periodic table

Table A (endpapers) displays the magnetic properties of the elements, distinguishing those that are paramagnetic, diamagnetic, ferromagnetic or antiferromagnetic at room temperature, and those that order magnetically at some lower temperature. Only sixteen elements have a magnetically ordered ground state, and all but oxygen belong to the $3d$ or $4f$ transition series. Besides iron, cobalt and nickel, only gadolinium can be ferromagnetic at room temperature, but that depends on the weather! The Curie temperature of gadolinium is just 292 K. Many other elements become superconducting at low enough temperature. The remainder are neither magnetic nor superconducting. No element manages to be both at the same time.

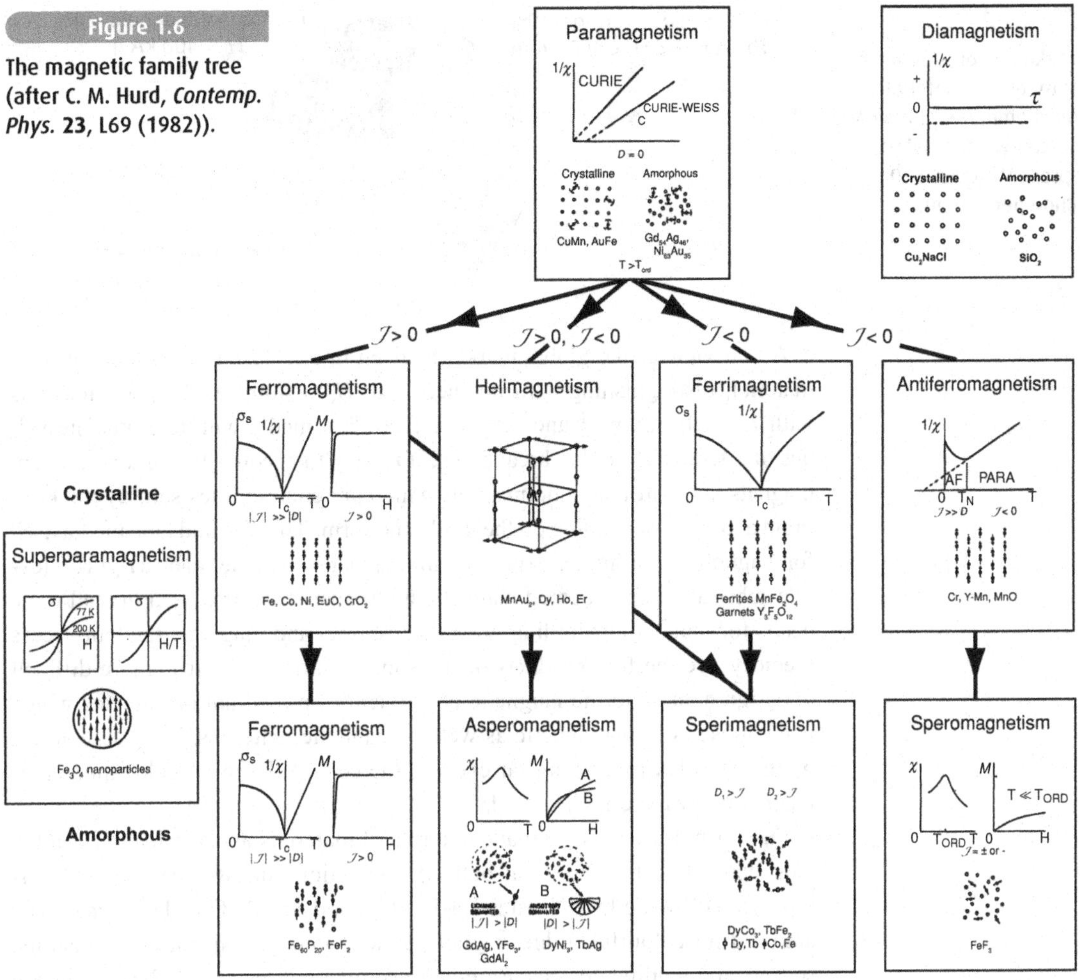

Figure 1.6

The magnetic family tree (after C. M. Hurd, *Contemp. Phys.* **23**, L69 (1982)).

1.3 Magnet applications

1.3.1 Overview of the world market

Applied magnetism. A Berlin postcard ca 1920.

Magnetic materials, recording media, heads and sensors constitute a market worth over \$30 billion per year. Since the population of the Earth is approaching 7 billion, this means an *average* of about \$5 per head. The world's goods are unevenly distributed. The richest billion, living mainly in North America, Europe and East Asia, consume the lion's share, but most people derive some benefit from magnetic technology, whether in the form of a cassette recorder, an electric pump in a tube well or a communal mobile phone.

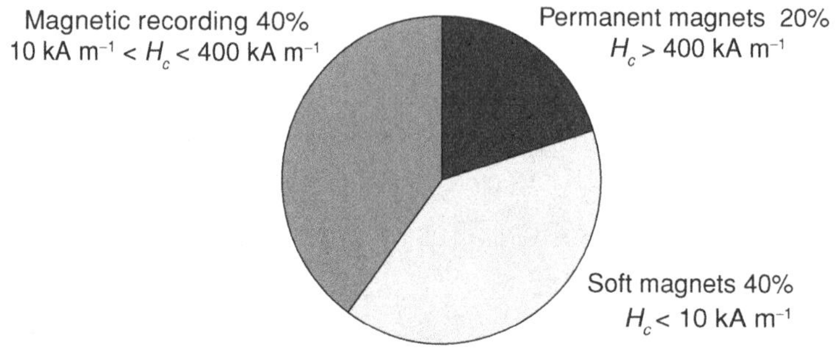

Figure 1.7

Breakdown of the market for magnetic materials, based on material type and coercivity. The total pie represents about $30 billion per annum.

A first view of the global market is given in Fig. 1.7. The breakdown is by material, distinguishing hard magnets with $H_c > 400$ kA m^{-1}, soft magnets with $H_c < 10$ kA m^{-1} and magnetic recording media with intermediate values of coercivity. In this breakdown, it is easy to account for bulk permanent magnets and soft magnetic magnets which are commodities sold by the kilogram at a price depending on the grade and form. The disc and tape media used for magnetic recording incorporate a film of magnetic material on a rigid or flexible substrate. Sophisticated magnetic multilayer structures used in read/write heads for magnetic recording, magnetic sensors and magnetic random-access memory, are the first products of the spin electronic age. It is more difficult to assign a value to the magnetic constituent of a medium or a device which is composed of nonmagnetic as well as magnetic materials. The value added by the complex processing far exceeds the cost of the minuscule amounts of magnetic raw material involved.

Further breakdowns are made in terms of materials and applications. In the hard magnet sector, the great bulk of production and over half the value is represented by the hard ferrites $Ba_2Fe_{12}O_{19}$ and $Sr_2Fe_{12}O_{19}$. These materials are used for colourful fridge magnets, as well as numerous motors, actuators, sensors and holding devices. Rare-earth compounds, especially $Nd_2Fe_{14}B$, are important in high-performance applications, and magnets based on Sm–Co alloys continue to be produced in smaller quantities.

Hard discs generally use thin films of a Co–Pt alloy. Thin film heads for magnetic recording typically use films of Fe–Ni or Fe–Co alloys in the writer and thin film stacks comprising Fe–Co and Mn-based alloys in the reader. These are soft magnetic films with a good high-frequency response, except for the Mn alloy which is an antiferromagnet. For flexible magnetic recording media, tapes and floppy discs, acicular fine particles of Fe, or Co-doped γ-Fe_2O_3, are commonly used.

Bulk soft magnetic materials are principally electrical steels. These Fe–Si alloys are produced in sheets about 300 μm thick for laminated temporary magnet cores in transformers and electrical machines. The better grades are grain-oriented with a specific crystalline texture. Soft ferrite is used for radiofrequency and microwave applications. Ferromagnetic metallic glasses, thin ribbons ($\approx$ 50 μm thick) of rapidly quenched amorphous Fe- or Co-based alloys are used in an intermediate frequency range (kHz–MHz).

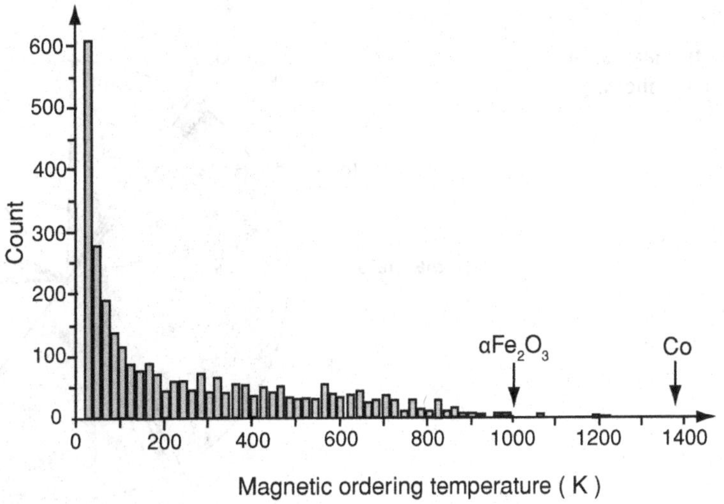

Figure 1.8

The distribution of magnetic ordering temperatures of ferromagnetic and antiferromagnetic materials (data are from T. F. Connolly and E. D. Copenhover (editors), Bibliography of Magnetic Materials, Oak Ridge National Laboratory, 1970).

To put everything in context, imagine a shopping basket with the average person's €5 worth of magnetic materials. It would include about 30 g of ferrite magnets, 1 g of rare-earth magnets, 1 m^2 of flexible recording media, an eighth of a hard disc, a quarter of a thin film head, 0.25 m^2 of electrical sheet steel, 30 g of soft ferrite and a few square centimetres of metallic glass.

Perhaps 95% of the market for magnetic materials is accounted for by barely a dozen different ferromagnetic and ferrimagnetic materials. That this is only a tiny fraction of the thousands that are known to order magnetically is testimony to the difficulty of developing new materials with the right combination of properties to bring to the market. The Curie temperature, for example, must be well above the maximum operating temperature for any practical magnetic material. A typical operating temperature range is −50 to 120 °C, so the Curie temperature needs to be at least 500 or 600 K. The distribution of magnetic ordering temperatures in Fig. 1.8 shows that only a small fraction of all known magnetic materials meet this requirement. Nevertheless, the magnetics industry has a far wider materials base than the semiconductor industry, with its overwhelming reliance on silicon.

Figure 1.9 is an attempt to break down the market by materials and applications. Magnetism is a pervasive and largely unnoticed component of the technology underpinning modern life. Our electricity is generated by movement of conductors in a magnetic field. Key components of audiovisual equipment, telephones, kitchen machines and the microwave oven are magnetic. Electrical consumer goods, where something moves when you switch on, invariably involve temporary or permanent magnets. Powerful medical imaging depends on magnetic resonance. Magnetic sensors offer contactless monitoring of position or velocity. Unimaginable amounts of information are stored and retrieved from magnetic discs in computers and servers throughout the world. Some non-volatile memory is magnetic. In 2008, consumers bought 500 million hard disk

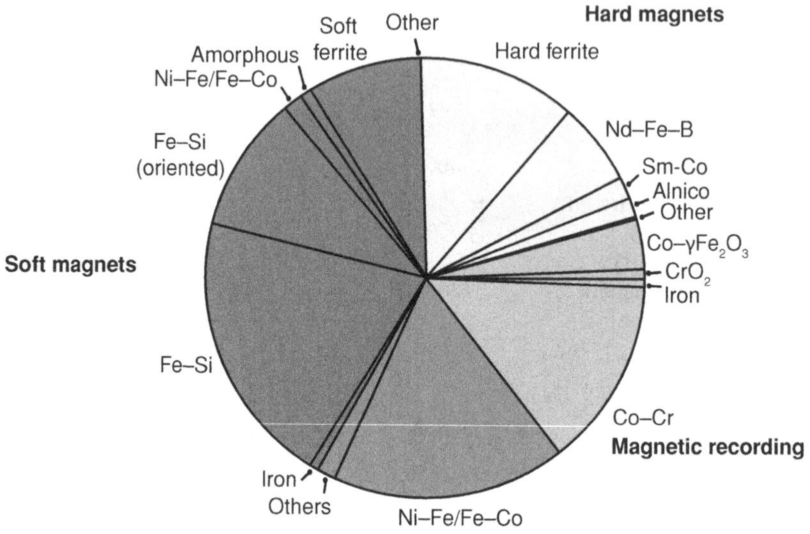

Figure 1.9
Magnetic materials and their applications.

drives and over a billion permanent-magnet motors. Magnetics is the partner of electronics in the global information revolution.

Ask any one of the wealthy billion how many magnets they own. A correct answer could be a couple of hundred or some billions depending on whether or not they possess a computer. On a hard disc drive every bit counts as an individually addressable magnet. Fifty years ago the answer might have been two or three. Fifty years hence, who knows?

1.3.2 Economics

Magnet applications depend critically on the cost and performance of ferromagnetic materials. For bulk magnets, the raw material cost may be significant. To a rough approximation, this cost is related to the abundance of the element in the Earth's crust. The composition of the crust is shown in Fig. 1.10. It can be expressed either as atom %, which emphasizes the light elements and relates to chemical formulae, or as weight %, which emphasizes the heavy elements. Note that abundances plotted in Fig. 1.10a) are in atomic % and those in Fig. 1.10b) are in weight %. Luckily, one ferromagnetic element, iron, features among the eight most abundant in the crust by either measure. Iron represents about 5% by weight of the crust, and it is the most abundant element overall when the composition of the entire globe is considered. In fact, it is 40 times as plentiful as all the other magnetic elements put together. We are truly fortunate that the cheapest metal is in many respects the best ferromagnet. Its rival, cobalt, is a thousand times scarcer, and about one hundred times more expensive. Some of the light rare-earth elements at the beginning of the $4f$ series have abundances comparable to cobalt (Fig. 1.11), but the heavy rare-earths live up to their name. Thulium, for example, sells for much more than gold or platinum. In thin-film devices, however, the cost of the element is

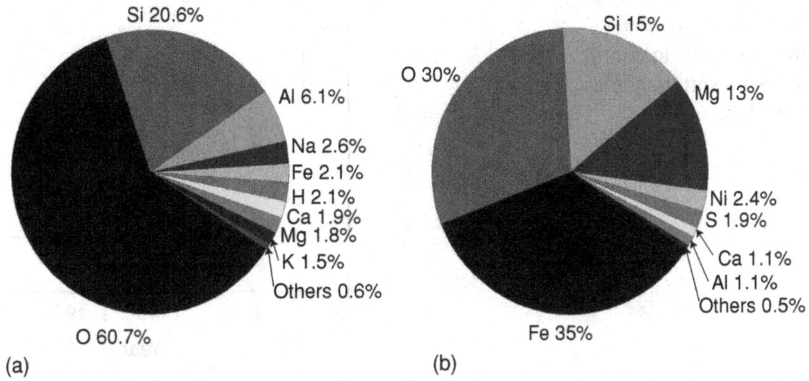

Figure 1.10

Elemental composition of: (a) the Earth's crust in atom %, and (b) the whole Earth in weight %.

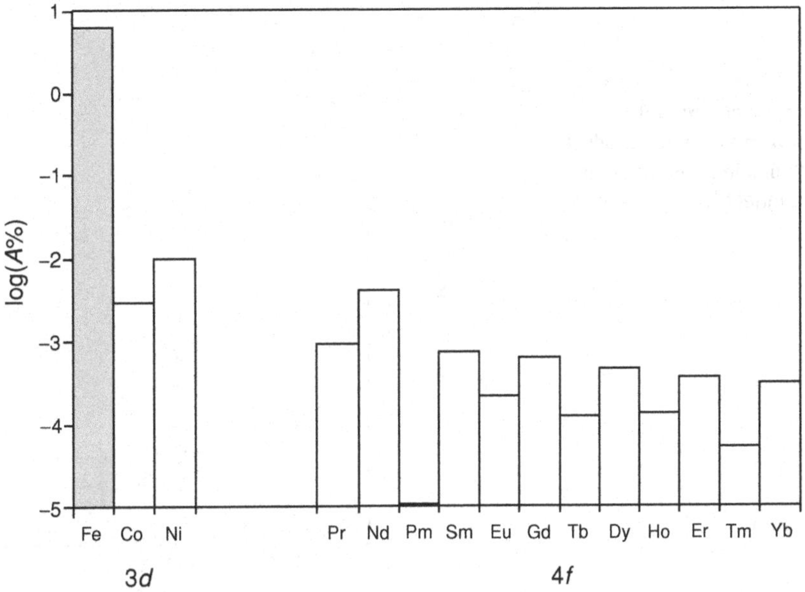

Figure 1.11

Crustal abundances of iron and other magnetic elements, shown on a logarithmic scale.

largely irrelevant because the mass of material used per device is measured in micrograms or less. Ruthenium, for example, is a rare metal used in spin-value stacks in layers a nanometre thick. A single sputtering target suffices to coat hundreds of wafers, with millions of devices.

The performance of magnetic materials improved by leaps and bounds in the twentieth century, although the records for magnetization and Curie temperature were not broken. Following from the progressive mastery of coercivity, and the relentless miniaturization in the feature size for magnetic recording, there has been an exponential improvement in properties in all three market segments.

In soft materials, the 60 Hz core losses halved every 15 years throughout much of the twentieth century (Fig. 1.12); the maximum available susceptibility doubled every 6 years over the first half. Further improvements here seem pointless, although there is a pressing need to improve the properties of temporary magnets at frequencies above 1 MHz.

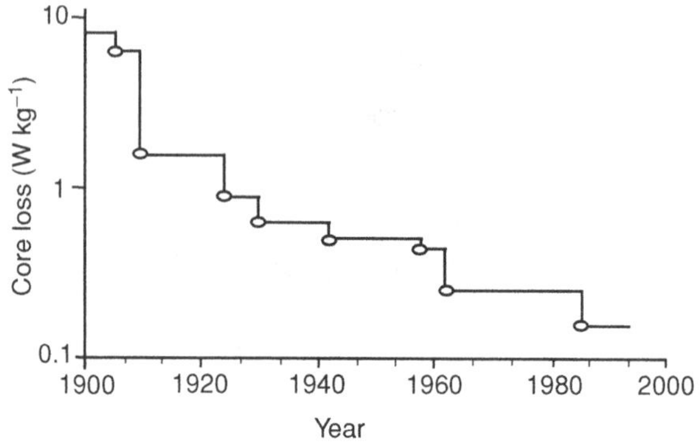

Figure 1.12

Improvements in energy losses for soft magnetic materials.

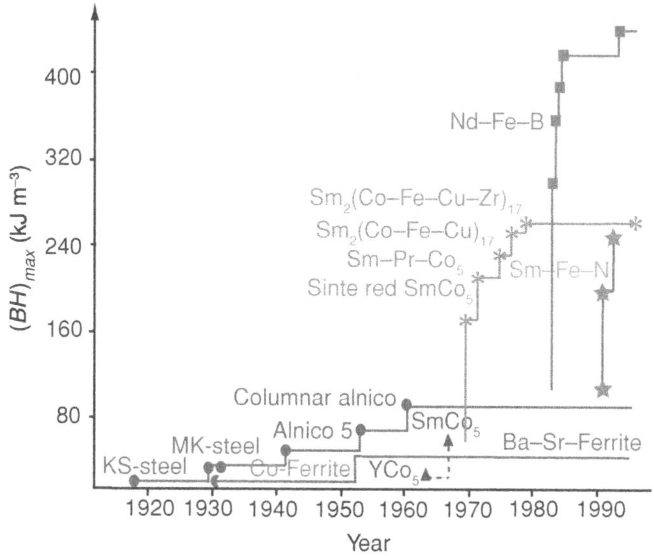

Figure 1.13

Improvement in the maximum energy product available for permanent magnets.[10]

An early eighteenth century lodestone, a ferrite magnet (right) and a Nd–Fe–B magnet (front), which all store about a joule of energy.

Hard magnets improved beyond recognition, breaking the shape barrier in 1950. This means that they could be made in any desired shape, without having to resort to horseshoes and bars to avoid self-demagnetization. The figure of merit here is the energy product, which is twice the energy stored in the magnetic field produced surrounding unit volume of an optimally shaped permanent magnet. Energy product doubled every 12 years up to 1990 (Fig. 1.13). The best permanent magnets have square hysteresis loops with $H_c > M_r/2$. The energy product cannot exceed $\mu_0 M_r^2/4$.

Most remarkable of all has been the progress of magnetic recording, illustrated in Fig. 1.14. For over a decade, the areal density doubled almost every

[10] In industry, units of MG Oe are used for energy product. 100 kJ m^{-3} = 12.57 MG Oe. Units are discussed in Appendix B.

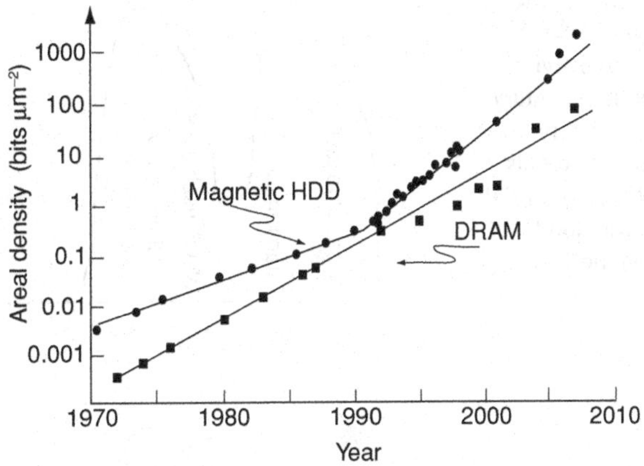

Figure 1.14

Moore's law for transistor density in semiconductors and for storage density in magnetic recording.[11]

year, the magnetic analogue of 'Moore's law' for semiconductors. Together, magnets and semiconductors have created the new era of big data where information is universally accessible, copies are free and privacy is moot.

The extraordinary technological progress epitomized by the three magnetic exponentials has vastly improved the quality of goods and services in ways that economics may fail to quantify. The consumer takes it all for granted, ignorant of the struggle of scientists and engineers to achieve their current mastery of nature, and of the science on which our civilization is based.

One thing is certain: exponential improvement cannot continue indefinitely. Permanent magnets are approaching the limit of energy product, 1200 J m^{-3}, defined by the remanence of permendur, the alloy of iron and cobalt having greatest room-temperature magnetization. Magnetic recording densities may saturate as magnetic instability is encounterd at densities well in excess of 1000 bits μm^{-2}, where each bit is so tiny that the corresponding volume of magnetic medium is too small to withstand thermal fluctuations.

Future improvements in the performance of bulk magnetic materials will probably focus on achieving desirable combinations of properties, e.g. a permanent magnet stable at 500 °C, a material combining low anisotropy and high magnetostriction, a multiferroic material that is both ferromagnetic and ferroelectric, all at the lowest possible cost. In devices, the trend is towards increasing integration of magnetic functionality with optics and electronics. As for the methods of investigation, intelligent experimentation will be supplemented increasingly by computer simulation and combinational synthesis.

1.4 Magnetism, the felicitous science

Magnetism is a wonderful example of how basic science, flowing from a magical natural phenomenon can become ubiquitous in our lives, thanks to the

[11] Industry prefers Gbits per square inch. 1 bit μm^{-2} = 0.645 Gbits per square inch.

Figure 1.15

An impression of activity over the twentieth century in fundamental theory, normal science, materials development and industrial production in relation to permanent magnetism.

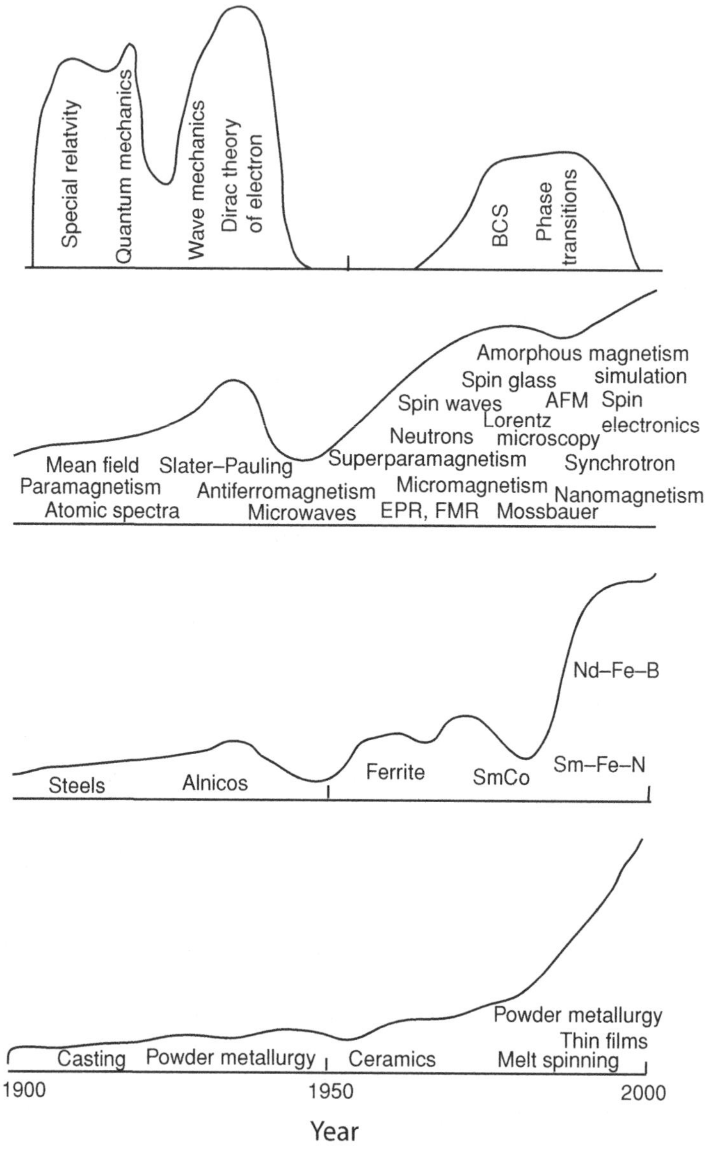

labours of successive generations of specialists. The twentieth century is a convenient time frame in which to trace how leaps in theoretical understanding and advances in experimental practice can relate to the emergence of a technology that creates wealth and facilitates our life. Figure 1.15 portrays the work in basic theory, normal science, materials development and industrial production (the latter two in relation to permanent magnetism). The four are interlinked, but cause-and-effect relations are not always obvious.

The context of magnetic research and development is schematized in Fig. 1.16. The activity employs roughly 30 000 people worldwide, at a cost

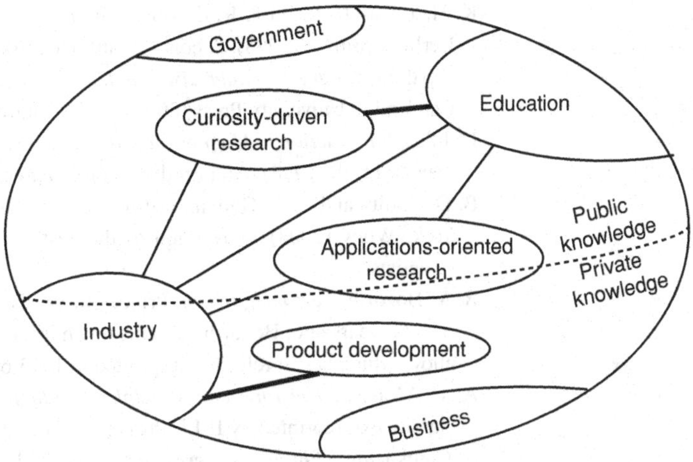

Figure 1.16

Relations between academic and industrial research and development.

of over €1 billion per year. It is instructive to contrast attitudes of academic and industrial scientists to new knowledge and technology. One seeks diffusion, the other ownership. Academic research is rewarded by peer recognition, industrial development by profit. Both share an understanding that by interrogating nature in a structured and rational way, trustworthy and objective knowledge of practical importance can be obtained. This is what unites the geomancer, the telegraph engineer, the PhD student struggling for data, the corporate scientist whose invention could trigger a billion dollar investment providing employment for thousands and the professor attempting to draw the strands together in a book on this felicitous science.

FURTHER READING

J. D. Livingston, *Driving Force*, Cambridge, MA: Harvard University Press (1996). An enjoyable account of magnetism for the general reader.

A. P. Guimares, *From Lodestone to Supermagnets*, Weinheim: Wiley-VCH (2005). Another popular account.

A. Kloss, *Geschichte des Magnetismus*, Berlin: VDE (1994). Packed with facts, an excellent one-volume history of magnetism from its beginnings to the early twentieth century.

D. C. Matthis, *Theory of Magnetism Made Simple*, Singapore: World Scientific (2006). An elegant introductory chapter outlines the history of ideas, especially in the nineteenth and twentieth centuries.

J. Needham, *Science and Civilisation in China*, Vol. IV:1, Cambridge: Cambridge University Press (1962). The definitive scholarly account of Chinese contributions to science and technology. This volume treats magnetism.

S. Blundell, *Magnetism in Condensed Matter*, Oxford: Oxford University Press (2001). A lively introduction for final-year undergraduates.

K. H. J. Buschow and F. R. de Boer, *Physics of Magnetism and Magnetic Materials*, Berlin: Springer (2003). A concise introduction to the principles and applications.

N. Spaldin, *Magnetic Materials, Fundaments and Device Applications*, Cambridge: Cambridge University Press (2003). A short introduction for undergraduates.

D. Jilles, *Introduction to Magnetism and Magnetic Material*, 2nd edition, London Chapman and Hall (1998). An introduction for engineers, in question and answer format.

B. D. Cullity and C. D. Graham, *Introduction to Magnetic Materials* 2nd edition, New York: Wiley (2008). A revision of the best-written book of its kind for materials scientists.

R. M Bozorth, *Ferromagnetism*, Princeton: van Nostrand (1951). Although published over 50 years ago, Bozorth contains much information, especially on 3*d* metals and alloys which is still relevant. Reprinted in 1993 by IEEE Press.

A. H. Morrish, *The Physical Principles of Magnetism*, New York: Wiley (1965). A classic text, reprinted by IEEE Press in 2001.

S. Chiukazumi, *Physics of Ferromagnetism*, 2nd edition, Oxford: Oxford University Press (1997). Another classic general text.

R. C. O'Handley, *Modern Magnetic Materials; Principles and Applications*, New York: Wiley (1999). A thorough treatment of modern materials and applications, including thin films.

H. Kronmuller and S. S. P. Parkin (editors), *Handbook of Magnetism and Advanced Magnetic Materials*, 5 volumes, Chichester: Wiley (2007). A modern multi-author reference work, with a focus on contemporary topics.

G. Rado and H. Suhl (editors), *Magnetism*, 5 volumes, New York: Academic Press (1960–73). A multiauthor treatise which details much of the modern understanding of the theory of magnetism.

K. H. J. Buschow and E. P. Wohlfarth (editors), *Handbook of Magnetic Materials*, 16 volumes, Amsterdam: North Holland/Elsevier (1980–). A continuing multiauthor series which is a mine of information on magnetic materials. New volumes appear every year or two.

EXERCISES

1.1 As a rule of thumb, a field 3 times greater than the spontaneous magnetization is needed to magnetize a permanent magnet. Given that the current in a lightning strike is 10^6 A, make an estimate of the time that will elapse before a particular rock outcrop of lodestone becomes magnetized.

1.2 Find one documented historical reference to magnetism, before 1200, from your own part of the world.

1.3 Estimate, and rank in decreasing order:
  (a) the magnetostatic energy stored in space around a 10 g permanent magnet;
  (b) the chemical energy stored in 10 g of cornflakes;
  (c) the gravitational potential energy in a 10 g pencil sitting on your desk;
  (d) the kinetic energy of a 10 g bullet moving at the speed of sound;
  (e) the mass energy released by fission of 10 g of ^{235}U.

1.4 When does the extrapolation of Fig. 1.14 'predict' that a bit will have dimensions smaller than an atom?

1.5 Given two apparently identical metal bars, one a temporary (soft) magnet, the other a permanent (hard) magnet and a piece of string, how could you distinguish between them?

1.6 Write two paragraphs on the area of magnetism which you think will have greatest commercial potential in 10 years time, explaining why. You may consult the last four chapters of the book.

1.7 Deduce Faraday's law of electromagnetic induction $\mathcal{E} = -\mathrm{d}\Phi/\mathrm{d}t$ from Maxwell's equation 1.1d where Φ, the magnetic flux through a circuit of area $\mathcal{A}$, is defined as $\int_s \boldsymbol{B}\cdot\mathrm{d}\,\mathcal{A}$ and $\mathcal{E}$ is the emf induced in the circuit.

2 Magnetostatics

Back to basics

The dipole moment $\mathfrak{m}$ is the elementary magnetic quantity, and magnetization $\boldsymbol{M}(\boldsymbol{r})$ is its mesoscopic volume average. The primary magnetic field $\boldsymbol{B}$ is related to the auxiliary magnetic field $\boldsymbol{H}$ and the magnetization by $\boldsymbol{B} = \mu_0(\boldsymbol{H} + \boldsymbol{M})$. Sources of magnetic field are electric currents and magnetized material. The field produced by a given distribution of magnetization can be calculated by integrating the dipole field due to each volume element $\boldsymbol{M}(\boldsymbol{r})\mathrm{d}V$, or using the equivalent distributions of electric currents or magnetic charge. Magnetic scalar and vector potentials φ_m and $\boldsymbol{A}$ are defined for $\boldsymbol{H}$ and $\boldsymbol{B}$, respectively. The internal, external and demagnetizing fields are distinguished. Internal field may be defined on a mesoscopic or a macroscopic scale, the latter in terms of the demagnetizing factor $\mathcal{N}$. Magnetic forces and energies are related to magnetization and external field.

We begin with magnetostatics, the classical physics of the magnetic fields, forces and energies associated with distributions of magnetic material and steady electric currents. The concepts presented here underpin the magnetism of solids. Magnetostatics refers to situations where there is no time dependence.

2.1 The magnetic dipole moment

The elementary quantity in solid-state magnetism is the magnetic moment $\mathfrak{m}$. On an atomic scale, intrinsic magnetic moments are associated with the spin of each electron and a further contribution is associated with its orbital motion around the nucleus. The nucleus itself may possess spin, but the corresponding nuclear moments are three orders of magnitude smaller than those associated with electrons, because the magnetic moment of a particle scales as 1/mass. We can often neglect them.

The spin and orbital moments of the atomic electrons add in ways governed by the laws of quantum mechanics, discussed in the next two chapters. Suffice it to say that most of them manage to cancel out, and only a few transition metal atoms or ions retain a resultant moment on the atomic scale in solids. Again these atomic moments sum to zero in the paramagnetic state, unless an external magnetic field is applied, but for a different reason – they are

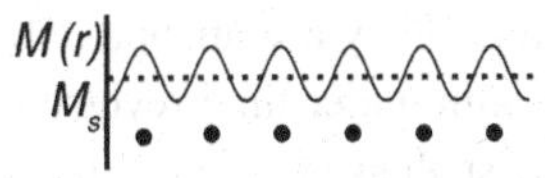

The local magnetization $M(r)$ fluctuates on an atomic scale – dots represent the atoms. The mesoscopic average, shown by the dashed line, is uniform.

disordered by thermal fluctuations. However, resultant moments do arise spontaneously within domains in the ferromagnetically ordered state.

We can therefore imagine a *local* magnetic moment density, the local magnetization $\boldsymbol{M}(\boldsymbol{r},t)$ which fluctuates wildly on a subnanometre scale, and also rapidly in time on a subnanosecond scale. But for our purposes, it is more useful to define a *mesoscopic average* $\langle \boldsymbol{M}(\boldsymbol{r},t)\rangle$ over a distance of order a few nanometres, and times of order a few microseconds to arrive at a steady, homogeneous, local magnetization $\boldsymbol{M}(\boldsymbol{r})$. The time-averaged magnetic moment $\delta\mathfrak{m}$ in a mesoscopic volume δV is

$$\delta\mathfrak{m} = \boldsymbol{M}\delta V. \tag{2.1}$$

This magnetization may be the spontaneous magnetization $\boldsymbol{M}_s$ within a ferromagnetic domain, or the uniform magnetization of a paramagnet or a diamagnet induced by an applied field. The representation of the magnetization of a solid by the quantity $\boldsymbol{M}(\boldsymbol{r})$ which varies smoothly on a mesoscopic scale is the continuous medium approximation. It is fundamental in magnetostatics.

The concept of magnetization of a ferromagnet is often extended to cover the *macroscopic average* over a sample:

$$\boldsymbol{M} = \sum_i \boldsymbol{M}_i V_i \Big/ \sum_i V_i, \tag{2.2}$$

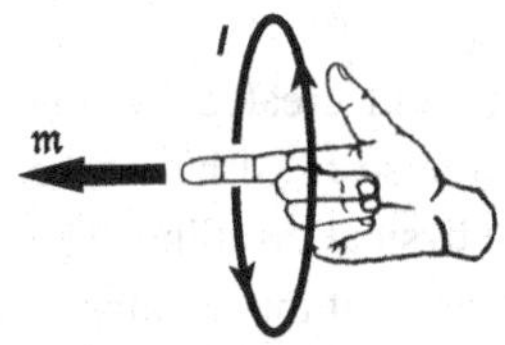

The corkscrew rule. When the tip of the right thumb turning in a clockwise sense traces the current, the index finger points along $\mathfrak{m}$.

where the sum is over all the domains, which have volume V_i. The sum $\sum_i V_i$ is the sample volume. Hysteresis loops are plots of the macroscopic average magnetization. For example, remanence $\boldsymbol{M}_r$ is a macroscopic average. Usually it will be clear from the context whether we are referring to a mesoscopic or a macroscopic average.

According to Ampère, a magnet is equivalent to a circulating electric current; the elementary magnetic moment $\mathfrak{m}$ can be represented by a tiny current loop. If the area of the loop is $\mathcal{A}$ square metres, and the circulating current is I amperes, then

$$\mathfrak{m} = I\mathcal{A}. \tag{2.3}$$

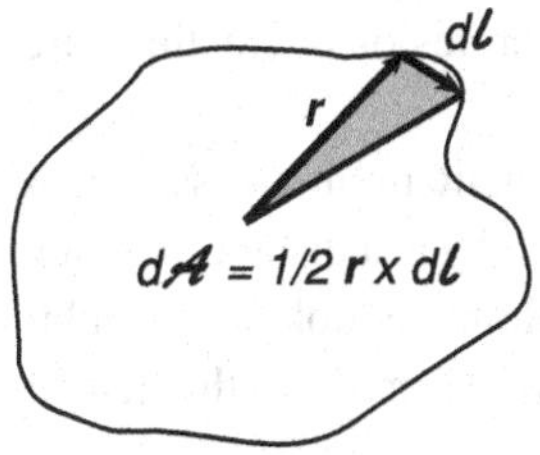

The area of a loop, obtained by summing the area of the elementary triangles, is $\frac{1}{2}\oint \boldsymbol{r} \times d\boldsymbol{l}$. The vector $d\mathcal{A}$ is into the plane of the paper.

The shape of the loop is unimportant provided the current flows in a plane. Units of $\mathfrak{m}$ are A m^2, and $\boldsymbol{M}$ from (2.1) has units of A m^{-1}. The relation between the directions of $\mathfrak{m}$ and I are given by the right-hand corkscrew rule.

A generalization of the relation (2.3) between the magnetic moment and the current is

$$\mathfrak{m} = \frac{1}{2}\int \boldsymbol{r} \times \boldsymbol{j}(\boldsymbol{r})\mathrm{d}^3 r, \tag{2.4}$$

where $\boldsymbol{j}(\boldsymbol{r})$ is now the *current density* at a point $\boldsymbol{r}$ in A m^{-2}, and $j = I/\mathfrak{a}$, where $\mathfrak{a}$ is the cross section in which the current flows. Considering an irregular plane loop, the current element is given by $I\delta l = j\delta V$. Hence $\frac{1}{2}\int \boldsymbol{r} \times \boldsymbol{j}(\boldsymbol{r})\mathrm{d}^3 r = \frac{1}{2}\oint \boldsymbol{r} \times I\mathrm{d}\boldsymbol{l} = I\int \mathrm{d}\mathcal{A} = \mathfrak{m}$.

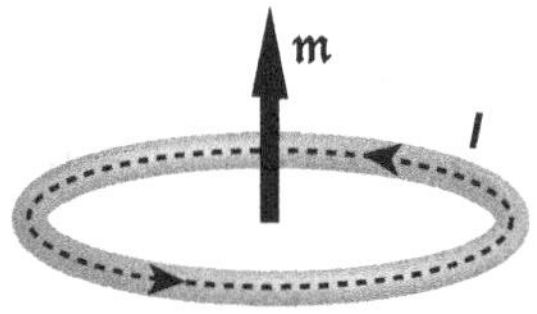

A magnetic moment m is equivalent to a current loop.

Magnetic moment and magnetization are axial vectors. They are unchanged under spatial inversion, $\boldsymbol{r} \rightarrow -\boldsymbol{r}$, but they do change sign under time reversal $t \rightarrow -t$. Axial vectors differ from normal polar vectors such as position, force, velocity and current density, which change sign on spatial inversion but not necessarily on time reversal. Strictly speaking, axial vectors are tensors; they can be written as a vector product of two polar vectors, as in (2.4), but their three independent components can be manipulated like those of a vector.

2.1.1 Fields due to electric currents and magnetic moments

In a steady state, the magnetic field $\delta \boldsymbol{B}$ created by a small current element $\boldsymbol{j}\delta V$ at a point P is given by the Biot–Savart law, which follows from Maxwell's equations (1.1a) and (1.1c):

$$\delta \boldsymbol{B} = -\frac{\mu_0}{4\pi}\frac{\boldsymbol{r} \times \boldsymbol{j}}{r^3}\delta V. \tag{2.5}$$

Written in terms of electric current I in a circuit element of length δl, where $I = \boldsymbol{j} \cdot \mathfrak{a}$ and $\delta V = \mathfrak{a} \cdot \delta \boldsymbol{l}$, the law of Biot and Savart is

$$\delta \boldsymbol{B} = -\frac{\mu_0}{4\pi} I \frac{\boldsymbol{r} \times \delta \boldsymbol{l}}{r^3}, \tag{2.6}$$

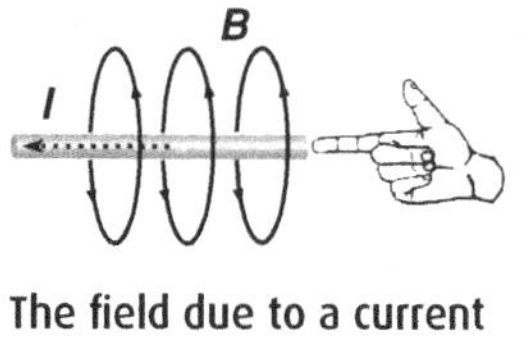

The field due to a current element. When the right index finger points along the current direction, the tip of the thumb traces the field as the right hand turns clockwise.

where the vector $\boldsymbol{r}$ goes from the current element to the point P.

Hence it is possible by integration to calculate the field created by any distribution of electric currents. Currents flow in circuits, so (2.6) has to be integrated over a complete circuit in order to acquire a physical meaning. The field falls as the *square* of the distance from the current element that creates it, with a constant of proportionality $\mu_0/4\pi$ which is exactly 10^{-7}, according to the definition of the ampere given in the next section. The units of the magnetic constant μ_0 are J A^{-2} m^{-1}, so the units of B are J A^{-1} m^{-2} or, equivalently, kg s^{-2} A^{-1}. In view of the importance of the $\boldsymbol{B}$-field in magnetism, the unit has its own special name, the tesla[1] (abbreviated to T). Equivalent units for μ_0 are T m A^{-1}. It is natural to think of the magnetic constant in these terms in expressions relating B with H or M, but when energies or interactions are involved, the form J A^{-2} m^{-1} is more convenient.

We use (2.6) to calculate the field created by a magnetic moment associated with a small current loop, firstly at the centre, and secondly at a distance r much greater than the size of the loop. For the first calculation, we look at a circular loop of radius a. Each element $Ia\delta\theta$ contributes $\mu_0 Ia\delta\theta/4\pi a^2$, so the field at the centre is

$$B_O = \frac{\mu_0 I}{2a}. \tag{2.7}$$

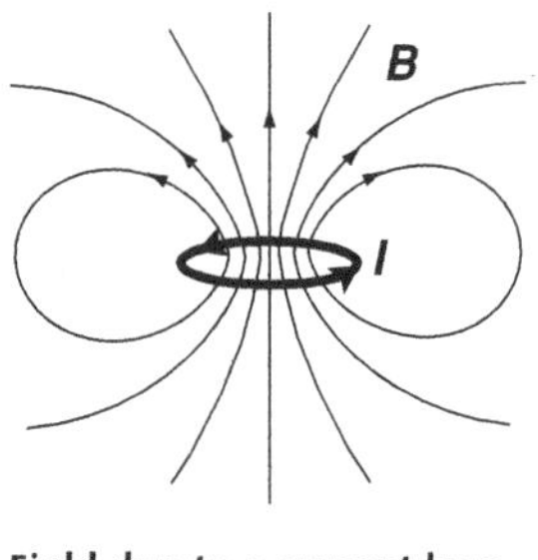

Field due to a current loop.

[1] The tesla is named for Nikola Tesla, the Serbian pioneer of electromagnetism.

Figure 2.1

Calculation of the dipolar field due to a magnetic moment $\mathfrak{m}$.

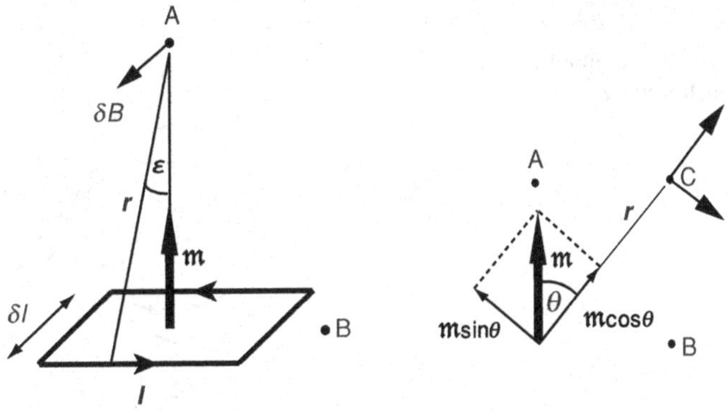

Nikola Tesla, 1856–1943.

To simplify the second calculation, we choose a square loop of side $\delta l \ll r$ and first evaluate the field at two special positions, point 'A' on the axis of the loop and point 'B' in the broadside position. At point 'A', there are four contributions like the one shown in Fig. 2.1, and their resultant is parallel to $\mathfrak{m}$ since the horizontal components cancel. Its magnitude is $B_A = 4\delta B \sin\epsilon = 4\mu_0 I\delta l \sin\epsilon/4\pi r^2$. Since $\sin\epsilon = \delta l/2r$, and $\mathfrak{m} = I(\delta l)^2$ according to (2.3), we obtain

$$B_A = 2\frac{\mu_0}{4\pi}\frac{\mathfrak{m}}{r^3}. \tag{2.8}$$

At point 'B' we must consider the contribution of the two sides of the square perpendicular to $\boldsymbol{r}$ as well as the two sides parallel to $\boldsymbol{r}$. Hence the magnitude

$$B_B = \frac{\mu_0}{4\pi} I\delta l\left\{\frac{1}{(r-\delta l/2)^2} - \frac{1}{(r+\delta l/2)^2} - \frac{2\sin\epsilon}{r^2}\right\}$$
$$\approx \frac{\mu_0}{4\pi}\frac{I\delta l}{r^2}\left\{\left(1+\frac{\delta l}{r}\right) - \left(1-\frac{\delta l}{r}\right) - \frac{\delta l}{r}\right\}:$$
$$B_B = \frac{\mu_0}{4\pi}\frac{\mathfrak{m}}{r^3}. \tag{2.9}$$

The field in the 'B' position is half as large as that in the 'A' position, and oppositely directed relative to $\mathfrak{m}$. To find the field at a general position 'C' far from the loop, we just resolve $\mathfrak{m}$ into two components, $\mathfrak{m}\cos\theta$ parallel to $\boldsymbol{r}$ and $\mathfrak{m}\sin\theta$ perpendicular to $\boldsymbol{r}$. The dipole field in polar coordinates is therefore

$$B_r = 2\left(\frac{\mu_0\mathfrak{m}}{4\pi r^3}\right)\cos\theta; \quad B_\theta = \left(\frac{\mu_0\mathfrak{m}}{4\pi r^3}\right)\sin\theta; \quad B_\phi = 0. \tag{2.10}$$

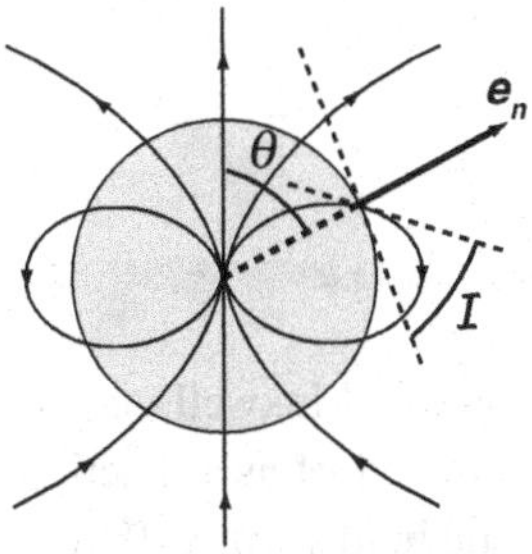

The angle of dip.

The field falls off rapidly, as the *cube* of the distance from the magnet. It has axial symmetry about $\mathfrak{m}$.

Faraday represented magnetic fields using lines of force. (The basic idea dates back to Descartes.) The lines provide a picture of the field by indicating its direction at any point; its magnitude is inversely proportional to the spacing of the lines. The direction of the field of a point dipole relative to the normal to $\boldsymbol{r}$ is known as 'dip'. The angle of dip $\mathcal{I}$ given by $\tan\mathcal{I} = B_r/B_\theta$ is related

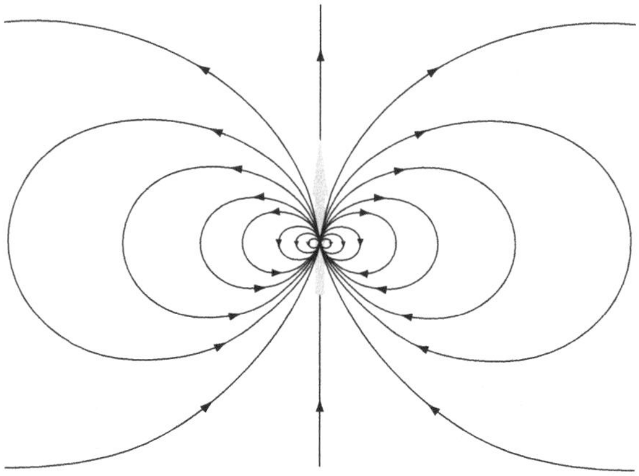

Figure 2.2

The field due to a magnetic dipole, illustrated by lines of force.

to θ by

$$\frac{B_r}{B_\theta} = \tan \mathcal{I} = 2\cot\theta. \tag{2.11}$$

Setting $\tan \mathcal{I} = \mathrm{d}r/r\mathrm{d}\theta$ gives the differential equation for the line of force $\mathrm{d}r/r\mathrm{d}\theta = 2\cot\theta$ and integration gives the parametric equation of a line of force, $r = c\sin^2\theta$, where c is a different constant for each line. The field of a small current loop or the equivalent magnetic moment is illustrated in Fig. 2.2.

The field of the magnetic moment has the same form as that of an electric dipole $\boldsymbol{p} = q\delta l$ formed of positive and negative charges $\pm q$ which are separated by a small distance δl. The vector $\boldsymbol{p}$ is directed from $-q$ to $+q$. Hence the magnetic moment $\mathfrak{m}$ may somehow be regarded as a *magnetic dipole*; its associated magnetic field is called the magnetic dipole field.

There are other ways of writing the dipole field which are equivalent to (2.10). One uses Cartesian coordinates, with $\mathfrak{m}$ in the z-direction:

$$\boldsymbol{B} = \frac{\mu_0 \mathfrak{m}}{4\pi r^5}\left[3xz\mathbf{e}_x + 3yz\mathbf{e}_y + (3z^2 - r^2)\mathbf{e}_z\right], \tag{2.12}$$

where $\mathbf{e}_i$ are unit vectors; another resolves the field into components parallel to $\mathbf{r}$ and $\mathfrak{m}$:

$$\boldsymbol{B} = \frac{\mu_0}{4\pi}\left[3\frac{(\mathfrak{m}\cdot\boldsymbol{r})\boldsymbol{r}}{r^5} - \frac{\mathfrak{m}}{r^3}\right]. \tag{2.13}$$

2.2 Magnetic fields

The magnetic field that appears in the Biot–Savart law and in Maxwell's equations in vacuum is $\boldsymbol{B}$, but the hysteresis loop of Fig. 1.3 traced $\boldsymbol{M}$ as a function of $\boldsymbol{H}$. It is time to explain why we need these two magnetic fields, with different units and dimensions.

2.2.1 The B-field

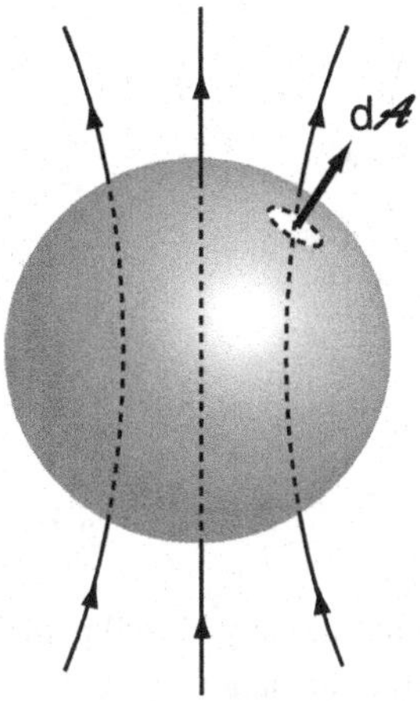

The B field is divergenceless, with no sources or sinks.

The existence of magnetic monopoles that could act as sources and sinks of magnetic field was originally suggested by Dirac as a means of explaining the quantization of electric charge. Their magnetic charge is $q_m \approx h/\mu_0 e$. Monopoles feature prominently in supersymmetric gauge theories, but they have never been observed in nature. They have nothing to do with the humble magnet. This truth is enshrined in Maxwell's equation (1.1a) which states that the magnetic field is divergenceless:[2]

$$\nabla \cdot \boldsymbol{B} = 0. \tag{2.14}$$

Fields with this property are said to be *solenoidal*; the lines of force all form continuous loops. Expressing this in integral form, and using the divergence theorem the equation requires that the flux $\boldsymbol{B}$ into any region enclosed by a surface S exactly balances the flux out of the same region. The net flux of $\boldsymbol{B}$ across any closed surface is zero, a result known as Gauss's theorem:

$$\int_S \boldsymbol{B} \cdot \mathrm{d}\mathcal{A} = 0. \tag{2.15}$$

In (2.15), the vector element of area $\mathrm{d}\mathcal{A}$ is defined to be in the direction of the outward normal at a point on the closed surface S, and $\mathrm{d}\Phi = \boldsymbol{B} \cdot \mathrm{d}\mathcal{A}$ is the element of magnetic flux flowing out of the surface S through area $\mathrm{d}\mathcal{A}$. Hence $B = \mathrm{d}\Phi/\mathrm{d}\mathcal{A}$, and an alternative name for the $\boldsymbol{B}$-field is magnetic flux density. Flux has a named (but little used) unit of its own, the weber, abbreviated to Wb. A unit equivalent to T is Wb m^{-2}. The other synonym for $\boldsymbol{B}$ is magnetic induction.

Magnetic flux is quantized in superconducting circuits, and the fundamental flux quantum $\Phi_0 = h/2e$ is equal to 2.068×10^{-15} Wb.

Sources of the $\boldsymbol{B}$-field are:

(i) electric currents flowing in conductors;
(ii) moving charges (which constitute an electric current); and
(iii) magnetic moments (which are equivalent to current loops).

Time-varying electric fields are also a source of magnetic fields, and vice versa, but we are restricting our attention to magnetostatics, which deals only with the magnetic fields created by steady currents $\boldsymbol{j}(\boldsymbol{r})$ and static distributions of magnetic moments $\boldsymbol{M}(\boldsymbol{r})$. The relation between the magnetic flux density $\boldsymbol{B}$ and the current density $\boldsymbol{j}$ can also be written in differential or integral form. In a steady state, the relation at any point is given by Maxwell's equation (1.1c):

$$\nabla \times \boldsymbol{B} = \mu_0 \boldsymbol{j}. \tag{2.16}$$

[2] Important definitions and results of vector calculus are summarized in Appendix C.

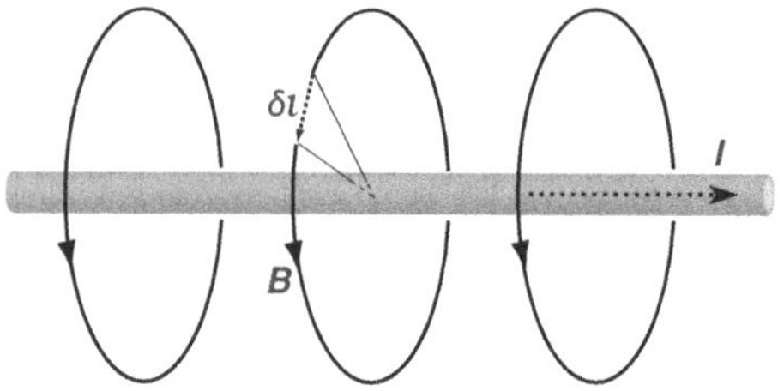

Figure 2.3

Ampère's law: the integral of $\boldsymbol{B}$ around any closed loop is proportional to the current threading the loop. The figure also illustrates the field due to a current in a long straight conductor.

Expressed in integral form, this becomes Ampère's law:

$$\oint \boldsymbol{B} \cdot \mathrm{d}l = \mu_0 I. \tag{2.17}$$

The integral, known as the circulation of $\boldsymbol{B}$, is taken around any closed path, and I is here the algebraic sum of the currents threading the path (Fig. 2.3). It is used to calculate the field due to highly symmetric current distributions. For example, a long straight wire produces a field that encircles the wire, as shown in Fig. 2.3. It is constant in magnitude at a distance r from the axis. Integrating (2.17) around such a circle of radius r gives

$$B(r) = \frac{\mu_0 I}{2\pi r}. \tag{2.18}$$

Electric and magnetic fields represent the interactions of electric charges, currents and magnetic moments located at different points in space. The fields provide the connection between charges, currents and moments, transferring information at the speed of light. All the sources of $\boldsymbol{B}$ are moving charges, but $\boldsymbol{B}$ itself interacts with charges only when they move. The fundamental relation between the fields and the *force* $\boldsymbol{f}$ exerted on a charged particle is the Lorentz expression (1.3)

$$\boldsymbol{f} = q(\boldsymbol{E} + \boldsymbol{v} \times \boldsymbol{B}). \tag{2.19}$$

The electric and magnetic fields can therefore be expressed in terms of the basic quantities of mass, length, time and current. The units of these four quantities in the *Système International* (SI) are the kilogram, metre, second and ampere. A coulomb is an ampere second. Hence the units of $\boldsymbol{E}$ and $\boldsymbol{B}$ are, respectively, newtons per coulomb (N C^{-1}), and N C^{-1} m^{-1} s. The latter reduces to kg s^{-2} A^{-1}, or tesla. Units and dimensions are discussed in Appendix A.

Equation (2.19) establishes the dimensions of $\boldsymbol{B}$, but the magnitude of the tesla depends on the definition of the ampere. According to (2.18) the field at a distance r from a long, straight conductor carrying a current I_1 is $B_1 = \mu_0 I_1/2\pi r$. It follows from (2.19) that the force per metre on another long, parallel conductor carrying current I_2 is $I_2 B_1$ or $\mu_0 I_1 I_2/2\pi r$. The ampere is then defined as the current flowing in conductors in vacuum which produces a force of precisely 2×10^{-7} N m^{-1} when the two conductors are 1 metre apart (Fig. 2.4). The force is attractive when the currents flow in the same sense. With this definition of the ampere, the constant μ_0 appearing in (2.5), (2.16)

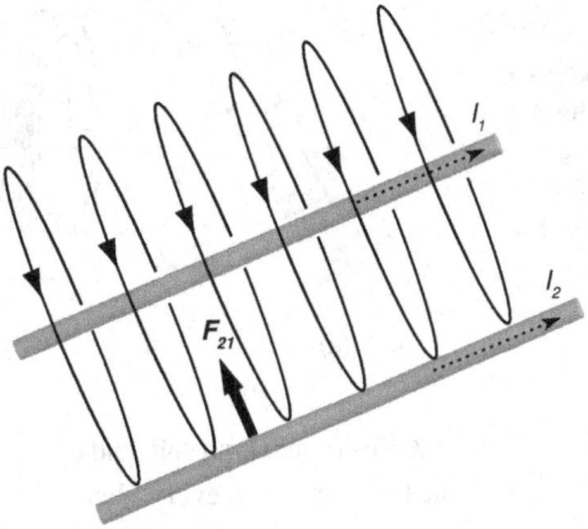

Figure 2.4

Force between two long straight conductors carrying currents I_1 and I_2. The field due to one of the conductors is shown.

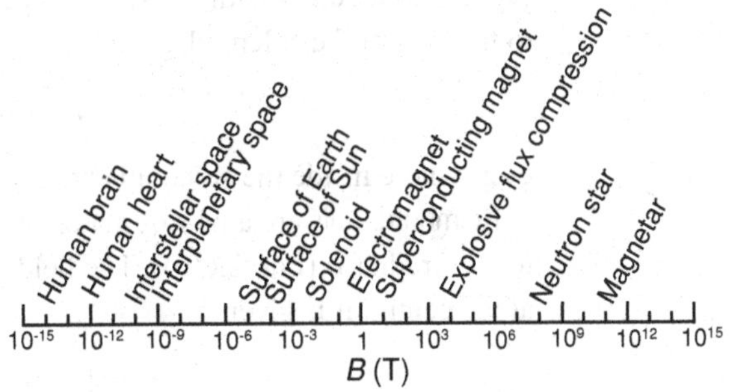

Figure 2.5

Magnitudes of some magnetic fields in tesla.

and (2.18) is precisely $4\pi \times 10^{-7}$ T m A^{-1}. The field at a distance of 1 metre from a long straight conductor carrying a current of one ampere is 2×10^{-7} T, or 0.2 μT. This cumbersome definition of the unit of electric current will be replaced in coming years by one based on a precisely defined number of electrons flowing per second in a high-frequency single-electron pump.

The tesla is a rather big unit. The largest continuous field ever produced in a laboratory is 45 T. Fields of order 1 T are created by electromagnets, and close to the surfaces of permanent magnets. The field a millimetre from a conductor carrying 5 A is only a millitesla. The field at the Earth's surface is a few tens of microtesla. The range of natural and man-made magnetic fields is illustrated further in Fig. 2.5.

2.2.2 Uniform magnetic fields

We have seen that the magnetic field far from flux sources that are confined in a limited region of space falls off in an anisotropic manner as $1/r^3$ (2.10). Nevertheless it is possible to devise structures of currents or magnets that create a field that is *uniform* in some limited volume; these are the magnetic analogues of the parallel-plate capacitor in electrostatics. We offer some examples.

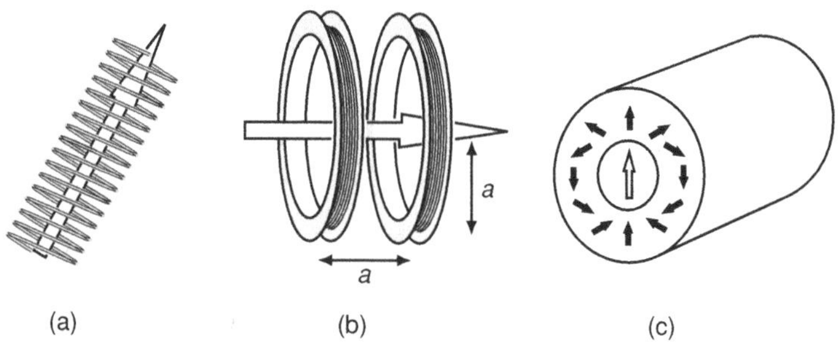

Figure 2.6

Structures that produce a uniform magnetic field in their bore: (a) a long solenoid, (b) Helmholtz coils and (c) a Halbach cylinder.

An infinitely long solenoid creates a uniform field that is parallel to its axis in the bore, and zero everywhere outside. If there are n turns per metre carrying a current I, application of Ampère's law (2.17) around the closed paths shown in Fig. 2.6(a) together with the symmetry argument that the field must lie parallel to the axis of the solenoid, gives

$$B = \mu_0 n I, \tag{2.20}$$

everywhere inside the solenoid and $B = 0$ outside.

Helmholtz coils are a pair of matched coaxial coils whose separation is equal to their radius a (Fig. 2.6(b)). The field is uniform, with zero second derivative at the centre. It is given by

$$B = [4/5]^{3/2}\mu_0 N I/a, \tag{2.21}$$

where N is the number of turns in each coil, and I is the current flowing in each. When the connections are reversed, the Helmholtz coils produce a uniform field gradient parallel to their axis.

The dipole ring or Halbach cylinder of Fig. 2.6(c) is an elongated cylindrical configuration of permanent magnets that creates a uniform field in the bore of the cylinder. The structure can be considered to be made up of long thin rods which are magnetized perpendicular to their axis, with a magnetic moment per unit length of λ A m. Integration of (2.10) or (2.12) shows that there is no difference in the magnitude of the field created in the A and B positions (Fig. 2.1) or at any other point at a distance r from the transversely magnetized long rod; $|B| = \mu_0\lambda/2\pi r^2$; the fields created by a long rod in cylindrical polar coordinates is

$$B_r = \left(\frac{\mu_0\lambda}{2\pi r^2}\right)\cos\theta, \quad B_\theta = \left(\frac{\mu_0\lambda}{2\pi r^2}\right)\sin\theta, \quad B_z = 0. \tag{2.22}$$

The structure of Fig. 2.6(c) is composed of many such elementary rods, all oriented so that the angle between their magnetization direction and the vertical y-axis is $\alpha = 2\theta$, so that the direction of the field created by each element is parallel to y. Further integration over these elementary rods yields the expression for the magnitude of the flux density in the bore,

$$B = \mu_0 M \ln(r_2/r_1), \tag{2.23}$$

and $B = 0$ outside. Here $\boldsymbol{M}$ is the magnetization of the permanent magnet, assumed to be uniform in magnitude, and r_1 and r_2 are the inner and outer radii, respectively.

2.2.3 The *H*-field

Now we come to the $\boldsymbol{H}$-field, also known as the magnetic field strength or magnetizing force. It is an indispensable auxiliary field whenever we have to deal with magnetic or superconducting material. The magnetization of a solid reflects the local value of $\boldsymbol{H}$.

The distinction between $\boldsymbol{B}$ and $\boldsymbol{H}$ is trivial in free space. They are simply related by the magnetic constant μ_0:

$$\boldsymbol{B} = \mu_0 \boldsymbol{H}. \tag{2.24}$$

The previous discussion of fields created by currents and magnetic moments, and Maxwell's equations themselves could as easily have been cast in terms of $\boldsymbol{H}$ in place of $\boldsymbol{B}$, provided we remain in free space. Equations (2.5)–(2.13) become equations for $\boldsymbol{H}$ if we drop μ_0. The problem arises in a material medium with (2.16), which leads to Ampère's law. The curl of $\boldsymbol{B}$ is related to the *total* current density

$$\nabla \times \boldsymbol{B} = \mu_0(\boldsymbol{j}_c + \boldsymbol{j}_m), \tag{2.25}$$

where $\boldsymbol{j}_c$ is the conduction current in electrical circuits and $\boldsymbol{j}_m$ is the Ampèrian magnetization current associated with the magnetized medium. The difficulty is that $\boldsymbol{j}_c$ can be measured, but $\boldsymbol{j}_m$ cannot. We have no experimental method, direct or indirect, of precisely determining the currents circulating within a solid that create its magnetization. Indeed the nature of these huge currents is quantum mechanical; mostly they represent the intrinsic spin of the electron.

The relation between $\boldsymbol{j}_m$ and $\boldsymbol{M}$ is simply[3]

$$\boldsymbol{j}_m = \nabla \times \boldsymbol{M}. \tag{2.26}$$

In order to retain Ampère's law in a practically useful form, we will *define* a new field

$$\boldsymbol{H} = \boldsymbol{B}/\mu_0 - \boldsymbol{M}, \tag{2.27}$$

so that $\nabla \times \boldsymbol{H} = \nabla \times \boldsymbol{B}/\mu_0 - \nabla \times \boldsymbol{M}$, and hence from (2.25) and (2.26)

$$\nabla \times \boldsymbol{H} = \boldsymbol{j}_c. \tag{2.28}$$

[3] This equation follows from the fact that the magnetization is associated with bound, atomic-scale magnetization currents; this is expressed by the result $\int_s \boldsymbol{j}_m \cdot \mathrm{d}\mathcal{A} = 0$ over any surface. This implies that $\boldsymbol{j}_m$ must be expressible as the curl of another vector $\boldsymbol{M}$. By Stokes's theorem, $\oint \boldsymbol{M} \cdot \mathrm{d}\boldsymbol{l} = \int_s (\nabla \times \boldsymbol{M}) \mathrm{d}\mathcal{A}$, hence by choosing the path to lie entirely outside the magnetized body, it follows that $\int_s (\nabla \times \boldsymbol{M}) \mathrm{d}\mathcal{A} = 0$ for any surface, so we can identify $\boldsymbol{j}_m$ with $\nabla \times \boldsymbol{M}$.

In integral form, Ampère's law for the H-field produced by conduction currents is

$$\oint \boldsymbol{H} \cdot \mathrm{d}\boldsymbol{l} = I_c, \tag{2.29}$$

where I_c is the total conduction current threading the path of the integral. The new field is no longer divergenceless, but has sources and sinks associated with nonuniformity of the magnetization. From (1.1a) and (2.27),

$$\nabla \cdot \boldsymbol{H} = -\nabla \cdot \boldsymbol{M}. \tag{2.30}$$

This equation is the basis of the Coulombian approach to magnetic field calculations presented in §2.4. In particular, the discontinuity at the surface of a piece of magnetized material constitutes a sheet of sources or sinks of $\boldsymbol{H}$. We can imagine that $\boldsymbol{H}$, like the electric field $\boldsymbol{E}$, arises from a distribution of positive and negative magnetic charge q_m. The field emanating from a single charge would be

$$\boldsymbol{H} = q_m \boldsymbol{r}/4\pi r^3. \tag{2.31}$$

Units of q_m are A m. The magnitude of the field falls off as $1/r^2$. The force on the magnetic charge in a field is

$$\boldsymbol{f}_m = \mu_0 q_m \boldsymbol{H}. \tag{2.32}$$

These fictitious charges are the fabled North and South poles[4] which used to feature so prominently in magnetism texts and remain, with the red horseshoe, embedded in popular imagination. The poles have no physical existence, but they have coloured our thinking about magnetism for centuries. The No*r*th seeking pole (positively charged) of magnets was painted *r*ed, the So*u*th seeking (negatively charged) pole was painted bl*u*e. Rather than using the letters N and S to denote the approximate positions of nonexistent poles, it is preferable to represent a magnet by an arrow denoting its direction of magnetization, a convention we adopt here. That said, magnetic charges do offer a mathematically convenient way of representing the $\boldsymbol{H}$-field, and some force and field calculations become much simpler if we make use of them. Charge avoidance is a useful principle in magnetostatics.

The relation (2.28) does not imply that $\boldsymbol{H}$ is *only* created by conduction currents. Any magnet will produce an $\boldsymbol{H}$-field both in the space around it and within its own volume. We can write the field as the sum of two contributions

$$\boldsymbol{H} = \boldsymbol{H}_c + \boldsymbol{H}_m,$$

[4] There are conflicting conventions for naming the poles. The 'North seeking' or positive pole of a bar magnet is the end that points North. The Earth's 'South' magnetic pole is therefore the one near the geographic North pole, as the magnetic field direction runs from 'N' to 'S'. Gilbert, and some modern charlatans, adopt the opposite convention.

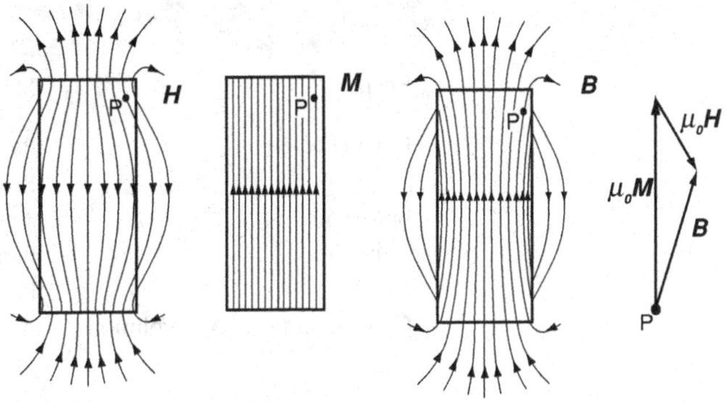

Figure 2.7
***B*, *M* and *H* for a magnet.**

where $\boldsymbol{H}_c$ is created by conduction currents and $\boldsymbol{H}_m$ is created by the magnetization distributions of other magnets and of the magnet itself. The second contribution is known as the stray field outside a magnet or as the demagnetizing field within it. It is represented by the symbol $\boldsymbol{H}_d$.

Equation (2.27) relating the fundamental field $\boldsymbol{B}$, the auxiliary field $\boldsymbol{H}$ and the magnetization $\boldsymbol{M}$ of the medium is normally rearranged as

$$\boldsymbol{B} = \mu_0(\boldsymbol{H} + \boldsymbol{M}). \tag{2.33}$$

In free space, $\boldsymbol{M} = 0$ and $\boldsymbol{B} = \mu_0 \boldsymbol{H}$. There $\boldsymbol{B}$ and $\boldsymbol{H}$ are indistinguishable, apart from the constant μ_0, which is so small that no confusion should ever arise between them. Units of $\boldsymbol{H}$, like those of $\boldsymbol{M}$, are A m^{-1}. One tesla is equivalent to 795 775 A m^{-1} (or approximately 800 kA m^{-1}). The quantities $\boldsymbol{B}$, $\boldsymbol{H}$ and $\boldsymbol{M}$ for a uniformly magnetized block of material in the absence of any external field are illustrated in Fig. 2.7, where the relation (2.33) between them is shown at a point ‘P’. Inside the magnet the $\boldsymbol{B}$-field and the $\boldsymbol{H}$-field are quite different, and oppositely directed. $\boldsymbol{H}$ is also oppositely directed to $\boldsymbol{M}$ inside the magnet, hence the name ‘demagnetizing field’. The field lines of $\boldsymbol{H}$ appear to originate on the horizontal surfaces of the magnet, where a magnetic charge of density $\sigma_m = \boldsymbol{M} \cdot \mathbf{e}_n$ resides; $\mathbf{e}_n$ is a unit vector normal to the surface. The $\boldsymbol{H}$-field is said to be *conservative* ($\nabla \times \boldsymbol{H} = \mathbf{0}$), whereas the $\boldsymbol{B}$-field, whose lines form continuous closed loops, is *solenoidal* ($\nabla \cdot \boldsymbol{B} = \mathbf{0}$).

When considering magnetization processes, $\boldsymbol{H}$ is chosen as the independent variable, $\boldsymbol{M}$ is plotted versus $\boldsymbol{H}$, and $\boldsymbol{B}$ is deduced from (2.33). The choice is justified because it is possible to specify $\boldsymbol{H}$ at points inside the material in terms of the demagnetizing field, acting together with the fields produced by external magnets and conduction currents.

2.2.4 The demagnetizing field

It turns out that in any uniformly magnetized sample having the form of an ellipsoid the demagnetizing field $\boldsymbol{H}_d$ is also uniform. The relation between $\boldsymbol{H}_d$

Table 2.1. Some demagnetizing factors for simple shapes

Shape	Magnetization direction	$\mathcal{N}$
Long needle	Parallel to axis	0
	Perpendicular to axis	1/2
Sphere	Any direction	1/3
Thin film	Parallel to plane	0
	Perpendicular to plane	1
General ellipsoid of revolution	$\mathcal{N}_c = (1 - 2\mathcal{N}_a)$	

and $\boldsymbol{M}$ is

$$H_{di} = -\mathcal{N}_{ij} M_j \qquad i, j = x, y, z, \tag{2.34}$$

where $\mathcal{N}_{ij}$ is the demagnetizing tensor, which is generally represented by a symmetric 3×3 matrix. A sum over the repeated index is implied. Along the principal axes of the ellipsoid, $\boldsymbol{H}_d$ and $\boldsymbol{M}$ are collinear and the principal components of $\mathcal{N}$ in diagonal form ($\mathcal{N}_x, \mathcal{N}_y, \mathcal{N}_z$) are known as demagnetizing factors. Only two of the three are independent because the demagnetizing tensor has unit trace:

$$\mathcal{N}_x + \mathcal{N}_y + \mathcal{N}_z = 1. \tag{2.35}$$

It is common practice to use a demagnetizing factor to obtain approximate internal fields, even in nonellipsoidal shapes such as cylinders and blocks where the demagnetizing field is not quite uniform. Demagnetizing factors for some simple shapes can be deduced by symmetry from (2.35). Examples are given in the Table 2.1.

Formulae for an ellipsoid of revolution having major axes (a, a, c) with $\alpha = c/a > 1$ and $\alpha = c/a < 1$ are

$$\mathcal{N}_c = \frac{1}{(\alpha^2 - 1)}\left[\frac{\alpha}{\sqrt{\alpha^2 - 1}}\cosh^{-1}(\alpha) - 1\right], \tag{2.36a}$$

$$\mathcal{N}_c = \frac{1}{(1 - \alpha^2)}\left[1 - \frac{\alpha}{\sqrt{1 - \alpha^2}}\cos^{-1}(\alpha)\right]. \tag{2.36b}$$

For nearly spherical shapes with $\alpha \approx 1$, $\mathcal{N}_c = \frac{1}{3} - \frac{1}{15}(\alpha - 1)$. Formulae valid in the limits $\alpha \gg 1$ and $\alpha \ll 1$ are $\mathcal{N}_c = (\ln 2\alpha - 1)/\alpha^2$ and $1 - \pi\alpha/2$, respectively. Figure 2.8(a) is a plot of the values of $\mathcal{N}_c$ for ellipsoids of revolution. Appendix D provides numerical values. From Table 2.1, the value of $\mathcal{N}_a$ is $\frac{1}{2}(1 - \mathcal{N}_c)$.

The fields in cylindrical and rectangular magnetized bodies are not uniform, even if the magnetization is assumed to be uniform. An effective demagnetizing factor can be obtained by representing the magnetization by a distribution of surface charge (§2.4) and calculating the field at the centre. Results for magnets of length $2c$ with circular and square cross sections of diameter or side $2a$

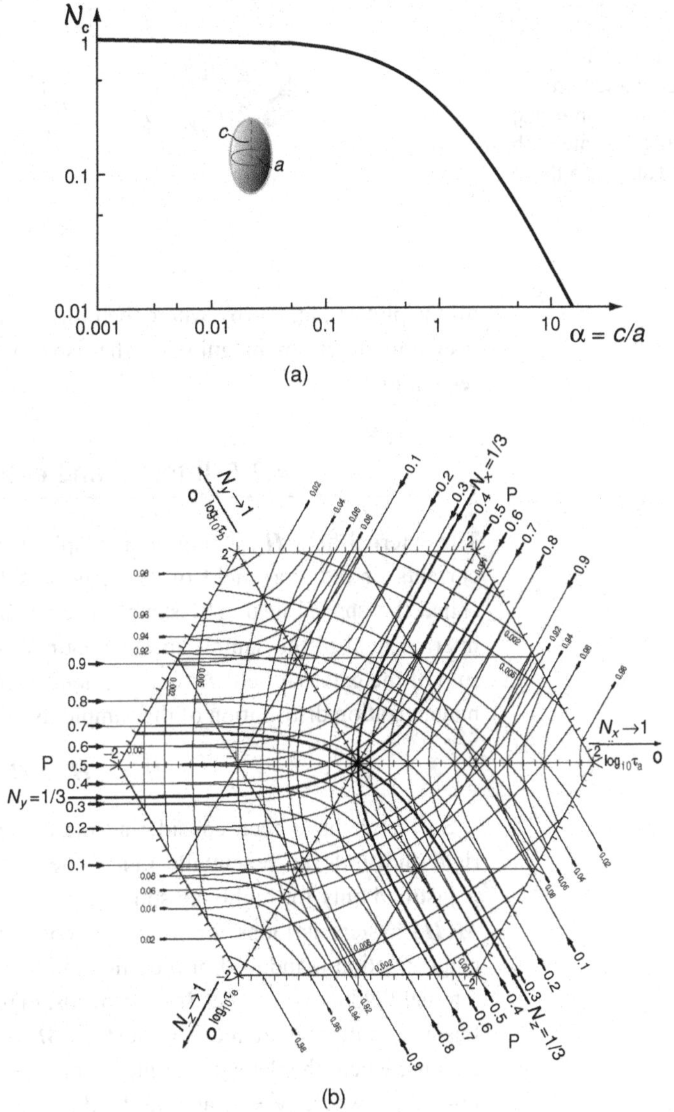

Figure 2.8

Demagnetizing factors for ellipsoids: (a) ellipsoids of revolution, (b) general ellipsoids. The letters O and P denote oblate and prolate ellipsoids. (M. Bellegia, M. de Graff and Y. Millev, *Phil. Mag.* 86, 2451 (2006)).

are $\mathcal{N}_c^{eff} = 1 - \alpha/\sqrt{\alpha^2+1}$ and $\mathcal{N}_c^{eff} = (2/\pi)\sin^{-1}[1/(1+\alpha)]$, respectively. Slightly different values are obtained from volume integrals over real, nonuniform magnetization distributions (e.g. Fig. 2.10(b)).

The general ellipsoid has major axes (a, b, c). Defining $\tau_a = a/c$, $\tau_b = b/c$ a general expression for $\mathcal{N}_c$ is

$$\mathcal{N}_c(\tau_a, \tau_b) = \frac{1}{2}\int_0^\infty \frac{1}{(1+u)^{3/2}(1+u\tau_a^2)^{1/2}(1+u\tau_b^2)^{1/2}}\mathrm{d}u.$$

The other principal components are obtained by rotation: $\mathcal{N}_a = \mathcal{N}_c(1/\tau_a, \tau_b/\tau_a)$, $\mathcal{N}_b = \mathcal{N}_c(\tau_a/\tau_b, 1/\tau_b)$. All three components can be read

Figure 2.9

Ways of measuring magnetization with no need for a demagnetizing correction: (a) a toroid, (b) a long rod and (c) a thin plate.

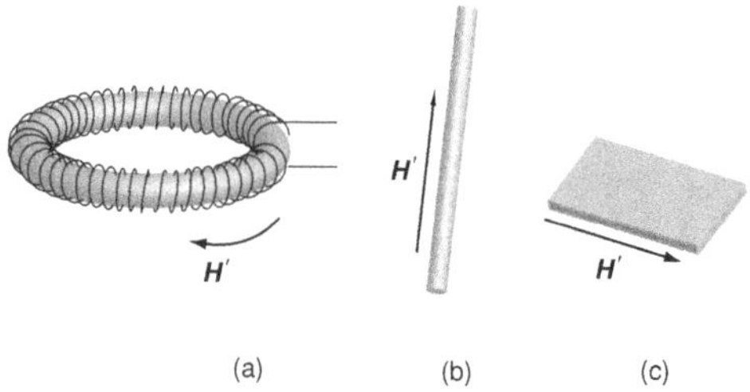

off the plot of Fig. 2.8(b), where the ratios τ are plotted on a logarithmic scale over four orders of magnitude. The axes in the plot represent ellipsoids of revolution.

2.2.5 Internal and external fields

The external field $\boldsymbol{H}'$, acting on a sample that is produced by steady electric currents or the stray field of magnets outside the sample volume, is often called the applied field. The sample itself makes no contribution to $\boldsymbol{H}'$. The internal field in the sample in our continuous medium approximation is the sum of the external field $\boldsymbol{H}'$ and the demagnetizing field $\boldsymbol{H}_d$ produced by the magnetization distribution of the sample itself:

$$\boldsymbol{H} = \boldsymbol{H}' + \boldsymbol{H}_d. \tag{2.37}$$

So far we have been considering the magnetization $\boldsymbol{M}$ of the material as rigid and uniform, essentially independent of the demagnetizing field. This is justified only for highly anisotropic permanent magnets having rectangular $\boldsymbol{M}(\boldsymbol{H})$ hysteresis loops for which the coercivity $\boldsymbol{H}_c > \boldsymbol{H}_d$. More generally, magnetization is induced or modified by the externally applied field $\boldsymbol{H}'$. The internal field in the magnet $\boldsymbol{H}(\boldsymbol{r})$ depends, in turn, on the magnetization $\boldsymbol{M}(\boldsymbol{r})$. Easiest to interpret are measurements of $\boldsymbol{M}(\boldsymbol{H})$ carried out in closed magnetic circuits, where the demagnetizing field is absent. An example is the toroid of Fig. 2.9(a) where $\mathcal{N} = 0$ and the field is that of a long solenoid $H = nI$. An alternative is to use a long bar or a thin film, and apply the field in the direction where $\mathcal{N} \approx 0$. If it is inconvenient to produce the sample in one of these forms, the best solution is to make it into a sphere for which the magnetization is uniform and the demagnetizing factor $\mathcal{N} = \frac{1}{3}$ is precisely known. Failing this, a cylindrical or block shape is used that can be approximately assimilated to an ellipsoid, and the applied field is corrected by the appropriate demagnetizing factor to obtain the internal field

$$\boldsymbol{H} \simeq \boldsymbol{H}' - \mathcal{N}\boldsymbol{M}. \tag{2.38}$$

The approximation is double: $\boldsymbol{H}$ is not uniform because the sample is not an ellipsoid, consequently $\boldsymbol{M}$ cannot be uniform either (Fig. 2.10).

Figure 2.10

Magnetization of (a) a ferromagnetic sphere and (b) a ferromagnetic cube in an applied field.

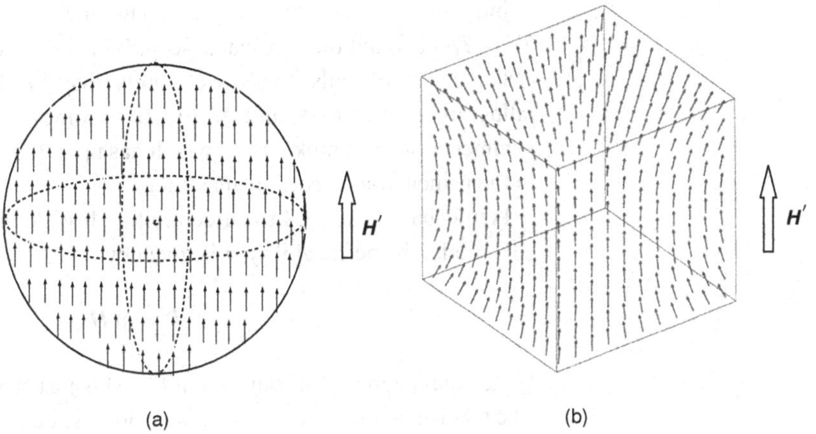

A common form of sample is a powder composed of irregularly shaped but roughly spherical particles. The demagnetizing field $\boldsymbol{H}_d(\boldsymbol{r})$ then fluctuates rapidly on the scale of the particle size. When the particles pack isotropically with a packing fraction $\mathfrak{f}$ into a sample holder having a shape with a demagnetizing factor $\mathcal{N}$, the effective demagnetizing factor for the powder sample is

$$\mathcal{N}_p \simeq \tfrac{1}{3} + \mathfrak{f}\left(\mathcal{N} - \tfrac{1}{3}\right). \tag{2.39}$$

Spatial fluctuations in $\boldsymbol{H}_d$ appear near protrusions and surface irregularities. The magnetization curves and hysteresis loops measured for any of these less-than-ideal shapes will evidently deviate from those determined for a fully dense, smooth toroid or sphere.

2.2.6 Susceptibility and permeability

The simplest materials are linear, isotropic and homogeneous (LIH). For magnetism, this means that the susceptibility or applied field is small and a small uniform magnetization is induced in the same direction as the external field:

$$\boldsymbol{M} = \chi' \boldsymbol{H}', \tag{2.40}$$

where χ' is a dimensionless scalar known as the external susceptibility. The magnetization is related to the internal field $\boldsymbol{H}$ (2.37) by

$$\boldsymbol{M} = \chi \boldsymbol{H}, \tag{2.41}$$

where χ is the internal susceptibility. It follows from (2.38) that

$$1/\chi = 1/\chi' - \mathcal{N}. \tag{2.42}$$

As long as χ is small, as it is for typical paramagnets and diamagnets ($\approx 10^{-5}$–10^{-2} and $\approx -10^{-5}$, respectively), the difference between χ' and χ can be safely neglected. Exceptions are superparamagnets and paramagnets close to the Curie point, and, of course, ferromagnets. The internal susceptibility χ diverges as $T \to T_C$ (1.8) but the external susceptibility χ' never exceeds $1/\mathcal{N}$.

For single crystals, the susceptibility may be different in different crystallographic directions, and $\boldsymbol{M} = \chi \boldsymbol{H}$ becomes a tensor relation with χ_{ij} a symmetric second-rank tensor, which has up to three independent components in the principal-axis coordinate system.

Permeability is related to susceptibility. It is defined in the internal field. In LIH media the permeability μ is given by

$$\boldsymbol{B} = \mu \boldsymbol{H}. \tag{2.43}$$

Hence the magnetic constant μ_0 in (2.24) is just the permeability of free space. The relative permeability μ_r is a dimensionless quantity defined as μ/μ_0. It follows from (2.33) and (2.43) that $\mu_r = 1 + \chi$. Permeability is usually discussed in relation to soft ferromagnetic materials, where μ_r can take very large values, up to 10^4 or more.

The relation between permeability of a soft magnetic material and magnetic field is akin to the relation between conductivity of a metal and electrical current. Flux density in a magnetic circuit is the analogue of current density in an electric circuit. A quantitative difference between magnetic and electric circuits, however, is that excellent insulators, including free space, exist to prevent current leakage, but free space is an imperfect magnetic insulator into which flux inevitably leaks. The magnetic equivalent of an insulator is a type I superconductor, where no flux penetrates, so its permeability is zero. The analogy between electric and magnetic circuits will be pursued in Chapter 13.

Consider the example of an isotropic, soft ferromagnetic sphere with high permeability and no hysteresis. The domains are supposed to be much smaller than the size of the sample, which has volume V. The macroscopic magnetization is averaged over a length scale greater than the domain size, so that the uniform average magnetization is $\boldsymbol{M} = \mathfrak{m}/V$, where $\mathfrak{m}$ is magnetic moment of the sphere. $\boldsymbol{M}$ is zero in the multidomain state that exists before the field is applied. The ideal soft material has $\chi \sim \infty$, hence the full spontaneous magnetization $\boldsymbol{M}_s$ is revealed in a very small internal field. Since $\mathcal{N} = \frac{1}{3}$ for a sphere, it follows from (2.42) that $\chi' = 3$, and an applied field $H' = \frac{1}{3}M_s$ is needed to achieve saturation, as shown in Fig. 2.11. During the magnetization process, $\boldsymbol{M}$ increases from zero to M_s in the external field, while the internal field remains zero.[5] The assumption of uniform macroscopic magnetization breaks down in small spheres which contain few domains.

The flux density induced by the applied field reaches the saturation value $B_s = \mu_0 M_s$ in an external field of $\frac{1}{3}M_s$ and an internal field of zero. Thereafter the $B(H')$ or $B(H)$ curve is linear with slope μ_0. Once the magnetization is

[5] An ideal soft ferromagnet behaves like a perfect conductor, where the external electric field $\boldsymbol{E}'$ is shielded by induced charges so that the internal field $\boldsymbol{E}$ is zero.

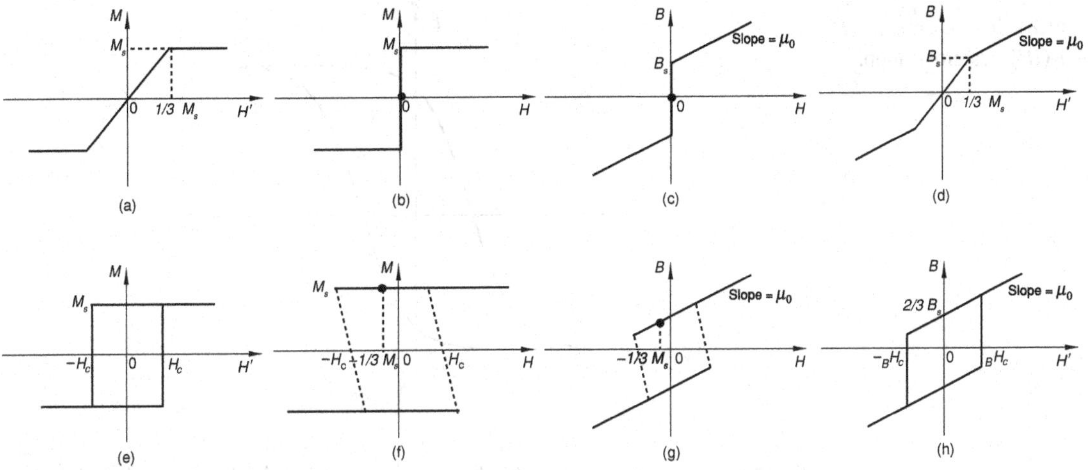

Figure 2.11

Magnetization, M, and induction, B, for a ferromagnetic sphere plotted as a function of the applied field, H', or the internal field, H. (a)–(d) are for a soft magnetic sphere and (e)–(h) are for a permanently magnetized sphere. The working point, where the sphere is subject only to its own demagnetizing field, is shown by the dot.

saturated, the magnetic medium becomes 'transparent', with the same permeability as free space.

The behaviour of a hard ferromagnetic sphere is quite different. It is permanently magnetized with $M = M_s$; in zero applied field, there is an internal field $H = -\frac{1}{3}M_s$ and the flux density is $B = \frac{2}{3}\mu_0 M_s = \frac{2}{3}B_s$ throughout the sphere. When a field is applied parallel to $\boldsymbol{M}$, $B(H)$ is linear, with slope μ_0. The working point of the magnetic sphere in its own demagnetizing field is marked in Fig. 2.11. A permanent magnet is one where the coercivity exceeds the demagnetizing field, giving a working point in the second quadrant.[6]

Generally, magnetic media are not linear, isotropic and homogeneous but nonlinear and hysteretic and often anisotropic and inhomogeneous as well! Then $\boldsymbol{B}$, like $\boldsymbol{M}$, is an irreversible and nonsingle-valued function of $\boldsymbol{H}$, represented by the $\boldsymbol{B}(\boldsymbol{H})$ hysteresis loop deduced from the $\boldsymbol{M}(\boldsymbol{H})$ loop using (2.33). A typical $\boldsymbol{B}(\boldsymbol{H}')$ loop is shown in Fig. 2.12. The coercivity on the $\boldsymbol{B}(\boldsymbol{H}')$ loop, denoted as ${}_BH_c$ is always less than or equal to H_c shown on the $\boldsymbol{M}(\boldsymbol{H}')$ loop in Fig. 1.3. The quantity H_c is sometimes (confusingly) called the 'intrinsic coercivity'. The switching for a macroscopic magnet is usually not the one-shot, square loop process assumed for the sphere in the previous example.

2.3 Maxwell's equations

Just as an auxiliary magnetic field is needed to account for a magnetically polarized medium, so an auxiliary electric field is needed to account for an

[6] Quadrants of a hysteresis loop are counted anticlockwise. The first is the one where M and H are both positive.

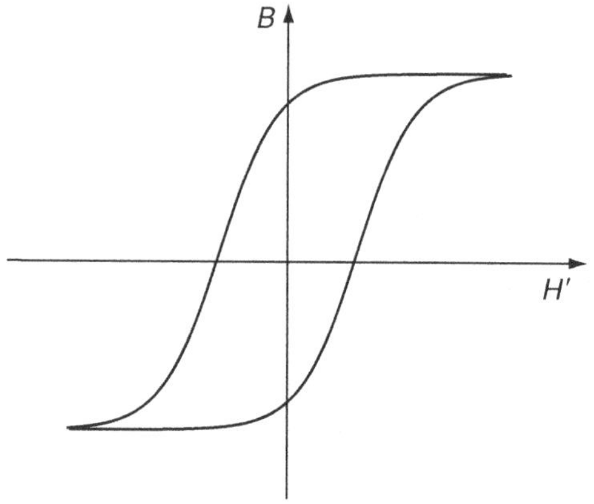

Figure 2.12
A $B(H')$ hysteresis loop.

electrically polarized one. This field is the electrical displacement $\boldsymbol{D}$, related to the electrical polarization $\boldsymbol{P}$ (electrical dipole moment per cubic metre) by

$$\boldsymbol{D} = \epsilon_0 \boldsymbol{E} + \boldsymbol{P}. \tag{2.44}$$

The constant ϵ_0, the permittivity of free space, is $1/\mu_0 c^2 = 8.854 \times 10^{-12}$ C V^{-1} m^{-1}. Engineers frequently brand the quantity $\boldsymbol{J} = \mu_0 \boldsymbol{M}$ as the magnetic polarization, where J, like B, is measured in tesla so that the relation between $\boldsymbol{B}$, $\boldsymbol{H}$ and $\boldsymbol{M}$ can be writted in a superficially similar form:

$$\boldsymbol{B} = \mu_0 \boldsymbol{H} + \boldsymbol{J}. \tag{2.45}$$

This is slightly misleading because the positions of the fundamental and auxiliary fields are reversed in the two equations. The form (2.27) $\boldsymbol{H} = \boldsymbol{B}/\mu_0 - \boldsymbol{M}$ is preferable insofar as it emphasizes the relation of the auxiliary magnetic field to the fundamental magnetic field and the magnetization of the material. Maxwell's equations in a material medium are expressed in terms of all four fields:

$$\nabla \cdot \boldsymbol{D} = \rho, \tag{2.46}$$

$$\nabla \cdot \boldsymbol{B} = 0, \tag{2.47}$$

$$\nabla \times \boldsymbol{E} = -\partial \boldsymbol{B}/\partial t, \tag{2.48}$$

$$\nabla \times \boldsymbol{H} = \boldsymbol{j} + \partial \boldsymbol{D}/\partial t. \tag{2.49}$$

Here ρ is the local electric charge density and $\partial \boldsymbol{D}/\partial t$ is the displacement current.

A consequence of writing the basic equations of electromagnetism in this neat and easy-to-remember form is that the constants μ_0, ϵ_0 and 4π are invisible. But they inevitably crop up elsewhere, for example in the Biot–Savart law (2.5). The fields naturally form two pairs: $\boldsymbol{B}$ and $\boldsymbol{E}$ are a pair, which appear in the Lorentz expression for the force on a charged particle (2.19), and $\boldsymbol{H}$ and $\boldsymbol{D}$ are the other pair, which are related to the field sources – free current density j and free charge density ρ, respectively.

In magnetostatics there is no time dependence of $\boldsymbol{B}$, $\boldsymbol{D}$ or ρ. We only have magnetic material and circulating conduction currents in a steady state. Conservation of electric charge is expressed by the equation,

$$\nabla \cdot \boldsymbol{j} = -\partial \rho / \partial t, \tag{2.50}$$

and in a steady state $\partial \rho / \partial t = 0$. The world of magnetostatics, which we explore further in Chapter 7, is therefore described by three simple equations:

$$\nabla \cdot \boldsymbol{j} = 0 \quad \nabla \cdot \boldsymbol{B} = 0 \quad \nabla \times \boldsymbol{H} = \boldsymbol{j}.$$

In order to solve problems in solid-state physics we need to know the response of the solid to the fields. The response is represented by the constitutive relations

$$\boldsymbol{M} = \boldsymbol{M}(\boldsymbol{H}), \quad \boldsymbol{P} = \boldsymbol{P}(\boldsymbol{E}), \; \boldsymbol{j} = \boldsymbol{j}(\boldsymbol{E}),$$

portrayed by the magnetic and electric hysteresis loops and the current–voltage characteristic. The solutions are simplified in LIH media, where $\boldsymbol{M} = \chi \boldsymbol{H}$, $\boldsymbol{P} = \epsilon_0 \chi_e \boldsymbol{E}$ and $\boldsymbol{j} = \sigma \boldsymbol{E}$ (Ohm's law), χ_e being the electrical susceptibility and σ the electrical conductivity. In terms of the fields appearing in Maxwell's equations, the linear constitutive relations are $\boldsymbol{B} = \mu \boldsymbol{H}$, $\boldsymbol{D} = \epsilon \boldsymbol{E}$, $\boldsymbol{j} = \sigma \boldsymbol{E}$, where $\mu = \mu_0(1 + \chi)$ and $\epsilon = \epsilon_0(1 + \chi_e)$. Of course, the LIH approximation is more-or-less irrelevant for ferromagnetic media, where $\boldsymbol{M} = \boldsymbol{M}(\boldsymbol{H})$ and $\boldsymbol{B} = \boldsymbol{B}(\boldsymbol{H})$ are the hysteresis loops in the internal field, which are related by (2.33).

2.4 Magnetic field calculations

In magnetostatics, the only sources of magnetic field are current-carrying conductors and magnetized material. The field from the currents at a point $\boldsymbol{r}$ in space is generally calculated from the Biot–Savart law (2.5). In a few situations with high symmetry such as the long straight conductor or the long solenoid, it is convenient to use Ampère's law (2.17) directly.

To calculate the field arising from a piece of magnetized material we are spoilt for choice. Alternative approaches are:

(i) calculate the dipole field directly by integrating over the volume distribution of magnetization $\boldsymbol{M}(\boldsymbol{r})$;
(ii) use the Ampèrian approach and replace the magnetization by an equivalent distribution of current density $\boldsymbol{j}_m$;
(iii) use the Coulombian approach and replace the magnetization by an equivalent distribution of magnetic charge q_m.

The three approaches are illustrated Fig. 2.13 for a cylinder uniformly magnetized along its axis. They yield identical results for the field in free space *outside* the magnetized material but not within it. The Amperian approach gives

Figure 2.13

Calculation of the magnetic field outside a uniformly magnetized cylinder by summing: (a) the fields produced by volume distribution of magnetic moments, (b) the fields produced by the distribution of currents, and (c) the fields produced by the distribution of magnetic charge.

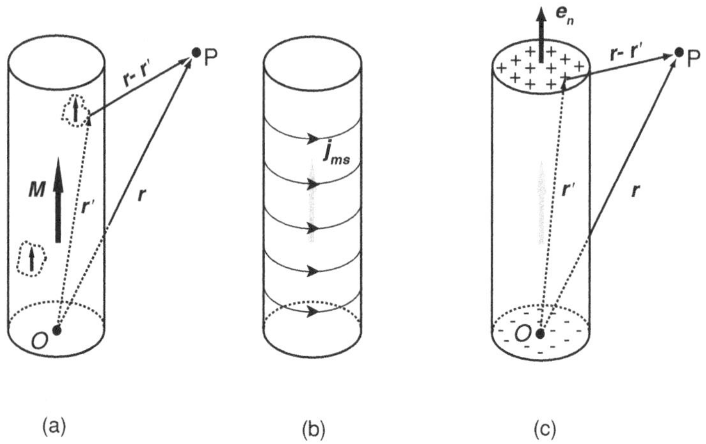

$\boldsymbol{B}$ correctly everywhere, whereas the Coulombian approach gives $\boldsymbol{H}$ correctly everywhere. The Coulombian approach is usually easiest from a computational point of view, especially when the field can be derived from a scalar potential, as explained in the next section.

The calculation for a distribution of magnetization $\boldsymbol{M}(\boldsymbol{r})$ proceeds by summing the fields produced by each magnetic dipole element $\boldsymbol{M}\mathrm{d}^3r$, given by (2.13). Hence

$$\boldsymbol{B}(\boldsymbol{r}) = \frac{\mu_0}{4\pi}\left[\int\left\{\frac{3\boldsymbol{M}(\boldsymbol{r}')\cdot(\boldsymbol{r}-\boldsymbol{r}')}{|\boldsymbol{r}-\boldsymbol{r}'|^5}(\boldsymbol{r}-\boldsymbol{r}') - \frac{\boldsymbol{M}(\boldsymbol{r}')}{|\boldsymbol{r}-\boldsymbol{r}'|^3} + \frac{2}{3}\mu_0\boldsymbol{M}(\boldsymbol{r}')\delta(\boldsymbol{r}-\boldsymbol{r}')\right\}\mathrm{d}^3r'\right]. \tag{2.51}$$

The last term is needed to take care of the divergence of the dipole field at the origin, $\boldsymbol{r}' = 0$. This approach yields the $\boldsymbol{B}$-field both inside and outside the solid medium.

The second approach considers the equivalent current distributions in the bulk and at the surface of the magnetized material:

$$\boldsymbol{j}_m = \nabla \times \boldsymbol{M} \text{ and } \boldsymbol{j}_{ms} = \boldsymbol{M} \times \mathbf{e}_n, \tag{2.52}$$

where $\mathbf{e}_n$ is the outward normal at a point on the surface. Using the Biot–Savart law (2.5) and adding the effects of the bulk and surface contributions to the current density,

$$\boldsymbol{B}(\boldsymbol{r}) = \frac{\mu_0}{4\pi}\left\{\int \frac{(\nabla' \times \boldsymbol{M}) \times (\boldsymbol{r}-\boldsymbol{r}')}{|\boldsymbol{r}-\boldsymbol{r}'|^3}\mathrm{d}^3r' + \int \frac{(\boldsymbol{M} \times \mathbf{e}_n) \times (\boldsymbol{r}-\boldsymbol{r}')}{|\boldsymbol{r}-\boldsymbol{r}'|^3}\mathrm{d}^2r'\right\}. \tag{2.53}$$

The first integral is zero for a uniform distribution of $\boldsymbol{M}$, because $\nabla' \times \boldsymbol{M}$ is then zero, and the calculation reduces to evaluating the integral of the surface current density. ∇' indicates differentiation with respect to $\boldsymbol{r}'$.

The third approach uses the equivalent distributions of magnetic charge in the bulk and at the surface of the magnetized material:

$$\rho_m = -\nabla \cdot \boldsymbol{M} \quad \text{and} \quad \sigma_m = \boldsymbol{M} \cdot \mathbf{e}_n. \tag{2.54}$$

From (2.31), the $\boldsymbol{H}$-field due to a small charged volume element δV is $\delta \boldsymbol{H} = (\rho_m \mathbf{r}/4\pi r^3)\delta V$. Hence

$$\boldsymbol{H}(\boldsymbol{r}) = \frac{1}{4\pi}\left\{-\int_V \frac{(\nabla' \cdot \boldsymbol{M})(\boldsymbol{r}-\boldsymbol{r}')}{|\boldsymbol{r}-\boldsymbol{r}'|^3}\mathrm{d}^3r' + \int_S \frac{\boldsymbol{M}\cdot\mathbf{e}_n(\boldsymbol{r}-\boldsymbol{r}')}{|\boldsymbol{r}-\boldsymbol{r}'|^3}\mathrm{d}^2r'\right\}. \tag{2.55}$$

Again, the first integral is zero for a uniform distribution of $\boldsymbol{M}$ because $\nabla \cdot \boldsymbol{M}$ is then zero. When the integrals are evaluated at a point $\boldsymbol{r}$ within the magnetized material, methods 1 and 2 yield $\boldsymbol{B}$, the solenoidal field, whereas method 3 yields $\boldsymbol{H}$, the conservative field.

Field calculations are frequently simplified by introducing a potential function, and taking the appropriate spatial derivative to obtain the field. Next we introduce the potentials corresponding to $\boldsymbol{B}$ and $\boldsymbol{H}$.

2.4.1 The magnetic potentials

Vector potential The flux density invariably satisfies $\nabla \cdot \boldsymbol{B} = 0$. Since the quantity $\nabla \cdot (\nabla \times \boldsymbol{A})$[7] is identically zero, we can always write

$$\boldsymbol{B} = \nabla \times \boldsymbol{A}, \tag{2.56}$$

where $\boldsymbol{A}$ is a magnetic vector potential. Units of A are T m. There is substantial latitude in the choice of vector potential for a given field. For example, a uniform field in the z-direction $(0, 0, B)$ may be represented by a vector potential $(0, xB, 0)$, or by $(-yB, 0, 0)$ or by $(-\frac{1}{2}yB, \frac{1}{2}xB, 0)$. Then $\boldsymbol{A}$ can be chosen so that its equipotentials are the lines of force. The definition of $\boldsymbol{A}$ is not unique, because it is permissible to add on the gradient of any arbitrary scalar function $f(\mathbf{r})$. Then if $\boldsymbol{A}' = \boldsymbol{A} + \nabla f(\boldsymbol{r})$, $\boldsymbol{B} = \nabla \times \boldsymbol{A}'$ because $\nabla \times \nabla f(\boldsymbol{r})$ is identically zero. $\boldsymbol{B}$ is unchanged by the transformation $\boldsymbol{A} \to \boldsymbol{A}'$, which is known as a gauge transformation. A useful gauge is the Coulomb gauge, where f is chosen so that $\nabla \cdot \boldsymbol{A} = 0$. A convenient expression for $\boldsymbol{A}$ in the Coulomb gauge is

$$\boldsymbol{A} = \tfrac{1}{2}\boldsymbol{B} \times \boldsymbol{r}. \tag{2.57}$$

It does not matter that the definition of $\boldsymbol{A}$ is not unique, because the observed effects depend on the magnetic field, not on the potential from which it is mathematically derived.

Since $(\boldsymbol{r}-\boldsymbol{r}')/|\boldsymbol{r}-\boldsymbol{r}'|^3 = -\nabla[1/|\boldsymbol{r}-\boldsymbol{r}'|]$, where ∇ refers to differentiation with respect to the field position $\boldsymbol{r}$, the expression for the field due to a distribution of currents obtained by integrating the Biot–Savart law (2.5) over the

[7] This, and other useful vector identities are collected in Appendix C.

variable $\boldsymbol{r}'$,

$$\boldsymbol{B}(\boldsymbol{r}) = \frac{\mu_0}{4\pi}\int \frac{(\boldsymbol{j}(\boldsymbol{r}') \times (\boldsymbol{r} - \boldsymbol{r}'))}{|\boldsymbol{r} - \boldsymbol{r}'|^3}\mathrm{d}^3r', \tag{2.58}$$

may be written in the form $\boldsymbol{B}(\boldsymbol{r}) = (\mu_0/4\pi)\nabla \times \int(\boldsymbol{j}(\boldsymbol{r}')/|\boldsymbol{r} - \boldsymbol{r}'|)\mathrm{d}^3r'$. Hence

$$\boldsymbol{A}(\boldsymbol{r}) = \frac{\mu_0}{4\pi}\int \frac{\boldsymbol{j}(\boldsymbol{r}')}{|\boldsymbol{r} - \boldsymbol{r}'|}\mathrm{d}^3r'. \tag{2.59}$$

In other words, a small contribution $\delta \boldsymbol{A}(r) = (\mu_0/4\pi)I\delta l/r$ to the vector potential is made by each small current element $I\delta l$ and $\boldsymbol{A}$, like $\boldsymbol{j}$ is a polar vector. Ampère's law $\nabla \times \boldsymbol{B} = \mu_0 \boldsymbol{j}$ can be written in terms of $\boldsymbol{A}$ as $\nabla \times (\nabla \times \boldsymbol{A}) = \mu_0 \boldsymbol{j}$. Since $\nabla \times (\nabla \times \boldsymbol{A}) = \nabla(\nabla \cdot \boldsymbol{A}) - \nabla^2 \boldsymbol{A}$, we see that in the Coulomb gauge, the vector potential satisfies Poisson's equation

$$\nabla^2 \boldsymbol{A} = -\mu_0 \boldsymbol{j}. \tag{2.60}$$

By expanding $1/|\boldsymbol{r} - \boldsymbol{r}'|$ as $(1/r)[1 + (r'/r)\cos\theta + \cdots]$, where θ is the angle between $\boldsymbol{r}$ and $\boldsymbol{r}'$, it follows that at large distances the vector potential for a magnetic moment $\mathfrak{m}$ equivalent to a current loop is

$$\boldsymbol{A}(\boldsymbol{r}) = \frac{\mu_0}{4\pi}\frac{\mathfrak{m} \times \boldsymbol{r}}{r^3}. \tag{2.61}$$

The expression for a distribution of magnetization $\boldsymbol{M}(\boldsymbol{r}')$ is

$$\boldsymbol{A}(\boldsymbol{r}) = \frac{\mu_0}{4\pi}\int \frac{\boldsymbol{M}(\boldsymbol{r}) \times (\boldsymbol{r} - \boldsymbol{r}')}{|\boldsymbol{r} - \boldsymbol{r}'|^3}\mathrm{d}^3r'. \tag{2.62}$$

Equations (2.59) and (2.62) specify the vector potential, and hence the magnetic flux density for any given distribution of magnetization or electric current. For example, the field due to a dipole is $\boldsymbol{B}(\boldsymbol{r}) = (\mu_0/4\pi)\nabla \times [(\mathfrak{m} \times \boldsymbol{r})/r^3]$ which can be shown to be equivalent to (2.13).

Scalar potential When the $\boldsymbol{H}$-field is produced only by magnets, and not conduction currents, it too can be expressed in terms of a potential. The field is then conservative, and Ampère's law (2.28) becomes $\nabla \times \boldsymbol{H} = 0$. Since $\nabla \times \nabla f(\boldsymbol{r}) = 0$ for any scalar $f(\boldsymbol{r})$, we can express H in terms of the magnetic scalar potential φ_m:

$$\boldsymbol{H} = -\nabla\varphi_m. \tag{2.63}$$

Units of φ_m are amperes. From (2.14) and (2.33), $\nabla \cdot (\boldsymbol{H} + \boldsymbol{M}) = 0$, so the scalar potential satisfies Poisson's equation:

$$\nabla^2\varphi_m = -\rho_m, \tag{2.64}$$

where the magnetic charge density $\rho_m = -\nabla \cdot \boldsymbol{M}$ in the bulk and $\sigma_m = \boldsymbol{M} \cdot \mathbf{e}_n$ at the surface. The potential due to a limited magnetized volume is

$$\varphi_m(r) = \frac{1}{4\pi}\left\{\int_v \frac{\rho_m}{|\boldsymbol{r} - \boldsymbol{r}'|}\mathrm{d}^3r' + \int_s \frac{\sigma_m}{|\boldsymbol{r} - \boldsymbol{r}'|}\mathrm{d}^2r'\right\}. \tag{2.65}$$

Figure 2.14

Boundary conditions of B and H at the interface between two media. $B_\perp$ and $H_\parallel$ are continuous.

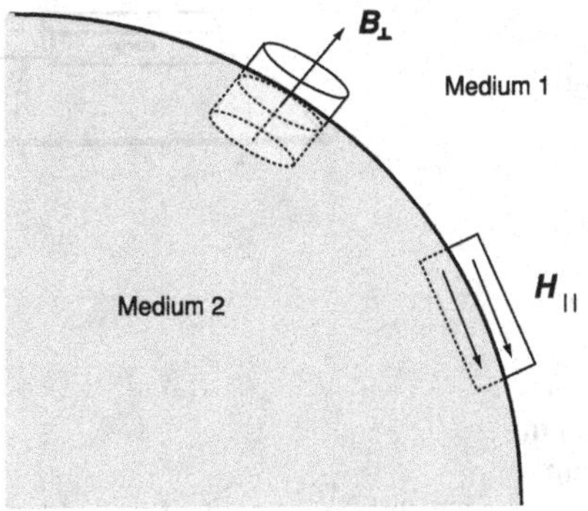

In other words, each small element of magnetic charge $\delta q_m = \rho_m \delta V$ makes a contribution $\delta\varphi_m = \rho_m \delta V / 4\pi r$ to the scalar potential. A small magnetic moment creates a potential $\mathfrak{m} \cdot \boldsymbol{r}/4\pi r^3$. Magnetostatic calculations are easier with the scalar potential, but it should be understood that it is only permissible to use it for problems where no conduction currents are present.

2.4.2 Boundary conditions

The two magnetic fields $\boldsymbol{B}$ and $\boldsymbol{H}$ satisfy different boundary conditions at an interface between two media. According to Gauss's law, (2.15), the integral of $\boldsymbol{B}$ over the surface of the thin flat 'pill box' in Fig. 2.14 is zero, hence

$$(\boldsymbol{B}_1 - \boldsymbol{B}_2) \cdot \mathbf{e}_n = 0. \tag{2.66}$$

The perpendicular component of $\boldsymbol{B}$ *is continuous.*

For $\boldsymbol{H}$, however, Ampère's law gives $\oint \boldsymbol{H} \cdot \mathrm{d}\boldsymbol{l} = 0$ for the circuit in Fig. 2.14. with two arms parallel to the boundary, provided there is no conduction current density at the surface. Hence

$$(\boldsymbol{H}_1 - \boldsymbol{H}_2) \times \mathbf{e}_n = 0. \tag{2.67}$$

The parallel component of $\boldsymbol{H}$ *is continuous.*

The corresponding boundary conditions on the potentials $\boldsymbol{A}$ and φ_m are obtained as follows. Since $\boldsymbol{B} = \nabla \times \boldsymbol{A}$, by Stokes's theorem the circulation of $\boldsymbol{A}$ is equal to the flux of $\boldsymbol{B}$; $\oint \boldsymbol{A} \cdot \mathrm{d}\boldsymbol{l} = \int_s \boldsymbol{B} \cdot \mathrm{d}\mathcal{A}$. Consider the small rectangular loop in Fig. 2.14, which is perpendicular to the surface; the flux of $\boldsymbol{B}$ though it is zero, hence the parallel component of $\boldsymbol{A}$ is continuous at the interface; hence

$$(\boldsymbol{A}_1 - \boldsymbol{A}_2) \times \mathbf{e}_n = 0. \tag{2.68}$$

The scalar potential is continuous at the interface:

$$\varphi_{m1} - \varphi_{m2} = 0. \tag{2.69}$$

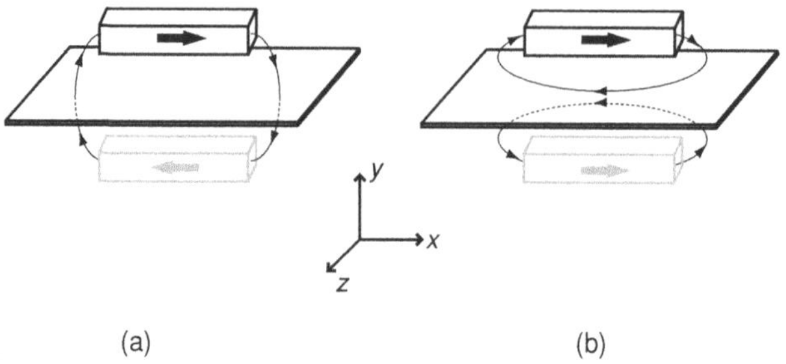

Figure 2.15

The image of a magnetic dipole in: (a) a soft ferromagnetic mirror $\mathfrak{m}_x \rightarrow -\mathfrak{m}_x$, $\mathfrak{m}_y \rightarrow -\mathfrak{m}_y$, $\mathfrak{m}_z \rightarrow \mathfrak{m}_z$ and (b) a superconducting mirror $\mathfrak{m}_x \rightarrow \mathfrak{m}_x$, $\mathfrak{m}_y \rightarrow \mathfrak{m}_y$, $\mathfrak{m}_z \rightarrow -\mathfrak{m}_z$. The magnetic field is perpendicular to the mirror surface in (a), but parallel to it in (b).

When we are dealing with LIH media where $\boldsymbol{B} = \mu_r \mu_0 \boldsymbol{H}$, the boundary condition (2.66) becomes $\boldsymbol{B}_1 \cdot \mathbf{e}_n = \boldsymbol{B}_2 \cdot \mathbf{e}_n$, or $\boldsymbol{H}_1 \cdot \mathbf{e}_n = (\mu_{r2}/\mu_{r1})\boldsymbol{H}_2 \cdot \mathbf{e}_n$. If medium 1 is air and medium 2 has a high permeability, this shows that the lines of magnetic field strength inside a highly permeable medium tend to lie parallel to the interface, whereas in air they tend to lie *perpendicular* to the interface. This is the reason why soft iron acts as a magnetic mirror as shown in Fig. 2.15. The reflected image of a magnetic moment is antiparallel to the magnet, and the magnet is attracted to its image in iron.

The situation is reversed if the iron is replaced by a sheet of superconductor. Ideally, the superconductor is a perfect diamagnet, with $\chi = -1$, into which flux does not penetrate. It follows that $\boldsymbol{B}$ is parallel to the surface, and the image is repulsive. Less than perfect diamagnets, like graphite for which $\chi \simeq -10^{-3}$, produce weakly repulsive images, which can nonetheless create a point where equilibrium in free space is stable, as in the levitation device illustrated in Fig. 15.4.

2.4.3 Local magnetic fields

Up to this point, we have worked in the continuous-medium approximation, assuming the magnetic material to be a homogeneous continuum with no atomic-scale structure. The local fluctuations of $\boldsymbol{B}$ and $\boldsymbol{M}$ on a 1 nm scale are smoothed out in the mesoscopic average (2.1). The $\boldsymbol{B}$- and $\boldsymbol{H}$-fields are supposed to be at most slowly varying within the solid. The minimum length-scale of the fluctuations in a ferromagnet is the exchange length, which is of order a few nanometres; this length scale is introduced in Chapter 7. In the example of a uniform, permanently magnetized sphere (Fig 2.10(a)), $\boldsymbol{B}$ and $\boldsymbol{H}$ are constant, equal to $(2/3)\mu_0\boldsymbol{M}$ and $-(1/3)\boldsymbol{M}$, respectively. In reality, solids are made up of atoms with a particular crystal structure, and we have the ability to probe the magnetic fields at the atomic nuclei experimentally via the hyperfine interactions. The question then arises: 'What is the value of the local magnetic field $\boldsymbol{H}_{loc}$ at a point in a solid?' The point may be an atomic site.

Figure 2.16

Calculation of the magnetism inside a solid. The sphere is the Lorentz cavity.

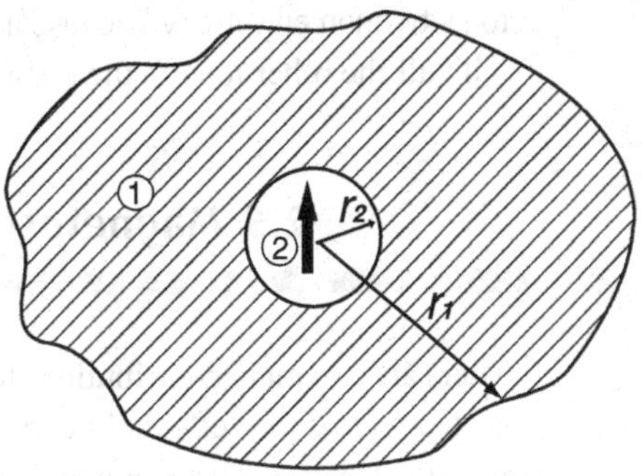

The calculation of $\boldsymbol{B}$ at any point $\boldsymbol{r}$ may be carried out in principle by replacing the integral (2.51) by a sum over atomic point dipoles $\mathfrak{m}_i$. Lorentz gave an argument to simplify the calculation. He divided the sample into two regions: region 1 which can be treated as a continuum, and the Lorentz cavity region 2, where the atomic-scale structure is taken into account (Fig. 2.16). Here $r_1 \gg r_2 \gg a$, where r_1 and r_2 are the average radii of the two regions, and a is the interatomic separation.

$$\boldsymbol{H}_{loc} = \boldsymbol{H}_1 + \boldsymbol{H}_2. \tag{2.70}$$

The Lorentz cavity is chosen to be spherical. The field due to region 1 can be evaluated from the distribution of surface charges $\sigma_m = \boldsymbol{M} \cdot \mathbf{e}_n$ on the inner and outer surfaces. Here we can use the continuum approximation to obtain $\boldsymbol{H}_1 = -\mathcal{N}\boldsymbol{M} + \frac{1}{3}\boldsymbol{M}$, where the second part

$$\boldsymbol{H}_L = \boldsymbol{M}/3, \tag{2.71}$$

is known as the Lorentz cavity field. The field $\boldsymbol{H}_2$ produced by the atoms contained within the cavity is evaluated as a dipole sum

$$\sum_i \left[\frac{3(\mathfrak{m}_i \cdot \mathbf{r}_i)\mathbf{r}_i}{r^5} - \frac{\mathfrak{m}_i}{r^3} \right].$$

The delta function in (2.51) is dropped unless we need the field acting at the nucleus. This atomic dipole sum can be expressed as $\boldsymbol{H}_{dip} = f_{dip}\boldsymbol{M}$, where f_{dip} is a geometric factor of order 1 which depends on the crystal structure of the lattice. In the particular case of a cubic latice, $f_{dip} = 0$, and the field $\boldsymbol{H}_2$ at the centre is zero. The contributions of all the dipoles on a cubic lattice exactly cancel out. When the sample itself is spherical, $\mathcal{N} = 1/3$ and $\boldsymbol{H}_1$ is also zero. There is therefore *no* local magnetic field at the centre of a uniformly-magnetized sphere of cubic material.

In general, f_{dip} is a tensor, so that $\boldsymbol{H}_{dip}$ depends on the direction of $\boldsymbol{M}$ relative to the crystal axes. This is a significant source of intrinsic magnetic anisotropy in non-cubic materials, since the interaction of $\boldsymbol{M}$ with $\boldsymbol{H}_{dip}$ defines certain easy directions of magnetization in the crystal. 'Intrinsic' here means that there is no dependence on the sample shape. This contribution is sometimes referred to as two-ion anisotropy because a central ion has a pairwise dipole interaction with all the other ions in the crystal.

2.5 Magnetostatic energy and forces

There are two main contributions to the energy of a ferromagnetic body: atomic-scale *electrostatic* effects like exchange or single-ion anisotropy, and *magnetostatic* effects. The magnetostatic effects, which involve the self-energy of interaction of the body with the field it creates by itself, as well as the interaction of the body with steady or slowly varying external magnetic fields, are considered here. Exchange and other electrostatic effects are the subject of Chapter 5.

The magnetostatic interactions are rather weak compared to the short-range exchange forces responsible for ferromagnetism, but they are important in ferromagnets nonetheless because the domain structure and magnetization process depends on them. It is the long-range nature of the dipole–dipole interaction, varying as r^{-3}, that allows these weak interactions to determine the magnetic microstructure. The magnetization of a typical ferromagnet is of order 1 MA m^{-1} and the demagnetizing fields $\boldsymbol{H}_d$ (2.37) though smaller, are similar in magnitude. Magnetostatic energies, $-\frac{1}{2}\mu_0 \boldsymbol{H}_d \cdot \boldsymbol{M}$, are therefore of order 10^6 J m^{-3}. An atomic volume is typically $(0.2\ \text{nm})^3$, so the corresponding energy per atom, 1×10^{-23} J, is equivalent to a temperature of about 1 K.

Generally, any product of a quantity with the dimensions of $\boldsymbol{B}$ with a quantity with the dimensions of $\boldsymbol{H}$ such as $\boldsymbol{B} \cdot \boldsymbol{H}$, $\boldsymbol{B} \cdot \boldsymbol{M}$, $\mu_0 H^2$ or $\mu_0 M^2$ has dimensions of energy per unit volume.

Magnetic fields do no work on electric currents or moving charges because the magnetic part of the Lorentz force $(\boldsymbol{j} \times \boldsymbol{B})$ per unit volume or $(\boldsymbol{v} \times \boldsymbol{B})$ per unit charge is always perpendicular to the motion. We cannot associate a potential energy function with the magnetic force. It is necessary to consider the work done by the transient electric fields when establishing a particular magnetic configuration $\boldsymbol{M}(\boldsymbol{r})$, or in setting up a particular current distribution $\boldsymbol{j}_c(\boldsymbol{r})$, in order to calculate the associated energy. Magnetostatic energy calculations can be rather subtle, and it is necessary to be clear about which energies are taken into account. For example, when a spanner is attracted towards an electromagnet, the energy of the spanner is lowered, but there are also modifications of the field both inside and outside the magnet, as well as the appearance of a demagnetizing field in the spanner. The energies associated with all these changes cancel out. Forces are nonetheless exerted, and damage may be done.

Let us first consider a small rigid magnetic dipole $\mathfrak{m}$ in a pre-existing steady field $\boldsymbol{B}$. We omit a prime, because the dipole has no internal structure. The

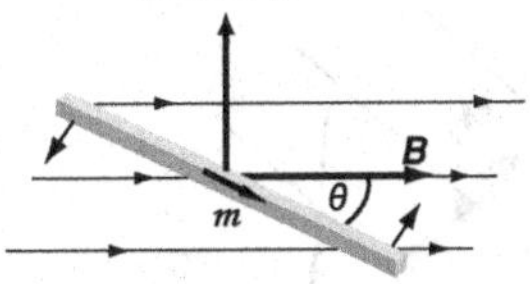

A magnet in a field. The energy is $-\mathrm{m}B\cos\theta$ and the torque is $\mathrm{m}B\sin\theta$.

moment experiences a torque $\boldsymbol{\Gamma}$ N m:

$$\boldsymbol{\Gamma} = \mathfrak{m} \times \boldsymbol{B}. \tag{2.72}$$

Taking the reference state at $\theta = 0$, where θ is the angle between $\mathfrak{m}$ and $\boldsymbol{B}$, and integrating the torque gives the 'potential energy' $\varepsilon_m = \int_0^\theta \Gamma \mathrm{d}\theta'$, which depends on *angle*. Apart from a constant,

$$\varepsilon_m = -\mathfrak{m} \cdot \boldsymbol{B}. \tag{2.73}$$

We assume that turning the magnetic dipole has no effect either on its moment or on the sources of $\boldsymbol{B}$. Equation (2.73) is the *Zeeman energy* of the magnetic moment in the external field. Although there is a torque, there is *no net force on a magnetic moment in a uniform field*; the 'potential energy' does not depend on position. However, if $\boldsymbol{B}$ is nonuniform, the energy of the dipole does depend on its position. The net force $\boldsymbol{f}_m = -\nabla\varepsilon_m$ resulting from (2.73) is

$$\boldsymbol{f}_m = \nabla(\mathfrak{m} \cdot \boldsymbol{B}). \tag{2.74}$$

The energy ε_m is minimized for a ferromagnet or paramagnet by this force tending to pull the material into a region where the field is greatest, but a diamagnet is pushed out to a region where the field is smallest.

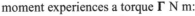

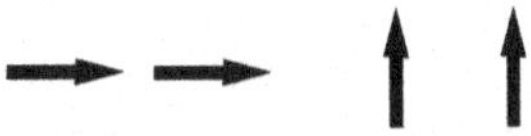

'Nose-to-tail' and 'broadside' configurations for a pair of magnetic dipoles; (a) is lower in energy and (b) is unstable.

Next we consider the mutual interaction of two parallel dipoles $\mathfrak{m}_1$ at $\mathbf{r}_1$ and $\mathfrak{m}_2$ at $\mathbf{r}_2$. The interaction between the pair ε_p may be considered as the energy of $\mathfrak{m}_1$ in the field $\boldsymbol{B}_{21}$ created by $\mathfrak{m}_2$ at $\mathbf{r}_1$ or vice versa, $\varepsilon_p = -\mathfrak{m}_1 \cdot \boldsymbol{B}_{21} = -\mathfrak{m}_2 \cdot \boldsymbol{B}_{12}$. Hence $\varepsilon_p = -\frac{1}{2}(\mathfrak{m}_1 \cdot \boldsymbol{B}_{21} + \mathfrak{m}_2 \cdot \boldsymbol{B}_{12})$.

The interaction of the parallel pair is anisotropic. From (2.10) and (2.73), $\varepsilon_p = -2\mu_0\mathfrak{m}^2/4\pi r^3$ in the 'nose-to-tail' configuration, whereas $\varepsilon_p = \mu_0\mathfrak{m}^2/4\pi r^3$ in the 'broadside' configuration. Here we have taken $\mathfrak{m}_1 = \mathfrak{m}_2$ and set $r = |\boldsymbol{r}_1 - \boldsymbol{r}_2|$. Hence freely suspended dipoles tend to aggregate in threads.

Reciprocity The two dipoles are an example of the reciprocity theorem, a useful result in magnetostatics, which states that the energy of interaction of two separate distributions of magnetization $\boldsymbol{M}_1$ and $\boldsymbol{M}_2$ producing fields $\boldsymbol{H}_1$ and $\boldsymbol{H}_2$ is

$$\varepsilon = -\mu_0 \int \boldsymbol{M}_1 \cdot \boldsymbol{H}_2 \mathrm{d}^3r = -\mu_0 \int \boldsymbol{M}_2 \cdot \boldsymbol{H}_1 \mathrm{d}^3r, \tag{2.75}$$

where ε is the interaction energy (Fig. 2.17). The reciprocity theorem is used to simplify magnetic energy calculations such as the interaction of a magnetic medium with a read head, for example.

2.5.1 Self-energy

These ideas can be extended to calculate the dipole–dipole interaction energy in a solid. We consider the energy of a body with magnetization $\boldsymbol{M}(\boldsymbol{r})$ in a

Figure 2.17

Mutual interaction of two distributions of magnetization.

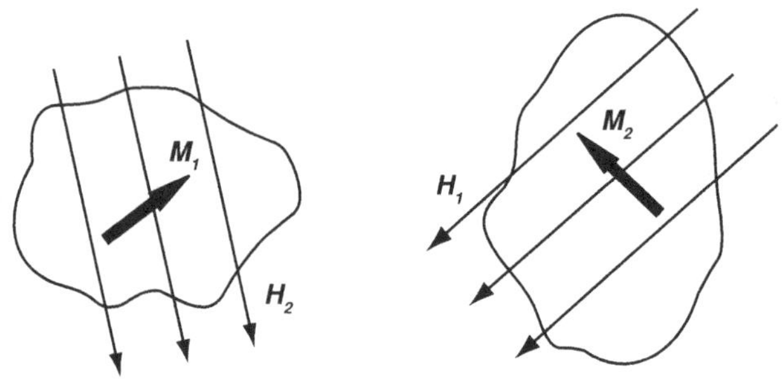

magnetic field. The result is different according to whether the field in question is an external field $\boldsymbol{H}'$ or the demagnetizing field $\boldsymbol{H}_d$ created by the body itself. Here we discuss the second case, considering first a small moment $\delta\mathfrak{m}$ at a point inside the macroscopic magnetized body. The energy needed to bring the small moment into position is $\delta\varepsilon = -\mu_0\delta\mathfrak{m}\cdot\boldsymbol{H}_{loc}$. We neglect $\boldsymbol{H}_2$ in (2.70) in the mesoscopic approximation, where

$$\boldsymbol{H}_{loc} = \boldsymbol{H}_d + \boldsymbol{H}_L. \tag{2.76}$$

Then, $\delta\varepsilon = -\mu_0\delta\mathfrak{m}\cdot(\boldsymbol{H}_d + \boldsymbol{H}_L)$. Since $\boldsymbol{H}_L = \frac{1}{3}\boldsymbol{M}$ (2.71), integration over the whole sample gives

$$\varepsilon = -\tfrac{1}{2}\int_V \mu_0\boldsymbol{H}_d\cdot\boldsymbol{M}\mathrm{d}^3r - \tfrac{1}{6}\int_V \mu_0 M^2\mathrm{d}^3r. \tag{2.77}$$

The factor of $\frac{1}{2}$ which always appears in expressions for the self-energy is needed to avoid double counting because each element $\delta\mathfrak{m}$ contributes as a field source and as a moment. This energy is plotted in Fig. 2.18 for a uniformly magnetized ellipsoid of revolution, for which $\varepsilon = \frac{1}{2}\mu_0 V(\mathcal{N} - \frac{1}{3})M^2$. The second term is actually unimportant; it tends to align the moments all in the same direction but it is much smaller than the exchange energy which has the same effect. The magnetostatic self-energy is conventionally defined as $\varepsilon_m = \varepsilon + \frac{1}{6}\int_v \mu_0 M^2\mathrm{d}^3r$ so that

$$\varepsilon_m = -\tfrac{1}{2}\int_V \mu_0\boldsymbol{H}_d\cdot\boldsymbol{M}\mathrm{d}^3r. \tag{2.78}$$

Since $\boldsymbol{M} = \boldsymbol{B}/\mu_0 - \boldsymbol{H}$, the integral may be written in an equivalent form:

$$\varepsilon_m = \tfrac{1}{2}\int \mu_0 H_d^2\mathrm{d}^3r, \tag{2.79}$$

where the integral is now over all space. We have used the handy result that for a magnet in its own field, when no currents are present,

$$\int \boldsymbol{B}\cdot\boldsymbol{H}_d\mathrm{d}^3r = 0, \tag{2.80}$$

where the integral is again over *all space*.[8]

[8] The proof is as follows: setting $\boldsymbol{B} = \nabla\times\boldsymbol{A}$ and using the vector identity $\boldsymbol{H}\cdot(\nabla\times\boldsymbol{A}) = \nabla\cdot(\boldsymbol{A}\times\boldsymbol{H}) + \boldsymbol{A}\cdot(\nabla\times\boldsymbol{H})$, where the second term is zero in the absence of conduction currents,

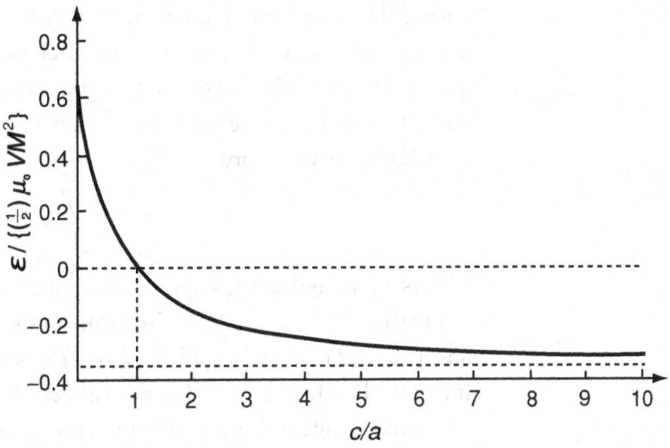

Figure 2.18

The dipolar self-energy for an ellipsoid of revolution.

This shows that the energy associated with a permanent magnet can be either associated with the integral of H_d^2 over all space (2.79), or with the integral of $-\boldsymbol{H}_d \cdot \boldsymbol{M}$ over the magnet (2.78), but not both. These are alternative ways of regarding the same energy term.

In the special case of a uniformly magnetized ellipsoid (2.78) gives

$$\varepsilon_m = \tfrac{1}{2}\mu_0 V \mathcal{N} M^2, \tag{2.81}$$

which is the same as the energy plotted in Fig. 2.18, apart from the constant term. The expression (2.78) assumes the magnetization is known in advance, so we can evaluate the magnetostatic energy in the field produced by the magnetization configuration. In practice, the magnetization tends to adopt a configuration which minimizes its self-energy. For an ellipsoid, there may be uniform magnetization along the axis where $\mathcal{N}$ is smallest.

2.5.2 Energy associated with a magnetic field

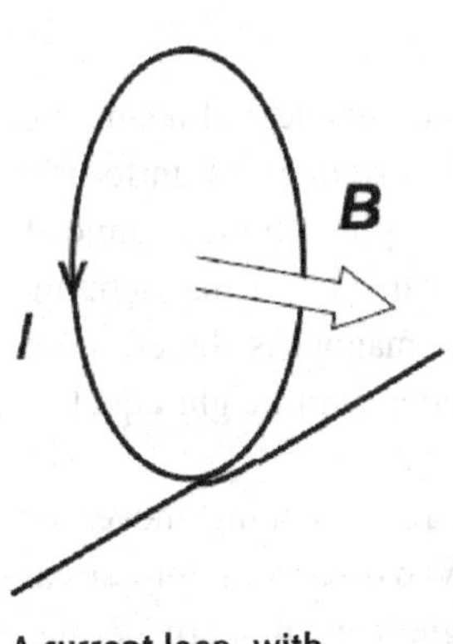

A current loop, with inductance L creates flux $\Phi = B\mathcal{A} = LI$.

An expression for the energy associated with a static magnetic field may be obtained by considering an inductor L consisting of a current loop which creates a flux $\Phi = LI$. By Faraday's law, $\mathcal{E} = -\mathrm{d}\Phi/\mathrm{d}t$ where $\mathcal{E}$ is the electromotive force (emf) developed in a circuit and Φ is the flux threading it, so the power needed to maintain a current I in the inductor is $-\mathcal{E}I = LI\mathrm{d}I/\mathrm{d}t$. Integrating from 0 to I gives an expression for the energy associated with the inductor: $\varepsilon = \frac{1}{2}LI^2 = \frac{1}{2}\Phi I$. The same energy can be associated with the field in space created by the current in the inductor. First the flux is expressed in terms of the vector potential $\boldsymbol{A}$ using Stokes's theorem $\int_s \boldsymbol{B} \cdot \mathrm{d}\mathcal{A} = \oint \boldsymbol{A} \cdot \mathrm{d}\boldsymbol{l}$ and the energy is written as $\varepsilon = \frac{1}{2}\oint I\boldsymbol{A} \cdot \mathrm{d}\boldsymbol{l}$. This can be generalized from a single

the integral $\int \boldsymbol{B} \cdot \boldsymbol{H}\mathrm{d}^3r = \int \boldsymbol{H} \cdot (\nabla \times \boldsymbol{A})\mathrm{d}^3r = \int \nabla \cdot (\boldsymbol{A} \times \boldsymbol{H})\mathrm{d}^3r$. By the divergence theorem, the volume integral can be converted to a surface integral $\int_s (\boldsymbol{A} \times \boldsymbol{H})\mathrm{d}^2r$. Far from the magnet, $A \sim 1/r^2$ and $H \sim 1/r^3$, so the integral over a surface of infinite radius is zero.

current loop to a continuous current density by replacing $I\mathrm{d}l$ by $\boldsymbol{j}\mathrm{d}^3r$ since $\mathrm{d}^3r = \mathrm{d}^2r\mathrm{d}l = \mathrm{d}\mathfrak{a} \cdot \mathrm{d}l$ and $\boldsymbol{j}$ is parallel to the area vector, $\mathfrak{a} \parallel \mathrm{d}\boldsymbol{l}$ therefore $\boldsymbol{j}\mathrm{d}^2r\mathfrak{a} \cdot \mathrm{d}\boldsymbol{l} = \boldsymbol{j} \cdot \mathfrak{a}\mathrm{d}^3r$ where $\boldsymbol{j}\mathrm{d}\mathfrak{a}$ is the current $\mathrm{d}I$ in a tube of cross section $\mathrm{d}\mathfrak{a}$. The general expression for the energy associated with a magnetic field distribution is therefore

$$\varepsilon = \tfrac{1}{2}\int \boldsymbol{j} \cdot \boldsymbol{A}\mathrm{d}^3r. \tag{2.82}$$

This is analogous to the expression for the electrostatic energy $\varepsilon = \frac{1}{2}\int \rho\varphi_e \mathrm{d}^3r$.

Finally, (2.82) is expressed in terms of the magnetic field. Since $\boldsymbol{j} = \nabla \times \boldsymbol{H}$ and $(\nabla \times \boldsymbol{H}) \cdot \boldsymbol{A} = \nabla \cdot (\boldsymbol{H} \times \boldsymbol{A}) + \boldsymbol{H} \cdot (\nabla \times \boldsymbol{A})$, there are two terms in the integral. The first is zero for a localized field source since by the divergence theorem the integral is equal to the flux of $\boldsymbol{H} \times \boldsymbol{A}$ through a surface at infinity. Hence, the only remaining term is $\varepsilon = \frac{1}{2}\int \boldsymbol{H} \cdot \boldsymbol{B}\mathrm{d}^3r$: in free space this becomes

$$\varepsilon = \tfrac{1}{2}\int \mu_0 H^2 \mathrm{d}^3r. \tag{2.83}$$

The local energy density associated with the magnetic field is $\frac{1}{2}\mu_0 H^2$. This is actually a general statement irrespective of whether the field is created by electric currents or magnetic material (2.79).

When designing magnetic circuits that include permanent magnets, the aim is usually to maximize the energy associated with the field created by the magnet in the space around it. From (2.78) and (2.79)

$$\tfrac{1}{2}\int \mu_0 H_d^2 \mathrm{d}^3r = -\tfrac{1}{2}\int_V \mu_0 \boldsymbol{H}_d \cdot \boldsymbol{M}\mathrm{d}^3r. \tag{2.84}$$

which may be rewritten as

$$\tfrac{1}{2}\int_o \mu_0 H_d^2 \mathrm{d}^3r = -\tfrac{1}{2}\int_i \mu_0 H_d^2 \mathrm{d}^3r - \tfrac{1}{2}\int_i \mu_0 \boldsymbol{M} \cdot \boldsymbol{H}_d \mathrm{d}^3r, \tag{2.85}$$

Ideal shape for a permanent magnet.

where the indices o and i indicate integrals over space outside and inside the magnet. The integral on the left is the one to be maximized. For a uniformly magnetized ellipsoid, the sum of the two integrals on the right over the volume of the magnet is $-\frac{1}{2}\mu_0 M^2(\mathcal{N}^2 - \mathcal{N})$; this is maximum when the demagnetizing factor $\mathcal{N}$ equals $\frac{1}{2}$. The ideal shape for a permanent magnet is therefore an ellipsoid of revolution with $c/a = 0.5$. A squat cylinder with height equal to radius is almost as good.

If you dismantle some consumer electronics where something moves or makes a noise, an earphone for example, you are likely to discover a cylindrical magnet of approximately this shape. Alas, the magnificent modern permanent magnet, shaped like a pill, has none of the iconic value of a horseshoe! From (2.33) and (2.85), the energy stored in the field outside the magnet, on the left-hand side of (2.85), is equal to $-\frac{1}{2}\int_i \mu_0 \boldsymbol{B} \cdot \boldsymbol{H}_d \mathrm{d}^3r$. The integral $-\int_i \mu_0 \boldsymbol{B} \cdot \boldsymbol{H}_d \mathrm{d}^3r$ is known as the energy product. It is *twice* the energy stored in the stray field of the magnet. Again, the usable energy may be associated

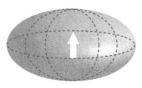

either with the stray field $\frac{1}{2}\int_o \mu_0 H_d^2 \mathrm{d}^3r$, or with the demagnetizing field in the magnet itself $-\frac{1}{2}\int_i \mu_0 \boldsymbol{B}\cdot\boldsymbol{H}_d \mathrm{d}^3r$, but not both. The two terms have the same absolute values, but their sum is zero. It is a matter of taste.

2.5.3 Energy in an external field

To extend the energy expression of a point dipole in an external field (2.73) to ferromagnetic materials, it is not adequate simply to fill the inductor with a medium of permeability μ, because ferromagnets are generally nonlinear; $\boldsymbol{B} \neq \mu\boldsymbol{H}$. They exhibit hysteresis, so that the energy needed to prepare a state described by $\boldsymbol{B}$ and $\boldsymbol{H}$ depends on the path followed. We can, however, evaluate the increment of work δw done to produce a small change in flux $\delta\Phi$; $\delta w = -\mathcal{E}I\delta t = I\delta\Phi$. But, by Ampère's law $I = \oint \boldsymbol{H}\cdot\mathrm{d}\boldsymbol{l}$ where the integral is taken around a closed loop. Hence $\delta w = \oint \delta\Phi\boldsymbol{H}\cdot\mathrm{d}\boldsymbol{l}$. This is generalized to the expression for magnetic work done on the system

$$\delta w = \int \boldsymbol{H}\cdot\delta\boldsymbol{B}\mathrm{d}^3r, \tag{2.86}$$

where the integral is over all space. When the magnetization is uniform, this expression becomes

$$\delta W = \boldsymbol{H}\cdot\delta\boldsymbol{B}, \tag{2.87}$$

where δW is the energy increment per unit volume.

More generally, we would love to have an expression for the energy of the magnetization distribution $\boldsymbol{M}(\boldsymbol{r})$ in the *external, applied field* $\boldsymbol{H}'$, which is supposedly undeformed by the presence of the magnetic material. It is rather unrealistic, even for good permanent magnets, to assume a perfectly uniform magnetization. $\boldsymbol{M}(\boldsymbol{r})$ is modified by the field applied during the magnetization process, and it depends on the value of $\boldsymbol{H}(\boldsymbol{r})$ throughout the body. We do not know what $\boldsymbol{H}(\boldsymbol{r})$ really looks like inside the magnet. It is not obvious that it is possible to find such an energy expression in terms of $\boldsymbol{H}'$. The real $\boldsymbol{H}$-field which is present throughout the material, and the one which features in Maxwell's equations, is (2.37)

$$\boldsymbol{H} = \boldsymbol{H}' + \boldsymbol{H}_d,$$

where $\boldsymbol{H}_d$ is the demagnetizing field produced by the material itself. The basic constitutive relation for the material is $\boldsymbol{M} = \boldsymbol{M}(\boldsymbol{H})$ not $\boldsymbol{M} = \boldsymbol{M}(\boldsymbol{H}')$, precisely because we did not want it to depend on extraneous features like sample shape.

The applied field $\boldsymbol{H}'$ is supposed to be created by some external current distribution $\boldsymbol{j}'$. It satisfies $\nabla\times\boldsymbol{H}' = \boldsymbol{j}'$ and $\nabla\cdot\boldsymbol{H}' = 0$ when there are no currents in the region of interest. The mesoscopic magnetostatic field $\boldsymbol{H}_d$ created by the body satisfies $\nabla\cdot\boldsymbol{H}_d = -\nabla\cdot\boldsymbol{M}$ and $\nabla\times\boldsymbol{H}_d = 0$. The corresponding induction is $\boldsymbol{B} = \mu_0(\boldsymbol{H}+\boldsymbol{M}) = \mu_0(\boldsymbol{H}'+\boldsymbol{H}_d+\boldsymbol{M})$. The general expression for the magnetic work done in changing the induction by $\delta\boldsymbol{B}$ is given by (2.86) with $\boldsymbol{H} = \boldsymbol{H}' + \boldsymbol{H}_d$. This expression includes a term associated with the

$\boldsymbol{H}$-field in empty space $\int \mu_0 \boldsymbol{H}' \delta \boldsymbol{H}' \mathrm{d}^3 r$, which is unrelated to the magnetic body, and should be subtracted to leave an expression for the work associated with the energy changes of the magnetized body alone. The relevant magnetic work is therefore

$$\delta w' = \int (\boldsymbol{H} \cdot \delta \boldsymbol{B} - \mu_0 \boldsymbol{H}' \cdot \delta \boldsymbol{H}') \mathrm{d}^3 r. \tag{2.88}$$

Using the expressions (2.37) and (2.33) for $\boldsymbol{H}$ and $\boldsymbol{B}$, $\boldsymbol{H} \cdot \delta \boldsymbol{B} = \mu_0(\boldsymbol{H}' + \boldsymbol{H}_d) \cdot (\delta \boldsymbol{H}' + \delta \boldsymbol{H}_d + \delta \boldsymbol{M})$. Hence

$$\delta w' = \mu_0 \left[\int \delta(\boldsymbol{H}' \cdot \boldsymbol{H}_d) \mathrm{d}^3 r + \int \boldsymbol{H}_d \delta \boldsymbol{H}_d \mathrm{d}^3 r + \int \boldsymbol{H} \cdot \delta \boldsymbol{M} \mathrm{d}^3 r \right]. \tag{2.89}$$

The first integral is zero for the same reason that $\int \boldsymbol{B} \cdot \boldsymbol{H}_d \mathrm{d}^3 r$ is zero over all space (2.80) ($\nabla \cdot \boldsymbol{H}' = 0$; $\nabla \times \boldsymbol{H}_d = 0$). The second integral is the contribution to the magnetostatic self-energy (2.79),

$$\delta w' = \delta \varepsilon_m + \mu_0 \int_V \boldsymbol{H} \cdot \delta \boldsymbol{M} \, \mathrm{d}^3 r,$$

where the integral is over the volume of the magnet as $\boldsymbol{M} = 0$ elsewhere. This expression relates the magnetic energy to the self-energy and the constitutive relation $\boldsymbol{M} = \boldsymbol{M}(\boldsymbol{H})$. From (2.78)

$$\delta \varepsilon_m = -\tfrac{1}{2} \mu_0 \int_V (\boldsymbol{H}_d \cdot \delta \boldsymbol{M} + \boldsymbol{M} \cdot \delta \boldsymbol{H}_d) \mathrm{d}^3 r, \tag{2.90}$$

$$\delta \varepsilon_m = -\mu_0 \int_V \boldsymbol{H}_d \cdot \delta \boldsymbol{M} \, \mathrm{d}^3 r, \tag{2.91}$$

by reciprocity. From (2.37), (2.89) and (2.91).

$$\delta w' = \mu_0 \int_V \boldsymbol{H}' \cdot \delta \boldsymbol{M} \, \mathrm{d}^3 r. \tag{2.92}$$

Equation (2.92) is the expression we sought for work done on the body by the applied field. The integral is over the body. Even though $\boldsymbol{H}'$ is not the field appearing in the constitutive relations, the work is expressed in terms of the quantities of direct physical interest, the applied field $\boldsymbol{H}'$ and the magnetization $\boldsymbol{M}$. This expression reduces to

$$\delta W' = \mu_0 \boldsymbol{H}' \cdot \delta \boldsymbol{M}, \tag{2.93}$$

where δW is the energy increment per unit volume.

If we consider a magnetic material with the $\boldsymbol{B}(\boldsymbol{H}')$ and $\boldsymbol{M}(\boldsymbol{H}')$ curves shown in Fig. 2.19, the integrals are as indicated in the figure caption. The energy $\int_0^M \mu_0 \boldsymbol{H}' \cdot \mathrm{d}\boldsymbol{M}$ expended to magnetize a sample is related to its anisotropy energy, including shape anisotropy, since the magnetization process in the external field depends on the orientation of the sample.

If the $\boldsymbol{M}(\boldsymbol{H}')$ relation is hysteretic, energy is expended on cycling the field $\boldsymbol{H}'$. For example, Fig. 2.19(b) shows the energy expended to magnetize the

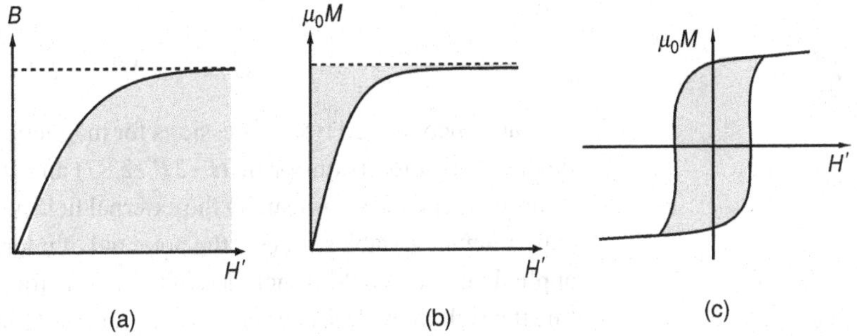

Figure 2.19

The energy associated with the magnetization curve. The shaded area in (a) indicates the energy associated with the applied field, whereas that in (b) represents the work done to magnetize the material, and that in (c) represents the hysteresis energy loss per cycle.

sample to remanence, and Fig. 2.19(c) shows the hysteresis loss on a complete cycle $\oint \mu_0 \boldsymbol{H}' \cdot \mathrm{d}\boldsymbol{M}$.

An expression for the energy needed to magnetize an LIH *paramagnetic* material in an external field may be deduced. Here the moment is induced by the field, according to (2.40). Hence, from (2.93)

$$W = \int_0^M \mu_0 H' \mathrm{d}M' = \tfrac{1}{2}\mu_0 M H'. \tag{2.94}$$

2.5.4 Thermodynamics of magnetic materials

In thermodynamics, the first law is written in terms of pairs of conjugate variables H_X and X, where H_X represents some external action on the system, and X is a state variable. The work done on the system is $H_X \mathrm{d}X$, and the first law is written

$$\mathrm{d}U = H_X \mathrm{d}X + \mathrm{d}Q, \tag{2.95}$$

where each term is an energy per unit volume. U refers to the internal energy of the system, and $\mathrm{d}Q$ is the heat absorbed by the system in the transformation. It is expressed in terms of entropy S, which for a reversible transformation is given by $\mathrm{d}Q = T\mathrm{d}S$. Here, T and S are the conjugate variables for thermal energy. The system is usually defined by fixing one variable in each of the (T, S) and (H_X, X) pairs. Four thermodynamic potentials can be defined by fixing two variables experimentally and leaving the other two variables free. The potentials are internal energy $U(X, S)$, enthalpy $E(H_X, S)$, Helmholtz free energy $F(X, T)$ and Gibbs free energy $G(H_X, T)$. All have units of J m^{-3} Thermodynamic equilibrium is reached when the appropriate potential reaches a minimum. When T is fixed, as is often the case, the relevant potentials are $F = U - TS$ and $G = F - H_X X$, for which

$$\mathrm{d}F = H_X \mathrm{d}X - S\mathrm{d}T \tag{2.96}$$

and

$$dG = -X dH_X - S dT. \tag{2.97}$$

We have encountered two expressions for magnetic work per unit volume when the magnetization is uniform, $\boldsymbol{H} \cdot \delta \boldsymbol{B}$ (2.87) and $\mu_0 \boldsymbol{H}' \cdot \delta \boldsymbol{M}$ (2.93). The latter is more practical, as it is usually the external field and temperature which are the experimental variables, so G is the potential of interest. The energy stored in the applied magnetic field is included in U. In transformations at constant temperature the Helmholtz free energy $F(M, T)$ and the Gibbs free energy $G(H', T)$ are related by

$$G = F - \mu_0 \boldsymbol{H}' \cdot \boldsymbol{M}. \tag{2.98}$$

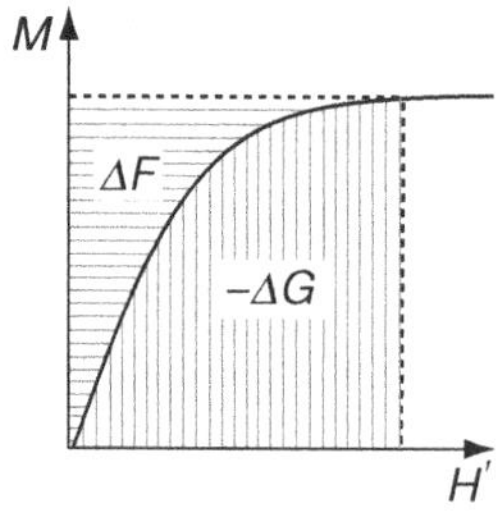

Changes in the thermodynamic free energies *F* and *G* associated with a reversible magnetization curve.

At thermodynamic equilibrium

$$dF = \mu_0 \boldsymbol{H}' \cdot d\boldsymbol{M} - S dT \tag{2.99}$$

and

$$dG = -\mu_0 \boldsymbol{M} \cdot d\boldsymbol{H}' - S dT \tag{2.100}$$

The changes in F and G at constant temperature are associated with the areas under the reversible $\boldsymbol{H}'(\boldsymbol{M})$ or $-\boldsymbol{M}(\boldsymbol{H}')$ curves. In the case of an LIH medium, the changes of Helmholtz free energy and Gibbs free energy on magnetizing the medium are $\frac{1}{2}\mu_0 M H'$ and $-\frac{1}{2}\mu_0 M H'$ respectively.

The spontaneous magnetization of a ferromagnet falls with increasing temperature. The fall becomes precipitous just below the Curie point, where the entropy of the spin system increases rapidly as the spin moments become disordered. In this temperature range, large entropy changes can be produced by modest applied fields. The entropy and magnetization of the ferromagnet are obtained as partial derivatives of the Gibbs free energy (2.100).

$$S = -\left(\frac{\partial G}{\partial T}\right)_{H'}, \qquad \mu_0 M = -\left(\frac{\partial G}{\partial H'}\right)_T. \tag{2.101}$$

Since $\delta Q = T\delta S$, the specific heat of magnetic origin C_m is equal to $-T(\partial^2 G/\partial T^2)_{H'}$.

Moreover, from the second derivatives of the four thermodynamic potentials, four Maxwell relations are obtained. For example, from the Gibbs free energy, using the fact that $(\partial^2 G/\partial H' \partial T) = (\partial^2 G/\partial T \partial H')$,

$$\left(\frac{\partial S}{\partial H'}\right)_T = \mu_0 \left(\frac{\partial M}{\partial T}\right)_{H'}. \tag{2.102}$$

According to the third law of thermodynamics, the entropy of the system tends to zero as $T \to 0$, regardless of magnetic field. It follows that $\partial S/\partial H' \to 0$ as $T \to 0$, hence $\partial M/\partial T \to 0$ as $T \to 0$. The temperature variation of the magnetization at $T = 0$ must have zero slope.

Another useful thermodynamic quantity in magnetism is the chemical potential, defined as $\mu = \partial G/\partial n$, where n is the number density of particles (electrons) in the system. The chemical potential is the increase of energy on adding an extra electron to the system. In metals, it is practically equivalent to the Fermi energy. The chemical potentials of spin-up and spin-down electrons will be different in an applied magnetic field. The spin-dependent chemical potential is used to analyse phenomena in spin electronics such as spin accumulation and giant magnetoresistance.

2.5.5 Magnetic forces

Forces in thermodynamics are related to the gradient of the free energy, which represents the ability of the system to do work. The Gibbs free energy when H' and T are the independent variables is $G = U - TS - \mu_0 \boldsymbol{H}' \cdot \boldsymbol{M}$. From (2.98), the force density due to a nonuniform field acting at constant temperature on a magnetized body $\boldsymbol{F}_m = -\nabla G$ is

$$\boldsymbol{F}_m = \nabla(\mu_0 \boldsymbol{H}' \cdot \boldsymbol{M}). \tag{2.103}$$

Using an identity for $\nabla(\boldsymbol{A} \cdot \boldsymbol{B})$ (Appendix C), the expression takes a simpler form when $\boldsymbol{M}$ is uniform and independent of $\boldsymbol{H}'$ ($\nabla \times \boldsymbol{M} = 0$), and no currents are present ($\nabla \times \boldsymbol{H}' = 0$). The first term on the right is zero because the curl is zero:

$$\nabla(\boldsymbol{H}' \cdot \boldsymbol{M}) = (\boldsymbol{H}' \cdot \nabla)\boldsymbol{M} + (\boldsymbol{M} \cdot \nabla)\boldsymbol{H}'$$

$$\boldsymbol{F}_m = \mu_0(\boldsymbol{M} \cdot \nabla)\boldsymbol{H}'. \tag{2.104}$$

This is known as the Kelvin force. When $\boldsymbol{M}$ is parallel to the z-direction and the change in $\boldsymbol{H}'$ is also in the z-direction, the expression is $\boldsymbol{F}_z = \mu_0 M(\partial H'/\partial z)$. In any case, the force is always in the direction of the gradient of the magnitude of the applied field. A general expression for the force density when $\boldsymbol{M}$ is not independent of $\boldsymbol{H}$ is

$$\boldsymbol{F}_m = -\mu_0 \nabla \left[\int_0^H \left(\frac{\partial M v}{\partial v} \right)_{H,T} \mathrm{d}H \right] + \mu_0 (\boldsymbol{M} \cdot \nabla)\boldsymbol{H}. \tag{2.105}$$

Note that H in this expression is the internal field, not the applied field H', and $v = 1/\mathrm{d}$, where d is the density. Hence $Mv = \sigma$, the specific magnetic moment per kg of sample. When this is independent of density, as it is for dilute solutions or suspensions of magnetic particles, the first term is zero and the force $\boldsymbol{F}_m$ is given by the Kelvin expression for a paramagnet with $H' = H$. The demagnetizing field is negligible in dilute paramagnetic solutions, but in more concentrated samples such as ferrofluids, the first term takes care of the dipole–dipole interactions.

FURTHER READING

J. D. Jackson, *Classical Electrodynamics*, third edition, New York: Wiley (1998). This classic textbook is now written in SI units.

G. Bertotti, *Hysteresis*, San Diego: Academic Press (2000). A monograph on all aspects of magnetostatics and the hysteresis loop.

E. S. Shire, *Classical Electricity and Magnetism*, London: Cambridge University Press (1960). A good basic account.

A. Rosencwaig, *Ferrohydrodynamics,* Mineola: Dover (1997). This includes a clear account of magnetic energy and forces, with particular reference to ferrofluids.

L. D. Landau and E. M. Lifschitz, *Electrodynamics of Continuous Media*, second edition, Oxford: Pergammon Press (1989). The definitive text.

EXERCISES

2.1 Use the Biot–Savart law to obtain an expression for the magnetic field at a point $\boldsymbol{B}(\boldsymbol{r})$ in terms of the integral of the local current density $\boldsymbol{j}(\boldsymbol{r}')$. Hence show that the divergence of $\boldsymbol{B}$ is zero.

2.2 Refer to the table of dimensions in Appendix B to show that T m A^{-1} and H m^{-1} are equivalent units for μ_0. Express the units in terms of another pair of SI quantities.

2.3 Show that (2.6) reduces to (a) (2.7) for a circular current loop and (b) (2.8) for a square current loop.

2.4 Show that the expressions for the dipolar field in (2.8), (2.9) and (2.10) are all equivalent.

2.5 Show that (2.10) is equivalent to (2.12). Repeat for (2.12) and (2.13).

2.6 (a) What is the flux density at a distance of 0.1 nm from an atom with moment $\mathrm{m} = 1\,\mu_B$?

(b) Use the expression for the dipolar field in Cartesian coordinates to show that the dipole field at any point on a cubic lattice is exactly zero.

2.7 Calculate the magnetic field due to electric currents I flowing in opposite directions in two parallel wires separated by a distance d, at a perpendicular distance R from the axis of the wires. Assume $R \gg d$.

2.8 (a) Using the expression for the magnetic field of a uniformly magnetized long rod, show that the Halbach cylinder of Fig. 2.6(c) produces a field $B = \mu_0 M_r \ln(r_2/r)$, where M is the magnetization of the permanent magnet.

(b) If the cylinder is made of a permanent magnet with remanence 1.5 T, what is the outer diameter of a cylinder which will produce a field of 2.0 T in a 25 mm bore? Make an estimate of the field near the ends of the cylinder. Is it possible to replace any of the permanent magnet by soft iron?

2.9 (a) A flake of graphite and (b) a bismuth needle are freely suspended in a uniform magnetic field. What happens?

2.10 A magnetic measurement on a sample of mass 5 mg gives a moment of $1.5\ 10^{-7}$ Am^2 in a field of 1.0 T and $3.0\ 10^{-7}$ Am^2 in 2.0 T. What is the dimensionless susceptibility of the material, if its density is 4500 kg m^{-3}? Convert this to mass susceptibility or molar susceptibility; the molecular weight of the material is 265 g mol^{-1}. Check your conversions using Table B (inside back cover).

2.11 Calculate the external susceptibility of a long, needle-shaped sample of gadolinium at the Curie point T_C if the magnetization is constrained to lie along the c-axis, which is oriented at 12° to the sample axis. (The internal susceptibility diverges at T_C.)

2.12 Show that the magnetic field of a point dipole $\mathfrak{m}$ can be derived from the vector potential $A = (\mu_0/4\pi r^2)\mathfrak{m} \times \mathbf{e}_r$.

2.13 (a) Use the expression (2.65) for the scalar potential to derive the expression (2.55) for the magnetic field.

(b) Use the expression (2.59) for the vector potential to derive the expression (2.53) for the magnetic field in terms of the current distribution.

2.14 Use Poisson's equation for the vector potential (2.60) to derive the Biot-Savert law for the magnetic field due to a current-carrying circuit $\boldsymbol{B}(\boldsymbol{r}) = -(\mu_0 I/4\pi)\oint[(\boldsymbol{r} - \boldsymbol{r}') \times d\boldsymbol{l}]/|\boldsymbol{r} - \boldsymbol{r}'|^3$.

2.15 Show that there is no magnetic force acting on a tube containing a paramagnetic solution with a concentration gradient $c(z)$ when it is placed in a uniform magnetic field.

2.16 Show that the energy density in a magnetic field is given by (2.83) for the special case of a long solenoid.

2.17 By considering the forces (2.32) on the surface charges of a small volume element, show that the force density on a uniformly magnetized material in a magnetic field gradient is given by (2.104).

2.18 Use Cartesian coordinates, and the vector identity $[\mathbf{A} \times (\nabla \times \mathbf{B})]_j = \sum_i [\mathbf{A}_i \nabla_j \mathbf{B}_i - \mathbf{A}_i \nabla_i \mathbf{B}_j]$ to show that (2.104) and $F_z = \mu_0 M(\partial H'/\partial z)$ follow from (2.103). Show that in dilute, linear, isotropic media, $\boldsymbol{F}_m = (\mu_0 \chi/2)\nabla H^2$.

2.19 Express the polarization of iron (2.15 T) as the equivalent magnetization in A m^{-1}, and also in the cgs units, gauss and emu. What is the specific magnetization of iron in SI and cgs units?

3 Magnetism of electrons

Could anything at first sight seem more impractical than a body which is so small that its mass is an insignificant fraction of the mass of an atom of hydrogen?

J. J. Thompson (*Atomic Physics*, film sound track (1934))

The magnetic moments in solids are associated with electrons. The microscopic theory of magnetism is based on the quantum mechanics of electronic angular momentum, which has two distinct sources – orbital motion and spin. They are coupled by the spin–orbit interaction. Free electrons follow cyclotron orbits in a magnetic field, whereas bound electrons undergo Larmor precession, which gives rise to orbital diamagnetism. The description of magnetism in solids is fundamentally different depending on whether the electrons are localized on ion cores, or delocalized in energy bands. A starting point for discussion of magnetism in metals is the free-electron model, which leads to temperature-independent Pauli paramagnetism and Landau diamagnetism. By contrast, localized noninteracting electrons exhibit Curie paramagnetism.

The nature of the electron as a tiny negatively charged particle was established at the outset of the twentieth century. Louis de Broglie proposed in 1924 that wave–particle duality extended to matter as well as light, and that the wavelength λ_e of an electron is related to its momentum by

$$p = h/\lambda_e. \tag{3.1}$$

This de Broglie relation, combined with Niels Bohr's postulate that the angular momentum of electrons in atoms was quantized in multiples of $\hbar$, $|\boldsymbol{r} \times \boldsymbol{p}| = n\hbar$, led to the idea that the allowed orbits of electrons in atoms were stationary states with an integral number of de Broglie wavelengths. This opened the door to the development of quantum physics.

Our understanding of the magnetism of the electron is rooted in quantum mechanics. Two basic approaches are wave mechanics, due to Schrödinger, and matrix mechanics, due to Heisenberg. In wave mechanics, the electron is represented by a complex wave function $\Psi(\boldsymbol{r})$ whose physical significance is that $\Psi^*(\boldsymbol{r})\Psi(\boldsymbol{r})\delta^3 r$ is the probability of finding an electron in a volume of $\delta^3 r$ at $\boldsymbol{r}$. Here Ψ^* denotes the complex conjugate of Ψ. The basic equation is the Schrödinger equation:

$$\mathcal{H}\Psi = \varepsilon\Psi, \tag{3.2}$$

Figure 3.1

Magnetic moments associated with (a) orbital and (b) spin angular momenta of an electron.

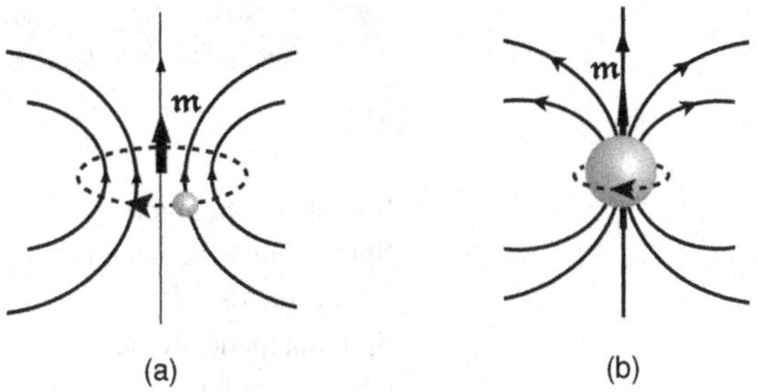

where $\mathcal{H}$ is the Hamiltonian operator. The solutions are the eigenstates or stationary states of the system $\Psi_i(\boldsymbol{r})$, and the eigenvalues are the energy levels ε_i, $i = 1, 2, \ldots$ The eigenfunctions are orthogonal, $\int \Psi_i^* \Psi_j \mathrm{d}^3 r = 0$, and form a basis of the system The Heisenberg formulation which is especially useful in magnetism when only a small number of eigenstates are relevant, uses an $n \times n$ matrix representation for the Hamiltonian. All physical observables can be represented by matrix operators. The eigenstates are $n \times 1$ column vectors, and the eigenvalues are real numbers. The procedure to determine them often involves diagonalizing a matrix to find its eigenvalues. Corrections due to small additional terms in the Hamiltonian are deduced from perturbation theory.

Equation (3.2) is used to find stationary states (energy eigenstates). When there is time dependence, $\Psi = \Psi(\boldsymbol{r}, t)$, the time-dependent Schrödinger equation must be used instead:

$$\mathcal{H}\Psi = \mathrm{i}\hbar \frac{\partial \Psi}{\partial t}. \tag{3.3}$$

For energy eigenstates, the solutions are of the form $\Psi \sim \mathrm{e}^{-\mathrm{i}\varepsilon t/\hbar}$, and (3.3) determines the evolution of any wave function.

3.1 Orbital and spin moments

Magnetism is intimately connected with *angular momentum of elementary particles*, so the quantum theory of magnetism is closely linked to the quantization of angular momentum. Protons, neutrons and electrons possess an intrinsic angular momentum $\frac{1}{2}\hbar$ known as spin, where $\hbar$ is Planck's constant h divided by 2π.

Nuclear spin creates much smaller magnetic moments than electronic spin because of the much greater nucleon mass, $\approx 1.67 \times 10^{-27}$ kg. In practice, the nuclear magnetism can often be neglected. Electrons are the main source of magnetic moments in solids. The electron is an elementary particle with charge $-e$ and mass m_e which has *two* distinct sources of angular momentum, one is associated with orbital motion around the nucleus, the other is spin (Fig. 3.1).

Table 3.1. Properties of the electron

Mass	m_e	9.109×10^{-31} kg
Charge	$-e$	-1.6022×10^{-19} C
Spin quantum number	s	1/2
Spin angular momentum	$\frac{1}{2}\hbar$	5.273×10^{-34} J s
Spin g-factor	g	2.0023
Spin magnetic moment	$\mathfrak{m}$	-9.285×10^{-24} A m^2
Classical radius $\mu_0 e^2/4\pi m_e$	r_e	2.818×10^{-15} m

3.1.1 Orbital moment

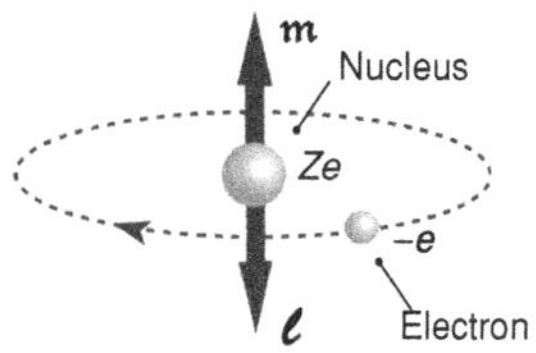

The Bohr atom. The electron moves in a circular orbit where its quantized angular momentum $\boldsymbol{\ell}$ and magnetic moment $\mathfrak{m}$ are oppositely directed.

The orbital moment can be introduced in terms of the Bohr model of the atom, where electrons revolve around a nucleus of charge Ze in circular orbits under the influence of the Coulomb potential $\varphi_e = -Ze/4\pi\epsilon_0 r$. An electron circulating in its orbit is equivalent to a current loop where the current direction is opposite to the sense of circulation because of the negative electronic charge. If the speed of the electron is v, its period of rotation is $\tau = 2\pi r/v$ and the equivalent current is $I = -e/\tau$. The magnetic moment associated with the current loop $\mathfrak{m} = I\mathcal{A}$ (2.3) is $-\frac{1}{2}e\boldsymbol{r} \times \boldsymbol{v}$, where the vector product shows the direction of $\mathfrak{m}$. In terms of angular momentum, $\boldsymbol{\ell} = m_e \boldsymbol{r} \times \boldsymbol{v}$, the moment is

$$\mathfrak{m} = -\frac{e}{2m_e}\boldsymbol{\ell}. \tag{3.4}$$

The proportionality between magnetic moment and angular momentum is a general result,

$$\mathfrak{m} = \gamma \boldsymbol{\ell}, \tag{3.5}$$

where the proportionality factor γ is known as the gyromagnetic ratio. For orbital motion of electrons, γ is $-(e/2m_e)$; the minus sign means that $\mathfrak{m}$ and $\boldsymbol{\ell}$ are oppositely directed because of the negative electron charge.

The orbital angular momentum is quantized in units of $\hbar$, in such a way that the component of $\mathfrak{m}$ in some particular direction, chosen as the z-direction, is

$$\mathfrak{m}_z = -\frac{e}{2m_e} m_\ell \hbar, \quad \text{where } m_\ell = 0, \pm 1, \pm 2, \ldots \tag{3.6}$$

Here m_ℓ is an orbital magnetic quantum number. The natural unit for electronic magnetism is therefore the Bohr magneton, defined as

$$\mu_B = \frac{e\hbar}{2m_e}; \tag{3.7}$$

$1\,\mu_B = 9.274 \times 10^{-24}$ A m^2. The z-component of the quantized orbital magnetic moment is an integral number of Bohr magnetons.

The remarkable difference between an electron in a quantum-mechanical stationary state and a classical charged particle is that the former can circulate indefinitely in its orbit as some sort of perpetual motion or electronic supercurrent – whereas the classical particle, or an electron in an unquantized orbit, must radiate energy on account of its continuous centripetal acceleration. Classical orbital motion will soon cease as a result of radiation loss.

The relation (3.5) is alternatively expressed in terms of a g-factor, which is defined as the ratio of the magnitude of the magnetic moment in units of μ_B to the magnitude of the angular momentum in units of $\hbar$: $(|\mathfrak{m}|/\mu_B) = (g|\ell|/\hbar)$. Hence g is exactly 1 for orbital motion.

The derivation of (3.4) can be generalized to noncircular orbits. From (2.3), $\mathfrak{m} = I\mathcal{A}$ for a planar loop of any shape and area $\mathcal{A}$. The angular momentum of an electron moving with angular velocity ω, $\ell = m_e r^2 \omega$, is a constant around the orbit. The current $I = -e/\tau = -(e\ell/m_e)\langle 1/2\pi r^2\rangle_{av} = -e\ell/2m_e\mathcal{A}$, where $\langle\cdots\rangle_{av}$ is the average over the orbit. Hence $\mathfrak{m} = -(e/2m_e)\ell$.

The Bohr model, a simplified version of the quantum mechanics of the atom, provides us with the natural units of length and energy for atomic physics. If $Z = 1$, Newton's second law for the centrally accelerated electron, $e^2/4\pi\epsilon_0 r^2 = m_e v^2/r$, and angular momentum quantization, $m_e v r = n\hbar$, give $r = n^2 a_0$, where the Bohr radius, a_0, is defined as

$$a_0 = \frac{4\pi\epsilon_0\hbar^2}{m_e e^2}. \tag{3.8}$$

The value of a_0 is 52.92 pm. The corresponding binding energy of the electron is $Z^2 R_0/n^2$, where $R_0 = (m_e/2\hbar^2)(e^2/4\pi\epsilon_0)^2$ is a constant known as the Rydberg. The value of R_0 is 2.180×10^{-18} J, which is equivalent to 13.606 eV.

3.1.2 Spin moment

The electron possesses intrinsic spin angular momentum with quantum number $s = \frac{1}{2}$. There is an associated intrinsic magnetic moment, unrelated to any orbital motion, which can only adopt one of two discreet orientations relative to a magnetic field. The electron is really a point particle, with radius $<10^{-20}$ m, much smaller than the classical radius (Table 3.1), so the image of a spinning ball of charge in Fig. 3.1 is ultimately misleading. The mysterious built-in angular momentum emerges as a consequence of relativistic quantum mechanics (§3.3). All fermions have spin and an associated magnetic moment. It turns out that the magnetic moment associated with the electron spin is not a half, but almost exactly one Bohr magneton. The gyromagnetic ratio γ is $-(e/m_e)$ and the g-factor is close to 2.

$$\mathfrak{m} = -\frac{e}{m_e}s. \tag{3.9}$$

The spin magnetic quantum number is $m_s = \pm\frac{1}{2}$, so there are only the two possible angular momentum states. The component of spin along any axis

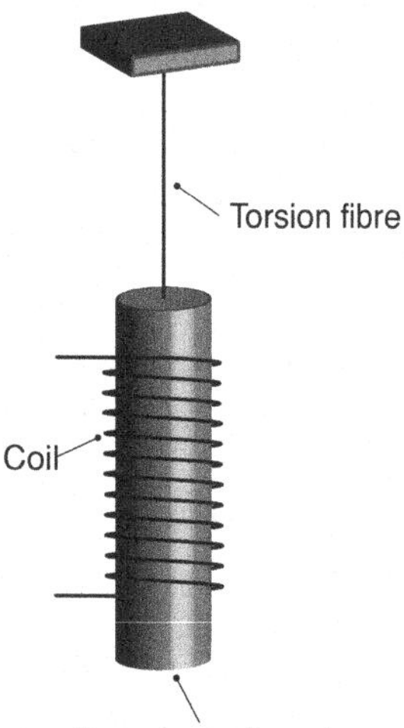

The Einstein–de Haas effect, demonstrating the relation between angular momentum and magnetic moment. A ferromagnetic rod is suspended on a fibre and the field in the solenoid is reversed, switching the direction of magnetization; the rod turns.

is $\pm\frac{1}{2}\hbar$:

$$\mathfrak{m}_z = -\frac{e}{m_e} m_s \hbar \text{ with } m_s = \pm\frac{1}{2}. \tag{3.10}$$

Spin angular momentum is therefore twice as efficient as orbital angular momentum at creating a magnetic moment. With higher-order corrections, g for the electron's intrinsic spin moment turns out to be 2.0023. The spin moment of the electron is 1.00116 μ_B. For practical purposes, this small correction can be ignored.

The reality of the link between magnetism and angular momentum, known as the Einstein–de Haas effect, was demonstrated in an experiment carried out by John Stewart in 1917. A ferromagnetic rod is suspended from a torsion fibre so that it can turn about its own axis. A vertical magnetic field created by a solenoid is sufficient to overcome the demagnetizing field and saturate the magnetization of the ferromagnet. The current in the solenoid is then reversed, switching the direction of magnetization of the rod, thereby delivering an angular impulse due to the reversal of the angular momentum of the electrons. The impulse causes the rod to rotate, and from the angle of rotation and the torsion constant of the fibre it is possible to deduce the change of angular momentum. In the case of iron, for which the spontaneous magnetization $M_s = 1710\,\mathrm{kA\,m^{-1}}$, the g-factor is found to be 2.09. This shows that the magnetization of iron is essentially due to electron spin. More surprising is the magnitude of the ferromagnetic moment, which works out at only $2.2\mu_B$ per atom.[1] The number of electrons per iron atom is equal to the atomic number, $Z = 26$, yet the ferromagnetic moment of iron corresponds to the spin moment of barely two of them. All the others form pairs with oppositely aligned spins, and contribute nothing.

3.1.3 Spin–orbit coupling

Generally, an atomic electron possesses both spin and orbital angular momentum. They may be coupled by spin–orbit interaction to create a total electronic angular momentum $\boldsymbol{j}$, with resultant magnetic moment

$$\mathfrak{m} = \gamma \boldsymbol{j}. \tag{3.11}$$

It is conventional to use lower-case letters ℓ, s, j to denote the the angular momentum quantum numbers of a single electron. Upper-case letters L, S, J are reserved for the multielectron atoms and ions discussed in the next chapter. The numbers may be integral or half-integral, except for ℓ and L, which can

[1] To calculate the magnetic moment $\mathfrak{m}$ in μ_B per molecular unit from σ the magnetic moment per kilogram, multiply σ by the molecular weight $\mathcal{M}$ and divide by 5585 ($\mathfrak{m} = \sigma\mathcal{M}/1000 N_A \mu_B$, where N_A is Avogadro's number). By chance, $\mathcal{M}/5585$ is exactly 1/100 for iron, which has atomic mass number = 55.85; $\sigma = M_s/\rho = 217$ A m^2 kg^{-1} for iron at 300 K, hence $\mathfrak{m} = 2.2\ \mu_B$. Other magnetization conversions can be found in Appendix E.

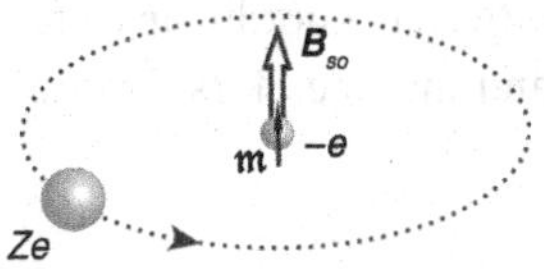

Spin–orbit interaction from the viewpoint of the electron.

only be integral. Bold symbols represent angular momentum vectors, which have units of $\hbar$.

From the electron's geocentric point of view, the nucleus revolves around it with speed v. The motion is equivalent to a current loop $I_n = Zev/2\pi r$, which creates a magnetic field $\mu_0 I_n/2r$ (2.7) at the centre. The spin–orbit interaction is due to this magnetic field, $B_{so} = \mu_0 Zev/4\pi r^2$, acting on the intrinsic magnetic moment of the electron. The electron's magnetic moments associated with $\boldsymbol{\ell}$ and $\boldsymbol{s}$ are oppositely aligned. The interaction energy (2.73) $\varepsilon_{so} = -\mu_B B_{so}$ can be written approximately in terms of the Bohr[2] magneton and the Bohr radius, since $r \simeq a_0/Z$ for an inner electron and $r = na_0$ for an outer electron, and $m_e vr \approx \hbar$. In the former case,

$$\varepsilon_{so} \approx -\frac{\mu_0 \mu_B^2 Z^4}{2\pi a_0^3}. \tag{3.12}$$

The Z variation means that the spin–orbit interaction, while weak for light elements becomes much more important for heavy elements and especially for inner shells. The associated magnetic field is of order 10 T for boron or carbon. The correct version of the spin–orbit interaction, resulting from a relativistic calculation, is given in §3.3.3. The expression (3.12) is modified by a factor 2. The interaction for a single electron is represented by the spin–orbit Hamiltonian

$$\mathcal{H}_{so} = \lambda \hat{\boldsymbol{l}} \cdot \hat{\boldsymbol{s}}, \tag{3.13}$$

where λ is the spin–orbit coupling energy. $\hat{\boldsymbol{l}}$ and $\hat{\boldsymbol{s}}$ are dimensionless operators – the $\hbar^2$ has been absorbed into λ, thus giving it dimensions of energy.

3.1.4 Quantum mechanics of angular momentum

The Bohr model is an oversimplification of the quantum theory of angular momentum. In quantum mechanics, physical observables are represented by differential operators or matrix operators, which we denote by bold symbols with a hat. For example, momentum is represented by $\hat{\boldsymbol{p}} = -i\hbar\nabla$ and kinetic energy by $\hat{\boldsymbol{p}}^2/2m = -\hbar^2\nabla^2/2m$. The allowed values of a physical observable are given by the eigenvalues, λ_i, of the equation $\hat{\boldsymbol{O}}\psi_i = \lambda_i\psi_i$, where $\hat{\boldsymbol{O}}$ is the operator and ψ_i are the eigenfunctions, which represent the possible observable states of the system. The eigenvalues are determined by solving the equation $|\hat{\boldsymbol{O}} - \lambda I| = 0$ where $|\cdots|$ denotes a determinant and I is the identity matrix.

The angular momentum operator is $\hat{\boldsymbol{l}} = \boldsymbol{r} \times \hat{\boldsymbol{p}}$, with components

$$\hat{\boldsymbol{l}} = -i\hbar(y\partial/\partial z - z\partial/\partial y)\mathbf{e}_x - i\hbar(z\partial/\partial x - x\partial/\partial z)\mathbf{e}_y - i\hbar(x\partial/\partial y - y\partial/\partial x)\mathbf{e}_z. \tag{3.14}$$

2 We usually approximate the spin moment of the electron as 1 μ_B. Strictly, this equation should be $\varepsilon_{so} = -g\mu_B m_s B_{so}$.

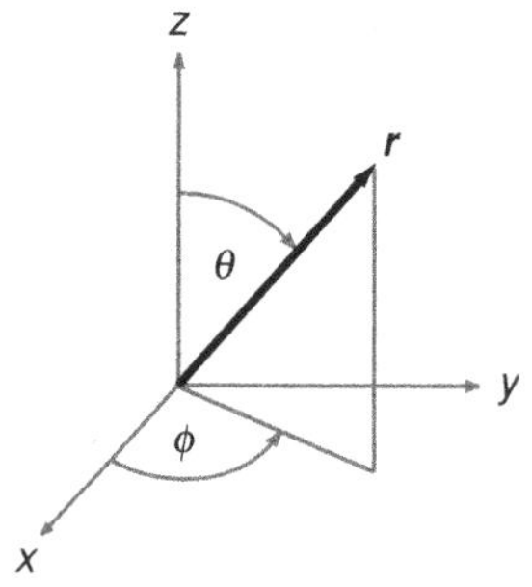

Spherical polar coordinates.

In terms of the spherical polar coordinates, the Cartesian coordinates are $x = r\sin\theta\cos\phi$, $y = r\sin\theta\sin\phi$ and $z = r\cos\theta$, and the operators for the components of the angular momentum become

$$\begin{aligned}\hat{l}_x &= \mathrm{i}\hbar(\sin\phi\partial/\partial\theta + \cot\theta\cos\phi\partial/\partial\phi),\\ \hat{l}_y &= \mathrm{i}\hbar(-\cos\phi\partial/\partial\theta + \cot\theta\sin\phi\partial/\partial\phi),\\ \hat{l}_z &= -\mathrm{i}\hbar(\partial/\partial\phi).\end{aligned} \tag{3.15}$$

The square of the total angular momentum is

$$\hat{l}^2 = \hat{l}_x^2 + \hat{l}_y^2 + \hat{l}_z^2 = -\hbar^2\left(\frac{\partial^2}{\partial\theta^2} + \cot\theta\frac{\partial}{\partial\theta} + \frac{1}{\sin^2\theta}\frac{\partial^2}{\partial\phi^2}\right). \tag{3.16}$$

An alternative way of representing angular momentum operators, which is invaluable when considering the spin of electrons, is with matrices. The magnetic systems have a small number ν of magnetic basis states, each denoted by a different magnetic quantum number m_i, and they can be represented by $\nu \times \nu$ square hermitian matrices.[3] For orbital angular momentum with quantum number ℓ, ν is $(2\ell + 1)$.

Similarly for spin, the electron with $s = \frac{1}{2}$ has just two basis states, denoted by $m_s = \pm\frac{1}{2}$. The spin angular momentum of the electron is represented by a 2×2 spin operator $\hat{\mathbf{s}}$. The three components of angular momentum are represented by the operators $\hat{s}_x, \hat{s}_y, \hat{s}_z$. Of these, $\hat{s}_z$ is conventionally chosen as the diagonal one and it has eigenvalues $\frac{1}{2}\hbar$ and $-\frac{1}{2}\hbar$, corresponding to $m_s = \pm\frac{1}{2}$. The two possible states of the electron are known as the $\downarrow$ or 'spin-up' and $\uparrow$ or 'spin-down' states.[4] Chemists refer to them as the β and α spin states. The eigenvectors corresponding to $|\downarrow\rangle$ and $|\uparrow\rangle$ are $\begin{bmatrix}1\\0\end{bmatrix}$ and $\begin{bmatrix}0\\1\end{bmatrix}$, so that $\hat{s}_z$ takes the matix form $\begin{bmatrix}1 & 0\\0 & -1\end{bmatrix}\frac{1}{2}\hbar$. A coordinate rotation about Oy by $\theta = \pi/2$ yields $\hat{s}_x = \begin{bmatrix}0 & 1\\1 & 0\end{bmatrix}\frac{1}{2}\hbar$ and a further rotation by $\phi = \frac{1}{2}\pi$ about Oz yields $\hat{s}_y = \begin{bmatrix}0 & -\mathrm{i}\\\mathrm{i} & 0\end{bmatrix}\frac{1}{2}\hbar$. The eigenvectors are known as spinors. The dimensionless operator $\hat{\boldsymbol{\sigma}}$ obtained by multiplying $\hat{\mathbf{s}} = (s_x, s_y, s_z)$ by $2/\hbar$ has components known as the Pauli spin matrices:

$$\hat{\boldsymbol{\sigma}} = \left(\begin{bmatrix}0 & 1\\1 & 0\end{bmatrix}, \begin{bmatrix}0 & -\mathrm{i}\\\mathrm{i} & 0\end{bmatrix}, \begin{bmatrix}1 & 0\\0 & -1\end{bmatrix}\right). \tag{3.17}$$

[3] A Hermitian matrix $[\mathrm{A}_{ij}]$ is one for which $\mathrm{A}_{ij} = \mathrm{A}^*_{ji}$, where * denotes the complex conjugate, obtained by replacing i by $-i$. Hermitian matrices have real eigenvalues, which correspond to physically observable quantities.

[4] The arrow indicates the direction of the magnetic moment. The negative charge of the electron means (somewhat confusingly) that $\uparrow$ is spin down and vice versa.

The fundamental property of angular momentum in quantum mechanics is that the operators representing the x, y, and z components satisfy the commutation rules

$$[\hat{s}_x, \hat{s}_y] = i\hbar\hat{s}_z, \quad [\hat{s}_y, \hat{s}_z] = i\hbar\hat{s}_x, \quad [\hat{s}_z, \hat{s}_x] = i\hbar\hat{s}_y. \tag{3.18}$$

The commutator is the square bracket, defined as $\hat{s}_x\hat{s}_y - \hat{s}_y\hat{s}_x$ etc. All three components $\hat{s}_x$, $\hat{s}_y$ and $\hat{s}_z$ satisfy the commutation rule, and all three of them have eigenvalues $\pm\frac{1}{2}\hbar$. These operators have to be Hermitian so that their eigenvalues are real. A neat way of summarizing the commutation relations is

$$\hat{\mathbf{s}} \times \hat{\mathbf{s}} = i\hbar\hat{\mathbf{s}}. \tag{3.19}$$

The differential operators for orbital angular momentum, (3.15), also obey these commutation rules, as all angular momentum operators must. Two operators are said to commute if their commutator is zero.

In quantum mechanics, only those physical quantities whose operators commute can be measured simultaneously. The three components of angular momentum do not commute and therefore cannot be measured at the same time. A precise measurement of the z component, for example, means that the x and y components are indeterminate. However, it is possible to measure the total angular momentum and any one of its components (but conventionally the z component) simultaneously. The square of the total spin angular momentum, $\hat{s}^2$ with eigenvalues $s(s+1)\hbar^2$ is proportional to the identity matrix and it is represented by

$$\hat{s}^2 = \hat{s}_x^2 + \hat{s}_y^2 + \hat{s}_z^2 = \begin{bmatrix} 1 & 0 \\ 0 & 1 \end{bmatrix} 3\hbar^2/4.$$

It commutes with $\hat{s}_x$, $\hat{s}_y$ and $\hat{s}_z$. The eigenvalue of the square of the total angular momentum $\langle\hat{s}^2\rangle = \langle i|\hat{s}^2|i\rangle$ is $3\hbar^2/4$ for both eigenstates $\begin{bmatrix} 1 \\ 0 \end{bmatrix}$ and $\begin{bmatrix} 0 \\ 1 \end{bmatrix}$. Both $\hat{s}_z$ and $\hat{s}^2$ are diagonal, and diagonal matrices always commute.[5] Hence the possiblity of measuring simultaneously both the square of the total angular momentum and its z-component.

Pictorially, the electronic angular momentum can be represented by a vector of length $\surd 3\hbar/2$ which precesses around Oz, and takes one of two orientations relative to Oz (Fig. 3.2). The component of spin angular momentum parallel to Oz can only take the values $m_s\hbar$, where $m_s = \pm\frac{1}{2}$. The two states with

[5] In the Dirac notation, $\langle i|\hat{a}|j\rangle$ is the i, j matrix element of the operator $\hat{a}$. The diagonal components are $\langle i|\hat{a}|i\rangle$. When the matrix has only diagonal terms, which can always be achieved by a suitable transformation if the matrix is Hermitian, the diagonal matrix elements are the eigenvalues. If $\hat{a}$ is the Hamiltonian, the eigenvalues are the energy levels of the system. $|i\rangle$, known as the 'ket', is the eigenfunction – a column vector in the matrix representation. $\langle i|$, known as the 'bra', is the complex conjugate row vector. Their product, the Dirac braket, $\langle i|i\rangle = 1$ for normalized eigenfunctions.

Figure 3.2

Vector model of electron spin. The total spin vector, of length $\sqrt{3}\hbar/2$, precesses around the applied field direction with the Larmor frequency. It has two possible projections $\pm\hbar/2$ along Oz, corresponding to the $|\uparrow\rangle$ and $|\downarrow\rangle$ states. The Zeeman splitting of the two magnetic energy levels in the field is shown.

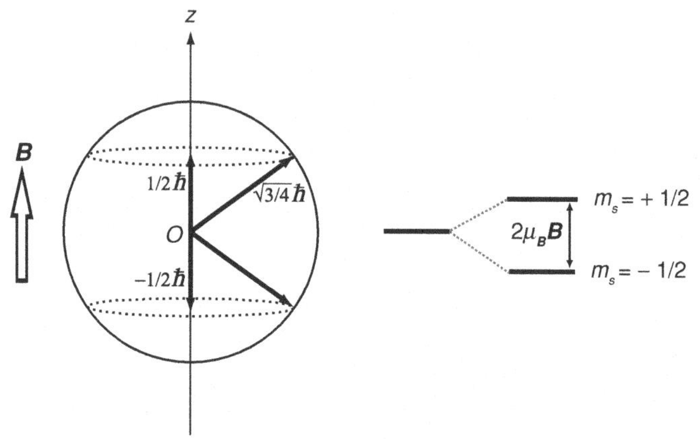

$m_s = \pm\frac{1}{2}$ have opposite magnetic moments and a Zeeman splitting of the two energy levels develops in a magnetic field $\boldsymbol{B}$. The Zeeman Hamiltonian $\mathcal{H}_Z = -\mathfrak{m}\cdot\boldsymbol{B} = (e/m_e)\hat{\boldsymbol{s}}\cdot\boldsymbol{B}$ has eigenvalues $g\mu_B m_s B \approx \pm\mu_B B$, which are the energy levels in the applied field. The spin splitting of the states is therefore $2\mu_B B$. We normally take g for electron spin to be exactly 2.

Two other operators, useful for manipulations in quantum mechanics are the ladder operators

$$\hat{s}^+ = \hat{s}_x + \mathrm{i}\hat{s}_y \quad \text{and} \quad \hat{s}^- = \hat{s}_x - \mathrm{i}\hat{s}_y.$$

They do not correspond to any measurable quantity, but they are helpful because they raise or lower m_s by unity while leaving s unchanged, which in the case of $s = \frac{1}{2}$ simply means that they transform $|\downarrow\rangle$ to $|\uparrow\rangle$ and vice versa. The operators $\hat{s}^+$ and $\hat{s}^-$ are represented by the non-Hermitian matrices $\begin{bmatrix} 0 & 1 \\ 0 & 0 \end{bmatrix}\hbar$, and $\begin{bmatrix} 0 & 0 \\ 1 & 0 \end{bmatrix}\hbar$, respectively. Hence $\hat{s}^+|\uparrow\rangle = \hbar|\downarrow\rangle$, $\hat{s}^+|\downarrow\rangle = |0\rangle$, $\hat{s}^-|\uparrow\rangle = |0\rangle$, $\hat{s}^-|\downarrow\rangle = \hbar|\uparrow\rangle$; $\hat{s}^2$ may be written in the form

$$\hat{s}^2 = \tfrac{1}{2}(\hat{s}^+\hat{s}^- + \hat{s}^-\hat{s}^+) + \hat{s}_z^2.$$

Equivalent versions are $\hat{s}^2 = (\hat{s}^+\hat{s}^- - \hbar\hat{s}_z + \hat{s}_z^2)$ and $\hat{s}^2 = (\hat{s}^-\hat{s}^+ + \hbar\hat{s}_z + \hat{s}_z^2)$. The commutation relations for the ladder operators are $[\hat{s}^2, \hat{s}^\pm] = 0$ and $[\hat{s}_z, \hat{s}^\pm] = \pm\hbar\hat{s}^\pm$.

To summarize, an electron with spin quantum number $s = \frac{1}{2}$ has total angular momentum $\sqrt{3}\hbar/2$. There are two spin states, $m_s = \pm\frac{1}{2}$ with a projection of the angular momentum along a specified direction Oz of $\pm\frac{1}{2}\hbar$. The states are degenerate in zero field, but split in a magnetic field. Alternative notations for the two spin states of the electron are $|\frac{1}{2}\rangle$ and $|-\frac{1}{2}\rangle$, $|\downarrow\rangle$ and $|\uparrow\rangle$, or α and β.

The magnetic moment operator of the electron $\hat{\mathfrak{m}}$ (in units of Bohr magnetons) is proportional to the associated angular momentum (in units of $\hbar$) and can be represented by a similar matrix, with the proportionality factor (g-factor) of 1 for orbital and 2 for spin moments. In fact, the meaning of (3.4) and (3.9)

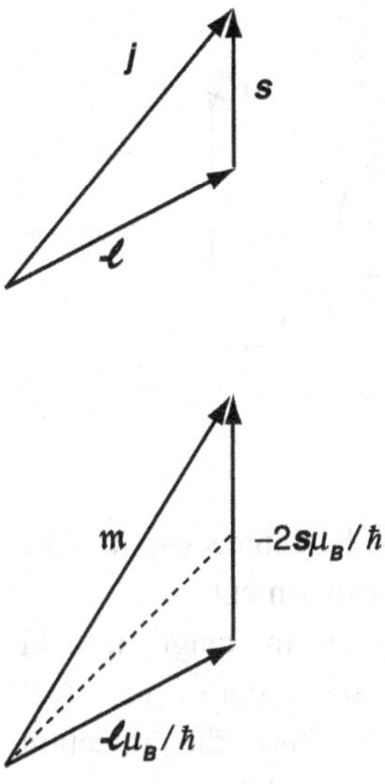

Addition of an electron's angular momenta and magnetic moments.

in quantum mechanics is that the matrix elements of the operators for $\hat{\mathfrak{m}}$ and $\hat{\boldsymbol{l}}$ or $\hat{\boldsymbol{s}}$ are proportional. The total magnetic moment of an electron is generally a vector sum of the spin and orbital magnetic moments:

$$\hat{\mathfrak{m}} = -(\mu_B/\hbar)(\hat{\boldsymbol{l}} + 2\hat{\boldsymbol{s}}). \tag{3.20}$$

The Zeeman interaction of these moments with an applied field $\boldsymbol{B}$ is represented by a term in the Hamiltonian:

$$\mathcal{H}_Z = (\mu_B/\hbar)(\hat{\boldsymbol{l}} + 2\hat{\boldsymbol{s}}) \cdot \boldsymbol{B}. \tag{3.21}$$

When $\boldsymbol{B}$ is in the z-direction, this Zeeman term is $(\mu_B/\hbar)(\hat{l}_z + 2\hat{s}_z)B$.

Polarization An electron in a general state has a wave function $|\psi\rangle = \alpha|\uparrow\rangle + \beta|\downarrow\rangle$, where α and β are complex numbers. If the wave function is normalized, $\langle\psi|\psi\rangle = 1$, and $\alpha^2 + \beta^2 = 1$. For example, the state with $\alpha = \beta = \frac{1}{\sqrt{2}}$, $\psi = \frac{1}{\sqrt{2}}\begin{bmatrix}1\\1\end{bmatrix}$ corresponds to the spin lying in the xy-plane. It is an equal superposition of $|\uparrow\rangle$ and $|\downarrow\rangle$ states. A measurement of s_z of such an electron will give $\hbar/2$ or $-\hbar/2$ with equal probability.

The polarization, P, of an ensemble of electrons is defined as

$$P = \frac{(n_\uparrow - n_\downarrow)}{(n_\uparrow + n_\downarrow)}, \tag{3.22}$$

where $n_\uparrow$ and $n_\downarrow$ are the densities of electrons in the two spin states. Units are m^{-3}. In the general case $P = (\alpha^2 - \beta^2)/(\alpha^2 + \beta^2)$. When they are all in the $|\uparrow\rangle$ state, the electrons are completely spin polarized, and $P = 1$. When they are equally likely to be in the $|\uparrow\rangle$ or the $|\downarrow\rangle$ state, the electrons are unpolarized, and $P = 0$.

The quantization axis for electrons Oz is defined by the local magnetic field direction. In principle, one way to separate $|\uparrow\rangle$ and $|\downarrow\rangle$ electrons and prepare an electron beam in a fully spin-polarized state is to pass an unpolarized beam through a region of space where it is subject to a nonuniform magnetic field. If the gradient is also in the z-direction, the field gradient force (2.74) is $\nabla(\mathfrak{m} \cdot \boldsymbol{B}) = \pm\mu_B(\mathrm{d}B_z/\mathrm{d}z)\boldsymbol{e}_z$ on electrons in the two spin states with moments of $\pm 1\mu_B$. The incident beam is split in two.

Otto Stern and Walther Gerlach were the first to observe bifurcation of this kind in 1921. Their experiment (Fig. 3.3) was performed, not with a beam of electrons which would be deflected by the Lorentz force (2.19), but on a beam of neutral silver atoms passing through the airgap of a specially shaped magnet where $\mathrm{d}B/\mathrm{d}z = 1000\ \mathrm{T\,m^{-1}}$. Atomic beams were expected to split into $(2L + 1)$ subbeams, where L is an integer. Silver has a $5s^1$ outer shell with no orbital moment, yet the silver beam splits in two, indicating a magnetic moment associated with half-integer angular momentum. It was originally to explain the fine structure in the spectrum of hydrogen in a magnetic field that Samuel

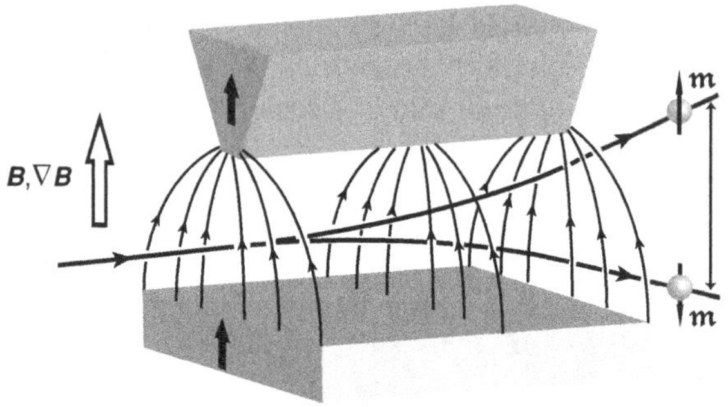

Figure 3.3
The Stern–Gerlach experiment. An unpolarized atom beam enters a region where there is a strong magnetic field gradient which splits the beam according to the projection of the magnetic moment along the field direction.

Goudsmit and George Uhlenbeck proposed the intrinsic half-integer spin of the electron. Their idea also explained the Stern–Gerlach experiment.

When a polarized electron beam enters a region where the magnetic field direction has changed, it is necessary to project the states onto a new Oz' direction in the xz-plane. If the angle between Oz and Oz' is θ, the projection of $\boldsymbol{s}$ on the new z axis $\boldsymbol{s} \cdot \boldsymbol{e}_{z'} = \boldsymbol{s}_z \cos\theta + \boldsymbol{s}_x \sin\theta$. The eigenvalues in the rotated frame are obtained by diagonalizing the matrix

$$\boldsymbol{s} \cdot \boldsymbol{e}_{z'} = \begin{bmatrix} \cos\theta & \sin\theta \\ \sin\theta & -\cos\theta \end{bmatrix} \frac{\hbar}{2}. \tag{3.23}$$

The diagonalization procedure for a matrix $\hat{\boldsymbol{M}}$ is to solve the eigenvalue equation $|\hat{\boldsymbol{M}} - \lambda \hat{\boldsymbol{I}}| = 0$, where $\hat{\boldsymbol{I}}$ is the unit matrix and $|\cdots|$ signifies the determinant of the matrix. Solving the equation $|\boldsymbol{s} \cdot \boldsymbol{e}_{z'} - (\hbar/2)\lambda \hat{\boldsymbol{I}}| = 0$ for λ gives $(\cos\theta - \lambda)(-\cos\theta - \lambda) - \sin^2\theta = 0$. Hence $\lambda^2 = 1$; the eigenvalues are $\pm\hbar/2$ as expected. The eigenvalues do not depend on the choice of coordinates. If $\begin{bmatrix} c_1 \\ c_2 \end{bmatrix}$ is an eigenvector, $\begin{bmatrix} c_1\cos\theta + c_2\sin\theta \\ c_1\sin\theta - c_2\cos\theta \end{bmatrix} = \lambda \begin{bmatrix} c_1 \\ c_2 \end{bmatrix}$. This yields the simultaneous equations

$$c_1(\cos\theta - \lambda) + c_2\sin\theta = 0, \quad c_1\sin\theta - c_2(\cos\theta + \lambda) = 0.$$

Hence $c_1/c_2 = -\sin\theta/(\cos\theta - \lambda)$. This leads to the expressions for the two eigenvectors in the original frame:

$$\begin{bmatrix} \cos\theta/2 \\ \sin\theta/2 \end{bmatrix} \text{ for } \lambda = 1 \text{ and } \begin{bmatrix} -\sin\theta/2 \\ \cos\theta/2 \end{bmatrix} \text{ for } \lambda = -1. \tag{3.24}$$

The transformation matrix $\hat{\mathrm{R}}_y(\theta)$ from the first frame, where $\hat{\boldsymbol{s}}_z$ is diagonal, to the second, where $\hat{\boldsymbol{s}}_{z'}$ is diagonal, is

$$\hat{\mathrm{R}}_y(\theta) = \begin{bmatrix} \cos\theta/2 & -\sin\theta/2 \\ \sin\theta/2 & \cos\theta/2 \end{bmatrix}. \tag{3.25}$$

Any observable A' in the new frame is related to the observable A in the old frame by $A' = \hat{R}^{-1} A \hat{R}$. The matrix (3.23) in the transformed frame is diagonal, and must be equal to $\begin{bmatrix} 1 & 0 \\ 0 & -1 \end{bmatrix} \hbar/2$. This can be checked by evaluating $\hat{R}_y^{-1} \begin{bmatrix} \cos\theta & \sin\theta \\ \sin\theta & -\cos\theta \end{bmatrix} \hat{R}_y$, where $\hat{R}_y^{-1}(\theta) = \begin{bmatrix} \cos\theta/2 & \sin\theta/2 \\ -\sin\theta/2 & \cos\theta/2 \end{bmatrix}$.

Rotation by θ about the other coordinate axes is represented by the matrices

$$\hat{R}_x(\theta) = \begin{bmatrix} \cos\theta/2 & -\mathrm{i}\sin\theta/2 \\ -\mathrm{i}\sin\theta/2 & \cos\theta/2 \end{bmatrix}, \qquad \hat{R}_z(\theta) = \begin{bmatrix} \mathrm{e}^{-\mathrm{i}\theta/2} & 0 \\ 0 & \mathrm{e}^{\mathrm{i}\theta/2} \end{bmatrix}.$$

These can be deduced from the general rotation operator for angular momentum $\hat{R}(\theta) = \exp(-\mathrm{i}\boldsymbol{\theta} \cdot \boldsymbol{S}/\hbar)$ and the Pauli spin matrices (3.17).

These rotations reveal a remarkable property of the electron, and other spin $\frac{1}{2}$ particles whose eigenvectors are spinors. Rotation through 2π is not equivalent to the identity matrix. A rotation by 4π is needed to turn the spinor into itself. Rotation of the quantization direction by 2π changes the phase of the spinor by π. This is an example of a Berry phase. Similarly, a rotation change in an angle ϕ about Oz introduces a Berry phase factor $\mathrm{e}^{\mathrm{i}\phi/2}$ into the spinors.

These ideas about angular momentum can be extended beyond the $s = \frac{1}{2}$ case, corresponding to the electron spin. For example, if an electron is in an orbital p-state with $\ell = 1$, there are three possible eigenstates for $\hat{l}_z$, corresponding to $m_\ell = 1, 0, -1$. The three states are represented by column vectors

$$\begin{bmatrix} 1 \\ 0 \\ 0 \end{bmatrix}, \quad \begin{bmatrix} 0 \\ 1 \\ 0 \end{bmatrix}, \quad \begin{bmatrix} 0 \\ 0 \\ 1 \end{bmatrix}.$$

The three components of the angular momentum $\hat{l}_x, \hat{l}_y, \hat{l}_z$ can be represented by the matrices

$$\begin{bmatrix} 0 & 1/\sqrt{2} & 0 \\ 1/\sqrt{2} & 0 & 1/\sqrt{2} \\ 0 & 1/\sqrt{2} & 0 \end{bmatrix} \hbar, \quad \begin{bmatrix} 0 & -i/\sqrt{2} & 0 \\ i/\sqrt{2} & 0 & -i/\sqrt{2} \\ 0 & i/\sqrt{2} & 0 \end{bmatrix} \hbar, \quad \begin{bmatrix} 1 & 0 & 0 \\ 0 & 0 & 0 \\ 0 & 0 & -1 \end{bmatrix} \hbar,$$

which have eigenvalues $\hbar$, 0, and $-\hbar$. The square of the total angular momentum $\hat{l}^2$, which has eigenvalue $\ell(\ell+1)\hbar^2$, equals $2\hbar^2$. It is represented by the matrix

$$\hat{l}^2 = \begin{bmatrix} 1 & 0 & 0 \\ 0 & 1 & 0 \\ 0 & 0 & 1 \end{bmatrix} 2\hbar^2,$$

the raising and lowering operators $\hat{l}^+$ and $\hat{l}^-$ are

$$\hat{l}^+ = \begin{bmatrix} 0 & \sqrt{2} & 0 \\ 0 & 0 & \sqrt{2} \\ 0 & 0 & 0 \end{bmatrix} \hbar \quad \text{and} \quad \hat{l}^- = \begin{bmatrix} 0 & 0 & 0 \\ \sqrt{2} & 0 & 0 \\ 0 & \sqrt{2} & 0 \end{bmatrix} \hbar.$$

These ideas can be generalized further to any integral or half-integral quantum number. The eigenvectors have $(2\ell+1)$ elements, and the Hermitian matrices have $(2\ell+1)$ rows and columns. The diagonal matrices for $\hat{\boldsymbol{l}}^2$ and $\hat{l}_z$ have elements $[\hat{\boldsymbol{l}}^2]_{pq} = \ell(\ell+1)\hbar^2\delta_{p,q}$ and $[\hat{l}_z]_{pq} = (\ell+1-p)\hbar\delta_{p,q}$, where $\delta_{p,q}$ is the Kronecker delta ($\delta_{p,q} = 1$ for $p = q$; $\delta_{p,q} = 0$ for $p \neq q$). The operator $\hat{l}^-$ has elements $[\hat{l}^-]_{pq} = \sqrt{p(2\ell+1-p)}\delta_{p,q-1}$, $[\hat{l}^+]$ is the reflection of $[\hat{l}^-]$ in the diagonal, and $\hat{l}_x = \frac{1}{2}(\hat{l}^+ + \hat{l}^-)$, $\hat{l}_y = \frac{-1}{2}i(\hat{l}^+ - \hat{l}^-)$.

3.2 Magnetic field effects

The effects of a magnetic field on an electron are to modify its linear or angular motion, and to induce some magnetization in the direction of the field, as a result of Boltzmann population of the energy levels obtained from (3.21). In this section, we discuss the effects of a magnetic field on the electron motion semiclassically.

3.2.1 Cyclotron orbits

If an electron travels with velocity $\boldsymbol{v}$ across a magnetic field $\boldsymbol{B}$, the Lorentz force $-e\boldsymbol{v} \times \boldsymbol{B}$ causes an acceleration perpendicular to the velocity, which produces circular motion. Newton's second law for the circularly accelerated motion gives $f = m_e v_\perp^2/r = ev_\perp B$, so the cyclotron frequency $f_c = v_\perp/2\pi r$ is proportional to the field:

$$f_c = \frac{eB}{2\pi m_e}. \tag{3.26}$$

The angular frequency ω_c equals $2\pi f_c$. Any component of the electron velocity parallel to the magnetic field is uninfluenced by the Lorentz force, so the trajectory is a helix around the field direction. Electrons which follow cyclotron orbits in vacuum radiate energy of frequency f_c. The cyclotron frequency is 28 GHz T^{-1}.

The cyclotron radius $r_c = m_e v_\perp/eB$ is of order a few micrometres for electrons in metals when $B \approx 1$ T. It is less for semiconductors and semimetals, where a smaller electron density leads to a lower Fermi velocity (§3.2.5) and the effective electron mass m^* may be less than m_e.

One example of the use of cyclotron radiation is the domestic microwave oven discussed in §13.3.2. Another is the synchrotron, where electrons accelerated to a large multiple γ_e of their rest energy and constrained to move in curved paths by bending magnets, emit linearly polarized white radiation in a narrow beam of width $1/\gamma_e$ radians. Synchrotron radiation is a valuable tool for probing the electronic structure of solids.

3.2.2 Larmor precession

Joseph Larmor, 1857–1942.

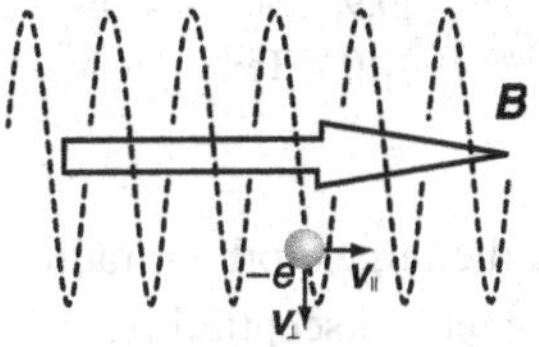

An electron moving freely in a magnetic field follows a helical path. The projection of the path in a plane normal to $\boldsymbol{B}$ is a circle where the electron circulates at the cyclotron frequency $eB/2\pi m_e$.

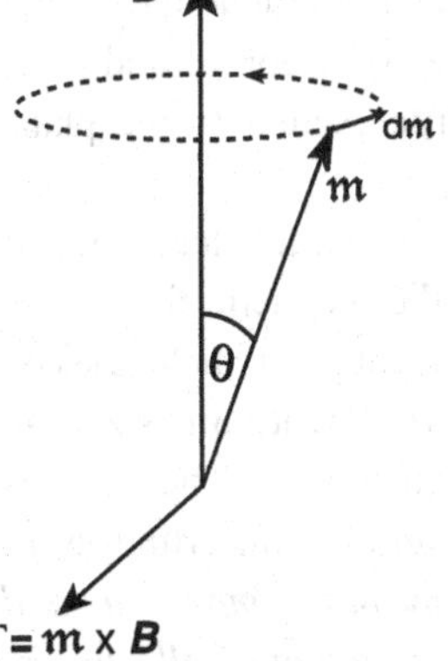

Torque on a magnetic moment in a magnetic field induces a precession around the field direction at the Larmor frequency. Note that γ is negative.

If an electron is constrained somehow to move in an orbit, it has an associated magnetic moment $\mathfrak{m} = \gamma \boldsymbol{\ell}$, where γ is the gyromagnetic ratio (3.5). The effect of the magnetic field is to exert a torque

$$\Gamma = \mathfrak{m} \times \boldsymbol{B} \tag{3.27}$$

on the current loop. Newton's law for angular momentum $\Gamma = \mathrm{d}\boldsymbol{\ell}/\mathrm{d}t$ gives

$$\frac{\mathrm{d}\mathfrak{m}}{\mathrm{d}t} = \gamma \mathfrak{m} \times \boldsymbol{B}. \tag{3.28}$$

When $\boldsymbol{B}$ is along the z-axis, the vector product in Cartesian coordinates gives

$$\frac{\mathrm{d}\mathfrak{m}_x}{\mathrm{d}t} = \gamma \mathfrak{m}_y B \qquad \frac{\mathrm{d}\mathfrak{m}_y}{\mathrm{d}t} = -\gamma \mathfrak{m}_x B \qquad \frac{\mathrm{d}\mathfrak{m}_z}{\mathrm{d}t} = 0. \tag{3.29}$$

The z component $\mathfrak{m}_z = \mathfrak{m}\cos\theta$ is independent of time, but the x and y components oscillate. The solution is $\mathfrak{m}(t) = (\mathfrak{m}\sin\theta\sin\omega_L t, \mathfrak{m}\sin\theta\cos\omega_L t, \mathfrak{m}\cos\theta)$, where $\omega_L = \gamma B$. The magnetic moment $\mathfrak{m}$ therefore *precesses* around the applied field direction at the Larmor frequency $f_L = \omega_L/2\pi$; here

$$f_L = \frac{\gamma B}{2\pi}. \tag{3.30}$$

The precession continues indefinitely if there is no way for the system to dissipate energy, and the angular momentum remains constant. Note that the Larmor precession frequency for an orbital moment ($\gamma = -e/2m_e$) is just half the cyclotron frequency, 28 GHz T^{-1}, whereas it is equal to the cyclotron frequency for a spin moment ($\gamma = -e/m_e$). The precession of the spin angular momentum around Oz occurs at the Larmor frequency.

3.2.3 Orbital diamagnetism

A semiclassical expression for the diamagnetic susceptibility of electrons with an orbital moment can be deduced from the Larmor precession. There is some angular momentum, and therefore a magnetic moment is associated with the precession of the electron orbit induced by the magnetic field. By Lenz's law, the induced moment is expected to oppose the applied field. The induced angular momentum is $m_e\omega_L\langle\rho^2\rangle$, where $\langle\rho^2\rangle = \langle x^2\rangle + \langle y^2\rangle$ is the mean square radius of the electron's orbit projected onto the plane perpendicular to $\boldsymbol{B}$. Since $\omega_L = \gamma B$, the induced moment is $-\gamma^2 m_e\langle\rho^2\rangle\boldsymbol{B}$, which gives a susceptibility $\mu_0 M/B$ of

$$\chi = -n\mu_0 e^2\langle r^2\rangle/6m_e, \tag{3.31}$$

where n is the number of electrons per cubic metre, $\gamma = -(e/2m_e)$ and $\langle r^2\rangle = (3/2)\langle\rho^2\rangle$ is the average squared radius of the electron orbit. In atoms, the effect

Table 3.2. Diamagnetic susceptibilities χ_m of common ions. Units are 10^{-9} m^3 kg^{-1}(after Sellwood, 1956)

H^+	0	Be^{2+}	0.6	Sc^{3+}	1.7	C^{4+}	0.1	V^{5+}	1.0	F^-	7.2
Li^+	1.1	Mg^{2+}	1.6	Y^{3+}	1.8	Si^{4+}	0.4	Nb^{5+}	1.0	OH^-	8.8
Na^+	2.7	Ca^{2+}	2.5	La^{3+}	1.8	Ge^{4+}	1.2	Ta^{5+}	1.0	Cl^-	9.2
K^+	4.2	Sr^{2+}	2.1	Lu^{3+}	1.2	Sn^{4+}	1.7			Br^-	5.6
Rb^+	2.9	Ba^{2+}	2.9			Pb^{4+}	1.4	Sb^{5+}	1.4	I^-	5.1
Cs^+	2.9			B^{3+}	0.2			Bi^{5+}	1.4		
Cu^+	2.4	Zn^{2+}	1.9	Al^{3+}	0.9	Ti^{4+}	1.3			O^{2-}	9.4
Ag^+	2.8	Cd^{2+}	2.5	Ga^{3+}	1.4	Zr^{4+}	1.4	Mo^{6+}	0.9	S^{2-}	14.8
Au^+	2.5	Hg^{2+}	2.3	In^{3+}	2.1	Hf^{4+}	1.1	W^{6+}	0.9	Se^{2-}	7.6
NH_4^+	8.0	Pb^{2+}	1.7			Th^{4+}	1.2	U^{6+}	1.0	Te^{2-}	6.8

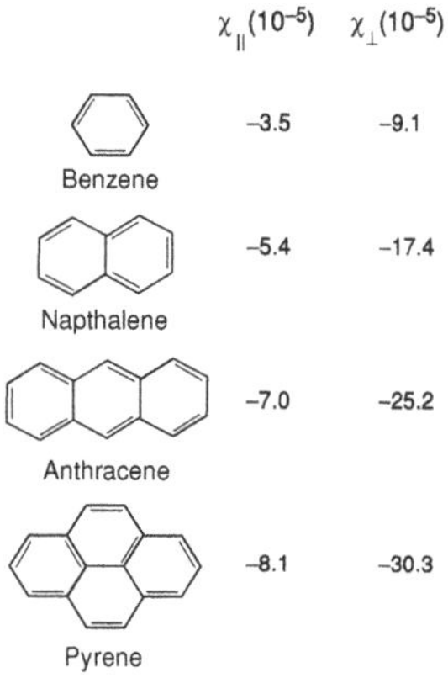

Susceptibility of some aromatic molecules, measured parallel and perpendicular to the plane of the molecular.

is dominated by the outer electron shells, which have the largest orbital radii. Negative ions therefore tend to have the largest diamagnetic susceptibility.

The order of magnitude of the orbital diamagnetic susceptibility χ for an element with $n \approx 6 \times 10^{28}$ atoms m^{-3} and $\sqrt{\langle r^2 \rangle} \approx 0.2$ nm is 10^{-5}. The corresponding mass susceptibility $\chi_m = \chi/d$, where d, the density, is of order 10^{-9} m^3 kg^{-1}. Orbital diamagnetism is a small effect, present to some extent for every element and molecule. It is the dominant susceptibility when there are no partially filled shells, which produce a larger paramagnetic contribution due to unpaired electron spins. Relatively large diamagnetic susceptibilities are observed for aromatic organic materials. Benzene rings, for example, have delocalized π electrons, where the induced currents can run around the carbon rings. The diamagnetic susceptibility is then quite anisotropic, and it is greatest in magnitude when the field is applied perpendicular to the plane of the ring.

Unfortunately, there is an underlying problem with classical calculations of the response of electrons to magnetic fields. Since the magnetic force $\boldsymbol{f} = -e(\boldsymbol{v} \times \boldsymbol{B})$ is perpendicular to the electron velocity, the magnetic field does no work on a moving electron, and cannot modify its energy. Hence $\delta w'$ is zero in (2.92), and it follows that there can be no change of magnetization. The idea was set out in the Bohr–van Leeuwen theorem, a famous and disconcerting result of classical statistical mechanics which states that *at any finite temperature and in all finite electric or magnetic fields, the net magnetization of a collection of electrons in thermal equilibrium vanishes identically.* Every sort of magnetism is impossible for electrons in classical physics! The semiclassical calculation of the orbital diamagnetism works only because we have *assumed* that there is a fixed magnetic moment associated with the orbit.

The orbital diamagnetism of common ions is tabulated in Table 3.2. The core diamagnetism of a compound is found by adding the individual ionic contributions. It is clear from Table A and Fig. 3.4 that the diamagnetic susceptibility of more than half the elements in the periodic table is

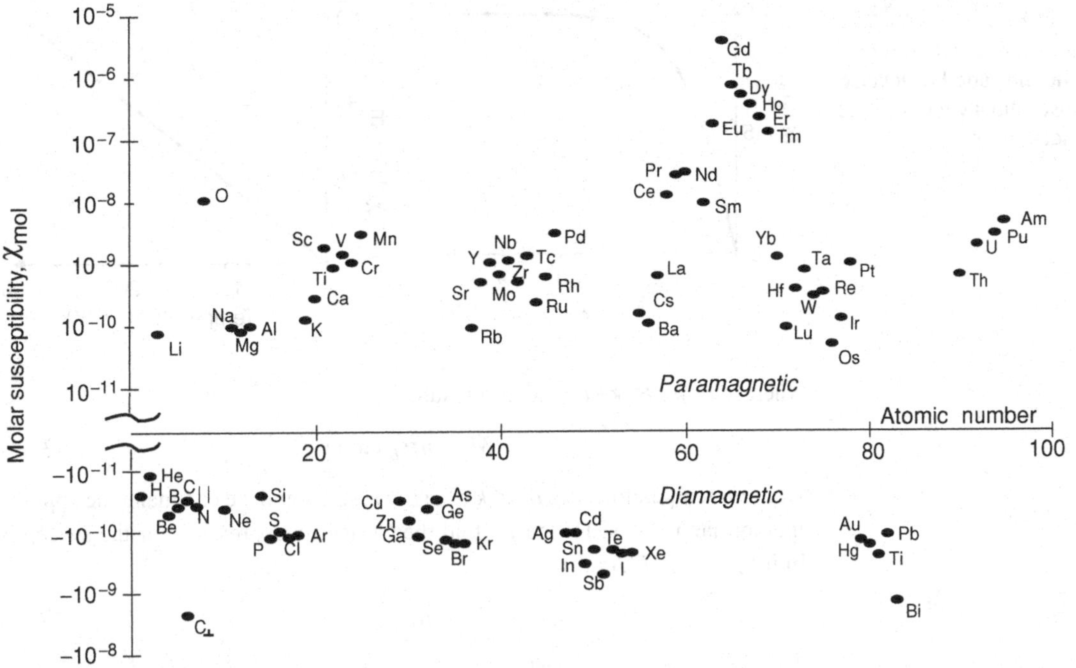

Figure 3.4

Molar magnetic susceptibility of the elements.

overwhelmed by a positive paramagnetic contribution. We now grant the electron its intrinsic spin moment, and examine how paramagnetic susceptibility arises in the two extreme models of magnetism, those of localized and delocalized electrons.

3.2.4 Curie-law paramagnetism

Considering an ensemble of single, *localized* electron spins in a magnetic field that is applied along Oz, the $|\uparrow\rangle$ and $|\downarrow\rangle$ states are split as shown in Fig. 3.2. The splitting is $\approx 2\mu_B B$. As the field increases, so does the Boltzmann population of the $|\uparrow\rangle$ state relative to the $|\downarrow\rangle$ state. If we have $n = (n^{\uparrow} + n^{\downarrow})$ electrons per unit volume, the induced magnetization along O_z is $(n^{\uparrow} - n^{\downarrow})\mu_B$, where $n^{\uparrow,\downarrow}$ are the Boltzmann populations of the two energy levels which are proportional to $\exp(\pm\mu_B B/k_B T)$. Hence

$$M = c\mu_B[(\exp(\mu_B B/k_B T) - \exp(-\mu_B B/k_B T)],$$

and

$$n = c[\exp(\mu_B B/k_B T) + \exp(-\mu_B B/k_B T)].$$

Hence the average z-component of the moment per atom $\langle \mathfrak{m}_z \rangle = (n^{\uparrow} - n^{\downarrow})\mu_B/(n^{\uparrow} + n^{\downarrow})$ is given by

$$\langle \mathfrak{m}_z \rangle = [\exp(x) - \exp(-x)]\mu_B/[(\exp(x) + \exp(-x)], \tag{3.32}$$

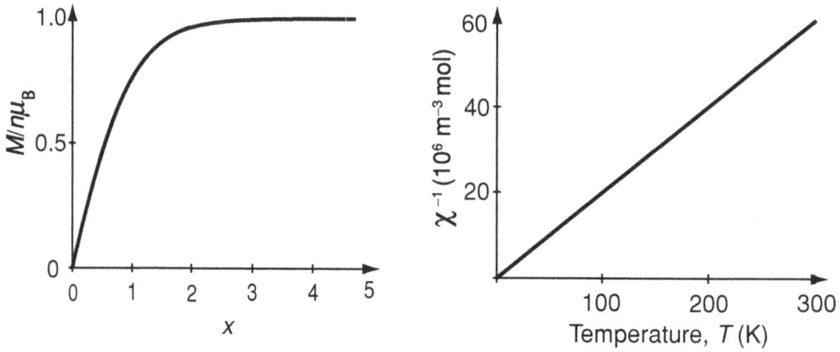

Figure 3.5
The magnetization curve and the Curie-law inverse susceptibility for localized electrons, $s = \frac{1}{2}$.

where $x = \mu_B B/k_B T$, and as a result,

$$M = n\mu_B \tanh x. \quad (3.33)$$

At room temperature, $\mu_B B \ll k_B T$, so x is small and we can make the approximation $\tanh x \approx x$. Hence we find the Curie-law expression for the susceptibility $\chi = \mu_0 M/B$:

$$\chi = n\mu_0\mu_B^2/k_B T. \quad (3.34)$$

The Curie law is often written

$$\chi = C/T, \quad (1.9)$$

where $C = n\mu_0\mu_B^2/k_B$ is the Curie constant (Fig. 3.5). A typical value of $n = 6 \times 10^{28}$ m^{-3} for one unpaired electron per atom gives $C = 0.5$ K and a susceptibility of 1.6×10^{-3} at room temperature. The susceptibility diverges as $T \longrightarrow 0$, according to the Curie law.

3.2.5 The free-electron model

In order to calculate the susceptibility in the opposite, *delocalized*, limit we introduce the simplest possible delocalized-electron model for a solid. The electrons are described as noninteracting waves confined in a box of dimension L. The Hamiltonian is the sum of terms representing the kinetic and potential energy:

$$\mathcal{H} = [(p^2/2m_e) + V(\boldsymbol{r})]. \quad (3.35)$$

When $V(\boldsymbol{r})$ is a constant, which can be set to zero, and $\boldsymbol{p}$ is replaced by the operator $-i\hbar\nabla$, Schrödinger's equation is

$$-(\hbar^2/2m_e)\nabla^2\psi = \varepsilon\psi. \quad (3.36)$$

Solutions are the free-electron waves $\psi = L^{-3/2}\exp(i\boldsymbol{k}\cdot\boldsymbol{r})$, where $\boldsymbol{k}$ is the electron wavevector ($k = 2\pi/\lambda_e$) and the $L^{-3/2}$ factor is required for normalization. The corresponding momentum and energy of the electron obtained

Table 3.3. Properties of the free-electron gas

Fermi wavevector	k_F	$(3\pi^2 n)^{1/3}$	1.2×10^{10}	m^{-1}
Fermi velocity	v_F	$\hbar k_F/m_e$	1.4×10^6	$m\ s^{-1}$
Fermi energy	ε_F	$(\hbar k_F)^2/2m_e$	9×10^{-19}	J
Fermi temperature	T_F	ε_F/k_B	6.5×10^4	K
Density of states	$\mathcal{D}_{\uparrow,\downarrow}(\varepsilon_F)$	$3n/4\varepsilon_F$	5×10^{46}	$m^{-3}\ J^{-1}$
Pauli susceptibility	χ_P	$3\mu_0\mu_B^2 n/2\varepsilon_F$	1.1×10^{-5}	
Hall coefficient	R_h	$1/ne$	1.0×10^{-10}	$m^3\ C^{-1}$

Numerical values are for $n = 6 \times 10^{28}$ m^{-3}. Density of states is for one spin.

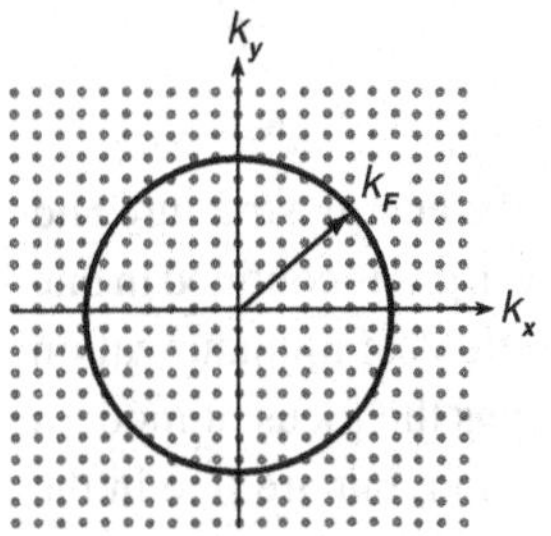

k-space. Each point represents a possible state for the electrons in a free-electron gas contained in a box of side L. Each state can accommodate a ↑ electron and a ↓ electron. The Fermi surface has radius k_F.

from the operators $\hat{\boldsymbol{p}} = -i\hbar\nabla$ and $\hat{\boldsymbol{p}}^2/2m_e$ are $\boldsymbol{p} = \hbar\boldsymbol{k}$ and $\varepsilon = \hbar^2k^2/2m_e$, respectively. The boundary conditions, which are periodic for free-electron waves, restrict the allowed values of $\boldsymbol{k}$ so that the components $k_i (i = x, y, z)$ are $\pm 2\pi n_i/L$, where n_i is an integer. Since indistinguishable electrons obey Fermi–Dirac statistics, each quantum state represented by the integers n_x, n_y, n_z can accommodate at most two electrons, one ↑, the other ↓. The allowed states form a simple cubic lattice in *k-space*, where the coordinates of a lattice point are (k_x, k_y, k_z). Each state occupies a volume $(2\pi/L)^3$, so the density of states for one spin in k-space is $(L^3/8\pi^3)$. Each state has two-fold spin degeneracy. At zero temperature the $N = nL^3$ electrons in the box occupy all the lowest available energy states, which fill a sphere of radius k_F, the Fermi wavevector. Since $\frac{4}{3}\pi k_F^3 = (N/2)(2\pi/L)^3$, it follows that

$$k_F = (3\pi^2 n)^{\frac{1}{3}}, \qquad (3.37)$$

and the corresponding energy, known as the Fermi energy, is

$$\varepsilon_F = (\hbar^2/2m_e)(3\pi^2 n)^{2/3}. \qquad (3.38)$$

The surface separating occupied and unoccupied states is the Fermi surface, which in the free electron model is a sphere. The Fermi velocity v_F is defined by $\hbar k_F = m_e v_F$, and the Fermi temperature T_F is defined by $\varepsilon_F = k_B T_F$. The density of states (states m^{-3} J^{-1}) $\mathcal{D}_{\uparrow,\downarrow}(\varepsilon) = \frac{1}{2}\mathrm{d}n/\mathrm{d}\varepsilon$ for *either* spin is

$$\mathcal{D}_{\uparrow,\downarrow}(\varepsilon) = (1/4\pi^2)(2m_e/\hbar^2)^{3/2}\varepsilon^{1/2}. \qquad (3.39)$$

Values of all these quantities for an electron density of 6 10^{28} m^{-3} are given in Table 3.3.

Using (3.37) and the result that $\varepsilon_F = \hbar^2k_F^2/2m_e$, the density of states at the Fermi level for our sample with n electrons per unit volume can be written

$$\mathcal{D}_{\uparrow,\downarrow}(\varepsilon_F) = 3n/4\varepsilon_F. \qquad (3.40)$$

The density of states of one of the bands is shown in Fig. 3.6. Units are states of one spin J^{-1} m^{-3}.

Provided the dimensions of the box are macroscopic, the electron states are very closely spaced in energy, and the expression for the density of states does

Figure 3.6

Density of ↑ or ↓ states in the free-electron model. The states are occupied with a probability given by the Fermi function (3.45), which is approximately a step function.

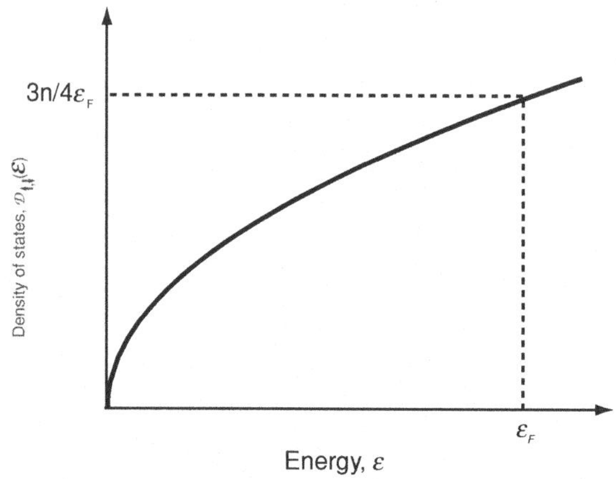

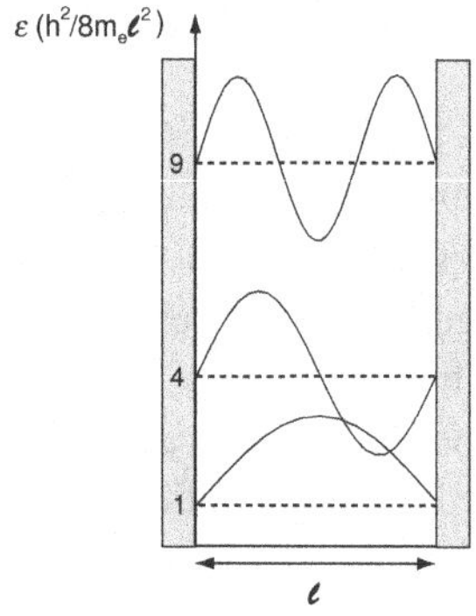

The lowest three modes for an electron confined in a well of width l. The energy levels are dashed, and energy is plotted on the vertical scale in units of $h^2/8m_el^2$.

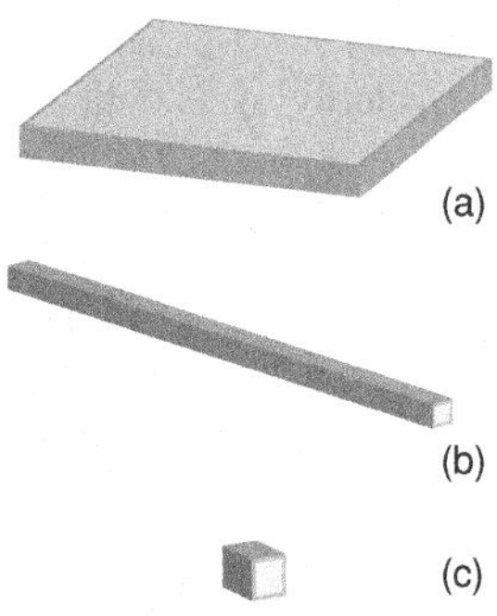

Confinement of the free-electron gas: (a) in two dimensions, (b) in one dimension – a quantum wire, and (c) in zero dimensions – a quantum dot.

not depend on L or on the shape of the box. However, the energy structure and density of states are drastically modified when the electron gas is confined in one or more directions on a nanometre length scale. Generations of miniaturization have made electron transport in such confined dimensions the focus of modern electronics. When electrons are confined in one dimension, but are free in the other two we have a quantum well. When they are confined in two dimensions, but free in just one we have a quantum wire. If all three dimensions are of order a nanometre, we are looking at a quantum dot, a sort of artificial atom. Confinement leads to a coarse-grained momentum and energy structure, which follows from the de Broglie relation (3.1).

Consider electrons confined in a well where the short dimension l is of order 1 nm in the z-direction. Localized boundary conditions on the electron wave function ψ_i are $\psi_i(0) = \psi_i(l) = 0$, so that an integral number n_i of half-wavelengths fits into the well, $l = n_i\lambda_e/2$, where the electron wavelength is given by (3.1). Hence $p_i = n_i h/2l$ and the corresponding energy levels are $\varepsilon_i = p_i^2/2m_e = (1/2m_e)(n_i h/2l)^2$. The first three quantized modes have energies $h^2/8l^2m_e$, $4h^2/8l^2m_e$ and $9h^2/8l^2m_e$. The separation of the first two when $l = 1$ nm is 1.1 eV, so only the lowest mode $n_i = 1$ will normally be occupied. We have a two-dimensional electron gas. The electron energy is

$$\varepsilon_i = \frac{\hbar^2}{2m_e}\left[k_x^2 + k_y^2 + \left(\frac{\pi n_i}{l}\right)^2\right]. \tag{3.41}$$

The free electrons occupy a Fermi circle with $\pi k_F^2 = (N/2)(2\pi/L)^2$ where $N = nL^2$. The Fermi energy ε_F equals $(\hbar^2/2m_e)2\pi n$, so the density of states $\mathcal{D}_{\uparrow,\downarrow}(\varepsilon_F) = (1/4\pi)(2m_e/\hbar^2)$ is a *constant*, independent of electron density. Likewise it can be shown that a quantum wire gives

$$\varepsilon_{ij} = \frac{\hbar^2}{2m_e}\left[k_x^2 + \left(\frac{\pi n_i}{l}\right)^2 + \left(\frac{\pi n_j}{l}\right)^2\right], \tag{3.42}$$

and a density of states $\mathcal{D}_{\uparrow,\downarrow}(\varepsilon_F) = (m_e/2\pi\hbar)(1/2m_e\varepsilon_F)^{1/2}$. Sharp singularities are a feature of the density of states of one-dimensional conductors.

Electrons moving in the lattice of a crystalline solid are subject to the periodic potential of the nuclei screened by their tightly bound ion cores. According to Bloch's theorem, the electronic states

$$\psi(\boldsymbol{r}) \approx \exp(i\boldsymbol{k} \cdot \boldsymbol{r}) u_k(\boldsymbol{r})$$

can be described as modified electron waves, still labelled by a wavevector $\boldsymbol{k}$. The function $u_k(\boldsymbol{r}) = u_k(\boldsymbol{r} + \boldsymbol{R})$ has the periodicity of the lattice, where $\boldsymbol{R} = p\boldsymbol{a}_1 + q\boldsymbol{a}_2 + r\boldsymbol{a}_3$ is a general lattice vector defined as a sum of the primitive lattice vectors $\boldsymbol{a}_1$, $\boldsymbol{a}_2$, $\boldsymbol{a}_3$. When the wavevector for an electron moving in some direction in k-space satisfies the Bragg condition,

$$2\boldsymbol{k} \cdot \boldsymbol{G} = G^2,$$

it will be reflected and a series of singularities will appear in the free-electron dispersion relation, which can lead to sharp structure and even gaps in the density of states. Here $\boldsymbol{G}$ is a lattice vector of the reciprocal lattice of the crystal in k-space, whose lattice points are $\boldsymbol{G} = h\boldsymbol{b}_1 + k\boldsymbol{b}_2 + l\boldsymbol{b}_3$, where $\boldsymbol{b}_1 = 2\pi(\boldsymbol{a}_2 \times \boldsymbol{a}_3)/(\boldsymbol{a}_1.(\boldsymbol{a}_2 \times \boldsymbol{a}_3))$ etc; h, k, l, like p, q, r, are integers. Compare the density of states for a metal such as iron (Fig. 5.13) with the free-electron density of states (Fig. 3.6). This structure in the density of states is critical for the appearance of ferromagnetism in metals.

3.2.6 Pauli susceptibility

The effect of an applied magnetic field $\boldsymbol{B}$ acting on the spin moment is to shift the two subbands by $\pm\mu_B B$, as shown in Fig. 3.7. Since $\mu_B B/k_B = 0.67$ K for a field of 1 T and $T_F \approx 65\,000$ K, the shifts induced by laboratory fields are minuscule compared with the Fermi energy. From the figure, it can be seen that the spin magnetization $M = (n^\uparrow - n^\downarrow)\mu_B$ is $2\mathcal{D}_{\uparrow,\downarrow}(\varepsilon_F)\mu_B^2 B$. Hence the susceptibility $\chi_P = \mu_0 M/B$ is given by

$$\chi_P = 2\mu_0\mu_B^2\mathcal{D}_{\uparrow,\downarrow}(\varepsilon_F). \quad (3.43)$$

This result is rather general. It is not restricted to the free-electron model as it depends only on the density of states at the Fermi level, but for this model, using (3.40) and setting $\varepsilon_F = k_B T_F$, the result can be written in the form

$$\chi_P = \frac{3n\mu_0\mu_B^2}{2k_B T_F}. \quad (3.44)$$

The Pauli susceptibility is temperature-independent to first order. Comparison with (3.34) shows that the Pauli susceptibilty is about two orders of magnitude smaller than the Curie susceptibility at room temperature. It is of order 10^{-5}.

At finite temperature, the occupancy of the states given by $\mathcal{D}(\varepsilon)$ is determined by the Fermi–Dirac distribution function:

$$f(\varepsilon) = \frac{1}{\{\exp[(\varepsilon - \mu)/k_B T] + 1\}}, \quad (3.45)$$

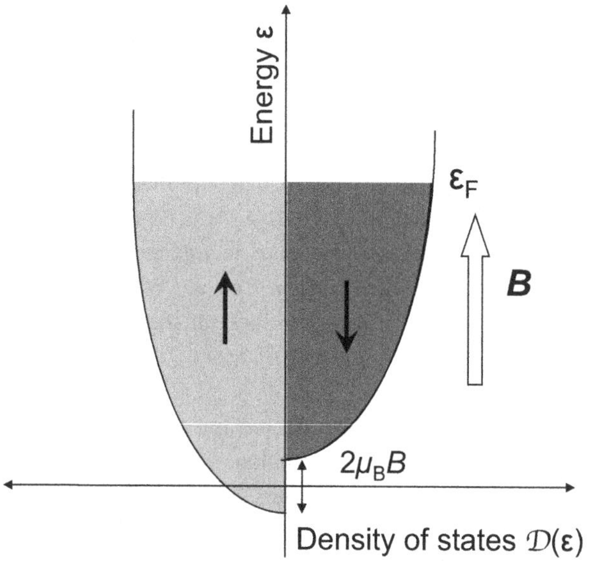

Figure 3.7

Spin splitting of the ↑ and ↓ densities of states in a magnetic field. A net moment results from the transfer of electrons at the Fermi level from the ↓ to the ↑ band.

where μ is the chemical potential. At $T = 0$ K, $\mu = \varepsilon_F$. The Fermi energy increases slightly at high temperature, as it is necessary to go to higher energy to accommodate the electrons, when some states below ε_F are unoccupied. The relation is

$$\mu = \varepsilon_F \left[1 - \frac{\pi^2}{12} \left(\frac{T}{T_F} \right)^2 + \cdots \right].$$

As a result of the decrease in the density of occupied states at ε_F there is a small temperature-dependent correction to (3.44), varying as T^2:

$$\chi_P = \frac{3n\mu_0\mu_B^2}{2k_B T_F} \left[1 - \frac{\pi^2}{12} \left(\frac{T}{T_F} \right)^2 + \cdots \right]. \tag{3.46}$$

Electron-electron correlation will increase the value of the Pauli susceptibility and give it more temperature dependence.

The electronic specific heat also depends on the density of state of the Fermi level, because only those electrons within $k_B T$ of ε_F can be excited thermally. The expression is

$$C_{el} = \gamma T = (4\pi^2/3)k_B^2 \mathcal{D}_{\uparrow,\downarrow}(\varepsilon_F)T$$

so from (3.43) the ratio γ/χ_P is a constant.

3.2.7 Electrical conduction

We will now consider electrical conduction in the free electron model of a metal, and how it is influenced by a magnetic field. When an electric field $\boldsymbol{E}$ is

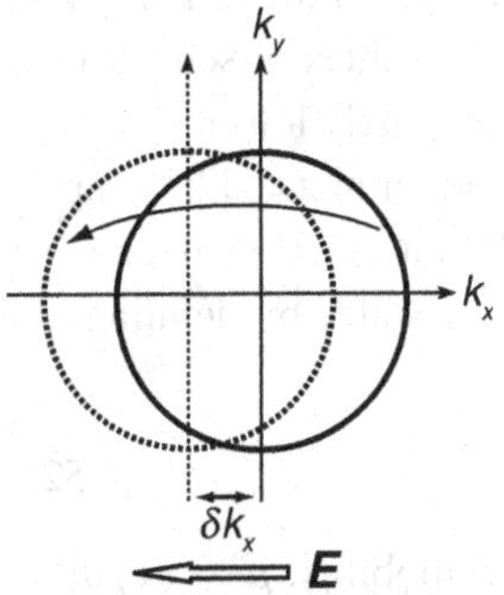

The conduction process in a metal. Electrons drift under the influence of the applied field, and are scattered into unoccupied states on the trailing edge of the Fermi surface. The shift $\delta k_x = m_e v_d/\hbar$ is greatly exaggerated in the drawing.

applied, a current density $\boldsymbol{j}$ flows in the same direction, given by Ohm's law:

$$\boldsymbol{j} = \sigma \boldsymbol{E}, \tag{3.47}$$

where σ is the electrical conductivity in S m^{-1}. An equivalent formulation is

$$\boldsymbol{E} = \varrho \boldsymbol{j},$$

where $\varrho = 1/\sigma$ is the resistivity in Ω m. The resistance R, in Ω, of a conductor of length l and uniform cross section a is $R = \varrho l/\mathrm{a}$. Since $E = V/l$, where V is the potential drop across the resistor, and $I = j\mathrm{a}$, (3.47) gives the familiar form of Ohm's law for a resistor, $V = IR$.

Ohm's law can be written in terms of the chemical potential, which is the change of energy when one extra electron is added to the metal. In an electric potential φ_e,

$$\mu = \mu_0 - e\varphi_e, \tag{3.48}$$

where μ_0 is the constant chemical potential in the absence of an electric field. Since $\boldsymbol{E} = -\nabla\varphi_e = \nabla(\mu/e)$,

$$\boldsymbol{j} = \frac{\sigma}{e}\nabla\mu. \tag{3.49}$$

A constant gradient of chemical potential is therefore associated with a flow of current in a conductor. The electrons are guided down the wire by a gradient of charge density at the surface of the conductor.

The entire Fermi surface is very slightly shifted in the direction of $\boldsymbol{E}$ as the electrons acquire a drift velocity v_d in the field direction. The electric current density $\boldsymbol{j}$ equals $-ne\boldsymbol{v}_d$. Mobility, defined as $\mu = v_d/E$, is a quantity with units m^2 V^{-1} s^{-1}, which is related to conductivity by

$$\sigma = ne\mu. \tag{3.50}$$

The conductivity of copper at room temperature $\sigma = 60 \times 10^6 \Omega$ m, and its electron density $n = 8.45 \times 10^{28}$ m^{-3}, give a mobility of 4×10^{-3} m^2 V^{-1} s^{-1}. A typical current density in copper of 1 A mm^{-2} corresponds to a drift velocity of only 0.07 mm s^{-1}. Electrons drift at the proverbial snail's pace, but their instantaneous Fermi velocity is an astonishing ten orders of magnitude greater.

The conduction process involves electrons being accelerated by the force $-e\boldsymbol{E}$ in a direction opposite to the field for a time τ, on average, before they are scattered across the Fermi surface into states where their velocity is randomized. Newton's second law gives $eE\tau = m_e v_d$, hence the expression for conductivity in terms of relaxation time is

$$\sigma = \frac{ne^2\tau}{m_e}. \tag{3.51}$$

The mean free path travelled by an electron in time τ between collisions is $\lambda = v_F\tau$. In our example of copper, the relaxation time is 2.5×10^{-14} s, and the mean free path is 40 nm. The conduction model breaks down when the mean free path is equal to the interatomic spacing $a \approx 0.2$ nm, which gives a minimum metallic conductivity $\sigma_{\min} = 0.3 \times 10^6\ \Omega$ m. Values in semimetals and semiconductors, where n is much less, are proportionately lower.

The free-electron model is quite a good approximation for metals like copper with a half-filled s-band and an almost-spherical Fermi surface. It can be extended to other metals with nonparabolic densities of states by defining an effective mass for the electrons as

$$m^* = \hbar^2(\partial^2\varepsilon/\partial k^2)^{-1}_{\varepsilon_F}. \quad (3.52)$$

Hence, narrow bands have high effective mass and low mobility, $\mu = e\tau/m^*$.

Generally, the conductivity or resistivity in Ohm's law (3.47) is a diagonal tensor, which reduces to the familiar scalar for cubic crystals or polycrystalline material. When a magnetic field is applied in the z-direction, the diagonal components $\varrho_i = \varrho_{xx}, \varrho_{yy}, \varrho_{zz}$ of the resistivity may change. **Magnetoresistance** is defined by

$$\Delta\varrho/\varrho = [\varrho_i(B) - \varrho_i(0)]/\varrho_i(0).$$

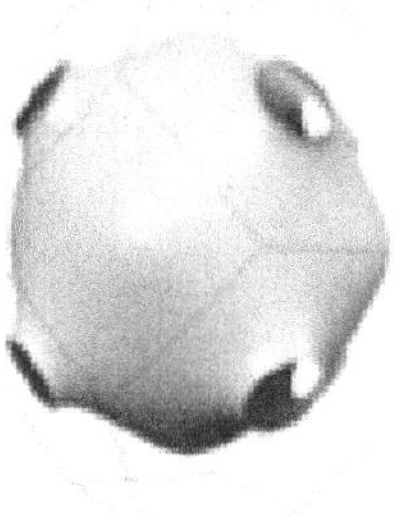

The Fermi surface of copper.

The resistance of a metal is inversely proportional to the mean free path. The change of resistance in an applied magnetic field results from the curtailing of the mean free path in the current direction when the electrons complete a significant fraction of a cyclotron orbit before they are scattered. Magnetoresistance effects associated with cyclotron motion can be significant when $\omega_c\tau \gtrsim 1$, where τ is the time between scattering events. The effect is initially quadratic in B. The longer the relaxation time, the greater the influence of magnetic field on the resistivity.[6] This magnetoresistance depends on interband scattering; and in fact it is strictly zero in the single-band free-electron model. The magnetoresistance is small ($\approx$1% in 1 T) in metals where scattering is strong, but it may be much larger in semimetals and semiconductors, where the electron mobility is high. Data on a bismuth film are shown in Fig. 3.8. Size effects can be observed in thin films, when the film thickness is comparable to r_c. Magnetoresistive sensors made from semiconducting InSb are useful for applications such as sensing the angular position of the rotor in a permanent magnet motor, where the fields are of order 0.1 T and a linear response is not required.

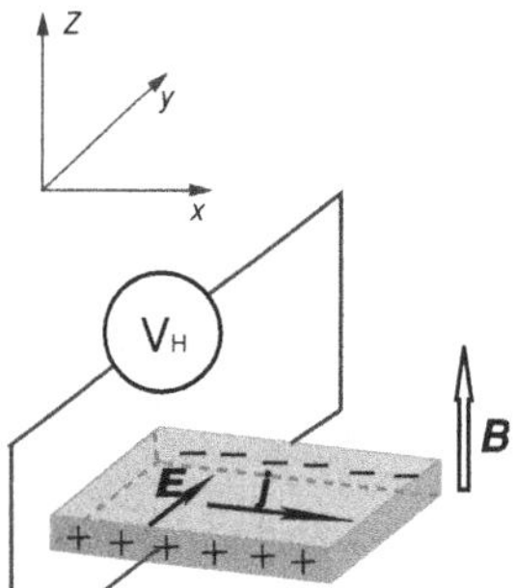

The Hall effect.

Furthermore, off-diagonal terms appear which are due to the Lorentz force. This leads to the **Hall effect**. When an electric current j_x of electrons moving with drift velocity v in the negative x-direction flows in a conductor, and a transverse magnetic field B_z is applied, the electrons are deflected and accumulate at the

[6] The magnetoresistance for a gas of free electrons is strictly zero because of the Hall effect. A compensating transverse electric field is set up in a solid, and there is no net force to deflect the electrons. However, magnetoresistance becomes possible whenever there is interband scattering of the electrons. This type of magnetoresistance is positive, in the sense that the applied field increases the resistance. It is called **ordinary magnetoresistance** (OMR), to distinguish it from effects encountered in ferromagnetic materials, which are often negative.

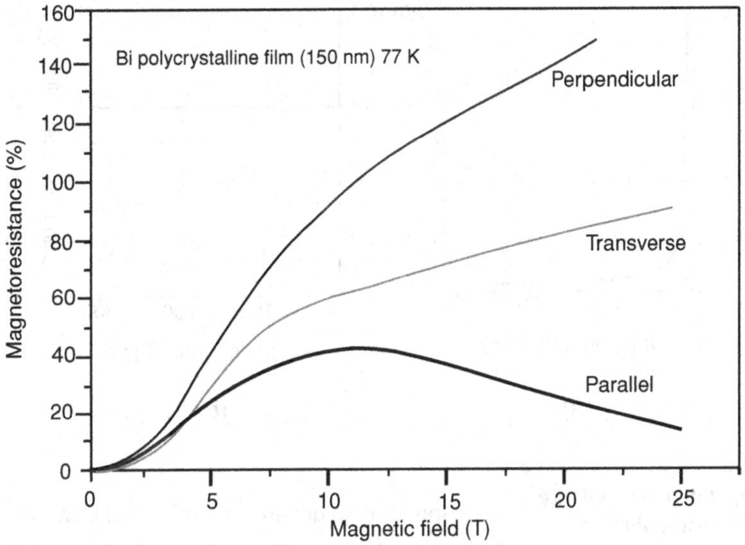

Figure 3.8

Magnetoresistance of a polycrystalline bismuth film 150 nm thick, measured with the current parallel, perpendicular or transverse to the applied field. The effects are large when B and j are perpendicular, and an influence of finite sample thickness is evident in the transverse geometry when $B > 10$ T. The variation in low fields is $\Delta\varrho \propto B^2$. (Data courtesy of J. McCauley)

edge of the sample until the electric field E_y they create is just sufficient to balance the Lorentz force. Thus $E_y = v_x B_z$. Since $j_x = -nev_x$,

$$E_y = -(1/ne)j_x B_z. \tag{3.53}$$

The quantity $R_h = -(1/ne)$ is known as the Hall coefficient, and the off-diagonal Hall resistivity ϱ_{xy} is $R_h B_z$. The Hall effect is inversely proportional to electron density, so it is large when n is small, as in semiconductors. The free-electron model is a starting point for conduction by ionized donors or acceptors, although for acceptors the conduction is due to free holes in the valence band rather than free electrons in the conduction band. The carriers are assigned an effective mass. Measurement of the Hall coefficient in a one-band solid will give their mobility; $\sigma = ne\mu$, hence $\mu = R_h/\varrho$.

In summary, the resistivity of an isotropic solid in a magnetic field is represented by the tensor

$$\hat{\boldsymbol{\varrho}} = \begin{bmatrix} \varrho_{xx} & -\varrho_{xy} & 0 \\ \varrho_{xy} & \varrho_{xx} & 0 \\ 0 & 0 & \varrho_{zz} \end{bmatrix},$$

$\varrho_{xy} \propto B$ and $\varrho_{xx} = \varrho_{zz} + \alpha B^2$. The form of the resistivity is determined by the Onsager principle, which requires that the off-diagonal terms in a response function satisfy $\sigma_{ij}(B) = -\sigma_{ji}(B) = \sigma_{ji}(-B)$.

Interesting transport effects arise when the free-electron gas is confined. The Hall resistance in the two-dimensional electron gas is quantized: $R_{xy} = h/\nu e^2$, where ν is an integer. The quantum Hall effect provides a precise standard for resistance; the quantum $h/e^2 = 25.813$ kΩ. Conductance in nanowires, where the electron gas is free only in one dimension, is confined to channels,

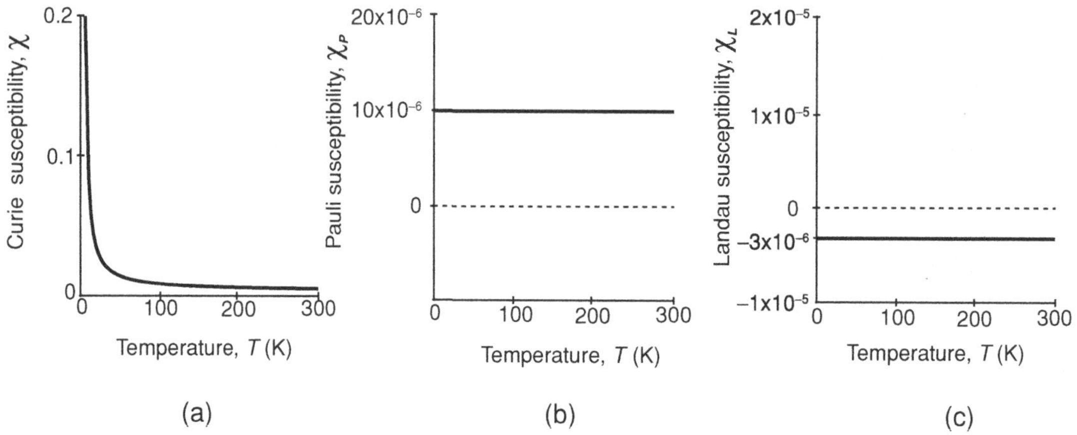

Figure 3.9

Temperature dependence of the susceptibility of electrons, comparing (a) the Curie-law susceptibility for localized electrons with (b) the Pauli and (c) the Landau contributions for free electrons.

corresponding to quantized modes in the cross section. The maximum conductance is G_0 per channel, assuming spin degeneracy, where $G_0 = e^2/h$. This is discussed further in Chapter 14.

3.2.8 Landau diamagnetism

The free-electron model was used by Landau to calculate the susceptibility due to orbital diamagnetism of the conduction electrons. The result is

$$\chi_L = -n\mu_0\mu_B^2/2k_BT_F, \tag{3.54}$$

which is exactly one third of the Pauli paramagnetism, but of opposite sign (Fig. 3.9). It looks as if the diamagnetism of the conduction electrons is just a correction to their paramagnetism, and never the dominant contribution. But this is not always correct if the real band structure of solids is approximately taken into account, by using the effective mass m^* (3.52) in the free-electron model. Then (3.54) is replaced by $\chi_L = -\frac{1}{3}(m_e/m^*)^2\chi_P$. For some semiconductors, and semimetals such as graphite or bismuth, $m^* \approx 0.01m_e$; despite the low electron densities, the diamagnetic susceptibility χ_L can be rather large ($\approx -10^{-4}$ for graphite). Quantum oscillations of the diamagnetic susceptibility are discussed in the next section.

A summary of the susceptibility of the elements is provided in Fig. 3.4. Some are gases at room temperature, so the definition $\chi = M/H$, where M is the induced magnetic moment per unit volume is not very useful. Instead, we plot the molar susceptibility $\chi_{mol} = \sigma\mathcal{M}/H$ in Fig. 3.4, where σ (in A m^2 kg^{-1}) is the induced magnetic moment per unit mass and $\mathcal{M}$ is the atomic weight in g mol^{-1}. Units of χ_{mol} are m^3 mol^{-1}. A mole of most solid elements occupies roughly 10 cm^3 = 10^{-5} m^3, so the molar susceptibility is roughly five orders of magnitude less than the dimensionless susceptibility χ. The mass susceptibility

Table 3.4. Mass susceptibilities χ_m of some paramagnetic and diamagnetic materials. Units are 10^{-9} m^3 kg^{-1}

MgO	−3.1	C(diamond)	−6.2	Cu	−1.1
Al_2O_3	−4.8	C(graphite)	$\chi_\parallel$ −6.3	Ag	−2.3
$LaAlO_3$	−2.7		$\chi_\perp$ −138.0	Au	−1.8
TiO_2	0.9	Si	−1.5	Al	7.9
$SrTiO_3$	−1.3	Ge	−1.4	Ta	10.7
ZnO	−6.2	NaCl	−6.4	Sc	88
ZrO_2	−1.1	ZnSe	−3.8	Zn	−2.2
HfO_2	−1.4	GaAs	−3.1	Pd	67.0
SiO_2	−7.1	GaN	−4.2	Pt	12.68(4)[a]
$MgAl_2O_4$	−4.2	InSb	−3.6	Ru	5.4
H_2O	−9.0	Perspex	−5.0	In	−7.0
D_2O	−8.1	DMSO	−6.6	Bi	−16.8

[a] NIST standard.

$\chi_m = \sigma/H$, with units m^3 kg^{-1} is three or four orders of magnitude less than χ. Since Avogadro's number $N_A = 6.022 \times 10^{23}$ mol^{-1}, the number density of atoms in solids is about 6×10^{28} m^{-3}. An atom occupies a volume of approximately (0.25 nm)^3.

Numerical values for the susceptibility of a selection of elements and compounds are provided in Table 3.4. Mass susceptibilities are listed here, because this is what is usually measured. The mass of a sample is much easier to measure than its volume.

Susceptibility units and conversions are a rich source of confusion; they are summarized in Appendix E and Table B, together with conversions to cgs units. A table of illustrative values of all the different susceptibilities is included for four representative materials.

3.3 Theory of electronic magnetism

Maxwell's equations (2.46)–(2.49) relate magnetic and electric fields to their sources. The other fundamental relation of electrodynamics is the Lorentz expression for the force on a moving particle with charge q:

$$\boldsymbol{f} = q(\boldsymbol{E} + \boldsymbol{v} \times \boldsymbol{B}). \tag{3.55}$$

The two terms are respectively the electric and magnetic forces. In a moving frame of reference, the separation of the electric and magnetic components may differ, but the relation between the total force and the fields measured in the same frame is invariant. The magnetic force produces the torque in (3.27).

When the charged particle is moving in a magnetic field, the momentum and energy are each the sum of kinetic and potential terms:

$$\hat{\boldsymbol{p}} = \hat{\boldsymbol{p}}_{kin} + q\boldsymbol{A}; \quad \mathcal{H} = (1/2m)\hat{\boldsymbol{p}}_{kin}^2 + q\varphi_e. \tag{3.56}$$

It is the canonical momentum $\hat{\boldsymbol{p}}$, including the vector potential term $q\boldsymbol{A}$, which is represented by the operator $-i\hbar\nabla$ in quantum mechanics. The vector potential $\boldsymbol{A}$ relates to the magnetic field $\boldsymbol{B}$ ($\boldsymbol{B} = \nabla \times \boldsymbol{A}$) and the electric scalar potential φ_e relates to the electric field $\boldsymbol{E}$ ($\boldsymbol{E} = -\nabla\varphi_e$). When the magnetic field is time-dependent $\boldsymbol{E} = -\nabla\varphi_e - \partial\boldsymbol{A}/\partial t$, hence the Hamiltonian of (3.56) for an electron becomes

$$\mathcal{H} = (1/2m_e)(\hat{\boldsymbol{p}} + e\boldsymbol{A})^2 + V(\boldsymbol{r}), \tag{3.57}$$

where $q = -e$, $V(\boldsymbol{r}) = -e\varphi_e$.

3.3.1 Orbital moment

The orbital paramagnetism and diamagnetism of the electron can be derived from (3.57). Making use of the Coulomb gauge, where $\boldsymbol{A}$ and $\hat{\boldsymbol{p}}$ commute, $(\hat{\boldsymbol{p}} + e\boldsymbol{A})^2$ is expanded as $\hat{\boldsymbol{p}}^2 + e^2\boldsymbol{A}^2 + 2e\boldsymbol{A}\cdot\hat{\boldsymbol{p}}$, so there are three terms in the Hamiltonian:

$$\mathcal{H} = [\hat{\boldsymbol{p}}^2/2m_e + V(\boldsymbol{r})] + (e/m_e)\boldsymbol{A}\cdot\hat{\boldsymbol{p}} + (e^2/2m_e)A^2, \tag{3.58}$$

$$\mathcal{H} = \mathcal{H}_0 + \mathcal{H}_1 + \mathcal{H}_2, \tag{3.59}$$

where $\mathcal{H}_0$ is the unperturbed Hamiltonian, $\mathcal{H}_1$ gives the paramagnetic response of the orbital moment and $\mathcal{H}_2$ describes its small diamagnetic response. Consider a uniform field $\boldsymbol{B}$ along Oz. The vector potential in component form can be taken as

$$\boldsymbol{A} = \tfrac{1}{2}(-yB, xB, 0),$$

so $\boldsymbol{B} = \nabla \times \boldsymbol{A} = \mathbf{e}_z(\partial A_y/\partial x - \partial A_x/\partial y) = \mathbf{e}_z B$.

More generally, $\boldsymbol{A} = \frac{1}{2}\boldsymbol{B} \times \boldsymbol{r}$, (2.57). Hence

$$(e/m_e)\boldsymbol{A}\cdot\hat{\boldsymbol{p}} = (e/2m_e)(\boldsymbol{B} \times \boldsymbol{r})\cdot\hat{\boldsymbol{p}} = (e/2m_e)\,\boldsymbol{B}\cdot(\boldsymbol{r} \times \hat{\boldsymbol{p}}) = (e/2m_e)\boldsymbol{B}\cdot\hat{\boldsymbol{l}},$$

since $\hat{\boldsymbol{l}} = \mathbf{r} \times \hat{\boldsymbol{p}}$. The angular momentum operator $\hat{\boldsymbol{l}}$ is therefore $\boldsymbol{r} \times (-i\hbar\nabla)$. Its z-component, for example, is $-i\hbar(x\partial/\partial y - y\partial/\partial x)$, or in polar coordinates $\hat{l}_z = -i\hbar\partial/\partial\phi$ (3.15).

The second term in the Hamiltonian (3.58) is therefore the Zeeman interaction for the orbital moment:

$$\mathcal{H}_1 = (\mu_B/\hbar)\hat{\boldsymbol{l}}_z B, \tag{3.60}$$

where the z axis has been chosen along the axis of $\boldsymbol{B}$. The eigenvalues of $\hat{\boldsymbol{l}}_z$ are $m_\ell\hbar$, where $m_\ell = -\ell, -\ell+1, \ldots, \ell$. The state with $m_\ell = -\ell$ is lowest in energy, due to the negative charge of the electron. The Hamiltonian (3.58) therefore gives an average orbital magnetic moment $\langle \mathfrak{m}_z \rangle = \sum_{-\ell}^{\ell} -m_\ell \mu_B \exp(-m_\ell \mu_B B/k_B T)/\sum_{-\ell}^{\ell} \exp(-m_\ell \mu_B B/kT)$, and an orbital susceptibility $n\langle \mathfrak{m}_z \rangle / H$, where n is the number of atoms per cubic metre.

The third term in (3.58) is $(e^2/8m_e)(\boldsymbol{B} \times \boldsymbol{r})^2 = (e^2/8m_e)B^2(x^2 + y^2)$. If the electron orbital is spherically symmetric, $\langle x^2 \rangle = \langle y^2 \rangle = \frac{1}{3}\langle r^2 \rangle$, so the energy corresponding to $\mathcal{H}_2$ is $\varepsilon = (e^2 B^2/12m_e)\langle r^2 \rangle$. This is the Gibbs free energy, because the Hamiltonian depends on the applied field $B = \mu_0 H$. Hence from (2.101) $\mathfrak{m} = -\partial\varepsilon/\partial B$. The diamagnetic susceptibility $\mu_0 n \mathfrak{m}/B$ is

$$\chi = -n\mu_0 e^2 \langle r^2 \rangle / 6m_e, \tag{3.61}$$

in agreement with the semiclassical expression (3.31).

3.3.2 Quantum oscillations

We now examine the diamagnetic response of the free-electron gas in more detail. The Hamiltonian of an electron in a magnetic field without the spin part is (3.57). Choosing a gauge $\boldsymbol{A} = (0, xB, 0)$ to represent the magnetic field, which is applied as usual in the z-direction, and setting $V(\boldsymbol{r}) = 0$ and $m_e = m^*$ we have Schrödinger's equation

$$\frac{1}{2m^*}\left[p_x^2 + (p_y + exB)^2 + p_z^2\right]\psi = \varepsilon\psi, \tag{3.62}$$

where $p_i = -\mathrm{i}\hbar\partial/\partial x_i$. It turns out that the y and z components of $\boldsymbol{p}$ commute with $\mathcal{H}$, so the solutions of this equation are plane waves in the y- and z-directions,with wave function $\psi(x)\mathrm{e}^{\mathrm{i}k_y y}\mathrm{e}^{\mathrm{i}k_z z}$. Substituting $\psi(x, y, z)$ back into Schrödinger's equation, we find

$$\left[-\frac{\hbar^2}{2m^*}\frac{d^2}{dx^2} + \frac{1}{2}m^*\omega_c^2(x - x_0)^2\right]\psi(x) = \varepsilon'\psi(x), \tag{3.63}$$

where $\omega_c = eB/m^*$ is the cyclotron frequency, $x_0 = -\hbar k_y/eB$ and $\varepsilon' = \varepsilon - (\hbar^2/2m)k_z^2$. Equation (3.63) is the equation of a one-dimensional harmonic oscillator, with motion centred at x_0. The oscillations are at the cyclotron frequency for a particle of mass m^*. The eigenvalues of the oscillator are $\varepsilon' = \varepsilon_n = (n + \frac{1}{2})\hbar\omega_c$ which are associated with the motion in the xy-plane, and the energy levels labelled by the quantum number n are known as Landau levels. The motion in the z-direction is unconstrained, so that

$$\varepsilon = \frac{\hbar^2 k_z^2}{2m^*} + \left(n + \frac{1}{2}\right)\hbar\omega_c. \tag{3.64}$$

The electron in the field, which classically follows the spiral trajectory, is represented in quantum mechanics by a plane wave along z and a one-dimensional

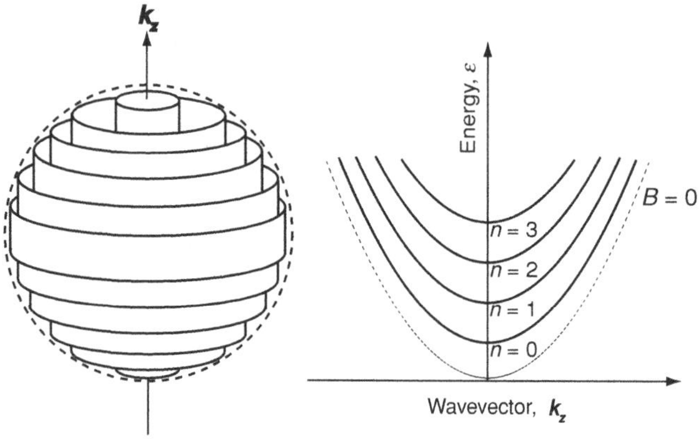

Figure 3.10
The states of the free-electron gas in a magnetic field coalesce onto a series of tubes. Each tube represents a Landau level. The dotted sphere is the $B = 0$ Fermi surface.

harmonic oscillator in the xy-plane. The system is like a magnetically confined quantum wire.

The states of the free-electron gas form a closely spaced lattice of points in k-space with spacing $2\pi/L$, where L is the dimension of the container. When the magnetic field is applied, these states collapse onto a series of tubes, as shown in Fig. 3.10. Each tube corresponds to a single Landau level, with index n. As B increases, the tubes expand and fewer of them are contained within the original Fermi sphere. The degeneracy of the Landau levels can be calculated from the area of k-space in the xy-plane between one level and the next

$$\pi\left(k_{n+1}^2 - k_n^2\right) = (2m^*\pi/\hbar^2)\left[\left(n + 1 + \tfrac{1}{2}\right)\hbar\omega_c - \left(n + \tfrac{1}{2}\right)\hbar\omega_c\right]$$
$$= 2m^*\pi\omega_c/\hbar,$$

where $k^2 = k_z^2 + k_n^2$. Each state occupies an area $(2\pi/L)^2$. The degeneracy of the Landau level is therefore

$$g_n = 4m^*\pi\omega_c/\hbar(2\pi/L)^2 = m^*L^2\omega_c/\pi\hbar.$$

The extra factor 2 accounts for the two electrons per level in the spin-degenerate case.

Oscillatory variations of the magnetization, conductivity and other properties with increasing field arise because of the periodic emptying of the uppermost Landau level as the field is increased. The Landau tubes expand with B, and an oscillatory variation of the energy as B^{-1} is observed. The oscillations of magnetic moment are known as the de Haas–van Alphen effect, whereas the corresponding oscillations of conductivity are known as the Shubnikov–de Haas effect. From the periodicity of the oscillations, it is possible to deduce the extremal areas of the Fermi surface normal to the direction of applied field. Hence the Fermi surface can be mapped.

3.3.3 Spin moment

The time-dependent Schrödinger equation

$$-\frac{\hbar^2}{2m}\nabla^2\psi + V\psi = i\hbar\frac{\partial\psi}{\partial t} \tag{3.65}$$

is not relativistically invariant because the derivatives $\partial/\partial t$ and $\partial/\partial x$ in the energy and momentum operators do not appear to the same power. A relativistically invariant version would use a 4-vector $X = (ct, x, y, z)$ with derivatives $\partial/\partial X_i$.

Dirac developed the relativistic quantum-mechanical theory of the electron, which involves the Pauli spin operators $\hat{\mathfrak{s}}_i$, and coupled equations for electrons and positrons. The nonrelativistic limit of his theory, including the interaction with a magnetic field represented by the vector potential $\boldsymbol{A}$, is represented by the Hamiltonian

$$\mathcal{H} = \left[\frac{\hbar^2}{2m}(\hat{\boldsymbol{p}} + e\boldsymbol{A})^2 + V(r)\right] - \frac{p^4}{8m_e^3c^2} + \frac{e}{m_e}(\nabla \times \boldsymbol{A}) \cdot \hat{\boldsymbol{s}} + \frac{1}{2m_e^2c^2r}\frac{\mathrm{d}V}{\mathrm{d}r}\hat{\boldsymbol{\ell}} \cdot \hat{\boldsymbol{s}} - \frac{1}{4m_e^2c^2}\frac{\mathrm{d}V}{\mathrm{d}r}. \tag{3.66}$$

- The *first term* is the nonrelativistic Hamiltonian (3.57).
- The *second term* is a higher-order correction to the kinetic energy.
- The *third term* is the interaction of the electron spin with the magnetic field. Taken together with (3.57), this gives the complete expression for the Zeeman interaction of the electron (3.21):

  $$\mathcal{H}_Z = (\mu_B/\hbar)(\hat{\boldsymbol{l}} + 2\hat{\boldsymbol{s}}) \cdot \boldsymbol{B}.$$

  The factor 2 is not quite exact. The correct value arising from quantum electrodynamics is $2(1 + \alpha/2\pi - \cdots) \approx 2.0023$, where $\alpha = e^2/4\pi\epsilon_0\hbar c \approx 1/137$ is the fine-structure constant.
- The *fourth term* is the spin–orbit interaction, which for a central potential $V(r) = -Ze^2/4\pi\epsilon_0 r$ with Ze as the nuclear charge becomes $-Ze^2\mu_0\hat{\boldsymbol{\ell}} \cdot \hat{\boldsymbol{s}}/8\pi m^2r^3$ since $\mu_0\epsilon_0 = 1/c^2$. The one-electron spin–orbit coupling is usually written as $\lambda\hat{\boldsymbol{\ell}} \cdot \hat{\boldsymbol{s}}$ (3.13). In an atom $\langle 1/r^3\rangle \sim (0.1\ \text{nm})^{-3}$ so the magnitude of the spin–orbit coupling λ is 8 K for lithium (Z = 3), 60 K for 3d elements (Z ≈ 25), and 160 K for actinides (Z ≈ 65). For the innermost electrons, $r \sim 1/Z$, which yields the Z^4 variation of (3.12).

  In a noncentral potential, the spin–orbit interaction is $(\hat{\boldsymbol{s}} \times \nabla V) \cdot \hat{\boldsymbol{p}}$.
- The *final term* just shifts the levels when $\ell = 0$.

3.3.4 Magnetism and relativity

The classification of interactions according to their relativistic character is based on the energy of a free particle, which is given by $\varepsilon^2 = m_e^2c^4 + p^2c^2$:

$$\varepsilon = \frac{m_ec^2}{\sqrt{[1 - (v^2/c^2)]}}. \tag{3.67}$$

The order of magnitude of the velocity of electrons in solids is αc, where c is the velocity of light and α is the fine-structure constant defined above. Expanding in powers of α shows the hierarchy of interactions:

$$\varepsilon = m_ec^2 + \tfrac{1}{2}\alpha^2 m_ec^2 - \tfrac{1}{8}\alpha^4 m_ec^2. \tag{3.68}$$

Here $m_ec^2 = 511$ keV; the second and third terms, which represent the order of magnitude of electrostatic and magnetostatic energies, are respectively 13.6 eV (the Rydberg, R_0) and 0.18 meV. Magnetic dipolar interactions are therefore of order 2 K.

Note that the Pauli susceptibility (3.43) can also be written in terms of the fine-structure constant as $\chi_P = \alpha^2 k_F a_0/\pi$, where k_F is the Fermi wavevector and a_0 is the Bohr radius. The force on a charged particle given by the Lorentz expression (2.19) does not change in a moving frame, but the relative contributions of the electric and magnetic fields to the force are modified. What looks like an electric field to a stationary electron acquires a magnetic field component in the frame of a moving electron, and vice versa.

From (2.19), the electric field acting on an electron subject to a magnetic field in a frame in which the electron is stationary is

$$\boldsymbol{E}^* = \boldsymbol{v} \times \boldsymbol{B}. \tag{3.69}$$

Conversely, the magnetic field acting on an electron subject to an electric field in a frame where the electron is stationary is

$$\boldsymbol{B}^* = -(1/c^2)\boldsymbol{v} \times \boldsymbol{E}. \tag{3.70}$$

This field leads to the Larmor precession of the spin of a moving electron in an electric field, which is known as the Rashba effect.

3.4 Magnetism of electrons in solids

The free-electron model provides a fair account of the outermost electrons in a metal or semiconductor. A better understanding of the magnetism of electrons in solids is achieved by considering first the situation for *free atoms*, summarized in the magnetic periodic table, Table A. The electronic moments are completely paired for some of the elements with even atomic number Z such as the alkaline earths or the noble gases, but most elements retain a magnetic

Figure 3.11

Formation of *d*- and *s*-bands in a metal. The broad *s*-bands have no moment. The *d*-bands may have one if they are sufficiently narrow; spin splitting is shown on the right. The interatomic spacing is r_0.

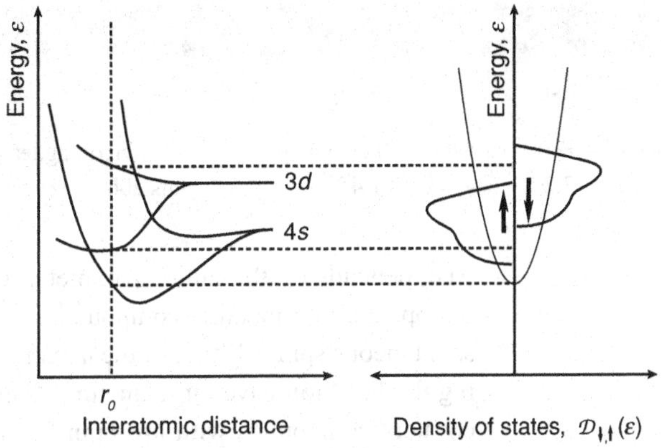

moment in the atomic state. These are the elements marked in bold type. The atomic moments are much less than $Z\mu_B$ for every element except hydrogen. The largest values of 10 μ_B are found for dysprosium and holmium. Electrons in filled shells have paired spins and no net orbital moment. Only unpaired spins in unfilled shells, usually the outermost one, contribute to the atomic moment.

Assembling the atoms together to form a *solid* is a traumatic process for the atomic moments. Magnetism tends to be destroyed by chemical interactions of the outermost electrons, which can occur in various ways:

- electron transfer to form filled shells in ionic compounds;
- covalent bond formation in semiconductors;
- band formation in metals.

Atomic iron, for example, has an electronic configuration (Ar)$3d^64s^2$. Four of the $3d$ electrons are unpaired, so the spin moment of the atom is 4 μ_B. When the iron atoms are brought together in a solid, as illustrated in Fig. 3.11, the outer $4s$ orbitals first overlap to form a broad $4s$-band, and then the smaller $3d$ orbitals follow suit to form a band that is considerably narrower. This results in $4s \rightarrow 3d$ charge transfer, producing an electronic configuration in iron metal of approximately (Ar)$3d^{7.4}4s^{0.6}$. The narrow $3d$ band has a tendency to split spontaneously to form a ferromagnetic state. This occurs below $T_C = 1044$ K in normal αFe, which has the bcc structure. The electrons occupy states with their magnetic moments either parallel (↑) or antiparallel (↓) to the ferromagnetic axis. A spin pair ↑↓ has no net moment. The electrons are perfectly paired in all the inner shells, and largely paired in the $4s$-band. The spin configuration of the $3d$-band is approximately $3d^{\uparrow 4.8}3d^{\downarrow 2.6}$ so the number of unpaired spins is 2.2. In Fig. 3.11, the ↑ and ↓ electrons are shown as occupying different, spin-split subbands.

It must be emphasized that the nature of the chemical bonding, and therefore the character of the magnetism, depends *critically* on crystal stucture and composition. The very existence of a moment in the other iron polymorph, fcc

Table 3.5. Atomic moments of iron in different crystalline environments, in units of μ_B

γ-Fe_2O_3	α-Fe	YFe_2	γ-Fe	YFe_2Si_2	FeS_2
Ferrimagnet 5.0	Ferromagnet 2.2	Ferromagnet 1.45	Antiferromagnet unstable	Pauli paramagnet 0	Diamagnet 0

γFe, depends on the lattice parameter. Compress it a little, and the moment disappears. Intermetallic compounds such as YFe_2Si_2 exist where there is no spontaneous spin splitting of the d-band. Insulating ionic compounds containing the Fe^{3+} ion have large unpaired spin moments of 5 μ_B per iron ion, but covalent compounds with low-spin Fe^{II} are nonmagnetic. A few examples to illustrate the range of properties of this most common magnetic element are given in Table 3.5.

3.4.1 Localized and delocalized electrons

It is a formidable task to make a physical theory that will adequately account for the behaviour of electrons in narrow bands, but two limits are accessible. One is the localized limit, where correlations of the electrons on the ion cores due to their mutual Coulomb interaction are strong, but transfer of electrons from one site to the next is negligible. The other is the delocalized limit where the electrons are confined in the solid, but Coulomb correlations among them and with the nuclear charges are relatively weak, of order 1 eV. Numerical methods for calculating the electronic structure, outlined in Chapter 5, then come into their own.

The localized model is best suited for $4f$ electrons in the rare-earth series, $R = \mathrm{Pr}, \ldots, \mathrm{Yb}$, which occupy orbitals belonging to an inner shell that barely participates in the bonding. The outer electrons have $5s$, $5p$, $5d$ or $6s$ character. There are usually two or three outermost $5d/6s$ valence electrons which either form a conduction band, as in the metals, or else are transferred to an electronegative ligand such as oxygen in the rare-earth oxides R_2O_3. There the rare-earth becomes an R^{3+} ion and the oxygen accepts two electrons to fill the $2p$ shell to become O^{2-}, with a stable closed-shell configuration[7] $2p^6$. Either way, the $4f$ shell is sufficiently well buried within the $5s$ and $5p$ shells to be uninvolved with chemical bonding, as shown in Fig. 3.12. It has an atom-like configuration with an integral number of electrons. There is a series of discrete energy levels for this strongly correlated electron shell, discussed in the next chapter. The electron orbitals describe localized states, and the $4f$ ions obey Boltzmann statistics. A similar picture applies to the heavy actinide ($5f$) elements, from Am onwards. Unfortunately, these elements become increasingly

[7] The closed shell configuration $2p^6$ is especially stable. Over 80% of the Earth's crust (Fig. 1.10) is made up of $2p^6$ ions.

Table 3.6. Summary of localized and delocalized magnetism

Localized magnetism	Delocalized magnetism
Integral number of $3d$ or $4f$ electrons	Nonintegral number of unpaired spins on the ion core
Integral number of unpaired spins per atom	
Discrete energy levels	Spin polarized energy bands with strong correlations
Ni^{2+} $3d^8$ $m = 2\,\mu_B$	Ni $3d^{9.4}4s^{0.6}$ $m = 0.6\,\mu_B$
$\Psi \approx \exp(-r/a_0)$	$\Psi \approx \exp(-i\boldsymbol{k}\cdot\boldsymbol{r})$
Boltzmann statistics	Fermi–Dirac statistics
$4f$ metals and compounds; some $3d$ compounds	$3d$ metals; some $3d$ compounds

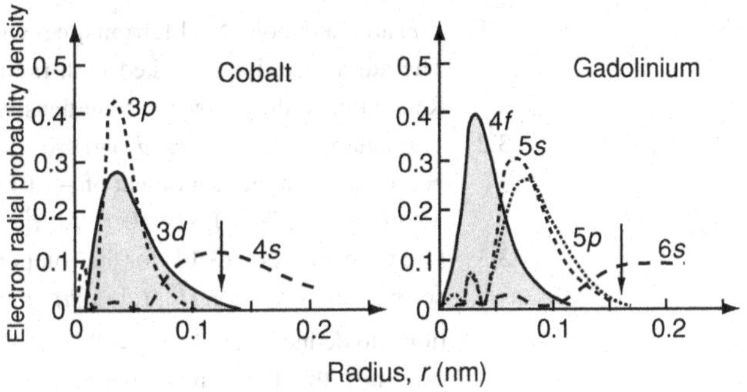

Figure 3.12 Comparison of the radial electron probability density in a $3d$ metal, Co, and a $4f$ metal, Gd. The arrow shows the interatomic spacing in each case.

radioactive the heavier they are, and magnetically interesting elements such as Bk or Cf are available only in milligram quantities, at best. The d or f electrons in compounds are usually more localized than they are in pure elements. The localized model is particularly suitable for insulating ionic compounds of $3d$ elements such as Fe_2O_3.

The delocalized model describes the magnetic electrons by wave-like extended states which form energy bands. The number of electrons in bands crossing the Fermi level is not integral, and the electrons obey Fermi–Dirac statistics. The delocalized model applies to the magnetism of $3d$ and $4d$ metals and alloys, and to their conducting compounds. It also applies at the very beginning of the $4f$ series (R = Ce) and to the light actinides from Th to Pu.

The differences between localized and delocalized magnetism are summarized in Table 3.6. Neither localized nor delocalized moments disappear above the Curie temperature; they just become disordered in the paramagnetic state, $T > T_C$.

FURTHER READING

R. H. Dicke and J. P. Wittke, *An Introduction to Quantum Mechanics*, Reading: Addison-Wesley (1960). A good basic treatment, including angular momentum.

J. Ziman, *Principles of the Theory of Solids*, London: Cambridge University Press (1965).

D. J. Craik, *Electricity, Relativity and Magnetism, A Unified Text*, Chichester: Wiley (1999).

S. Datta, *Quantum Transport: Atom to Transistor*, Cambridge: Cambridge University Press (2005).

P. W. Selwood, *Magnetochemistry* (second edition), New York: Interscience (1956). A classic monograph, focussed on diamagnetic and paramagnetic susceptibilities of atoms and molecules.

EXERCISES

3.1 Consider a hypothetical ferromagnet with an orbital moment of 1 μ_B per atom. The atoms are dense-packed with radius 0.1 nm. What is the magnetization? Relate this to the surface current density.

3.2 The neutron, an *uncharged* particle with angular momentum $\frac{1}{2}\hbar$, nevertheless possesses a magnetic moment of −1.913 nuclear magnetons. (The nuclear magneton $e\hbar/2m_p = 5.051 \times 10^{-27}$ A m^2). Why is this?

3.3 Use the expression (3.14) for the components of angular momentum to prove the commutation relation $[\hat{l}_x, \hat{l}_y] = i\hbar\hat{l}_z$ then use this relation, and cyclic permutations, to deduce that $[\hat{l}^2, \hat{l}_z] = 0$.

3.4 (a) Show that the spin angular momentum operators $\hat{s}_x$ and $\hat{s}_y$ satisfy the commutation rule $[\hat{s}_x, \hat{s}_y] = i\hbar\hat{s}_z$ and that each of them have eigenvalues $\pm\frac{1}{2}\hbar$.

(b) Use these commutation relations to show that the operators $\hat{s}^2$ and $\hat{s}_z$ commute, and hence that it is possible to measure $\hat{s}_z$ and $(\hat{s}_x^2 + \hat{s}_y^2)$ simultaneously.

3.5 If $\hat{s} = (s_x, s_y, s_z)$ represents an angular momentum vector, use (3.19) to show that $-\hat{s} = (-s_x, -s_y, -s_z)$ is *not* an angular momentum vector.

3.6 Show that the ladder operators $\hat{s}^+$ and $\hat{s}^-$ satisfy the commutation relations $[\hat{s}^2, \hat{s}^\pm] = 0$, and $[\hat{s}_z, \hat{s}^\pm] = \pm\hbar\hat{s}^\pm$.

3.7 Use the rotation matrix (3.25) to transform the spin matrix $\hat{s}_z$ into $\hat{s}_x$.

3.8 Given that $\mathcal{D}_{\uparrow,\downarrow}(\varepsilon) \propto \varepsilon^{1/2}$ show by integration that (3.40) follows.

3.9 (a) Derive (3.43) for the Pauli susceptibility of the free-electron gas.

(b) Express the Pauli susceptibility in terms of the fine structure constant α, the Bohr radius a_0 and the Fermi wave vector k_F.

(c) Use data in Tables 3.2 and 3.4 to deduce the Pauli susceptibility of copper. Compare with the result in (a).

3.10 Deduce the expression for the density of states in a quantum wire, which follows from the energy expression (3.42).

3.11 By considering the reduction of the mean free path due to cyclotron motion in a transverse magnetic field, deduce an expression for the B^2 magnetoresistance.

3.12 Consider a free electron in a semiconductor with $n = 6 \times 10^{22}$ m^{-3} moving ballistically with the Fermi velocity, which is subject to an electric field of 10^8 V m^{-1}. Initially its spin is parallel to its velocity. How far does it travel before the spin is reversed?

4 Magnetism of localized electrons on the atom

More matter, with less art.
Hamlet II ii

The quantum mechanics of a single electron in a central potential leads to classification of the one-electron states in terms of four quantum numbers. The individual electrons' spin and orbital angular momenta are coupled in an isolated many-electron ion to give total spin and orbital quantum numbers S and L. Spin-orbit coupling then operates to split the energy levels into a series of J-multiplets, the lowest of which is specified by Hund's rules. Curie-law susceptibility, $\chi = C/T$, is calculated for a general value of J. When placed in a solid, the ion experiences a crystal field due to the charge environment which disturbs the spin-orbit coupling, leaving either S or J as the appropriate quantum number. The crystal field modifies the structure of the lowest M_S or M_J magnetic sublevels which are split by the Zeeman interaction and it introduces single-ion anisotropy.

Atomic physics is concerned with the energy levels of an atom or ion and the possible transitions between them, which are usually in the optical or ultraviolet energy range (1–10 eV). Magnetism is concerned with the energy levels that are occupied at ambient temperature, which usually means only the *ground state*, and its sublevels resulting from interactions with magnetic or electric fields, which produce a splitting of less than 0.1 eV. At ambient temperature k_BT is about 25 meV.

4.1 The hydrogenic atom and angular momentum

The problem of a single electron in a central potential is treated in numerous texts on quantum mechanics and atomic physics. We summarize the results here, because we need to understand the symmetry of the one-electron states – mainly d-states – which are important in magnetism. A hydrogenic atom is composed of a nucleus of charge Ze at the origin and an electron with position labelled by r, θ, ϕ. First, consider a single electron in a central potential $\varphi_e = Ze/4\pi\epsilon_0 r$. The Hamiltonian is

$$\mathcal{H} = -\frac{\hbar^2}{2m_e}\nabla^2 - \frac{Ze^2}{4\pi\epsilon_0 r}. \tag{4.1}$$

Spherical polar coordinates are appropriate for a problem with this symmetry. The operator ∇^2 is then (Appendix C)

$$\nabla^2 = \frac{1}{r^2 \sin\theta}\left[\sin\theta\frac{\partial}{\partial r}\left(r^2\frac{\partial}{\partial r}\right) + \frac{\partial}{\partial\theta}\left(\sin\theta\frac{\partial}{\partial\theta}\right) + \frac{1}{\sin\theta}\frac{\partial^2}{\partial\phi^2}\right]. \quad (4.2)$$

This is rearranged as

$$\nabla^2 = \frac{\partial^2}{\partial r^2} + \frac{2}{r}\frac{\partial}{\partial r} + \frac{1}{r^2}\left(\frac{\partial^2}{\partial\theta^2} + \cot\theta\frac{\partial}{\partial\theta} + \frac{1}{\sin^2\theta}\frac{\partial^2}{\partial\varphi^2}\right). \quad (4.3)$$

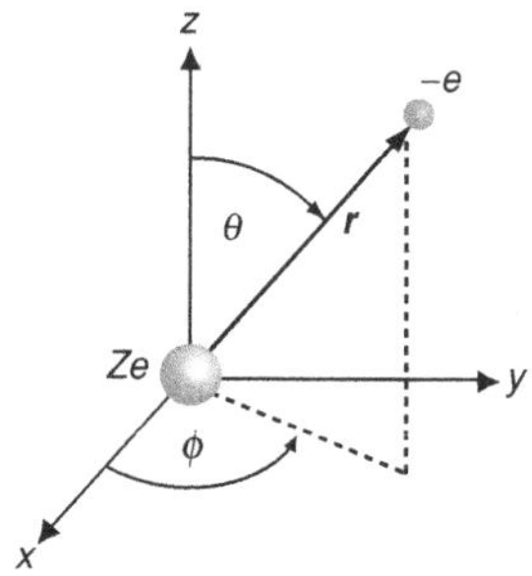

Spherical polar coordinates.

The term in parentheses which takes account of all the angular variation may be recognized as $-\hat{\boldsymbol{l}}^2/\hbar^2$ (3.15), where $\hat{\boldsymbol{l}}$ is the orbital angular momentum operator.

Schrödinger's equation for the energy levels of the atom is

$$\mathcal{H}\psi_i = \varepsilon_i\psi_i,$$

where ε_i are the energy eigenvalues and ψ_i are the corresponding wave functions. The wave function ψ determines the probability of finding the electron in a small volume $\mathrm{d}V$ at $\boldsymbol{r}$ as $\psi^*(\boldsymbol{r})\psi(\boldsymbol{r})\mathrm{d}V$, where ψ^* is the complex conjugate of ψ[1]. The Schrödinger equation is therefore:

$$\left[-\frac{\hbar^2}{2m_e}\left(\frac{\partial^2}{\partial r^2} + \frac{2}{r}\frac{\partial}{\partial r} - \frac{1}{\hbar^2 r^2}\hat{\boldsymbol{l}}^2\right) - \frac{Ze^2}{4\pi\epsilon_0 r}\right]\psi_i = \varepsilon_i\psi_i. \quad (4.4)$$

Solutions of partial differential equations like (4.4) are obtained by separation of variables; they may be written in the form

$$\psi(r,\theta,\phi) = R(r)\Theta(\theta)\Phi(\phi).$$

Each factor depends on only one of the variables.

The azimuthal (ϕ-dependent) part of the solution is an eigenfunction of $\hat{l}_z = -\mathrm{i}\hbar\partial/\partial\phi$ (3.15). The eigenvalues are $m_\ell\hbar$, with $m_\ell = 0, \pm1, \pm2, \ldots$ and the corresponding eigenfunctions are $\Phi(\phi) = \exp(\mathrm{i}m_\ell\phi)$. The polar ($\theta$-dependent) part of the solution is an associated Legendre polynomial $\Theta(\theta) = P_\ell^{m_\ell}(\theta)$, which depends on the angular momentum quantum number ℓ and also on m_ℓ. The quantum number $\ell \geq |m_\ell|$, and $\ell = 0, 1, 2, 3, \ldots$ while $m_\ell = 0, \pm1, \pm2, \ldots, \pm\ell$. There are $(2\ell+1)$ different values of m_ℓ for a given ℓ.

The product of the azimuthal and polar parts is a spherical harmonic, which depends on two integers ℓ and m_ℓ, such that $\ell \geq 0$ and $|m_\ell| \leq \ell$:

$$Y_\ell^{m_\ell}(\theta,\phi) \propto P_\ell^{m_\ell}(\theta)\mathrm{e}^{\mathrm{i}m_\ell\phi}.$$

The normalized spherical harmonics are listed in Table 4.1. The square of the orbital angular momentum $\hat{\boldsymbol{l}}^2$ has eigenvalues $\ell(\ell+1)\hbar^2$. Thus the orbital angular momentum has magnitude $\sqrt{\ell(\ell+1)}\hbar$, and its projection along Oz

[1] The complex conjugate of a complex number is obtained by replacing i by −i, where $\mathrm{i} = \sqrt{-1}$.

Table 4.1. The normalized spherical harmonics

s	$Y_0^0 = \sqrt{1/4\pi}$	
p	$Y_1^0 = \sqrt{3/4\pi}\cos\theta$	$Y_1^{\pm 1} = \mp\sqrt{3/8\pi}\sin\theta\exp(\pm i\phi)$
d	$Y_2^0 = \sqrt{5/16\pi}(3\cos^2\theta - 1)$	$Y_2^{\pm 1} = \mp\sqrt{15/8\pi}\sin\theta\cos\theta\exp(\pm i\phi)$
f	$Y_3^0 = \sqrt{7/16\pi}(5\cos^3\theta - 3\cos\theta)$	$Y_3^{\pm 1} = \mp\sqrt{21/64\pi}(5\cos^2\theta - 1)\sin\theta\exp(\pm i\phi)$
s		
p		
d	$Y_2^{\pm 2} = \sqrt{15/32\pi}\sin^2\theta\exp(\pm 2i\phi)$	
f	$Y_3^{\pm 2} = \sqrt{105/32\pi}\sin^2\theta\cos\theta\exp(\pm 2i\phi)$	$Y_3^{\pm 3} = \mp\sqrt{35/64\pi}\sin^3\theta\exp(\pm 3i\phi)$

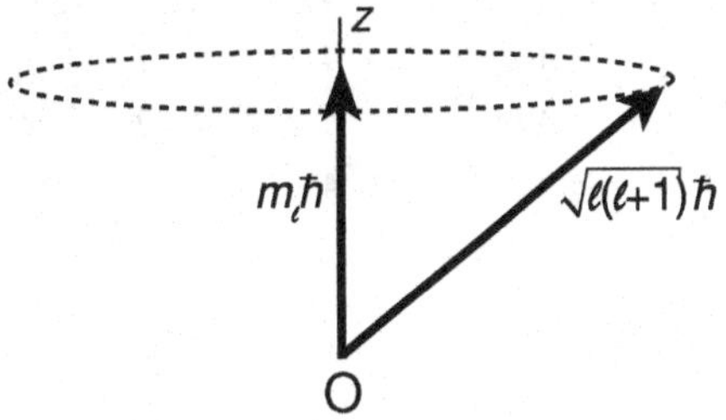

Figure 4.1

The total angular momentum and its z component. The former precesses around Oz.

can take any of $(2\ell + 1)$ values from $-\ell\hbar$ to $+\ell\hbar$ (Fig. 4.1). As explained in §3.1.4, the quantities $\boldsymbol{\ell}_z$ and $\boldsymbol{\ell}^2$ can be measured simultaneously because their operators are diagonal matrices which commute.

The *radial part* of the wavefunction $R(r)$ depends on ℓ and n. The latter also takes only integer values, and is known as the principal quantum number; $n > \ell$; hence $\ell = 0, 1, \ldots, (n-1)$. Thus there are n different l values for a given n.

$$R(r) = V_n^\ell(Zr/na_0)\exp[-(Zr/na_0)]$$

where V_n^ℓ are related to the associated Laguerre polynomials; the first one, V_1^0, equals 1. Here $a_0 = 4\pi\epsilon_0\hbar^2/m_e e^2$ is the Bohr radius (3.8), the basic length scale in atomic physics: $a_0 = 52.92$ pm.

The energy levels of the one-electron atom with a central Coulomb potential $V(r)$ depend on n, but *not* on ℓ or m_ℓ:

$$\varepsilon_n = \frac{-Z^2 me^4}{8\epsilon_0^2 h^2 n^2} = \frac{-Z^2 R_0}{n^2}. \tag{4.5}$$

Here $R_0 = me^4/8\epsilon_0^2 h^2$ is the Rydberg, the basic energy scale in atomic physics; $R_0 = 13.61$ eV.

The three quantum numbers n, ℓ, m_ℓ specify a wave function $\psi_i(r, \theta, \phi)$ with a characteristic spatial distribution of electronic charge known as an *orbital*. For historical reasons related to the appearance of lines in atomic spectra, the notation for orbitals is

$$nx_{m_\ell}, \tag{4.6}$$

Table 4.2. The hydrogenic orbitals. The number of states per orbital is $2(2\ell + 1)$

	n	ℓ	m_ℓ	m_s	States
1s	1	0	0	$\pm\frac{1}{2}$	2
2s	2	0	0	$\pm\frac{1}{2}$	2
2p	2	1	$0, \pm1$	$\pm\frac{1}{2}$	6
3s	3	0	0	$\pm\frac{1}{2}$	2
3p	3	1	$0, \pm1$	$\pm\frac{1}{2}$	6
3d	3	2	$0, \pm1, \pm2$	$\pm\frac{1}{2}$	10
4s	4	0	0	$\pm\frac{1}{2}$	2
4p	4	1	$0, \pm1$	$\pm\frac{1}{2}$	6
4d	4	2	$0, \pm1, \pm2$	$\pm\frac{1}{2}$	10
4f	4	3	$0, \pm1, \pm2, \pm3$	$\pm\frac{1}{2}$	14

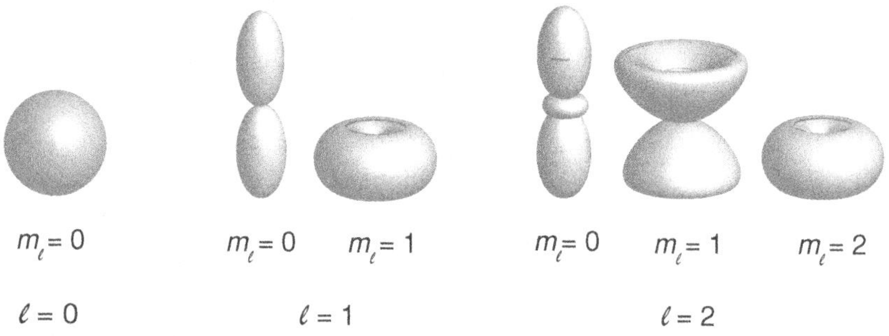

Figure 4.2

Electron densities in hydrogenic orbitals specified by their ℓ and m_ℓ values.

where $x = s, p, d, f$ stands for $\ell = 0, 1, 2, 3$. Each orbital can accommodate up to two electrons with spin $m_s = \pm\frac{1}{2}$. No two electrons can be in a state with the same four quantum numbers, which is a way of stating the Pauli exclusion principle. It is forbidden for two electrons to occupy the same quantum state because they are fermions.

The possible hydrogenic orbitals are enumerated in Table 4.2. The angular parts of the wave functions are the spherical harmonics illustrated in Fig. 4.2, and the radial functions are plotted in Fig. 4.3. Note that the radial functions have between 0 and $n - 1$ nodes, depending on the value of ℓ. The number of different orbitals for an ℓ-shell is $2\ell + 1$; the s, p, d, f shells having $\ell = 0, 1, 2, 3$ contain $1, 3, 5, 7$ orbitals respectively.

4.2 The many-electron atom

When more than one electron is present, mutual Coulomb repulsion terms between the electrons $e^2/4\pi\epsilon_0 r_{ij}$ must be added into (4.1). The Hamiltonian

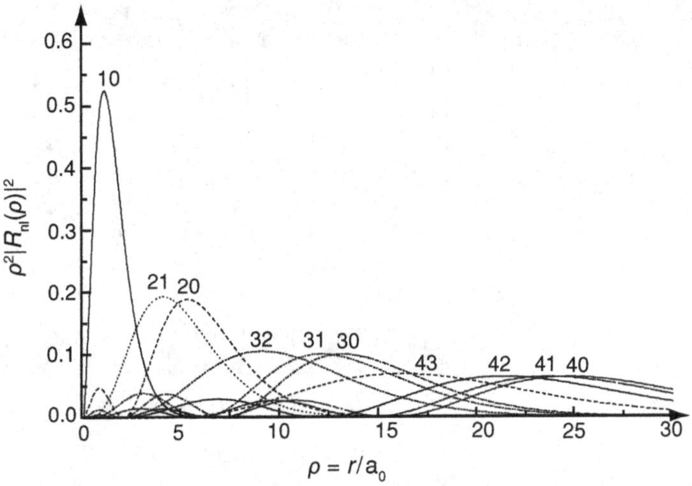

Figure 4.3

Radial probability distributions deduced from the radial parts of the wave function $R(r)$. The curves are labelled with the values of n and ℓ.

becomes

$$\mathcal{H}_0 = \sum_i [-(\hbar^2/2m_e)\nabla^2 - Ze^2/4\pi\epsilon_0 r_i] + \sum_{i<j} e^2/4\pi\epsilon_0 r_{ij}. \quad (4.7)$$

The sum over many pairs of interacting particles makes this an intractable analytical problem.[2] An appropriate way of dealing with the extra Coulomb interactions is to suppose that each electron experiences the central potential of some different, spherically symmetric charge distribution. The potential with many electrons is no longer a simple Coulomb well and the degeneracy of the energy of electrons with different ℓ-values but the same principal quantum number n is lifted. The $4s$ shell, for example, then turns out to be lower in energy than the $3d$ shell, the energy change depends on filling. The sense in which the shells are filled defines the shape of the periodic table. The potentials can be determined self-consistently. This is known as the Hartree–Foch approximation.

n	1	2	3	4	5	6
	1s	2s	3s	4s	5s	6s
		2p	3p	4p	5p	6p
			3d	4d	5d	6d
				4f	5f	6f
					5g	6g

Sequence of shell filling for a many electron atom.

When several electrons are present on the same atom, at most two of them, with opposite spin, can occupy the same orbital. The ions of interest in magnetism generally follow the L-S coupling scheme, where individual spin and orbital angular momenta add[3] to give resultant quantum numbers (here S and $L \geq 0$):

$$S = \sum s_i, \quad M_S = \sum m_{si}, \quad L = \sum \ell_i, \quad M_L = \sum m_{\ell i}.$$

[2] There is a very complicated analytical solution to the three-body problem, but there are usually many more than three particles involved in atoms.

[3] When spin-orbit coupling is very strong, as it is in the actinides, it is appropriate to first couple l_i and s_i for each electron, to form j_i, and then to couple these total angular momenta. This is the $j-j$ coupling scheme.

Table 4.3. Example of the six-electron carbon atom; $1s^2 2s^2 2p^2$

$1s$	↑↓	↑↓	↑↓	↑↓	↑↓	↑↓	↑↓	↑↓	↑↓	↑↓	↑↓	↑↓	↑↓	↑↓	↑↓
$2s$	↑↓	↑↓	↑↓	↑↓	↑↓	↑↓	↑↓	↑↓	↑↓	↑↓	↑↓	↑↓	↑↓	↑↓	↑↓
$2p_{-1}$	↑↓	↑	↑	↓	↓	↑	↑		↓	↓					
$2p_0$		↑	↓	↑	↓			↑↓			↓	↑	↓	↑	
$2p_1$						↑	↓		↑	↓	↓	↓	↑	↑	↑↓
M_L	2	1	1	1	1	0	0	0	0	0	−1	−1	−1	−1	−2
M_S	0	−1	0	0	+1	−1	0	0	0	+1	+1	0	0	−1	0

Table 4.4. Terms for the carbon atom.

Term	L	S	(M_L, M_S)
^{1}S	0	0	(0,0)
^{3}P	1	1	(1, 1)(1, 0)(1, −1)(0, 1)(0, 0)(0, −1)(−1, 1)(−1, 0)(−1, −1)
^{1}D	2	0	(2, 0)(1, 0)(0, 0)(−1, 0)(−2, 0)

Consider, for example, the carbon atom with its six electrons. The electronic configuration is $1s^2 2s^2 2p^2$. Each s-shell has no alternative but to accommodate the pair of electrons with opposite spin. However, there are 15 ways of accommodating two electrons in the $2p$-orbitals. The possibilities are listed in Table 4.3, a subscript denotes the m_l value.

The 15 states are grouped into three terms. This is done by considering the number of states with a particular combination of M_L and M_S, and decomposing them into blocks, as follows:

$M_S \backslash M_L$	−2	−1	0	1	2		$L=1, S=1$				$L=2, S=0$						$L=S=0$
−1	–	1	1	1	–		1	1	1								
0	1	2	3	2	1	=	1	1	1	+	1	1	1	1	1	+	1
1	–	1	1	1	–		1	1	1								

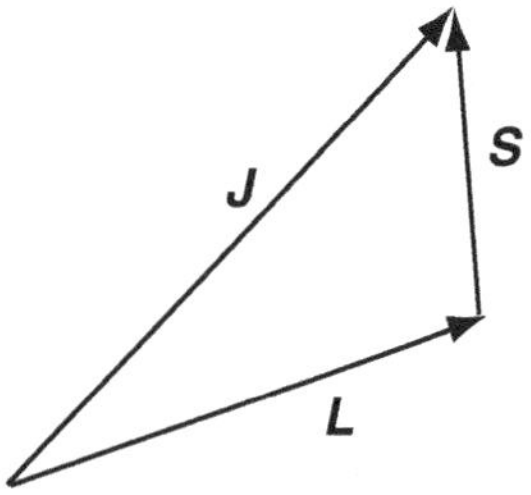

Addition of spin and orbital angular momentum in the vector model: $\boldsymbol{J} = \boldsymbol{L} + \boldsymbol{S}$.

and identifying each block as a term. Terms for carbon are listed in Table 4.4. The convention is to denote $L = 0, 1, 2, 3, 4, 5$ by S, P, D, F, G, H, and then indicate the spin multiplicity $2S+1$ by a left superscript ^{2S+1}X. The energy splitting of the terms is of order 1 eV, so the allowed $\Delta L = \pm 1$ transitions to excited states require optical excitation. The final step is to introduce spin-orbit interaction which couples L and S together to form J. Each term gives rise to several J-states, with $|L - S| \leqslant J \leqslant L + S$. There are either $(2S+1)$ or $(2L+1)$ of them, depending on whether L or S is smaller. The different J-states of a term are known as multiplets. The general notation for multiplets shows the J-value as a subscript on the term, thus

$$^{2S+1}X_J.$$

The ground state is normally the only one of the $(2J+1)$ M_J-states occupied at room temperature; it determines the magnetic properties of the atom.

Hund provided an *empirical* prescription for determining the lowest-energy state of a multielectron atom or ion ; Hund's rules are:

(1) First maximize S for the configuration.
(2) Then maximize L consistent with S.
(3) Finally couple L and S to form J: $J = L - S$ if the shell is less than half full, and $J = L + S$ if the shell is more than half full. When the shell is exactly half full, $L = 0$ and $J = S$.

The justification for the first rule is that electrons minimize their Coulomb interaction by avoiding each other, which is best achieved if they can occupy different orbitals. To this end, intra-atomic exchange tends to align the spins of different orbitals parallel. The second rule means that the electrons orbit in the same sense whenever possible. The third and weakest rule is a consequence of the sign of the spin-orbit coupling. Hund's rules only predict the *ground state*; they tell us nothing about the position and order of the excited states.

In our example of carbon, Hund's rules give $S = 1, L = 1$ and $J = 0$, so $M_J = 0$, and the ground state of the carbon atom is actually *nonmagnetic*, thanks to spin-orbit coupling. Free atoms of other elements in the same column of the magnetic periodic table are likewise nonmagnetic, as well as s^2 atoms, p^6 atoms and a few others (Table A).

More examples of the use of Hund's rules to determine the ground-state multiplet of some common magnetic ions are:

Fe^{3+}	$3d^5$	↑↑↑↑↑ |ooooo		
$S = 5/2$		$L = 0$	$J = 5/2$	$^6S_{5/2}$
Ni^{2+}	$3d^8$	↑↑↑↑↑ | ↓↓↓oo		
$S = 1$		$L = 3$	$J = 4$	$^3F_4$
Nd^{3+}	$4f^3$	↑↑↑oooo|ooooooo		
$S = 3/2$		$L = 6$	$J = 9/2$	$^4I_{9/2}$
Dy^{3+}	$4f^9$	↑↑↑↑↑↑↑ | ↓↓ooooo		
$S = 5/2$		$L = 5$	$J = 15/2$	$^6H_{15/2}$

The symbols ↑, ↓ and o show whether a $3d$- or $4f$-orbital is occupied by a ↑ or a ↓ electron, or is unoccupied. Values of L, S and J for the whole series of $3d^n$ and $4f^n$ ions are plotted in Fig. 4.4. We saw already in Table A that most free atoms possess a magnetic moment.

4.2.1 Spin-orbit coupling

We return to the relatively weak relativistic interaction responsible for Hund's third rule. The spin-orbit interaction is the origin of many of the most

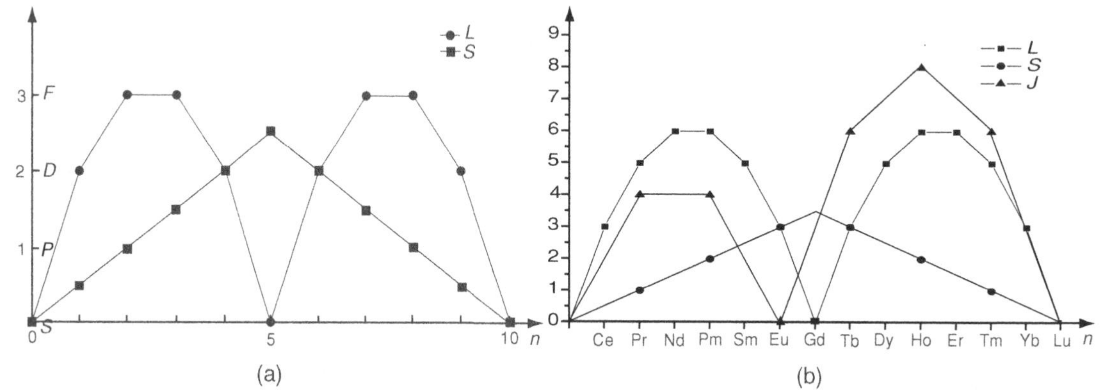

Figure 4.4

(a) L and S for the series of $3d^n$ ions; (b) L, S and J for the series of trivalent $4f$ ions.

interesting phenomena in magnetism, including magnetocrystalline anisotropy, magnetostriction, anisotropic magnetoresistance and the anomalous planar and spin Hall effect.

In the multielectron atom, the single-electron spin-orbit coupling we encountered in §3.1.3 becomes

$$\mathcal{H}_{so} = (\Lambda/\hbar^2)\hat{\boldsymbol{L}} \cdot \hat{\boldsymbol{S}}.$$

As Hund's third rule testifies, Λ is positive for the first half of the $3d$ or $4f$ series and negative for the second half. The coupling becomes large in heavy elements where the energy of the orbiting electrons is a significant fraction of their rest-mass energy $m_e c^2$. The constant Λ is related to the one-electron coupling constant λ of (3.13) by $\Lambda = \pm\lambda/2S$ for the first and second halves of the series (Table 4.5). Since $\boldsymbol{J} = \boldsymbol{L} + \boldsymbol{S}$, the identity $\boldsymbol{J}^2 = \boldsymbol{L}^2 + \boldsymbol{S}^2 + 2\boldsymbol{L} \cdot \boldsymbol{S}$ can be used to evaluate $\mathcal{H}_{so}$, as the eigenvalues of $\boldsymbol{J}^2$, $\boldsymbol{L}^2$ and $\boldsymbol{S}^2$ are all known. Turning on the spin-orbit coupling, the atomic states in the $L-S$ coupling scheme are labelled by (L, S, J, M_J),where M_J is the total magnetic quantum number.

4.2.2 The Zeeman interaction

By analogy with (3.20), the magnetic moment of an atom is represented by the operator

$$\hat{\mathfrak{m}} = -(\mu_B/\hbar)(\hat{\boldsymbol{L}} + 2\hat{\boldsymbol{S}}). \tag{4.8}$$

The Zeeman Hamiltonian for the magnetic moment in a field $\boldsymbol{B}$ is

$$\mathcal{H}_Z = (\mu_B/\hbar)(\hat{\boldsymbol{L}} + 2\hat{\boldsymbol{S}}) \cdot \boldsymbol{B}. \tag{4.9}$$

When $\boldsymbol{B}$ is applied along Oz, the Hamiltonian becomes $(\mu_B/\hbar)(\hat{\boldsymbol{L}}_z + 2\hat{\boldsymbol{S}}_z)B$.

Next we define the Landé g-factor for the multielectron atom or ion as the ratio of the component of magnetic moment along $\boldsymbol{J}$ in units of μ_B to the magnitude

Table 4.5. Spin-orbit coupling constants for ions in the $3d$ and $4f$ series, in kelvin. $\Delta\varepsilon$ is the energy of the first excited multiplet

	Ion	λ	Λ	$\Delta\varepsilon$
$3d^1$	Ti^{3+}	229	229	573
$3d^2$	V^{3+}	305	153	458
$3d^3$	Cr^{3+}	393	131	328
$3d^4$	Mn^{3+}	500	125	125
$3d^6$	Fe^{2+}	656	−164	656
$3d^7$	Co^{2+}	818	−272	1224
$3d^8$	Ni^{2+}	987	−494	3948
$4f^1$	Ce^{3+}	920	920	3220
$4f^2$	Pr^{3+}	1080	540	2700
$4f^3$	Nd^{3+}	1290	430	2365
$4f^4$	Pm^{3+}	1540	380	1900
$4f^5$	Sm^{3+}	1730	350	1225
$4f^6$	Eu^{3+}	1950	330	330
$4f^8$	Tb^{3+}	2450	−410	2460
$4f^9$	Dy^{3+}	2730	−550	4125
$4f^{10}$	Ho^{3+}	3110	−780	6240
$4f^{11}$	Er^{3+}	3510	−1170	8775
$4f^{12}$	Tm^{3+}	3800	−1900	11400
$4f^{13}$	Yb^{3+}	4140	−4140	14490

of the angular momentum, in units of $\hbar$. The g-factor is a dimensionless version of the gyromagnetic ratio γ (3.5).

Thus, in vector notation

$$g = -(\mathfrak{m} \cdot \boldsymbol{J}/\mu_B)/(|\boldsymbol{J}|^2/\hbar) = -\mathfrak{m} \cdot \boldsymbol{J}/J(J+1)\mu_B\hbar,$$

but

$$\begin{aligned} \mathfrak{m} \cdot \boldsymbol{J} &= -(\mu_B/\hbar)[(\boldsymbol{L}+2\boldsymbol{S}) \cdot (\boldsymbol{L}+\boldsymbol{S})] \\ &= -(\mu_B/\hbar)[(\boldsymbol{L}^2 + 3\boldsymbol{L}\cdot\boldsymbol{S} + 2\boldsymbol{S}^2)] \text{ since } \boldsymbol{L} \text{ and } \boldsymbol{S} \text{ commute} \\ &= -(\mu_B/\hbar)[\boldsymbol{L}^2 + 2\boldsymbol{S}^2 + (3/2)(\boldsymbol{J}^2 - \boldsymbol{L}^2 - \boldsymbol{S}^2)] \\ &= -(\mu_B\hbar)[(3/2)J(J+1) - (1/2)L(L+1) + (1/2)S(S+1)]. \end{aligned}$$

The expression for the g-factor is therefore

$$g = \tfrac{3}{2} + \{S(S+1) - L(L+1)\}/2J(J+1). \qquad (4.10)$$

The Landé g-factor is also the ratio of the z components of magnetic moment (in units of μ_B) and angular momentum (in units of $\hbar$). From the vector model, Fig. 4.5, where the magnetic moment precesses rapidly about $\boldsymbol{J}$, the ratio

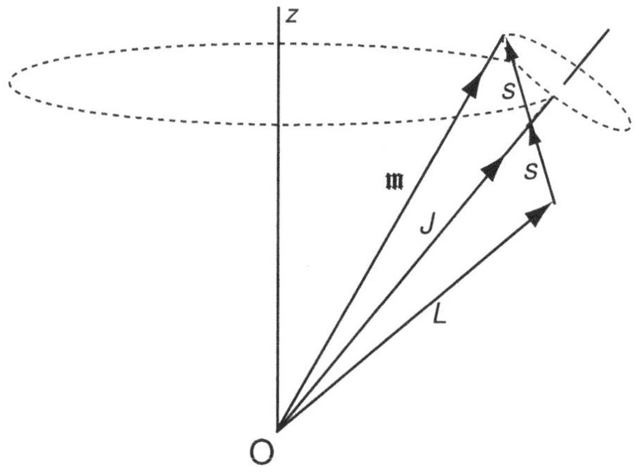

Figure 4.5

The vector model of the atom, showing both the magnetic moment and the angular momentum. $\mathfrak{m}$ precesses rapidly around J, and the time-averaged moment then precesses around Oz.

$\mathfrak{m}_z/J_z = \mathfrak{m} \cdot \boldsymbol{J}/J^2 = -g\mu_B/\hbar$, so we can write the projected/average $\mathfrak{m}$ in terms of the g-factor as

$$\mathfrak{m} = -(g\mu_B/\hbar)\boldsymbol{J}.$$

The level with $M_J = -J$ is lowest.

The Zeeman energy in a magnetic field $\boldsymbol{B}$ applied along Oz is $\varepsilon_Z = -\mathfrak{m}_z B$. This is $-(\mathfrak{m}_z/J_z)J_z B = (g\mu_B/\hbar)J_z B$. Hence the Zeeman energy is

$$\varepsilon_Z = g\mu_B M_J B.$$

Note the magnitude of the energies involved here. The splitting of two adjacent Zeeman energy levels is $g\mu_B B$, which is of order 1 K for a field of 1 T. Hundreds of tesla would be needed to establish a significant population surplus in the $M_J = -J$ ground state at room temperature. No such steady field has ever been produced in the laboratory. Matters are easier in the liquid helium temperature range, where normal laboratory fields of a few tesla will suffice to establish a preponderant population of the ground state, and saturate the magnetization of the ion. The entropy S of the ion is $k_B \ln \Omega$, where Ω is the number of configurations. When the $2J+1$ M_J configurations are equally populated, $S = k_B \ln(2J+1)$, but if only one M_J sublevel is occupied, $S = 0$.

An example to summarize the discussion up to this point is given in Fig. 4.6, where the energy levels of a free ion, $Co^{2+}3d^7$, are shown. According to Hund's rules, the $J = \frac{9}{2}$ multiplet is the ground state.

4.3 Paramagnetism

We now examine in more detail the response of a localized magnetic moment $\mathfrak{m}$ to an applied magnetic field $\boldsymbol{H} = \boldsymbol{B}/\mu_0$ in free space. The $'$ is dropped because

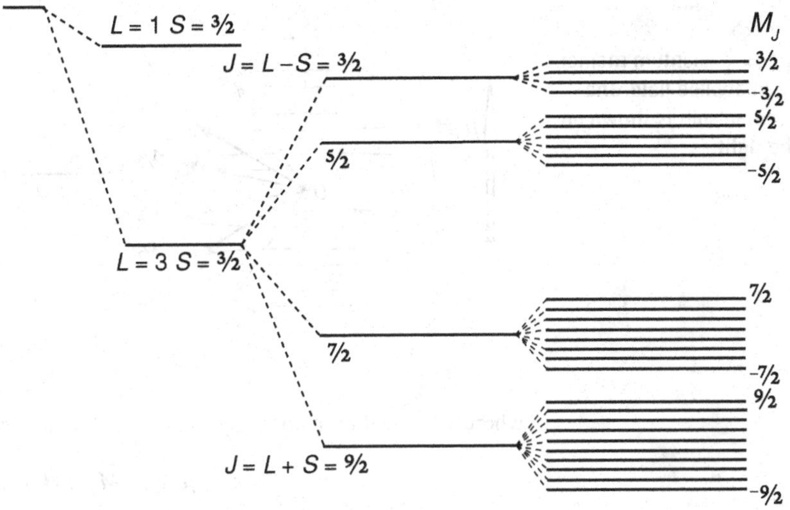

Figure 4.6

The energy levels for a free ion with electronic configuration $3d^7$: Co^{2+} $S = \frac{3}{2}, L = 3, J = \frac{9}{2}; g = \frac{4}{3}$.

there is no distinction between applied field and internal field for a single atom or particle. It is instructive to look first at the general quantum case where the moment $\mathfrak{m} = -g\mu_B \boldsymbol{J}/\hbar$ can make any of $2J + 1$ polar angles with the applied field, and then at the theory for classical magnetic moments which can adopt any orientation whatsoever with respect to the applied field. We have already presented the theory for a spin-$\frac{1}{2}$ quantum system in §3.2.4, where the moment $\mathfrak{m} = -g\mu_B \boldsymbol{S}/\hbar$ can have only one of two projections along the applied field. The classical and extreme quantum limits correspond to $J \longrightarrow \infty$ and $J = \frac{1}{2}$, respectively.

4.3.1 Brillouin theory

The general expression for the thermodynamic average value of any quantity q is $\langle q \rangle = \mathrm{Tr}[q_i \exp(-\varepsilon_i/k_BT)]/\mathcal{Z}$, where $\mathcal{Z} = \mathrm{Tr}\exp(-\varepsilon_i/k_BT)$ is the partition function. The trace (Tr) is the sum over the i energy states, and q_i is the value of q in the ith one. This amounts to taking an average over the i states, weighted by their Boltzmann populations. $\mathcal{Z}$ is the normalization factor. Evaluating the thermodynamic average $\langle \mathfrak{m} \rangle$,

$$\langle \mathfrak{m} \rangle = \frac{\sum_i \mathfrak{m}_i \exp(-\varepsilon_i/k_BT)}{\sum_i \exp(-\varepsilon_i/k_BT)}. \tag{4.11}$$

The extreme quantum limit is the case $J = 1/2$. Usually this arises when $S = 1/2$, $L = 0$. There are only two energy levels and two orientations of the moment relative to the applied field (which, as usual, is along Oz), corresponding to the $|\uparrow\rangle$ and $|\downarrow\rangle$ states. Then (4.11) for the z component of $\mathfrak{m}$ reduces to

$$\langle \mathfrak{m}_z \rangle = g\mu_B J \tanh x, \tag{4.12}$$

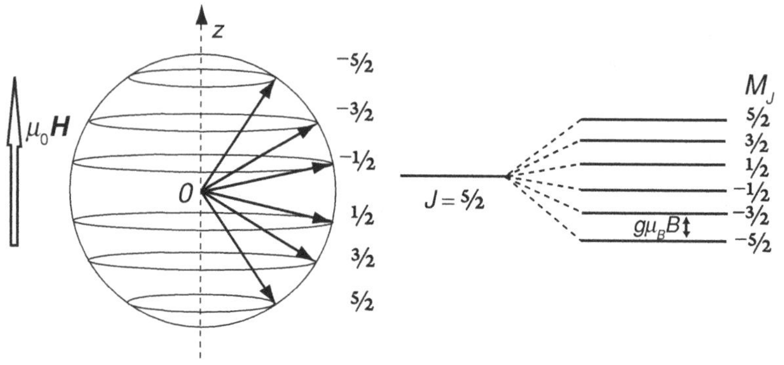

Figure 4.7

A $J = \frac{5}{2}$ quantum moment in an applied field. The Zeeman shift is shown on the right.

where x is the dimensionless ratio of the Zeeman energy to the thermal energy.

$$x = \mu_0 g \mu_B M_j H / k_B T. \tag{4.13}$$

When $J = S = \frac{1}{2}$, $g = 2$, this becomes $\langle \mathfrak{m}_z \rangle = \mathfrak{m}_z \tanh x$, with $\mathfrak{m}_z = \mu_B$ and $x = \mu_0 \mu_B H / k_B T$, as seen in §3.2.4. In small fields, $\tanh x \approx x$, and if there are n atoms per unit volume, the susceptibility $n\langle \mathfrak{m}_z \rangle / H$ is

$$\chi = \mu_0 n g^2 \mu_B^2 J^2 / k_B T. \tag{4.14}$$

Equation (4.14) reduces to (3.34) for $J = 1/2$, $g = 2$. Many two-level systems in physics are treated using this theory, by assigning them a pseudospin $S = \frac{1}{2}$; the two levels correspond to $M_s = \pm\frac{1}{2}$, and their separation is $2\mu_0 \mu_B H$.

The general quantum case was treated by Léon Brillouin; $\mathfrak{m}$ is $-g\mu_B \boldsymbol{J}/\hbar$, and x is $\mu_0 g \mu_B M_J H / k_B T$. There are now $2J + 1$ energy levels $\varepsilon_i = +\mu_0 g \mu_B M_J H$, with moments $\mathfrak{m}_{zi} = -g\mu_B M_J$, where $M_J = J$, $J - 1, J - 2, \ldots, -J$. The sums in (4.11) each have $2J + 1$ terms. The example of $J = \frac{5}{2}$ is illustrated in Fig. 4.7.

To calculate the susceptibility, we take the limit $x \ll 1$. Susceptibility is the initial slope of the magnetization curve. The exponentials in (4.11) are expanded as $\exp(x) \approx 1 + x + \cdots$. Hence, keeping only the leading terms,

$$\langle \mathfrak{m}_z \rangle = \frac{\sum_{-J}^{J} -g\mu_B M_J (1 - \mu_0 g \mu_B M_J H / k_B T)}{\sum_{-J}^{J} (1 - \mu_0 g \mu_B M_J H / k_B T)}.$$

Using the identities

$$\sum_{-J}^{J} 1 = 2J + 1, \qquad \sum_{-J}^{J} M_J = 0, \qquad \sum_{-J}^{J} M_J^2 = J(J+1)(2J+1)/3,$$

we find $\langle \mathfrak{m}_z \rangle = \mu_0 g^2 \mu_B^2 J(J+1) H / 3 k_B T$. The susceptibility is $n\langle \mathfrak{m} \rangle / H$, thus the general form of Curie's law is $\chi = C/T$, where

$$C = \frac{\mu_0 n g^2 \mu_B^2 J(J+1)}{3 k_B}. \tag{4.15}$$

To make a connection with (3.34), the susceptibility is written in terms of an effective moment $\mathfrak{m}_{eff} = g\mu_B\sqrt{J(J+1)}$ (or an effective Bohr magneton number $p_{eff} = g\sqrt{J(J+1)}$) as

$$\chi = \frac{\mu_0 n \mathfrak{m}_{eff}^2}{3k_B T}.$$

The susceptibility depends on the square of the magnitude or 'length' of $\mathfrak{m}$, and not on its projection $\mathfrak{m}_z$. The Curie constant is

$$C = \mu_0 n \mathfrak{m}_{eff}^2 / 3k_B. \tag{4.16}$$

A typical value of C obtained for $n = 6 \times 10^{28}$ m^{-3}, $g = 2$ and $J = 1$ is 1.3 K. The molar Curie constant C_{mol}, obtained by setting $n = N_A$, is a useful quantity. Its numerical value is

$$C_{mol} = 1.571 \times 10^{-6} p_{eff}^2.$$

In order to calculate the complete magnetization curve, we set $y = \mu_0 g\mu_B H/k_B T$ and then use (4.11) and the result $\mathrm{d}(\ln z)/\mathrm{d}y = (1/z)\mathrm{d}z/\mathrm{d}y$ to write the thermodynamic average $\langle \mathfrak{m}_z \rangle = \langle -g\mu_B M_J \rangle$ as

$$\langle \mathfrak{m}_z \rangle = g\mu_B \frac{\partial}{\partial y}\left[\ln \sum_{-J}^{J} \exp(-M_J y)\right].$$

The sum over the energy levels must be evaluated; it can be written as $\exp(Jy)[1 + r + r^2 + \cdots + r^{2J}]$, where $r = \exp(-y)$. The sum of the geometric progression is $(r^{2J+1} - 1)/(r - 1)$. Therefore, multiplying top and bottom by $\exp(y/2)$,

$$\begin{aligned}\sum_{-J}^{J} \exp(M_J y) &= \{\exp[-(2J+1)y] - 1\}\exp(Jy)/[\exp(-y) - 1] \\ &= \sinh[(2J+1)y/2]/\sinh(y/2).\end{aligned}$$

Hence

$$\begin{aligned}\langle \mathfrak{m}_z \rangle &= g\mu_B \partial \ln[\sinh((2J+1)y/2)]/[\sinh(y/2)]]/\partial y \\ &= (g\mu_B/2)[(2J+1)\coth((2J+1)y/2) - \coth(y/2)].\end{aligned}$$

Setting $x = Jy$ we obtain finally

$$\langle \mathfrak{m}_z \rangle = \mathfrak{m}_0 \left\{\frac{2J+1}{2J}\coth\frac{2J+1}{2J}x - \frac{1}{2J}\coth\frac{x}{2J}\right\}, \tag{4.17}$$

where $\mathfrak{m}_0 = g\mu_B J$ is the maximum magnitude of the moment and the quantity in braces is the Brillouin function $\mathcal{B}_J(x)$:

$$\langle \mathfrak{m}_z \rangle = \mathfrak{m}_0 \mathcal{B}_J(x). \tag{4.18}$$

It reduces to the Langevin function in the limit $J \longrightarrow \infty$ and to $\tanh x$ when $J = 1/2$ and $g = 2$ (4.12), Fig. 4.8.

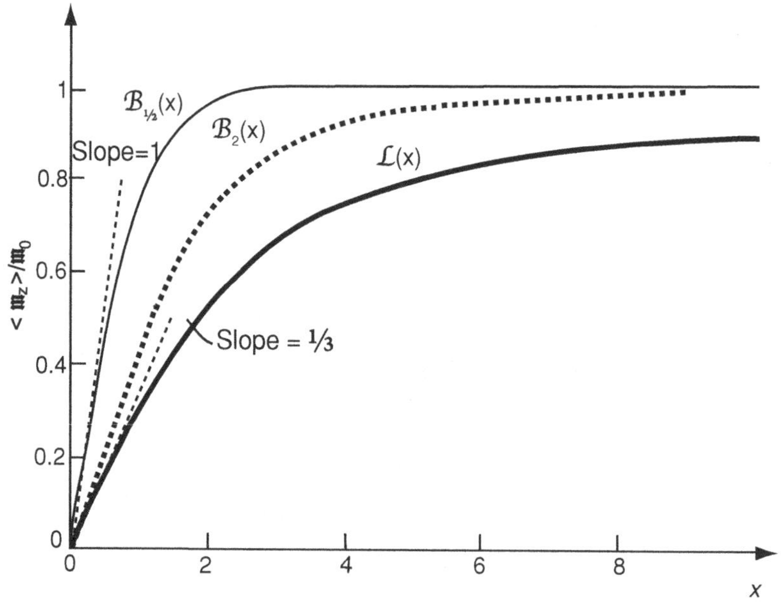

Figure 4.8

Comparison of the Brillouin functions for $J = \frac{1}{2}$ and $J = 2$ with the Langevin function, which is the classical limit. The slope at the origin of the Langevin function is $\frac{1}{3}$.

In the small-x limit,

$$\mathcal{B}_J(x) \approx \frac{(J+1)}{3J}x - \frac{[(J+1)^2 + J^2](J+1)}{90J^2}x^3 + \cdots. \tag{4.19}$$

The leading term gives the Curie law susceptibility $\chi = g^2\mu_0 nJ(J+1)\mu_B^2/3k_BT = C/T$.

The theory of localized magnetism gives an excellent account of magnetically dilute $3d$ and $4f$ salts, where the magnetic moments do not interact with each other, for example in alums such as $KCr(SO_4)_2 \cdot 12H_2O$, where Cr^{3+} ions are well separated by sulphate ions and water of crystallization. Except in large fields or at very low temperatures, the $M(H)$ response is linear. The excellence of the theory is illustrated by the fact that all the data for quite different temperatures superimpose on a single Brillouin curve when plotted as a function of $x \sim H/T$, as shown in Fig. 4.9.

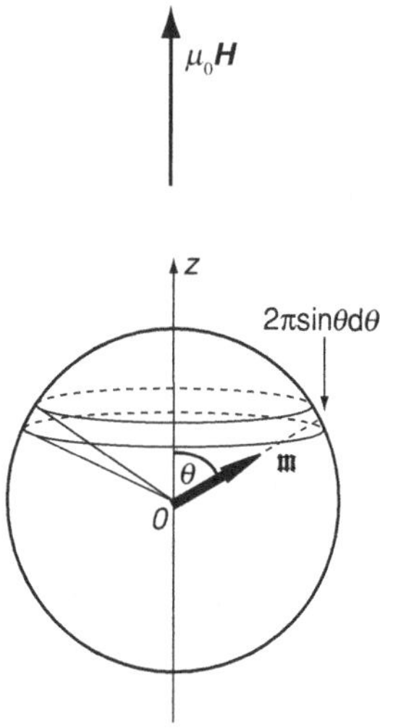

A classical moment in an applied field.

4.3.2 Langevin theory

The classical theory of paramagnetism, which is the limit of quantum theory when $J \longrightarrow \infty$, was worked out in 1905 by Paul Langevin. Colloidal ferromagnetic nanoparticles suspended in a liquid or small grains of ferromagnetic minerals, usually magnetite, dispersed in a rock are examples of systems where the classical theory is expected to apply. Each atom or particle has a macroscopic moment $\mathfrak{m}$ which can take any orientation relative to the field applied in

Figure 4.9

Reduced magnetization curves for three paramagnetic salts, compared with Brillouin theory predictions. (W. E. Henry, *Phys. Rev.* 88, 559 (1952)).

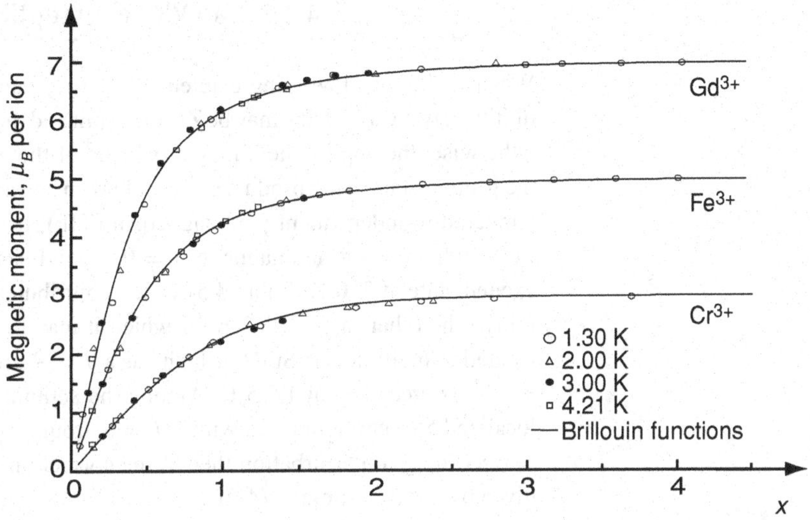

the z-direction $\boldsymbol{H} = \boldsymbol{B}/\mu_0$. The Zeeman energy (2.73) $\varepsilon = -\mathfrak{m} \cdot \boldsymbol{B}$ gives

$$\varepsilon(\theta) = -\mu_0 \mathfrak{m} H \cos\theta.$$

The probability $P(\theta)$ that the moment makes an angle θ with Oz is the product of a Boltzmann factor $\exp[-\varepsilon(\theta)/k_B T]$ and a geometric factor $2\pi \sin\theta$. Hence $P(\theta) = \kappa 2\pi \sin\theta \exp(\mu_0 \mathfrak{m} H \cos\theta / k_B T)$, where the constant κ is determined by the normalization condition $\int_0^\pi P(\theta) d\theta = N$. Here N is the number of atoms or particles in the system. Then

$$\langle \mathfrak{m}_z \rangle = \frac{\int_0^\pi \mathfrak{m} \cos\theta \, P(\theta) d\theta}{\int_0^\pi P(\theta) d\theta}. \tag{4.20}$$

To evaluate the integrals, let $a = \cos\theta$, $da = -\sin\theta d\theta$ and define the dimensionless ratio of magnetic to thermal energy $x = \mu_0 \mathfrak{m} H / k_B T$. The result is $\langle \mathfrak{m}_z \rangle = \mathfrak{m} \mathcal{L}(x)$, where

$$\mathcal{L}(x) = (\coth x - 1/x). \tag{4.21}$$

The quantity $\mathcal{L}(x)$, known as the Langevin function, is plotted in Fig. 4.8. The leading terms in the expansion when x is small are $\mathcal{L}(x) = x/3 - x^3/45 + 2x^5/945 - \cdots$. At low fields or high temperatures, $\mathcal{L}(x) \approx x/3$, which is just $\frac{1}{3}$ of the value in extreme quantum case $J = \frac{1}{2}$. The susceptibility of an ensemble of n moments per cubic metre is $\chi = n\langle \mathfrak{m}_z \rangle / H$, hence

$$\chi = \mu_0 n \mathfrak{m}^2 / 3 k_B T. \tag{4.22}$$

This is the classical form of the famous Curie law $\chi = C/T$, where $C = \mu_0 n \mathfrak{m}^2 / 3 k_B$ is the Curie constant.

At high fields, $x \gg 1$, the moments are aligned and the magnetization saturates. The high-field limit of the Langevin function is $\mathcal{L}(x) \approx 1 - 1/x$.

4.3.3 Van Vleck susceptibility

When an ion has low-lying excited states, they can contribute to the susceptibility in two ways. They may be Zeeman split and thermally populated as usual. Otherwise, the applied field may mix some of the excited-state character into the ground state. This produces a term known as van Vleck paramagnetism or temperature-independent paramagnetism (TIP). It is especially important for Eu^{3+}, which has a nonmagnetic $J = 0$ ground state, but a low-lying $J = 1$ excited state at 330 K (Table 4.5) which contributes a paramagnetic susceptibility. The other rare-earth ion for which it may be necessary to consider the excited J-multiplets is Sm^{3+}, which has a $J = \frac{5}{2}$ ground state and a low-lying $J = \frac{7}{2}$ excited state at 1225 K. Among the actinides, Bk^{5+} and Cm^{4+} have a localized $5f^6$ configuration, with a $J = 0$ ground state.

According to perturbation theory, the correction to the energy of the ion is given by the Hamiltonian (4.9)

$$\varepsilon_1 = \left[\frac{\mu_B}{\hbar}\right]^2 \frac{|\langle g|(\hat{\boldsymbol{L}} + 2\hat{\boldsymbol{S}}) \cdot \boldsymbol{B}|e\rangle|^2}{\Delta\varepsilon}, \tag{4.23}$$

where g and e represent the ground and excited states, respectively, and $\Delta\varepsilon$ is their energy separation. From the expression for the Gibbs free energy of a paramagnet in an applied field, $-\frac{1}{2}BM$ with $M = \chi B/\mu_0$, it follows that $\chi = 2n\mu_0\varepsilon_1/B^2$, where n is the number of ions per unit volume. Hence the susceptibility depends on the off-diagonal matrix elements of the magnetic moment:

$$\chi = 2n\mu_0 \left[\frac{\mu_B}{\hbar}\right]^2 \frac{|\langle g|\hat{\boldsymbol{L}}_z + 2\hat{\boldsymbol{S}}_z|e\rangle|^2}{\Delta\varepsilon}. \tag{4.24}$$

4.3.4 Adiabatic demagnetization

A remarkable application of the Curie-law paramagnetism of dilute salts is the achievement of temperatures in the millikelvin range by adiabatic diamagnetization.

In zero applied field, the energy levels of each ion are $(2J + 1)$-fold degenerate. The number of possible configurations for N ions in the sample is $\Omega = (2J + 1)^N$. The corresponding entropy is $S = k_B \ln \Omega$ or

$$S = R \ln(2J + 1), \tag{4.25}$$

per mole, where $R = N_A k_B$ is the gas constant, 8.315 J mol^{-1}. The entropy is reduced by applying a magnetic field, which leads to preferential occupation of the energy levels with more negative M_J values, according to the Boltzmann probability $\exp(-g\mu_B M_J \mu_0 H/k_B T)$. At zero temperature, all the ions are in the $M_J = -J$ state. There is only one configuration for each one, so $\Omega = 1^N$ and $S = 0$. Increasing temperature leads to absorption of heat by the

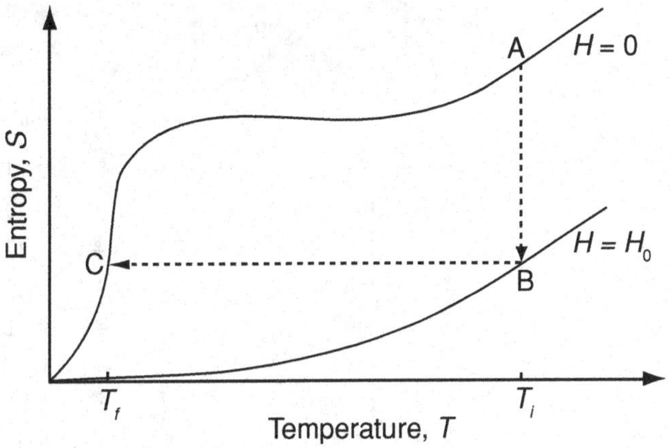

Figure 4.10

Cooling by adiabatic demagnetization. The two curves are $S(T)$ for zero field, $H = 0$, and an applied field, $H = H_0$. The material is first magnetized isothermally at T_i by the applied field H_0 (AB), and then cooled adiabatically to T_f when the field H is reduced almost to zero (BC).

system and population of the higher Zeeman levels, which increases the entropy. Increasing field favours preferential population of the lower Zeeman levels, which decreases the entropy. Since the population of the energy levels depends solely on the Boltzmann factor, entropy is a monotonically decreasing function of H/T. The entropy can be calculated from the Helmholtz free energy $F = -Nk_BT \ln \mathcal{Z}$, where $\mathcal{Z}$ is the partition function, using the thermodynamic relation $S = -(\partial F/\partial T)_H$.

An adiabatic process is one where there is no exchange of heat between a system and its surroundings. Since $\delta Q = T\delta S$, entropy is conserved in such a process. Adiabatic demagnetization is usually carried out in a cryostat, where, for example, a pill of cerium magnesium nitrate is initially in thermal contact via helium exchange gas with a bath of pumped liquid helium at 1.2 K or liquid ^{3}He at 0.3 K. The first step (Fig. 4.10) is isothermal magnetization in a field of several teslas, generated by a superconducting magnet. The second step is to cut the thermal contact with the bath by evacuating the exchange gas, and reduce the field to zero. During the adiabatic demagnetization, there is no change of entropy $S(H/T)$ so it follows that

$$\frac{H_i}{T_i} = \frac{H_f}{T_f}, \tag{4.26}$$

where i and f denote the initial and final states. If H_f were really zero, the procedure would attain the absolute zero of temperature! Unfortunately, there is always some residual field due to the dipole–dipole interaction of the ionic moments. These stray fields are of order a millitesla in dilute salts, so the best temperatures that can be reached are in the millikelvin range.

Microkelvin temperatures are attainable by adding a second stage where adiabatic demagnetization is applied to the nuclei. Nuclear magnetic moments are three orders of magnitude smaller than electronic ones, so the stray fields are proportionately less, but the angular momentum $I\hbar$ and the entropy in the disordered state, $R \ln(2I + 1)$ per mole, are similar. Copper, which has two

Table 4.6. The $4f$ ions. The paramagnetic moment m_{eff} and the saturation moment m_0 are in units of μ_B

$4f^n$		S	L	J	g	$m_0 = gJ$	$m_{eff} = g\sqrt{J(J+1)}$	m_{eff}^{exp}
1	Ce^{3+}	$\frac{1}{2}$	3	$\frac{5}{2}$	$\frac{6}{7}$	2.14	2.54	2.5
2	Pr^{3+}	1	5	4	$\frac{4}{5}$	3.20	3.58	3.5
3	Nd^{3+}	$\frac{3}{2}$	6	$\frac{9}{2}$	$\frac{8}{11}$	3.27	3.52	3.4
4	Pm^{3+}	2	6	4	$\frac{3}{5}$	2.40	2.68	
5	Sm^{3+}	$\frac{5}{2}$	5	$\frac{5}{2}$	$\frac{2}{7}$	0.71	0.85	1.7
6	Eu^{3+}	3	3	0	0	0	0	3.4
7	Gd^{3+}	$\frac{7}{2}$	0	$\frac{7}{2}$	2	7.0	7.94	8.9
8	Tb^{3+}	3	3	6	$\frac{3}{2}$	9.0	9.72	9.8
9	Dy^{3+}	$\frac{5}{2}$	5	$\frac{15}{2}$	$\frac{4}{3}$	10.0	10.65	10.6
10	Ho^{3+}	2	6	8	$\frac{5}{4}$	10.0	10.61	10.4
11	Er^{3+}	$\frac{3}{2}$	6	$\frac{15}{2}$	$\frac{6}{5}$	9.0	9.58	9.5
12	Tm^{3+}	1	5	6	$\frac{7}{6}$	7.0	7.56	7.6
13	Yb^{3+}	$\frac{1}{2}$	3	$\frac{7}{2}$	$\frac{8}{7}$	4.0	4.53	4.5

isotopes each with $I = 3/2$, is often used. It is first cooled to the millikelvin range where the nuclei are polarized in an applied field of a few tesla, and then demagnetized adiabatically.

4.4 Ions in solids; crystal-field interactions

In practice, it should be possible to deduce m_0, the maximum value of m_z, by saturating the magnetization at low temperature, thereby obtaining the value of $gJ\mu_B$. The magnitude of the moment can be deduced from the paramagnetic susceptibility, which yields $m_{eff}^2 = g^2\mu_B^2 J(J+1)$. In practice, these measurements are made on magnetically dilute salts, not free ions. Results of these measurements are summarized in the last columns of Tables 4.6 and 4.7, where m_{eff}, listed in units of Bohr magnetons, is just the effective Bohr magneton number p_{eff}.

From data on the $4f$ ions in Table 4.6 it is seen that the experimentally measured m_{eff} corresponds to $g\mu_B\sqrt{J(J+1)}$, except for radioactive Pm, for which it has never been measured, and Sm^{3+} and Eu^{3+}, where there is an additional van Vleck contribution to the susceptibility from excited multiplets. J is the good quantum number for the rare-earth series. However, Table 4.7 tells a different story for the $3d$ ions. There is a large discrepancy between $gJ\mu_B$ and m_0 for all except the $3d^5$ ions, for which Hund's second rule gives $L = 0$ and $J = S$. In fact, $m_0 \approx gS\mu_B$ and $m_{eff} \approx g\mu_B\sqrt{S(S+1)}$. It appears that the magnetism of the $3d$ series is due to the spin moment, with little or no orbital contribution. S is the good quantum number for the $3d$ series. An exception

Table 4.7. The $3d$ ions. $\mathfrak{m}_{eff}$ is in units of μ_B

$3d^n$		S	L	J	g	$m_{eff} = g\sqrt{J(J+1)}$	$m_{eff} = g\sqrt{S(S+1)}$	m_{eff}^{exp}
1	Ti^{3+}, V^{4+}	$\frac{1}{2}$	2	$\frac{3}{2}$	$\frac{4}{5}$	1.55	1.73	1.7
2	Ti^{2+}, V^{3+}	1	3	2	$\frac{2}{3}$	1.63	2.83	2.8
3	V^{2+}, Cr^{3+}	$\frac{3}{2}$	3	$\frac{3}{2}$	$\frac{2}{5}$	0.78	3.87	3.8
4	Cr^{2+}, Mn^{3+}	2	2	0			4.90	4.9
5	Mn^{2+}, Fe^{3+}	$\frac{5}{2}$	0	$\frac{5}{2}$	2	5.92	5.92	5.9
6	Fe^{2+}, Co^{3+}	2	2	4	$\frac{3}{2}$	6.71	4.90	5.4
7	Co^{2+}, Ni^{3+}	$\frac{3}{2}$	3	$\frac{9}{2}$	$\frac{4}{3}$	6.63	3.87	4.8
8	Ni^{3+}	1	3	4	$\frac{5}{4}$	5.59	2.83	3.2
9	Cu^{2+}	$\frac{1}{2}$	2	$\frac{5}{2}$	$\frac{6}{5}$	3.55	1.73	1.9

seems to be Co^{2+}, where an orbital contribution raises $\mathfrak{m}_{eff}$ significantly above the spin only value.

Summarizing, for free ions:

- Filled electronic shells are not magnetic. A $\uparrow$ electron is paired with a $\downarrow$ electron in each orbital.
- Only partly filled shells may possess a magnetic moment.
- The magnetic moment is related to the angular momentum by $\mathfrak{m} = -g(\mu_B/\hbar)\boldsymbol{J}$, where the quantum number J denotes the total angular momentum. For a given configuration, the values of J and g follow from Hund's rules and (4.10).

The third point in the summary has to be modified for ions in solids.

- Orbital angular momentum for $3d$ ions is quenched. The magnitude of the spin-only moment is $\mathfrak{m} \approx g\mu_B S$, with $g = 2$.
- Certain crystallographic directions become *easy axes* of magnetization.

These two effects result from electrostatic fields present in the crystal.

4.4.1 Crystal fields

When an ion or atom is embedded in a solid, the Coulomb interaction of its electronic charge distribution $\rho_0(\boldsymbol{r})$ with the surrounding charges in the crystal must be considered. This is the crystal-field interaction. Both the quenching of orbital angular momentum and the development of single-ion anisotropy are due to the crystal electric field.

The potential $\varphi_{cf}(\boldsymbol{r})$ produced by the distribution of charge $\rho(\boldsymbol{r}')$ outside the ion is

$$\varphi_{cf}(\boldsymbol{r}) = \int \frac{\rho(\mathbf{r}')}{4\pi\epsilon_0|\boldsymbol{r} - \boldsymbol{r}'|} \mathrm{d}^3 r'. \tag{4.27}$$

Table 4.8. Typical magnitudes of energy terms (in K) for $3d$ and $4f$ ions in solids

	$\mathcal{H}_0$	$\mathcal{H}_{so}$	$\mathcal{H}_{cf}$	$\mathcal{H}_Z$
$3d$	$1{-}5 \times 10^4$	$10^2{-}10^3$	$10^4{-}10^5$	1
$4f$	$1{-}6 \times 10^5$	$1{-}5 \times 10^3$	$\approx 3 \times 10^2$	1

$\mathcal{H}_Z$ is for 1 T

Here $1/|\boldsymbol{r} - \boldsymbol{r}'|$ can be expanded in spherical harmonics using spherical polar coordinates $\boldsymbol{r} = (r, \theta, \phi)$ and $\boldsymbol{r}' = (r', \theta', \phi')$:

$$\frac{1}{|\boldsymbol{r} - \boldsymbol{r}'|} = \frac{1}{r'} \sum_{n=0}^{\infty} \frac{4\pi}{(2n+1)} \left(\frac{r}{r'}\right)^n \sum_{m=-n}^{n} (-1)^m Y_n^{-m}(\theta', \phi') Y_n^m(\theta, \phi).$$

Hence

$$\varphi_{cf}(r, \theta, \phi) = \sum_{n=0}^{\infty} \sum_{m=-n}^{n} r^n \gamma_{nm} Y_n^m(\theta, \phi), \tag{4.28}$$

where

$$\gamma_{nm} = \frac{4\pi}{(2n+1)} \int \frac{\rho(\boldsymbol{r}')(-1)^m Y_n^{-m}(\theta', \phi')}{r'^{n+1}} \mathrm{d}^3 r'.$$

The complete Hamiltonian of an ion in a solid has four terms:

$$\mathcal{H} = \mathcal{H}_0 + \mathcal{H}_{so} + \mathcal{H}_{cf} + \mathcal{H}_Z.$$

$\mathcal{H}_0$ takes account of the Coulomb interactions among the electrons and between the electrons and the nucleus, giving rise to the total spin and orbital angular momenta $\boldsymbol{L}$ and $\boldsymbol{S}$. The terms $\mathcal{H}_{so}$, $\mathcal{H}_{cf}$ and $\mathcal{H}_Z$ are the spin-orbit, crystal-field and Zeeman terms, respectively. The crystal-field Hamiltonian is

$$\mathcal{H}_{cf} = \int \rho_0(\boldsymbol{r}) \varphi_{cf}(\boldsymbol{r}) \mathrm{d}^3 r. \tag{4.29}$$

Relative magnitudes of these interactions are given in Table 4.8.

$\mathcal{H}_{cf}$ is relatively weak in the rare-earths because the $4f$ shell is buried deep inside the atom (Fig. 3.12) and the crystal-field potential φ_{cf} is shielded by the outer electrons. $\mathcal{H}_{so}$ must be considered before $\mathcal{H}_{cf}$ in any perturbation scheme for evaluating the energies of $4f$ ions. J is a good quantum number, and the $|J, M_J\rangle$-states form a basis. The crystal field is treated as a perturbation. The converse is true for $3d$ ions, where the $3d$ shell is outermost. The crystal field acts on the states $|L, M_L, S, M_S\rangle$ resulting from the Hamiltonian $\mathcal{H}_0$ for the $3d$ transition metal series. The orbital ground states for the $3d$ ions can only have $L = 0, 2$ or 3, that is an S, D or F state (Fig. 4.4(a)).

4.4.2 One-electron states

Generally, $4f$ electrons are localized in solids of any description, and $3d$ electrons are delocalized in metals, but usually localized in oxides and other ionic

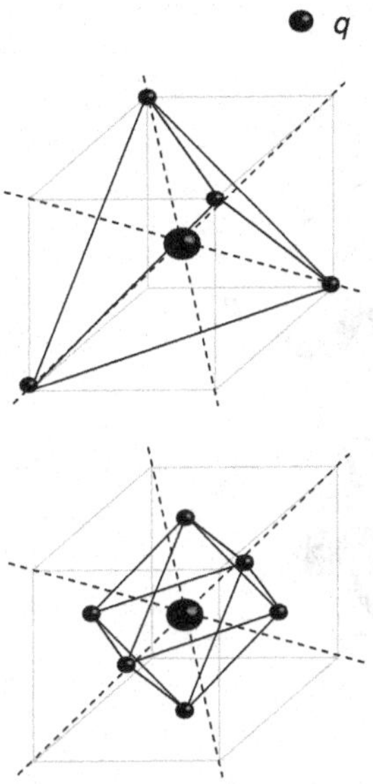

Tetrahedral and octahedral coordination. The central on site has cubic symmetry in each case.

compounds, where their most common coordination is six-fold (octahedral) or four-fold (tetrahedral). If undistorted, both of these sites have cubic symmetry. The essential feature of cubic symmetry is four three-fold symmetry axes oriented along the cube's diagonals.

We begin by considering the effect of the crystal field on one-electron p- and d-states, and generalize to multielectron $3d$ ions in the next section. To demonstrate quenching of orbital angular momentum, we take the example of the p-states with $\ell = 1$ and $m_\ell = 0, \pm 1$. These orbital states, obtained from (4.1), and illustrated in Fig. 4.2, are

$$\psi_0 = |0\rangle = R(r)\cos\theta,$$
$$\psi_{\pm 1} = |\pm 1\rangle = (1\sqrt{2})R(r)\sin\theta\exp(\pm i\phi).$$

Next we suppose that a p^1 ion is surrounded by six other ions that form an octahedron. These could be oxygen O^{2-} anions. In the point-charge approximation the anions may be represented by point charges q at the corners of the octahedron. The crystal-field Hamiltonian $\mathcal{H}_{cf}$ depends on the Coulomb potential

$$V_{cf} = D_c(x^4 + y^4 + z^4 - 3y^2z^2 - 3z^2x^2 - 3x^2y^2), \tag{4.30}$$

where $D_c = 7q/8\pi\epsilon_0 a^5$ and the six charges are placed at $\pm a$ along each of the x, y and z axes. However, the orbital states $|\pm 1\rangle$ are *not* eigenstates of $\mathcal{H}_{cf}$. In other words, the matrix elements $\langle i|\,\mathcal{H}_{cf}\,|j\rangle \not\propto eV_{cf}\delta_{ij}$ (δ_{ij} is the Kronecker delta-function; $\delta_{ij} = 0$ for $i \neq j$ and $\delta_{ii} = 1$); the 3×3 matrix which represents the crystal-field term in the Hamiltonian is not diagonal. This can be easily seen by inspecting the integrals over the wave functions, which are the matrix elements $\langle i|eV_{cf}|j\rangle = \int \psi_i^* eV_{cf}\psi_j \mathrm{d}^3r$. The integral for $i = 1, j = -1$, for instance, involves $\sin^2\theta$ which is even in θ and will not integrate to zero. We need to find linear combinations of the hydrogenic orbital states that are eigenfunctions of $\mathcal{H}_{cf}$, namely

$$\psi_0 = R(r)\cos\theta = zR(r)/r = p_z,$$
$$(1/\sqrt{2})(\psi_1 + \psi_{-1}) = R(r)\sin\theta\cos\phi = xR(r)/r = p_x,$$
$$(i/\sqrt{2})(\psi_1 - \psi_{-1}) = R(r)\sin\theta\sin\phi = yR(r)/r = p_y.$$

These new wave functions are the p_x, p_y and p_z orbitals shown in Fig. 4.11. The expectation value of the z-component of angular momentum; $\ell_z = -i\hbar\partial/\partial\phi$ is zero for all three wave functions. The orbital angular momentum is said to be *quenched* by the crystal field. Unlike the rings of charge density which allow for clockwise or anticlockwise circulation of electrons in the xy-plane in ψ_1 and ψ_{-1} orbits of the free ion, no such circulation is possible in the p_x or p_y orbitals in the solid.

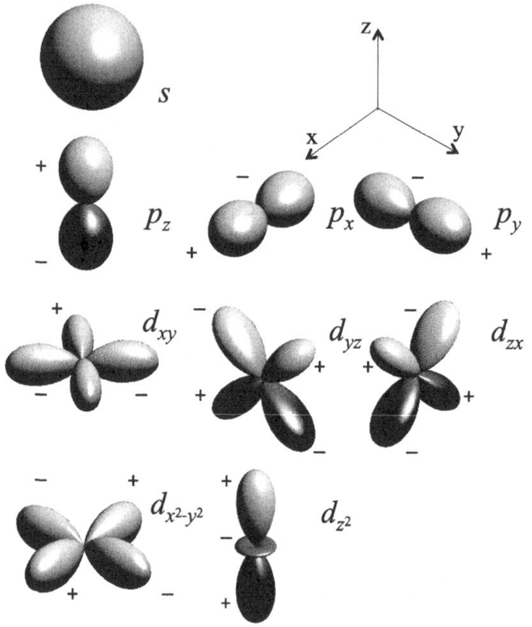

Figure 4.11
Electron densities for the *s*-, *p*- and *d*-orbitals, represented by boundary surfaces showing the angular distribution probability for electrons in each orbital. The sign of the wave function is indicated.

Similar arguments apply for the $3d$-orbitals; the eigenfunctions there are three t_{2g}, and two e_g, orbitals[4]

t_{2g} orbitals:

$$\begin{aligned}
d_{xy} &= -(i/\sqrt{2})(\psi_2 - \psi_{-2}) = R'(r)\sin^2\theta\sin 2\phi \approx xyR(r)/r^2\\
d_{yz} &= (i/\sqrt{2})(\psi_1 + \psi_{-1}) = R'(r)\sin\theta\cos\theta\sin\phi \approx yzR(r)/r^2\\
d_{zx} &= -(1/\sqrt{2})(\psi_1 - \psi_{-1}) = R'(r)\sin\theta\cos\theta\cos\phi \approx zxR(r)/r^2
\end{aligned}$$

e_g orbitals:

$$\begin{aligned}
d_{x^2-y^2} &= (1/\sqrt{2})(\psi_2 + \psi_{-2}) = R'(r)\sin^2\theta\cos 2\phi \approx (x^2 - y^2)R(r)/r^2\\
d_{3z^2-r^2} &= \psi_0 = R(r)(3\cos^2\theta - 1) \approx (3z^2 - r^2)R(r)/r^2
\end{aligned}$$

Crystal fields and ligand fields The splitting of the $3d$-orbitals in octahedral oxygen coordination, for example, is regarded in crystal-field theory as an electrostatic effect of the neighbouring oxygen anions, which are treated as point charges. This is an oversimplification. The $3d$- and oxygen $2p$-orbitals overlap, and they form a partially covalent bond. The oxygens bonding to the

[4] *Notation:* Conventionally, *a* and *b* denote nondegenerate electron orbitals, *e* a two-fold degenerate orbital and *t* a threefold degenerate orbital. Lower-case letters refer to single-electron states, capital letters refer to multielectron states. *a*, *A* are nondegenerate and symmetric with respect to the principal axis of symmetry (the sign of the wave function is unchanged); *b*, *B* are antisymmetric with respect to the principal axis (the sign of the wave function changes). Subscripts *g* and *u* indicate whether the wave function is symmetric or antisymmetric under inversion. Subscript 1 refers to mirror planes parallel to a symmetry axis, 2 refers to diagonal mirror planes.

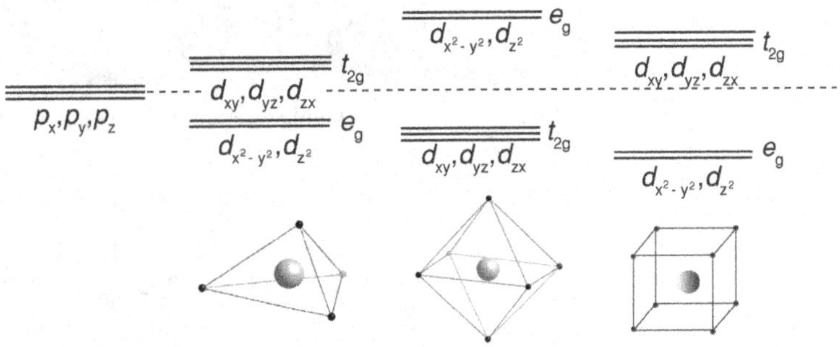

Figure 4.12

Splitting of one-electron 3*d* energy levels in different cubic crystal fields; 2*p* levels are unsplit.

3*d* metal are the ligands. On account of their orientation, the overlap is greater for e_g- than t_{2g}, orbitals. The overlap leads to mixed wave functions, producing bonding and antibonding orbitals, whose splitting increases with overlap. The hybridized orbitals are

$$\phi = \alpha\psi_{2p} + \beta\psi_{3d},$$

where $\alpha^2 + \beta^2 = 1$, for normalization. The splitting between the mainly $t_{2g}\,\pi^*$ and the mainly $e_g\sigma^*$-orbitals for 3*d* ions in octahedral coordination[5] is usually 1–2 eV . The ionic and covalent contributions to the splitting Δ are actually of comparable magnitude.

The sequence of ligands in order of increasing effectiveness at producing crystal/ligand field splitting Δ is known as the spectrochemical series. The ligands that concern us here, in a sequence which reflects both ionic charge and the tendency to bond covalently are

$$Br^- < Cl^- < F^- < OH^- < CO_3^{2-} < O^{2-} < H_2O < NH_3 < SO_3^{2-}$$
$$< NO_2^- < S^{2-} < CN^-.$$

The bond is mostly ionic at the beginning of the series, and mostly covalent at the end.

Covalency is stronger in tetrahedral coordination, on account of the shorter metal-oxygen distances, but the crystal-field splitting Δ_{tet} is $\frac{4}{9}\Delta_{oct}$. The relative splittings of the 3*d* crystal-field levels in different sites with cubic symmetry are illustrated in Fig. 4.12 for the one-electron states. The symmetry of the one-electron states is preserved even in metals, where the coordination number is 8 or 12, and the neighbours appear negatively charged.

As the site symmetry is reduced, the degeneracy of the one-electron energy levels is lifted. For example, a tetragonal extension of the octahedron along the

[5] σ-bonds have charge density mainly around the line joining the atoms, whereas π-bonds have charge density above or below this line. The * denotes an antibonding orbital. The 3*d*- states usually lie higher in energy than the 2*p*- states in oxides, so the hybridized orbitals with mainly 2*p* character are bonding and those with mainly 3*d* character are antibonding.

Table 4.9. The splitting of the one-electron levels in different symmetry

	ℓ	Cubic	Tetragonal	Trigonal	Rhombohedral
s	0	1	1	1	1
p	1	3	1, 2	1, 2	1, 1, 1
d	2	2, 3	1, 1, 1, 2	1, 2, 2	1, 1, 1, 1, 1
f	3	1, 3, 3	1, 1, 1, 2, 2	1, 1, 1, 2, 2	1, 1, 1, 1, 1, 1, 1

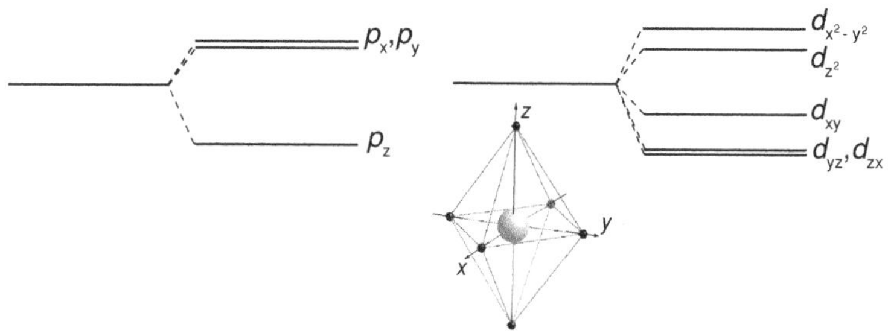

Figure 4.13
The effect of a tetragonal distortion of octahedral symmetry on the one-electron energy levels.

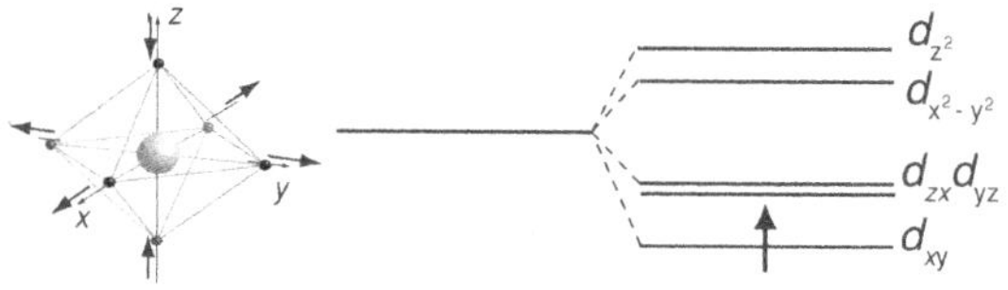

Figure 4.14
Jahn–Teller distortion of an octahedral site containing a d^1 ion.

z axis will lower p_z and raise p_x and p_y. The effect on the d-states is shown in Fig. 4.13. The degeneracy of the d-orbitals in different symmetries is shown in Table 4.9.

A system with a single electron (or hole) in a degenerate level will tend to distort spontaneously as shown in Fig. 4.14. This is the Jahn–Teller effect. The effect is particularly strong for d^4 and d^9 ions in octahedral symmetry (Mn^{3+}, Cu^{2+}), which can lower their energy by distorting the crystal environment. If the local strain is ϵ, the energy change $\delta\varepsilon = -A\epsilon + B\epsilon^2$, where the first term is the crystal-field stabilization energy and the second term is the increased elastic energy. Here A and B are constants. The minimum is at $\epsilon = A/2B$. The crystal field plays an important part in stabilizing ionic structures. The stabilization energy for a d^1 ion is $0.4\Delta_{oct}$ in an octahedral site, and it can be further increased by a uniaxial distortion that splits the t_{2g} levels. The splittings conserve the centre of gravity of the levels, so the e_g levels are raised by $0.6\Delta_{oct}$ while the t_{2g} levels are lowered by $-0.4\Delta_{oct}$ in an octahedral crystal field. The d^5 ions Fe^{2+} or Mn^{2+}, which have all orbits singly occupied, exhibit no crystal-field stabilization energy.

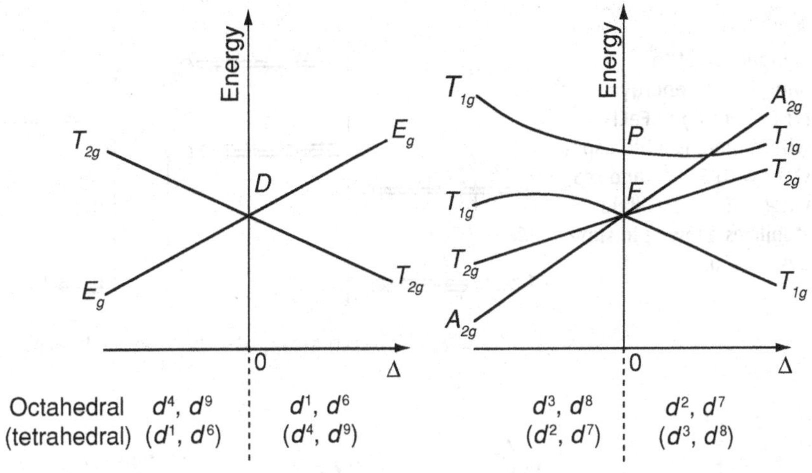

Figure 4.15

Orgel diagrams for D and F terms. The corresponding d^n ion configuration is shown.

4.4.3 Many-electron states

The one-electron picture is modified by the strong Coulomb interactions among the electrons. There are only three distinct types of energy-level diagrams for $3d$ ions, corresponding to the D, F or S ground-state terms. The D-states map directly onto the one-electron levels because there is just one electron or hole in an otherwise empty, half-filled or filled d-shell. These are the d^1, d^4, d^6 and d^9 configurations, with $L = 2$ according to Hund's second rule. The F term is found for the d^2, d^3, d^7 and d^8 configurations, with $L = 3$. The S term is for d^5, where $L = 0$. Orgel diagrams (Fig. 4.15) show the splitting of the ground-state D or F level in a cubic crystal field. The S term does not split, on account of the spherical symmetry of the half-filled shell.

The hierarchy of interactions $\mathcal{H}_0 > \mathcal{H}_{cf} > \mathcal{H}_{so}$ is not always maintained for $3d$ ions. Some ligands create such intense electrostatic fields that they manage to overturn not only Hund's second rule, but also his first rule as well. The crystal field can drive the ion into a low-spin state. The on-site Coulomb interaction has the effect of raising the energy of doubly occupied orbitals by U, compared with singly occupied orbitals. This is the electrostatic penalty for double occupancy. When U exceeds the crystal-field splitting of the one-electron levels, Hund's first rule applies as advertised, but when U $< \Delta$ the t_{2g}-orbitals for an octahedral site, for example, will tend to be doubly occupied. Figure 4.16 illustrates the high-spin and low-spin states for Fe^{2+}.

In some materials, where the low-spin state lies only slightly lower in energy than the high-spin state, a spin crossover may occur as a function of temperature at a phase transition, driven by magnetic entropy $R\ln(2S+1)$.

Tanabe–Sugano diagrams show the splitting of both ground-state and higher terms in the crystal field. They are drawn with the ground-state energy set to zero. Some diagrams for $3d$ ions in octahedral coordination are shown in Fig. 4.17. The stabilization of low-spin states in strong crystal fields is evident

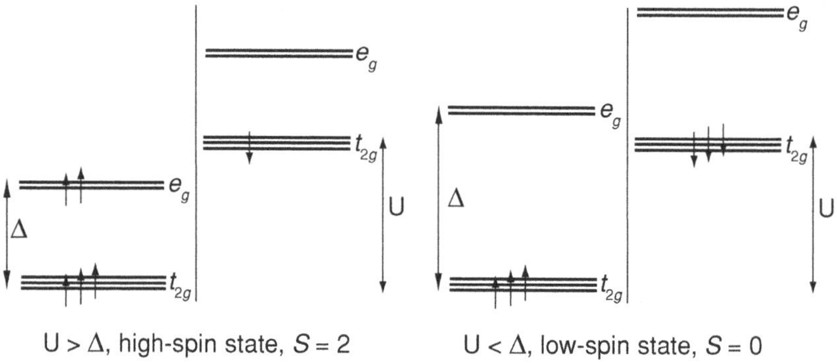

Figure 4.16

Comparison of the one-electron energy levels for Fe^{2+} ($3d^6$) in $FeCl_2$, where there is a high-spin state with $S = 2$, and FeS_2 where the crystal field stabilizes a low-spin state with $S = 0$.

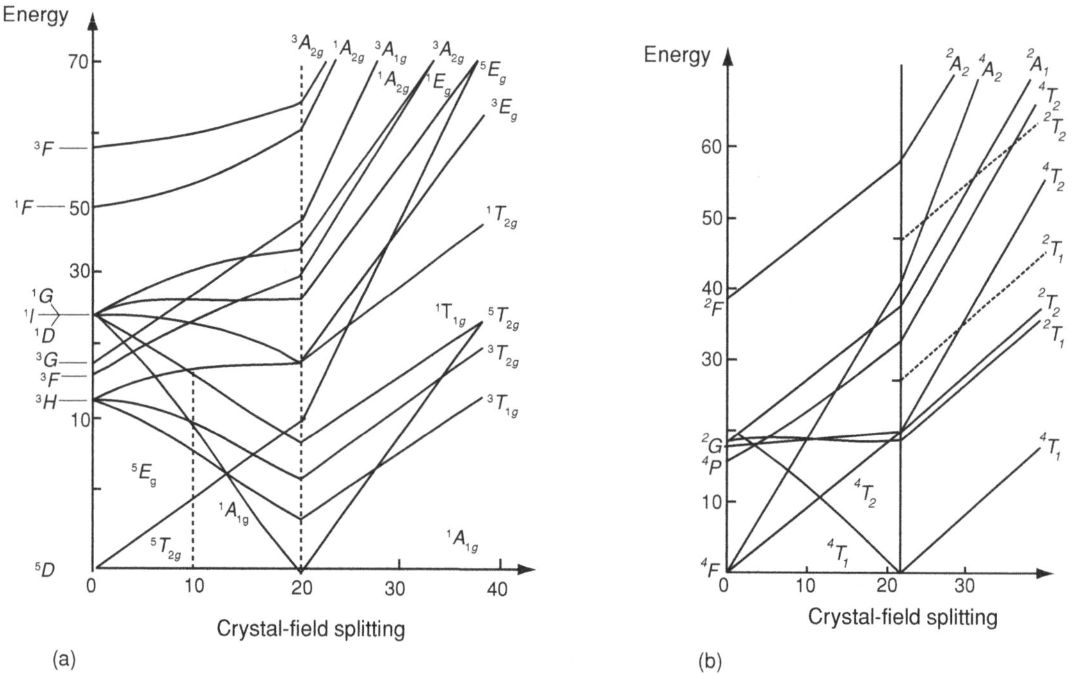

Figure 4.17

Tanabe–Sugano diagrams for: (a) a $3d^6$ ion and (b) a $3d^7$ ion. The high-spin–low-spin crossover is indicated by the vertical line.

from these diagrams. The crystal-field parameters, and especially the cubic crystal-field parameter Δ (also known as $10Dq$) can be deduced from the wavelengths of optical transitions from the ground state to different excited states.

4.4.4 Single-ion anisotropy

Single-ion anisotropy is due to the electrostatic interaction (4.31) of charge in orbitals containing the magnetic electron distribution $\rho_0(\boldsymbol{r})$ with the potential $\varphi_{cf}(\boldsymbol{r})$ created at the atomic site by the rest of the crystal. As we have seen in Figs. 4.14–4.17, the crystal-field interaction tends to stabilize particular orbitals.

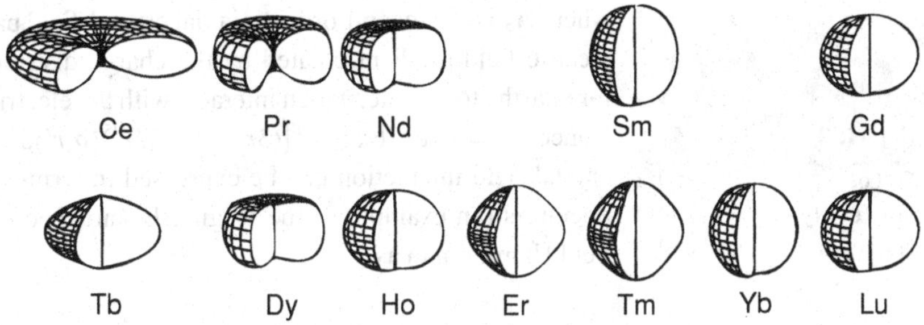

Figure 4.18

Charge density distributions for the trivalent rare-earth ions at zero temperature. The deviations from spherical symmetry are exaggerated.

Spin-orbit interaction $\Lambda \boldsymbol{L} \cdot \boldsymbol{S}$, then leads to magnetic moment alignment in specific directions in the crystal.

Crystal-field anisotropy is easiest to treat in the case of the rare-earths. With these ions it is possible to cleanly divide the charge density $e\rho(\boldsymbol{r}')$ producing the crystal field, from the charge $e\rho_{4f}(\boldsymbol{r})$ in the $4f$ shell on which it acts. If the electrostatic potential due to the surroundings is $\varphi_{cf}(\boldsymbol{r})$ and the $4f$ electron density is ρ_{4f}, the interaction energy is

$$\varepsilon_a = \int e\rho_{4f}(\boldsymbol{r})\varphi_{cf}(\boldsymbol{r})\mathrm{d}^3r. \qquad (4.31)$$

It is appropriate to expand the $4f$ charge density in spherical harmonics and express the electrostatic interaction in terms of the 2^n-pole moments of the charge distribution, where n is even. The $n = 2$ term is the quadrupole moment:

$$Q_2 = \int \rho_{4f}(r)(3\cos^2\theta - 1)r^2\mathrm{d}^3r.$$

The sign of Q_2 reflects the shape of the $4f$ electron cloud whether prolate (elongated, $Q_2 > 0$) or oblate (flattened, $Q_2 < 0$), as shown in Fig. 4.18. Units are m^2.

The $n = 4$ and $n = 6$, hexadecapole and 64-pole moments are defined as

$$Q_4 = \int \rho_{4f}(r)(35\cos^4\theta - 30\cos^2\theta + 3)r^4\mathrm{d}^3r,$$

$$Q_6 = \int \rho_{4f}(r)(231\cos^6\theta - 315\cos^4\theta + 105\cos^2\theta - 5)r^6\mathrm{d}^3r.$$

These higher-order moments decide, for instance, whether a prolate charge distribution resembles a bone ($Q_4 < 0$) or a snake which has just eaten a rabbit ($Q_4 > 0$). Units are m^4 and m^6, respectively.

In a few cases it is possible to represent the multipole moments Q_m in terms of closed expressions. For example, in the $J = J_z$ ground state of the heavy rare-earths where $\hat{\mathbf{O}}_2^0 = 2J^2 - J$, quadrupole moments are given by

$$Q_2 = -(1/45)(14 - n)(7 - n)(21 - 2n)\langle r_{4f}^2 \rangle,$$

where n is the number of $4f$ electrons. The leading term in (4.31) is

$$\varepsilon_a = (1/2)Q_2A_2^0(3\cos^2\theta - 1), \qquad (4.32)$$

Values of A_2^0 at the rare earth site in some intermetallic compounds (units $\mathrm{K}a_0^{-2}$)	
$SmCo_5$	−380
$Sm_2Fe_{17}N_3$	−240
$Nd_2Fe_{14}B$	310

where A_2^0 is the second-order uniaxial crystal-field parameter that describes the electric field gradient created by the charge distribution of the crystal at the rare-earth atomic site, which interacts with the electric quadrupole moment Q_2. Hence $A_2^0 = -(e/16\pi\epsilon_0)\int[(3r_z'^2 - r'^2)/r'^5]\rho(r')\mathrm{d}^3r'$. Stevens showed that the crystal-field interaction can be expressed in terms of the angular momentum operators (an example of the Wigner-Eckart theorem). He wrote the crystal-field Hamiltonian as

$$\mathcal{H}_{cf} = \sum_{n=0,2,4,6} \sum_{m=-n,\dots,n} B_n^m \hat{\mathbf{O}}_n^m, \tag{4.33}$$

where $B_n^m = \theta_n \langle r_{4f}^n \rangle A_n^m$, and θ_n is a constant, different for each rare-earth, which is proportional to the 2^n-pole moment: $Q_2 = 2\theta_2\langle r_{4f}^2\rangle$, and $Q_4 = 8\theta_4\langle r_{4f}^4\rangle$, $Q_6 = 16\theta_6\langle r_{4f}^6\rangle$, and $\hat{\mathbf{O}}_n^m$ are the Stevens operators, tabulated in Appendix H and by Hutchings (1965). For example,

$$\hat{\mathbf{O}}_2^0 = [3\hat{\boldsymbol{J}}_z^2 - J(J+1)].$$

Here we have dropped the $\hbar$ in the angular momentum operators. The temperature dependence of the multipole moments Q_n can be evaluated from the thermodynamic averages $\langle \hat{\mathbf{O}}_n^0 \rangle$.

In rare-earths, a limited number of terms in the expansion (4.33) are needed, depending on the symmetry of the site. The crystal-field coefficients B_n^m or A_n^m are then parameters to be determined experimentally. At low temperatures, second-, fourth- and sixth-order terms may all be important, but at room temperature it is often enough to consider only the leading, second-order terms

$$\mathcal{H}_{cf} = \theta_2 \langle r_{4f}^2 \rangle \left[A_2^0 \hat{\mathbf{O}}_2^0 + A_2^2 \hat{\mathbf{O}}_2^2\right], \tag{4.34}$$

which, in the case of axial symmetry when there is no off-diagonal term, is further simplified to

$$\mathcal{H}_{cf} = \mathrm{D} J_z^2, \tag{4.35}$$

where D is the uniaxial crystal field parameter.

Another commonly encountered expression, for cubic anisotropy, is

$$\mathcal{H}_{cf} = \theta_4 \langle r_{4f}^4 \rangle \left[A_4^0 \hat{\mathbf{O}}_4^0 + 5A_4^4 \hat{\mathbf{O}}_4^4\right] + \theta_6 \langle r_{4f}^6 \rangle \left[A_6^0 \hat{\mathbf{O}}_6^0 - 21A_6^4 \hat{\mathbf{O}}_6^4\right]. \tag{4.36}$$

Only the fourth-order term exists for $3d$ ions, with r_{3d} taking the place of r_{4f}.

A general expression for ε_a is given in Appendix H.

Kramer's theorem, which is a consequence of time reversal symmetry, states that energy levels for systems with an odd number of electrons, and therefore half-integral quantum numbers, are always at least two-fold degenerate in the absence of a magnetic field. When J is half-integral, the second-order crystal field produces a series of *Kramer's doublets* $|\pm M_J\rangle$; but when J is integral there is a singlet $|0\rangle$ and a series of doublets. If D is negative, the $|\pm J\rangle$ state is

Table 4.10. Data on the rare-earth ions. The operators $\hat{O}_n^0$ are evaluated at $T = 0\,K$, where $J_z = J$

	J	θ_2 (10^{-2})	$\theta_2\langle r^2\rangle O_2^0$ (a_0^2)	θ_4 (10^{-4})	$\theta_4\langle r^4\rangle O_4^0$ (a_0^4)	θ_6 (10^{-6})	$\theta_6\langle r^6\rangle O_6^0$ (a_0^6)	G	γ_s	C_{mol}
Ce^{3+}	5/2	−5.714	−0.748	63.5	1.51			0.18	−1/3	8.0
Pr^{3+}	4	−2.101	−0.713	−7.346	−2.12	60.99	5.89	0.80	−1/2	16.0
Nd^{3+}	9/2	−0.643	−0.258	−2.911	−1.28	−37.99	−8.63	1.84	−1/4	16.4
Sm^{3+}	5/2	4.127	0.398	25.012	0.34			4.46	−5	0.9
Gd^{3+}	7/2				–			15.75	1	78.8
Tb^{3+}	6	−1.010	−0.548	1.224	1.20	−1.12	−1.28	10.50	2/3	118.2
Dy^{3+}	15/2	−0.635	−0.521	−0.592	−1.46	1.04	5.64	7.08	1/2	141.7
Ho^{3+}	8	−0.222	−0.199	−0.333	−1.00	−1.29	−10.03	4.50	2/5	140.7
Er^{3+}	15/2	0.254	0.190	0.444	0.92	2.07	8.98	2.55	1/3	114.8
Tm^{3+}	6	1.010	0.454	1.633	1.14	−5.61	−4.05	1.17	2/7	71.5
Yb^{3+}	7/2	3.175	0.435	−17.316	−0.79	1.48	0.73	0.29	1/4	25.7

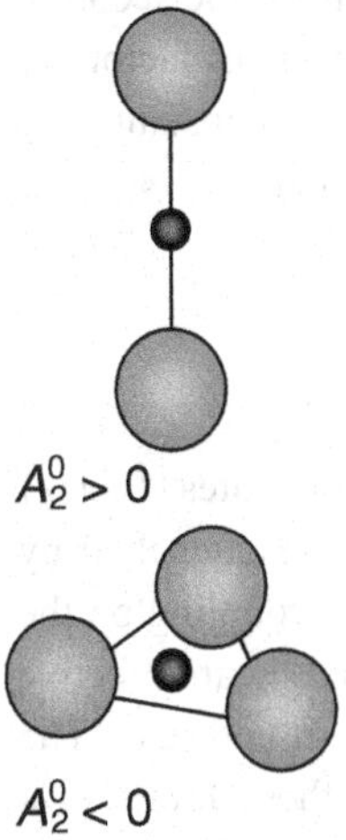

Examples of atomic configurations that produce positive and negative electric field gradients at the central site.

lowest, and the susceptibility in a magnetic field applied along the crystal-field axis is given by (4.14). If D is positive, the $M_J = 0$ singlet is the ground state. The crystal field appears to have destroyed the moment. The initial susceptibility is zero, although the induced moment increases quadratically with field. However, when the magnetic field is applied perpendicular to the crystal-field axis, there is a large susceptibility. What this means is that there is an easy axis of magnetization in the first case, but an easy plane in the second.

Single-ion anisotropy is the major source of anisotropy in hard ferromagnetic materials. The tendency for the magnetic moment to lie along specific crystal axes generally makes the susceptibility a tensor rather than a scalar quantity. There are three principal axes, for which the induced moment lies parallel to the applied magnetic field.

In general, the details of the response of the system to an applied field are obtained by diagonalizing the Hamiltonian $\mathcal{H}_{cf} + \mathcal{H}_z$, and evaluating $\langle \mathfrak{m} \rangle$ from (4.11). In a series of isostructural rare-earth metals or compounds, A_n^m is roughly constant. The leading term in the anisotropy just depends on the product of A_2^0 and $\theta_2\langle r_{4f}^2\rangle J_z^2$ for the rare-earth in question. Both θ_2, and the quadrupole moment change sign at each quarter-shell filling (Fig. 4.18, Table 4.10). When A_2^0 is positive, ions such as Nd^{3+}, which have a negative quadrupole moment and an oblate charge distribution, will exhibit easy axis anisotropy, whereas ions such as Sm^{3+}, which have a positive quadrupole moment and a prolate charge distribution, will exhibit hard axis anisotropy.

Consider a rare-earth, for which the spin-orbit interaction stabilizes a $(2J+1)$-fold degenerate multiplet $|J, M_J\rangle$. The crystal field defines the z axis and splits out the M_J states, subject to Kramers theorem.

In the figure, Sm^{3+} has a $|\pm 5/2\rangle$ ground state, so the moment has its maximum positive or negative projection along Oz, which precesses as indicated by the vector model. The average moment lies along the z-direction, and application of a field in this direction splits out the $\pm\frac{5}{2}$ states, leading to a susceptibility $\chi = C/T$ given by (4.14) with $J = \frac{5}{2}$. When the magnetic field is applied in the x-direction, the susceptibility is much smaller.

The situation is different for Pr^{3+}. The crystal field in this case makes $|0\rangle$ the ground state, with no projection of the moment along Oz. The ground state is nonmagnetic in the vector model, the moment in the $M_J = 0$ state lies in an indeterminate direction in the xy-plane. The susceptibility is zero at low temperature in a small magnetic field applied along Oz. However, in large fields, the Zeeman splitting of the excited states $|\pm 1\rangle$, $|\pm 2\rangle$, $|\pm 3\rangle$ and $|\pm 4\rangle$ produces a series of level crossings and the magnetization in high fields is ultimately that of the $|\pm 4\rangle$ state. However, if the field is applied along Ox, the situation is quite different. There is immediately a high susceptibility associated with the $|\pm 4\rangle$ state. The Pr^{3+} in this example has hard axis–easy plane anisotropy.

Data on the rare-earth ions are collected in Table 4.10. The fourth, sixth and eighth columns respectively indicate the relative strengths of the second-, fourth- and sixth-order crystal-field interactions; G is the de Gennes factor, γ_s is the ratio of spin to total moment and C_{mol} is the molar Curie constant. The crystal-field parameters A_n^m or B_n^m are determined by experiment.

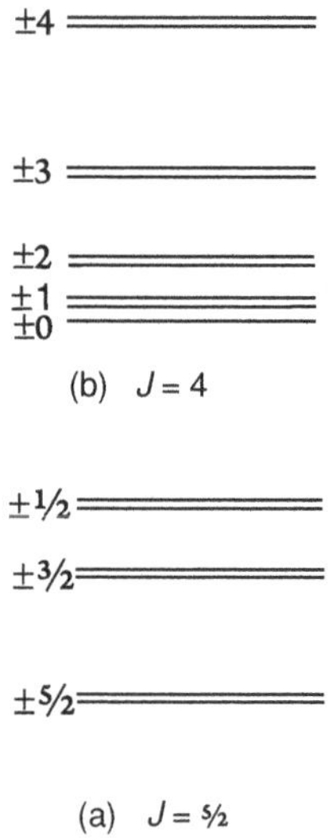

Crystal field splitting for ions in an axial field with $A_2^0 < 0$: (a) Sm^{3+}; (b) Pr^{3+}.

4.4.5 The spin Hamiltonian

The orbital states of a multielectron $3d$ ion are represented by the states $|L, M_L\rangle$ and the spin states by $|S, M_S\rangle$. Since the orbital moments are quenched by the crystal field, the effects of the crystal field and spin-orbit coupling on the magnetic energy levels of an ion are conveniently represented by an effective spin Hamiltonian based on the $2S + 1$ $|M_S\rangle$ sublevels of the ground state. The magnetic properties of the ion depend on the splitting of these levels in a magnetic field.

Following (4.35), the simplest spin Hamiltonian, used for sites with uniaxial symmetry, is

$$\mathcal{H}_{spin} = \mathsf{D}S_z^2. \tag{4.37}$$

An orthorhombic distortion adds a term $\mathsf{E}(S_x^2 - S_y^2)$. The next term in tetragonal symmetry is $\mathsf{F}S_z^4$. Cubic symmetry gives a term $\mathsf{D}_c(S_x^4 + S_y^4 + S_z^4)$. The splitting of the levels usually depends on the orientation of the applied field, which is reproduced by an anisotropic $\hat{\mathbf{g}}$-tensor, so the Zeeman term is written as $\mu_B/\hbar(\boldsymbol{B} \cdot \hat{\mathbf{g}} \cdot \boldsymbol{S}) = (\mu_B/\hbar)\sum_{i,j} g_{ij} B_i S_j$. The spin Hamiltonian approach is widely used in electron paramagnetic resonance, discussed in Chapter 9, where the spectra are parameterized in terms of D, E, ..., $g_{\parallel}$, $g_{\perp}$. The ability to diagonalize large Hamiltonian matrices which include crystal-field, spin-orbit and Zeeman terms by computer has reduced the need for these approximation methods.

FURTHER READING

S. McMurry, *Quantum Mechanics*, Wokingham: Addison Wesley (1994). An excellent final-year undergraduate text.

J. J. Sakurai, *Modern Quantum Mechanics*, New York: Addison Wesley (1985). A clearly written graduate text.

R. G. Burns, *Mineralogical Applications of Crystal Field Theory* (second edition), Cambridge: Cambridge University Press (1993). A detailed account of crystal-field interactions and spectroscopy of transition-metal oxides.

C. Ballhausen, *Introduction to Ligand Field Theory*, New York: McGraw Hill (1962). An introduction to crystal-field theory for $3d$, $4d$ and $5d$ ions in molecules and solids, with many examples.

M. T. Hutchings, *Point-Charge Calculations of Energy Levels of Magnetic Ions in Crystalline Electric Fields*, Solid State Physics, Vol. 16, p. 227–273, New York: Academic Press (1964). The standard account of crystal field theory for both $3d$ and $4f$ ions, with comprehensive tables of the Stevens operators.

EXERCISES

4.1 Calculate the multiplet splitting in terms of the spin-orbit coupling constant Λ for an ion with $L = 3$, $S = \frac{1}{2}$.

4.2 Deduce the expression for the Langevin function from (4.20).

4.3 Calculate the Van Vleck susceptibility for a mole of Eu^{3+} ions.

4.4 Plot the crystal field stabilization energy for high-spin $3d^n$ ions for $1 \leq n \leq 9$ in (a) octahedral and (b) tetrahedral sites. Use the one-electron levels shown in Fig. 4.12 and take $\Delta_{tet} = -(4/9)\Delta_{oct}$.

4.5 Make a table of the spin moments of $3d$ ions in six-fold, octahedral coordination in high- and low-spin states. Repeat for eight-fold cubic coordination and four-fold tetrahedral coordination.

4.6 At zero temperature in a compound containing octahedrally coordinated Fe^{2+}, the low-spin energy level lies just 50 meV below the high-spin energy level. At what temperature would you expect spin crossover to occur? What is the order of the transition?

4.7 Show that the low-temperature specific heat is linear in temperature for a sample containing Kramers ions which are subject to a uniform magnetic field when the local crystal field axes are oriented at random with respect to the field direction.

4.8 An ion is in a site where the spin Hamiltonian is $\mathsf{D}S_z^2$. If $\mathsf{D} = 6$ K, $S = 1$ and $g = 2$. Sketch the magnetization as a function of field when the field is applied (a) along Oz and (b) along Ox at $0K$ and $4K$. Make use of the angular momentum operators for $l = 1$, given in §3.1.4.

4.9 Show that for an ion in an $S = 1$ state, described by the spin Hamiltonian $\mathcal{H} = g_{||}\mu_B B_z S_z + g_{\perp}\mu_B(B_x S_x + B_y S_y) + \mathsf{D}S_z^2$, the energy eigenstate ε can be deduced from $\varepsilon = (\varepsilon - \mathsf{D})^2 - \varepsilon(g_{||}\mu_B B\cos\theta)^2 - (\varepsilon - \mathsf{D})(g_{\perp}\mu_B \sin\theta)^2$. Find the solutions for $\theta = 0$ and $\theta = \pi/2$, where θ is the angle between $\boldsymbol{B}$ and the crystal field axis O_z. Make use of the angular momentum operators for $L = 1$.

5 Ferromagnetism and exchange

Order, order, order!

Ferromagnetism and the Curie temperature were explained by Weiss in terms of a huge internal 'molecular field' proportional to the magnetization. The theory is applicable both to localized and delocalized electrons. No such magnetic field really exists, but it is a useful way of approximating the effect of the interatomic Coulomb interaction in quantum mechanics, which Heisenberg described by the Hamiltonian $\mathcal{H} = -2\mathcal{J}\boldsymbol{S}_1 \cdot \boldsymbol{S}_2$, where $\boldsymbol{S}_1$ and $\boldsymbol{S}_2$ are operators describing the localized spins on two adjacent atoms. When $\mathcal{J} > 0$, ferromagnetic exchange leads to ferromagnetic order in three dimensions. Spin waves are the low-energy excitations of the exchange-coupled magnetic lattice. In the delocalized electron picture, a ferromagnet has spontaneously spin-split energy bands. The density of $\uparrow$ and $\downarrow$ states is calculated using spin-dependent density functional theory. Important physical phenomena associated with ferromagnetism are discussed in this chapter, including magnetic anisotropy and, magnetoelastic, magneto-optic and magnetotransport effects.

The characteristic feature of a ferromagnet is its spontaneous magnetization M_s, which is due to alignment of the magnetic moments located on an atomic lattice. The magnetization tends to lie along easy directions determined by crystal structure, atomic-scale texture or sample shape. Heating above a critical temperature known as the Curie point, which ranges from less than 1 K for magnetically dilute salts to almost 1400 K for cobalt, leads to a reversible collapse of the spontaneous magnetization. Although there is no reason in principle why uniform ferromagnetic liquids should not exist, it seems that there are none. Ferrofluids, while ferromagnetic and liquid, are actually colloidal suspensions of solid ferromagnetic particles.

Important modifications of the electronic, thermal, elastic and optical properties are associated with magnetic order, whether ferromagnetic, or one of the more complex multisublattice or noncollinear ordered magnetic structures presented in the next chapter.

5.1 Mean field theory

5.1.1 Molecular field theory

The first modern theory of ferromagnetism, and one that remains useful today, was proposed by Pierre Weiss in 1906. Weiss's original theory was based on the classical paramagnetism of Langevin, but it was soon extended to the more general Brillouin theory of localized magnetic moments. His idea was that there is an internal 'molecular field' which is proportional to the magnetization of the ferromagnet. If n_W is the constant of proportionality, this adds to the internal contribution of any externally applied field:

$$\boldsymbol{H}^i = n_W \boldsymbol{M} + \boldsymbol{H}. \tag{5.1}$$

$\boldsymbol{H}^i$ has to be enormous to induce a spontaneous magnetization at room temperature; the Weiss coefficient n_W is approximately 100. The magnetization is given by the Brillouin function (4.17) with $M_0 = n\mathfrak{m}_0 = ng\mu_B J$, where n is the number of magnetic atoms per unit volume,

$$M = M_0 \mathcal{B}_J(x), \tag{5.2}$$

but now

$$x = \mu_0 \mathfrak{m}_0 (n_W M + H)/k_B T. \tag{5.3}$$

In zero external field, $\boldsymbol{M}$ is the spontaneous magnetization $\boldsymbol{M}_s$ so we have

$$M_s/M_0 = \mathcal{B}_J(x_0), \tag{5.4}$$

where $x_0 = \mu_0 \mathfrak{m}_0 n_W M_s / k_B T$. Combining x_0 with $M_0 = n\mathfrak{m}_0$, we find $M_s/M_0 = (nk_B T/\mu_0 M_0^2 n_W)x_0$, which is conveniently written in terms of the Curie constant C (4.16) as

$$M_s/M_0 = [T(J+1)/3JCn_W]x_0. \tag{5.5}$$

The simultaneous solution of (5.4) and (5.5) is found graphically as indicated in Fig. 5.1. Otherwise the equations can be solved numerically. Results for M_s/M_0 versus T/T_C are plotted in Fig. 5.2 for some values of J, including the classical limit $J \longrightarrow \infty$ where (5.4) is replaced by the Langevin function (4.21). In the Brillouin theory, the magnetization approaches zero temperature with horizontal slope, as required by thermodynamics (§2.5.4). Numerical values of the reduced spontaneous magnetization M_s/M_0 versus reduced temperature T/T_C are listed in Appendix G for different values of J. When S is the good quantum number, J in these formulae is replaced by S. Theory and experiment for nickel are compared in Fig. 5.3.

Weiss's molecular field theory was the first mean field theory of a phase transition. The moments are completely disordered at and above T_C, where the

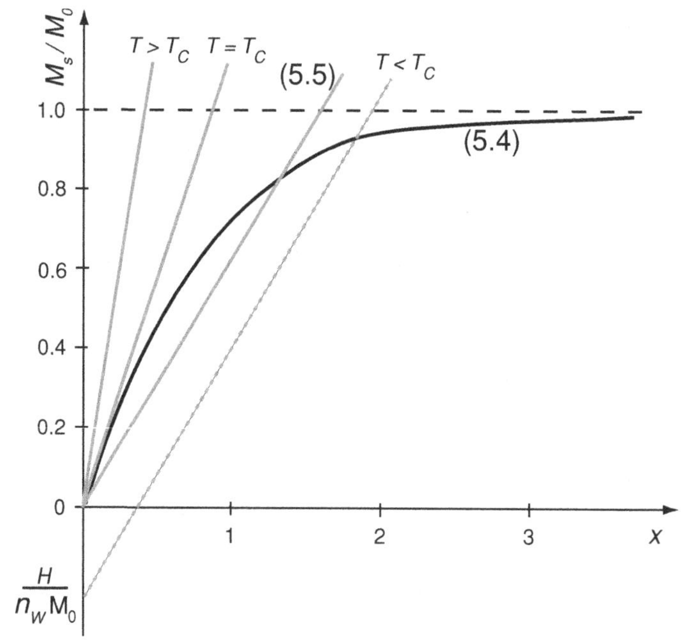

Figure 5.1

Graphical solution of (5.4) and (5.5) for $J = \frac{1}{2}$ to find the spontaneous magnetization M_s when $T < T_C$. Equation (5.5) is also plotted for $T = T_C$ and $T > T_C$. The effect of an external field is to offset (5.5), as shown by the dotted line.

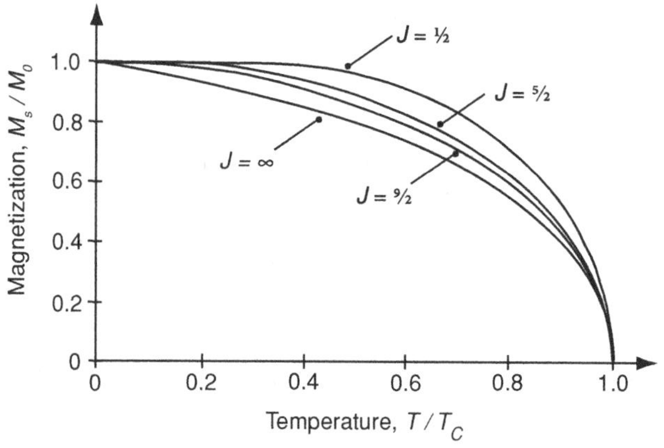

Figure 5.2

Spontaneous magnetization as a function of temperature calculated from molecular field theory, based on the Brillouin function for different values of J. The classical limit $J = \infty$ is based on the Langevin function.

$2J + 1$ energetically degenerate M_J levels are equally populated. The magnetic entropy (4.25) then is $R \ln(2J + 1)$ per mole, where $R = N_A k_B$ is the gas constant, 8.315 J mol^{-1}. Below T_C, and especially just below, there is a specific heat of magnetic origin, as energy is absorbed to disorder the moments when the system is heated. A discontinuity in specific heat appears at T_C.

On a plot of M_s/M_0 versus x, the slope of (5.5) precisely at the Curie temperature is equal to the slope at the origin of the Brillouin function. For small x (4.19) $\mathcal{B}_J(x) \approx [(J + 1)/3J]x$, hence there is a direct relation between Curie constant and Curie temperature:

$$T_C = n_W C. \tag{5.6}$$

Figure 5.3

The spontaneous magnetization for nickel, together with the theoretical curve for $J = \frac{1}{2}$ from the molecular field theory. Note that the theoretical curve is scaled to give correct values at each end.

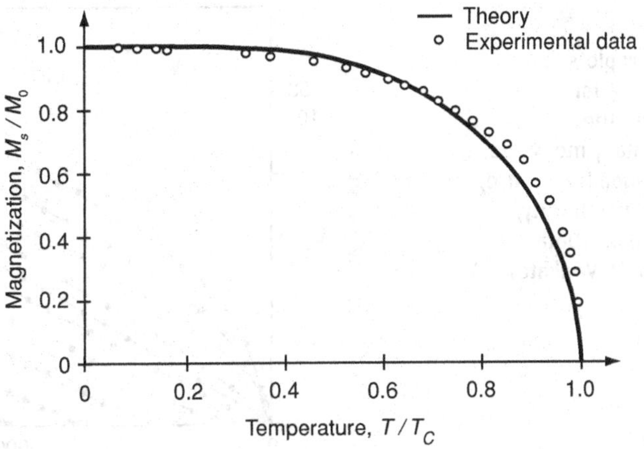

In practice, T_C is used to determine n_W. Taking gadolinium as an example: $T_C = 292$ K, $J = S = 7/2; g = 2; n = 3.0 \times 10^{28}\ \mathrm{m}^{-3}$. Hence

$$C = \mu_0 n g^2 \mu_B^2 J(J+1)/3k_B \tag{4.16}$$

is 4.9 K, and the Weiss coefficient works out as $n_W = 59$.

The paramagnetic susceptibility above T_C is obtained from (4.19), (5.3) and (5.4) in the small-x limit. The result is the Curie–Weiss law

$$\chi = C/(T - \theta_p), \tag{5.7}$$

where

$$\theta_p = T_C = \mu_0 n_W n g^2 \mu_B^2 J(J+1)/3k_B. \tag{5.8}$$

The Curie constant C is often written in terms of the effective moment $\mathfrak{m}_{eff}$ as $C = \mu_0 n \mathfrak{m}_{eff}^2/3k_B$, where $\mathfrak{m}_{eff} = g\sqrt{J(J+1)}\mu_B$. In this theory, the paramagnetic Curie temperature θ_p is equal to the Curie temperature T_C, which is the point where the susceptibility diverges.

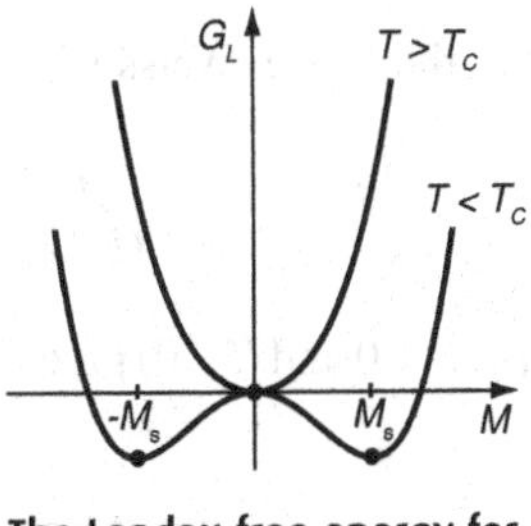

The Landau free energy for a ferromagnet at temperatures close to the Curie temperature. There are two energy minima at $\pm M_s$ for $T < T_C$, but a single minimum at $M = 0$ for $T > T_C$.

5.1.2 Landau theory

An approach that is equivalent to molecular field theory close to T_C, where M is small and aligned with any field external H', is to expand the free energy G_L in even powers of M. Only even powers are permitted in the series, because time reversal symmetry requires that the energy is unchanged on reversing the magnetization, $G_L(M) = G_L(-M)$ in the absence of the external field:

$$G_L = \mathsf{A}M^2 + \mathsf{B}M^4 + \cdots - \mu_0 H' M. \tag{5.9}$$

The coefficients A and B depend on temperature. There is a difference between the Landau free energy $G_L = f(M,T) - \mu_0 H' M$ and the Gibbs free energy $G(H',T) = F(M,T) - \mu_0 H' M$ (§2.5.4), where M is expressed as a function

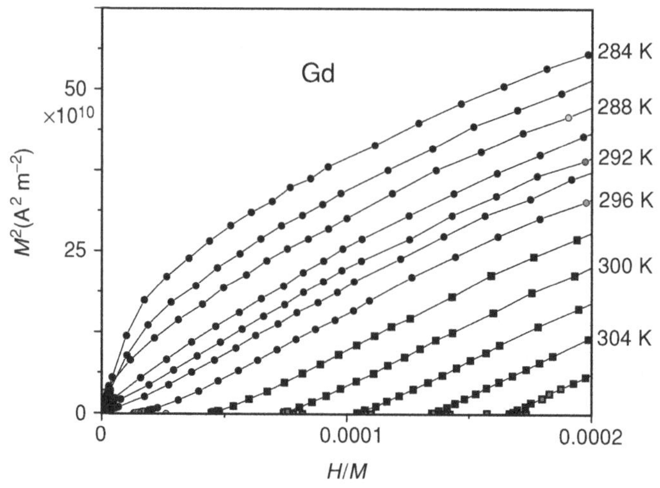

Figure 5.4

Arrott Belov plots to determine T_C for gadolinium. The experimentally measured magnetization is $\sigma = M$ d, where d is the density, rather than M. (Data courtesy of M. Venkatesan.)

Lev Landau 1908–1968.

of the variables H', T via the equation of state $M = M(H', T)$; G_L is the energy of the state when M is forced to adopt a particular value, as if it were an external constraint. G_L is minimized in a local energy minimum with that value of M, which makes the approach useful for treating problems of hysteresis.

For $T < T_C$, energy minima at $M = \pm M_s$ imply $\mathsf{A} < 0$ and $\mathsf{B} > 0$. For $T > T_C$ an energy minimum at $M = 0$ implies $\mathsf{A} > 0$ and $\mathsf{B} > 0$. It follows that A must change sign at T_C. It has the form $a(T - T_C)$, where a is a constant independent of temperature, $a > 0$. The equilibrium magnetization minimizes G_L with respect to M; $\partial G_L/\partial M = 0$ implies

$$2\mathsf{A}M + 4\mathsf{B}M^3 = \mu_0 H'. \tag{5.10}$$

Close to T_C, in zero field, $M_s^2 = -\mathsf{A}/2\mathsf{B}$, hence

$$M_s \approx \sqrt{a/2\mathsf{B}}(T_C - T)^{\frac{1}{2}}, \tag{5.11}$$

as shown in Fig. 5.2. Ignoring the demagnetizing field, the Curie–Weiss susceptibility M/H' is given by (5.10) as $\mu_0/2\mathsf{A}$;

$$\chi \approx (\mu_0/2a)(T - T_C)^{-1}. \tag{5.12}$$

When the system is at a temperature exactly equal to T_C, $\mathsf{A} = 0$ and (5.10) gives the critical isotherm

$$M = (\mu_0/4\mathsf{B})^{1/3} H'^{1/3}, \tag{5.13}$$

whereas in the vicinity of T_C (5.10) gives

$$M^2 = (\mu_0/4\mathsf{B})H'/M - (a/2\mathsf{B})(T - T_C). \tag{5.14}$$

This last equation is the basis of Arrott–Belov plots used for precise determination of the Curie temperature. The $M(H)$ curves at different temperatures are plotted

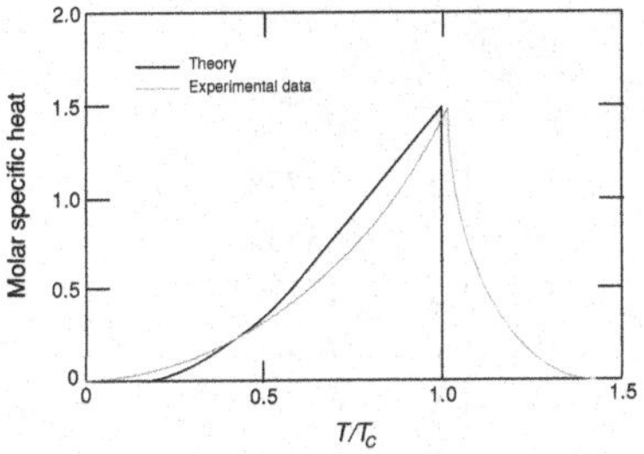

Comparison of the measured magnetic specific heat of a ferromagnet near T_C (dotted line) with the prediction of mean field theory (solid line).

Table 5.1. The critical exponents of a mean field ferromagnet

Specific heat	$C_m \sim \|T_C - T\|^\alpha$	$\alpha = 0$
Magnetization	$M_s \sim (T_C - T)^\beta$	$\beta = \frac{1}{2}$
Susceptibility	$\chi \sim (T - T_C)^{-\gamma}$	$\gamma = 1$
Critical isotherm	$M_s \sim H^{1/\delta}$	$\delta = 3$

as M^2 versus H'/M, and the isotherm that extrapolates to zero is the one at T_C (Fig. 5.4).

The magnetic specific heat C_m can also be calculated from Landau theory using $C_m = -T(\partial^2 G_L/\partial T^2)$. Results from (5.9) and (5.14) are $C_m = Ta^2/2B$ when $T = T_C^-$ and $C_m = 0$ when $T = T_C^+$. There is a stepwise discontinuity at T_C, with no magnetic specific heat above the transition where $M = 0$.

The Landau theory can be adapted to any continuous or discontinuous phase transition. M is the order parameter for the ferromagnet, H' is the conjugate field and the relation between them is the generalized susceptibility χ. Whatever the interpretation of these parameters in different physical systems, the power laws describing their variations with T near T_C are exactly the same. The same powers are obtained from Landau theory and from Weiss's molecular field theory. This can be verified by expanding the Brillouin function to order x^3 (4.19), which gives an expression equivalent to (5.9). Both are mean field theories of ferromagnetism. Other terms may be added to the free energy to include additional fields such as pressure or stress. It is remarkable how many relations can be established between different measurable physical quantities from an expansion of the free energy in powers of the order parameter.

The power law variations of the physical properties in the vicinity of T_C are summarized in Table 5.1. The values of the static critical exponents $\alpha, \beta, \gamma, \delta$ are common to all mean field theories.

Experimentally, the properties of ferromagnets do show power law behaviour in $(T - T_C)$ provided measurements are made sufficiently close to the Curie point, but the critical exponents are somewhat different from those predicted by mean field theory. For example, ferromagnets usually show a λ-type anomaly in their specific heat at T_C, rather than a stepwise discontinuity. The divergence is described by a critical exponent $\alpha \approx 0.1$, rather than zero. The residual magnetic specific heat above T_C is witness to the persistence of short-range order, which is not predicted by the theory. Above T_C, the susceptibility follows a power law $\chi \sim (T - T_C)^{-\gamma}$, where γ is about 1.3, whereas in mean field theory γ is 1 (the Curie–Weiss law). The critcal exponents $\alpha, \beta, \gamma, \delta$ for nickel, for example, are 0.10, 0.42, 1.32 and 4.5, respectively. We return to this topic at the end of Chapter 6.

The other place where significant deviations from the mean field theory are found is at low temperatures, where the spin-wave excitations discussed later in the chapter are important.

Table 5.2. Dimensionless susceptibility of some metals at 298 K (units: 10^{-6})

Li	14	Sc	263	Cu	−10	Ce	1778
K	6	Y	121	Zn	−16	Nd	3433
Be	−24	Ti	182	Au	−34	Eu	15570
Ca	22	Nb	237	Al	21	Gd	476300
Ba	7	Mo	123	Sn	−29	Dy	68400
		Pd	805	Bi	−164	Tm	17710
		Pt	279				

5.1.3 Stoner criterion

The starting point for a discussion of ferromagnetism in metals is the band paramagnetism introduced in §3.2.6. The Pauli susceptibility is a small, positive quantity, practically independent of temperature because delocalized electrons obey Fermi–Dirac statistics; only the small fraction of them with energy close to ε_F are able to respond to a change in temperature or magnetic field.

In the three-dimensional free-electron model, the density of states $\mathcal{D}(\varepsilon)$ (states m^{-3} J^{-1}) varies as $\sqrt{\varepsilon}$ (3.39), and the ↑ and ↓ bands shift by $\mp\mu_0 H\mu_B$ in the field as shown in Fig. 3.7. The resulting susceptibility (3.43) can be written

$$\chi_P = \mu_0\mu_B^2\mathcal{D}(\varepsilon_F), \tag{5.15}$$

where $\mathcal{D}(\varepsilon_F)$ is the density of states at the Fermi level for both spins, which is double the density of states for one spin $\mathcal{D}_{\uparrow,\downarrow}(\varepsilon_F)$. The Pauli susceptibility is approximately 10^{-5} for many metals, but it approaches 10^{-3} for the 4d metal Pd (Table 5.2). Narrower bands tend to have higher susceptibility, because the density of states at ε_F scales as the inverse of the bandwidth. When the density of states is high enough, it becomes energetically favourable for the bands to split, and the metal becomes spontaneously ferromagnetic.

Stoner applied Weiss's molecular field idea to the free-electron gas. Assuming the linear variation of internal field with magnetization has a coefficient n_S:

$$\boldsymbol{H}^i = n_S\boldsymbol{M} + \boldsymbol{H}, \tag{5.16}$$

the Pauli susceptibility in the internal field is $\chi_P = M/(n_S M + H)$. Hence, the response to the field H

$$\chi = M/H = \chi_P/(1 - n_S\chi_P) \tag{5.17}$$

is a susceptibility that is enhanced when $n_S\chi_P < 1$ and diverges when $n_S\chi_P = 1$. Stoner expressed this condition in terms of the local density of states at the Fermi level, $\mathcal{D}(\varepsilon_F)$. Writing the exchange energy (in J m^{-3}) $-\frac{1}{2}\mu_0 H^i M = -\frac{1}{2}\mu_0 n_S M^2$ as $-(\mathcal{I}/4)(n^\uparrow - n^\downarrow)^2/n$, where $M = (n^\uparrow - n^\downarrow)\mu_B$

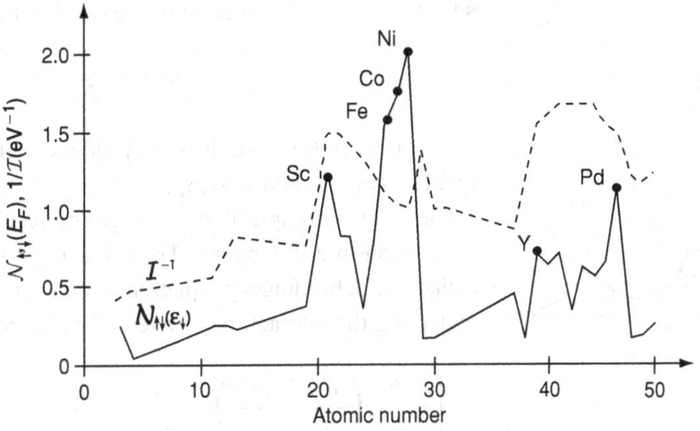

Figure 5.5

Comparision of $\mathcal{N}_{\uparrow,\downarrow}(\varepsilon_F)$ with $1/\mathcal{I}$ for metallic elements.

and n is the number of atoms per unit volume, it follows from (5.15) that $n_S\chi_P = \mathcal{I}\mathcal{D}(\varepsilon_F)/2n$. The metal becomes spontaneously ferromagnetic when the susceptibility diverges spontaneously; in other words when

$$\mathcal{I}\mathcal{N}_{\uparrow,\downarrow}(\varepsilon_F) > 1, \tag{5.18}$$

where $\mathcal{N}_{\uparrow,\downarrow}(\varepsilon) = \mathcal{D}(\varepsilon)/2n$ is the density of states per atom for each spin state. This is the famous Stoner criterion. The Stoner exchange parameter $\mathcal{I}$ is roughly 1 eV for the $3d$ ferromagnets, and $n_S \gtrsim 10^3$ for spontaneous band splitting. The exchange parameter has to be comparable to the width of the band for spontaneous splitting to be observed. Ferromagnetic metals have narrow bands and a peak in the density of states $\mathcal{N}(\varepsilon)$ at or near ε_F. The data in Fig. 5.5 show that only Fe, Co and Ni meet the Stoner criterion. Pd comes close.

5.2 Exchange interactions

The origin of the effective field $\boldsymbol{H}^i$ is the exchange interaction, which reflects the Coulomb repulsion of two nearby electrons, usually on neighbouring atoms, acting in conjunction with the Pauli principle, which forbids the two electrons to enter the same quantum state. Electrons cannot be in the same place if they have the same spin. There is an energy difference between the $\uparrow_i\uparrow_j$ and $\uparrow_i\downarrow_j$ configurations of the spins of neighbouring atoms i, j. Interatomic exchange in insulators is usually one or two orders of magnitude weaker than the ferromagnetic intra-atomic exchange between electrons on the same atom, which leads to Hund's first rule.

As stated in §4.1, the Pauli principle forbids more than one electron to enter a quantum state, denoted by a particular set of quantum numbers. Electrons are indistinguishable, so exchange of two electrons must give the same electron density $|\Psi(1, 2)|^2 = |\Psi(2, 1)|^2$. Since electrons are fermions, the only solution is for the total wave function of the two electrons to be antisymmetric

$$\Psi(1,2) = -\Psi(2,1). \tag{5.19}$$

The total wave function Ψ is the product of functions of space and spin coordinates $\phi(\boldsymbol{r}_1, \boldsymbol{r}_2)$ and $\chi(s_1, s_2)$.

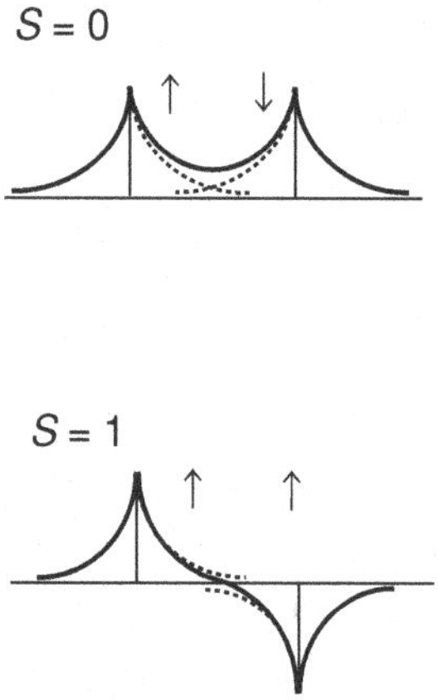

The spatially symmetric and antisymmetric wave functions for the H_2 molecule.

The simple example of the hydrogen molecule H_2 with two atoms each having an electron in a hydrogenic 1s-orbital $\psi_i(\boldsymbol{r}_i)$ gives an idea of the physics of exchange. Schrödinger's equation is $\mathcal{H}(\boldsymbol{r}_1, \boldsymbol{r}_2)\Psi(\boldsymbol{r}_1, \boldsymbol{r}_2) = \varepsilon\Psi(\boldsymbol{r}_1, \boldsymbol{r}_2)$ where, neglecting the interactions between the electrons,

$$\left[-\frac{\hbar^2}{2m}\left(\frac{\partial^2}{\partial r_1^2} + \frac{\partial^2}{\partial r_2^2}\right) - \frac{e^2}{4\pi\epsilon_0}\left(\frac{1}{r_1} + \frac{1}{r_2}\right)\right]\Psi(\boldsymbol{r}_1, \boldsymbol{r}_2) = \varepsilon\Psi(\boldsymbol{r}_1, \boldsymbol{r}_2). \tag{5.20}$$

There are two molecular orbits, a spatially symmetric bonding orbital ϕ_s, with electronic charge piled up between the atoms, and a spatially antisymmetric antibonding orbital ϕ_a having a nodal plane with no charge midway between them. Chemical bonds which involve hybridized wave functions of electrons of neighbouring atoms are generally classified in this way:

$$\phi_s = (1/\sqrt{2})(\psi_1 + \psi_2) \qquad \phi_a = (1/\sqrt{2})(\psi_1 - \psi_2). \tag{5.21}$$

ψ_1 and ψ_2 are the spatial components of the individual wave functions of electrons 1 and 2 respectively. The wave functions $\psi_1(\boldsymbol{r}_1)$ and $\psi_2(\boldsymbol{r}_2)$ are the solutions of Schrödinger's equation for each individual atom.

The symmetric and antisymmetric spin functions are the spin triplet and spin singlet states:

$S = 1; \quad M_S = 1, 0, -1$

$\chi_s = |\uparrow_1, \uparrow_2\rangle; \quad (1/\sqrt{2})[|\uparrow_1, \downarrow_2\rangle + |\downarrow_1, \uparrow_2\rangle]; \quad |\downarrow_1, \downarrow_2\rangle$

$S = 0; \quad M_S = 0$

$\chi_a = (1/\sqrt{2})[\uparrow_1, \downarrow_2\rangle - |\downarrow_1, \uparrow_2\rangle]$

According to (5.19), the symmetric space function must multiply the antisymmetric spin function, and vice versa. Hence the total antisymmetric wave functions are

$$\Psi_I = \phi_s(1,2)\chi_a(1,2),$$
$$\Psi_{II} = \phi_a(1,2)\chi_s(1,2).$$

When the two electrons are in a spin triplet state, there can be no chance of finding them at the same point of space. Electrons with parallel spins avoid each other. But if the electrons are in the spin singlet state, with antiparallel spins, there is some probability of finding them in the same place, because the spatial part of the wave function is symmetric under exchange of the electrons.

The energies of the two states can be evaluated from the Hamiltonian $\mathcal{H}(\mathbf{r}_1, \mathbf{r}_2)$ in (5.20):

$$\varepsilon_{I,II} = \int \phi^*_{s,a}(\mathbf{r}_1, \mathbf{r}_2)\mathcal{H}(\mathbf{r}_1, \mathbf{r}_2)\phi_{s,a}(\mathbf{r}_1, \mathbf{r}_2)\mathrm{d}r_1^3\mathrm{d}r_2^3.$$

For the hydrogen molecule, ε_I is lower than ε_{II}. In other words, the bonding orbital/spin singlet state lies below the antibonding orbital/spin triplet state because of the spatial constraint on the triplet. Setting the exchange integral $\mathcal{J} = (\varepsilon_I - \varepsilon_{II})/2$, we can write the energy in the form

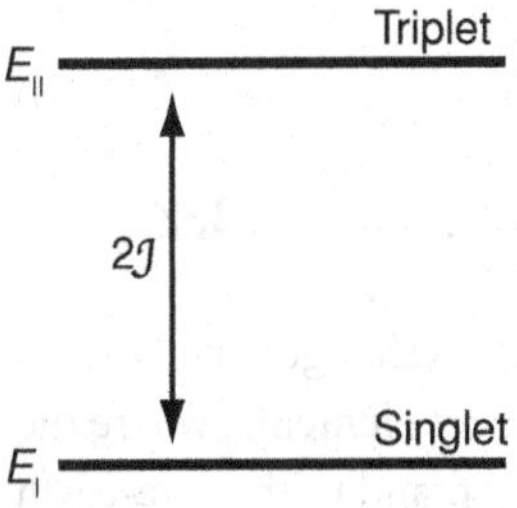

Splitting of the spin singlet and spin triplet states for the H_2 molecule. The exchange integral $\mathcal{J}$ is negative, so the singlet is lower.

$$\varepsilon = -2(\mathcal{J}/\hbar^2)\mathbf{s}_1 \cdot \mathbf{s}_2, \tag{5.22}$$

where the product $\mathbf{s}_1 \cdot \mathbf{s}_2$ is $\frac{1}{2}[(\mathbf{s}_1 + \mathbf{s}_2)^2 - \mathbf{s}_1^2 - \mathbf{s}_2^2]$. According to whether the spin quantum number $S = s_1 + s_2$ is 0 or 1, the eigenvalues are $-\frac{3}{4}\hbar^2$ or $+\frac{1}{4}\hbar^2$. The energy splitting between the singlet state Ψ_I and the triplet state Ψ_{II} is $2\mathcal{J}$. Here $\mathcal{J}$ is the exchange integral

$$\mathcal{J} = \int \psi_1^*(r')\psi_2^*(r)\mathcal{H}(r, r')\psi_1(r)\psi_2(r')\mathrm{d}r^3\mathrm{d}^3r'.$$

In the H_2 molecule, the spin singlet state is lower, so the integral is negative. In an atom, however, the orbitals are orthogonal and $\mathcal{J}$ is positive.

Heisenberg generalized (5.22) to many-electron atomic spins $\mathbf{S}_1$ and $\mathbf{S}_2$, writing his famous Hamiltonian

$$\mathcal{H} = -2\mathcal{J}\hat{\mathbf{S}}_1 \cdot \hat{\mathbf{S}}_2, \tag{5.23}$$

where $\hat{\mathbf{S}}_1$ and $\hat{\mathbf{S}}_2$ are dimensionless spin operators, like the Pauli spin matrices in (3.17). The $\hbar^2$ has been absorbed into the exchange constant $\mathcal{J}$, which has units of energy. From now on we will adopt this convention, in order to avoid writing $\hbar$ everywhere. We also drop the hat on the spin operators, $\hat{\mathbf{S}}_i$. The exchange integral $\mathcal{J}$ then has dimensions of energy, and it is often expressed in kelvins by dividing it by k_B, Boltzmann's constant. $\mathcal{J} > 0$ indicates a ferromagnetic interaction, which tends to align the two spins parallel; $\mathcal{J} < 0$ indicates an antiferromagnetic interaction, which tends to align the two spins antiparallel.

When there is a lattice, the Hamiltonian[1] is generalized to a sum over all pairs of atoms on lattice sites i, j:

$$\mathcal{H} = -2\sum_{i>j} \mathcal{J}_{ij}\mathbf{S}_i \cdot \mathbf{S}_j. \tag{5.24}$$

This is simplified to a sum with a single exchange constant $\mathcal{J}$ if only nearest-neighbour interactions count. The interatomic exchange coupling described by the Heisenberg Hamiltonian can only be ferromagnetic or antiferromagnetic.

The Heisenberg exchange constant $\mathcal{J}$ can be related to the Weiss constant n_W of the molecular field theory. Suppose that a moment $g\mu_B S_i$ interacts with an effective field $H^i = n_W M = n_W n g \mu_B S$, and that in the Heisenberg model

[1] Other conventions exist, omitting the 2 and/or counting each pair in the sum twice.

Junjiro Kanamori, 1930–2012.

only the nearest neighours of S_i interact appreciably with it. Then the site Hamiltonian is

$$\mathcal{H}_i = -2\left[\sum_j \mathcal{J}\boldsymbol{S}_j\right]\cdot\boldsymbol{S}_i \approx -\mu_0 H^i g\mu_B S_i. \tag{5.25}$$

The molecular field approximation amounts to averaging out the local correlations between $\boldsymbol{S}_i$ and $\boldsymbol{S}_j$. If Z is the number of nearest neighbours in the sum, then $\mathcal{J} = \mu_0 n_W n g^2 \mu_B^2/2Z$. Hence, from (5.8)

$$T_C = \frac{2Z\mathcal{J}S(S+1)}{3k_B}. \tag{5.26}$$

Taking the example of gadolinium again, where $T_C = 292$ K, $S = 7/2$, $Z = 12$, we find $\mathcal{J}/k_B = 2.3$ K.

The Heisenberg Hamiltonian (5.23) indicates that exchange interactions couple the atomic *spins*. It can be applied directly to the $3d$ elements, where the crystal field ensures that spin is a good quantum number, and to the rare-earth ions Eu^{2+} and Gd^{3+}, which have no orbital moment. However, J is the good quantum number for the other rare-earths, so S must be projected onto J, as explained below. Ions with a $J = 0$ ground-state multiplet, Sm^{2+} and Eu^{3+}, cannot order magnetically despite their large spin quantum number, $S = 3$.

Generally, the energy of any electronic system is lowered as the wave functions spread out. This follows from the uncertainty principle $\Delta p \Delta x \approx \hbar$. When many more-or-less delocalized electrons are present in different orbitals, the calculation of exchange is a delicate matter. Orbital degeneracy, absent in the H_2 molecule, opens the possibility that triplet states may be lower in energy than singlets. The energies involved are only $\approx$1 meV, compared with bandwidths of order 1–10 eV. Competing exchange interactions may coexist with different signs of coupling. It is therefore best to describe exchange phenomenologically, and determine the exchange interactions experimentally.

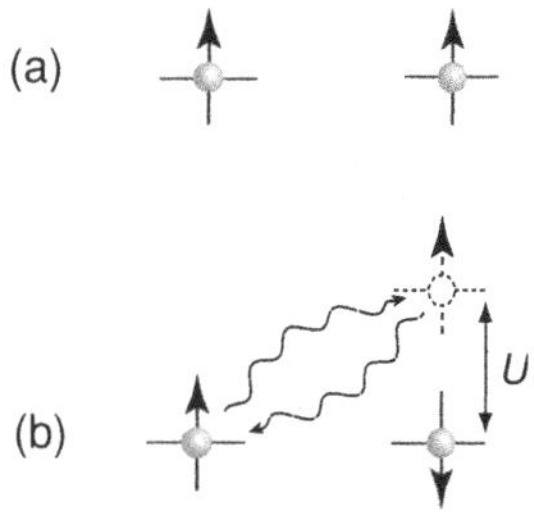

The antiferromagnetic superexchange interaction. Two neighbouring sites with singly occupied orbitals are shown with (a) parallel or (b) antiparallel spin alignment. Hopping is forbidden by the Pauli principle in the parallel case. There is an energy gain due to virtual hopping in the antiparallel case.

5.2.1 Exchange in insulators

Superexchange The electrons in insulators are localized. Oxides are a good example. There is little direct $3d$–$3d$ overlap in transition-metal oxides, but the $3d$-orbitals are hybridized with the oxygen $2p$-orbitals; $\phi_{3d} = \alpha\psi_{3d} + \beta\psi_{2p}$ with $|\alpha|^2 + |\beta|^2 = 1$. The oxygen bridges transmit a 'superexchange' interaction, which can be described by the Heisenberg Hamiltonian.

Figure 5.6 shows a typical superexchange bond. In the case of a singly occupied $3d$-orbital or a half-filled d shell (Fe^{3+}, Mn^{2+}), configuration (b) is lower in energy than configuration (a) because both electrons in an oxygen $2p$-orbital can then spread out into unoccupied $3d$-orbitals. The superexchange interaction $\mathcal{J}$ involves simultaneous virtual transfer of two electrons with the instantaneous formation of a $3d^{n+1}2p^5$ excited state; the interaction is of order $-2t^2/U$, where t is the p–d transfer integral and U is the on-site $3d$ Coulomb

Figure 5.6

A typical superexchange bond. Configuration (b) is lower in energy than configuration (a).

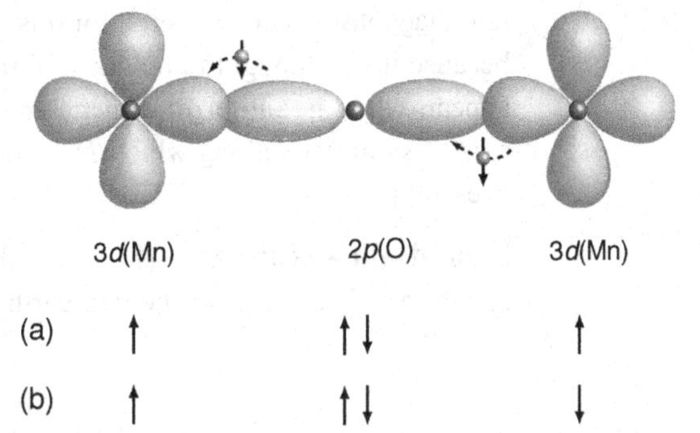

John B. Goodenough, 1922–.

interaction. The transfer integral is of order 0.1 eV and the on-site Coulomb interaction is in the range 3–5 eV. $\mathcal{J}$ depends sensitively on the interatomic separation, but also on the M–O–M bond angle, varying as $\cos^2\theta_{12}$.

The occupancy and orbital degeneracy of the 3*d* states is the critical factor in determining the strength and sign of superexchange. There are many possible cases to consider and the results were summarized in the Goodenough-Kanamori rules. The rules were reformulated by Anderson, in a simpler way that makes it unnecessary to consider the oxygen.

(i) When two cations have lobes of singly occupied 3*d*-orbitals which point towards each other giving large overlap and hopping integrals, the exchange is strong and *antiferromagnetic* ($\mathcal{J} < 0$). This is the usual case, for 120–180° M–O–M bonds.
(ii) When two cations have an overlap integral between singly occupied 3*d*-orbitals which is zero by symmetry, the exchange is *ferromagnetic* and relatively weak. This is the case for ~90° M–O–M bonds.
(iii) When two cations have an overlap between singly occupied 3*d* orbitals and empty or doubly occupied orbitals of the same type, the exchange is also *ferromagnetic*, and relatively weak.

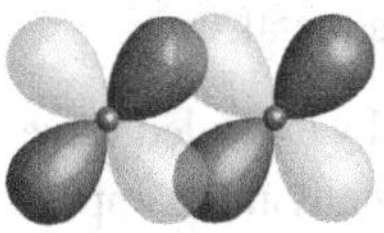

(a)

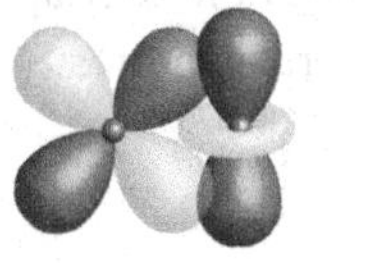

(b)

Overlapping *d*-orbitals characterized by (a) nonzero and (b) zero overlap integrals. Dark and light shading denotes positive and negative sign of the wave function.

Superexchange is more commonly antiferromagnetic than ferromagnetic, because the overlap integrals are more likely to be large than zero.

Antisymmetric exchange A few materials with low symmetry exhibit a weak antisymmetric coupling, the Dzyaloshinski-Moriya interaction. This is represented by the Hamiltonian

$$\mathcal{H} = -\mathcal{D}\cdot(\boldsymbol{S}_i \times \boldsymbol{S}_j), \tag{5.27}$$

where $\mathcal{D}$ is a vector which lies along a high-symmetry axis, so the tendency is to couple the two spins *perpendicularly*. This is a higher-order effect, ocurring between ions already coupled by superexchange; $|\mathcal{D}/\mathcal{J}| \approx 10^{-2}$. In an antiferromagnet, the spins may be canted away from the antiferromagnetic axis by about 1°. Antisymmetric exchange is the reason why antiferromagnets with a uniaxial crystal structure such as MnF_2, $MnCO_3$ and αFe_2O_3 may exhibit a weak ferromagnetic moment. In the older literature the term parasitic

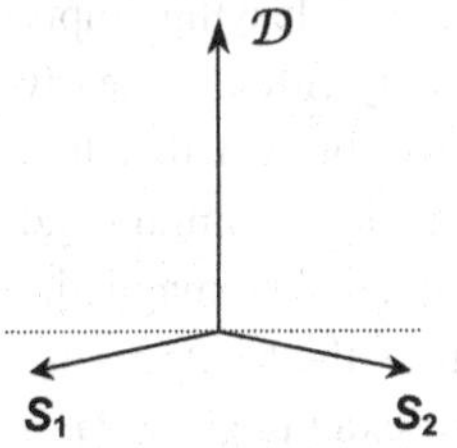

Canted antiferromagnetism due to the Dzyaloshinski-Moriya interaction.

ferromagnetism is encountered for this kind of intrinsic weak ferromagnetism, because it was thought to be due to ferromagnetic impurities. A moment only appears when the antiferromagnetic axis is perpendicular to the crystallographic axis of symmetry, along which $\mathcal{D}$ is constrained to lie. It disappears when the axes are parallel.

Biquadratic exchange This is another weak, higher-order effect which is sometimes detectable for the rare-earths. It is represented by the Hamiltonian

$$\mathcal{H} = -\mathcal{B}(\boldsymbol{S}_i \cdot \boldsymbol{S}_j)^2. \tag{5.28}$$

5.2.2 Exchange in metals

The principal exchange mechanism in ferromagnetic and antiferromagnetic metals involves overlap of the partly localized atomic orbitals of adjacent atoms. Other exchange mechanisms involve the interaction of purely delocalized electrons or of localized and delocalized electrons in the metal.

Direct exchange In $3d$ metals, the electrons are described by extended wave functions and a spin-polarized local density of states. It is usually more appropriate to describe them by the one-electron d wave functions of §4.4.2, rather than the free-electron waves of §3.2.5. In the tight-binding model the overlap of the one-electron wave functions is small and the electrons remain mostly localized on the atoms. The model Hamiltonian is

$$\mathcal{H} = \sum_{ij} t_{ij} c_i^\dagger c_j,$$

where the sum represents the conduction band in terms of the electron creation and annihilation operators[2] $c^\dagger$ and c. Usually only nearest-neighbour interactions are important and the interatomic transfer integral $t_{ij} = t$. The bandwidth in the tight-binding model is $W = 2Zt$, where Z is the number of nearest neighbours. In $3d$ metals $t \approx 0.1$ eV and $Z = 8 - 12$, so the d bands are a few eV wide. Exchange in a roughly half-filled band is antiferromagnetic, because the energy gain associated with letting the wave functions expand onto neighbouring sites is only achieved when the neighbours are antiparallel, leaving empty $\uparrow$ orbitals on the neighbouring sites to transfer into. Nearly filled or nearly empty bands tend to be ferromagnetic (Fig. 5.7) because electrons can then hop into empty states with the same spin. This helps to explain why chromium and manganese are antiferromagnetic, but iron, cobalt and nickel are ferromagnetic.

Bandwidth is the enemy of exchange. As t becomes large, the electrons are delocalized regardless of their spin. The alkali metals, for example, are Pauli

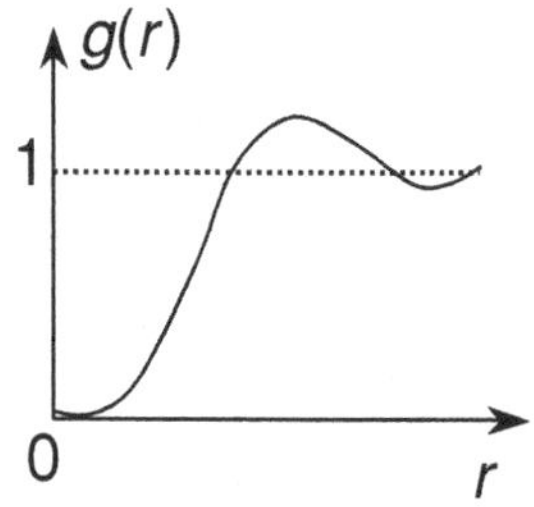

The exchange hole: the normalized probability of finding two electrons with the same spin a distance r apart.

[2] c_j is an operator that destroys an electron on site j, while $c_i^\dagger$ is an operator that creates an electron on site i. The product $c_i^\dagger c_j$ therefore transfers an electron from site j to site i.

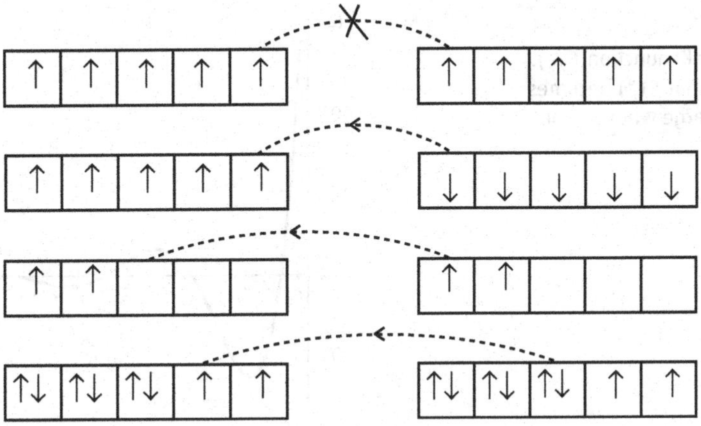

Figure 5.7

Electron delocalization in *d* bands which are half-full, almost empty or almost full.

paramagnets described by a free-electron model with one electron per atom. The early $3d$ metals scandium, titanium and vanadium are not ferromagnetic because t is too big. Scandium comes close. If it were possible to dilate the lattice to reduce t a little, scandium would become ferromagnetic.

The sign of the direct exchange depends principally on band occupancy, and then on the interatomic spacing, with ferromagnetic exchange favoured at larger spacing. The exchange is greatest just after the critical condition for the appearance of magnetism, $U/W > (U/W)_{crit}$, where U is the on-site Coulomb interaction, and W is the bandwidth.

***s–d* model** Coupling of the spins $\boldsymbol{s}$ of the conduction electrons with core spins $\boldsymbol{S}$ in a metal is generally represented by a Hamiltonian including the term

$$-\mathcal{J}_{sd}\Omega\,|\psi|^2\,\boldsymbol{S}\cdot\boldsymbol{s}, \tag{5.29}$$

where Ω is the volume of the core d shell and $|\psi|^2$ is the s-electron probability density. The s–d coupling is an on-site interaction, so the coupling constant is large, $\mathcal{J}_{sd} \approx 1$ eV. This interaction may lead to long-range ferromagnetic coupling between the core spins, regardless of whether $\mathcal{J}_{sd}$ is positive or negative. The host conduction band is supposed to be uniformly spin-polarized parallel or antiparallel to the core spins.

RKKY interaction The 's–d' model applies as well to rare-earths, where the core spins are not $3d$, but $4f$. The localized moments in the $4f$ shell interact via electrons in the $5d/6s$ conduction band. The on-site interaction between a core spin $\boldsymbol{S}$ and a conduction electron spin $\boldsymbol{s}$ is $-\mathcal{J}_{sf}\boldsymbol{S}\cdot\boldsymbol{s}$, where $\mathcal{J}_{sf} \approx 0.2$ eV. Ruderman, Kittel, Kasuya and Yosida showed that a single magnetic impurity actually creates a nonuniform, oscillating spin polarization in the conduction band which falls off as r^{-3}. This spin polarization is related to the Friedel oscillations of charge density around the impurity which have wavelength π/k_F. It leads to long-range oscillatory coupling between core spins. For free

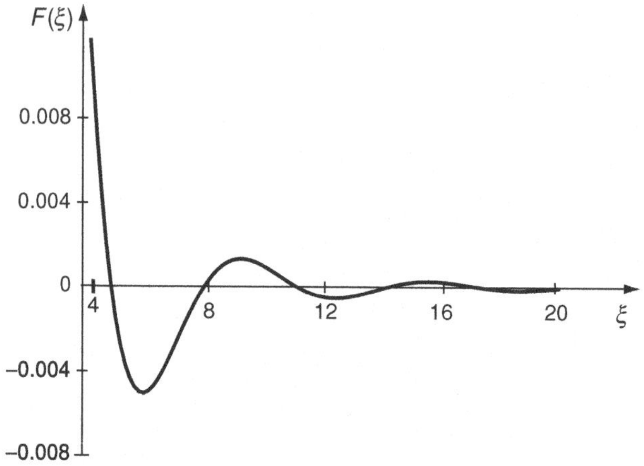

Figure 5.8

The RKKY function $F(\xi)$. Note that $F(\xi)$ becomes very large when $\xi < 4$.

electrons, the polarization is proportional to the RKKY function

$$F(\xi) = (\sin\xi - \xi\cos\xi)/\xi^4,$$

where $\xi = 2k_F r$, k_F being the Fermi wavevector (Fig. 5.8). This oscillating spin polarization results from the different potential seen by the ↑ and ↓ conduction electrons at the local moment site. The first zero of $F(\xi)$ is at $\xi = 4.5$. The effective coupling between two localized spins is

$$\mathcal{J}_{eff} \approx \frac{9\pi \mathcal{J}_{sf}^2 \nu^2 F(\xi)}{64\mathcal{E}_F}, \tag{5.30}$$

where ν is the number of conduction electrons per atom and $\mathcal{E}_F$ is the Fermi energy. Since the Fermi wavevector is about 0.1 nm^{-1} (Table 3.3), the sign of $\mathcal{J}_{eff}$ fluctuates on a scale of nanometres. When only ferromagnetic nearest-neighbour coupling is important, the Curie temperature can be deduced from (5.26). The RKKY interaction in the low-electron-density limit is equivalent to the s–d model with ferromagnetic coupling. Analogous oscillatory exchange is found in ferromagnetic multilayers with nonmagnetic spacer layers.

Among the rare-earth metals, only gadolinium has S as a good quantum number. The others have J as their quantum number, yet the exchange interaction couples spins. We therefore need to project S onto J when calculating the exchange coupling, whether direct or indirect. Since $\boldsymbol{L} + 2\boldsymbol{S} = g\boldsymbol{J}$ and $\boldsymbol{J} = \boldsymbol{L} + \boldsymbol{S}$, $\boldsymbol{S} = (g-1)\boldsymbol{J}$. This introduces a factor $(g-1)^2 J(J+1)$ into the exchange coupling. The factor is squared, because the spin enters the exchange interaction between two rare-earths twice. The effective coupling is

$$\mathcal{J}_{RKKY} = G\mathcal{J}_{eff}.$$

where $G = (g-1)^2 J(J+1)$ is the de Gennes factor. Magnetic ordering temperatures for any series of rare-earth metals or compounds with the same conduction-band structure and similar lattice spacings should scale with G,

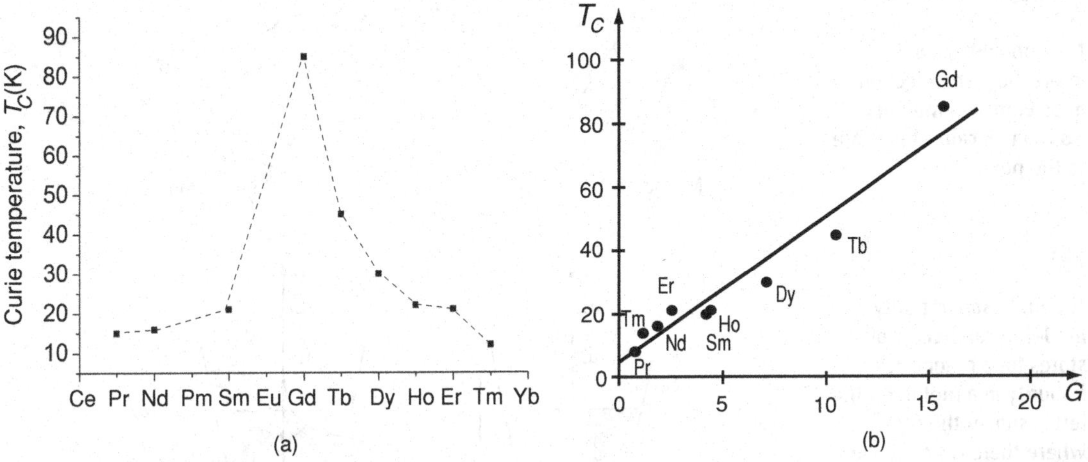

Figure 5.9

(a) Curie temperatures of ferromagnetic RNi_2 compounds; (b) a plot of T_C versus the de Gennes factor G.

and show a maximum for gadolinium. The de Gennes factor was included in Table 4.10. Figure 5.9 displays data for the Curie temperature of a series of ferromagnetic RNi_2 compounds. When plotted versus G, the data follow a straight line. The nickel in this series of intermetallic compounds is nonmagnetic.

Double exchange This interaction arises between 3*d* ions which have both localized and delocalized *d* electrons. Unlike ferromagnetic superexchange, mixed valence configurations are required for double exchange, as they are in any metal, but unlike a normal metal, the number of configurations is restricted to just two. In copper, for example, with its one electron in a broad 4*s* band, the instantaneous atomic configurations are s^0, s^1 and s^2. Electronic correlations are weak in broad bands, so the three configurations will appear with probabilities of $\frac{1}{4}$, $\frac{1}{2}$ and $\frac{1}{4}$. By contrast, a double-exchange material, such as the manganite $(La_{0.7}Ca_{0.3})MnO_3$, has both $Mn^{4+}$and Mn^{3+} ions (d^3 and d^4) present on octahedral sites. The two Mn valence states are imposed by the charge states of the other ions in the compound, La^{3+}, Ca^{2+} and O^{2-}. The d^3 core electrons for both octahedrally coordinated ions are localized in a narrow $t_{2g}^{\uparrow}$ band, but the fourth d electron inhabits a broader $e_g^{\uparrow}$ band, hybridized with oxygen, where it can hop from one d^3 core to another, Fig. 5.10. The configurations $d_i^3 d_j^4$ and $d_i^4 d_j^3$ on adjacent sites i and j are practically degenerate. On each site, there is strong on-site Hund's rule exchange coupling $\mathcal{J}_H \approx 2$ eV between t_{2g} and e_g electrons. Electrons can hop freely if the core spins are parallel, but when they are antiparallel there is a large energy barrier due to the Hund's rule interaction. If the quantization axes of adjacent sites are misaligned by an angle θ, the eigenvector of a $\uparrow$ electron in the rotated frame is $\begin{vmatrix}\cos\theta/2\\ \sin\theta/2\end{vmatrix}$ (3.24). The transfer integral t therefore varies as $\cos(\theta/2)$. Double exchange is ferromagnetic because the transfer is zero when the ions on adjacent sites are antiparallel, $\theta = \pi$.

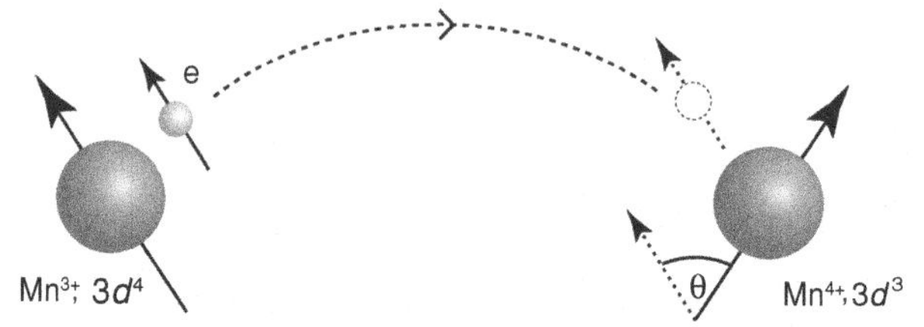

Figure 5.10

The double-exchange interaction. The electron hops with spin memory from one localized ion core to the next.

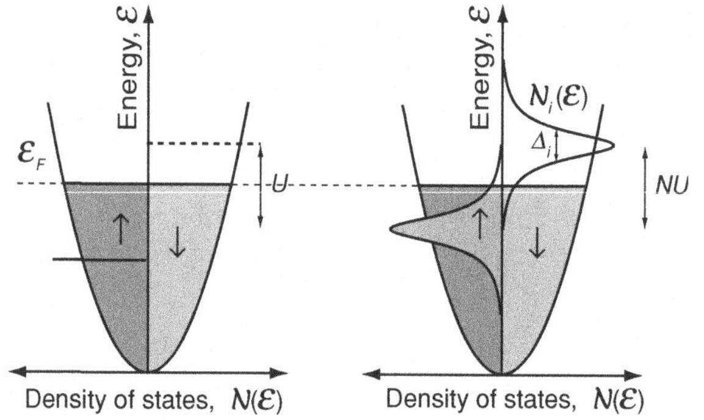

Figure 5.11

The Anderson impurity model. Local density of states for a magnetic impurity in a metal. On the left is shown the case where there is no mixing of the wave functions of the impurity level with the conduction electrons, on the right is the result of such hybridization.

Another common double-exchange pair is Fe^{3+} and Fe^{2+}, which are d^5 and d^6 ions respectively. The d^5 configuration is a half-filled ↑ shell, and the sixth d electron occupies the bottom of a $t_{2g}^{\downarrow}$ band when the ion is octahedrally coordinated by oxygen where it can hop directly from one d^5 core to another.

5.3 Band magnetism

5.3.1 Magnetic impurities in nonmagnetic metals

The above discussion of exchange between localized moments and conduction electrons in a metal begs the question of whether a magnetic impurity can really retain its moment when diluted in a nonmagnetic matrix. Does a single atom of cobalt, for example, still have a moment when it is diluted in copper? The magnetic impurity problem engrossed the magnetism community in the 1960s and 1970s. The $3d$ electrons of cobalt will hybridize with the $4s$ electrons of copper, broadening the local atomic level into a Lorentzian-like feature in the density of states. Figure 5.11 shows the energy level of a singly occupied d orbital before hybridization, with the doubly occupied orbital higher in energy by the on-site Coulomb repulsion energy U. Hybridization with the conduction band states broadens the impurity level, giving it a width Δ_i. Anderson showed that a moment, albeit one reduced by hybridization, is stable provided $U > \Delta_i$. Further broadening of the local density of states destroys the moment

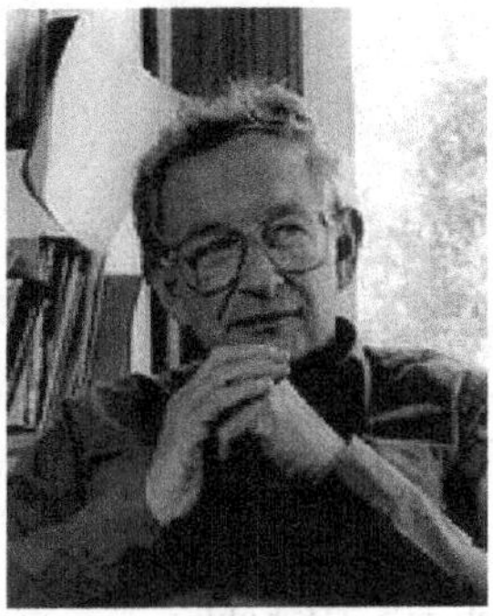

P. W. Anderson 1923–.

completely. The broadening is more effective for p-metals than for s-metals, because more electrons are available to hybridize with the impurity d-orbitals. Cobalt keeps its moment in copper, but loses it in aluminium.

The number of unpaired impurity electrons $N = N^{\uparrow} - N^{\downarrow}$ is

$$N = \nu(\varepsilon_F^+) - \nu(\varepsilon_F^-), \tag{5.31}$$

where $\varepsilon_F^{\pm} = \varepsilon_F \pm \frac{1}{2}NU$ and $\nu(\varepsilon)$, is the integral of the impurity density of states, $\int_o^{\varepsilon} \mathcal{N}_i(\varepsilon')\mathrm{d}\varepsilon'$. Expanding this expression as a power series for small N, we find $N = NU\mathcal{N}_i(\varepsilon_F) + \frac{1}{24}(NU)^3\mathcal{N}''(\varepsilon_F)$, where the second derivative $\mathcal{N}_i''(\varepsilon) = \mathrm{d}^2\mathcal{N}_i(\varepsilon)/\mathrm{d}\varepsilon^2$ is negative. Hence $N^2 = [24(1 - U\mathcal{N}_i(\varepsilon_F))/\mathcal{N}_i''(\varepsilon_F)U^3]$. A moment will form spontaneously at the impurity provided

$$U\mathcal{N}_i(\varepsilon_F) > 1. \tag{5.32}$$

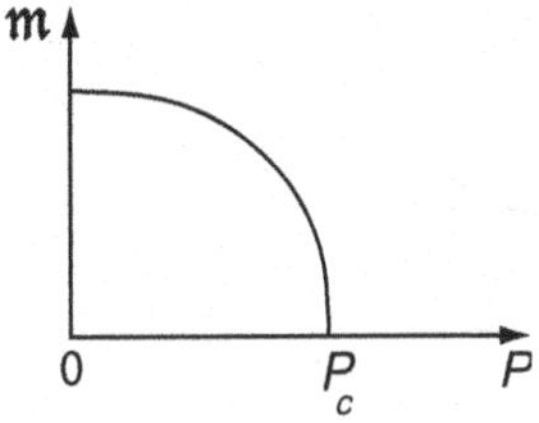

Destruction of an impurity magnetic moment at a critical pressure P_c in the Anderson model.

Since $\mathcal{N}_i(\varepsilon_F)$ is approximately $1/\Delta_i$, we find the Anderson criterion $U \gtrsim \Delta_i$ for magnetism of the impurity. It may be compared with the Stoner criterion for ferromagnetism (5.18). Strong correlations favour magnetism, strong mixing destroys it. If $\mathcal{N}_i(\varepsilon_F)$ varies *smoothly* with some parameter x such as pressure or concentration, then the magnetic moment on the impurity just below the critical value x_c where it disappears will vary as $(x - x_c)^{\frac{1}{2}}$.

The existence of a moment on an atomic site in an alloy may depend sensitively on the local environment. For example, Fe carries a moment when dilute in Mo, but not in Nb. The Fe–Nb hybridization is more effective than Fe–Mo hybridization at broadening the local iron density of states. Iron impurities in $Nb_{1-x}Mo_x$ alloys are nonmagnetic when surrounded by less than seven Mo atoms and magnetic when there are seven or more Mo nearest neighbours. In alloys with $x \simeq 0.6$, magnetic and nonmagnetic iron impurities coexist on different sites with different atomic environments. The model where magnetic moments in alloys are governed by the local chemical environment is known as the Jaccarino–Walker model.

Granted a local moment, the s–d Hamiltonian, also known as the Kondo Hamiltonian when $\mathcal{J}_{sd}$ is negative, is written as

$$\mathcal{H} = \sum_{i,j} t_{ij} c_i^{\dagger} c_j - \sum_{k,l} \mathcal{J}_{sd} \boldsymbol{S}_k \cdot \boldsymbol{s}_l. \tag{5.33}$$

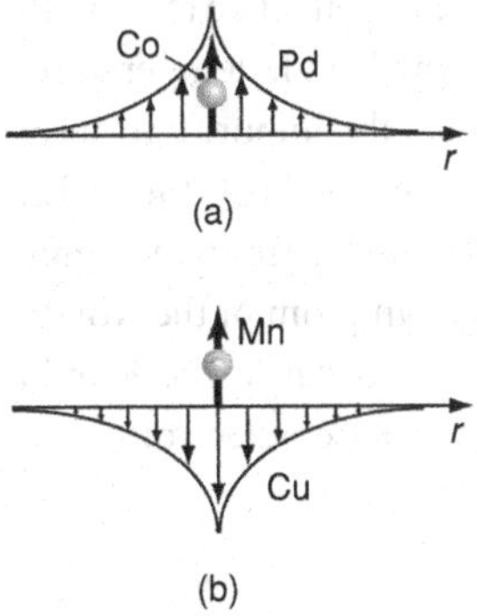

(a) A giant moment and (b) a Kondo singlet showing local spin polarization of the host conduction band.

Possible consequences of the interaction between the magnetic impurity and the conduction electrons are the formation of a giant moment when the s–d exchange is ferromagnetic, $\mathcal{J}_{sd} > 0$, or the Kondo effect when the s–d exchange is antiferromagnetic, $\mathcal{J}_{sd} < 0$. The giant moment is due to a cloud of positively polarized electron density surrounding the impurity site. When the paramagnetic susceptibility of the host (5.17) is enhanced beyond the Pauli susceptibility expected from the bare density of states yet not quite sufficiently to meet the Stoner criterion, the dressed local moment can be very large. Cobalt impurities in a palladium host have associated moments of several tens

Table 5.3. Kondo temperatures (in kelvin). The host metal is indicated in bold type.

	Cr	Mn	Fe	Co	Ni
Cu	1.0	0.01	25	2000	5000
Ag	.02	.04	3		
Au	0.01	0.01	0.3	200	
Zn	3	1.0	90		
Al	1200	530	5000		

From D. L. Wohlleben and B. R. Coles in Magnetism 5, (H. Suhl, editor), New York: Academic Press, 1973.

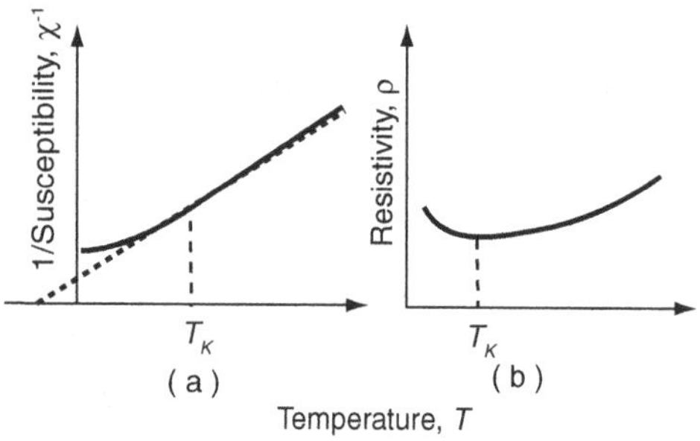

Figure 5.12

Experimental signs of the Kondo effect: (a) inverse susceptibility of a Kondo alloy and (b) the temperature dependence of the resistivity.

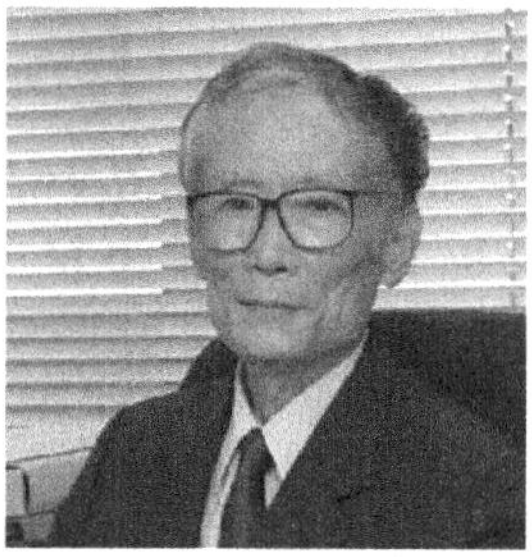

Jun Kondo 1930–.

of Bohr magnetons. There is a threshold beyond which the entire matrix turns ferromagnetic; for Co in Pd, the threshold is only 1.5 at% (see Fig. 10.13(c)).

When the exchange coupling between the impurity moment and the conduction electrons $\mathcal{J}_{sd}$ is negative, there is a possibility of forming a nonmagnetic spin singlet state from the impurity and the surrounding cloud of negatively polarized conduction electrons. A good example is iron in copper. The susceptibility is ambiguous; it shows Curie–Weiss temperature dependence (5.7) with negative θ_p above a certain temperature T_K, known as the Kondo temperature, but it becomes temperature-independent below T_K, when the impurity forms a nonmagnetic singlet state with the conduction electrons of the host, Fig. 5.12. According to the system, the Kondo temperature can lie anywhere in the range 1–1000 K. Some values are given in Table 5.3. Another symptom of the Kondo effect is a shallow minimum in the resistivity near T_K, because the Kondo singlets provide an additional channel for scattering conduction electrons. The Kondo temperature is

$$T_K \approx (\Delta_i / k_B) \exp[\Delta_i / 2\mathcal{J}_{sd}]$$

and the excess resistivity due to the Kondo scattering varies logarithmically with temperature; it is proportional to $(\mathcal{J}_{sd}^2/\Delta_i)S(S+1)[1 + (2\mathcal{J}_{sd}/\Delta_i)\ln(\Delta_i/k_BT)]$.

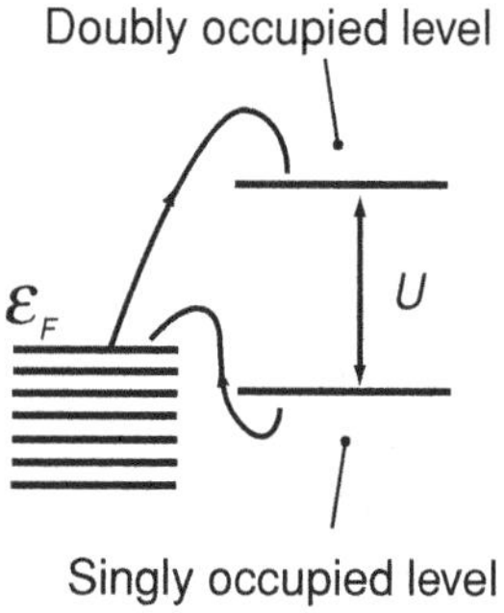

Kondo scattering. A singlet state is formed between the impurity spin and the conduction electrons.

Figure 5.13

Densities of states for some metallic elements in the paramagnetic state. Calculations by courtesy of Chaitania Das.

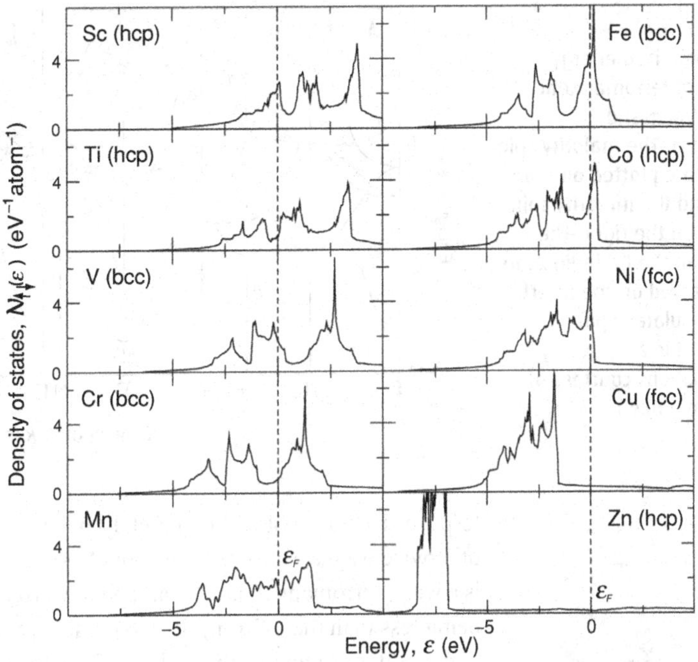

5.3.2 Ferromagnetic metals

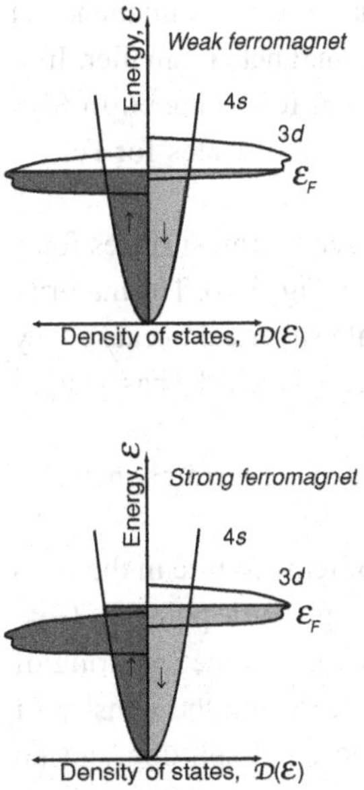

Schematic densities of states for a strong and a weak ferromagnet. The 3$d^{\uparrow}$-band is full for the strong ferromagnet.

The calculated densities of states for some paramagnetic metals are illustrated in Fig. 5.13. A highly structured 3d-band is superposed on a much broader band of 4s character. The structure of the d-bands reflects the crystal-field splitting of the t_{2g}- and e_g-bands in 8-fold or 12-fold coordination (Fig. 4.12) in the body-centred cubic (bcc) or face-centred cubic (fcc) structures, as well as the bonding/antibonding splitting between the states near the bottom or the top of the bands, and singularities that appear when the bands cross the boundary of the Brillouin zone.

The band diagrams show the dispersion relations $\varepsilon(k)$ of the five d-bands along different directions in k-space in the first Brillouin zone. The Brillouin zone is a primitive unit cell of reciprocal space defined by the Wigner–Seitz procedure, which involves forming the perpendicular bisector planes of the vectors from the origin to neighbouring reciprocal lattice points. The example of iron metal is illustrated in Fig. 5.14. The spin-up and spin-down bands are shown on separate panels. The broad parabolic free-electron-like s-bands, starting at −4 V are barely spin polarized; they hybridize with the d states between −3 and 2 eV for spin-down. The flatter, spin-up d-bands are filled. Two spin-down d bands lie mainly above Fermi level.

In the Stoner picture of metallic ferromagnetism, the bands split spontaneously provided the criterion (5.18) is satisfied. If the splitting is sufficient

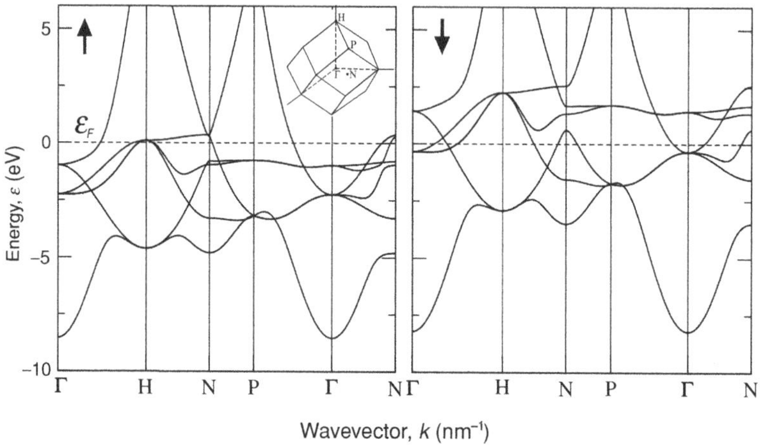

Figure 5.14

Spin-polarized energy bands of ferromagnetic α-iron for $\uparrow$ and $\downarrow$ electrons. The majority spin bands are plotted on the left, and the minority spin bands on the right. The points in the Brillouin zone are marked on the insert. The calculated spin moment is 2.2 μ_B. (Calculations courtesy of Chaitania Das.)

Edmund Stoner, 1899–1968.

to push the $\uparrow$ d-subband completely below ε_F we have a strong ferromagnet, otherwise we have a weak ferromagnet. Among the ferromagnetic elements, Fe is a weak ferromagnet, but Co and Ni are strong (despite their atomic moments being less than that of iron). In each case there are approximately 0.6 electrons at the bottom of an unsplit sp-band. The $3d$ levels lie above the bottom of the $4s$-band thanks to the term in the Schrödinger equation for the multielectron atom (4.7) which is identified with orbital kinetic energy. The spin moments of Ni and Co are 0.6 μ_B and 1.6 μ_B, respectively. Co has a residual unquenched orbital moment of 0.14 μ_B. But that of the other ferromagnets is smaller. Iron would have a spin moment of 2.6 μ_B if it were a strong ferromagnet. In fact, its moment is 2.2 μ_B. The calculated spin-split densities of states for Fe, Co and Ni are shown in Fig. 5.15.

The different filling of $\uparrow$ and $\downarrow$ bands leads to different Fermi surfaces for $\uparrow$ and $\downarrow$ electrons. They are illustrated for Fe, Co and Ni in Fig. 5.16. The majority spin surfaces for Co and Ni are quite small and roughly spherical because they contain only electrons with predominantly $4s$ character, whereas Fe has a larger $\uparrow$ Fermi surface. All three have large $\downarrow$ Fermi surfaces.

Table 5.4 summarizes the most important properties of the ferromagnetic $3d$ metals.

Stoner calculated the magnetization as a function of temperature in the free-electron model. His calculation gave an unrealistically high Curie point, $k_B T_C \approx \varepsilon_F$ because the only effect of temperature he considered was the smearing of the Fermi–Dirac occupancy function (3.45) which decreases the density of states near ε_F when $k_B T_C \approx \varepsilon_F$. There should be no band splitting and no moment above T_C. The temperature dependence of the susceptibility above T_C is that given by (3.46). However, in most metallic ferromagnets, T_C is at least an order of magnitude less than predicted by the Stoner theory, and there is a substantial Curie–Weiss-like variation of the susceptibility above T_C. On-site electronic correlations sustain an atomic-scale moment which does not

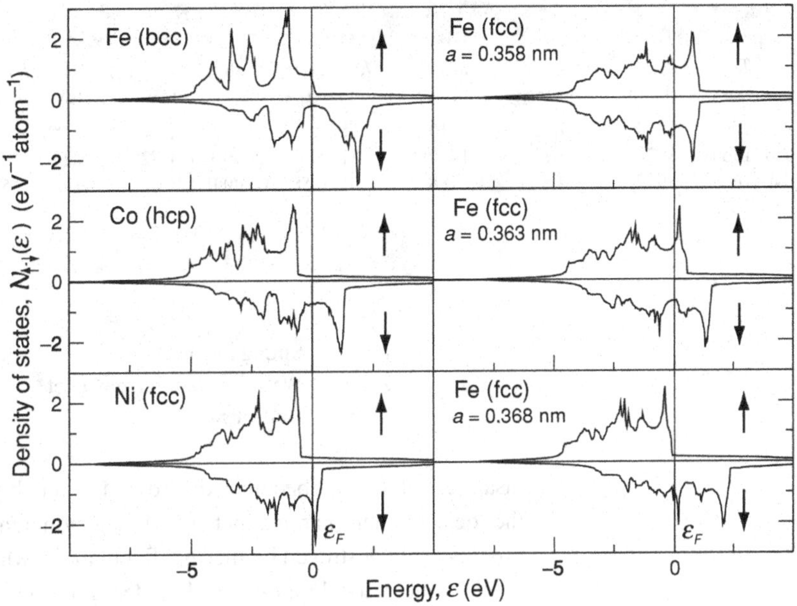

Figure 5.15

Densities of states for some elements in the ferromagnetic state. Fe is a weak ferromagnet, Co and Ni are strong. Results for γFe with different lattice parameters illustrate the sensitivity of the Fe moment to lattice parameter in a dense-packed structure. (Calculations courtesy of Ivan Rungger.)

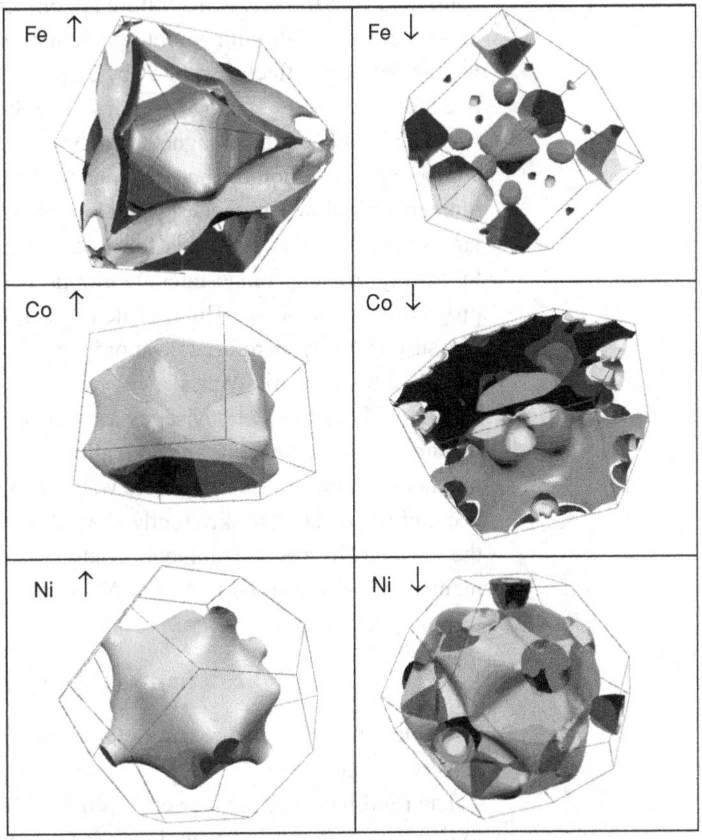

Figure 5.16

The Fermi surfaces of Fe, Co and Ni for ↑ and ↓ electrons.

Table 5.4. Intrinsic properties of the ferromagnetic 3d elements at room temperature

	T_C (K)	d (kg m^{-3})	σ_s (A m^2 kg^{-1})	M_s (kA m^{-1})	J_s (T)	$\mathfrak{m}$ (spin/orbit) (μ_B)	$\mathcal{N}_{\uparrow,\downarrow}(\varepsilon_F)$ (eV^{-1})	$\mathcal{I}$ (eV)	K_1 (J m^{-3})	λ_s (10^{-6})	g	D_{sw} (10^{-40} J m^2)
Fe	1044	7874	217	1710	2.15	2.17 (2.09/0.08)	1.54	0.93	48	−7	2.08	4.5
Co	1360(ε)	8920	162(ε)	1440	1.81	1.71 (1.57/0.14)	1.72	0.99	410	−60	2.17	8.0
Ni	628	8902	54.8	488	0.61	0.58 (0.53/0.05)	2.02	1.01	−5	−35	2.18	6.3

Table 5.5. Moments in metallic ferromagnets

		$\mathfrak{m}_{eff}$	$\mathfrak{m}_0$	T_c
Ni	Strong ferromagnet	1.0	0.6	628
$ZrZn_2$	Weak itinerant ferromagnet	1.8	0.2	25
CrO_2	Half-metal	2.4	2.0	396

disappear at T_C, but becomes disordered in much the same way as it does for the local-moment paramagnet (§4.3) or ferromagnet (§5.1.1). The moment is progressively destroyed by thermal fluctuations when $T \gg T_C$.

Whenever a local moment is disordered but stable in temperature, the effective moment $\mathfrak{m}_{eff}$ deduced from the susceptibility above T_C using (5.7) should be consistent with the zero-temperature ferromagnetic moment $\mathfrak{m}_0$ in the sense of Table 5.5. For metals with nonintegral numbers of unpaired electrons per atom, we can define an effective spin S^* by $2\mu_B S^* = \mathfrak{m}_0$, and a corresponding effective moment $\mathfrak{m}_{eff}$ as $2\sqrt{S^*(S^*+1)}\mu_B$. The Stoner model applies best to some very weak itinerant ferromagnets with $S^* \ll \frac{1}{2}$, such as $ZrZn_2$, an intermetallic compound of two nonmagnetic elements which exhibits a small ferromagnetic moment and a low Fermi energy. For a weak itinerant ferromagnet, $\mathfrak{m}_{eff}$ is even larger than expected from this formula. The effect of temperature is not just to destroy the long-range intersite atomic correlations, but also to eliminate progressively the on-site Hund's rule correlations that sustain a local moment. The susceptibility therefore falls more rapidly with increasing temperature than predicted by the Curie–Weiss law.

The rigid-band model envisages a fixed, spin-split density of states for the ferromagnetic 3d elements and their alloys, which is filled up with the necessary number of electrons as if they were water being poured into a jug. The jugs for bcc and fcc metals are differently shaped. Ignoring the small contribution of the 4s-band, the average moment per atom is $\langle m \rangle \approx (N_{3d}^{\uparrow} - N_{3d}^{\downarrow})\mu_B$. The total number of 3$d$ electrons is $N_{3d} = N_{3d}^{\uparrow} + N_{3d}^{\downarrow}$, where $N_{3d}^{\uparrow} = 5$ for the strong ferromagnets. Hence

$$\langle \mathfrak{m} \rangle \approx (10 - N_{3d})\mu_B. \tag{5.34}$$

This relation applies to any strong ferromagnet regardless of the details of the density of states.

The rigid-band picture is oversimplified. Nevertheless, the model has merit. In Cu_xNi_{1-x} alloys, for example, each Cu atom brings an extra electron. The

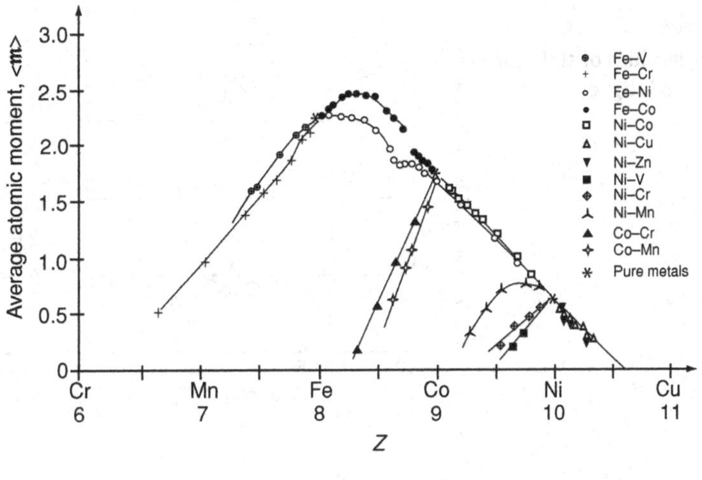

Figure 5.17

The Slater-Pauling curve. The average atomic moment is plotted against the number of valence ($3d + 4s$) electrons.

d-band of Ni has 0.6 holes and the ferromagnetism of Cu_xNi_{1-x} disappears when the d-band is full up at $x = 0.6$, as predicted.

Bonding states with delocalized singlet-like wave functions are found near the bottom of a metallic band and antibonding states with more localized triplet-like states are near the top. Since the total wave function must be antisymmetric, this helps to explain why $3d$ elements near the end of the series tend to be ferromagnetic, while those at the beginning of the series are not. The binding energy of $3d$ electrons increases by about 5 eV across the series, which is comparable to the $3d$ bandwidth. The band narrowing resulting from the increased nuclear charge is sufficient to offset the broadening resulting from reduced metal–metal distances as we move across the $3d$ series.

The famous Slater–Pauling curve, Fig. 5.17, is a plot of the magnetic moment per atom for binary alloys of $3d$ elements plotted against Z, the total number of $3d$ and $4s$ electrons per atom. There is inevitably some mixture of $4p$ character in the $4s$-band. The alloys on the right-hand side of Fig. 5.17 are strong ferromagnets. The slope of the branch on the right is -1. The multiple branches with slope ≈ 1, as expected for rigid bands, are for alloys of late $3d$ elements with early $3d$ elements for which the $3d$-states lie well above the Fermi level of the ferromagnetic host. The assumption of a common band in the rigid-band model really applies only when the charge difference of the constituent atoms is small, $\Delta Z \lesssim 2$. Otherwise a split band with a joint density of states reflects the densities of states of the constituents. The partial densities of states of FeV and $FeNi_3$, for example, are compared in Fig. 5.18.

The magnetic valence model is a more general formulation of these ideas which allows us to estimate the average atomic moment per atom of any alloy of a $3d$ element, provided it is a strong ferromagnet. The valence of an atom is given by $\mathcal{Z} = N^{\uparrow} + N^{\downarrow}$, where $N^{\uparrow}$ and $N^{\downarrow}$ are the numbers of $\uparrow$ and $\downarrow$ valence electrons per atom. The magnetic moment is given by $\mathrm{m} = (N^{\uparrow} - N^{\downarrow})\mu_B = (2N^{\uparrow} - \mathcal{Z})\mu_B$. Now the value of $N_d^{\uparrow}$ is exactly 5 for strong ferromagnetic

Figure 5.18

Partial densities of states of (a) FeV and (b) $FeNi_3$.

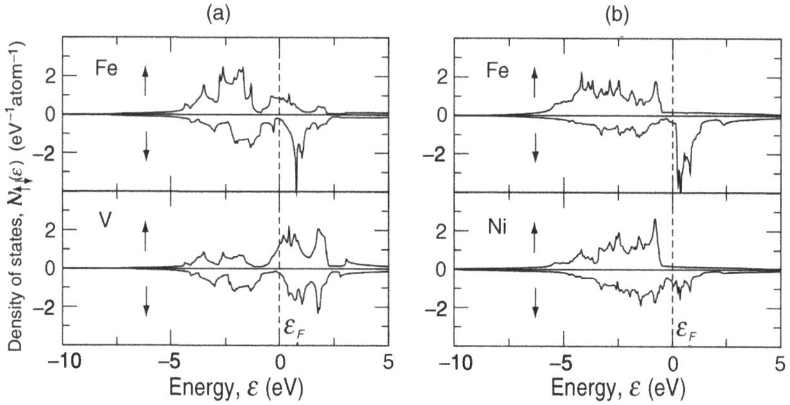

elements, and 0 for main group elements which have no d electrons. The magnetic valence of an element $\mathcal{Z}_m$, defined as

$$\mathcal{Z}_m = 2N_d^{\uparrow} - \mathcal{Z}, \tag{5.35}$$

is an integer. Its moment is $\mathfrak{m} = (\mathcal{Z}_m + 2N_s^{\uparrow})\mu_B$, where $2N_s^{\uparrow} \approx 0.6 - 0.7$ is the number of electrons in the unpolarized $4sp$-band. The average moment per atom in an alloy is obtained by replacing $\mathcal{Z}_m$ by its weighted average value over all atoms present in the alloy:

$$\langle \mathfrak{m} \rangle = (\langle \mathcal{Z}_m \rangle + 2N_s^{\uparrow})\mu_B. \tag{5.36}$$

In this way it is possible to estimate the magnetization of any strong ferromagnetic alloy based on iron, cobalt or nickel. Some magnetic valences are $\mathcal{Z}_m = -3$ for B, Y, La and all rare-earths, -4 for C, Si, Ti, -5 for V, P, -6 for Cr, but 2 for Fe, 1 for Co and 0 for Ni. Taking YFe_2 as an example, the average moment is $[\frac{1}{3}(-3 + 0.6) + \frac{2}{3}(2 + 0.6)] = 0.93$ μ_B/atom or 2.8 μ_B/(formula unit, fu). We can consider that the yttrium has reduced the iron moment from 2.2 μ_B to 1.4 μ_B. Adding more yttrium to the alloy will eventually destroy the magnetism entirely (Exercise 5.7(c)). Moments per atom for rare-earth–iron alloys are shown in Fig. 5.19.

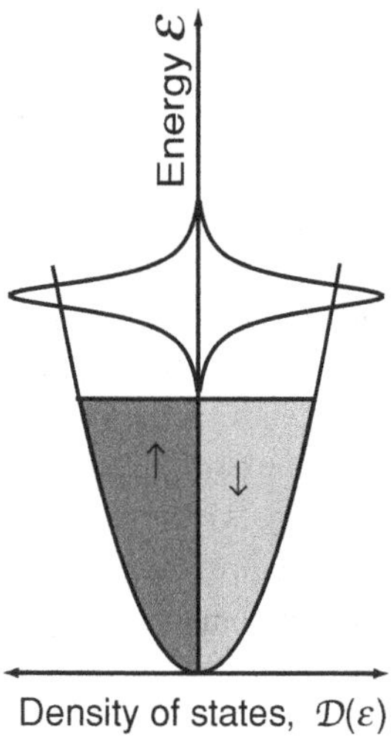

A nonmagnetic virtual bound state.

5.3.3 Impurities in ferromagnets

The converse of the problem considered in §5.3.1, the behaviour of a single impurity atom in a ferromagnetic host, is also interesting. If the impurity is a much lighter $3d$ element than the host, like V in Ni (Fig. 5.20) its d levels lie above the Fermi level in the $4s$ conduction band. If V_{kd} is the hopping integral from the d level to the conduction band, the level acquires a width

$$\Delta_i = \pi \mathcal{N}_{4s}(\varepsilon_F)V_{kd}^2, \tag{5.37}$$

Figure 5.19

Variation of the magnetic moment per atom as a function of magnetic valence for some rare-earth iron alloys.

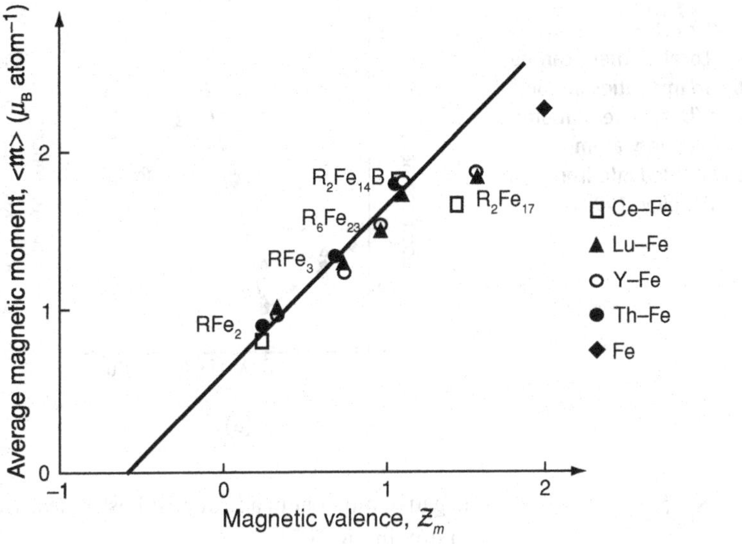

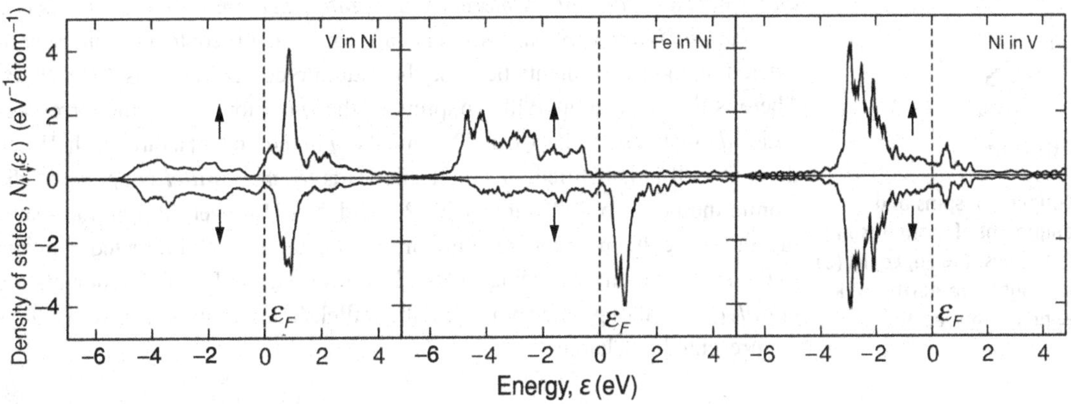

Figure 5.20

The local density of states for V in Ni, Fe in Ni and Ni in V. (Calculations by courtesy of Nadjib Baadji).

which may be of order 1 eV. The impurity level is then known as a virtual bound state. The width is inversely related to the time an electron dwells on the impurity site.

When the virtual bound state lies entirely above ε_F, the $3d$ impurity electrons are emptied into the $3d$-band of the host. If the host $3d^{\uparrow}$-band is full, there will be a moment reduction of N_{3d}^i Bohr magnetons, where N_{3d}^i is the number of impurity $3d$ electrons. In addition, the moment of one host atom is suppressed at the site of the substitution. For example, when a V impurity ($Z = 5$, $N_{3d}^i \approx 4$) is substituted in a Ni host, the moment reduction is drastic, $4 + 0.6 = 4.6\ \mu_B/\mathrm{V}$.

There will inevitably be some hybridization of the impurity and host $3d$-states, which will be more effective for the host $3d^{\downarrow}$ electrons, because they lie closer to the Fermi level. A light $3d$ element will therefore acquire a small

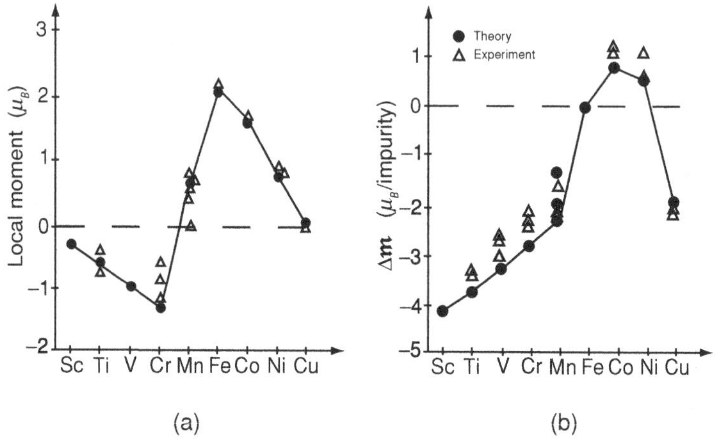

Figure 5.21

(a) Local moment carried by 3d impurities in iron, and (b) change of moment per impurity atom substituted into iron (O'Handley, 1999).

(a) $\overset{\longleftarrow\;\longmapsto}{S_R\;\;S_T}$

$\overset{\longrightarrow\;\;\longrightarrow}{m_R\;\;m_T}$

(b) $\overset{\longleftarrow\;\longmapsto}{S_R\;\;S_T}$

$\overset{\longleftarrow\;\longmapsto}{m_R\;\;m_T}$

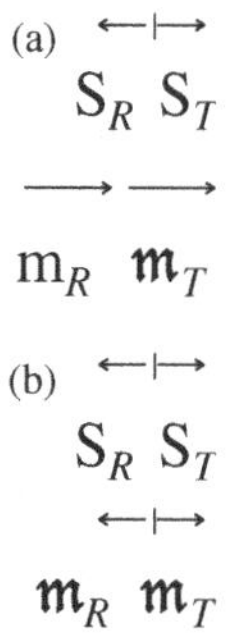

Coupling of spins and alignment of moments in R–T alloys, T = Fe, Co, Ni (a) R = light rare-earth, (b) R = heavy rare-earth.

negative moment in a heavy 3d host. These trends are illustrated for impurities in iron in Fig. 5.21.

This is an example of the more general rule that *exchange coupling between atoms with d shells that are more than half full with atoms whose d shell is less than half full is antiferromagnetic.* The rare-earths in this context should be considered as light d elements because their atomic configuration is $4f^n5d^16s^2$. There is therefore antiparallel coupling of the *spin* moments of the ferromagnetic 3d elements T = Fe, Co and Ni, and the *spin* moment of a rare-earth. When the 4f shell is half-filled, or more, this leads to *antiparallel* coupling of the atomic moments in R–T alloys with R = Gd–Yb. However, in light rare-earth metals where the moment is mainly orbital in character, and directed opposite to the spin moment according to Hund's third rule, the R and T moments are *parallel*, even though the spins are antiparallel. Many examples of R–T alloys are presented in Chapter 11.

5.3.4 Half-metals

These oddly named materials are ferromagnets with electrons of only one spin polarization at the Fermi level. Cobalt and nickel are *not* half-metals because of the presence of the 4s electrons at ε_F which are not fully spin-polarized. Indeed, no ferromagnetic element is a half-metal. It is necessary to form a compound where the 4s electrons can be removed from the vicinity of the Fermi energy by charge transfer or hybridization. Examples include the oxide CrO_2 and the ordered intermetallic compound MnNiSb, which are both discussed in Chapter 11. The characteristic feature of a half-metal is a spin gap in the ↑ or ↓ density of states at ε_F. Furthermore, the spin moment per formula unit in a stoichiometric half-metallic compound is an *integral number of Bohr magnetons*. This is because there are an integral number $N^\uparrow + N^\downarrow$ of electrons per formula unit, and the band with the spin gap must contain an integral number of $N^{\uparrow}$ (or $N^{\downarrow}$) electrons, hence $N^{\uparrow} - N^{\downarrow}$ is also an integer.

Spin-orbit interaction tends to destroy half-metallicity by mixing $\uparrow$ and $\downarrow$ states, as explained in §5.6.4.

5.3.5 The two-electron model

Further insight into the physical interactions of importance in metals is provided by a diatomic two-electron model. Although highly simplified, the model contains most of the ingredients of the physics of the many-electron problem, except orbital degeneracy. Important quantities are the on-site Coulomb repulsion U, the transfer or 'hopping' integral t which gives rise to the bandwidth W and the direct exchange $\mathcal{J}_d$.

The Hamiltonian $\mathcal{H}(\boldsymbol{r}_1, \boldsymbol{r}_2)$ is that in the Schrödinger equation of (5.20), with an additional term $e^2/4\epsilon_0\,|\boldsymbol{r}_1 - \boldsymbol{r}_2|$ to take account of the Coulomb interaction of the two electrons with each other. The spatially symmetric and antisymmetric wave functions

$$\phi_s = (1/\sqrt{2})(\psi_1 + \psi_2), \qquad \phi_a = (1/\sqrt{2})(\psi_1 - \psi_2), \tag{5.21}$$

may be regarded as embryonic Bloch functions (electron waves) for the metal with $k = 0$ and $k = \pi/d$, where d is the interatomic spacing. We can replace ϕ_s and ϕ_a by embryonic Wannier functions which are mostly localized on the left and right atoms:

$$\phi_l = (1/N)(\psi_1 + a\psi_2), \qquad \phi_r = (1/N)(a\psi_1 + \psi_2), \tag{5.38}$$

with

$$a = \frac{-1 + \sqrt{1 - \mathcal{S}^2}}{\mathcal{S}},$$

where $\mathcal{S}$ is the overlap integral $\int \psi_1^*(r)\psi_2(r)\mathrm{d}^3r$ and N is a normalization factor. These Wannier functions, Fig. 5.22, differ from ψ_1 and ψ_2, the eigenfunctions of the one-electron problem, in that they are supposed to be orthogonal $\int \phi_l^*(r)\phi_r(r)\mathrm{d}^3r = 0$.

There are now four possible two-electron wave functions $\Psi_i(\boldsymbol{r}, \boldsymbol{r}')$:

$$\Psi_1 = \phi_l(\boldsymbol{r})\phi_l(\boldsymbol{r}'); \qquad \Psi_2 = \phi_l(\boldsymbol{r})\phi_r(\boldsymbol{r}');$$
$$\Psi_3 = \phi_r(\boldsymbol{r})\phi_l(\boldsymbol{r}'); \qquad \Psi_4 = \phi_r(\boldsymbol{r})\phi_r(\boldsymbol{r}').$$

The functions Ψ_1 and Ψ_4 represent doubly occupied states. The interaction matrix is

$$\begin{pmatrix} U & t & t & \mathcal{J}_d \\ t & 0 & \mathcal{J}_d & t \\ t & \mathcal{J}_d & 0 & t \\ \mathcal{J}_d & t & t & U \end{pmatrix}. \tag{5.39}$$

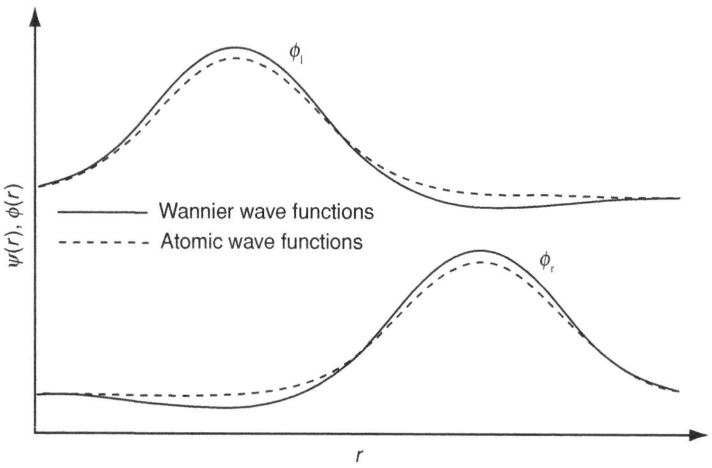

Figure 5.22

Wannier functions and atomic wave functions for the diatomic molecule.

The Coulomb interaction U is the energy penalty when two electrons are put into the same orbital. It is several electron volts:

$$U = \int \phi_l^*(\boldsymbol{r})\phi_l^*(\boldsymbol{r}')\mathcal{H}(\boldsymbol{r}, \boldsymbol{r}')\phi_l(\boldsymbol{r})\phi_l(\boldsymbol{r}')\mathrm{d}^3r\mathrm{d}^3r'.$$

The transfer or hopping integral t is also positive, and is $\lesssim 1$ eV. It represents the bandwidth. More generally, in the tight-binding approximation, the bandwidth is $2Z_nt$, where Z_n is the number of nearest neighbours.

$$t \approx \int \phi_R^*(\boldsymbol{r})\phi_l^*(\boldsymbol{r}')\mathcal{H}(\boldsymbol{r}, \boldsymbol{r}')\phi_l(\boldsymbol{r})\phi_r(\boldsymbol{r}')\mathrm{d}^3r\mathrm{d}^3r'$$

The direct exchange between doubly occupied sites is smaller, and of order 0.1 eV:

$$\mathcal{J}_d = \int \phi_l^*(\boldsymbol{r})\phi_l^*(\boldsymbol{r}')\mathcal{H}(\boldsymbol{r}, \boldsymbol{r}')\phi_r(\boldsymbol{r})\phi_r(\boldsymbol{r}')\mathrm{d}^3r\mathrm{d}^3r'.$$

The interaction matrix (5.39) can be diagonalized directly. Two doubly occupied states have eigenvalues of order U, and are therefore neglected. The other states, which are much lower in energy, are: (i) a delocalized ferromagnetic state (the spatial part of the wave function is antisymmetric) with eigenvalue $\varepsilon_{FM} = -\mathcal{J}_d$

$$\Psi_{FM} = (1/\sqrt{2})[\phi_l(\boldsymbol{r})\phi_r(\boldsymbol{r}') - \phi_r(\boldsymbol{r})\phi_l(\boldsymbol{r}')];$$

and (ii) an antiferromagnetic state (the spatial part of the wave function is symmetric)

$$\Psi_{AF} = (\sin\chi/\sqrt{2})[\phi_l(\boldsymbol{r})\phi_l(\boldsymbol{r}') + \phi_r(\boldsymbol{r})\phi_r(\boldsymbol{r}')] + (\cos\chi/\sqrt{2})[\phi_l(\boldsymbol{r})\phi_r(\boldsymbol{r}') + \phi_r(\boldsymbol{r})\phi_l(\boldsymbol{r}')],$$

where $\tan\chi = 4t/U$. The associated energy is $\varepsilon_{AF} = U/2 + \mathcal{J}_d - \sqrt{4t^2 + U^2/4}$.

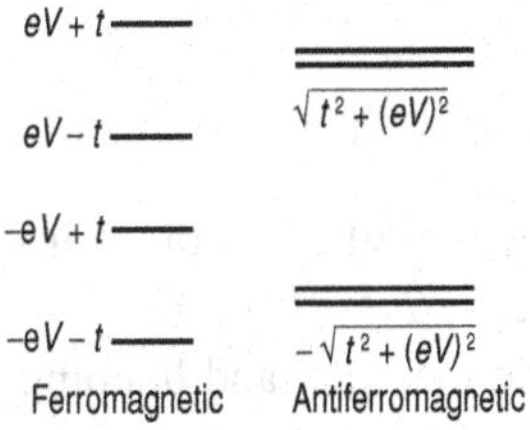

Energy levels of the ferromagnetic and antiferromagnetic states for the two-atom/four-state problem.

The effective exchange is $\mathcal{J}_{eff} = \frac{1}{2}(\varepsilon_{AF} - \varepsilon_{FM})$,

$$\mathcal{J}_{eff} = \mathcal{J}_d + U/4 - \sqrt{(t^2 + U^2/16)}. \quad (5.40)$$

$\mathcal{J}_{eff} > 0$ indicates a 'ferromagnetic' ground state and $\mathcal{J}_{eff} < 0$ indicates an 'antiferromagnetic' ground state. Direct exchange favours ferromagnetism, but strong interatomic hopping t favours antiferromagnetism. When $U \gg t$, and $\mathcal{J}_d = 0$, the exchange is antiferromagnetic, as illustrated in §5.2.2:

$$\mathcal{J}_{eff} = -2t^2/U. \quad (5.41)$$

Our rudimentary model can also illustrate how exchange depends on band filling. We consider the ferromagnetic and the antiferromagnetic states, for which the one-electron Hamiltonians are,

$$\mathcal{H}_F = \begin{bmatrix} \pm eV & t \\ t & \pm eV \end{bmatrix},$$

$$\mathcal{H}_{AF} = \begin{bmatrix} \pm eV & t \\ t & \mp eV \end{bmatrix},$$

where V is the local exchange potential experienced by an electron on site 1 or 2 and t is the interatomic hopping integral. When $t = 0$, there is one-electron exchange splitting of the states. Diagonalizing the matrices to find the eigenvalues involves solving the determinant $|\mathcal{H} - \lambda\mathbf{I}| = 0$. For the ferromagnetic state, the energy levels are $\pm eV + t$ and $\pm eV - t$, whereas for the antiferromagnetic state, they are doubly degenerate $\pm\sqrt{t^2 + (eV)^2}$. It can be seen that a single electron or three electrons (quarter-filled or three-quarters-filled band) go into a ferromagnetic state, but two electrons (half-filled band) prefer the antiferromagnetic state.

5.3.6 The Hubbard model

A famous model Hamiltonian which represents electron correlation in the tight-binding model for an array of one-electron atoms is

$$\mathcal{H} = -\sum_{i,j} t c_i^\dagger c_j + U \sum_i N_i^\uparrow N_i^\downarrow,$$

where $N_i^{\uparrow,\downarrow}$ are the numbers of spin-up and spin-down electrons, respectively, on the ith atom. The first term is the transfer term that creates the band of width $W = 2Zt$; the second is the Coulomb energy penalty involved in placing two electrons on the same atom.

Electrons are localized when $U/W > 1$ because there are then no states available to accommodate the double-occupancy charge fluctuations that are indispensable for electronic conduction. Compounds with an integral number of electrons per atom which satisfy this condition are known as Mott–Hubbard insulators.

The second term can be rewritten using

$$UN_i^{\uparrow}N_i^{\downarrow} = U[(N^{\uparrow} + N^{\downarrow})^2/4 - (N^{\uparrow} - N^{\downarrow})^2/4].$$

The Stoner interaction $-(\mathcal{I}/4)(N^{\uparrow} - N^{\downarrow})^2$ is thereby identified as the spin-dependent part of the on-site Coulomb interaction; hence $\mathcal{I} \approx U$. In the Hubbard model, the on-site correlations create a magnetic moment, and hopping between adjacent nondegenerate singly occupied orbitals provides an antiferromagnetic interaction.

A variant of the Hubbard model is the t–$\mathcal{J}$ model, where the second term is replaced by $-2\mathcal{J}\sum_{i>j} \boldsymbol{S}_i \cdot \boldsymbol{S}_j$ with $\mathcal{J} = -2t^2/U$.

5.3.7 Electronic structure calculations

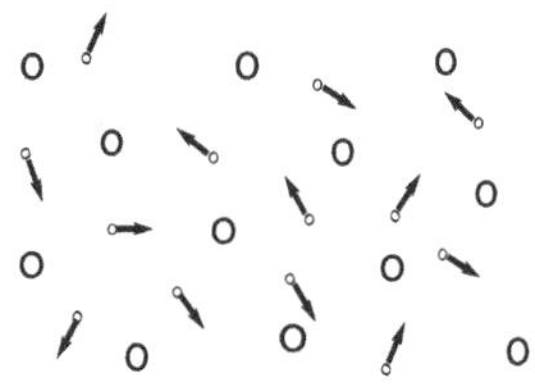

The Born–Oppenheimer approximation: the electrons (marked with arrows) move in a background of frozen nuclei (open circles).

A solid is a system of N electrons at positions $\{\boldsymbol{r}_i\}$ and N' nuclei, usually centred on a periodic lattice $\{\boldsymbol{R}_I\}$. Inner electrons occupy tightly bound, localized core orbitals around the nuclei. Outer orbitals with binding energies of a few electron volts or less are the home of the valence and conduction electrons, which determine the electronic character of the solid, be it metal, semiconductor, insulator, ferromagnet, antiferromagnet, superconductor, . . . The instantaneous velocity of these outer electrons is of order the Fermi velocity, $v_F \approx 10^6$ m s^{-1}. Ion cores vibrate at phonon frequencies that are of order 10^{14} Hz, with an amplitude of about 10 pm, which means that their velocity is of order 10^3 m s^{-1}. We are therefore justified in thinking that the electron sees the potential of a set of nuclei instantaneously frozen in position – the Born–Oppenheimer approximation. Furthermore, we will ignore the atomic displacements, which lead to electron scattering, and assume that the ion cores are localized at the lattice sites. The electrons experience Coulomb interactions with the nuclei, and with each other. The Hamiltonian, with factors of $\frac{1}{2}$ to avoid double counting, takes the form

$$\mathcal{H} = -\sum_i \frac{\hbar^2\nabla_i^2}{2m_e} - \sum_{i,I} \frac{Ze^2}{4\pi\epsilon_0 R_{Ii}} + \frac{1}{2}\sum_{i,j} \frac{e^2}{4\pi\epsilon_0 r_{ij}}. \tag{5.42}$$

The problem is to solve Schrödinger's equation $\mathcal{H}\Psi = \varepsilon\Psi$, where $\Psi(\{R_I\}, \{r_i\})$ is a wave function for the huge number of electrons and nuclei in the system.

An abbreviated notation for $\mathcal{H}$ is

$$\mathcal{H} = \mathcal{T} + \mathcal{V} + \mathcal{U}, \tag{5.43}$$

where $\mathcal{T}$ and $\mathcal{V}$ are the terms corresponding to the one-electron kinetic and potential energy and $\mathcal{U}$ represents the two-electron interactions that capture the complexity of the physics.

Many first-principles methods for solving the many-electron Schrödinger equation use wave functions based on Slater determinants. The idea is to build in the antisymmetry of the wave function under exchange of any two electrons

John Hubbard, 1931–1980

with given space and spin coordinates x, y, z, σ denoted as $\boldsymbol{x}_i$ and $\boldsymbol{x}_j$. In the case of just two electrons, for example, $\Psi = \psi_1(\boldsymbol{x}_1)\psi_2(\boldsymbol{x}_2)$ is not antisymmetric, but $(1/\sqrt{2})[\psi_1(\boldsymbol{x}_1)\psi_2(\boldsymbol{x}_2) - \psi_1(\boldsymbol{x}_2)\psi_2(\boldsymbol{x}_1)]$ is a suitable wave function. It can be written in the form of a determinant:

$$\Psi(\boldsymbol{x}_1, \boldsymbol{x}_2) = \frac{1}{\sqrt{2}} \begin{vmatrix} \psi_1(\boldsymbol{x}_1) & \psi_2(\boldsymbol{x}_1) \\ \psi_1(\boldsymbol{x}_2) & \psi_2(\boldsymbol{x}_2) \end{vmatrix}. \tag{5.44}$$

Placing two electrons in the same orbit $\psi_1 = \psi_2$ gives $\Psi(\boldsymbol{x}_1, \boldsymbol{x}_2) = 0$, as required by the Pauli principle. Slater generalized this idea to N electrons, writing the wave function as

$$\Psi(\boldsymbol{x}_1, \boldsymbol{x}_2, \ldots, \boldsymbol{x}_N) = \frac{1}{\sqrt{N}} \begin{vmatrix} \psi_1(\boldsymbol{x}_1) & \psi_2(\boldsymbol{x}_1) & \cdots & \psi_N(\boldsymbol{x}_1) \\ \psi_1(\boldsymbol{x}_2) & \psi_2(\boldsymbol{x}_2) & \cdots & \psi_N(\boldsymbol{x}_2) \\ \vdots & \vdots & \ddots & \vdots \\ \psi_1(\boldsymbol{x}_N) & \psi_2(\boldsymbol{x}_N) & \cdots & \psi_N(\boldsymbol{x}_N) \end{vmatrix}. \tag{5.45}$$

A compact way to denote a Slater determinant is as a ket, $|1, 2, \ldots, N\rangle$.

The Hartree–Fock method assumes that the exact N-electron wave function of the system can be approximated as a single Slater determinant. A variational solution is then based on a linear combination of these one-electron wave functions with coefficients chosen to minimize the energy. The method completely neglects electron correlations, but takes perfect account of the exchange. Each electron is surrounded by an exchange hole, from which any other electron with the same spin is excluded.

An alternative approach to Hartree–Fock calculations is density functional theory (DFT), which provides an approximate solution for both exchange and correlation energies. It succeeds in mapping a many-electron problem with $\mathcal{U}$ onto a one-electron problem without $\mathcal{U}$. The theory is based on two theorems, proved by Hohenberg and Kohn in the mid 1960s. The first is that *the density* $n(\boldsymbol{r})$ *of a system of* N *electrons determines all the ground-state electronic properties*. The ground-state wave function Ψ_0 is a unique functional of electron density $\Psi_0[n(\boldsymbol{r})]$. (A functional is just a function which has another function as its argument.) Other physical properties can be derived from the wave function. In particular, the ground-state energy is

$$\varepsilon_0[n(\boldsymbol{r})] = \langle \Psi_0 | \mathcal{T} + \mathcal{V} + \mathcal{U} | \Psi_0 \rangle. \tag{5.46}$$

The second is that *the energy functional* $\varepsilon_0[n(\boldsymbol{r})]$ *is lower in energy for the ground-state density* $n_0(\boldsymbol{r})$ *than any other state.* The term in the energy which needs to be minimized in (5.46) is the one that depends on $\{R_I\}$, $\mathcal{V}[n(\boldsymbol{r})] = -e \int V(\mathbf{r}) n(\mathbf{r}) \mathrm{d}^3 r$. The significance of these theorems is that it is immensely easier to base a calculation on the density, which depends on only three variables, x, y, z, or four when we include the spin σ, than it is on the wave functions of an N-electron system which depend on $4N$ variables. The problem is that the correct density functional is unknown, and it must be arrived at by inspired approximation. Furthermore, the method applies to the ground state, and it does not give the structure of excited states, although a time-dependent variant of the theory may remedy this defect.

Density functional theory is usually implemented using the Kohn–Sham method, where the problem of strongly interacting electrons moving in the potential of the nuclei is reduced to the more tractable problem of noninteracting electrons in an effective potential which somehow takes care of the inter-electronic Coulomb interaction, as well as exchange and correlation effects. The energy of (5.46) is rewritten as

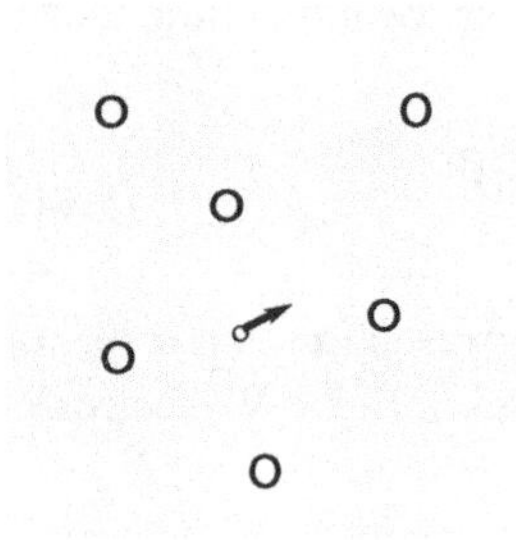

The Kohn–Sham formalism. Each electron moves independently in an effective potential created by the others.

$$\varepsilon_0[n(\boldsymbol{r})] = T_s[n(\boldsymbol{r})] + \varepsilon_V[n(\boldsymbol{r})], \tag{5.47}$$

where T_s is the noninteracting kinetic energy and ε_V is the total potential energy. The Kohn–Sham equations for this noninteracting system are just a set of effective, single-particle Schrödinger equations

$$\left[-\frac{\hbar^2\nabla_i^2}{2m_e} + \mathcal{V}_s(\boldsymbol{r})\right]\phi_i(\boldsymbol{r}) = \varepsilon_i\phi_i(\boldsymbol{r}) \tag{5.48}$$

which yield a set of orbitals ϕ_i which are approximate wave functions for the real system of electrons that reproduce the density of the original many-electron system $n(\boldsymbol{r}) = \sum_i |\phi_i^2|$. The effective single-particle potential is usually written as

$$\mathcal{V}_s = \mathcal{V} + \frac{1}{2}\int \frac{e^2 n(\boldsymbol{r}')}{4\pi\epsilon_0\,|\boldsymbol{r}-\boldsymbol{r}'|}\mathrm{d}^3r' + \mathcal{V}_{xc}[n(\boldsymbol{r})]. \tag{5.49}$$

The first term is the Coulomb interaction of the electron with the nuclei, the second term is the Hartree term $\mathcal{V}_H$ describing the electron–electron Coulomb repulsion, and the key term is the third one, the exchange-correlation potential, which includes all the many-electron correlations. The Kohn–Sham equations are solved by following an iterative procedure. Taking an initial guess for $[n(\boldsymbol{r})]$, $\mathcal{V}_s$ is calculated and the equations are solved for $\phi_i(\boldsymbol{r})$, from which a new density is calculated, and the process is repeated. The local density approximation (LDA) assumes that the exchange-correlation functional $\mathcal{V}_{xc}[n\{\boldsymbol{r}\}]$ depends only on the density at the point where the function is evaluated $\mathcal{V}_{xc}[n(\boldsymbol{r})]$. A variant is the general gradient approximation, where $\mathcal{V}_{xc}$ depends also on the density gradient $\mathcal{V}_{xc}[n(\boldsymbol{r}), \nabla n(\boldsymbol{r})]$. In systems of more ionic character, a U term can be added to reduce double orbital occupancy.

All this can be generalized to the spin-dependent case. In the local spin density approximation (LSDA) two densities must be taken into account, the scalar electron density $n(\boldsymbol{r})$ and the vector magnetization density $\boldsymbol{m}(\boldsymbol{r}) = \mu_B[n^\uparrow(\boldsymbol{r}) - n^\downarrow(\boldsymbol{r})]\mathbf{e}_z$. They are both incorporated in a 2×2 density matrix $\hat{\boldsymbol{n}}(\boldsymbol{r}) = \frac{1}{2}[n(\boldsymbol{r})\hat{\mathbf{I}} + \hat{\boldsymbol{\sigma}}\cdot\boldsymbol{s}(\boldsymbol{r})]$, where $\hat{\mathbf{I}}$ is the identity matrix $\begin{bmatrix}1 & 0\\ 0 & 1\end{bmatrix}$, $\hat{\boldsymbol{\sigma}}$ are the Pauli spin matrices (3.17) and $\boldsymbol{s}(\boldsymbol{r})$ is the local spin density $\boldsymbol{m}(\boldsymbol{r})/\mu_B$. For a collinear spin configuration, the density matrix is diagonal:

$$\hat{\boldsymbol{n}}(\boldsymbol{r}) = \begin{bmatrix} n^{\uparrow}(\boldsymbol{r}) & 0 \\ 0 & n^{\downarrow}(\boldsymbol{r}) \end{bmatrix}, \tag{5.50}$$

so that $n(\boldsymbol{r}) = n^{\uparrow}(\boldsymbol{r}) + n^{\downarrow}(\boldsymbol{r})$ and $m(\boldsymbol{r}) = \mu_B[n^{\uparrow}(\boldsymbol{r}) - n^{\downarrow}(\boldsymbol{r})]$. The potential matix is given by $\hat{\mathcal{V}} = \mathcal{V}\hat{\mathbf{I}} + \mu_B \boldsymbol{m}(\boldsymbol{r}) \cdot \boldsymbol{B}$, where $\boldsymbol{B}$ is the magnetic field. The spin-polarized version of the Kohn–Sham equations is

$$\left[\left(-\frac{\hbar^2 \nabla_i^2}{2m_e} + \mathcal{V}_H\right)\hat{\mathbf{I}} + \mathcal{V} + \mathcal{V}_{\mathbf{xc}}\right]\begin{bmatrix} \phi_i^{\uparrow}(\boldsymbol{r}) \\ \phi_i^{\downarrow}(\boldsymbol{r}) \end{bmatrix} = \varepsilon_i \begin{bmatrix} \phi_i^{\uparrow}(\boldsymbol{r}) \\ \phi_i^{\downarrow}(\boldsymbol{r}) \end{bmatrix}. \tag{5.51}$$

The three forces in magnetism: theory, experiment, and simulation. (Courtesy Wiebke Drenckhan).

The exchange-correlation matrix depends on both $n(\boldsymbol{r})$and $\boldsymbol{m}(\boldsymbol{r})$,

$$\mathcal{V}_{xc} = \mathcal{V}_{xc}[n(\boldsymbol{r}), \boldsymbol{m}(\boldsymbol{r})], \tag{5.52}$$

for which a suitable approximation must be found. Equations (5.51) yield the wave functions $\phi_i^{\uparrow,\downarrow}(\boldsymbol{r})$ from which the density matrix $\hat{\boldsymbol{n}}(r)$ is deduced. A self-consistent solution is obtained, as in the the nonmagnetic case. A number of computer codes are available to do the job. If the density matrix is diagonal, the magnetic structure is collinear, but the general formalism allows for noncollinear structures. DFT is an accurate method for calculating magnetic moments and spin-polarized band structures, especially in metallic systems.

Exponential growth of computer power, and especially multiprocessor computer clusters, has enabled computer simulation to establish itself as a third force, alongside experiment and theory, for investigations in magnetism. Not only in electronic structure calculations, where it is becoming possible to investigate the crystal structure and magnetic order of a new compound without ever actually having to make it in the laboratory, but also in the areas of electronic transport properties and micromagnetism, are computational methods making their mark. Large numbers of atoms, of order 1000 or more, can be handled using current DFT codes, which makes it possible to investigate spin-dependent transport in a molecule, or to study the appearance of a magnetic moment on different types of lattice defects in a solid. It is immensely more convenient to create a specific complex lattice defect on a computer than it is in the laboratory.

5.4 Collective excitations

The comparison of magnetization data on nickel with the predictions of molecular field theory for $J = \frac{1}{2}$ in Fig. 5.3 shows discrepancies both at low temperature, and in the vicinity of T_C. Actually the discrepancies are worse than they appear because T_C is used to determine n_W, so the model is constrained to return the right Curie temperature, and the correct value of m_0.

Experimental methods discussed in Chapter 10, exist to determine the exchange constants directly, so it is possible to make a more telling comparison

Figure 5.23

Spontaneous magnetization and inverse susceptibility as a function of temperature. The molecular field predictions (grey) are compared with typical experimental data (black).

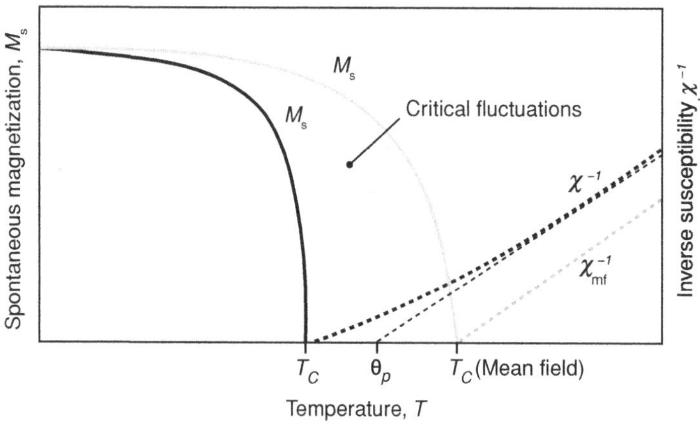

Figure 5.24

Illustration of a spin wave.

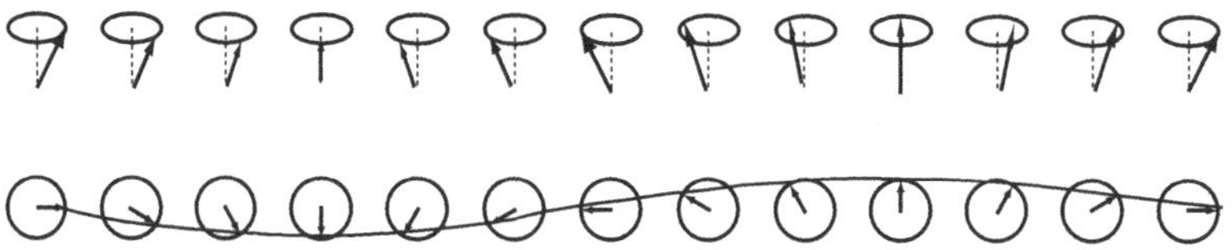

between theory and experiment, as indicated in Fig. 5.23. Here it is evident that ferromagnetism is considerably less stable at elevated temperature than molecular field theory would have us believe; it overestimates T_C by as much as a factor of 2, depending on the dimensions and the lattice type. The spontaneous magnetization is diminished at low temperatures by spin-wave excitations. Near T_C the critical fluctuations destroy it.

5.4.1 Spin waves

The total exchange energy in the ferromagnetic ground state is $-2Z\mathcal{J}S^2$ per site, where Z is the number of magnetic nearest neighbours and $\mathcal{J}$ is the nearest-neighbour exchange interaction. The elementary excitations from the ferromagnetic ground state are not, as might be imagined, flips of individual spins that reduce an atomic moment from a state with $M_s = S$ to one with $M_s = (S-1)$. A single localized spin reversal in an $S = \frac{1}{2}$ chain $\uparrow\uparrow\uparrow\uparrow\downarrow\uparrow\uparrow\uparrow\uparrow$ costs $8\mathcal{J}S^2$ or $2\mathcal{J}$ when $S = \frac{1}{2}$, which is twice as large as k_BT_C for the chain treated in the molecular field approximation (5.26); $Z = 2$ for a chain, so $k_BT_C = 2\mathcal{J}ZS(S+1)/3 = \mathcal{J}$. Such expensive excitations cannot occur at low temperature. Instead, all the atoms share out the spin reversal, with periodic oscillation of their transverse spin orientation. The spin deviations spread over the whole lattice in a propagating spin wave with wave vector $\boldsymbol{q}$ and energy $\varepsilon_q = \hbar\omega_q$, as illustrated in Fig. 5.24. Spin waves exist as classical excitations, but the extended, quantized spin deviations in solids are known as magnons by analogy with phonons, the quantized lattice waves. Think of spin waves as oscillations in the relative orientations of spins on a lattice, whereas lattice waves are oscillations of the relative positions of atoms on a lattice.

The relation between the wavevector $q = 2\pi/\lambda$ and frequency ω_q of the spin wave can be calculated classically, or from quantum mechanics. The classical approach considers the spin angular momentum of the atom at site j, $\hbar \boldsymbol{S}_j$ and equates the torque exerted by the molecular field to the rate of change of angular momentum, thus

$$\hbar \frac{\mathrm{d}\boldsymbol{S}_j}{\mathrm{d}t} = \mu_0 g \mu_B \boldsymbol{S}_j \times \boldsymbol{H}^j. \tag{5.53}$$

In a chain, the molecular field $\boldsymbol{H}^j$ at site j is due to the neighbours at sites $j \pm 1$. From (5.25), $\boldsymbol{H}^j = 2\mathcal{J}(\boldsymbol{S}_{j-1} + \boldsymbol{S}_{j+1})/\mu_0 g \mu_B$, hence $\hbar \mathrm{d}\boldsymbol{S}_j/\mathrm{d}t = 2\mathcal{J}\boldsymbol{S}_j \times (\boldsymbol{S}_{j-1} + \boldsymbol{S}_{j+1})$. This can be written in Cartesian coordinates:

$$\hbar \frac{\mathrm{d}S_j^x}{\mathrm{d}t} = 2\mathcal{J}\left[S_j^y(S_{j-1}^z + S_{j+1}^z) - S_j^z(S_{j-1}^y + S_{j+1}^y)\right]$$

plus cyclic permutations. For small deviations, we can approximate $S_j^z = S_j = S$ and neglect terms like $S_i^x S_j^y$. Hence

$$\begin{aligned}
\hbar \frac{\mathrm{d}S_j^x}{\mathrm{d}t} &= 2\mathcal{J}S\left[2S_j^y - S_{j-1}^y - S_{j+1}^y\right], \\
-\hbar \frac{\mathrm{d}S_j^y}{\mathrm{d}t} &= 2\mathcal{J}S\left[2S_j^x - S_{j-1}^x - S_{j+1}^x\right], \\
\hbar \frac{\mathrm{d}S_j^z}{\mathrm{d}t} &= 0.
\end{aligned} \tag{5.54}$$

Solutions are of the form $S_j^x = uS \exp[\mathrm{i}(jqa - \omega_q t)]$, $S_j^y = vS \exp[\mathrm{i}(jqa - \omega_q t)]$, where q is the wavevector and a is the interatomic spacing. Substituting back into (5.54) gives $-\mathrm{i}\hbar\omega_q u = 4\mathcal{J}S(1 - \cos qa)v$, $\mathrm{i}\hbar\omega_q v = 4\mathcal{J}S(1 - \cos qa)u$. Multiplying these results for a one-dimensional chain of isotropic spins, gives

$$\hbar\omega_q = 4\mathcal{J}S(1 - \cos qa). \tag{5.55}$$

In the limit of small wavevectors, the spin-wave dispersion relation becomes

$$\varepsilon_q \approx D_{sw} q^2, \tag{5.56}$$

where $\varepsilon_q = \hbar\omega_q$ and the spin-wave stiffness parameter is $D_{sw} = 2\mathcal{J}Sa^2$. It takes a vanishingly small energy to create a long-wavelength magnetic excitation. The generalization to a three-dimensional cubic lattice with nearest-neighbour interactions is

$$\hbar\omega_q = 2\mathcal{J}S\left[Z - \sum_{\delta} \cos \boldsymbol{q} \cdot \boldsymbol{\delta}\right],$$

where the sum is over the Z vectors $\boldsymbol{\delta}$ connecting the central atom to its nearest neighbours. The same dispersion relation with $D_{sw} = 2\mathcal{J}Sa_0^2$ is found in *any* of

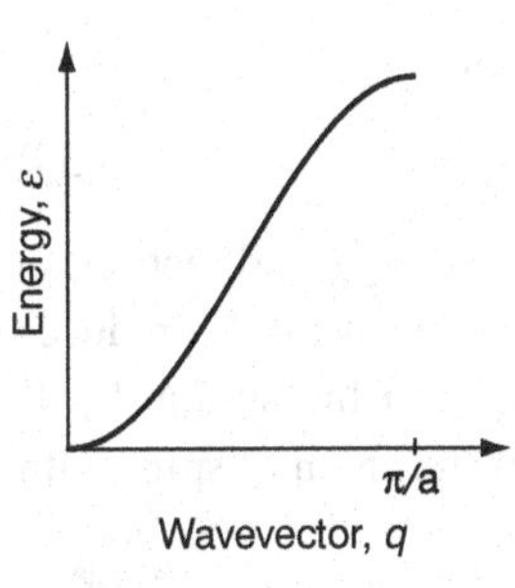

The spin-wave dispersion relation for a chain of atoms.

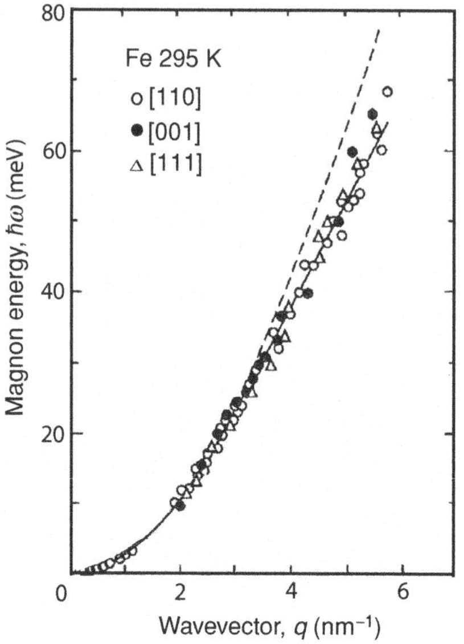

Figure 5.25

The magnon dispersion relations for iron measured in different directions in the unit cell. The dashed line corresponds to $D_{sw} = 4.5 \times 10^{-40}$ J m^2. (G. Shirane *et al.*, Journal of Applied Physics, **39**, 383 (1968))

the three basic cubic lattices, where a_0 is now the lattice parameter. Dispersion of magnons differs from that of phonons, $\varepsilon_q \approx c_0 q$ where c_0 is the velocity of sound, which is linear in the small-q limit. The value of D_{sw} for cobalt, for example, is 8.0×10^{-40} J m^2 (500 meV Å^2). It is smaller for other ferromagnets (Table 5.4). The dispersion relation for iron is shown in Fig. 5.25.

Equation (5.55) can be derived quantum mechanically from the Heisenberg Hamiltonian (5.24) where the sum is over nearest-neighbour pairs i, j. The Hamiltonian

$$\boldsymbol{S}_i \cdot \boldsymbol{S}_j = S_i^x S_j^x + S_i^y S_j^y + S_i^z S_j^z \tag{5.57}$$

is written in terms of the raising and lowering operators.

$$\boldsymbol{S}_i \cdot \boldsymbol{S}_j = S_i^z S_j^z + \tfrac{1}{2}(S_i^+ S_j^- + S_i^- S_j^+). \tag{5.58}$$

The ground state of the system $|\Phi\rangle$ has all the spins aligned in the z-direction, so that $\mathcal{H}|\Phi\rangle = -2\mathcal{J}(N-1)S^2|\Phi\rangle$. Flipping a spin $\frac{1}{2}$ at site i using S_i^- reduces M_S^i from S to $S-1$; $|i\rangle = S_i^-|\Phi\rangle$ lowers the total spin of the system by 1. However, $|i\rangle$ is not an eigenstate of the Hamiltonian of the chain of spins with nearest-neighbour interactions

$$\mathcal{H} = -2\mathcal{J}\sum_{i=1}^{N-1}\left[S_i^z S_{i+1}^z + \tfrac{1}{2}(S_i^+ S_{i+1}^- + S_i^- S_{i+1}^+)\right]$$

because $\mathcal{H}|i\rangle = 2\mathcal{J}[-(N-1)S^2 + 2S|i\rangle - S|i+1\rangle - S|i-1\rangle]$. It is necessary to form linear combinations like

$$|\boldsymbol{q}\rangle = \frac{1}{\sqrt{N}}\sum_i \mathrm{e}^{\mathrm{i}\mathbf{q}\cdot\mathbf{r}_i}|i\rangle \tag{5.59}$$

Figure 5.26

Magnon dispersion relation for terbium (J. Jonsen and A. R. Mackintosh, *Rare Earth Magnetism*, Oxford University Press 1991).

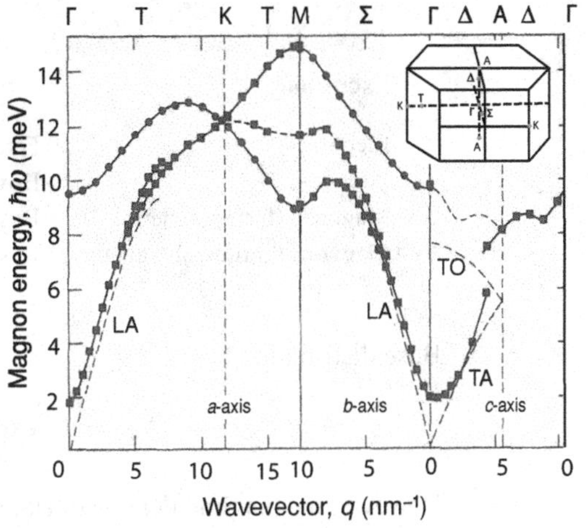

This state is a magnon, a spin flip delocalized on the chain, with wavevector $\boldsymbol{q}$. Then

$$\mathcal{H}|\boldsymbol{q}\rangle = \frac{2\mathcal{J}}{\sqrt{N}} \sum_{i=1}^{N-1} e^{i\mathbf{q}\cdot\mathbf{r}_i}[-(N-1)S^2 + 2S|i\rangle - S|i+1\rangle - S|i-1\rangle]$$
$$= [-2\mathcal{J}(N-1)S^2 + 4\mathcal{J}S(1-\cos qa)]|\boldsymbol{q}\rangle. \quad (5.60)$$

Dropping the first term, which is constant, we have $\varepsilon(q) = 4\mathcal{J}S(1-\cos qa)$, as before.

Dispersion relations are best measured by inelastic neutron scattering, which is discussed in Chapter 10. There are multiple magnon branches when the unit cell is noncubic, or if it contains more than one magnetic atom. The energy in a mode of frequency ω_q containing N_q magnons is $(N_q + \frac{1}{2})\hbar\omega_q$. Excitation of magnons is responsible for the fall of magnetization with increasing T. They also contribute to resistivity and magnetic specific heat. By analysing spin-wave dispersion relations measured across the Brillouin zone, it is possible to deduce the exchange interactions $\mathcal{J}(\boldsymbol{r}_{ij})$ for different atom pairs. Alternatively, the wavevector-dependent exchange $\mathcal{J}(\boldsymbol{q})$ can be fitted to the data. When the minimum of $\mathcal{J}(\boldsymbol{q})$ does not fall at $q = 0$, a spatially modulated magnetic structure is stable (§6.3).

Figure 5.26 shows the spin-wave dispersion relation for terbium. There is an energy gap at $q = 0$ due to the single-ion anisotropy of this rare-earth metal (§4.4.4). Excitation of spin waves can be suppressed at very low temperatures by the anisotropy. The energy gap at $q = 0$ is K_1/n, where n is the number of atoms per unit volume. In hexagonal close packed (hcp) cobalt, for example, $n = 9 \times 10^{28}$ m^{-3}, $K_1 = 500$ kJ m^{-3}, the spin-wave gap is 0.4 K.

Magnons behave like bosons; each magnon corresponds to the reversal of one spin $\frac{1}{2}$ over the whole sample, or a change $\Delta M_S = 1$ for the whole system. Hence the average number of quantized spin waves in a mode q is given by the

Table 5.6. Comparison of excitations in solids

Excitation		Dispersion	Specific heat
Electrons	Fermions	$\varepsilon_k \approx (\hbar/2m)k^2$	γT
Phonons	Bosons	$\varepsilon_q \approx c_0 q$	T^3
Magnons (ferromagnetic)	Bosons	$\varepsilon_q \approx D_{sw} q^2$	$T^{3/2}$
Magnons (antiferromagnetic)	Bosons	$\varepsilon_q \approx D_{af} q$	T^3

Bose distribution

$$\langle N_q \rangle = 1/[\exp(\hbar\omega_q/k_B T) - 1].$$

However, the magnon density of states $\mathcal{N}(\omega_q) \propto \omega_q^{\frac{1}{2}}$, just like that of electrons which have a similar dispersion relation $\varepsilon_k = \hbar^2 k^2/2m$. Dispersion relations and the corresponding low-temperature specific heats are summarized in Table 5.6. It can be shown that the reduction in magnetization at low temperatures due to the excitation of magnons is

$$\Delta M/M_0 = (0.0587/\nu)(k_B T/2S\mathcal{J})^{3/2}. \tag{5.61}$$

This is the Bloch $T^{3/2}$ power law. The integer ν equals 1, 2 or 4 for a simple cubic, bcc or fcc lattice. Specific heat follows the same power law at low temperature. A consequence of spin-wave excitation is that Curie temperatures are much lower than expected from molecular field theory, given the exchange constant $\mathcal{J}$ (Fig. 5.23).

For electrons in ferromagnetic metals, there is an additional scattering process, in addition to scattering from defects, phonons and other electrons. The electron can be inelastically spin-flip scattered, with the creation or annihilation of a magnon $(\omega_q, \boldsymbol{q})$. This leads to a term in resistivity varying as T^2.

Our discussion of spin waves has been based on localized spins and Heisenberg exchange coupling, but the idea is more general; any ferromagnetic continuum with exchange stiffness will exhibit spin-wave excitations.

5.4.2 Stoner excitations

Besides spin waves, another type of excitation in a metal can reduce its magnetization. Electrons at the Fermi level can be excited from filled states in the majority-spin band to empty states in the minority-spin band. If the initial state has wavevector $\boldsymbol{k}$ and the final state has wavevector $\boldsymbol{k} - \boldsymbol{q}$, an excitation of wavevector $\boldsymbol{q}$ is produced. The energy of the excitation is given by

$$\hbar\omega_q = \varepsilon_k - \varepsilon_{k-q} + \Delta_{ex}.$$

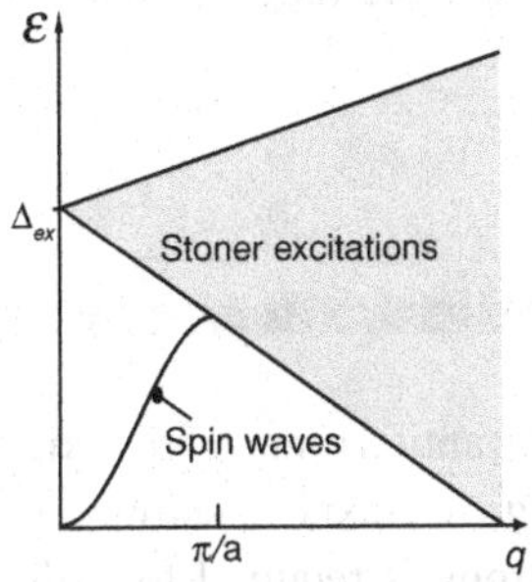

Spin waves and Stoner excitations (shaded) in a ferromagnet.

At $q = 0$, these excitations require the full exchange splitting Δ_{ex}, but at finite q there is a broad continuum of Stoner excitations.

5.4.3 Mermin–Wagner theorem

The derivation of the spin-wave dispersion has been based on the existence of a ferromagnetic state in an isotropic chain, or a three-dimensional lattice. The assumptions warrant scrutiny. The number of magnons excited at a temperature T is given by

$$\mathfrak{n}_m = \int_0^\infty \frac{\mathcal{N}(\omega_q)\mathrm{d}\omega_q}{\mathrm{e}^{\hbar\omega_q/kT} - 1},$$

where the density of states for magnons $\mathcal{N}(\omega_q)$ in one, two and three dimensions varies as $\omega_q^{-1/2}$, $\omega_q^o =$ constant and $\omega_q^{1/2}$, respectively. The argument is similar to that for the electron gas, given in §3.2.5, which has similar dispersion relations. Setting $x = \hbar\omega_q/k_BT$, the integral in three dimensions varies as $(k_BT/\hbar)^{3/2}\int_0^\infty x^{1/2}\mathrm{d}^3x/(\mathrm{e}^x - 1)$, whence comes the Bloch $T^{3/2}$ law (5.61). However, the integrals *diverge* at finite temperature in one and two dimensions. The ferromagnetically ordered state should be unstable in dimensions lower than 3. This is the Mermin–Wagner theorem. Magnetic order is possible in the Heisenberg model in three dimensions, but not in one or two. The linear chain, our example of spin-wave dispersion, cannot order except at $T = 0$ K.

The consequences of this theorem are not as catastrophic as they seem at first sight. The divergence is avoided if there is some anisotropy in the system, which creates a gap in the spin-wave spectrum at $\boldsymbol{q} = 0$; the lower limit of integration is then greater than zero and the divergence is avoided. Some anisotropy is always caused by crystal field or dipolar interactions. Two-dimensional ferromagnetic layers do exist in reality, thanks to anisotropy (§8.1).

5.4.4 Critical behaviour

Not only at low temperatures does the mean-field theory fail to account properly for the temperature dependence of the magnetization of a ferromagnet. There is a discrepancy in the critical region, close to T_C where the variation of M with temperature is as $(T - T_C)^\beta$ with $\beta \approx 0.34$, rather than $\frac{1}{2}$ (5.14). If the magnetization is calculated from molecular field theory using the measured exchange parameters $\mathcal{J}(r_{ij})$ derived from spin-wave dispersion relations, the discrepancy in the predicted and measured values of T_C may be as great as 60%. This is partly due to spin waves, but the exchange parameters are renormalized to lower values on increasing temperature because of critical fluctuations. These remarkable fluctuations, which are self-similar over a wide range of length scales, are discussed in Chapter 6.

5.5 Anisotropy

Magnetic anisotropy means that the ferromagnetic or antiferromagnetic axis of a sample lies along some fixed direction(s). Strong easy-axis anisotropy is a prerequisite for hard magnetism. Near-zero anisotropy is required for soft magnets. Generally, the tendency for magnetization to lie along an easy axis is represented by the energy density term introduced in Chapter 1:

$$E_a = K_1 \sin^2 \theta, \tag{5.62}$$

where θ is the angle between $\boldsymbol{M}$ and the anisotropy axis. K_1 has units J m^{-3}. Values may range from less than 1 kJ m^{-3} to more than 20 MJ m^{-3}. The anisotropy depends on temperature, and must tend to zero at T_C if there is no applied field. Three main sources of anisotropy are related to sample shape, crystal structure and atomic or micro-scale texture.

Shape anisotropy derives from the demagnetizing field discussed in §2.2.4. The energy of a sample in its demagnetizing field H_d gives a contribution to the self-energy (§2.5.1), which depends on the direction of magnetization in the sample. Obviously, this cannot be an intrinsic property of the material, as it depends on the sample shape.

Magnetocrystalline anisotropy is an intrinsic property. The magnetization process is different when the field is applied along different crystallographic directions, and the anisotropy reflects the crystal symmetry. Its origin is in the crystal-field interaction and spin-orbit coupling, or else the interatomic dipole–dipole interaction.

Induced anisotropy arises when an easy direction of magnetization is created by applied stress, or by depositing or annealing a disordered alloy in a magnetic field to create some atomic-scale texture. The texture is often subtle, and difficult to discern by conventional X-ray or electron scattering methods. Coarser-scale texture may be associated with mesoscopic fluctuations of the compositions, such as spinodal decomposition.

We first consider the phenomenological expressions for E_a, and then look deeper into its microscopic origins.

5.5.1 Shape anisotropy

The magnetostatic energy of a ferromagnetic ellipsoid with magnetization M_s is

$$\varepsilon_m = \tfrac{1}{2}\mu_0 V \mathcal{N} M_s^2. \tag{2.81}$$

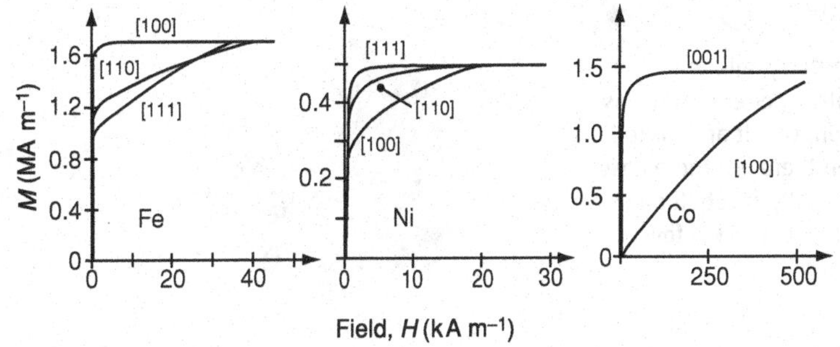

Figure 5.27

Magnetization of single crystals of iron, cobalt and nickel.

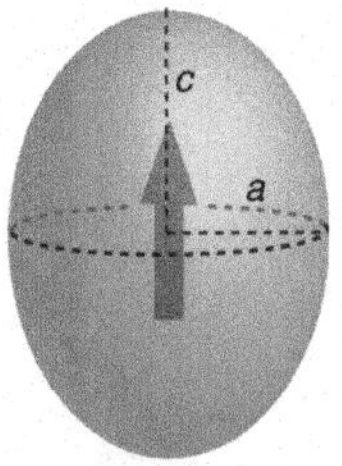

Magnetization of a prolate ellipsoid of revolution with $c > a$ and no magnetocrystalline anisotropy. The c axis is the easy direction of magnetization.

Other simple shapes can be approximated to ellipsoids. The anisotropy energy is related to the difference in energy $\Delta\varepsilon$ when the ellipsoid is magnetized along its hard and easy directions. $\mathcal{N}$ is the demagnetizing factor for the easy direction; $\mathcal{N}' = \frac{1}{2}(1 - \mathcal{N})$ is the demagnetizing factor for the perpendicular, hard, directions (Table 2.1). Hence $\Delta\varepsilon_m = \frac{1}{2}\mu_0 V M_s^2[\frac{1}{2}(1 - \mathcal{N}) - \mathcal{N}]$, which gives for a prolate ellipsoid

$$K_{sh} = \frac{1}{4}\mu_0 M_s^2(1 - 3\mathcal{N}). \tag{5.63}$$

This is zero for a sphere ($\mathcal{N} = \frac{1}{3}$), as expected. Shape anisotropy is only fully effective in samples which are so small that they do not break up into domains. Non-ellipsoidal shapes are approximately described by an effective demagnetizing factor. In a multidomain state, each domain creates its own demagnetizing field, and is subject to the stray field of the other domains. The order of magnitude of shape anisotropy for a ferromagnet with $\mu_0 M_s \approx 1$ T is 200 kJ m^{-3} when $\mathcal{N} = 0$.

5.5.2 Magnetocrystalline anisotropy

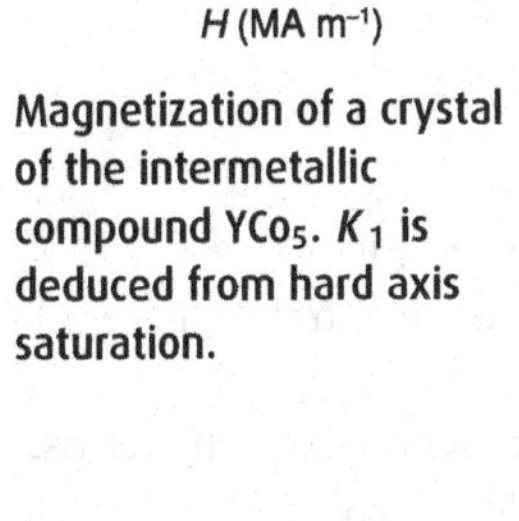

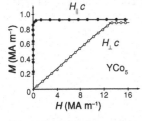

Magnetization of a crystal of the intermetallic compound YCo_5. K_1 is deduced from hard axis saturation.

The magnetization curves of single crystals of the three 3d ferromagnetic elements, corrected for the demagnetizing field, show a different approach to saturation when magnetized in different directions (Fig. 5.27). For iron, the cube edges ⟨100⟩ are easy directions and the cube diagonals[3] ⟨111⟩ are hard directions. For nickel it is the other way around. Cobalt, and many intermetallic compounds such as YCo_5 have the hexagonal axis [001] as the unique easy direction. It is much more difficult to saturate YCo_5, for example, than Co in a hard direction, perpendicular to [001]. Uniaxial anisotropy is a prerequisite for permanent magnetism.

The three-dimensional anisotropy surfaces for the three ferromagnetic elements are shown in Fig. 5.28.

[3] [] denotes a single direction and ⟨⟩ denotes a set of equivalent directions. Similarly () denotes a plane and {} a set of equivalent planes.

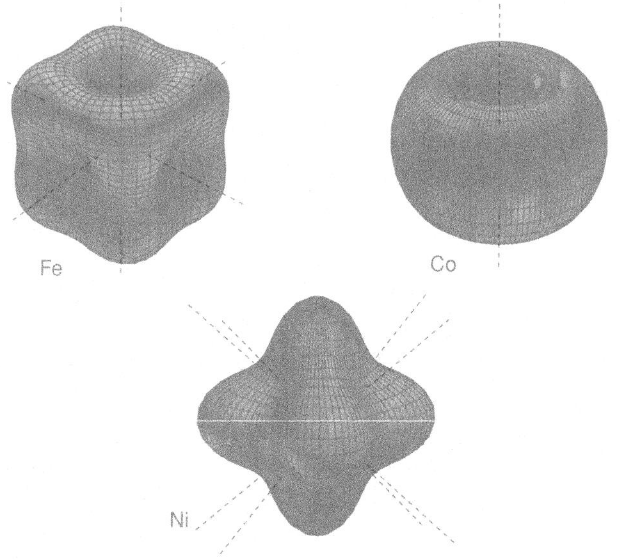

Figure 5.28

Magnetocrystalline anisotropy energy surfaces for iron, cobalt and nickel. In iron there are three easy axes ⟨100⟩, in cobalt one [001] and in nickel four ⟨111⟩.

Conventional expressions for the anisotropy energy in different symmetries are:

$$\begin{aligned}
&\textit{Hexagonal:} && E_a = K_1 \sin^2\theta + K_2 \sin^4\theta + K_3 \sin^6\theta + K_3' \sin^6\theta \sin 6\phi,\\
&\textit{Tetragonal:} && E_a = K_1 \sin^2\theta + K_2 \sin^4\theta + K_2' \sin^4\theta \cos 4\phi + K_3 \sin^6\theta\\
& && \quad + K_3' \sin^6\theta \sin 4\phi,\\
&\textit{Cubic:} && E_a = K_{1c}\left(\alpha_1^2\alpha_2^2 + a_2^2\alpha_3^2 + \alpha_3^2\alpha_1^2\right) + K_{2c}\left(\alpha_1^2\alpha_2^2\alpha_3^2\right),
\end{aligned}$$

where α_i are the direction cosines of the magnetization.

The K_{1c} term is equivalent to $K_{1c}(\sin^4\theta\cos^2\phi\sin^2\phi + \cos^2\theta\sin^2\theta)$. When $\theta \approx 0$, $\phi = 0$ this reduces to $K_{1c}\sin^2\theta$, so the leading term in every case is of the form (5.62).

An alternative way of writing the anisotropy expressions is in terms of a set of orthonormal spherical harmonics with anisotropy coefficients κ_l^m, and the crystal-field coefficients A_l^m introduced in §4.4:

$$E_a = \sum_{l=2,4,6} \kappa_l^m A_l^m Y_l^m(\theta,\phi). \tag{5.64}$$

For example,

$$\begin{aligned}
&\textit{Hexagonal:} && E_a^{hex} = \kappa_0 + \kappa_2^0\left(\alpha^2 - \tfrac{1}{3}\right) + \kappa_4^0\left(\alpha^4 - \tfrac{6}{7}\alpha^2 + \tfrac{3}{35}\right) + \cdots\\
& && \quad \text{with } \alpha = \cos\theta.\\
&\textit{Cubic:} && E_a^{cubic} = \kappa_0 + \kappa_4^4\left(\alpha_1^2\alpha_2^2 + \alpha_2^2\alpha_3^2 + \alpha_3^2\alpha_1^2 - \tfrac{1}{5}\right)\\
& && \quad + \kappa_6^4\left(\alpha_1^2\alpha_2^2\alpha_3^2 - \tfrac{1}{11}\left(\alpha_1^2\alpha_2^2 + \alpha_2^2\alpha_3^2 + \alpha_3^2\alpha_1^2 - \tfrac{1}{5}\right) - \tfrac{1}{105}\right)
\end{aligned}$$

This shows clearly that K_{1c} actually relates to fourth-order anisotropy terms, whereas K_1 in uniaxial structures relates to second-order anisotropy.

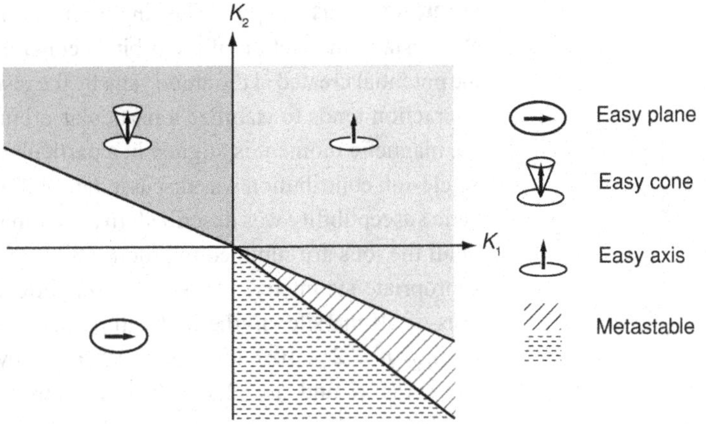

Figure 5.29

Magnetic phase diagram for uniaxial magnets. In the metastable regions, there are two energy minima, at $\theta = 0$ and $\theta = \pi/2$.

An interesting magnetic phase diagram (Fig. 5.29) arises for a uniaxial magnet when both K_1 and K_2 are taken into account; minimizing the anisotropy energy $E_a = K_1 \sin^2\theta + K_2 \sin^4\theta$, an easy cone phase appears when $K_1 < 0$ and $K_2 > -K_1/2$. The cone angle is $\sin^{-1}\surd(|K_1|/2K_2)$. In practice, when $K_1 > 0$, the two anisotropy constants K_1 and K_2 can be deduced from a plot of the hard axis magnetization curve as H/M versus M^2. This is the Sucksmith–Thomson plot.

The anisotropy field H_a is defined as the field needed to saturate the magnetization of a uniaxial crystal in a hard direction:

$$E = K_u \sin^2\theta - \mu_0 M_s H \cos(\pi/2 - \theta).$$

Minimizing E, $\partial E/\partial\theta = 0$ and setting $\theta = \pi/2$,

$$H_a = 2K_u/\mu_0 M_s. \tag{5.65}$$

Since $\mu_0 M_s \approx 1$ T for a typical ferromagnet, H_a can range from <2 kA m^{-1} to >20 MA m^{-1}, with typical values for shape anisotropy for $\mathcal{N} = 1$ of 200 kA m^{-1}. An equivalent definition of H_a is the applied field along the easy axis which would reproduce the change in energy due to a small deviation of the magnetization from this axis. This gives $K_u \sin^2\delta\theta = \mu_0 H_a M_s(1 - \cos\delta\theta)$, which returns the same result as (5.65).

Beware of taking the idea of anisotropy field too literally. Except at small angles, the energy variation in a field is not the same as the leading term in the anisotropy. A magnetic field defines an easy direction, not an easy axis.

5.5.3 Origin of magnetocrystalline anisotropy

There are two distinct sources of magnetocrystalline anisotropy:

- single-ion contributions;
- two-ion contributions.

Single-ion anisotropy The single-ion contribution is essentially due to the electrostatic interaction of the orbitals containing the magnetic electrons with the potential created at the atomic site by the rest of the crystal. This crystal-field interaction tends to stabilize a particular orbital, and by spin-orbit interaction the magnetic moment is aligned in a particular crystallographic direction. The single-ion contribution was discussed in §4.4, where its effect on the paramagnetic susceptibility was described. In a ferromagnetic crystal, the contributions of all the ions are summed to produce a set of macroscopic energy terms with appropriate symmetry. The sum is straightforward when the local anisotropy axes of all the sites in the unit cell coincide. For example, a uniaxial crystal having $n = 2 \times 10^{28}$ ions m^{-3}, described by a spin Hamiltonian $\mathrm{D}S_z^2$ with $\mathrm{D}/k_B = 1$ K and $S = 2$ will have anisotropy constant $K_1 = n\mathrm{D}S^2 = 1.1 \times 10^6$ J m^{-3}.

Broadside and head-to-tail configurations for a pair of ferromagnetically coupled magnetic moments. The latter is lower in energy.

Two-ion anisotropy The two-ion contribution often reflects the anisotropy of the dipole–dipole interaction. Comparing the broadside and head-to-tail configurations of two dipoles, each with moment $\mathfrak{m}$, it can be seen from (2.10) and (2.73) that the head-to-tail configuration is lower in energy by $3\mu_0\mathfrak{m}^2/4\pi r^3$. Magnets tend to align head-to-tail. This anisotropy is of order 1 K per atom, or 100 kJ m^{-3}. However, the dipole sum has to be extended over the entire lattice, and it vanishes for certain lattices (including all the cubic lattices). In noncubic lattices, the dipole interaction is an appreciable source of ferromagnetic anisotropy.

Another source of two-ion anisotropy is anisotropic exchange. The Heisenberg Hamiltonian is perfectly isotropic, but there are higher-order corrections involving spin-orbit coupling that lead to preferred orientations of the exchange-coupled pairs.

5.5.4 Induced anisotropy

One way of inducing uniaxial anisotropy is to anneal certain alloys in a magnetic field, Fig. 5.30. A good example is permalloy, $Ni_{80}Fe_{20}$, which is an fcc alloy with the property that $K_1 \approx 0$. When annealed at $\sim$800 K, atomic diffusion proceeds in the sense of favouring head-to-tail pairs of iron atoms, which have a larger moment than nickel. The iron pairs tend to align with the field, Fig. 5.31. This built-in texture produces a weak anisotropy. Amorphous ferromagnets may likewise acquire a uniaxial anisotropy due to pairwise texture when annealed in a magnetic field. Similar texture can be achieved by atomic deposition of thin films in a magnetic field.

Another way to induce uniaxial anisotropy in a ferromagnetic solid is to apply uniaxial stress σ. The magnitude of the stress-induced anisotropy is $K_{u\sigma} = \frac{3}{2}\sigma\lambda_s$, where λ_s is the saturation magnetostriction discussed in the next

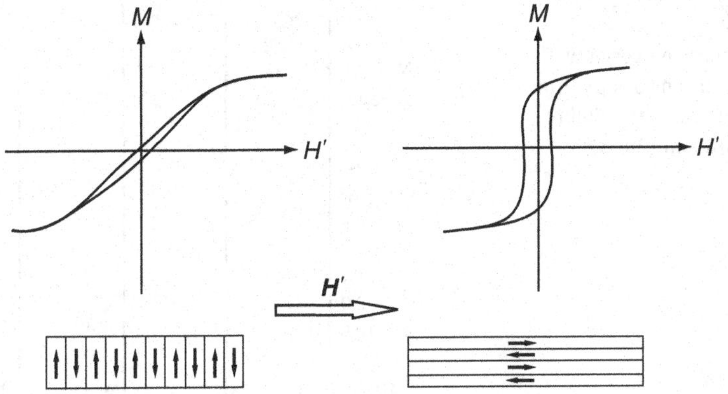

Figure 5.30

Magnetization of a thin film with induced anisotropy created by annealing in a magnetic field. The 'sheared' loop on the left is found when the measuring field H' is applied perpendicular to the annealing field direction. The open loop on the right is obtained when the two directions are parallel.

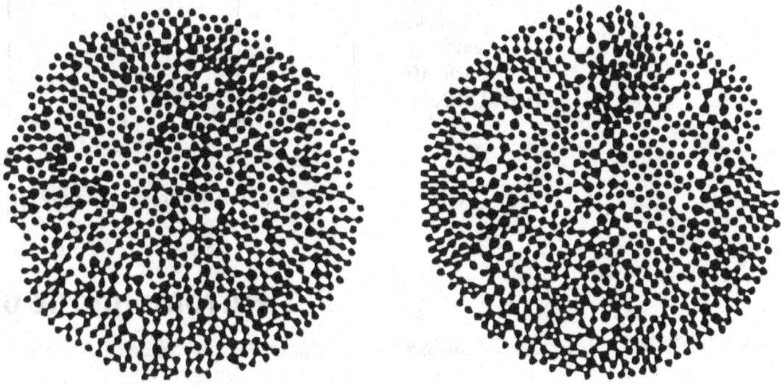

Figure 5.31

Pairwise texture induced by magnetic annealing. Pairs of atoms represented by the larger dots tend to be aligned vertically (G. S. Cargill and T. Mizoguchi, *J. Appl. Phys.* 49 1753 (1978)).

section. Both the single-ion and two-ion anisotropy contribute to the stress-induced effect.

The largest values of uniaxial anisotropy are found in hexagonal and other uniaxial crystals. Smallest values are found in certain cubic alloys and amorphous ferromagnets. The magnitudes of different anisotropy contributions are summarized in Fig. 5.32.

5.5.5 Temperature dependence

Any anisotropy originating from the magnetic dipole interaction will vary with temperature as M^2 because the dipole field $H_d \propto M_s$ and the energy density $E_a = -\frac{1}{2}\mu_0 \boldsymbol{M}_s \cdot \boldsymbol{H}_d \propto M_s^2$.

Single-ion anisotropy due to the crystal field exhibits a $j(j+1)/2$ power law at low temperatures for the second-, fourth- and sixth-order terms, which vary respectively as the third, tenth and twenty-first powers of the magnetization; $K_j(T)/K_j(0) = M^{j(j+1)/2}$. Close to T_C, these turn into M^j power laws. Here is a rare example of an effect whose temperature dependence is determined by symmetry alone.

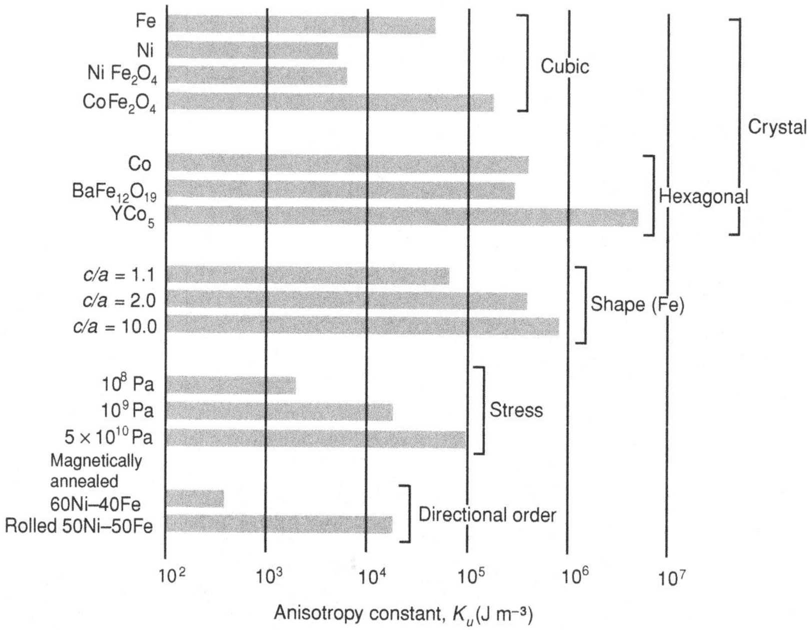

Figure 5.32

Quantitative overview of magnetic anisotropy, from all sources. (After Cullity and Graham (2008))

5.6 Ferromagnetic phenomena

Magnetic order in a crystal can influence the lattice parameters and the elastic moduli. These effects reflect the volume or strain dependence of bandwidth, exchange energy, magnetic dipole energy or crystal-field interaction. Some effects, like linear magnetostriction, are small, $\lambda_s \approx 10^{-5}$. Others like the ΔE effect, the change of Young's modulus associated with domain alignment, can be as large as 90%. Even very small magnetoelastic effects may be critical when optimizing the properties of soft magnets, because of the sensitivity of coercivity to stress-induced anisotropy. Ferromagnetism also influences the thermal, electrical and optical properties of solids. These *intrinsic* effects in ferromagnets are discussed in the following paragraphs.

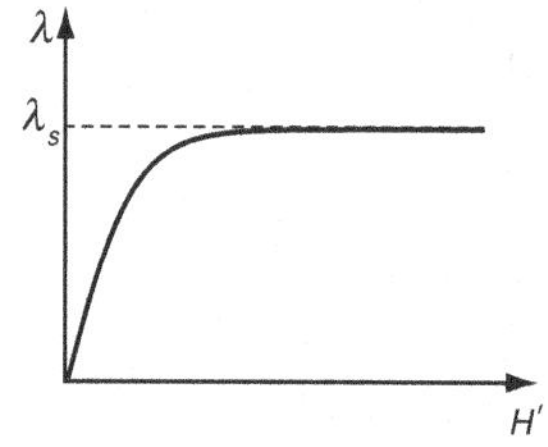

Development of magnetostriction of a polycrystalline material as a function of applied magnetic field.

5.6.1 Magnetostriction

Spontaneous volume magnetostriction ω_s is the fractional change of volume of an isotropic crystal due to magnetic order; the effect is proportional to M^2. The sign may be positive or negative but the magnitude does not normally exceed 1%. Iron-rich fcc alloys, for example, have $\omega_s > 0$, reflecting the sensitivity of exchange to interatomic spacing (Fig. 5.15). By expanding slightly, the iron-rich alloy increases the ferromagnetic exchange interaction, and lowers its energy.

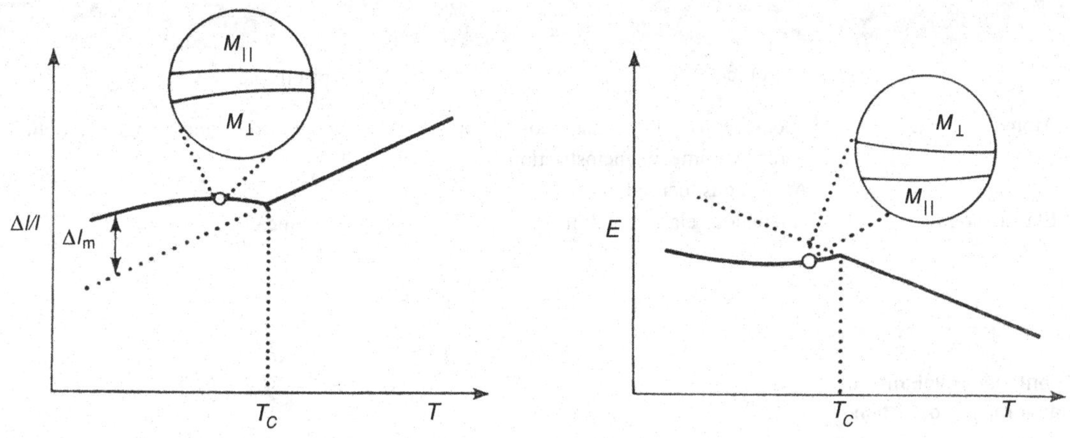

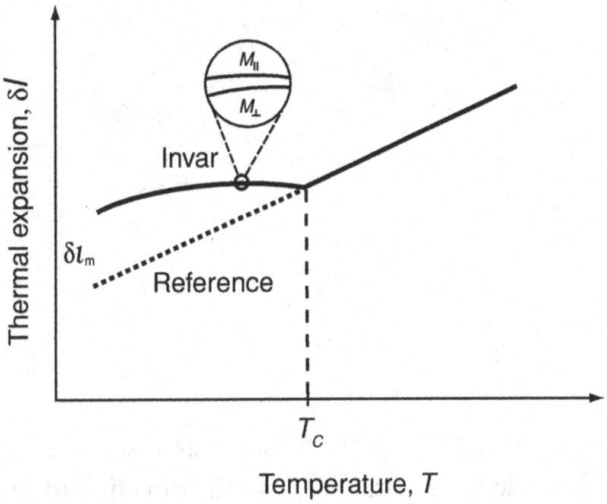

Figure 5.33

The invar effect; magnetostriction, Young's modulus and thermal expansion of an iron-based ferromagnetic alloy. The dotted line shows the thermal expansion of a reference material, while the thermal expansion of an invar alloy is shown by the solid line. The difference is the magnetovolume anomaly.

Invar is an fcc iron–nickel alloy of composition $Fe_{65}Ni_{35}$, discovered by Charles Guillaume in 1896. The alloy has zero thermal expansion because of a positive magnetovolume anomaly. Its discovery was important at the time because of the practical uses for alloys that are dimensionally stable in precision instruments. By varying the Fe:Ni ratio slightly, a match can be achieved with the thermal expansion of other materials, such as Si or quartz. The spontaneous volume magnetostriction of a ferromagnet is inferred by comparison with the thermal expansion of a nonmagnetic reference material, as shown in Fig. 5.33.

In weak ferromagnets, there is also a forced volume magnetostriction in high magnetic fields, as the moment is slightly increased by additional field-induced band splitting.

Table 5.7. Summary of magnetoelastic effects

	Isotropic	Anisotropic
Atomic spacing	Spontaneous volume magnetostriction Forced volume magnetostriction Anomalous thermal expansion	Spontaneous linear magnetostriction
Elastic moduli	Anomalous elastic moduli	Morphic effect, ΔE effect

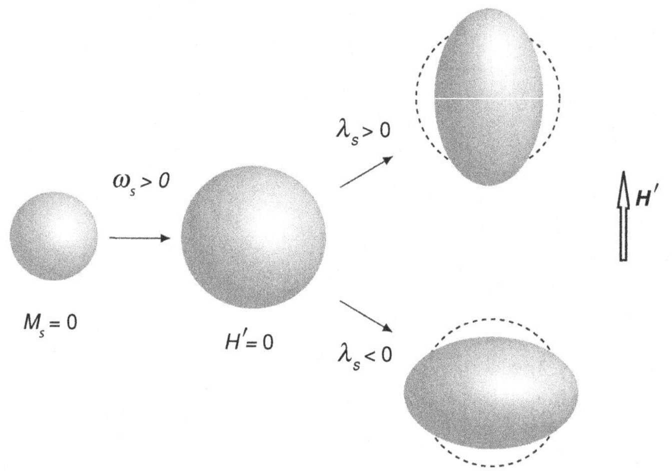

Figure 5.34
Spontaneous volume and linear magnetostriction.

James Joule 1818–1889.

Linear magnetostriction was discovered by James Joule in nickel in 1842. The linear strain $\delta l/l$ in the direction of magnetization is associated with the magnetization process. Saturation magnetostriction is denoted as λ_s. Iron has $\lambda_s = -7 \times 10^{-6}$ (the dimensionless unit of 10^{-6} is known as a *microstrain*), so it contracts by 8 ppm along the magnetic axis as the magnetization is saturated. Volume is conserved, which means that it also expands by 4 ppm in the perpendicular direction and $\lambda_{\parallel} + 2\lambda_{\perp} = 0$. Thanks to magnetostriction, no ferromagnetically ordered crystal can ever be strictly cubic! Linear magnetostriction is the reason why transformers hum.

In general, if θ is the angle between the magnetization and the easy axis,

$$\lambda(\theta) = \lambda_s(3\cos^2\theta - 1)/2.$$

The main magnetostrictive effects, summarized in Table 5.7, are greatly exaggerated in Fig. 5.34.

Linear magnetostriction depends on the direction of magnetization relative to the crystal axes, although, not of course, on the sense of its orientation along a particular direction. A change of strain is associated with rotation of domains as the magnetization approaches saturation, not with motion of 180° domain

Table 5.8. Summary of linear magnetostrictive effects

Joule effect	Villari effect
Field-induced strain	*Strain-induced* anisotropy
Wiedemann effect	Matteuchi effect
Helical-*field-induced* torque	*Torque-induced* anisotropy

walls. Magnetization reversal within a domain produces no strain. A single parameter is sufficient to describe the linear magnetostriction of an isotropic polycrystalline or amorphous material, but magnetostriction is really a third-rank tensor, since it relates strain ϵ_{ij} (a second-rank tensor) and magnetization M_k (a vector – first-rank tensor). In cubic crystals, there are two principal values. For example, in iron $\lambda_{100} = 15 \times 10^{-6}$ and $\lambda_{111} = -21 \times 10^{-6}$. The isotropic average,

$$\lambda_s = \tfrac{2}{5}\lambda_{100} + \tfrac{3}{5}\lambda_{111},$$

is -7×10^{-6}. Larger magnetostriction is found in nickel, where Joule first saw the effect, and in cobalt (Table 5.4). Some mixed rare-earth–iron alloys exhibit λ_s values of up to 2000×10^{-6}. The converse effect, where mechanical stress applied to a demagnetized multidomain ferromagnet modifies its easy axes and changes the initial susceptibility is known as the Villari effect.

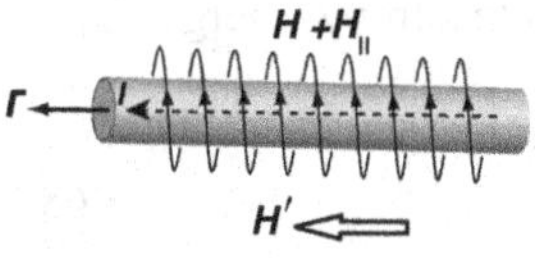

The Wiedemann effect. The sense of the twist shown is for $\lambda_s > 0$.

A variant of linear magnetostriction is the Wiedemann effect. A ferromagnetic rod subject to a transverse helicoidal field tends to twist. Torque can therefore be generated by the field produced by passing a current through a ferromagnetic wire while applying a field along its axis. The twist angle ϕ is proportional to the length of the wire l, the current density j and the magnetostriction λ_s:

$$\phi = \tfrac{3}{2}\lambda_s j l H_{\parallel}. \tag{5.66}$$

The effect can be used to measure magnetostriction.

The converse of the Wiedemann effect, known as the Matteuchi effect, is the modification of the susceptibility of a ferromagnetic wire by torque. The Joule and Wiedemann effects find applications in various types of magnetic actuators, whereas the Villari and Matteuchi effects are exploited in electromechanical sensors. The linear magnetostrictive effects are summarized in Table 5.8.

5.6.2 Other magnetoelastic effects

The effect of imposing a uniaxial stress σ N m^{-2} on a ferromagnetic material is to create strain-induced anisotropy. The elastic energy density for a polycrystalline material is

$$E_{ms} = -\lambda_s(E/2)(3\cos^2\theta - 1)\epsilon + (1/2)E\epsilon^2,$$

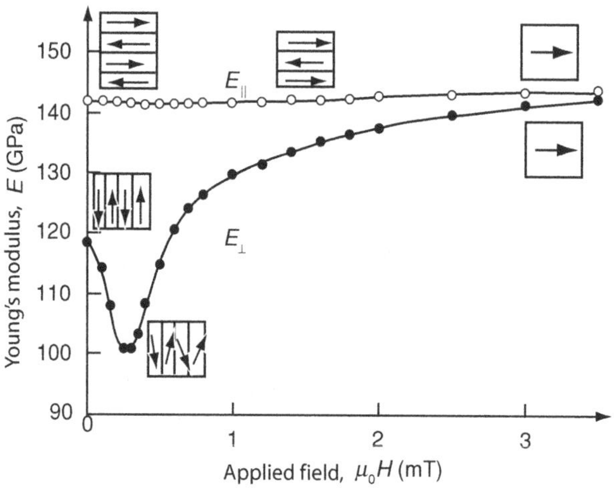

Figure 5.35

The ΔE effect in a soft, magnetostrictive ferromagnet. E is Young's modulus. The inserts show the evolution of the domain structure when stress is exerted in an applied field. (after B. S. Berry and W. C. Pritchett, *Phys Rev. Lett* **34**, 1022 (1975))

where ϵ is the strain, E is Young's modulus and θ is the angle between the magnetization and the strain axis. Minimizing the energy with respect to ϵ gives $\epsilon = \lambda_s/2(3\cos^2\theta - 1)$. If no stress is applied, the equilibrium strain $\epsilon = \lambda_s$. Since $\sigma = \epsilon E$, the angular-dependent term is

$$E_{ms} = -\lambda_s(\sigma/2)(3\cos^2\theta - 1). \tag{5.67}$$

Comparing with the usual expression for uniaxial anisotropy energy $E_a = K_u \sin^2\theta$ (5.62), we see that

$$K_u = \tfrac{3}{2}\lambda_s\sigma, \tag{5.68}$$

as mentioned in §5.5.4. In iron $\lambda_s \approx -7 \times 10^{-6}$ and $E \approx 200$ GPa, hence a stress of 1.4 MPa (14 bars) can be created by the magnetostriction.

The coupling of strain and magnetostriction is manifest in a dramatic way in the ΔE effect. The magnetization directions of domains in an unsaturated ferromagnet will be modified by an applied stress, which in turn leads to a magnetostrictive strain in the direction of the stress. This looks like a change in Young's modulus E, hence the name. Regardless of the sign of λ, the effect is always to reduce E, which is defined as the ratio of linear stress to strain (σ/ϵ). If a tensile stress is applied to a material with positive λ, the magnetization directions align with the stress, and there is an elongation, which increases ϵ and reduces E. If the tensile stress is applied to a material with negative λ, the magnetization directions align perpendicular to the stress, producing an elongation which is half as large, but again reduces E. The effect disappears as soon as the magnetization is saturated along or perpendicular to the stress direction. Some applications of the ΔE effect (Fig. 5.35) in ferromagnetic thin films are presented in Chapter 12.

The morphic effect is the dependence of the elastic constants on the direction of magnetization of a ferromagnet.

Table 5.9. Summary of uniaxial anisotropy

Anisotropy	Energy	K_u	kJ m^{-3}
Magnetocrystalline	Crystal field Magnetic dipole	K_1	1–10^4
Shape	Magnetostatic (magnetic dipole)	$\frac{1}{4}\mu_0 M_s^2[3\mathcal{N}-1]$	1–500
Stress	Magnetoelastic	$3\lambda_s\sigma/2$	0–100

Table 5.10. Energy contributions in a ferromagnet, in kJ m^{-3}

Exchange	$-2n\mathcal{J}ZS^2$	10^3–10^5
Anisotropy	K_u, K_i	10^{-1}–10^4
Magnetic self-energy	E_d	0–2 × 10^3
External field energy (in 1 T)	$B_0 M_s$	10^2–10^3
External stress energy (in 1 GPa)	$\sigma\lambda$	1–10^2
Magnetostrictive self-energy	$c\lambda^2$	0–1

10^3 kJ m^{-3} is roughly equivalent to 1 K or 0.1 meV atom^{-1}

The sources of uniaxial anisotropy are summarized in Table 5.9. A summary of the magnitudes of the energies involved in ferromagnetism is provided in Table 5.10.

5.6.3 Magnetocaloric effect

We have seen that a specific heat of magnetic origin is associated with the progressive disordering of atomic moments in a magnetically ordered material on increasing the temperature. There is a λ anomaly at the Curie point. The contributions to the total specific heat of nickel are illustrated in Fig. 5.36. The exchange energy, E_{ex} in the molecular field model is $-\frac{1}{2}\mu_0 H^i M_s$, where $H^i = n_W M_s$. Hence the specific heat of magnetic origin, $C_m = \mathrm{d}E_{ex}/\mathrm{d}T = (\mathrm{d}E_{ex}/\mathrm{d}M_s)(\mathrm{d}M_s/\mathrm{d}T)$, giving

$$C_m = -\mu_0 n_W M_s \frac{\mathrm{d}M_s}{\mathrm{d}T}. \tag{5.69}$$

In this model, there is no magnetic contribution above T_C where $M_s = 0$. However, short-range spin correlations persist above T_C. A more accurate expression for the specific heat, deduced from (5.24) for nearest-neighbour exchange coupling in terms of the spin correlation function $\langle \boldsymbol{S}_i \cdot \boldsymbol{S}_j \rangle$, is

$$C_m = -2nZ\mathcal{J}\frac{\partial \langle \boldsymbol{S}_i \cdot \boldsymbol{S}_j \rangle}{\partial T}.$$

where n is the number of atoms per unit volume.

The persistence of magnetic specific heat above T_C is evident in Fig. 5.36.

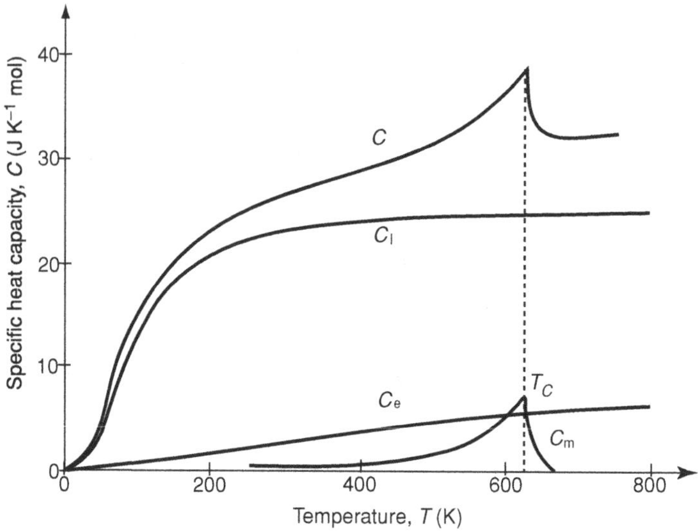

Figure 5.36

Heat capacity of nickel, showing electronic, lattice and magnetic contributions, and their sum *C*.

The effect of an applied magnetic field is to reduce magnetic disorder, and decrease magnetic entropy. If this is done adiabatically, with no exchange of heat with the surroundings, there has to be an increase of temperature of the sample. Conversely, if the thermally isolated sample is magnetized in an applied field, which is then reduced to zero, there must be a decrease of temperature in the sample due to adiabatic demagnetization, as described for paramagnets in §4.3.4. The entropy change is related to the temperature derivative of the Magnetization by Eq. 2.102.

The work done by the field to produce an increase of magnetization δM is $\delta W = \mu_0 H' \delta M$ (2.93), so the heat resulting from the change of magnetization, which is the difference between this work and the change of exchange energy $-\mu_0 n_W M \delta M$, is

$$\delta Q = \mu_0 (H' + n_W M) \delta M. \tag{5.70}$$

In the paramagnetic temperature range above T_C, $\chi = C/(T - T_C)$, which from (5.6) is $T_C/[n_W(T - T_C)]$, hence

$$M = \frac{H' T_C}{n_W(T - T_C)}. \tag{5.71}$$

Substituting (5.71) into (5.70), it follows that

$$\delta Q = \frac{\mu_0}{2n_W} \frac{T T_C}{(T - T_C)^2} \delta(H^2).$$

The corresponding change of temperature is $\delta T = \delta Q / C_M$, where C_M is the specific heat at constant magnetization (in units of J K^{-1} m^{-3}). Expressed in terms of the change of magnetization, we find from (5.71) for $T > T_C$:

$$\delta T = \frac{\mu_0 n_W}{2C_M} \frac{T}{T_C} \delta(M^2).$$

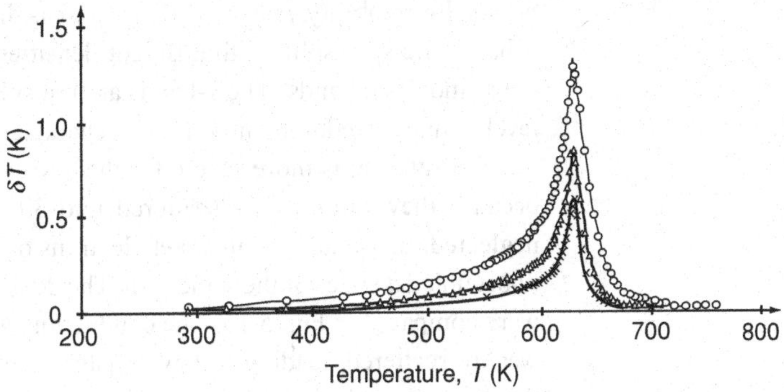

Figure 5.37

The magnetocaloric effect for Ni in different applied fields. It peaks at T_C.

The result for $T < T_C$ is almost the same; it is obtained by neglecting the applied field H' relative to the molecular field $n_W M$ in the expression for δQ

$$\delta T = \frac{\mu_0 n_W}{2C_M}\delta(M^2).$$

The magnetocaloric effect for Ni in a field of 2 T is illustrated in Fig. 5.37. The temperature change at T_c is of order 2 K in a field of 1 T. Gd-based alloys with T_c around 320 K may be used for magnetic refrigeration.

5.6.4 Magnetotransport

Here we consider *intrinsic* magnetotransport effects which do not depend on the shape or form of a ferromagnetic sample. Several types of magnetoresistive and Hall effects fall into this category. Magnetoresistance may be defined as

$$MR = [\varrho(B) - \varrho(0)]/\varrho(0). \quad (5.72)$$

The ordinary positive B^2 magnetoresistance of a nonmagnetic metal due to cyclotron motion of the electrons was mentioned in §3.2.7.

There are other magnetoresistance phenomena intrinsic to ferromagnets, notably anisotropic magnetoresistance (AMR), a small effect depending on the relative orientation of current and magnetization, and colossal magnetoresistance (CMR) a much larger negative effect which appears in the vicinity of the Curie point of materials where the exchange coupling is by double exchange, and the electron transfer is linked to nearest-neighbour spin alignment. A further source of negative magnetoresistance in ferromagnets is spin disorder scattering in the vicinity of T_C. Ferromagnets also exhibit an additional term in the Hall effect which is proportional to the magnetization.

Conduction electrons in $3d$ ferromagnets and their alloys are in s-like or d-like states, which coexist at the Fermi level. The s-electrons resemble free electrons; they have high mobility and carry most of the current. The d-electrons are relatively ineffective current carriers on account of their high effective mass

m^* and low mobility, $\mu = e\tau/m^*$, where τ is the scattering relaxation time. The d-bands are spin-split, with different densities of states at the Fermi level for the $\uparrow$ and $\downarrow$ subbands. The s-bands are not spin-split, but s states at the Fermi level acquire predominantly $\uparrow$ or $\downarrow$ character by hybridization with d states there. Scattering is more severe for the $\downarrow$ s-electrons in a strong ferromagnet, because they can only be scattered into $3d^{\downarrow}$ states if spin-flip scattering is neglected. In a weak ferromagnet electrons of either spin can be scattered into unoccupied d states of the same spin. The densities of state of copper and nickel were compared in Fig. 5.15. The conduction electrons in copper are s-like and weakly scattered, leading to low resistivity. In nickel, a strong ferrromagnet, the $\uparrow$ states at the Fermi level conduct like copper, but the $\downarrow$ states are very strongly scattered on account of the large density of d states at the Fermi level into which they can scatter.

The situation in $4f$ metals and alloys is different. The $4f$-band is usually very narrow because of the limited overlap of $4f$ wave functions on adjacent sites. There is integral $4f^n$ electron occupancy, so the band does not conduct. The Mott criterion for conduction in a narrow band is $W > U$, where W is the bandwidth and U is the screened Coulomb energy of an extra electron on the atomic site. When $W < U$, the charged configurations $4f^{n\pm1}$ necessary for metallic conduction cannot be accommodated in the band, and the material is a Mott–Hubbard insulator. Conduction in the rare-earth metals and alloys is therefore due to the partially occupied $5d$- and $6s$-orbitals at the Fermi level. The $5d$ states are exchange-split by on-site $4f$–$5d$ exchange coupling. The electronic configuration of a rare-earth is roughly $4f^n5d^26s$, so they behave in some ways like light d-elements.

The $5f$-band of ferromagnetic compounds of the early actinides is broad and conducting, as in the $3d$ metals, but for the heavy actinides there is integral $5f$ occupancy, and the f electrons are localized, as in the rare-earths.

For ionic compounds, such as oxides, the $3d$ band is narrow and the electrons tend to be localized. There are a few exceptions such as CrO_2 or $SrRuO_3$ where the band is broad enough for $3d^n$ and $3d^{n\pm1}$ configurations to coexist. Thanks to the on-site exchange interaction, these d states are spin polarized and $3d^{n+1} \rightleftharpoons 3d^n$ electron hopping produces ferromagnetism, via the double exchange interaction. Mobility is low, and the velocity of the electrons may be slow enough in oxides for the Born–Oppenheimer approximation to be invalid. The electrons drag a local lattice distortion along with them, forming a polaron, which further enhances their effective mass. When the conduction electrons are spin polarized, they drag a cloud of spin polarization of the $3d$ or $4f$ ion cores along with them; then we have a spin polaron.

The $\uparrow$ or $\downarrow$ electrons in a ferromagnet may undergo spin-flip scattering events, associated with excitation or absorption of a magnon ($\Delta S = 1$). The spin-orbit interaction also mixes $\uparrow$ and $\downarrow$ channels, because $\boldsymbol{L} \cdot \boldsymbol{S} = L_xS_x + L_yS_y + L_zS_z$ can be written in terms of raising and lowering operators as $\frac{1}{2}(L^+S^- + L^-S^+) + L_zS_z$. It therefore mixes $|\frac{1}{2}, m_\ell\rangle$ with $|-\frac{1}{2}, m_\ell + 1\rangle$

Table 5.11. Room-temperature resistivity of metals (10^{-8} Ω m)

Metal	Orbitals	Magnetization	$\varrho\uparrow$	$\varrho\downarrow$	ϱ	α
Cu	*s*-band	Paramagnet	4	4	2	1
Ni	*d*-band	Strong ferromagnet	13	65	11	5
Co	*d*-band	Strong ferromagnet	8	120	7	15
Fe	*d*-band	Weak ferromagnet	32	28	15	0.9
V	*d*-band	Paramagnet	52	52	26	1
$aFe_{80}B_{20}$	*d*-band	Amorphous ferromagnet	320	320	120	1
Gd	*f*-level; 5*d*/5*s*-band	Ferromagnet	270	270	130	1

states. Nevertheless, these spin flipping events are relatively rare compared with normal momentum scattering events. An electron in a ferromagnetic 3*d* metal may undergo 100 or more scattering events, before it experiences a spin flip. This led Mott to propose his two-current model in 1936, whereby the ↑ and ↓ conduction channels are regarded as independent, and conduct in parallel. Hence $\sigma = \sigma_\uparrow + \sigma_\downarrow$, or in terms of resistivity

$$\varrho = \frac{\varrho_\uparrow \varrho_\downarrow}{(\varrho_\uparrow + \varrho_\downarrow)}. \tag{5.73}$$

The quantity α is defined as the ratio of conductivity in the ↑ and ↓ channels: $\alpha = \sigma_\uparrow/\sigma_\downarrow = \varrho_\downarrow/\varrho_\uparrow$. The conductivity of a metal is always greater than the conductivity of either spin channel – double when the metal is paramagnetic and $\sigma_\uparrow = \sigma_\downarrow$. Values of $\varrho_{\uparrow,\downarrow}$ and α for a few metals are given in Table 5.11.

A useful result for the field dependence of the resistivity of a metal where the resistivity can be described by a single scattering time is Kohler's rule:

$$\frac{\varrho(H)}{\varrho(0)} = f\left(\frac{H}{\varrho(0)}\right)^2. \tag{5.74}$$

It follows that $\Delta\varrho/\varrho \approx (H/\varrho)^2$.

Spin-disorder scattering Electrons in ferromagnetic solids experience a spin-dependent potential. When the material is perfectly ordered, there is no magnetic scattering. But as the temperature approaches T_C the electrons see a potential that fluctuates between values differing by the on-site exchange interaction. This random potential contributes to the resistivity as shown in Fig. 5.38.

The normal temperature variation of the resistivity of a metal is described by Matthiesen's rule, a statement that the resistivity is the sum of a temperature-independent term ϱ_0 due to impurity scattering, which is of order 10 nΩ m per % of impurity, and a temperature-dependent term $\varrho_{ph}(T)$ due to phonon scattering, which varies as T^n, where $n = 1$ at temperatures well above the

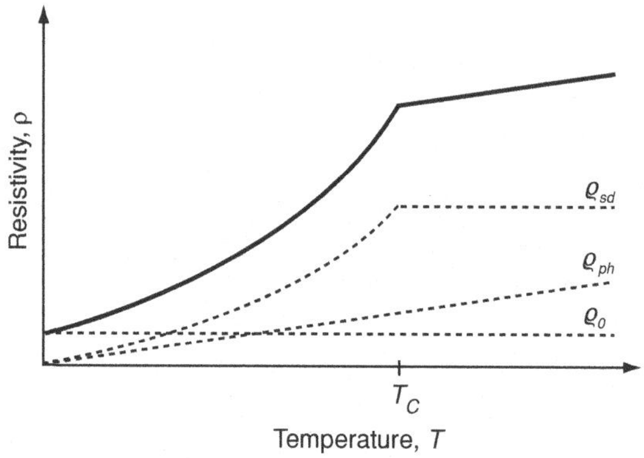

Figure 5.38

The resistivity of a normal metal showing the impurity contribution ϱ_0 and the phonon contribution ϱ_{ph}. An extra term due to spin-disorder scattering ϱ_{sd} appears in a ferromagnet.

Debye temperature and $n = 3$–5 far below it. In ferromagnets, there is an extra impurity-like term due to disordered spins. When majority-spin electrons, which carry most of the current, encounter an atom with reduced spin moment, they are strongly scattered because of the increased local exchange potential. The spin disorder scattering term in the paramagnetic state varies as

$$\varrho_{para} \sim \frac{k_F m_e^2 \mathcal{J}_{sd}^2}{e^2 \hbar^2} s(s+1), \tag{5.75}$$

where $\mathcal{J}_{sd}$ is the exchange between localized and conduction electrons. This leads to an expression for the spin disorder resistivity in the ferromagnetic state

$$\varrho_{ferro} = \varrho_{para}\{1 - [M_s(T)/M_s(0)]^2\}. \tag{5.76}$$

Spin-disorder scattering is pronounced in rare-earth metals and alloys, and in half-metals. Applying a magnetic field close to T_C where the susceptibility is high, increases M, and therefore produces a negative magnetoresistance.

An alternative way to describe spin mixing is to introduce a spin mixing resistivity $\varrho_{\uparrow\downarrow}$ which is zero at low temperature, but large above T_C. The expression for resistivity is then

$$\varrho = \frac{\varrho_\uparrow \varrho_\downarrow + \varrho_{\uparrow\downarrow}(\varrho_\uparrow + \varrho_\downarrow)}{\varrho_\uparrow + \varrho_\downarrow + 4\varrho_{\uparrow\downarrow}}. \tag{5.77}$$

This has the correct behaviour in the high-temperature limit, where it tends to $(\varrho_\uparrow + \varrho_\downarrow)/4 = \varrho_{para}$, and in the low-temperature limit where it reduces to (5.73).

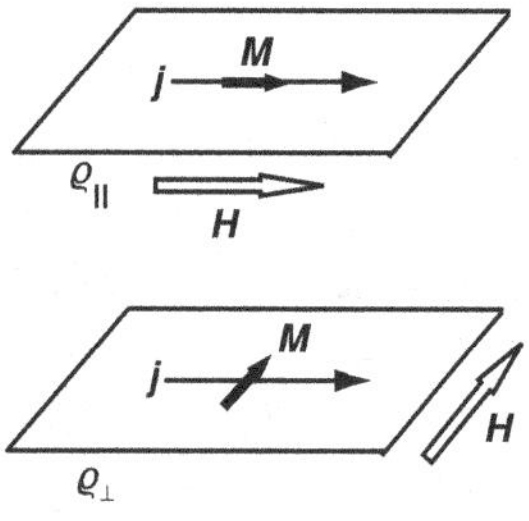

Measurement of AMR for a thin film.

Anisotropic magnetoresistance AMR is a much smaller change, but it is useful for sensors because the resistance change can be achieved in thin films with magnetic fields of a millitesla or less. William Thompson discovered in 1857 that the resistivity of nickel varies slightly with the direction of current flow, relative to the direction of magnetization. The resistance is a few per cent higher when the current flows parallel to an applied field than when it is in

Figure 5.39

AMR of a permalloy film, and its angular dependence.

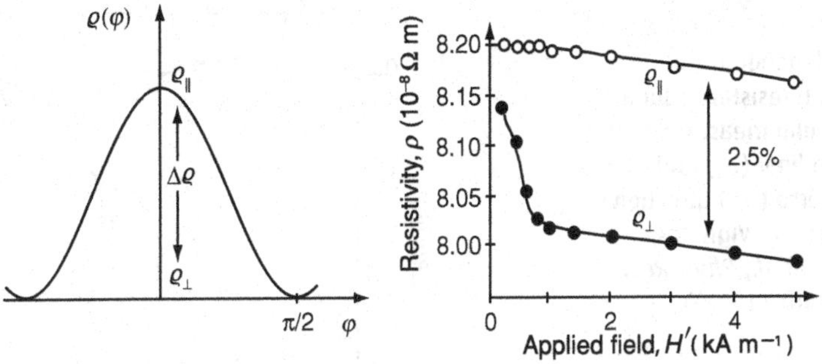

the transverse direction. Anisotropy in this sense is usual for ferromagnetic metals. The maximum AMR may be defined as $\Delta\varrho/\varrho_\perp$, where $\Delta\varrho = (\varrho_\parallel - \varrho_\perp)$. The magnitude of the effect is usually no more than about 3%. The angular dependence of the saturated anisotropic magnetoresistance is

$$\varrho(\varphi) = \varrho_\perp + (\varrho_\parallel - \varrho_\perp)\cos^2\varphi, \tag{5.78}$$

where φ is the angle between the current density $\boldsymbol{j}$ and the magnetization $\boldsymbol{M}_s$. Results for permalloy are shown in Fig. 5.39.

Note that the greatest sensitivity of the resistance to a change in φ, determined by the condition $d^2\varrho/d^2\varphi = 0$, arises when $\varphi = \pi/4$, a fact that is exploited in AMR sensors, discussed in §14.3.

When the current flows in the x-direction, an easy-anisotropy direction – this could be determined by the shape anisotropy of a wire, for example – with anisotropy constant K_u and the field applied in the perpendicular y-direction, then φ is obtained by minimizing

$$E = M_s H' \sin\varphi + K_u \cos^2\varphi \tag{5.79}$$

with respect to φ, which gives $\sin\varphi = M_s H/2K_u$. Hence, from (5.78)

$$\varrho(H) = \varrho_\perp \left[1 + \frac{\Delta\varrho}{\varrho_\perp}\left(1 - \frac{M_s^2 H^2}{4K_u^2}\right)\right] \tag{5.80}$$

or $\varrho(H) = \varrho(0) - \Delta\varrho(M_s^2 H^2/4K_u^2)$, provided $M_s H/2K_u < 1$. Note the H^2 variation, as predicted by Kohler's rule. This result applies for anhysteretic magnetization rotation. Magnetization processes that only involve motion of domain walls in a direction perpendicular to the easy axis have no effect on the magnetoresistance, whatever the angle between the current and applied field.

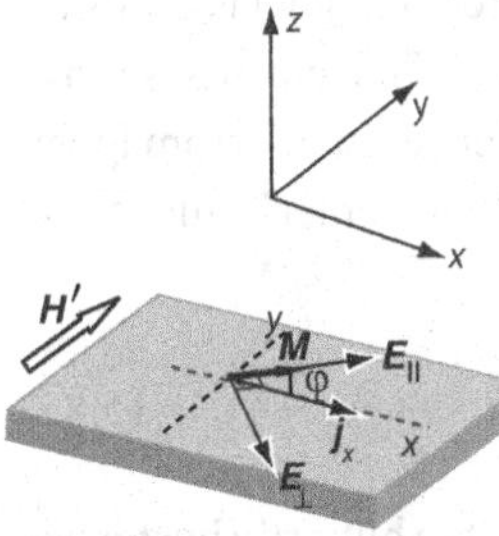

Calculation of the planar Hall effect.

The reason why AMR is usually positive ($\varrho_\parallel > \varrho_\perp$) is related to the spin-orbit interaction, which tends to mix the ↑ and ↓ channels. Spin-flip scattering of the mobile ↑ electrons is permitted, and the scattering is effective when the plane of the orbit lies in the plane of the current, but not when it lies in the perpendicular direction.

A phenomenon related to AMR, which depends on the direction of magnetization of a conducting ferromagnet, is the planar Hall effect. Amagnetic field is

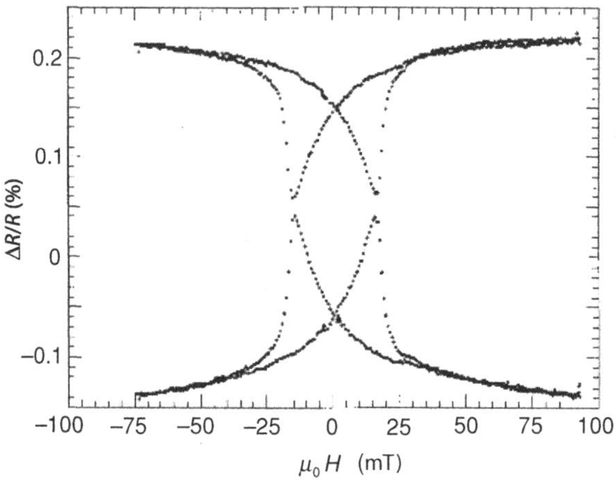

Figure 5.40
Field dependence of magnetoresistance for a nickel film measured in the longitudinal ($\rho_{\|}$) and transverse ($\rho_{\perp}$) direction (M. Viret, D. Vignoles, D. Cole *et al.*, *Phys. Rev.* B53, 8464 (1996)).

applied in the plane of a thin sample, transverse to the direction of the current j_x, which is in the x-direction. If the magnetization lies in-plane at an angle φ to the current direction, a transverse electric field E_y appears, just as in the normal Hall effect. The anisotropic resistivity is related to the direction of $\boldsymbol{M}$. Resolving the electric field $\boldsymbol{E}$ into components parallel and perpendicular to M, we have

$$E_{\|} = \varrho_{\|} j_x \cos\varphi; \quad E_{\perp} = \varrho_{\perp} j_x \sin\varphi.$$

Now the components of the electric field parallel and perpendicular to the current are $E_x = E_{\|}\cos\phi + E_{\perp}\sin\phi$ and $E_y = E_{\|}\sin\varphi - E_{\perp}\cos\varphi$, hence

$$E_x = j_x(\varrho_{\perp} + \Delta\varrho\cos^2\varphi) \text{ and } E_y = j_x\Delta\varrho\sin\varphi\cos\varphi.$$

The planar Hall voltage is therefore

$$V_{pH} = j_x w \Delta\varrho \sin\varphi\cos\varphi, \tag{5.81}$$

where w is the width of the film. The planar Hall resistivity is $\varrho_{xy} = \frac{1}{2}\Delta\rho\sin 2\varphi$. The greatest change is found when the magnetization switches from $\varphi = 45°$ to $\varphi = 135°$.

So far, we have assumed that the magnetization of the ferromagnet is uniform, and that the current flows at some fixed angle φ with M. When electrons move in a solid they are influenced by the $\boldsymbol{B}$-field they experience. In a ferromagnet, the magnetization contributes to electron scattering, so that Kohler's rule (5.74) should be generalized to read

$$\frac{\Delta\varrho}{\varrho} = a(H/\varrho)^2 + b(M/\varrho)^2. \tag{5.82}$$

Data for nickel in Fig. 5.40 show that the resistance exhibits a butterfly hysteresis loop that reflects the hysteresis of the magnetization, although it is an even function of M, as expected.

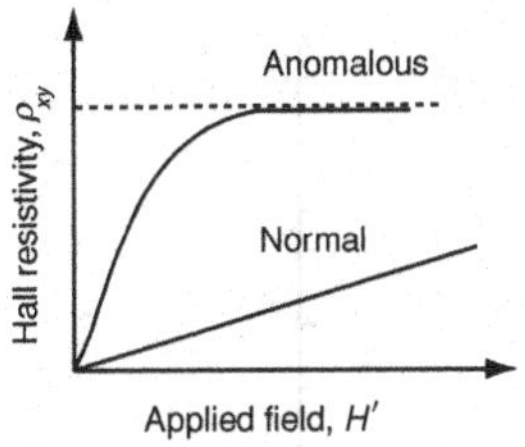

Anomalous Hall effect.

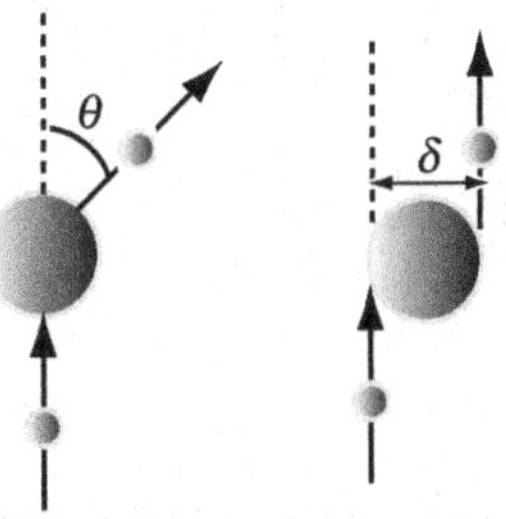

(i) Skew scattering and (ii) side jump scattering which both contribute to the anomalous Hall effect.

Anomalous Hall effect There is a term in the Hall resistivity of a ferromagnet when the field is applied in the z-direction, perpendicular to the plane of the film, in addition to the normal Hall effect (3.53). This is the anomalous Hall effect, which varies with the magnitude of the magnetization M:

$$\varrho_{xy} = \mu_0(R_h H' + R_s M). \tag{5.83}$$

The anomalous Hall effect is yet another consequence of spin-orbit coupling. The symmetry of the radial component of the Lorentz force $\boldsymbol{j} \times \boldsymbol{B}$ which produces the normal Hall effect is the same as the symmetry of the spin-orbit interaction $\boldsymbol{L} \cdot \boldsymbol{S}$ since $\boldsymbol{L} = \boldsymbol{r} \times \boldsymbol{p}$, $\boldsymbol{p} \propto \boldsymbol{j}$, $\boldsymbol{S} \propto \mu_0 \boldsymbol{M}$.

In a ferromagnet the anomalous Hall effect varies as the macroscopic average magnetization. Generally there are contributions which vary as ϱ_{xx} and as ϱ_{xx}^2, which are associated with different scattering mechanisms. Deviation of the electron trajectories due to spin-orbit interaction is known as skew scattering.

Writing $\varrho_m = \mu_0 R H'$, the Hall angle ϕ_H is defined as ϱ_m/ϱ_{xx}. Thus $\phi_H = \alpha + \beta\varrho_{xx}$; α is the skew scattering angle. The second term is often larger. It is associated with the side-jump mechanism due to impurity scattering. If $\delta \approx 0.1$ nm is the side jump, the Hall angle here is δ/λ, which is proportional to ϱ_{xx}. Here λ is the mean free path.

The anomalous Hall effect is especially large in disordered ferromagnets with strong impurity scattering, and in magnetic glasses. There is a small effect in paramagnets, which follows a Curie–Weiss law. It is most useful for measuring hysteresis in films when their Magnetization lies perpendicular to the plane.

Colossal magnetoresistance CMR is the substantial fall in resistance observed in the vicinity of the Curie temperature of double-exchange materials such as $La_{0.7}Ca_{0.3}MnO_3$ when they are subjected to a large applied magnetic field. The field aligns the partly disordered manganese moments, which promotes electron hopping (§5.2), and in turn enhances the exchange. Such ferromagnetic mixed valence oxides, where double exchange is the dominant ferromagnetic coupling mechanism, show enhanced negative magnetoresistance in the vicinity of T_C. With our definition (5.72) the limit is -100%, but an alternative, optimistic definition places $\varrho(H)$ in the denominator:

$$MR = [\varrho(H) - \varrho(0)]/\varrho(H),$$

producing a bigger number for the same change. For instance, if $\varrho(H) = 0.1\varrho(0)$, a 90% *MR* becomes a 900% effect with the new definition. There is some justification for this, when the primary effect of magnetic field is to increase conductance, since the definition is equivalent to

$$MR = [\sigma(H) - \sigma(0)]/\sigma(0).$$

This is the order of magnitude that can be observed in materials such as $La_{0.7}Sr_{0.3}MnO_3$ in a large field, which merits the epithet 'colossal'. Since

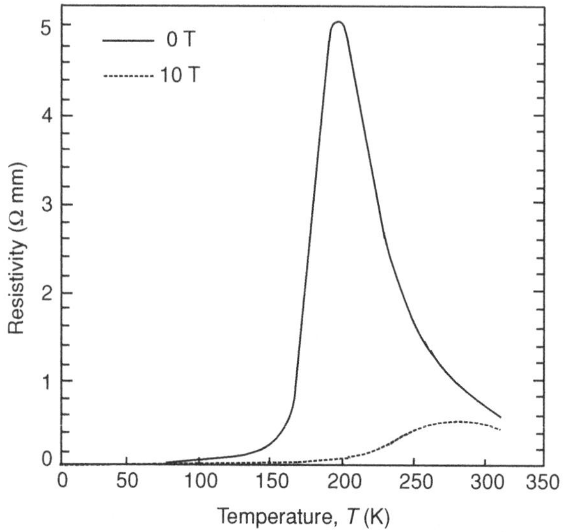

Figure 5.41

Magnetoresistance of $La_{0.7}Ca_{0.3}MnO_3$ in a field of 10 T.

electron hopping (Fig 5.10) in a mixed-valence half-metal depends on $\cos(\theta_{ij}/2)$ the resistance depends critically on the nearest-neighbour correlations $\langle \boldsymbol{S}_i \cdot \boldsymbol{S}_j \rangle \approx \langle 1 - \theta_{ij}^2 \rangle$. Applying the field has the double effect of reducing θ_{ij} thereby increasing the ferromagnetic exchange and reducing θ_{ij} further. Some typical CMR data are shown in Fig. 5.41.

Despite the impressive magnitude of the resistance change, CMR is of limited use because of the huge fields needed to create it. If the ratio of the resistance change to the field needed to produce it is taken as the figure of merit, one is better served by an AMR of 1% in a few milliteslas, than a CMR of 90% in several teslas. The temperature dependence of resistivity near T_C in CMR materials has found some use in bolometers.

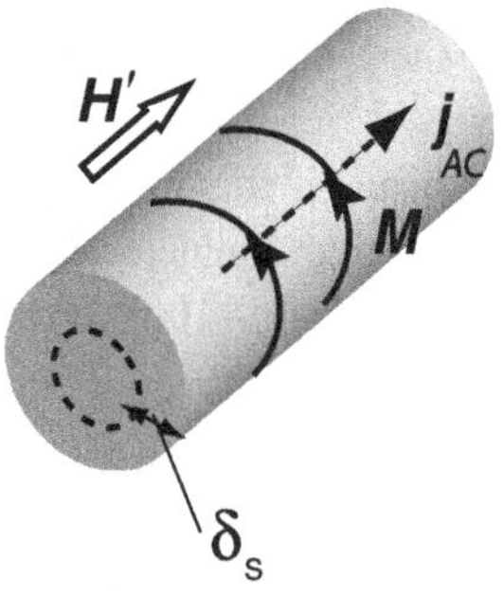

Magnetoimpedance. The AC field created by j_{AC} penetrates the wire to a depth δ_s, the skin depth. The permeability of the circumferentially magnetized wire is saturated in the applied field H'.

Magnetoimpedance The total impedance Z of a ferromagnetic component changes when it is exposed to a static magnetic field. The effect relates to the inductance L:

$$Z = R + i\omega L,$$

where $L = \Phi/I$ is the ratio of magnetic flux to current. Magnetoimpedance is conventionally defined as

$$MI = \frac{Z(H) - Z_0}{Z_0}, \tag{5.84}$$

where Z_0 is the impedance in a field that is large enough to saturate the effect. In soft ferromagnets, the effect may be hundreds of per cent at megahertz frequencies. Hence the 'giant' epithet again.

The effect is related to the skin depth δ_s (12.2), which is the depth an AC field penetrates into a metal. In a ferromagnet, δ_s varies as $(\mu_r\omega)^{-1/2}$.

Figure 5.42

Frequency response of the impedance of an amorphous $Fe_{4.3}Co_{68.2}Si_{12.5}B_{15}$ wire, 30 μm in diameter, in zero field, and in a field of 800 A m^{-1} applied parallel to the wire (after L. V. Panina and K. Mohri, *Appl. Phys. Lett.* **65** 1189 (1994)).

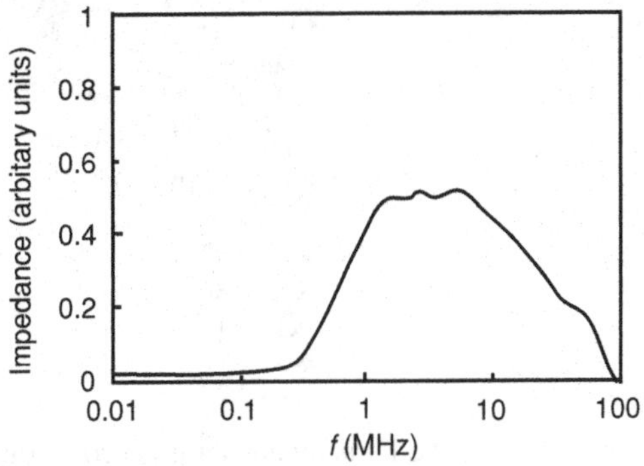

Magnetoimpedance is best explained for a soft ferromagnetic wire carrying an AC current $I = I_0 \exp(i\omega t)$. The current creates a circumferential magnetic field around the wire that penetrates to a depth δ_s. Circumferential anisotropy K_c is created if the wire has a small positive magnetostriction. The anisotropy field $H_a = 2K_c/\mu_0 M_s$ controls the response of the induced magnetization to a DC field H' applied parallel to the wire, and hence the permeability in the skin depth. For good results, H_a should be an order of magnitude greater than the field H' it is desired to measure. A typical response of a soft amorphous ferromagnetic wire is shown in Fig. 5.42.

5.6.5 Magneto-optics

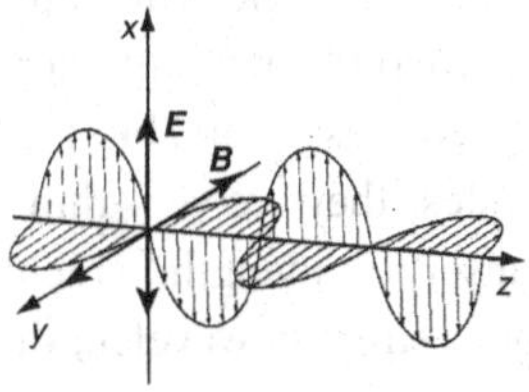

An electromagnetic wave.

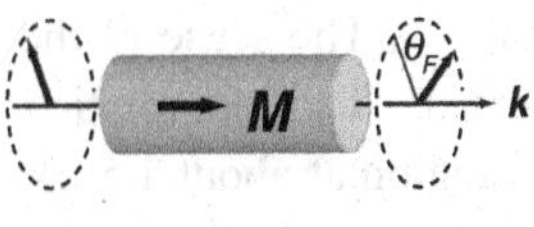

The magneto-optic Faraday effect. The plane of polarization of incident light is rotated through an angle θ_F. In a magneto-optic isolator, $\theta_F = \pi/4$, and polarizers offset by this angle are placed at each end.

Michael Faraday's 1845 discovery of a connection between magnetism and light was an advance comparable in importance to Hans-Christian Oersted's discovery of the connection between magnetism and electricity. He found that the plane of polarization of light rotates as it passes through a glass plate when a magnetic field is applied parallel to the direction $\mathbf{e}_K$ of propagation of the light. Light is an electromagnetic wave, where the $\boldsymbol{E}$-field and the $\boldsymbol{H}$-field are perpendicular to each other, and transverse to the propagation vector $\boldsymbol{K} = (2\pi/\Lambda)\mathbf{e}_K$; Λ is the wavelength of light. The direction of polarization is defined by $\boldsymbol{E}$.

An $\boldsymbol{H}$-field has no influence on the electromagnetic wave; the Faraday effect is due to the interaction of light with the induced magnetization $\boldsymbol{M}$ of the solid medium. Faraday's original demonstration was for paramagnetic glass, but Faraday rotation is also observed when light traverses a diamagnet or a transparent, spontaneously magnetized ferromagnet or ferrimagnet. The Faraday rotation in water, for example, is 3.9 rad $T^{-1}m^{-1}$. The sense of rotation reverses when the magnetization direction is switched. Conventionally, the angle of rotation θ_F is positive if the plane of polarization of the light turns clockwise when the observer faces the source.

Table 5.12. Faraday rotation in ferromagnetic metals and ferrimagnetic insulators

	θ_F/t at 830 nm (° μm^{-1})		θ_F/t at 1.06 μm (° mm^{-1})
Fe	35	$Y_3Fe_5O_{12}$	28
Co	36	$Tb_3Fe_5O_{12}$	54
Ni	10		

The Faraday constant is proportional to the optical path length

$$\theta_F = k_V \int \mu_0 \boldsymbol{M} \cdot \mathrm{d}\boldsymbol{l} \tag{5.85}$$

where $k_V(\Lambda)$ is the Verdet constant of the material, which depends on wavelength. Since the $\boldsymbol{H}$-field has no influence, (5.85) may be equally well written as $\theta_F = k_V Bl$ for a uniformly magnetized sample. The Faraday effect is nonreciprocal in the sense that the rotation (clockwise or counterclockwise) depends on whether the wavevector $\boldsymbol{K}$ of light is parallel or antiparallel to $\boldsymbol{M}$. Hence Faraday rotation is cumulative; as light is reflected back along its path in a magnetic medium, the resultant rotation is twice that for a single pass, not zero. This property is exploited in magneto-optic isolators. If the length l is chosen so that $\theta_F = \pi/4$, light entering through a polarizer passes through the isolator in one direction, but reflected light is blocked. Some values of θ_F per unit path length are given in Table 5.12. The Verdet constants are 1000 times greater for ferromagnetic metals than for transparent ferrimagnetic insulators, but the comparison is misleading because metals are opaque unless they are very thin. A better figure of merit is the product of k_V and the absorption length, which is a few degrees per tesla for both. The Faraday rotation is strongly wavelength dependent, and it is enhanced near an absorption edge.

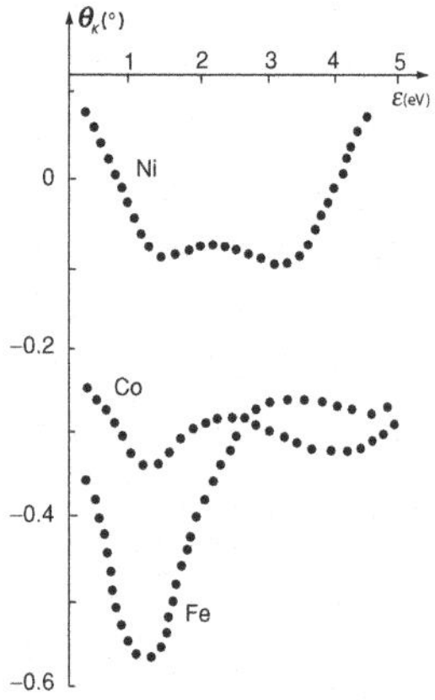

Kerr rotation as a function of photon energy for Fe, Co, Ni.

A related effect for reflected light was discovered by John Kerr in 1877. He noticed that the plane of polarization of light reflected from the polished iron pole face of an electromagnet was rotated by less than 1°. The sense of the rotation again switched when the magnetization reversed. The Kerr rotation for $3d$ ferromagnets shows a maximum in the optical region at about 1.5 eV (Table 5.13).

The polar configuration is one of three where a magneto-optic Kerr effect (MOKE) can be observed. In the other two, the magnetization lies in-plane, either transverse to the plane containing the incident and reflected rays, or longitudinal within that plane (Fig. 5.43). In the longitudinal configuration, there is also a rotation of the polarization of the incident ray by less than a degree, and a slight ellipticity of the reflected beam, but in the transverse configuration there is just a difference in reflectivity for light polarized in-plane or perpendicular to the plane of incidence, which depends on the magnetization of the sample.

Table 5.13. Kerr rotation in ferromagnetic metals at 830 nm (1.5 eV)

	θ_K (°)		θ_K (°)
Fe	−0.53	CoPd	−0.17
FePt	−0.39	CoPt	−0.36
FeCo	−0.60	Ni	−0.09
Co	−0.36	PtMnSb	−1.3

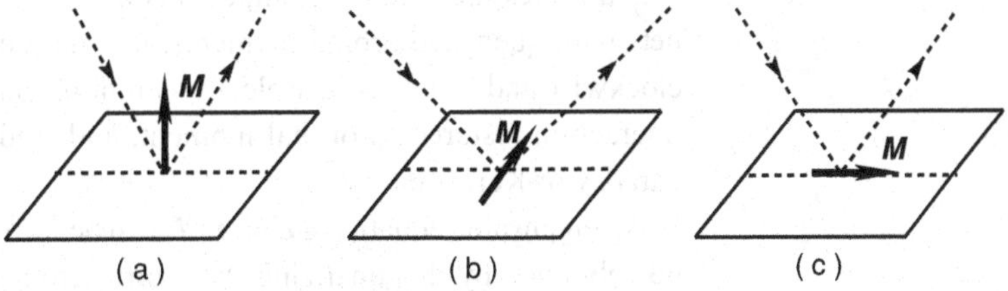

Figure 5.43

Three configurations in which the magneto-optic Kerr or Faraday effects can be observed: (a) polar, (b) transverse, (c) longitudinal.

For the polar and longitudinal effects, there is a small, magnetization-sensitive component k_K at right angles to the reflected intensity r_K which leads to the Kerr rotation θ_K and ellipticity e_K. When $k_K \ll r_K$, θ_K and e_K are given by the expressions

$$\theta_K = \psi_K \cos\phi_K \quad \text{and} \quad e_K = \psi_K \sin\phi_K,$$

where $\psi_K = |K/r|$, and ϕ_K is the phase difference between K and r. The polar Kerr effect finds applications in measuring the hysteresis of thin films, domain imaging and magneto-optic recording.

Both Faraday and Kerr effects are related to spin-orbit coupling. They may be enhanced by incorporating heavy elements such as bismuth into the crystal lattice. Magnetic permeability is not involved because the permeability is frequency-independent and equal to μ_0 at optical frequencies, which are orders of magnitude greater than the Larmor precession or ferromagnetic resonance frequency. The speed of light in a medium, v, just depends on its electrical permittivity ϵ: $v = (\mu_0\epsilon)^{-1/2}$, where $\epsilon = \epsilon_0\epsilon_r$.

The permittivity of a crystal is generally a symmetric tensor defined by $D_i = \epsilon_{ij}E_j$. It is simplified in the principal axis system to one with only diagonal elements, $\epsilon_{xx}, \epsilon_{yy}, \epsilon_{zz}$. When the crystal is isotropic, all three are equal. However, if the crystal is magnetized along Oz the direction of propagation of the light, the magneto-optic permittivity tensor becomes a Hermitian, skew-symmetric matrix if spin-orbit coupling is present:

$$\hat{\epsilon} = \epsilon_0\epsilon_r \begin{pmatrix} 1 & \mathrm{i}Q & 0 \\ -\mathrm{i}Q & 1 & 0 \\ 0 & 0 & 1 \end{pmatrix}. \tag{5.86}$$

This is an example of Onsager's principle for the symmetry of a response function $\epsilon_{ij}(M) = -\epsilon_{ji}(M) = \epsilon_{ji}(-M)$. Both the index of refraction $n = \epsilon_r^{1/2}$ and the magneto-optic parameter Q have real and imaginary parts:

$$n = n' - \mathrm{i}n'' \quad \text{and} \quad Q = Q' - \mathrm{i}Q'',$$

which are complex constants of the material that define its magneto-optic properties.

An interpretation of the off-diagonal matrix elements is that the magnetic field $\boldsymbol{H}$ in the electromagnetic wave gives rise to a Lorentz force $-\mu_0 e(\boldsymbol{v} \times \boldsymbol{H})$ which mixes the x and y components of the electron's motion. In ferromagnets with quenched orbital moments, it is not possible to distinguish between clockwise and anticlockwise electron motion, and $\epsilon_{xy} = 0$. However, spin-orbit interaction restores an orbital moment, and leads to magnetic dichroism and Faraday or Kerr rotation.

An eigenvalue equation $\hat{\epsilon}\boldsymbol{E} = n_0^2\boldsymbol{E}$ can be formed from (5.86). The eigenvalues obtained by diagonalizing the matrix are $n = \epsilon_r^{1/2}(1 \pm Q)^{1/2} \simeq \epsilon_r^{1/2}(1 \pm \frac{1}{2}Q)$ since Q is a small quantity, of order 10^{-3} or 10^{-4}. The two eigenmodes are $\begin{bmatrix} 1 \\ \mathrm{i} \end{bmatrix}$ and $\begin{bmatrix} 1 \\ -\mathrm{i} \end{bmatrix}$ or $E_x \pm \mathrm{i}E_y$, which corresopond to left and right circular polarization of the electric field. These modes, denoted σ^+ and σ^-, experience slightly different refractive indices, and therefore propagate at slightly different speeds $c/n = c\epsilon_r^{-1/2}(1 \pm \frac{1}{2}Q)^{-1}$. This is an effect known as circular birefringence and the resulting phase difference between the modes causes the Faraday rotation.

The two fundamental modes σ^+ and σ^- correspond to the quanta of electromagnetic radiation, known as photons, which are massless particles with $\ell = 1$ and $m_\ell = \pm 1$ (but not zero). Photons carry angular momentum, which can take two values, $\pm\hbar$, when projected along the direction of propagation. The polarization of the left- and right-circularly polarized beams rotate clockwise and anticlockwise respectively, when viewing the source. When the light is absorbed by the electron system of a solid, the photons can transmit their angular momentum to the electrons, thereby changing the projection of the magnetic moment along Oz.

When $\boldsymbol{K} \| \boldsymbol{M}$, the orbital contribution to the magnetization involves electrons circulating in the same sense as $\boldsymbol{E}$ for one mode and in the opposite sense for the other. The two modes see slightly different relative permittivities $\epsilon_r^{+,-}$ and refractive indices $n^{+,-}$, where $\epsilon_r = n^2$. The absorption is also slightly different for the two modes, an effect known as circular dichroism which leads to an elliptical polarization of the rectilinearly polarized incident light.

A summary of the four main magneto-optic phenomena is provided in Table 5.14. Besides the Faraday and Kerr effects, linear in M, which depend on spin-orbit coupling, there are also second-order magneto-optic effects varying as M^2 which depend on whether the plane of polarization of the light lies parallel or perpendicular to the magnetization, which is in the xy-plane.

Table 5.14. Magneto-optic phenomena

Geometry	Physical effect	Phenomenon	Condition	M dependence
$\boldsymbol{M} \parallel \boldsymbol{K}$	Circular birefringence	Faraday/Kerr rotation	$n^+ \neq n^-$	$\sim M$
	Circular dichroism	Magnetic circular dichroism	$\alpha^+ \neq \alpha^-$	$\sim M$
$\boldsymbol{M} \perp \boldsymbol{K}$	Linear birefringence	Cotton-Mouton effect	$n^\parallel \neq n^\perp$	$\sim M^2$
	Linear dichroism	Magnetic linear dichroism	$\alpha^\parallel \neq \alpha^\perp$	$\sim M^2$

Magneto-optic effects are not restricted to the optical frequency range, but exist also in the microwave, ultraviolet and X-ray ranges. Magnetic permeability is important at microwave frequencies. Magnetic circular dichroism (MCD) is enhanced near X-ray absorption edges, especially the L-edges of transition metals. It is a useful technique for investigating magnetic materials because it is element specific, and capable of separating the spin and orbital contributions to the magnetic moment (see §10.4.2).

Finally, we mention nonlinear optical effects which appear as a result of the intense electric fields in pulsed laser beams. These effects are forbidden in centrosymmetric crystals. At surfaces, there is no centre of symmetry, and second harmonic generation is a particularly sensitive method for selectively studying the magnetization of ferromagnetic surfaces and interfaces in reflection.

FURTHER READING

J. S. Smart, *Effective Field Theories of Magnetism,* Philadelphia: Saunders (1966). A concise summary of molecular field theory of ferromagnets, antiferromagnets and ferrimagnets.

D. C. Matthis, *The Theory of Magnetism Made Simple,* Singapore: World Scientific (2006). The retitled new edition of an accessible theoretical text, first published in 1965.

A. K. Zvedin and V. A. Kotov, *Modern Magnetoptics and Magneto-optical Materials,* Bristol: IOP (1997).

A. M. Tishin and Y. I. Spichkin, *Magnetocaloric Effect and Its Applications*, Bristol: IOP (2004).

P. Mohn, *Magnetism in the Solid State; Introduction,* Berlin: Springer (2005). Focussed on itinerant electron ferromagnetism.

E. du Tremolet de Lachesserie, *Magnetostriction; Theory and Applications of Magnetoelasticity*, Boston: CRC Press (1993). Detailed theory of magnetostriction.

A. del Morel, *Handbook of Magnetostriction and Magnetostrictive Materials,* 2 volumes, Zaragoza: del Moral (2008). A new and exhaustive treatment of the subject.

R. C. O'Handley, *Modern Magnetic Materials; Principles and Applications*, New York: Wiley (1999). A thorough treatment of modern materials and applications, especially metal, alloys and thin films.

M. Knober, M. Vásquez, and L. Kraus, Giant magnetoimpedance, in *Handbook of Magnetic Materials* (K. H. J. Buschow, editor). Amsterdam: North Holland, Vol 15, p. 497–564.

EXERCISES

5.1 Calculate the heat capacity for a ferromagnet near T_C in Landau theory.

5.2 Estimate the Curie temperature of gadolinium from the RKKY model. Take $S = 7/2$ and $n = 3 \times 10^{28}$ m^{-3}.

5.3 Use the magnetic valence model to estimate:
(a) the average moment/atom in an $Fe_{40}Ni_{30}B_{20}$ alloy and in $Co_{80}Cr_{20}$;
(b) the critical concentration for the appearance of magnetism in amorphous $Y_{1-x}Fe$.

5.4 Diagonalize the matrix (5.39). Rank the eigenvalues in order and identify the states to which they correspond.

5.5 Use (5.24) and (5.57) to deduce the result that the ground state for a chain of spins has eigenvalue $-2(N-1)\mathcal{J}S^2$.

5.6 Show that the low-temperature heat capacity due to excitations which obey a dispersion relation $\varepsilon = Dq^2$ varies as $T^{3/2}$ for a three-dimensional solid. How is the heat capacity modified by an anisotropy gap Δ at $q = 0$?

5.7 By minimizing the energy of a uniaxial ferromagnet with two anisotropy constants K_1 and K_2, show how to deduce the two constants from the magnetization curve in the hard direction (Sucksmith–Thompson plot).

5.8 Explain why it is impossible to produce a permanent magnet of arbitrary shape using shape anisotropy alone.

5.9 Show explicitly that the quadrupole moment of a rare-earth ion vanishes at high temperatures. Is it possible to observe anisotropy above the Curie point?

5.10 Deduce the $\kappa_j(T)/\kappa_j(0) = M^{j(j+1)/2}$ law for $j = 2, 4$ (see H. B. Callen and E. Callen; *J. Phys. Chem. Solids*, **27** 1271 (1966)).

5.11 Show that the spontaneous volume magnetostriction varies as M_s^2. The associated energy is $\frac{1}{2}K\omega_s^2$, where K is the bulk modulus.

5.12 Discuss the effect on (a) the magnetization and (b) the resistivity of introducing 5% of Cr into a Ni host.

5.13 Devise a mechanical experiment to show that light carries angular momentum, making an estimate of the magnitude of the effect that might be observed.

5.14 The magnetization calculated from Langevin theory has nonzero slope at $T = 0$ K. According to (2.102) this implies that the entropy is finite, which is incompatible with the third law of thermodynamics. Discuss.

6 Antiferromagnetism and other magnetic order

Je prefere explorer les forêts vierges que cultiver un jardin de curé.
Louis Néel

Negative exchange $\mathcal{J} < 0$ leads to magnetic order that depends on lattice topology. Structures with more than one magnetic sublattice include antiferromagnets and ferrimagnets. An antiferromagnet has two equal but oppositely directed sublattices, where the sublattice magnetization disappears above the Néel point T_N. Two unequal oppositely directed magnetic lattices constitute a ferrimagnet. The molecular field theory is extended to cover these cases. A wealth of more complex noncollinear magnetic structures exist. The subtle effects of a noncrystalline structure are manifest in amorphous magnets, where spins sometimes freeze in random orientations. Magnetic model systems highlight the influence of some particular feature on collective magnetic order, such as reduced space or spin dimensionality, a particular distribution of exchange interactions, special topology or lack of crystal structure. Examples include the two-dimensional Ising model, frustrated antiferromagnets and canonical spin glasses.

Antiferromagnetism is an occult magnetic order. A crystal lattice is subdivided into two or more atomic sublattices which order in such a way that their net magnetization is zero. Louis Néel, who was a student of Pierre Weiss, first discussed this possibility in 1936 for two equal and oppositely aligned sublattices. The antiferromagnetic ordering transition, known as the Néel point, is marked by a small peak in the magnetic susceptibility, and a substantial specific heat anomaly, similar to that found at the Curie point of a ferromagnet (Fig. 5.36). It was only with the advent of neutron scattering in the 1950s that it became possible to measure the sublattice magnetization M_α directly. Sublattice magnetization is the order parameter, and its conjugate field (which we cannot produce in the laboratory, although we can measure the generalized susceptibility $\chi(\boldsymbol{K}, \Omega)$ by neutron scattering) is a staggered field which alternates in direction from one atomic site to the next, in at least one direction.

Data on a few well-known antiferromagnets are collected in Table 6.1. Transition-metal oxides and fluorides are frequently antiferromagnetic, as are Cr, Mn and many of their alloys.

Table 6.1. Some common antiferromagnets

	Structure	T_N (K)	θ_p (K)	$\mu_0 M_\alpha$ (T)
Cr	sdw	311		0.20
Mn	Complex	96	~ -2000	0.20
NiO	Néel	524	-1310	0.54
αFe_2O_3	Canted	958	-2000	0.92
MnF_2	Néel	67	-80	0.78
FeMn	Néel	510		0.53
$IrMn_3$	Néel	690		0.50

sdw – spin density wave; Néel – two collinear sublattices.

6.1 Molecular field theory of antiferromagnetism

Néel antiferromagnets have two equal and oppositely directed magnetic sublattices, designated 'A' and 'B', with sublattice magnetizations $\boldsymbol{M}_A = -\boldsymbol{M}_B$. The negative Weiss coefficient n_{AB} represents the intersublattice molecular field coupling, and a further coefficient n_{AA} may be included to account for intrasublattice molecular field interactions.

The 'molecular' fields acting on each sublattice are

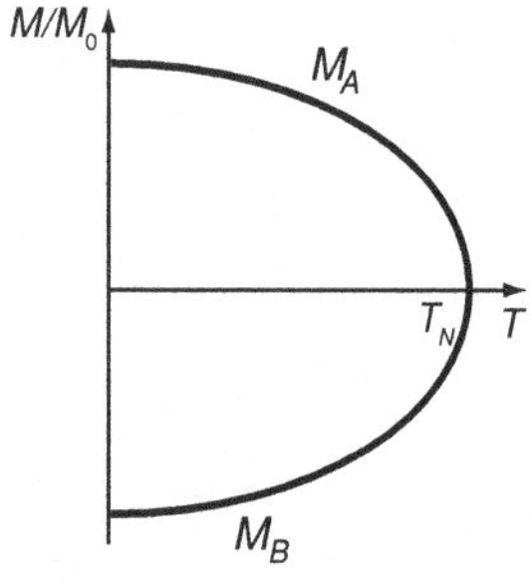

Sublattice magnetization of an antiferromagnet. T_N is the Néel temperature.

$$\begin{aligned} \boldsymbol{H}^i_A &= n_{AA}\boldsymbol{M}_A + n_{AB}\boldsymbol{M}_B + \boldsymbol{H}, \\ \boldsymbol{H}^i_B &= n_{BA}\boldsymbol{M}_A + n_{BB}\boldsymbol{M}_B + \boldsymbol{H}, \end{aligned} \tag{6.1}$$

where $n_{AA} = n_{BB}$, $n_{AB} = n_{BA}$ and $\boldsymbol{H}$ is the contribution from an externally applied field.

The net magnetization $\boldsymbol{M} = \boldsymbol{M}_A + \boldsymbol{M}_B$ is zero when $\boldsymbol{H} = 0$. The magnetization of each sublattice falls to zero at the Néel point T_N, and the spontaneous magnetization of each of them is represented by a Brillouin function (4.17):

$$M_\alpha = M_{\alpha 0}\mathcal{B}_J(x_\alpha), \tag{6.2}$$

where $\alpha = A, B$ and $x_\alpha = \mu_0 \mathfrak{m}|H^i_\alpha|/k_B T$. Here $M_{A0} = -M_{B0} = (n/2)g\mu_B J = (n/2)\mathfrak{m}$. The number of magnetic ions per unit volume is n, with $n/2$ on each sublattice.

In the paramagnetic region above T_N, $M_\alpha = \chi H^i_\alpha$, where $\chi = C'/T$ with $C' = \mu_0(n/2)\mathfrak{m}^2_{eff}/3k_B$. Hence

$$\begin{aligned} \boldsymbol{M}_A &= (C'/T)(n_{AA}\boldsymbol{M}_A + n_{AB}\boldsymbol{M}_B + \boldsymbol{H}), \\ \boldsymbol{M}_B &= (C'/T)(n_{BA}\boldsymbol{M}_A + n_{BB}\boldsymbol{M}_B + \boldsymbol{H}). \end{aligned} \tag{6.3}$$

The condition for the appearance of spontaneous sublattice magnetization is that these equations have a nonzero solution in zero applied field. The determinant of the coefficients of M_A and M_B must be zero, hence

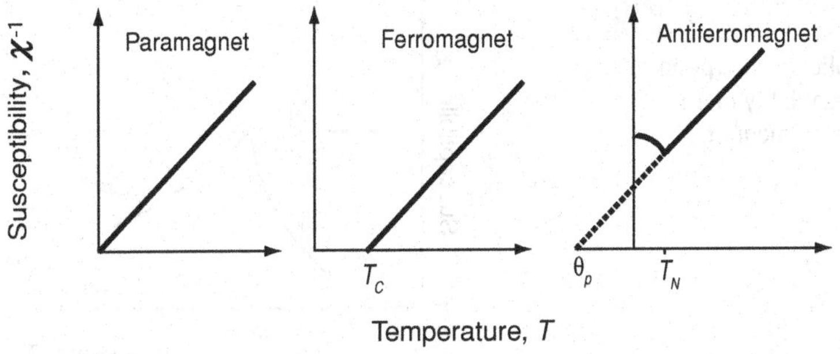

Figure 6.1

Comparison of the inverse susceptibility of a paramagnet, a ferromagnet and an antiferromagnet.

$[(C'/T)n_{AA} - 1]^2 - [(C'/T)n_{AB}]^2 = 0$, which yields the Néel temperature:

$$T_N = C'(n_{AA} - n_{AB}). \tag{6.4}$$

n_{AB} is negative and T_N is positive, so the negative sign is chosen in the square root. To calculate the susceptibility above T_N we evaluate $\chi = (M_A + M_B)/H$. Adding Eq. (6.3) gives the Curie–Weiss law

$$\chi = C/(T - \theta_p), \tag{6.5}$$

where $C = 2C'$ and the paramagnetic Curie temperature is given by

$$\theta_p = C'(n_{AA} + n_{AB}). \tag{6.6}$$

The inverse susceptibility of an antiferromagnet is shown schematically in Fig. 6.1. In the two-sublattice model, we can therefore evaluate both n_{AB} and n_{AA} from a knowledge of T_N and θ_p. Since $n_{AB} < 0$, it follows that $\theta_p < T_N$; the paramagnetic Curie temperature is usually negative. Normally $1/\chi$ is plotted versus T to determine θ_p by extrapolation, and m_{eff} is obtained from the slope using (4.16). T_N is marked only by a small cusp in the susceptibility.

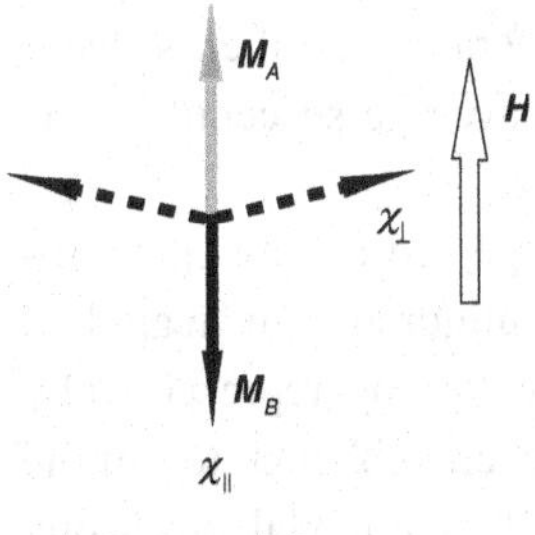

Calculation of the susceptibility of an antiferromagnet well below T_N. The dashed lines show the configuration after a spin flop.

The antiferromagnetic axis along which the two sublattice magnetizations lie is determined by magnetocrystalline anisotropy, and the magnetic response below T_N depends on the direction of $\boldsymbol{H}$ relative to this axis. There is no shape anisotropy in an antiferromagnet, because there can be no demagnetizing field when $M = 0$. It might be thought that there would be no antiferromagnetic domains for the same reason, but entropy can drive domain formation at finite temperature.

If a small field is applied *parallel* to the antiferromagnetic axis, we can calculate the parallel susceptibility $\chi_\parallel$ by expanding the Brillouin functions about their argument x_0 in zero applied field in terms of the derivative $\mathcal{B}'_J(x) = \partial\mathcal{B}_J(x)/\partial x$: $\mathcal{B}_J(x_0 + \delta x) = \mathcal{B}_J(x_0) + \delta x \mathcal{B}'_J(x_0)$. For simplicity we set $n_{AA} = 0$. The result for $\chi_\parallel = [M_A(H) + M_B(H)]/H$ is

$$\chi_\parallel = \frac{2C'[3J/(J+1)]\mathcal{B}'_J(x_0)}{T - n_{AB}C'[3J/(J+1)]\mathcal{B}'_J(x_0)}, \tag{6.7}$$

where x_0 is $-\mu_0 \mathrm{m} n_{AB} M_\alpha / k_B T$, $|\boldsymbol{M}_A| = |\mathbf{M}_B| = ng\mu_B J/2$ and $\mathrm{m}^2 = g^2\mu_B^2 J(J+1)$. As $T \to 0$ K, it can be seen from (6.7) that $\chi_\parallel \longrightarrow 0$, so

Figure 6.2

Parallel and perpendicular susceptibility of an antiferromagnet.

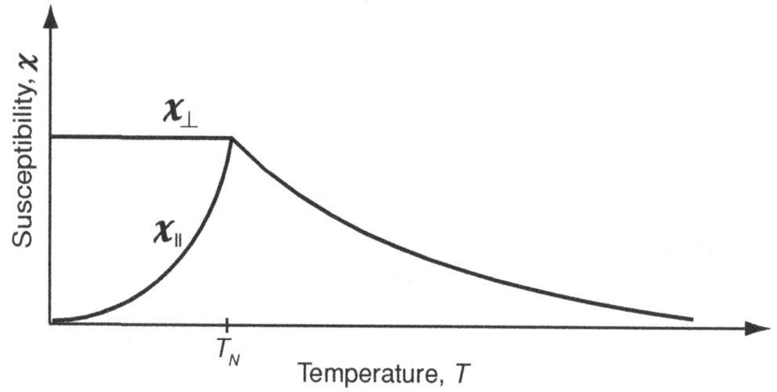

the field has no effect because the two sublattices are saturated, $\mathcal{B}'_J(x_0) = 0$. As T increases, M_A and M_B are reduced by thermal fluctuations, and the susceptibility rises from zero at $T = 0$ K to reach the paramagnetic value (6.5) at T_N. When $M_\alpha = 0$ and $\mathcal{B}'_J(0) = (J+1)/3J$ the susceptibility adopts the Curie–Weiss form, since $T_N = -\theta_p = -C' n_{AB}$. The susceptibility reaches its greatest value $\chi_{\parallel} = -1/n_{AB}$ at T_N. The stronger the antiferromagnetic interaction, the lower the maximum susceptibility.

The *perpendicular* susceptibility can be calculated assuming the sublattices are canted by a small angle δ. In equilibrium the *torque* on each one is zero, hence $M_A H = M_A n_{AB} M_B \sin 2\delta$. Since $M_\perp = 2M_\alpha \sin\delta$,

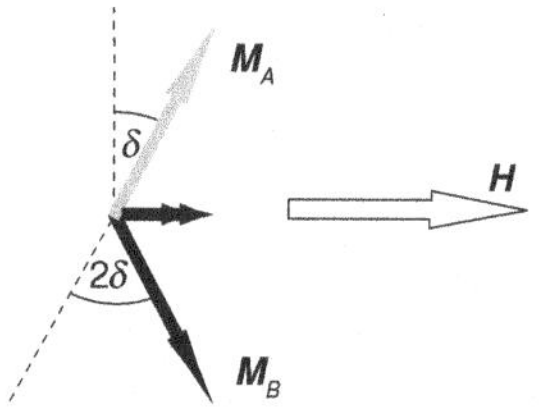

Calculation of perpendicular susceptibility.

$$\chi_\perp = -1/n_{AB}. \tag{6.8}$$

The perpendicular susceptibility is therefore constant, independent of temperature up to T_N, Fig. 6.2. The exchange parameter $|n_{AB}|$ is greater than 100 for materials that are antiferromagnetic at room temperature, so their susceptibility is typically of order 10^{-2}. For a powder, the average susceptibility is $\frac{1}{3}\chi_{\parallel} + \frac{2}{3}\chi_\perp$, or $2/(3n_{AB})$ at low temperatures.

Since $\chi_\perp > \chi_{\parallel}$ for all $T < T_N$, one might have expected that an antiferromagnet would *always* adopt the transverse, flopped configuration in an applied field. That it does not is due to magnetocrystalline anisotropy, represented by an effective anisotropy field (§5.5.2), which acts on each sublattice to pin the magnetization along the easy antiferromagnetic axis. If the uniaxial anisotropy constant is K_1, the anisotropy field H_a is equal to $K_1/\mu_0 M_\alpha$. When $\boldsymbol{H}$ is applied parallel to $\boldsymbol{M}_\alpha$ a spin flop occurs in the field where the energies of the parallel and perpendicular configurations are equal:

$$-2M_\alpha H_a - \tfrac{1}{2}\chi_{\parallel} H_{sf}^2 = -\tfrac{1}{2}\chi_\perp H_{sf}^2$$

$$H_{sf} = [4M_\alpha H_a/(\chi_\perp - \chi_{\parallel})]^{\frac{1}{2}}. \tag{6.9}$$

When $T \ll T_N$, from (6.2), (6.7) and (6.8) this reduces to $H_{sf} = 2(H_a H_\alpha^i)^{\frac{1}{2}}$. Taking as orders of magnitude for the anisotropy field $\mu_0 H_a$ and the molecular field $\mu_0 H^i$, 1 T and 100 T respectively, the spin-flop field $\mu_0 H_{sf}$ is of order 10 T. Further increase of the applied field leads to saturation when $H = H_\alpha^i$. Low-temperature antiferromagnets with weak intrasublattice interactions which

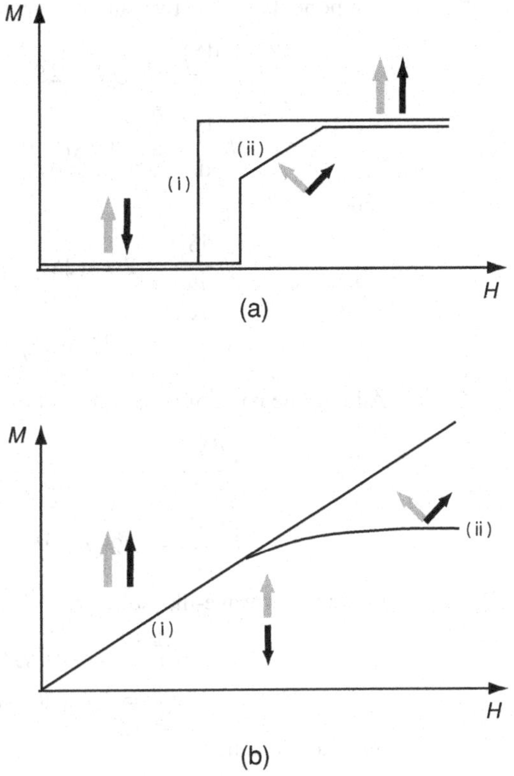

Figure 6.3

(a) Magnetization of an antiferromagnet as a function of applied magnetic field, showing: (i) a metamagnetic transition and (ii) a spin-flop transition. (b) The phase diagram for the antiferromagnet.

undergo a transition directly to the saturated ferromagnetic state when $H > H_a$ are known as metamagnets, Fig. 6.3.

6.1.1 Artificial antiferromagnets

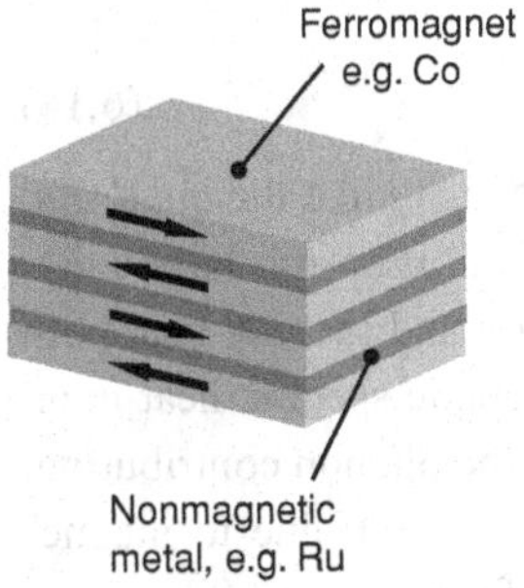

An artificial antiferromagnet. The nonmagnetic coupling layer is shaded dark grey. The simplest structure consists of just three layers.

Artificial antiferromagnets are thin-film stacks with alternating layers of a ferromagnet such as cobalt and a nonmagnetic metal such as copper or ruthenium which serves to provide magnetic coupling from one ferromagnetic layer to the next. The structure can be reduced to a three-layer sandwich. Now M_α is the magnetization of the ferromagnetic layer, and $\mu_0 H^i_\alpha$ represents the interlayer coupling. Typically $\mu_0 H^i_\alpha$ and $\mu_0 H_a$ are of order 1 T, so the spin-flop field $\mu_0 H_{sf}$ is similar in magnitude. Artificial antiferromagnets are used in spin-valve structures, discussed in Chapter 8.

6.1.2 Spin waves

The spin waves in an antiferromagnet can be calculated by considering a two-sublattice chain where the atoms are numbered sequentially and writing the equations of motion for each of the two sublattices. Following the procedure of §5.4.1, we obtain the following equations of motion for the transverse spin components for the two sublattices:

$$\hbar \frac{\mathrm{d}S_j^x}{\mathrm{d}t} = 2\mathcal{J}S(-2S_j^y - S_{j-1}^y - S_{j+1}^y) \qquad j \text{ odd} \tag{6.10}$$

$$\hbar \frac{\mathrm{d}S_j^y}{\mathrm{d}t} = -2\mathcal{J}S(-2S_j^x - S_{j-1}^x - S_{j+1}^x) \tag{6.11}$$

and

$$\hbar \frac{\mathrm{d}S_j^x}{\mathrm{d}t} = 2\mathcal{J}S(2S_j^y + S_{j-1}^y + S_{j+1}^y) \qquad j \text{ even} \tag{6.12}$$

$$\hbar \frac{\mathrm{d}S_j^y}{\mathrm{d}t} = -2\mathcal{J}S(2S_j^x + S_{j-1}^x + S_{j+1}^x). \tag{6.13}$$

Adding the pairs of equations and using the definitions of $S^{\pm}$ (§3.1.4),

$$\hbar \frac{\mathrm{d}S_j^+}{\mathrm{d}t} = 2\mathrm{i}\mathcal{J}S[2S_j^+ + S_{j-1}^+ + S_{j+1}^+] \quad \text{for } j \text{ odd},$$

$$\hbar \frac{\mathrm{d}S_j^+}{\mathrm{d}t} = -2\mathrm{i}\mathcal{J}S[2S_j^+ + S_{j-1}^+ + S_{j+1}^+] \text{ for } j \text{ even}.$$

Looking for wave-like solutions

$$S_j^+ = A \exp \mathrm{i}(qja - \omega_q t) \quad \text{for } j \text{ odd},$$

$$S_j^+ = B \exp \mathrm{i}(qja - \omega_q t) \quad \text{for } j \text{ even}$$

leads to the equations

$$-\mathrm{i}\hbar\omega_q A = 4\mathrm{i}\mathcal{J}S[A + B\cos qa],$$

$$-\mathrm{i}\hbar\omega_q B = -4\mathrm{i}\mathcal{J}S[A\cos qa + B].$$

Equating the determinant of the coefficients to zero then yields the antiferromagnetic spin-wave dispersion relation:

$$\hbar\omega_q = 4\mathcal{J}S \sin qa. \tag{6.14}$$

The dispersion is *linear* for small values of q, $\hbar\omega_q \approx q$, unlike the *quadratic* dispersion for ferromagnetic magnons (5.56). In general, if bosons have a dispersion relation $\hbar\omega_q \approx q^n$ the sublattice magnetization and specific heat both vary as $T^{3/n}$ (Exercise 6.3). The T^3 variation of magnetic specific heat in an antiferromagnet is practically indistinguishable from the phonon contribution. As with ferromagnetic magnons, there is a gap Δ at $q = 0$ due to magnetocrystalline anisotropy. Magnon dispersion relations for haematite, a common antiferromagnet, are shown in Fig. 6.4.

6.2 Ferrimagnets

A ferrimagnet may be regarded as an antiferromagnet with two unequal sublattices. Most oxides which possess a net ordered magnetic moment are

Table 6.2. Some common ferrimagnets

	Sublattices	T_c (K)	T_{comp} (K)
Fe_3O_4	$8a;16d$	856	
YFe_5O_{12}	$16a;24d$	560	
$BaFe_{12}O_{19}$	$2a,2b,4f1,4f2,12k$	740	
$TbFe_2$	$8a,16d$	698	
$GdCo_5$	$1a,2c,3g$	1014	287

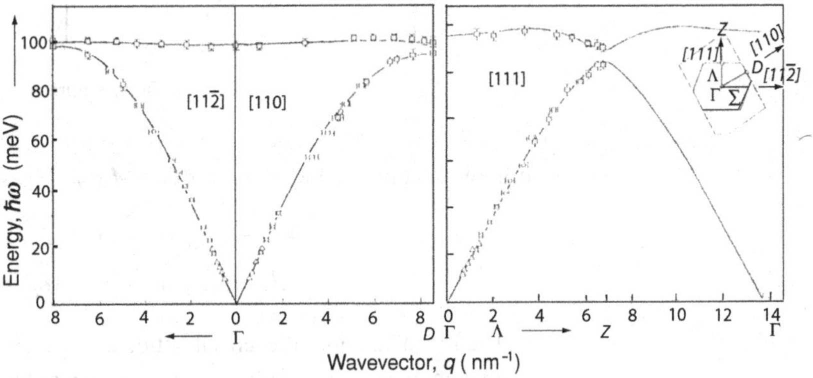

Figure 6.4

Antiferromagnetic spin-wave dispersion relations in αFe_2O_3 along different directions in the Brillouin zone, determined by inelastic neutron scattering. (E. J. Samuelsen and G. Shirane, *Phys. Stat. Sol* b42, 241 (1970).)

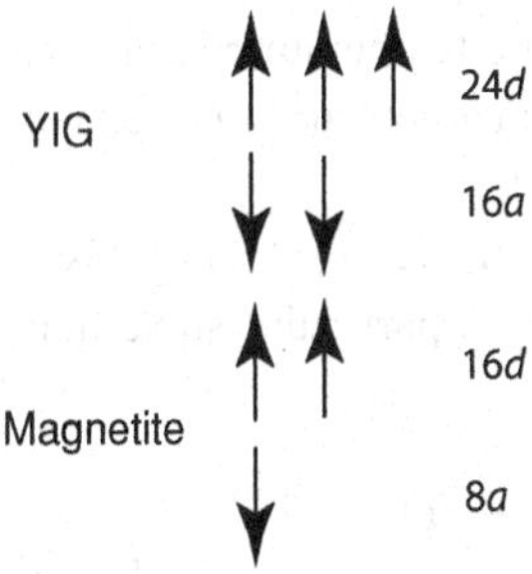

Ferrimagnetic configurations of YIG and magnetite.

ferrimagnets. Some of them are listed in Table 6.2. An example is yttrium–iron garnet (YIG) $Y_3Fe_5O_{12}$ (§11.6.6). The iron in YIG is ferric (Fe^{3+}, $3d^5$) and occupies two different crystallographic sites, one (*16a*) octahedrally coordinated by oxygen, the other (*24d*) tetrahedrally coordinated by oxygen. Neighbouring sites share a common oxygen ligand, and there is a strong antiferromagnetic *a–d* interaction. The ferrimagnetic configuration leads to a moment of $5\mu_B$ per formula at $T = 0$ K, due to the single uncompensated $3d^5$ ion per formula unit.

The most famous ferrimagnet of them all is magnetite (§11.6.4), the archetypical permanent magnet. Magnetite has the spinel structure with formula Fe_3O_4. Again there are two cation sites with different multiplicities in the unit cell, an *8a* tetrahedral site (A site) and a *16d* octahedral site (B site). The iron in magnetite is a 2:1 ferric:ferrous mixture to ensure charge neutrality with the O^{2-}. The A site is occupied by ferric iron, Fe^{3+}, and the B site by an equal mixture of ferric and ferrous, Fe^{3+} and Fe^{2+}. The ferrimagnetic configuration leads to an uncompensated ferrous (Fe^{2+}, $3d^4$) spin-only moment of $4\mu_B$ per formula at $T = 0$ K.

Labelling the two unequal and oppositely directed magnetic sublattices 'A' and 'B', the net magnetization $\boldsymbol{M} = \boldsymbol{M}_A + \boldsymbol{M}_B$ is nonzero. Three different Weiss coefficients n_{AA}, n_{BB} and n_{AB} are needed to represent the inter- and intrasublattice interactions. The essential interaction, n_{AB}, is negative. The

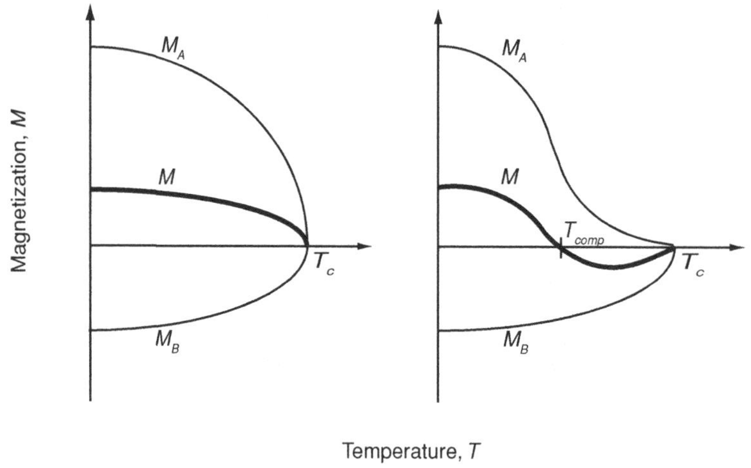

Figure 6.5

Sublattice magnetization of a ferrimagnet. T_c is the ferrimagnetic Néel temperature. On the left, $|n_{AB}| \gg |n_{AA}|, |n_{BB}|$, and on the right, $|n_{BB}| \gg |n_{AB}|, |n_{AA}|$. T_{comp} is the compensation temperature where the net magnetization is zero.

difference compared with (6.1) is that $\boldsymbol{M}_A \neq \boldsymbol{M}_B$ and $n_{AA} \neq n_{BB}$:

$$\begin{aligned} \boldsymbol{H}^i_A &= n_{AA}\boldsymbol{M}_A + n_{AB}\boldsymbol{M}_B + \boldsymbol{H}, \\ \boldsymbol{H}^i_B &= n_{BA}\boldsymbol{M}_A + n_{BB}\boldsymbol{M}_B + \boldsymbol{H}. \end{aligned} \tag{6.15}$$

The magnetization of each sublattice is represented by a Brillouin function, and when $H = 0$ each falls to zero at a critical temperature T_c, known as the ferrimagnetic Néel temperature. In some circumstances (Fig. 6.5) it is possible for the two sublattice magnetizations to cancel exactly at a temperature known as the compensation temperature, T_{comp}. The sublattice magnetization M_α equals $M_{\alpha 0}\mathcal{B}_J(x_\alpha)$, where $\alpha = A, B$ and $x_\alpha = \mu_0 \mathfrak{m}_\alpha H^i_\alpha / k_B T$.

Above T_c, $\boldsymbol{M}_\alpha = \chi_\alpha \boldsymbol{H}^i_\alpha$, where $\chi_\alpha = C_\alpha / T$ with $C_\alpha = \mu_0 n_\alpha m^2_{eff}/3k_B$. Here n_α is the number of atoms per cubic metre on a particular sublattice. Hence

$$\begin{aligned} \boldsymbol{M}_A &= (C_A/T)(n_{AA}\boldsymbol{M}_A + n_{AB}\boldsymbol{M}_B + \boldsymbol{H}), \\ \boldsymbol{M}_B &= (C_B/T)(n_{AB}\boldsymbol{M}_A + n_{BB}\boldsymbol{M}_B + \boldsymbol{H}). \end{aligned} \tag{6.16}$$

The condition for the appearance of spontaneous sublattice magnetization is that these equations have a nonzero solution in zero applied field. The determinant of the coefficients is zero, hence $[(C_A/T)n_{AA} - 1][(C_B/T)n_{BB} - 1] - (C_A C_B/T^2)n^2_{AB} = 0$, which yields

$$T_c = \tfrac{1}{2}\left[(C_A n_{AA} + C_B n_{BB}) + \sqrt{(C_A n_{AA} - C_B n_{BB})^2 + 4C_A C_B n^2_{AB}}\right]. \tag{6.17}$$

The expression for the susceptibility above T_c is obtained from (6.16):

$$\frac{1}{\chi} = \frac{T - \theta}{C_A + C_B} - \frac{C''}{T - \theta'}, \tag{6.18}$$

Figure 6.6

Inverse susceptibility of a ferrimagnet above its ferrimagnetic Néel point.

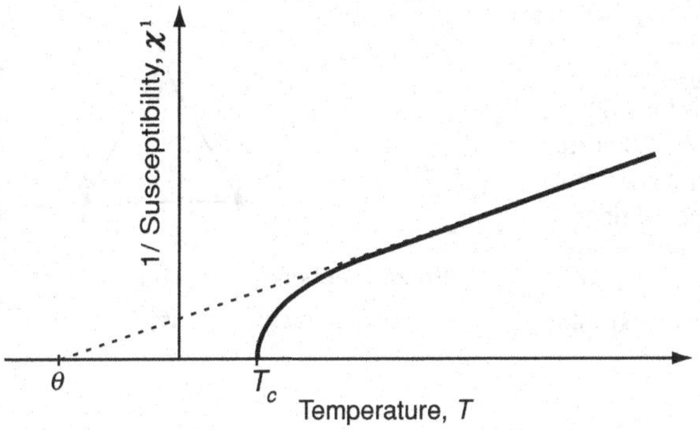

where

$$C'' = \left(\frac{C_A C_B n_{AB}^2}{C_A + C_B}\right)[C_A(1 + n_{AA}) - C_B(1 + n_{BB})]^2,$$

$$\theta = \left(\frac{C_A C_B n_{AB}}{C_A + C_B}\right)\left[n_{AA}\frac{C_A}{C_B} - n_{BB}\frac{C_B}{C_A} - 2\right],$$

$$\theta' = \left(\frac{C_A C_B n_{AB}}{C_A + C_B}\right)[n_{AA} + n_{BB} + 2].$$

Equation (6.18) is the equation of an hyperbola, as shown in Fig. 6.6.

6.3 Frustration

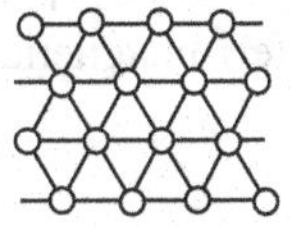

Triangular lattice

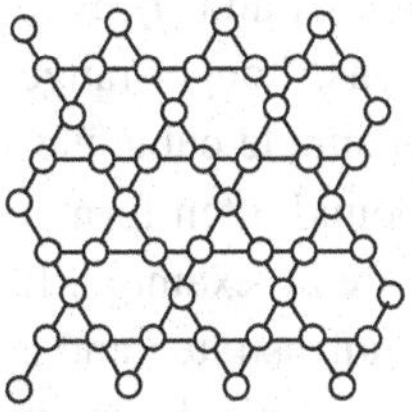

Kagomé lattice

Some frustrated two-dimensional antiferromagnetic lattices – the triangular and kagomé lattices.

Ferromagnetic interactions are satisfied by a parallel alignment of the atomic moments. Antiferromagnetic interactions may not be so easily appeased. In structures with odd-membered rings it is impossible to satisfy all the antiferromagnetic interactions simultaneously. A consequence is that $T_N \ll |\theta_p|$. Examples of crystal lattices where nearest-neighbour interactions are naturally frustrated include the triangular, kagomé, fcc and tetrahedral lattices. The 16*d* sites of the spinel structure form a tetrahedral lattice; each tetrahedron has four triangular faces.

As an illustration of frustration, consider the lowest-energy configurations for three-, four- and five-membered rings in Fig. 6.7. The exchange is supposed to be of the form $-2\mathcal{J}\boldsymbol{S}_i \cdot \boldsymbol{S}_j$ with either Ising spins, which have only one component ($\boldsymbol{S} = (0, \pm 1)$) or vector spins ($\boldsymbol{S} = (S_x, S_y)$, with $S_x^2 + S_y^2 = 1$).

Exchange in the odd-membered rings is frustrated, which means that the total exchange energy divided by the number of bonds is less than $\mathcal{J}$. Besides a low ordering temperature, the hallmarks of frustration are increased degeneracy of the ground state and a tendency to form noncollinear spin structures, epitomized by the three- and five-membered rings in Fig. 6.7(b).

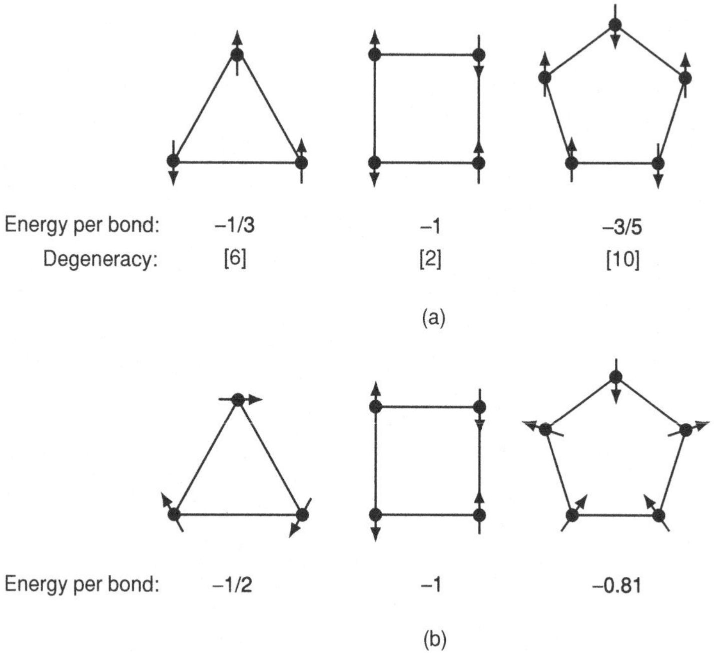

Figure 6.7

Lowest-energy configurations for (a) Ising and (b) classical 2D vector spins forming three-, four- or five-membered rings. Energy per bond in units of $\mathcal{J}$ and the degree of degeneracy for Ising spins are shown.

6.3.1 Cubic antiferromagnets

A feature of many crystal lattices is that there can be several different ways of constituting the two equal antiferromagnetic sublattices. Different spin configurations have different topology. The orientation of the spins relative to the crystal axes is a separate issue, determined by magnetocrystalline anisotropy. Heisenberg exchange is isotropic, so it imposes no particular antiferromagnetic axis.

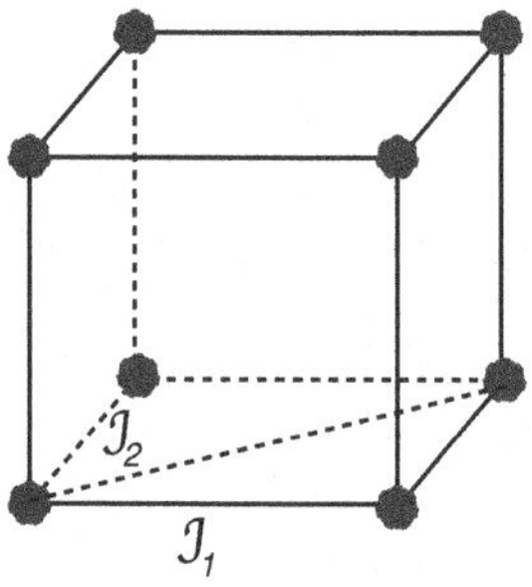

Magnetic interactions in a simple cubic lattice.

Simple cubic Four possible antiferromagnetic modes for a simple cubic lattice are shown in Fig. 6.8. Two possible superexchange paths $\mathcal{J}_1$ and $\mathcal{J}_2$ exist for nearest-neighbour and next-nearest-neighbour interactions. Longer-range interactions are possible, especially in metals. Although no simple cubic elements exist, except polonium, the magnetic ions in a compound often form a simple-cubic sublattice: the 'B' sites in perovskite, ABO_3 are an example. If $\mathcal{J}_1$ is the only antiferromagnetic interaction, the bonds are unfrustrated and a G mode is adopted, where all the six nearest neighbours of a particular atom lie on the opposite sublattice. The structure is composed of alternate ferromagnetic [111] planes. If $\mathcal{J}_2$ is the only antiferromagnetic interaction, it becomes impossible to satisfy all twelve next-nearest-neighbour bonds simultaneously, and the best solutions have eight of them on the opposite sublattice and four on the same sublattice, as in the A and C modes. There are four magnetic sublattices, and the equation for T_N involves a fourth-order determinant. The

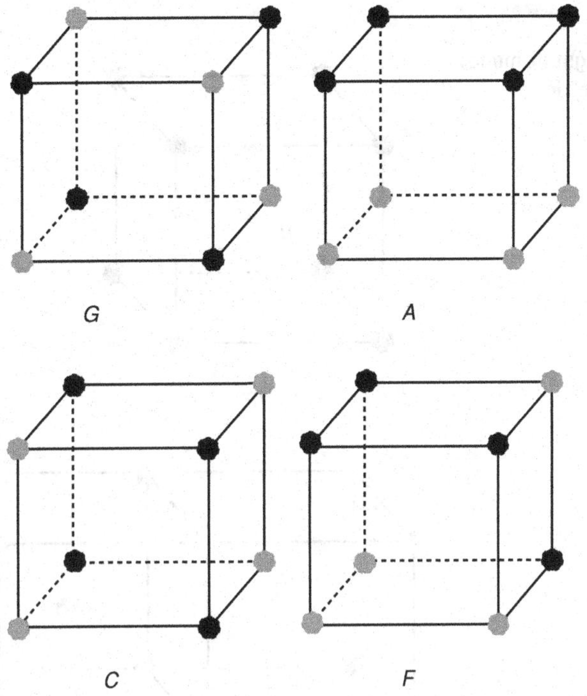

Figure 6.8

Antiferromagnetic modes for the simple cubic lattice. The two sublattices are represented by light and dark shading.

fourth, F, mode is ferromagnetic. More general magnetic structures can be generated by combining different modes with components along the Cartesian axes. Small distortions of the lattice may favour particular orbital occupancy for Jahn-Teller ions, which modifies orbital overlap, and hence the exchange.

Body-centred cubic Here, there exist unfrustrated structures where either nearest-neighbour interactions $\mathcal{J}_1$ or next-nearest-neighbour interactions $\mathcal{J}_2$ can be completely satisfied. In the latter case, there are two completely decoupled simple-cubic antiferromagnetic structures. The structures are known as type I and type II bcc order, respectively. If both antiferromagnetic interactions are present, conflicts arise which may have to be accommodated in a partly frustrated, compromise ground state. If the two exchange interactions $\mathcal{J}_1$ and $\mathcal{J}_2$ are represented by molecular field coefficients, the regions of stability of the ferromagnetic and antiferromagnetic structures are as shown in Fig. 6.9.

Face-centred cubic The best-known frustrated antiferromagnet has an fcc lattice. Here the nearest-neighbour exchange $\mathcal{J}_1$ is always frustrated, just like $\mathcal{J}_2$ in the simple-cubic lattice. The fcc lattice divides into four simple-cubic sublattices with magnetizations $\boldsymbol{M}_A$, $\boldsymbol{M}_B$, $\boldsymbol{M}_C$ and $\boldsymbol{M}_D$. Each atom has four nearest neighbours on each of the other three sublattices. At best two out of three neighbours of a given spin can be aligned antiparallel. In zero applied field, the molecular field equations are $\boldsymbol{M}_A = \boldsymbol{M}_0\mathcal{B}_J(x_A)$, where $x_A = \mu_0 m_A[n_{AA}\boldsymbol{M}_A + n_{AB}(\boldsymbol{M}_B + \boldsymbol{M}_C + \boldsymbol{M}_D) + \boldsymbol{H}]/k_BT$, etc.

Figure 6.9

Antiferromagnetic modes for the bcc lattice.

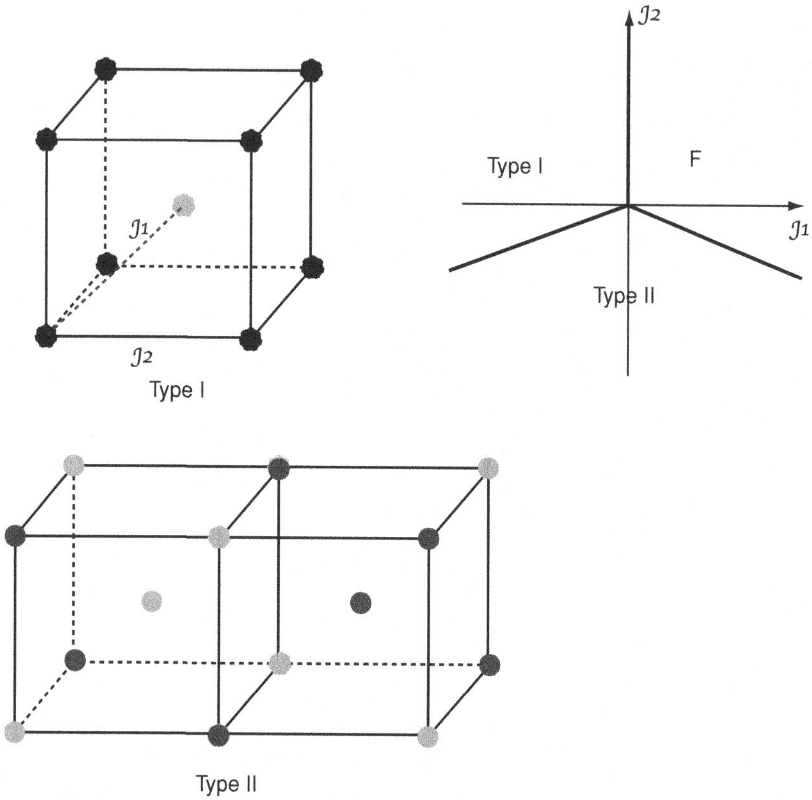

Considering only the intersublattice interaction n_{AB}, it follows that

$$T_N = C' n_{AB}, \quad \theta_p = 3C' n_{AB}, \tag{6.19}$$

where $C' = \mu_0 N_A \mathrm{m}_{\mathit{eff}}^2 / 3k_B$. Note that $\theta_p = 3T_N$. We can read the strength of the individual exchange interactions in extrapolating to zero the susceptibility measured at high temperature. The Néel temperature, however, reflects the extent to which it is possible to satisfy all the exchange bonds simultaneously. When the magnetic structure is known, it may be possible to deduce the two molecular field coefficients, and hence the exchange interactions $\mathcal{J}_1$ and $\mathcal{J}_2$ from knowledge of T_N and θ_p.

There are three possible magnetic modes for the fcc lattice, illustrated in Fig. 6.10. The structure with alternating ferromagnetic [001] planes is type I. That with a structure of alternating ferromagnetic [111] planes is type II. Type III is made up of alternating antiferromagnetic [001] planes. The transition-metal monoxides MnO, FeO, CoO and NiO all have type II order. $MnTe_2$ and MnS_2 are examples of type I and type III order, respectively. The regions of stability for different values of the first and second neighbour molecular field coefficients are shown in Fig. 6.10. A fourth type of order, consisting

Figure 6.10

Antiferromagnetic modes for the fcc lattice.

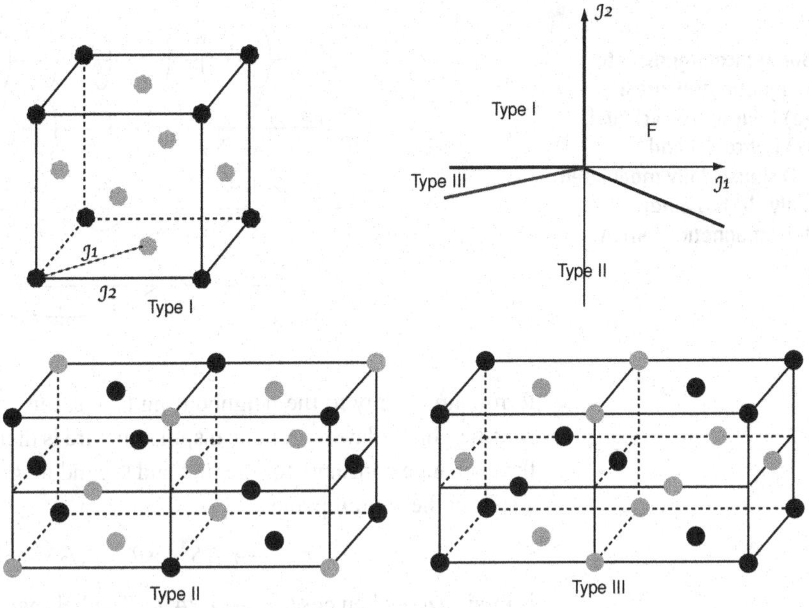

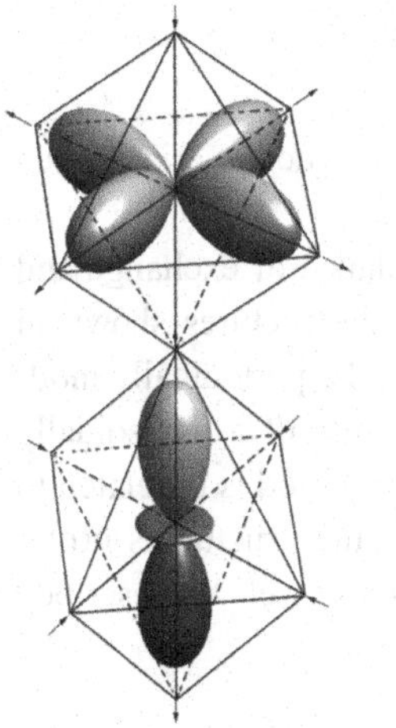

Correlated distortions of oxygen octahedra around d^4 or d^9 ions. The arrows indicate displacements of the oxygen ions.

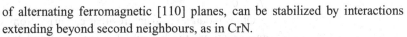

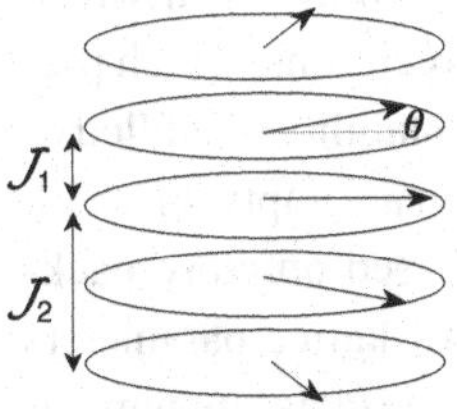

A planar helimagnet.

of alternating ferromagnetic [110] planes, can be stabilized by interactions extending beyond second neighbours, as in CrN.

Just as there can be no strictly cubic ferromagnet, because of magnetostriction, so there can be no strictly cubic antiferromagnet for the same reason. Magnetostriction produces a deformation of the structure along the antiferromagnetic axis. A slight tetragonal or rhombohedral distortion arises, depending on whether the axis is [100] or [111].

6.3.2 Orbital order

Cations with d-electrons in orbitally degenerate energy levels may exhibit electronic ordering effects that are formally equivalent to antiferromagnetism. Consider the example of Cu^{2+}, a $3d^9$ cation, octahedrally coordinated by oxygen. This cation has its ninth electron in an $e_g^{\downarrow}$ level, where the d_{z^2} and $d_{x^2-y^2}$ orbitals are degenerate. However, d^9 is a Jahn–Teller ion, which tends to elongate or flatten the octahedron in order to stabilize one orbital or the other. It may be energetically favourable to alternate the configurations, so that the distortions of neighbouring octahedra match. There is then G-type *orbital* order, where there are two sublattices, composed of ions with electrons in d_{z^2} or $d_{x^2-y^2}$ orbitals.

6.3.3 Helimagnets

It is possible for ferromagnetic and antiferromagnetic interactions to be in conflict. A good example is a layer structure where ferromagnetic layers couple

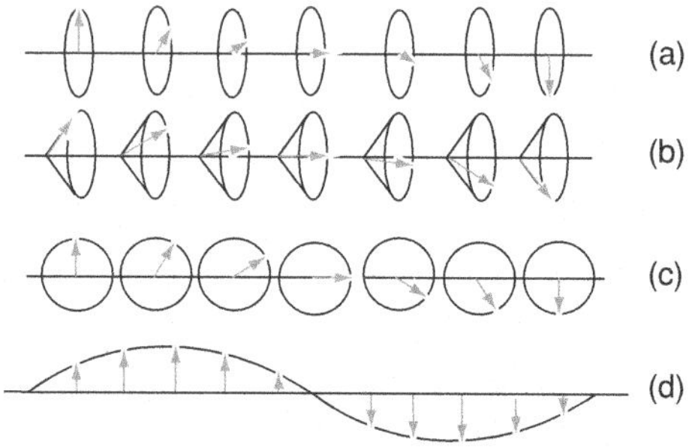

Figure 6.11

Some incommensurate magnetic structures: (a) helical, (b) cycloidal, (c) helicoidal and (d) sinusoidally modulated. Only (b) has a net ferromagnetic moment.

ferromagnetically to the neighbouring layers, but antiferromagnetically to the next-neighbour layers. A helical spin structure is then possible. If $\mathcal{J}_1$ and $\mathcal{J}_2$ are the exchange constants for the first and second neighbour planes, the energy of a spin in the central plane,

$$\varepsilon = -4\mathcal{J}_1 S^2 \cos\theta - 4\mathcal{J}_2 S^2 \cos 2\theta,$$

is minimized when $\cos\theta = -\mathcal{J}_1/4\mathcal{J}_2$. The helimagnetic structure arises when $\mathcal{J}_1 > 0$, $\mathcal{J}_1 < -4\mathcal{J}_2$.

Other modulated structures that can arise from a balance of exchange and anisotropy are the helicoidal (easy cone) and cycloidal structures shown in Fig. 6.11. The magnitude of the moment can sometimes be periodically modulated, rather than its direction. The best-known example of a sinusoidally modulated structure is chromium. When the modulation period is unrelated to the underlying periodicity of the lattice, as it is in pure Cr, the structure is incommensurate. Any commensurate magnetic order can equally well be described as a multisublattice structure in an enlarged unit cell.

6.3.4 Rare-earth metals

The $4f$ metals from cerium to ytterbium, and their nonmagnetic analogues yttrium, lanthanum and lutecium, are a playground for researchers in magnetism. The metals are chemically similar, but they possess widely differing magnetic moments and single-ion anisotropy. The magnetic properties are remarkably diverse. Rare-earths usually adopt a trivalent configuration of the $4f$ shell in metals $4f^n(5d6s)^3$, and in insulators they lose their three outermost electrons to form trivalent R^{3+} $4f^n$ ions. Yb and Eu are exceptions, which prefer a divalent configuration to benefit from the particularly stability of a filled or half-filled $4f$ shell (Fig. 11.2). Ce can be quadrivalent, with an empty $4f$ shell.

The crystal structures of the trivalent rare-earths are based on close packing of two-dimensional hexagonal sheets. The decrease of lattice parameters along the series is known as the lanthanide contraction, and the quantum number J varies as shown in Fig. 4.4. Effective exchange coupling scales as the de Gennes factor G (Table 4.10). The extraordinary variety and complexity of

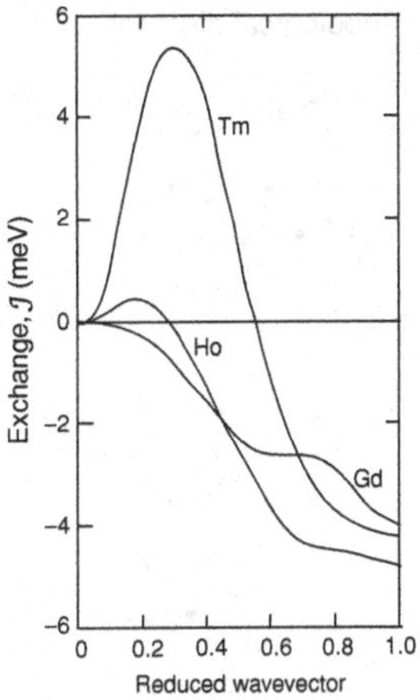

The wavevector dependence of the exchange interaction $\mathcal{J}(q)$ for heavy rare-earth metals. Gd is ferromagnetic, the others are helimagnets.

rare-earth magnetism, which embraces ferromagnetic, antiferromagnetic, helical and periodically modulated structures, sometimes in the same metal at different temperatures, reflects the interplay of anisotropy and exchange. Exchange is long range, of the RKKY type, with many shells of interacting neighbours having positive or negative coupling (§5.2.2). In these circumstances, it is convenient to define a wavevector-dependent exchange interaction

$$\mathcal{J}(q) = \sum_j \mathcal{J}(r_{ij})e^{-iqr_{ij}}, \tag{6.20}$$

which is the Fourier transform of the real-space exchange interaction $\mathcal{J}(r_{ij})$. The period of the helical magnetic structure is determined by the wavevector for which $\mathcal{J}(q)$ is minimum. If the minimum is at $q = 0$, the structure is ferromagnetic.

Anisotropy at high temperatures is predominantly second-order – easy-axis or easy-plane – with a sign determined by the product of the rare-earth quadrupole moment and the electric field gradient at the rare-earth site, as discussed in §4.4.4. At low temperatures, fourth- and sixth-order terms, proportional to $\langle J_z^n \rangle$, come into play and modify the magnetic structure to suit themselves. The crystal field can play tricks, such as stabilizing a nonmagnetic $|M_J = 0\rangle$ state for Pr on some sites in the structure, but not in others. We return to the rare-earths and rare earth intermetallics in Chapter 11.

The rare earths form extended families of isostructural intermetallic compounds and pseudobinary solid solutions with the 3d elements such as RFe_2 or R_2Co_{17}, which are stable across the 4f series. These families include many of the most interesting and useful magnetic materials; examples are $SmCo_5$ (huge anisotropy), $(Tb,Dy)Fe_2$ (giant magnetostriction), and Gd_5 (Si, Sn)$_4$ or $La(Fe_{11}Si)$ (large magnetocaloric effect).

6.4 Amorphous magnets

Amorphous solids have no crystal lattice. The atoms are in a frozen, liquid-like state. It is practically impossible to retain pure metallic elements in such a state at room temperature, so noncrystalline materials of interest in magnetism are alloys or compounds of the 3d or 4f elements, sometimes in the form of very thin films. Amorphous alloys are usually prepared by quenching from the liquid or vapour phase. Generally a mixture of the chemical (site), bond and topological disorder, illustrated in Fig. 6.12, is present in the structure.

The structure of an amorphous solid is represented in an average way by the radial distribution function $\mathcal{G}(r)$, where $\mathcal{G}(r)dr$ is the number of atoms between r and $r + dr$ from an arbitrary central atom, averaged over all positions of the central atom, Fig. 6.13. At large distances $\mathcal{G}(r)$ tends to a parabola because the atomic density is uniform when averaged over a sufficient volume, but at short distances the radial distribution function shows a few peaks corresponding to

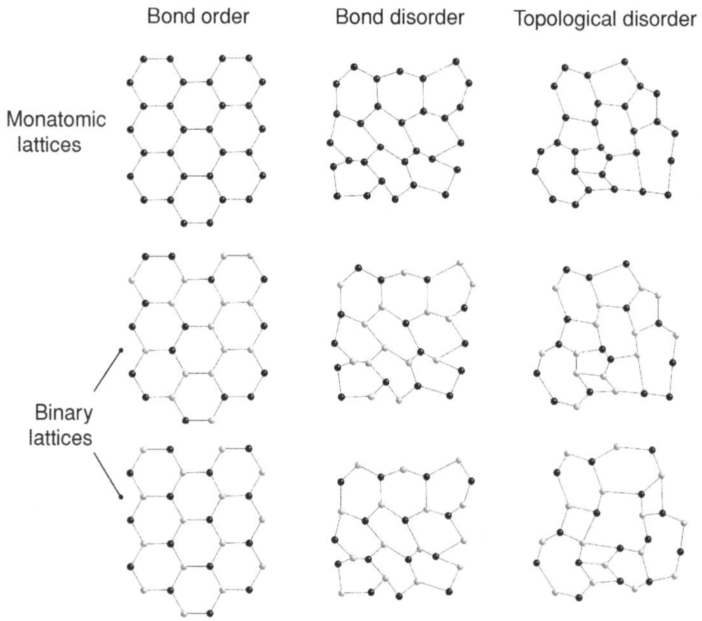

Figure 6.12

Types of disorder on two-dimensional monatomic and diatomic networks.

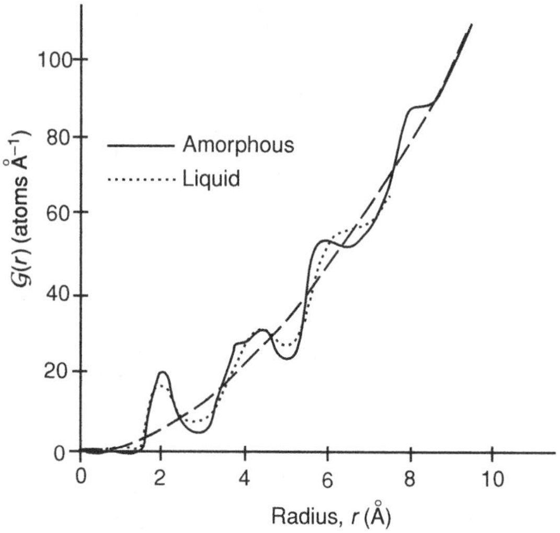

Figure 6.13

The radial distribution functions for liquid and amorphous cobalt. The dashed parabola is the large-r limit.

the first atomic coodination shells. The short-range order is like that of a frozen liquid. $\mathcal{G}(r)$ can be derived experimentally from the Fourier transform of the diffraction pattern $\mathcal{I}(k)$.

The prefix 'a-' will be used for amorphous materials. A binary a-AB alloy needs three partial radial distribution functions $\mathcal{G}_{AA}$, $\mathcal{G}_{AB}$ and $\mathcal{G}_{BB}$ just to provide this averaged description of the structure. $\mathcal{G}_{AB}(r)\mathrm{d}r$ denotes the number of B atoms between r and $r + \mathrm{d}r$ from a central A atom.

J. D. Bernal, 1901–1971.

Further scraps of structural information are obtained from techniques such as NMR and Mössbauer spectroscopy which probe the local electric field gradients, and other methods that are sensitive to three-centre correlations such as bond angle. Local coordination numbers can be deduced from the X-ray absorption edge fine structure (EXAFS). Nevertheless, modelling is the single most powerful technique for learning about the structure of noncrystalline materials. The test of a model is that it must reproduce experimental observations, especially X-ray or neutron diffraction patterns.

The packing fraction f of an atomic structure is defined as the volume fraction of space occupied by the atoms, regarded as hard spheres. Amorphous metals often adopt the random-dense-packed Bernal structure, which has $f = 0.64$, compared to 0.74 for cubic or hexagonal close packing and 0.68 for the bcc lattice, which is not close-packed. Locally there are dense clusters of 13 atoms (filled dodecahedra) in the random close-packing, but these cannot fill space without leaving large interstitial holes. The ratio of atoms to holes is about 4:1. The holes can be blocked by smaller atoms, which stabilize the noncrystalline structure. A typical composition is a-$Fe_{80}B_{20}$, where the covalent boron plugs the holes in the Bernal structure of iron. Another example is a-$Gd_{80}Au_{20}$.

The Bernal structure can be generated in the laboratory by packing peas into a jar, or ball-bearings into a football bladder, where they are fixed with wax, and then removed one-by-one to examine their coordination and bond lengths. Such tedious but instructive procedures now tend to be conducted by computer.

Eutectic melts with depressed freezing temperatures T_m are good glass formers. A glass is defined as a noncrystalline solid obtained by simply quenching the melt to below a temperature known as the glass transition temperature T_g where diffusive motion ceases on the timescale of the measurement. This is easiest near a eutectic point on a binary, or pseudobinary phase diagram, where $(T_m - T_g)/T_m$ is small. Metallic, semiconducting and insulating glasses can be formed in this way.

Insulating dielectric glasses, like those in a window, are typically mixtures of several different oxides. One of them acts as a glass former, for example, SiO_2, which forms a tetrahedral continuous random network, and the others, whose cations occupy sites with tetrahedral, octahedral or greater oxygen coordination, act as network modifiers. Magnetic $3d$ cations normally occupy tetrahedral or octahedral sites in the oxygen lattice. They are network modifiers rather than glass formers, and their concentration is usually insufficient to allow percolation of the nearest-neighbour exchange interactions. The percolation threshold x_p is the fraction of magnetic cations beyond which continuous exchange paths form throughout the structure. It is approximately $2/Z_c$, where Z_c is the cation–cation coordination number; you need at least two bonds to form a network. Covalently bonded amorphous structures such as the four-fold tetrahedral random network structure of a-Si are not favoured by magnetic atoms or ions. Nevertheless, a few concentrated amorphous ionic compounds do exist, notably amorphous

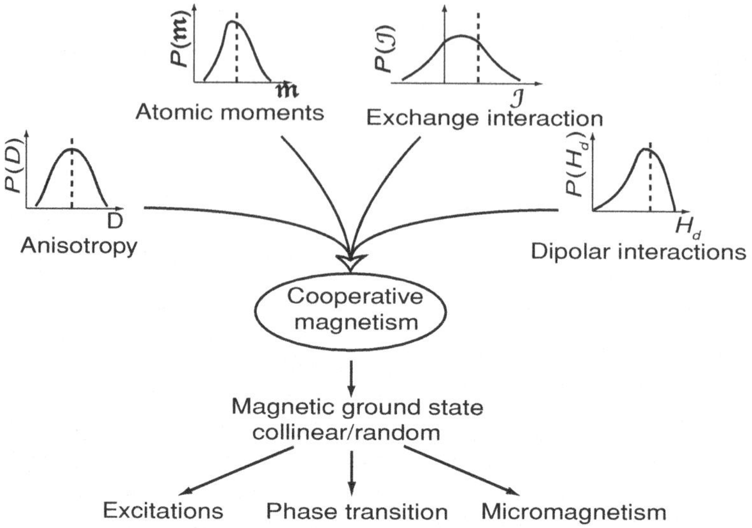

Figure 6.14

Summary of the ingredients of cooperative magnetism, showing how unique values of $\mathfrak{m}$, D, $\mathcal{J}$ and H_d are replaced by probability distributions in a noncrystalline solid.

FeF_3, which has an octahedral continuous random network structure which preserves six-fold coordination of iron by fluorine, and two-fold coordination of fluorine by iron. Examples of magnetic glasses are presented in Chapter 11.

Any amorphous structure entails a *distribution* of nearest-neighbour environments and bond lengths for a given magnetic atom, described by the radial distribution function and higher-order correlation functions. These distributions lead to a distribution of site moments, exchange interactions, dipolar and crystal fields, all of which influence the nature of the magnetic order, as summarized in Fig. 6.14.

6.4.1 One-network structures

These are structures with a single magnetic network. For example, the iron atoms in a-$Fe_{80}B_{20}$ or a-FeF_3 constitute a single magnetic network. In this discussion of magnetic order in amorphous materials, we will ignore the magnetostatic, dipole–dipole interactions which lead to domain formation in structures with a net ferromagnetic moment, focussing rather on the consequences of distributions of exchange and single-ion anisotropy.

$\mathcal{J} > 0$ When exchange is ferromagnetic and local single-ion anisotropy is negligible, the magnetism is straightforwardly ferromagnetic. There is a Curie point, and differences compared with a crystalline ferromagnet are minor. Obviously, there can be no magnetocrystalline anisotropy K_1, because there is no crystal lattice, and there is no overall preferred direction of magnetization in the bulk unless it is due to shape or an anisotropic texture somehow imparted to the amorphous solid (by annealing near the glass transition in a magnetic field, for example). Nevertheless local magnetic anisotropy does exist at an *atomic* scale. Linear magnetostriction is present in amorphous alloys, since

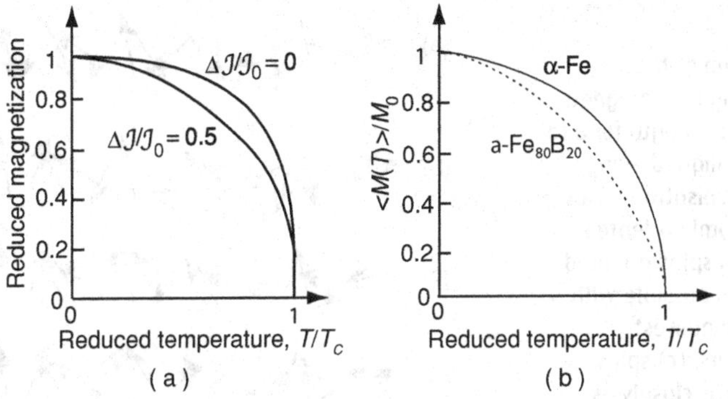

Figure 6.15

(a) Reduced magnetization versus temperature for crystalline and amorphous ferromagnets, according to the Handrich molecular field model (R. Handrich, *Phys Stat Sol* **32**, K55 (1969)). (b) Reduced magnetization data for crystalline Fe and amorphous $Fe_{80}B_{20}$.

magnetostriction depends on spin-orbit coupling, which is an atomic-scale interaction. Volume magnetostriction may also be present as it depends on the distance dependence of interatomic exchange. It is possible to find amorphous alloys that exhibit zero magnetostriction – the invar effect. Indeed, it is easier to tune the properties by modifying the composition in the amorphous state than it is in a crystal, because the amorphous dense-packed structures evolve continuously with composition. For example, the Fe : Co ratio can be varied in all proportions in a-$Fe_{80-x}Co_xB_{20}$, whereas in $Fe_{100-x}Co_x$ there is a change from bcc to fcc at $x = 20$.

The amorphous structure is accounted for in molecular field theory by admitting a distribution in $\mathcal{J}$, such that

$$\mathcal{J} = \mathcal{J}_0 + \Delta\mathcal{J}, \tag{6.21}$$

where $\Delta\mathcal{J}$ is a random quantity with a symmetric gaussian distribution. This is the Handrich model. In the absence of an external field, the Brillouin function is replaced by

$$\langle M(T)\rangle = (M_0/2)\{\mathcal{B}_S[x(1+\delta)] + \mathcal{B}_S[x(1-\delta)]\}, \tag{6.22}$$

where δ is the normalized root-mean-square exchange fluctuation: δ is defined by the expression

$$\delta^2 = \left\langle \left(\sum\nolimits_j \Delta\mathcal{J}\right)^2 \right\rangle \Big/ \left(\sum\nolimits_j \langle\mathcal{J}\rangle\right)^2.$$

As a result, the magnetization declines more rapidly with temperature than it would otherwise. T_C is usually reduced by the spatial fluctuations of $\mathcal{J}$, as shown in Fig. 6.15.

Although there is no net magnetic anisotropy, there are local easy directions on the scale of an atom or a nanoscale volume with dimensions of a few interatomic spacings. There is always an electrostatic field at any atomic site, although there is no precise site symmetry in an amorphous solid. A differently oriented electric field gradient tensor leads to a different easy axis $\boldsymbol{e}_i$ at every site. The leading term in the electrostatic interaction is the second-order term (4.32). This random local anisotropy may be sufficient to pin the local magnetization

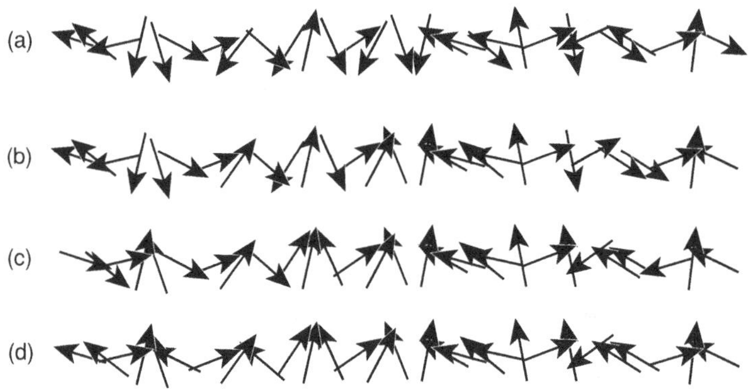

Figure 6.16

Illustration of the increasing ferromagnetic correlation length for a row of spins aligned along random anisotropy axes: (a) randomly oriented spins, (b) spins oriented as closely as possible with their two nearest neighbours, (c) spins oriented as closely as possible with eight neighbours and (d) spins oriented as closely as possible with all their neighbours.

direction in $4f$ alloys, but it is generally insufficient to pin it in $3d$ alloys. In some sense, however, the random anisotropy always destroys the ferromagnetic order.

The situation can be represented by a Hamiltonian known as the Harris–Plischke–Zuckermann model, which is based on (4.33):

$$\mathcal{H}_{HPZ} = -2\sum_{ij} \mathcal{J}_{ij} \boldsymbol{S}_i \cdot \boldsymbol{S}_j - \sum_i \mathsf{D}_i(\boldsymbol{e}_i \cdot \boldsymbol{S}_i)^2 - \sum_i \mu_0 g \mu_B(\boldsymbol{S}_i \cdot \boldsymbol{H}_z). \tag{6.23}$$

This can be simplified if exchange is restricted to nearest neighbours, $\mathcal{J}_{ij} = \mathcal{J}$, and if the random anisotropy has constant magnitude, but random direction, $\mathsf{D}_i = \mathsf{D}$. The key parameter is the ratio $\alpha = \mathsf{D}/\mathcal{J}$. If $\alpha \gg 1$, the random anisotropy dominates the exchange, and destroys the ferromagnetic order. The length scale over which the ferromagnetism is scrambled can be estimated by considering the direction of the exchange field at any site. This is the resultant of a z-component proportional to $\frac{1}{2}Z$, and a transverse component in the xy-plane proportional to $(\pi/4)\sqrt{Z}$. Here Z is the number of interacting neighbours, and $\frac{1}{2}$ and $(\pi/4)$ are the average values of a unit vector, randomly oriented within a hemisphere, in a direction parallel and perpendicular to Oz. Therefore the exchange field is misoriented by an angle $\zeta = \tan^{-1}(\pi/2\sqrt{Z})$ on average. (If $Z = 12$, $\zeta = 24°$.) These misorientations accumulate in a random way on moving out from the central atom in any direction, so that, approximating $\tan\zeta \approx \zeta$, beyond a distance $(\pi/2\zeta)^2 a$, where a is the interatomic spacing, memory of the original z-direction is lost. The ferromagnetic correlation length is $\approx Za$. The increase of correlation length with the number of interacting neighbours is illustrated in one dimension in Fig. 6.16. The local ferromagnetic axis wanders over a few nearest-neighbour distances when there are only a few interacting neighbours, but as $Z \to \infty$, the magnetization directions are distributed at random within a single hemisphere centred on the z axis, and the magnetization $M_s = \frac{1}{2}n\mathfrak{m}$ because the integral of $\mathfrak{m}\cos\theta$ over a hemisphere

$$\langle \mathfrak{m} \rangle = \int_0^{\pi/2} \mathfrak{m} \cos\theta 2\pi r^2 \sin\theta d\theta / \int_0^{\pi/2} 2\pi r^2 \sin\theta d\theta = \tfrac{1}{2}\mathfrak{m}.$$

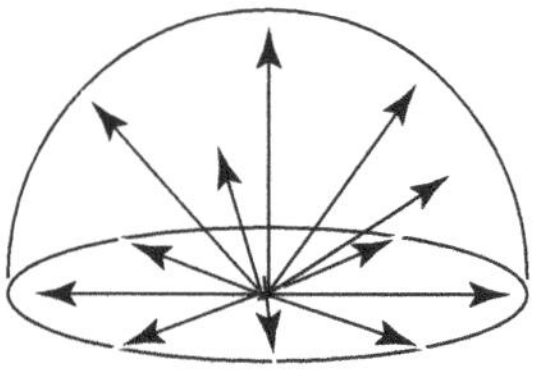

A uniform dispersion of atomic moment orientations within a hemisphere gives an average magnetization $\langle \mathfrak{m} \rangle = \frac{1}{2}\mathfrak{m}$.

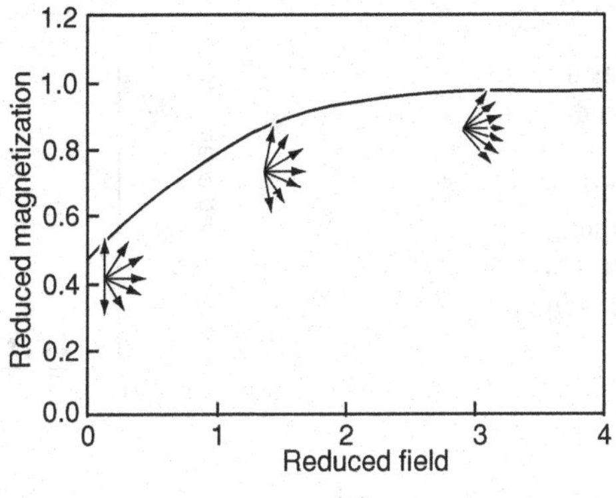

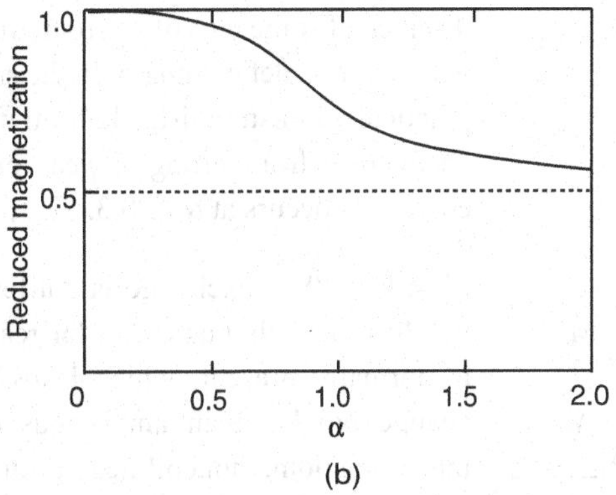

Figure 6.17

(a) Reduced magnetization curve at $T = 0$ K for the random-anisotropy model with negligible exchange. h is the reduced applied field $[g\mu_B/DS]\mu_0 H$. (b) Reduced remanence as a function of $\alpha = D/\mathcal{J}$, the ratio of anisotropy to exchange energy. Calculations are for classical spins.

In an applied field, the magnetization continuously approaches saturation, as shown for classical spins in Fig. 6.17. The remanence in the same limit falls from 1 to 0.5 as α increases from 0 to ∞.

Turning to the limit $\alpha \ll 1$, which is more realistic for amorphous 3d ferromagnets where the local anisotropy may be 100 times less than the exchange, we again find a wandering ferromagnetic axis, but on a much longer length scale. Admitting that the spin directions are correlated in a region of dimension L containing $N = (L/a)^3$ atoms, the anisotropy produces local deviations from parallel alignment of order α. Statistical fluctuations within the region will ensure that some particular direction of magnetization is preferred. The average anisotropy energy per atom ε_a will be $DS^2/\sqrt{N}$. The ferromagnetic axis changes its orientation by $\pi/2$ on passing from one region to the next, so the average increase in exchange ε_{ex} is $\mathcal{J}S^2(\pi a/2L)^2$. Minimizing the sum of these two energies gives

$$L = (1/9\alpha^2)\pi^4 a. \tag{6.24}$$

The argument shows that random anisotropy never averages away completely. A pure ferromagnetic state will always be destroyed by random anisotropy, however small. But if α is small to begin with, the effect can be neglected. In

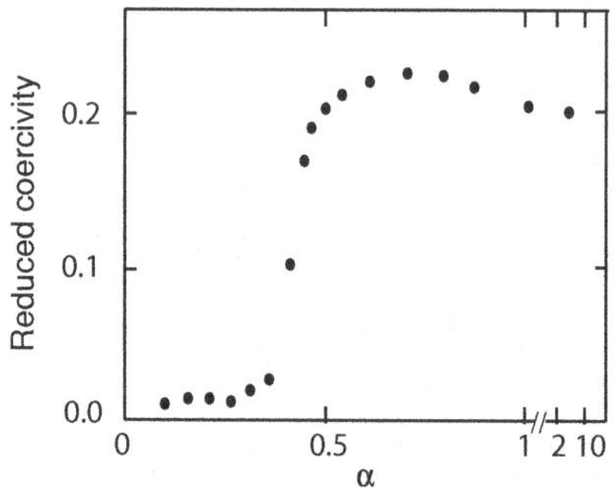

Figure 6.18

Reduced coercivity as a function of α, deduced from Monte-Carlo simulations for a block of 996 spins in the random anisotropy model. (C. Chi and R. Alben, *J. Appl. Phys.* 48, 2487 (1977).)

our example, $\alpha = 1/100$, $L \approx 10^5 a$. The ferromagnetic axis wanders over a distance of some tens of micrometres, which is comparable to the size of a normal ferromagnetic domain. In the small-α limit, the amorphous ferromagnet is practically indistinguishable from its soft crystalline counterpart. A qualitative changeover from strong to weak pinning, accompanied by the appearance of coercivity occurs at $\alpha \approx 0.3$, Fig. 6.18.

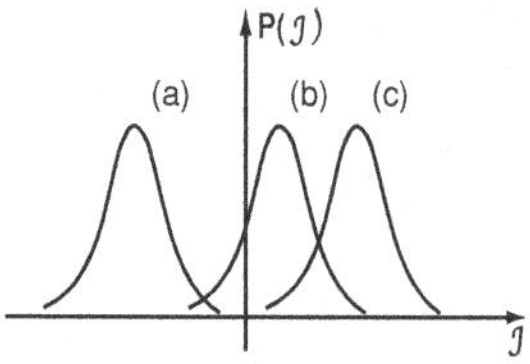

Distributions of exchange interactions giving rise to (a) speromagnetism, (b) asperomagnetism and (c) ferromagnetism.

$\mathcal{J} < 0$ When exchange is antiferromagnetic, the absence of a crystal lattice has more dramatic and far-reaching consequences. Topological disorder is normally present, which leads to frustration of the individual superexchange bonds in an amorphous oxide or halide. Spins ultimately freeze into a random, noncollinear ground state with a high degree of degeneracy. Many different spin configurations of the system have almost the same energy. The freezing temperature T_f is much less than the magnitude of the paramagnetic Curie temperature θ_p, which reflects the averaged strength of the individual antiferromagnetic interactions. Site-averaged spin correlations $\langle \boldsymbol{S}_i(0) \cdot \boldsymbol{S}_j(r) \rangle$ are negative at the nearest-neighbour distance, but rapidly average to zero at longer distances. This random spin freezing is known as speromagnetism.

An intermediate situation arises when the exchange distribution is broad, but biased towards a net positive value. Locally, there is a net magnetization, but the ferromagnetic axis wanders under the influence of the local balance of exchange. This type of random spin freezing is called asperomagnetism. The same random magnetic structure arose when $\mathcal{J} > 0$ under the influence of strong random local uniaxial anisotropy.

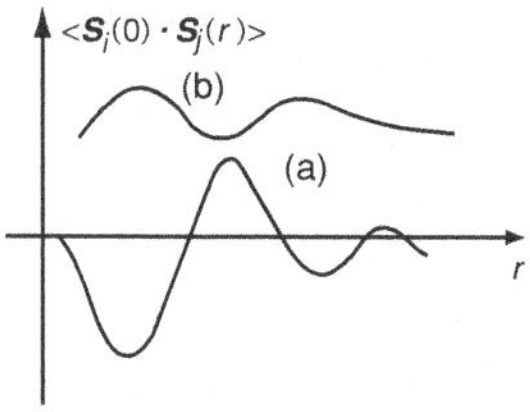

Averaged spin correlations in (a) a speromagnet and (b) an asperomagnet.

Speromagnetism and asperomagnetism are distinguished by the length scale over which the spin correlations average to zero. In a speromagnet, this is at most a couple of interatomic spacings (the nearest-neighbour correlations are antiferromagnetic), whereas in asperomagnets it is much longer, and the integrated correlations on a mesoscopic scale are ferromagnetic.

The different sorts of one-sublattice magnetic order in amorphous materials, sketched in Fig. 6.19, are best distinguished by their magnetization curves, which are compared in Fig. 6.20.

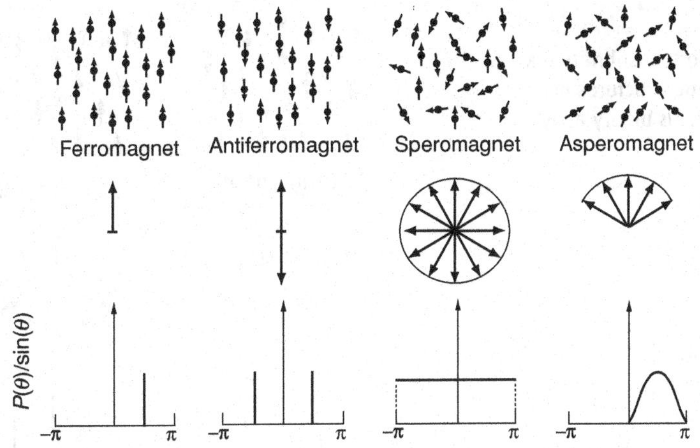

Figure 6.19

Possible one network magnetic structures in amorphous solids.

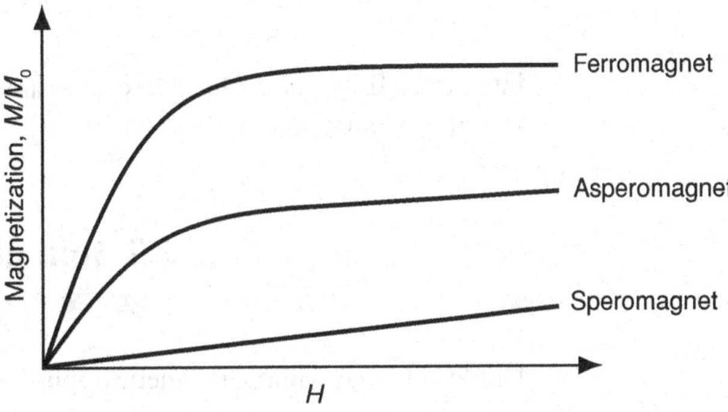

Figure 6.20

Magnetization curves at low temperature for a ferromagnet, a speromagnet and an asperomagnet.

6.4.2 Two-network structures

One might imagine a situation where two amorphous sublattices could be defined *topologically*, as sketched in Fig. 6.12, but it seems unlikely that this ever arises in practice.

There remains, however, the possibility of distinguishing two magnetic subnetworks in amorphous solids on a *chemical* basis. Generally, these are the sublattices composed of $3d$ and $4f$ atoms, The d–d exchange is strongly ferromagnetic, defining a ferromagnetic $3d$ subnetwork, and the $3d$–$4f$ interactions then tend to align the $4f$ subnetwork *spins* antiparallel to the $3d$ subnetwork. Hence the subnetwork moments are aligned parallel for heavy rare-earths, but antiparallel for the light rare-earths. We can therefore have an amorphous ferrimagnet where the A and B subsystems are defined chemically. An example is a-$Gd_{25}Co_{75}$. There may be a compensation point at the temperature where $M_{4f} > M_{3d}$, just as in crystalline ferrimagnets.

For rare-earths with strong uniaxial anisotropy and weak exchange coupling to the $3d$ subnetwork, their local easy axes are defined by the local crystal-field

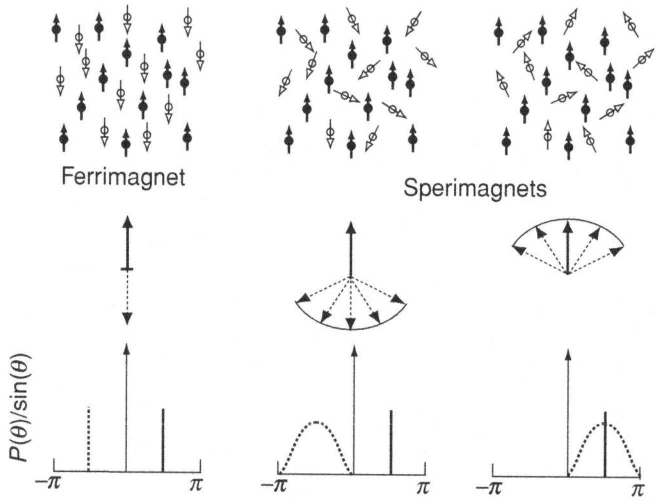

Figure 6.21
Possible two-subnetwork magnetic structures in amorphous binary alloys.

interaction $D_i J_{zi}^2$. These local easy axes are random, leading to the sperimagnetic structures illustrated in Fig. 6.21.[1]

6.5 Spin glasses

Dilute alloys containing magnetic impurity atoms which retain their moments in a nonmagnetic matrix, where the positions of the magnetic atoms are essentially random are known as canonical spin glasses. Examples are **Cu**Mn and **Au**Fe (the metal in bold type is the host, and the alloy composition is $\mathbf{M}_{100-x}\mathrm{T}_x$). When the impurity concentration x is below 1%, the discrete nature of the lattice is imperceptible and the impurities couple magnetically via the long-range RKKY interaction

$$\mathcal{H} = -2\sum_{i>j} \mathcal{J}_{RKKY} \boldsymbol{S}_i \cdot \boldsymbol{S}_j.$$

$\mathcal{J}_{RKKY}$ is given by (5.30) $(9\pi\nu^2/64\varepsilon_F)\mathcal{J}_{sd}{}^2F(\xi)$, where ν is the number of conduction electrons per atom and $F(\xi)$ is the RKKY function, which falls off as $1/r_{ij}^3$ in the limit of large separation in three dimensions. $\mathcal{J}_{sd}$ is the coupling between an impurity moment and the conduction electrons. The distances between impurities become random in the dilute limit. Coupling varies as V_0/r^3 with V_0 equally likely to be positive or negative. The average strength of an exchange bond is zero in this case. A symmetric exchange distribution $P(\mathcal{J})$

[1] The spero/speri root in the names for these random structures comes from the greek $\delta\iota\alpha\sigma\pi\varepsilon\iota\rho\varepsilon\iota\nu$, meaning to scatter in all directions. Other words with this root are dispersion and diaspora.

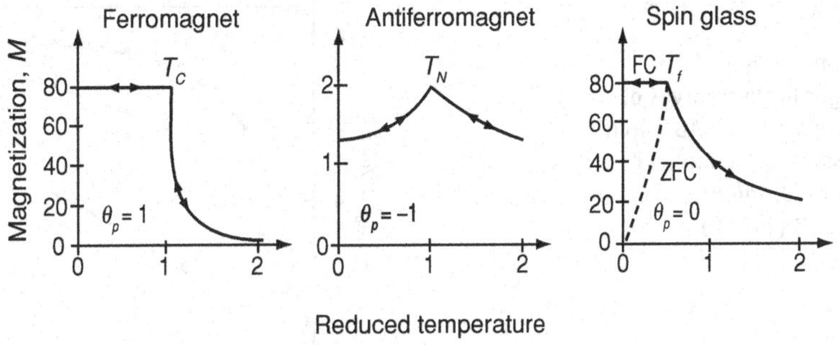

Figure 6.22

Comparison of the magnetization measured in a small applied field for a ferromagnet, an antiferromagnet and a spin glass. The dashed line indicates zero-field-cooled behaviour (ZFC), which evolves with time. The field-cooled line (FC) is reversible.

of width $\Delta\mathcal{J}$, centred at zero, is the hallmark of a canonical spin glass. Since the average exchange interaction is zero, the inverse susceptibility follows a Curie law, not a Curie–Weiss law.

The characteristic experimental feature of a spin glass is a cusp in susceptibility at a temperature $T_f < \Delta\mathcal{J}/k_B$, known as the spin freezing temperature. The spins freeze in essentially random directions below T_f. There is a myriad of nearly degenerate magnetic ground-state configurations for the system, and an applied field picks out the one which has the biggest net moment in the direction of the field. Unlike a ferromagnet, where the response to a small applied field is reversible provided there is no coercivity, the field-cooled and zero-field-cooled responses of the spin glass are quite different, with a small remanent magnetization being frozen in below T_f in the former case, Fig. 6.22. The dynamics of the response to any change in field there are very protracted, varying as the logarithm of time. Another experimental signature of canonical spin glasses is a specific heat of magnetic origin which is linear in temperature and independent of the concentration of magnetic impurities below T_f, yet there is no specific heat anomaly at the freezing temperature.

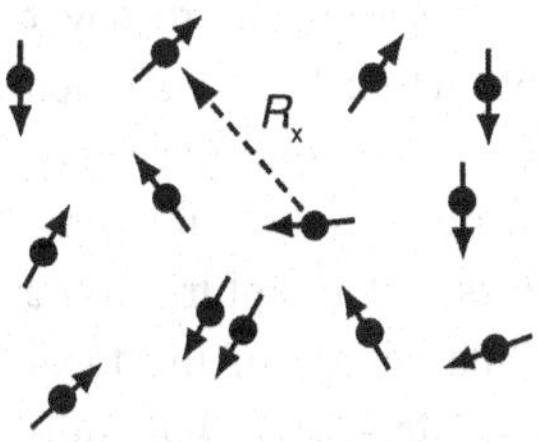

A picture which can be used to represent a dilute alloy, regardless of concentration (scaling).

Some insight into the behaviour of dilute-alloy spin glasses is achieved by picturing the positions of randomly placed impurities. It is impossible to infer the concentration from the picture. If a length scale R_x related to the distance between two impurities is chosen, the volume R_x^3 contains a constant number of impurities proportional, on average, to xR_x^3. Any thermodynamic property $\mathcal{P}$ of the system can be derived from the partition function $\mathcal{Z} = \mathrm{Tr}\exp(-\mathcal{H}/k_BT)$. If $\mathcal{H}$ and T are scaled by the same constant, the properties of the system are unchanged. Dividing $\mathcal{H}_{RKKY}$ (5.30) by x makes it a function of xr_{ij}^{-3}, which is independent of concentration. If a Zeeman term is included, the applied field H must also be divided by x. Hence the thermodynamic properties $\mathcal{P}$ obey scaling laws of the form $\mathcal{P}/x = f(T/x, H/x)$. Specifically, the specific heat and magnetization can be written as $C = xf_C(T/x, H/x)$ and $M = xf_M(T/x, H/x)$, Fig. 6.23.

Spin-glass behaviour is not restricted to the dilute-alloy spin glasses, where $T_f \propto x$. It is much more general, being found in a wide range of dilute and concentrated magnetic materials, amorphous or crystalline. Essential

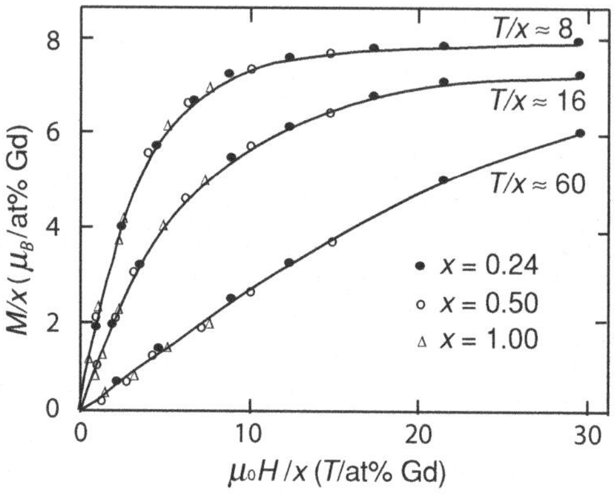

Figure 6.23
Scaling of the magnetization curves of dilute $Gd_xLa_{80-x}Au_{20}$ spin glasses. (S. J. Poon and J. Durand, *Solid State Comm* **21**, 793 (1977).)

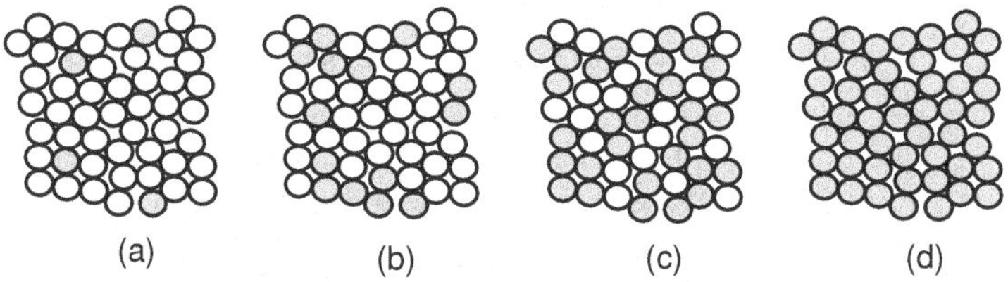

Figure 6.24
Different concentration regimes for an amorphous magnetic alloy: (a) is the dilute limit, where scaling laws apply; (b) is below the percolation threshold, where nearest-neighbour clusters form; (c) is at the percolation threshold; and (d) is the concentrated limit.

ingredients are disorder and frustration of the interactions. As concentration increases, the scaling breaks down, and magnetic atoms begin to form nearest-neighbour clusters. Beyond the percolation threshold, a 'bulk cluster' appears, which means that there is a group of connected atoms of infinite extent (Fig. 6.24(c)). This can lead to ferromagnetic long-range order throughout the material if the nearest-neighbour coupling is ferromagnetic. The magnetic phase diagrams for many of these systems with ferromagnetic nearest-neighbour interactions resemble that shown in Fig. 6.25. There is a crossover from spin-glass to ferromagnetic order near the percolation threshold, but when $x \gtrsim x_p$ there is a re-entrant transition at a temperature T_{xy} to a spin-glass-like phase which is characterized by a difference between field-cooled and zero-field-cooled magnetization, a softening of the spin waves and an upturn in the hyperfine field, which is proportional to the average magnitude of the local moment $\langle \mathrm{m}_i \rangle$. The longitudinal component of the spins orders at T_C, but additional freezing of the transverse components in random directions occurs at T_{xy}, which is a ferromagnetic $\rightarrow$ asperomagnetic transition.

In a broad sense, all speromagnets and asperomagnets can be considered as variants of spin glass. The essential feature of all these systems is that some random frustration of the exchange interactions leads to many nearly degenerate ground states. Differing degrees of magnetic short-range order can be inbuilt, in the form of ferromagnetic clusters or antiferromagnetic nearest-neighbour correlations, but the site-average correlations $\langle \mathrm{m}_i \cdot \mathrm{m}_j \rangle$ tend to zero as r_{ij} extends much beyond a few interatomic spacings.

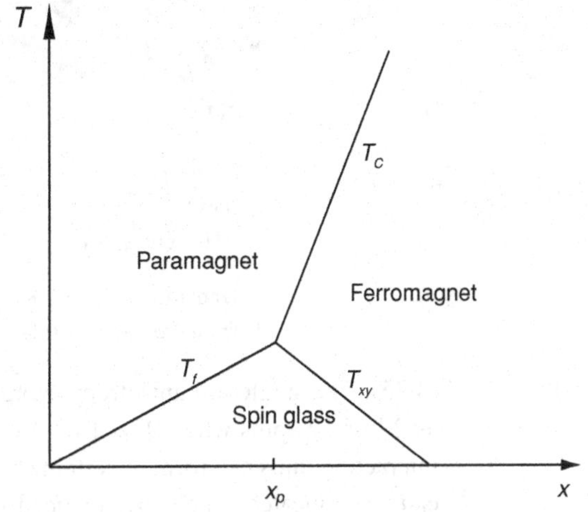

Figure 6.25

Phase diagram for a metallic spin-glass system as a function of concentration.

6.6 Magnetic models

A magnetic model system is one with a simple lattice structure, populated by spins with one, two or three components, which is susceptible to precise theoretical analysis. Although every real sample has a finite, defective, three-dimensional lattice, populated by atoms or electrons with three spin components, as always in physics there is much insight to be gained by calculating the properties of simplified models. Furthermore, solid-state chemists have been remarkably adept at devising compounds with localized magnetic moments that are good approximations to the ideal theoretical models.

6.6.1 Heisenberg, xy and Ising models

The interaction between two spins in the Heisenberg model $-2\mathcal{J}\boldsymbol{S}_1 \cdot \boldsymbol{S}_2$ may be written out in terms of their x, y and z components:

$$\mathcal{H} = -2\mathcal{J}(S_1^x S_2^x + S_1^y S_2^y + S_1^z S_2^z). \tag{6.25}$$

One can imagine two-dimensional spins which are confined to a plane – the xy model:

$$\mathcal{H} = -2\mathcal{J}(S_1^x S_2^x + S_1^y S_2^y), \tag{6.26}$$

or even one-dimensional spins which have their single component along the z axis – the Ising model:

$$\mathcal{H} = -2\mathcal{J} S_1^z S_2^z. \tag{6.27}$$

Experimentally, it is possible to approximate the latter models with reduced spin dimensionality by using ions with a large uniaxial anisotropy constant D

Table 6.3. Summary of magnetic model systems which exhibit a phase transition

$d\backslash D$	1	2	3
1(Ising)	×[a]	√	√
2(xy)	×	√[b]	√
3(Heisenberg)	×	×[a]	√

[a] Transition at $T = 0$ K.
[b] Transition at $T > 0$ K to a state with $M = 0$.

(4.35). The single-ion anisotropy energy is written as $\mathsf{D}S_z^2$, so the ions behave as $2d$ or $1d$ spins when $\mathsf{D} \gg 1$ or $\mathsf{D} \ll 1$. Magnetically decoupled or weakly interacting units can form in certain crystal lattices where magnetic sublattices can be delineated which have reduced spatial dimensionality D.

Chains are one-dimensional, while layer compounds have two-dimensional character. The space and spin dimensionality $\{D, d\}$ together define a model class. Some of these models, such as the two-dimensional Ising model {2,1}, can be solved exactly. Others must be solved numerically. It is convenient in this section to set $S = 1$, so the energy of a pair of nearest neighbours with parallel spins is $-2\mathcal{J}$.

Magnetic ordering at a finite phase transition temperature becomes more likely for lower spin dimensionality d, because the spins then have fewer degrees of freedom, and in higher spatial dimension D, because fluctuations are then less likely to destroy the order, as indicated in Table 6.3.

6.6.2 Critical behaviour

Thermal fluctuations characterize the behaviour of a magnetic system in the vicinity of a continuous, second-order phase transition, a zone known as the critical region. There is a mathematical singularity in the Gibbs free energy at the transition temperature T_C, where the order parameter drops continuously to zero, and there is a discontinuity in its temperature derivative. The critical fluctuations are absent at a first-order transition, where the order parameter is discontinuous. Strictly speaking, a system must be infinite in extent for the free energy to exhibit any singularity, so a perfectly sharp phase transition is a fiction realized only in an infinite system. In practice, rounding of the transition only becomes evident in submicrometre-sized samples.

An important result of statistical thermodynamics is the fluctuation dissipation theorem that relates the susceptibility of a ferromagnet in thermal equilibrium to fluctuations of the magnetization by

$$\chi = \frac{\mu_0}{k_B T}(\langle M^2 \rangle - \langle M \rangle^2). \tag{6.28}$$

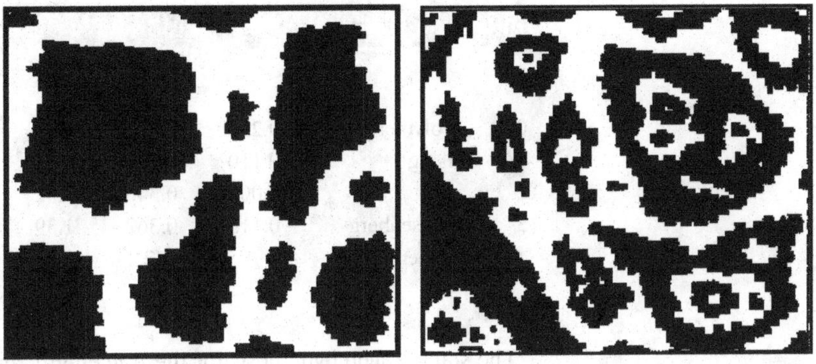

Figure 6.26

Illustration of critical fluctuations at two temperatures just above the Curie point in the two-dimensional Ising model. The black and white squares represent ↑ and ↓ atomic moments. The correlation length diverges at T_c, but the magnetization there is zero.

The fluctuations diverge at T_C, where the correlation length also diverges. Analogous expressions relate the specific heat to fluctuations in the enthalpy and compressibility to fluctuations in the density. These results allow physical observables to be deduced from computer simulations of model systems of spins, for example. The pair correlation function $\Gamma(r)$ between two spins i and j is defined as

$$\Gamma(r_{ij}) = \langle \boldsymbol{S}_i \cdot \boldsymbol{S}_j \rangle \; - \langle \boldsymbol{S}_i \rangle \cdot \langle \boldsymbol{S} j \rangle .$$

The correlations decay exponentially, and the correlation length ξ is defined by $\lim_{r\to\infty} \Gamma(r) \sim \exp(-r/\xi)$.

The pair of numbers representing the space and spin dimensionality $\{D, d\}$ specify a universality class for critical behaviour. Examples are the two-dimensional Ising model $\{2.1\}$ or the three-dimensional Heisenberg model $\{3{,}3\}$. Close to T_C is a region where critical fluctuations are important. There is a different type of critical behaviour for each universality class, but all the materials within a class behave similarly, regardless of their composition or lattice structure. A reduced temperature is defined, $\varepsilon = (1 - T/T_C)$, and the critical region can be considered as the zone where $\varepsilon < 10^{-2}$. The critical exponents were introduced in §5.1.2 in the context of the mean field model. When ε is small, $M \approx \varepsilon^{\beta}(\varepsilon \gtrsim 0)$, $M \approx H^{1/\delta}(\varepsilon = 0)$, $\chi \approx |\varepsilon|^{\gamma}(\varepsilon \approx 0)$ and $C \approx |\varepsilon|^{-\alpha}(\varepsilon \approx 0)$. Here M is the order parameter, H is the conjugate field and χ is the susceptibility $\mathrm{d}M/\mathrm{d}H$. In an antiferromagnet, the order parameter is the sublattice magnetization M_α and the conjugate field is a staggered field. In the spin glass there is an order parameter $\tilde{q}$ for which the conjugate field is a random field.

Two more critical exponents ν and η describe the correlation length ξ and the correlation function at T_C. They are defined by

$$\xi \approx |\varepsilon|^{-\nu}(\varepsilon \approx 0) \quad \text{and} \quad \Gamma(r) \approx |r|^{-(D-2+\eta)}(\varepsilon = 0).$$

Critical fluctuations are shown for a plane of Ising spins in Fig. 6.26. Fluctuations are self-similar on different length scales. The figure illustrates how it is possible for the correlation length to diverge, while the magnetization remains vanishingly small.

Table 6.4. Critical exponents for the three-dimensional d-vector models

d		α	β	γ	δ	ν	η
0	Polymer	0.236	0.302	1.16	4.85	0.588	0.03
1	Ising	0.110	0.324	1.24	4.82	0.630	0.03
2	xy	−0.007	0.346	1.32	4.81	0.669	0.03
3	Heisenberg	−0.115	0.362	1.39	4.82	0.705	0.03
∞	Spherical	−1	1/2	2	5	1	0

The static scaling hypothesis for the free energy and the correlation function implies that only two of the exponents are actually independent. They are related by equalities such as

$$\begin{aligned} 2 &= \alpha + 2\beta + \gamma, \\ \gamma &= \beta(\delta - 1), \\ \alpha &= 2 - \nu D, \\ (2 - \eta)\nu &= \gamma. \end{aligned}$$

The mean field exponents, are $\alpha = 0$, $\beta = \frac{1}{2}$, $\gamma = 1$, $\delta = 3$, $\nu = \frac{1}{2}$ and $\eta = 0$. The mean field theory of a ferromagnet or an antiferromagnet does not account properly for the real critical fluctuations that are observed when $D = 3$, but when $D = 4$, according to the equalities, the theory could be exact! The dimension where the mean field theory is exact is known as the upper critical dimension. Generally in the critical region, close to T_C, the equation of state can be written (5.14)

$$(H/M)^{\gamma} = a(T - T_C) - bM^{1/\beta}. \tag{6.29}$$

The critical exponents have been calculated numerically when there is no analytical solution, using the renormalization-group method developed by Kenneth Wilson, Leo Kadanoff and others. The properties of the original lattice are compared with those of a lattice expanded by a scaling factor. It turns out that iterative scaling preserves the physics of the critical region. Values for the practically important three-dimensional Heisenberg model are included in Table 6.4. Critical components for the Ising model, including the exact Onsager solution in two dimensions, are collected in Table 6.5.

The value of the critical temperature, the Curie or Néel point, is not independent of lattice structure. It too can be calculated numerically; it increases with D and coordination number Z, as seen in Table 6.6, and also with spin dimension d. For the three-dimensional Heisenberg model, the ratio $k_B T_C / Z\mathcal{J}$ is 0.61, 0.66 and 0.70 for the simple cubic, bcc and fcc lattices, respectively.

Table 6.5. Some Ising-model critical exponents; $D \geq 4$ is the mean field case

D	α	β	γ	δ	ν	η
2	0	1/8	7/4	15	1	1/4
3[a]	1/8	5/16	5/4	5	5/8	0
≥4	0	1/2	1	3	1/2	0

[a] Approximate values.

Table 6.6. The ratio $k_B T_C / Z \mathcal{J}$ for Ising spins on different lattices

Lattice	D	Z	
Chain	1	2	0
Honeycomb	2	3	0.506
Square	2	4	0.567
Triangular	2	6	0.607
Diamond	3	4	0.676
Simple cubic	3	6	0.752
Body-centred cubic	3	8	0.794
Face-centred cubic	3	12	0.916

6.6.3 Spin-glass theory

Returning to spin glasses, a much-debated theoretical question has been 'Is there a phase transition at T_f, or do the spin dynamics evolve continuously, but exponentially with temperature as the spins progressively freeze?' In other words, is the freezing of the spins simply analogous to the freezing out of long-range diffusive motion in a glass at its glass transition (as the name 'spin glass' would suggest), or is there some sort of collective behaviour producing a singularity in the free energy or its derivatives at T_f, just as there is at the Curie point.

If there is a phase transition, it should be possible to identify the order parameter that plays the role of magnetization in a ferromagnet or sublattice magnetization in an antiferromagnet, and goes to zero at T_f. The local magnetic moment $\mathfrak{m}_i$ at the ith site averaged over all sites $\langle \mathfrak{m}_i \rangle$ is *not* a possible choice, because it is zero at all temperatures. It is better to take the projection of the spins onto a particular random configuration, or replica of the system. There is an energy landscape where different spin configurations occupy different, mutually inaccessible energy minima. An order parameter was defined by Edwards and Anderson as the mean-square spontaneous magnetization in a single minimum α, averaged over all possible minima:

$$\tilde{q} = \sum P_\alpha \langle \mathfrak{m}_{i\alpha}^2 \rangle, \tag{6.30}$$

Figure 6.27

Theoretical phase diagrams calculated in mean field theory for (a) an Ising spin glass by D. Sherrington and S. Kirkpatrick (*Phys. Rev. Letters* **35**, 1792 (1975)) and (b) for vector spins by M. Gabay and G. Toulouse (*Phys. Rev. Letters* **47**, 201 (1981)). There is a distribution of exchange of width $\Delta\mathcal{J}$ and average value $\mathcal{J}_0$.

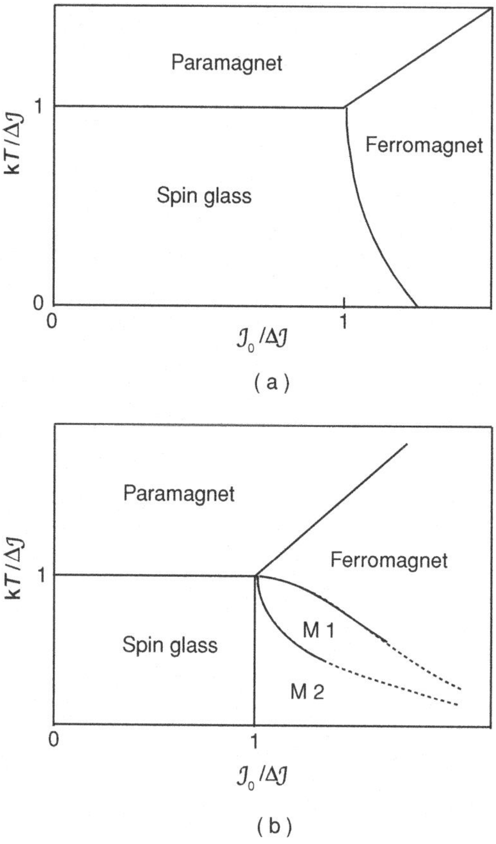

where $P_\alpha = \exp(-\varepsilon_\alpha/k_BT)/\sum \exp(-\varepsilon_\alpha/k_BT)$. Associated with the order parameter is a conjugate field. This is not the uniform field accessible in the laboratory, but a different randomly staggered field for each configuration. The corresponding susceptibility is $\chi_{\tilde{q}}$. Fortunately, it turns out that $\chi_{\tilde{q}}$ is not inaccessible, because the nonlinear susceptibility χ_{nl}, defined by

$$M = \chi H - \chi_{nl} H^3, \tag{6.31}$$

is proportional to $\chi_{\tilde{q}}$.

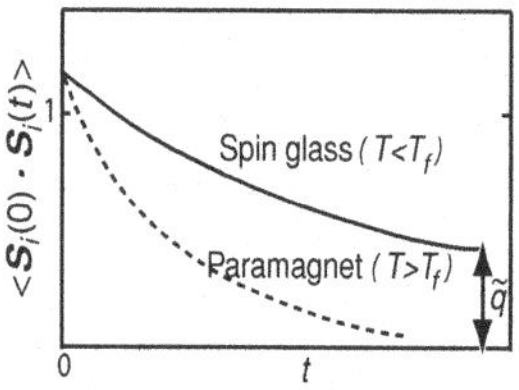

Time dependence of the autocorrelation function $\langle S_i(0) \cdot S_i(t) \rangle$ for a paramagnet and a spin glass.

The question of whether or not there is a phase transition at T_f turns out to be unexpectedly subtle. It is unclear whether the system ever really reaches equilibrium, as the relaxation is logarithmic in time. A solution of the model, where there is a Gaussian distribution of exchange interactions of width $\Delta\mathcal{J}$, centred at $\mathcal{J}_0$, has been given for an Ising spin glass in the mean field approximation (Fig. 6.27(a)). It shows a re-entrant transition to uniform spin glass. The diagram for a mean field solution of the Heisenberg model shows a transition T_{xy}, where the transverse spin components freeze, and another transition at a lower temperature where irreversibility sets in.

Table 6.7. The lower critical dimensionality in a uniform crystal (first column) compared with materials with random anisotropy and random field

		D_i	ΔJ
Ising	2	1	2
xy	2[a]	3	4[a]
Heisenberg	3	4	4

[a] There is a transition to a state where $M = 0$.

The molecular field approximation exaggerates the tendency towards magnetic order. More sophisticated calculations use the renormalization group method. They indicate the lowest spatial dimension where a phase transition can be expected, Table 6.7, which is believed to be 4 for a spin glass with Heisenberg spins but 2 for Ising spins. Theoretically, there should be no phase transition in three dimensions unless the spin dimensionality is reduced from 3 by local anisotropy. Mean field theory for a spin glass is thought to be accurate in $D = 6$ dimensions, compared with four dimensions for a ferromagnet.

6.6.4 Chains, ladders and sheets

Chains Isolated magnetic chains never order magnetically. The most favourable case would be the one-dimensional Ising chain {1,1}, but a simple energetic argument shows that order is impossible. Consider the cost of reversing a block of spins, as shown below:

$$\uparrow\uparrow\uparrow\uparrow\uparrow\uparrow\uparrow\uparrow\downarrow\downarrow\downarrow\downarrow\downarrow\downarrow\downarrow\downarrow\downarrow\downarrow\downarrow\uparrow\uparrow\uparrow\uparrow\uparrow\uparrow\uparrow$$

Reversal of the segment increases the energy of the chain by $8\mathcal{J}$, but this can always be recovered from the entropy term in the free energy, $F = U - TS$, which is $-k_B T \ln N$, by making the N-member chain sufficiently long. The ordering temperature is then $8\mathcal{J}/k_B \ln N$, which tends to zero as $N \longrightarrow \infty$.

If interchain interactions $\mathcal{J}'$ are not completely negligible, they easily transform the magnetic dimensionality of the system from $D = 1$ to $D = 3$. The three-dimensional magnetic ordering temperature varies as $\sqrt{\mathcal{J}\mathcal{J}'}$, so even if $\mathcal{J}/\mathcal{J}' = 100$, the weak interchain coupling can induce magnetic order at a fairly high Curie temperature.

Although there is no long-range magnetic order in a chain, there can be magnetic excitations. In the Ising chain, the excitation is the flip of a block of spins. Once created, the excitation can expand, or move along the chain at no extra cost. The dispersion relation is flat. The behaviour is the same whether the chain is ferromagnetic or antiferromagnetic.

The antiferromagetic Heisenberg spin chain {1,3} is quite different. The excitations there are known as spinons. They have a dispersion relation of the form

$$\hbar\omega_q = \pi \, |\mathcal{J} \sin qa| \,, \tag{6.32}$$

which differs only by a factor $\pi/2$ from that of antiferromagnetic magnons with $S = \frac{1}{2}$ (6.14). The main difference is that the magnon is an excitation with spin-1, corresponding to a delocalized spin deviation (or spin flip in an $S = \frac{1}{2}$ system), whereas the spinon is a spin-$\frac{1}{2}$ excitation. Magnon dispersion can be probed directly by inelastic neutron scattering, because the neutron is a spin-$\frac{1}{2}$ particle which undergoes a change $\Delta S = 1$ in a magnon scattering event. However, the neutron must excite spinons two by two, and there is a continuum of excitation energies. Furthermore, the excitations of the linear Heisenberg chain differ according to whether the ions forming the chain have integral or half-integral spin. Half-integral spins have no gap at $q = 0$, as indicated by (6.32) but integer spins exhibit a gap, known as the Haldane gap. The difference is a consequence of the different behaviour of fermions and bosons under exchange.

The half-integral spin chains can develop a gap if they dimerize to form alternate pairs of strong and weak exchange bonds. This is analogous to the mechanical instability – the Peierls distortion – that appears in a one-dimensional atomic chain which spontaneously deforms to create alternating long and short bonds. The magnetic version is the spin-Peierls effect. An example is $CuGeO_3$, which has $T_{sp} = 14$ K.

Ladders Solid-state chemists have produced materials with structures corresponding to many of the magnetic models that theoreticians conceive. There are materials having structures with single chains, or double or multiple-leg ladders. The exchange is usually different along the chain $\mathcal{J}_{\parallel}$ and between chains $\mathcal{J}_{\perp}$. An example of a two-leg ladder is $SrCu_2O_3$, and the series of compounds $Sr_{n-1}Cu_{n+1}O_{2n}$ when n is odd is composed of [$(N+1)/2$]-leg ladders. The coupling $\mathcal{J}_{\perp}$ creates a gap in the excitation spectrum if the number of legs is even, but not if it is odd.

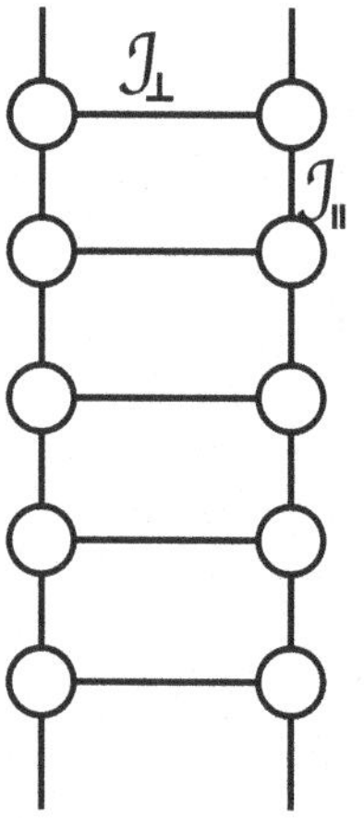

A magnetic ladder.

Sheets Crystal structures with planes of magnetic ions are model two-dimensional systems. These may be layer-like structures, like the clay minerals, with thick blocks of nonmagnetic material intercalated between the magnetic planes, or else they can be three-dimensional structures, like K_2NiF_4, where the antiferromagnetic Ni^{2+} planes are stacked in such a way that there is no net coupling between adjacent planes when they are antiferromagnetically ordered. Again, the effective spin dimensionality d may be 3, 2 or 1 depending on the single-ion anisotropy.

The two-dimensional Heisenberg model does not order magnetically at any finite temperature, acording to the Mermin–Wagner theorem, but there is evidence that there are regions of correlated short-range order and that the

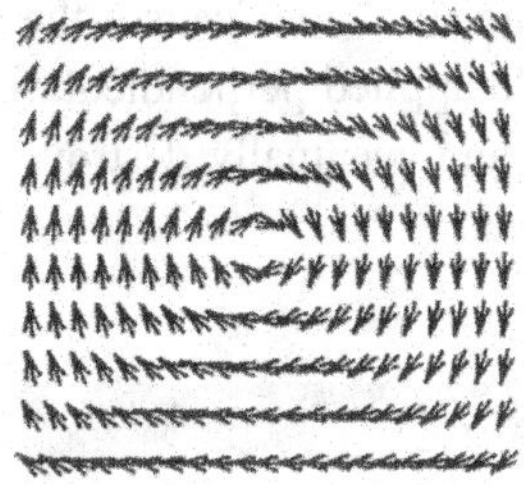

A magnetic vortex in the xy model.

correlation length diverges exponentially as $T \longrightarrow 0$. The system behaves as if it had a phase transition at zero temperature.

The two-dimensional xy model is even stranger. There is a tendency to form a vortex in a plane of xy spins. The exchange energy associated with the vortex is approximately $\pi \mathcal{J} \ln(R/a)$, where R is the size of the system; it diverges slowly as $R \longrightarrow \infty$. However, the centre of the vortex can be at any of the $(R/a)^2$ lattice sites, so the entropy of the vortex is $k_B \ln(R/a)^2$. The free energy $F = [\pi \mathcal{J} - 2Tk_B]\ln(R/a)$ becomes negative below a temperature T_{TK}, the Thouless–Kosterlitz transition:

$$T_{TK} = \frac{\pi \mathcal{J}}{2k_B}.$$

It is an unusual transition, because the low-temperature phase (like that of a spin glass) has no long-range magnetic order, and there is not the spontaneous symmetry breaking that usually accompanies a phase transition. There is a large energy barrier between degenerate states of opposite chirality.

A related defect in the three-dimensional Heisenberg ferromagnet is the *hedgehog*. Again there are two degenerate versions, one inward and one outward, with a large energy barrier between them. A similar argument gives $T_H = \pi \mathcal{J}/3k_B$, which is lower than the ferromagnetic ordering temperature $T_C = 3.66\mathcal{J}/k_B$.The ferromagnetic state is more stable than the hedgehog configuration. However, it may be possible to stabilize the hedgehog in a small particle by surface anisotropy.

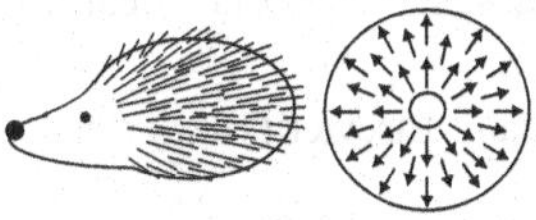

A hedgehog.

Finally, the two-dimensional Ising model {2,1} has a famous exact solution, worked out by Onsager in 1944. The solution for $S_z = \pm 1$ on a square lattice – a benchmark in the theory of phase transitions – is

$$\langle S \rangle = [1 - \sinh^4(2\mathcal{J}/k_B T)]^{1/8} \tag{6.33}$$

and the Curie temperature is $T_C = 2\mathcal{J}/k_B \ln(1 - \sqrt{2})$.

6.6.5 Quantum phase transitions

Quantum phase transitions take place at 0 K, unlike normal phase transitions which occur as a function of temperature, driven by the greater entropy of the high-temperature phase. An electronic phase transition as a function of composition x or gate voltage, or a magnetic phase transition as a function of $J_0/\Delta J$ (Fig. 6.27) may be examples of a quantum phase transition. Magnetic field or pressure are the easiest variables to control in the laboratory. In any case, by tuning some variable g one may enter a region where two states compete to be the ground state of the system. In quantum systems, fluctuations like those described by the uncertainty principle are always present.

An example is the compound $LiHoF_4$, where the Ho^{3+} ion sits in a site with uniaxial anisotropy which stabilizes an $M_J = \pm 8$ doublet, giving the ion an Ising-like character. Weak coupling between the Ho^{3+} ions causes the compound to order ferromagnetically at 1.6 K. A field applied perpendicular to the axis causes tunnelling between the two states, and eventually destroys the ferromagnetic order, creating a quantum paramagnet.

FURTHER READING

J. S. Smart, *Effective Field Theories of Magnetism*, Philadelphia: J. B. Saunders (1966). A succinct account of molecular field theory.

A Herpin, *Théorie du Magnétisme*, Paris: Presses Universitaires de France (1968). A detailed account of molecular field theory, with applications to numerous materials.

A. H. Morrish, *Physical Principles of Magnetism*, New York: Wiley (1965). A comprehensive treatment of localized-electron magnetism.

J. A. Mydosh, *Spin Glasses*, London: Taylor and Francis (1993). An experimental perspective.

K. H. Fisher and J. A. Hertz, *Spin Glasses*, Cambridge: Cambridge University Press (1993). A theoretical perspective.

D. H. Ryan (editor), *Recent Progress in Random Magnets*, Singapore: World Scientific (1992). Articles on spin freezing by experimentalists.

K. Moorjani and J. M. D. Coey, *Magnetic Glasses*, Amsterdam: Elsevier (1985). Magnetism in noncrystalline solids.

EXERCISES

6.1 Derive the susceptibility at $T = T_N$ from (6.5).

6.2 Derive (6.7). Calculate χ when $n_{AA} = n_{BB} \neq 0$ and show that T_N defined by the condition $\chi_{\parallel} = \chi_{\perp}$ is given by (6.4).

6.3 Sketch the magnetic phase diagram for an antiferromagnet at $T = 0$ K, taking H/H^i and H_a/H^i as the x and y axes.

6.4 Deduce the expressions for T_N and θ_p for the fcc lattice, considering two interactions, n_1 and n_2. Hence derive the exchange constants $\mathcal{J}_1$ and $\mathcal{J}_2$ for the inter- and intrasublattice interactions for the NaCl-structure monoxides.

6.5 Show that if magnetic excitations in a lattice of dimension d obey a dispersion relation $\omega_q = \alpha q^n$, the specific heat varies as $T^{d/n}$.

6.6 Writing the Hamiltonian of a spin glass as the sum of terms due to exchange (RKKY, in the limit of large r) and interactions with an external field, show that any thermodynamic property P that can be derived from the partition function follows a scaling law $P/x = f(T/x, H/x)$.

6.7 Find a solution for the temperature dependence of the reduced magnetization of an amorphous magnet with $S = 1$ and $\delta = 0.5$, by solving (5.3) and (6.22) graphically.

7 Micromagnetism, domains and hysteresis

Get into the loop

The domain structure of ferromagnets and ferrimagnets is a result of minimizing the free energy, which includes a self-energy term due to the dipole field $\boldsymbol{H}_d(\boldsymbol{r})$. Free energy in micromagnetic theory is expressed in the continuum approximation, where atomic structure is averaged away and $\boldsymbol{M}(\boldsymbol{r})$ is a smoothly varying function of constant magnitude. Domain formation helps to minimize the energy in most cases. The Stoner–Wohlfarth model is an exactly soluble model for coercivity based on the simplification of coherent reversal in single-domain particles. The concepts of domain-wall pinning and nucleation of reverse domains are central to the explanation of coercivity in real materials. The magnetization processes of a ferromagnet are related to the modification, and eventual elimination of the domain structure with increasing applied magnetic field.

The basic premise of micromagnetism is that a magnet is a mesoscopic continuous medium where atomic-scale structure can be ignored (§2.1): $\boldsymbol{M}(\boldsymbol{r})$ and $\boldsymbol{H}_d(\boldsymbol{r})$ are generally nonuniform, but continuously varying functions of $\boldsymbol{r}$. $\boldsymbol{M}(\boldsymbol{r})$ varies in direction only: its magnitude is the spontaneous magnetization M_s. Domains tend to form in the lowest-energy state of all but submicrometre-sized ferrromagetic or ferrimagnetic samples, Fig. 7.1, because the system wants to minimize its total self-energy, which can be written as a volume integral of the energy density E_d, in terms of the demagnetizing field (2.78):

$$\varepsilon_d = -\tfrac{1}{2}\int \mu_0 \boldsymbol{H}_d \cdot \boldsymbol{M}\,\mathrm{d}^3r. \tag{7.1}$$

Energy minimization is subject to constraints imposed by exchange, anisotropy and magnetostriction. The domain structure is eliminated by a large-enough applied field, and the underlying spontaneous magnetization of the ferromagnet is then revealed. On reducing the field, a new domain structure forms and hysteresis appears as shown in Fig. 1.3. The hysteretic response of a ferromagnet, like the behaviour of a person, depends not only on current circumstances but on what has gone before. Magnets have memory.

Demagnetizing fields and stray fields arise whenever the magnetization has a component normal to an external or internal surface. They also arise whenever the magnetization is nonuniform in such a way that $\nabla \cdot \boldsymbol{M} \neq 0$. The direction of magnetization in a domain is mainly governed by magnetocrystalline anisotropy, so a stray field associated with surface 'charge' density $\sigma = \boldsymbol{M} \cdot \mathbf{e}_n$

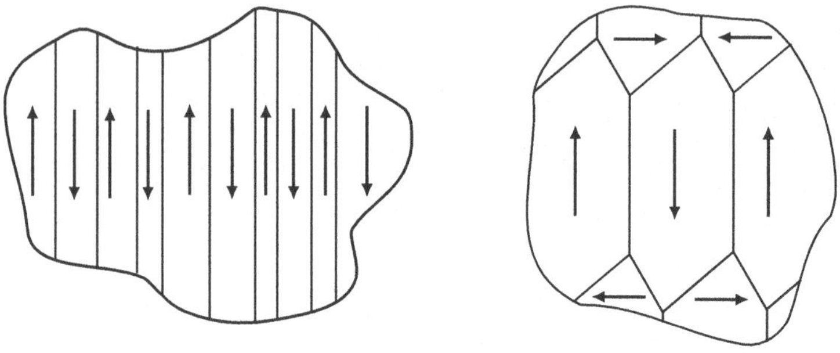

Figure 7.1

Schematic domain structures for hexagonal and cubic crystals.

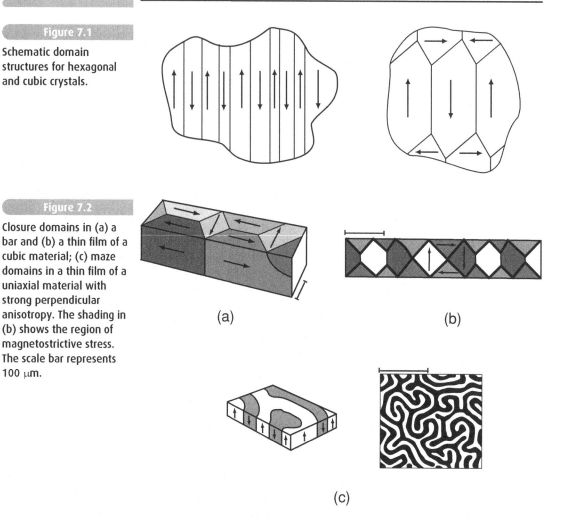

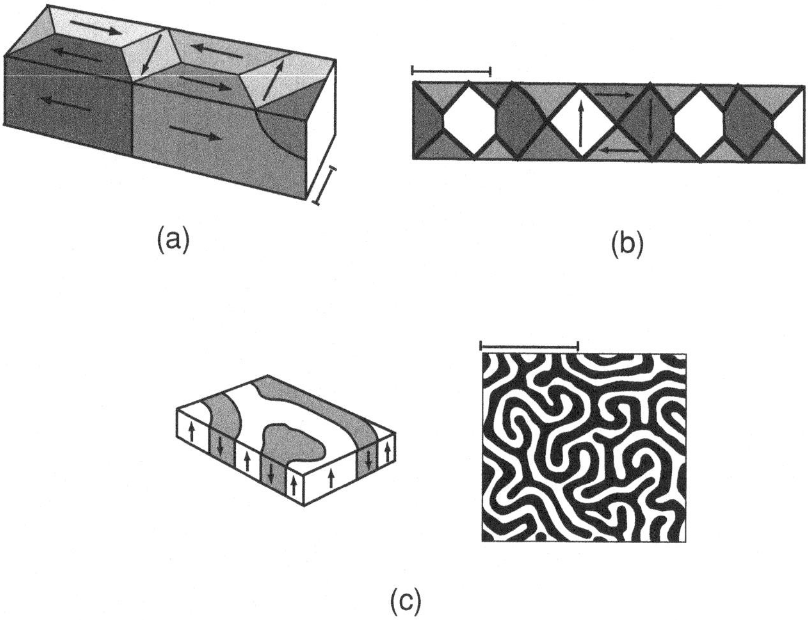

Figure 7.2

Closure domains in (a) a bar and (b) a thin film of a cubic material; (c) maze domains in a thin film of a uniaxial material with strong perpendicular anisotropy. The shading in (b) shows the region of magnetostrictive stress. The scale bar represents 100 μm.

may be created at surfaces which do not lie parallel to the easy axis of uniaxial magnets (2.54). Magnetic charge q_m, measured in A m, is a perennially useful concept in micromagnetism, a typical value of surface magnetic charge density is 10^6 A m^{-1}. In cubic materials with $\langle 100 \rangle$ anisotropy, closure domains like those in Fig. 7.2 eliminate the stray field due to surface charge. The internal walls at 45° to the magnetization bear no net charge either. Cubic materials with $K_{1c} > 0$ exhibit these 90° walls, whereas materials with $\langle 111 \rangle$ anisotropy and $K_{1c} < 0$ exhibit 71° and 109° walls. The formation of closure domains, and walls where the magnetization directions in adjacent domains are not antiparallel, is inhibited by magnetostriction. The magnetostrictive strains λ_s are then incompatible, and there is an additional magnetoelastic energy associated with a non-180° domain wall. The formation of domain walls is a consequence of magnetocrystalline anisotropy. A perfectly soft ferromagnet would tend to minimize its energy by adopting the most gradual possible variation of

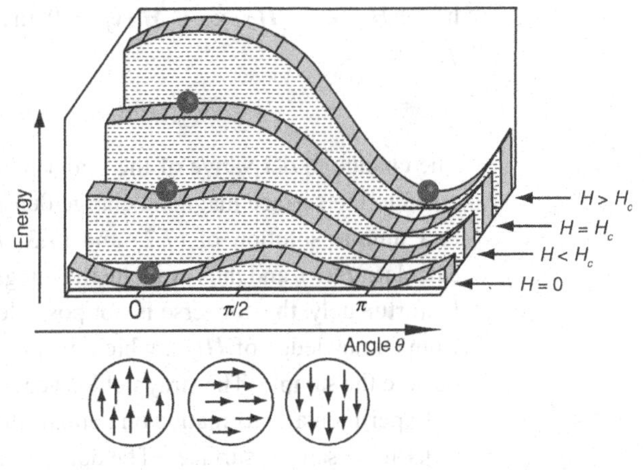

Figure 7.3

An energy landscape with metastable minima gives rise to remanence and coercivity.

magnetization direction with $\boldsymbol{M}$ everywhere parallel to the surface, so as to create no surface charge.

Coercivity and hysteresis are related to a multivalley energy landscape in magnetic configuration space, Fig. 7.3, which exists even for a uniformly magnetized single-domain particle. There are energy barriers between different magnetization configurations $\boldsymbol{M}(\boldsymbol{r})$ and jumps from one configuration to another, driven by the external field, are ireversible. The magnetization reversal process in real materials is horribly complicated, generally involving coherent and incoherent reversal processes, as well as nucleation of reverse domains and movement of domain walls. A coherent process is one where the direction of $\boldsymbol{M}$ remains everywhere the same during reversal, independent of $\boldsymbol{r}$ whereas an incoherent reversal process involves an intermediate state which is not uniformly magnetized. In all but the very smallest, nanometer-sized particles, the reversal process is either incoherent, or it involves an intermediate, multidomain state. Early progress in understanding coercivity largely relied on simplified models and phenomenological explanations. More recently large-scale computer simulations with software like OOMMF are providing direct insight into the complexity of magnetization reversal.

In the magnetostatic limit (Chapter 2), no time dependence or conduction currents are involved, and Maxwell's equations are then

$$\nabla \times \boldsymbol{H} = 0, \quad \nabla \cdot \boldsymbol{B} = 0.$$

Using $\boldsymbol{B} = \mu_0(\boldsymbol{H} + \boldsymbol{M})$, $\boldsymbol{H} = -\nabla\varphi_m$, it follows that $-\nabla^2\varphi_m + \nabla \cdot \boldsymbol{M} = 0$, hence the magnetic scalar potential obeys Poisson's equation, where the volume magnetic charge density ρ_m is $-\nabla \cdot \boldsymbol{M}$:

$$\nabla^2\varphi_m = -\rho_m. \tag{7.2}$$

The boundary condition for $\boldsymbol{B}$ at the surface of a ferromagnetic material (subscript 1) in air (subscript 2) is

$$B_1^{\perp} = \mu_0(H_1^{\perp} + M^{\perp}) = B_2^{\perp},$$

hence $\boldsymbol{H}_1 \cdot \mathbf{e}_n - \boldsymbol{H}_2 \cdot \mathbf{e}_n + \boldsymbol{M} \cdot \mathbf{e}_n = 0$. In terms of the magnetic scalar potential φ_m,

$$\partial\varphi_{m1}/\partial r_n - \partial\varphi_{m2}/\partial r_n = \boldsymbol{M} \cdot \mathbf{e}_n. \tag{7.3}$$

The change in derivative of the scalar potential φ_m at the surface is therefore equal to the surface magnetic charge density $\sigma_m = \boldsymbol{M} \cdot \mathbf{e}_n$. Given $\boldsymbol{M}(\boldsymbol{r})$, there is a unique solution of (7.2) and (7.3) for φ_m. If we know $\boldsymbol{M}(\boldsymbol{r})$ we can calculate $\boldsymbol{H}(\boldsymbol{r})$, and thus evaluate the integral of (7.1) over the whole specimen. Unfortunately, the converse is not possible; $\boldsymbol{M}(\boldsymbol{r})$ cannot be inferred uniquely from a knowledge of $\boldsymbol{H}(\boldsymbol{r})$, which, in any case, we can normally measure only outside the sample. The stray field is accessible, the demagnetizing field is not.

Experimental information on domain structures comes mainly from observations at the sample surface.[1] The domain structure of the interior is inscrutable, and it is practically impossible to infer unambiguously the arrangement of domains from the stray field measured outside, except in the case of thin films. Domain investigations often depend on models or numerical simulations which are validated by their predictions at the surface.

7.1 Micromagnetic energy

The domain structure is a result of minimizing the *total* free energy, and it reflects either a local or an absolute energy minimum. There are five other terms besides ε_d that may have to be considered:

$$\varepsilon_{tot} = \varepsilon_{ex} + \varepsilon_a + \varepsilon_d + \varepsilon_Z + \varepsilon_{stress} + \varepsilon_{ms}. \tag{7.4}$$

The first three terms due to exchange, magnetocrystalline anisotropy and the demagnetizing field are always present to some extent in a ferromagnet. The fourth is the response to an applied field, and it defines the magnetization process and hysteresis loop. The last terms are due to applied stress and magnetostriction. We neglect them at first, because the associated energies are small (Table 5.10).

The free energy is then written as a volume integral over the sample, where $\boldsymbol{M} = \boldsymbol{M}(\boldsymbol{r})$ and $\boldsymbol{H} = \boldsymbol{H}(\boldsymbol{r})$:

$$\varepsilon_{tot} = \int \{A(\nabla \boldsymbol{M}/M_s)^2 - K_1 \sin^2\theta - \cdots - \tfrac{1}{2}\mu_0 \boldsymbol{M} \cdot \boldsymbol{H}_d - \mu_0 \boldsymbol{M} \cdot \boldsymbol{H}\}\mathrm{d}^3r. \tag{7.5}$$

The meaning of $(\nabla \boldsymbol{M}/M_s)^2$ is the sum of the squares of the gradient of the three components $(\nabla M_x/M_s)^2 + (\nabla M_y/M_s)^2 + (\nabla M_z/M_s)^2$. The exchange

[1] It is possible, in principle, to build up a three-dimensional image of domains in a solid by neutron scattering, electron holography or NMR tomography, but these methods are impracticable for most real specimens.

stiffness A and anisotropy constants $K_1, \ldots,$ may also be position-dependent, but the magnitude of the magnetization M_s is supposed to remain constant everywhere. Only its direction, represented by the unit vector $\mathbf{e}_M$, wanders from point to point. The reference z axis is generally taken as the anisotropy axis. We now examine the terms in (7.5) one by one.

7.1.1 Exchange

The first term in (7.5) is the exchange energy in the continuum picture

$$\varepsilon_{ex} = \int A(\nabla \mathbf{e}_M)^2 \mathrm{d}^3 r,$$

where $\mathbf{e}_M = \boldsymbol{M}(\boldsymbol{r})/M_s$ is a unit vector in the local direction of magnetization (θ, ϕ) relative to the z axis (usually defined by the leading term in the anisotropy). By writing the unit vector $\mathbf{e}_M$ in Cartesian coordinates $(\sin\theta\cos\phi, \sin\theta\sin\phi, \cos\theta)$, it can be shown that an equivalent form for this term is

$$\varepsilon_{ex} = \int A[(\nabla\theta)^2 + \sin^2\theta(\nabla\phi)^2]\mathrm{d}^3 r. \tag{7.6}$$

A simplification arises if the magnetization twist is confined to a plane ϕ =constant; the exchange term is then $A(\nabla\theta)^2$, which reduces further to $A(\partial\theta/\partial x)^2$ when the magnetization varies along a single direction, as in a Bloch wall. However we choose to write it, the sense of the exchange term is to maintain the smoothest possible variation of $\mathbf{e}_M$ in all directions. Rapid fluctuations of θ and ϕ incur an exchange energy penalty.

The exchange stiffness A is related to the Curie temperature T_C: A is roughly $k_B T_C/2a_0$, where a_0 is the lattice parameter in a simple structure. It is also proportional to the exchange constant $\mathcal{J}$. The relation is

$$A \approx \mathcal{J} S^2 Z_c/a_0, \tag{7.7}$$

where Z_c is the number of atoms per unit cell – one for simple cubic, two for bcc and four for fcc. For hcp, $A = 2\sqrt{2}\mathcal{J} S^2/a$. However, the best way to derive A is from the spin-wave stiffness D_{sw} in the low-energy magnon dispersion relation (5.56), since the energy of the long-wavelength spin waves is associated with a gradual twist of the magnetization. The relation is

$$A(T) = \frac{M_s(T)D_{sw}}{2g\mu_B}.$$

A typical value of A for a ferromagnet having a Curie temperature well above room temperature is 10 pJ m^{-1}. Cobalt and permalloy have $A = 31$ pJ m^{-1} and 10 pJ m^{-1}, respectively.

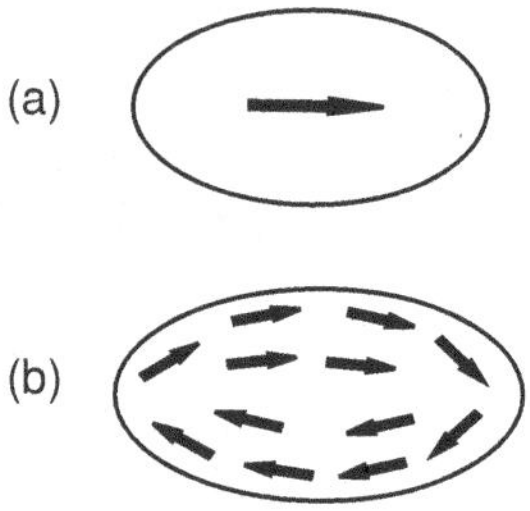

Stable ferromagnetic configurations in a soft spheroidal particle: (a) without and (b) with the effect of the demagnetizing field.

The competition between the exchange energy ε_{ex} and the dipolar energy ε_d is characterized by the exchange length:[2]

$$l_{ex} = \sqrt{\frac{A}{\mu_0 M_s^2}}. \tag{7.8}$$

A typical value of l_{ex} is 3 nm (Table 7.1). This is the shortest scale on which the magnetization can be twisted in order to minimize the dipolar interaction.

If ε_{ex} were the only term to consider, there would be no incentive for $\mathbf{e}_M$ to vary, and the magnetization would remain uniform. But ε_d is ever-present, and its tendency is to reduce the net moment of an isotropic sample to zero by continuous rotation of the magnetization. There will be an energy cost of order $\mathcal{J}S^2 \ln(R/a)$ associated with the vortex, where R is the radius of the particle. For micrometre-sized particles, $\ln(R/a) \approx 10$. This cost can be easily compensated by the reduction in demagnetizing energy $(V/6)\mu_0 M_s^2$ for all but the smallest spherical particles of volume V. Soft magnetic particles with negligible anisotropy tend to adopt a curling or vortex state when they are larger than about 10 nm.

7.1.2 Anisotropy

The anisotropy energy E_a is the single-ion or two-ion magnetocrystalline anisotropy, the leading term of which is $K_1 \sin^2\theta$ with an anisotropy constant K_1 ranging from 0.1 to 10^4 kJ m^{-3}. The anisotropy energy is usually expressed in terms of the polar angles (θ, ϕ) of the magnetization direction $\boldsymbol{e}_M$, but it is sometimes useful to make an expansion in terms of the operators $\hat{\boldsymbol{O}}_n^m$ of the crystal-field Hamiltonian. Shape anisotropy which is related to the demagnetizing field is included in ε_d. In each crystallite there are easy directions $\mathbf{e}_n$, which will vary with position in a polycrystalline sample. The balance of exchange and anisotropy usually leads to a structure in domains where the magnetization lies along an easy axis, separated by narrow domain walls, where the magnetization rotates from one easy direction to another.

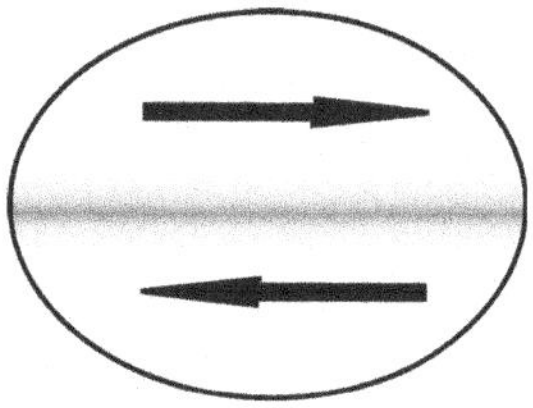

A ferromagnetic domain state resulting from the interplay of uniaxial anisotropy and demagnetizing field. The domain wall is the shaded region.

For thin films and nanoparticles, we have to take into account an additional surface anisotropy term

$$\varepsilon_{as} = \int K_s[1 - (\mathbf{e}_M \cdot \mathbf{e}_n)^2]\mathrm{d}^2 r,$$

where $\mathbf{e}_n$ is the surface normal, which defines the z axis, and the integral is over the sample surface. Typical values of surface anisotropy K_s are 0.1–1 mJ m^{-2}.

Exchange-related anisotropy may be present when there is coupling at a ferromagnetic–antiferromagnetic or soft–hard interface. Unlike the other forms

[2] Other definitions of exchange length can be found in the literature, such as $\sqrt{2A/\mu_0 M_s^2}$ or $\sqrt{A/K_{eff}}$, where K_{eff} is an effective anisotropy constant.

of anisotropy, this one is unidirectional,

$$\varepsilon_{ea} = -\int K_{ex} \cos\theta \mathrm{d}^2 r,$$

where θ is the angle between $\mathbf{e}_M$ and the preferred anisotropy direction.

7.1.3 Demagnetizing field

When no external field is present, it follows from $\boldsymbol{B} = \mu_0(\boldsymbol{H} + \boldsymbol{M})$ and $\nabla \cdot \boldsymbol{B} = 0$ that $\nabla \cdot \boldsymbol{H}_d = -\nabla \cdot \boldsymbol{M}$. Using the result (2.80) $\int \boldsymbol{B} \cdot \boldsymbol{H} \mathrm{d}^3 r = 0$, where the integral is over all space, it follows that the demagnetizing energy (7.1) can be written in two equivalent ways:

$$\varepsilon_d = -\tfrac{1}{2} \int \mu_0 \boldsymbol{H}_d \cdot \boldsymbol{M} \mathrm{d}^3 r \quad \text{(integral over the volume of the magnet)}$$
$$\varepsilon_d = \tfrac{1}{2} \int \mu_0 H_d^2 \mathrm{d}^3 r \quad \text{(integral over all space).} \tag{7.9}$$

If $\boldsymbol{H}_d$ is expressed in terms of the scalar potential φ_m, the integral in the second expression for the energy can be evaluated from the volume and surface charge distributions, $-\nabla \cdot \boldsymbol{M}$ and $\boldsymbol{M} \cdot \mathbf{e}_n$ (2.54), using Gauss's theorem for a vector field. In the case of a uniformly magnetized ellipsoid, the volume integral is zero and the surface contribution is that of the uniform demagnetizing field $-\mathcal{N}\boldsymbol{M}$, namely $-\frac{1}{2}\mu_0 \mathcal{N} M_s^2$. Values of E_d range up to 2000 kJ m^{-3}.

7.1.4 Strain

An external stress σ_{ij} applied to a sample introduces a strain term in the energy:

$$\varepsilon_{stress} = -\sum_{i,j} \sigma_{ij} \epsilon_{ij}, \tag{7.10}$$

where $\epsilon_{ij} = \hat{m}_{ijkl} H_k H_l$ is the magnetoelastic strain tensor. For an isotropic material with uniaxial stress σ along Oz, the contribution depending on $\boldsymbol{M}$ is $-\frac{1}{2}\lambda_s \sigma (3\cos^2\theta - 1)$ (5.67), where λ_s is the spontaneous magnetostriction. This is equivalent to a uniaxial anisotropy with $K_u = (3/2)\lambda_s \sigma$.

7.1.5 Magnetostriction

Local stress may also be generated by the magnetostriction of the ferromagnetic material itself. Stresses only arise if the magnetostrictive strains are frustrated. The corresponding strain energy is

$$\varepsilon_{ms} = \tfrac{1}{2} \int (\boldsymbol{p}_e - \boldsymbol{\epsilon}) \cdot \boldsymbol{c} \cdot (\boldsymbol{p}_e - \boldsymbol{\epsilon}) \mathrm{d}^3 r. \tag{7.11}$$

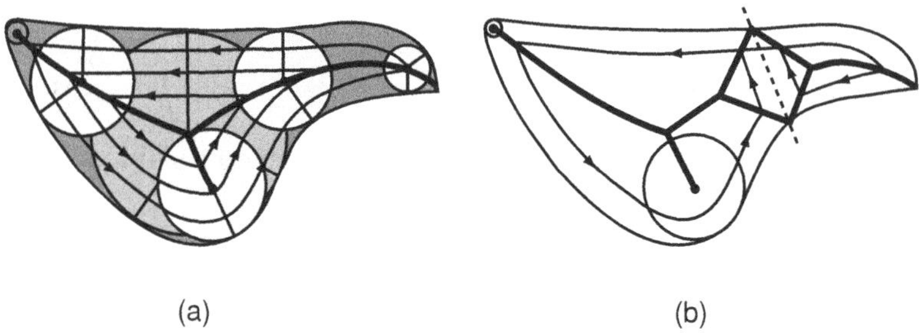

Figure 7.4
(a) The van den Berg construction to obtain the domain structure of an ideally soft two-dimensional element. In (b) extra domains are introduced by a virtual cut shown by the dotted line.

Here $\boldsymbol{p}_e$ is the deviation from the nonmagnetostrictive state, $\boldsymbol{\epsilon}$ is the strain in the freely deformed state and $\mathbf{c}$ is the tensor of elastic constants. $\boldsymbol{c} \cdot (\boldsymbol{p}_e - \boldsymbol{\epsilon})$ is the magnetostrictive stress $\boldsymbol{\sigma}$. Thankfully this term is usually small, < 1 kJ m^{-3}.

7.1.6 Charge avoidance

The task of understanding how nature minimizes the sum of all six terms in the micromagnetic free energy of a ferromagnetic specimen to arrive at a stable configuration is a formidable one. Vortex states tend to be stable in soft ferromagnets, multidomain states in hard ferromagnets and single-domain states in the smallest ferromagnetic elements. A rough but useful guide for minimizing the energy in a multidomain body is the charge avoidance principle. Surface charge $\boldsymbol{M} \cdot \mathbf{e}_n$ is loathe to form, because of the high energy cost of the stray field H_d which it creates (2.79). Charges of the same sign tend to avoid each other as far as possible for the same reason. The magnetization in the bulk should have as little divergence as possible to avoid creating volume charge $\nabla \cdot \boldsymbol{M}$. Complete freedom of the magnetization to rotate in order to maximize charge avoidance implies an absence of anisotropy that is characteristic of very soft magnetic material. The cost in exchange energy of forming the domain walls needed to achieve the 'pole-free' configurations is overlooked, but charge avoidance provides us with an idea of the magnetic configurations which would arise from the influence of dipole interactions alone.

An elegant construction due to van den Berg produces 'pole-free' configurations in thin-film elements that depend only on the shape of the element. His method ensures that the magnetization is everywhere parallel to the surface. The domain walls are the locus of points corresponding to the centre of a circle which touches the edge of the element at two points at least, but does not cut it anywhere. The magnetization pattern so obtained is not only parallel to the surface, but also divergenceless in the interior. Walls end at singular points within the element, or else at sharp corners. The construction is illustrated in Fig. 7.4. Virtual cuts can be introduced anywhere in the element to create new domains, as shown in Fig. 7.4(b). The van den Berg construction shows that domains can arise from the dipole interaction (7.1) alone. It is best applied

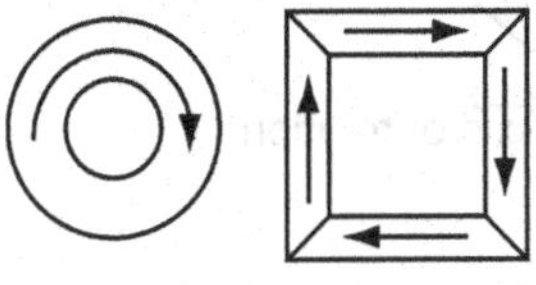

Configurations that create no stray field: (a) ring and (b) picture frame.

to soft magnetic films with little anisotropy and dimensions which exceed the domain wall width.

Applying the van den Berg construction to a ring-shaped or rectangular frame gives a configuration with a single wall running around the centre. In fact, a lower 'pole-free' configuration exists without any wall. These stray-field-free configurations are useful for measuring internal susceptibility (§2.2.6). The picture-frame configuration with one or more domain walls is compatible with $\langle 100 \rangle$ cubic anisotropy of a single crystal. Closure domains with 90 ° domain walls (Fig. 7.2) are very effective at reducing the stray field in soft ferromagnets such as permalloy, as there is no surface charge.

A permanent magnet is the antithesis of charge avoidance. Here the purpose is to generate as much stray field as possible in surrounding space. Permanent magnets are characterized by lots of surface charge and deep metastable minima in their energy landscape which make configurations with close-to-saturated magnetization stable almost indefinitely. The anisotropy term in (7.5) is the key to achieving this.

Generally, the minimization of (7.4) or (7.5) involves finding a solution where the magnetization direction is stable at every point, subject to boundary conditions. If $\boldsymbol{H}_{eff}$ is the local effective field, which makes an angle ϑ with the local magnetization direction $\mathbf{e}_M$, the condition $\partial E_{tot}/\partial\vartheta = 0$ means that $-MH_{eff}\sin\vartheta = 0$. This is equivalent to the condition that no torque acts anywhere on the magnetization; $\boldsymbol{\Gamma} = \boldsymbol{M} \times \boldsymbol{H}_{eff} = 0$, so

$$\mathbf{e}_M \times \boldsymbol{H}_{eff} = 0. \tag{7.12}$$

William Fuller Brown, 1904–1983.

The local effective field, in a vector notation, is

$$\boldsymbol{H}_{eff} = \frac{2A}{\mu_0 M_s}\nabla^2\mathbf{e}_M - \frac{1}{\mu_0 M_s}\frac{\partial E_a}{\partial \mathbf{e}_M} + H_d + H'. \tag{7.13}$$

The notation $\nabla^2\mathbf{e}_M$ indicates a vector whose Cartesian coordinates are $M_s^{-1}[(\partial^2 M_x/\partial x^2 + \partial^2 M_x/\partial y^2 + \partial^2 M_x/\partial z^2), \ldots]$, and $\partial E_a/\partial\mathbf{e}_M$ is a vector whose Cartesian coordinates are $[\partial E_a/\partial\mathbf{e}_{Mx}, \partial E_a/\partial\mathbf{e}_{My}, \partial E_a/\partial\mathbf{e}_{Mz}]$. These two are Brown's micromagnetic equations, which have to be solved numerically, subject to the surface boundary condition

$$\mathbf{e}_M \times \left[2A(\mathbf{e}_n \cdot \nabla)\mathbf{e}_M + \frac{\partial E_{as}}{\partial \mathbf{e}_M}\right] = 0.$$

Their meaning is that in equilibrium the magnetization lies everywhere parallel to $\boldsymbol{H}_{eff}$ given by (7.13).

7.2 Domain theory

The micromagnetic approach is capable, in principle, of predicting the equilibrium magnetic configurations of any system where the exchange stiffness $A(\boldsymbol{r})$

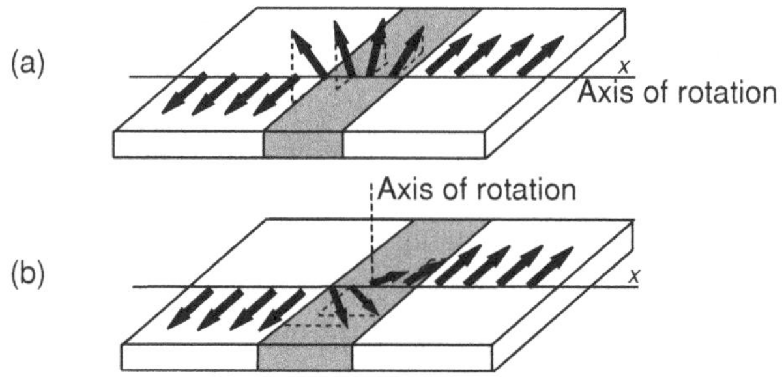

Figure 7.5
(a) A Bloch wall, and (b) a Néel wall.

and anisotropy $E_a(\boldsymbol{r})$ can be specified throughout. Hysteresis may be deduced, knowing the magnetic history in an applied field $\boldsymbol{H}'(t)$. No account is taken of temperature. It is impractical to implement micromagnetic theory in any but idealized situations. The problem is mathematically complex, and real materials contain local defects and disorder which cannot be specified precisely, but which nonetheless tend to dominate the magnetization process.

Domain theory is an attempt to reduce this complexity to manageable proportions. It postulates the existence of large regions of uniform magnetization in a macroscopic sample, which are separated by planar regions – the domain walls – where the magnetization rotates from one easy direction to another. Domain observations support the model. If domains exist, so must domain walls. An applied field changes the net magnetization of the sample, either by causing the walls to move or by making the magnetization in the domains rotate towards the applied field direction. The magnetostatic energy depends on the wall positions and the domain orientations.

Domain theory breaks down in very soft magnetic materials, especially in thin film elements where the demagnetizing field is small. There, instead of domains, states with continuous rotation of magnetization tend to form.

Now we look more carefully into the structure of the domain walls. A flip of magnetization from one plane of atoms to the next would be prohibitively expensive, costing $4\mathcal{J}S^2/a^2 \approx 2A/a$ or about 0.1 J m^{-2}. Magnetization rotates continuously over many interatomic distances under the combined influence of exchange and anisotropy. Dimensional analysis gives the wall width $\delta_w \approx \sqrt{A/K_1}$, which is of order 10–100 nm, and the wall energy $\gamma_w \approx \sqrt{AK_1}$, which is of order 1 mJ m^{-2}. The association of energy with the wall area means that the domain wall behaves like an elastic membrane or soap film. Two common types of domain wall are illustrated in Fig. 7.5.

7.2.1 Bloch wall

The commonest is the 180° Bloch wall illustrated in more detail in Fig. 7.6, where the magnetization rotates in the plane of the wall. The Bloch wall has the remarkable property that it creates no divergence of the magnetization. Each of

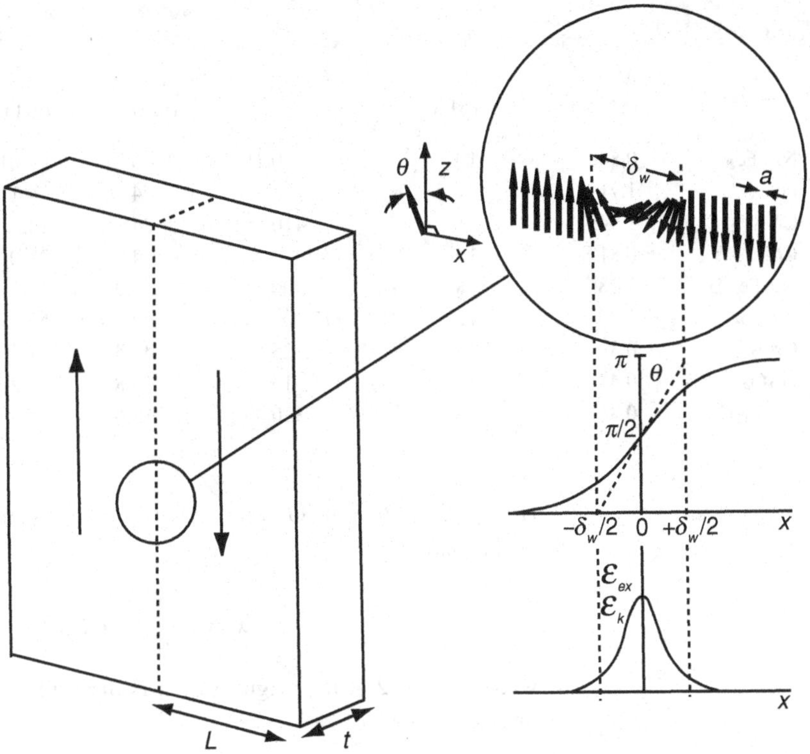

Figure 7.6

Detail of the 180° Bloch wall.

the three terms in $\nabla \cdot \boldsymbol{M} = \partial M_x/\partial x + \partial M_y/\partial y + \partial M_z/\partial z$ is zero; there is no component of magnetization in the x-direction, and the spins in each yz-plane are parallel to each other. Since $\nabla \cdot \boldsymbol{M} = 0$, there is no magnetic charge and no source of demagnetizing field in the wall.

To calculate the form of the Bloch wall, we minimize the free energy (7.5) for clockwise (or anticlockwise) rotation. Ignoring the magnetostatic energy originating at the sample surface, and considering only the leading anisotropy term

$$\varepsilon_{tot} = \varepsilon_{ex} + \varepsilon_k = \int [A(\partial\theta/\partial x)^2 + K \sin^2\theta]\mathrm{d}x. \qquad (7.14)$$

Here $K = K_1$ is assumed to be positive, so that $\theta = 0$ and $\theta = \pi$ are equivalent easy directions. Magnetization is confined to the plane $\phi = \pi/2$, and the only variation of θ is along the x axis. Additional anisotropy terms are neglected. There is no need to consider demagnetizing energy because there are no sources of demagnetizing field in the wall. Minimizing the integral, which is of the form $\int F(x, \theta(x), \theta'(x))\mathrm{d}x$, is equivalent to solving the Euler equation $\partial F/\partial\theta - (\mathrm{d}/\mathrm{d}x)(\partial F/\partial\theta') = 0$ where $\theta' = \partial\theta/\partial x$:

$$\partial(K \sin^2\theta)/\partial\theta - 2A\partial^2\theta/\partial x^2 = 0. \qquad (7.15)$$

Table 7.1. Domain wall parameters for some ferromagnetic materials

	M_s (MA m^{-1})	A (pJ m^{-1})	K_1 (kJ m^{-3})	δ_w (nm)	γ_w (mJ m^{-2})	κ	l_{ex} (nm)
$Ni_{80}Fe_{20}$	0.84	10	0.15	2000	0.01	0.01	3.4
Fe	1.71	21	48	64	4.1	0.12	2.4
Co	1.44	31	410	24	14.3	0.45	3.4
CoPt	0.81	10	4900	4.5	28.0	2.47	3.5
$Nd_2Fe_{14}B$	1.28	8	4900	3.9	25	1.54	1.9
$SmCo_5$	0.86	12	17 200	2.6	57.5	4.30	3.6
CrO_2	0.39	4	25	44.4	1.1	0.36	4.4
Fe_3O_4	0.48	7	−13	72.8	1.2	0.21	4.9
$BaFe_{12}O_{19}$	0.38	6	330	13.6	5.6	1.35	5.8

Hence $K\sin^2\theta = A(\partial\theta/\partial x)^2$, $\partial\theta/\partial x = \sqrt{K/A}\sin\theta$, yielding the domain-wall equation

$$x = \sqrt{\frac{A}{K}}\ln\tan(\theta/2), \tag{7.16}$$

where $\theta = \pi/2$ at the origin, which is the wall centre. The equation can be inverted to read

$$\theta(x) = \tan^{-1}[\sin k(\pi x/\delta_w)] + \pi/2,$$

where

$$\delta_w = \pi\sqrt{\frac{A}{K}}. \tag{7.17}$$

The domain wall does not have a precisely defined width, since the direction of magnetization only approaches 0 or π asymptotically. But it is usual to define the extrapolated width from the tangent at the origin to (7.16). An alternative is to define the width as the distance between points where some fraction, say 90%, of the rotation has taken place. In that case $\delta_w \approx 4\sqrt{(A/K)}$. Since $K\sin^2\theta = A(\partial\theta/\partial x)^2$; the two terms in the integral (7.14) for the energy are equal to each other at every point in the wall. The energy per unit domain wall area is

$$\gamma_w = 4\sqrt{AK}. \tag{7.18}$$

If K were negative, and there were no terms other than $K\sin^2\theta$ in the expansion of the anisotropy energy, the magnetization would lie in the plane $\theta = \pi/2$. The domain wall $\boldsymbol{M}(\phi)$ between regions with $\phi = 0$ and $\phi = \pi$ would be infintely extended because no anisotropy cost is incurred. Anisotropy of some sort is necessary for a finite wall width. Domain wall parameters for different materials are collected in Table 7.1.

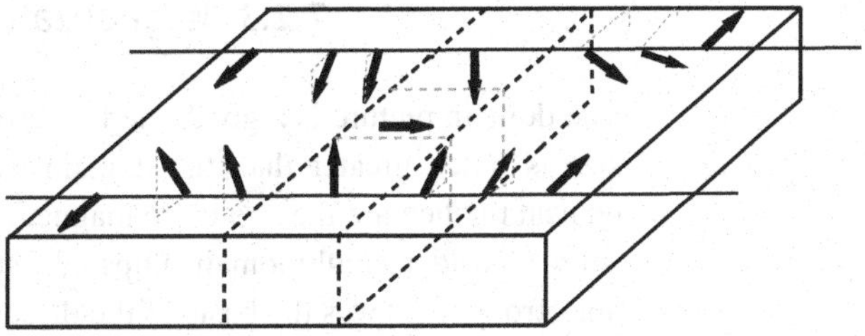

Figure 7.7
Néel lines in a Bloch wall.

Néel lines are topological defects between segments of Bloch walls where the magnetization rotates in opposite senses, Fig. 7.7. Néel lines are rather stable magnetic defects where the magnetization rotates in the plane of the domain magnetization; they can only be eliminated by completely saturating the magnetization.

Expressions for wall width and energy are different for cubic anisotropy. For example, the energy for a 90° wall, which forms in iron and other materials with $\langle 100 \rangle$ easy axes, is $\gamma_{100} = \sqrt{AK_{1c}}$. When $\langle 111 \rangle$ axes are easy, as in nickel, 71°, 109° and 180° walls may form, with energy $c\sqrt{AK_{1c}}$ with $c = 0.5, 1.5$ and 2.2, respectively.

7.2.2 Néel wall

The Néel wall, where the magnetization rotates within the plane of the domain magnetization, is normally higher in energy than the Bloch wall because of the stray field created by the nonzero divergence of $\boldsymbol{M}$. But unlike the Bloch wall, the Néel wall creates no surface charge, and there is no associated stray field. Néel lines are actually strips of Néel wall. When two regions of Néel wall with opposite chirality meet, there forms a strip of Bloch wall, which we call a Bloch line.[3]

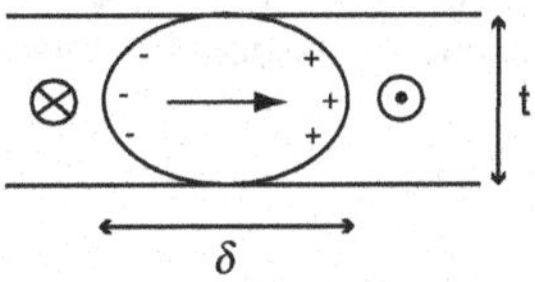

Explanation of formation of Néel walls in thin films.

Néel walls are only stable in films thinner than the wall width. To explain the formation of his walls Néel represented the wall by an elliptical cylinder of cross section t × δ where t is the film thickness. For a bulk sample the demagnetizing factor for the Bloch wall is zero, while that for the Néel wall is 1. When t $< \delta$ the Néel wall has the lower magnetostatic energy. However, the Bloch wall width itself is much reduced in thin films in order to minimize the stray field. The Bloch–Néel crossover thickness in a film of permalloy ($Ni_{80}Fe_{20}$) is at 60 nm.

A cross-tie wall is a Néel wall in which the magnetization rotates in opposite directions in adjacent sections. Transitions between domains in thicker films may have vortex structures, mixtures with Néel caps at the surface and Bloch character at the centre. There is rich fauna in the micromagnetic jungle.

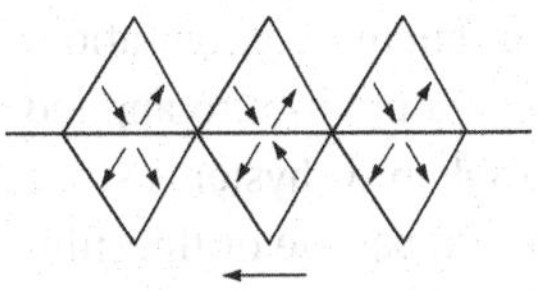

A cross-tie wall.

[3] Often in the literature, the chirality-related line defect in a Bloch wall is called a Bloch line. It seems more reasonable to use this name for the chirality-related defect in a Néel wall where the magnetization rotates around a line perpendicular to the wall.

7.2.3 Magnetization processes

The domain picture is a good one for ferromagnetic solids, when the domain size is much greater than the domain wall width. Domain wall motion and domain rotation are the two basic magnetization processes in any multidomain solid. Consider a multidomain ellipsoidal specimen with no magnetocrystalline anisotropy. This was the basis of the discussion of the demagnetizing fields in §2.2.4. When the field is applied along the easy axis, the effect is to move the domain walls in such a way as to grow the parallel domains and shrink the antiparallel domains. The average magnetization $M_{av} = M[(V_p - V_{ap})/(V_p + V_{ap})]$ increases with increasing applied field in such a way that $H = H' - \mathcal{N} M_{av} = 0$. The external susceptibility due to domain wall motion is $\chi_{ext} = M_{av}/H' = 1/\mathcal{N}$.

The other magnetization process, magnetization rotation, applies when the applied field $\boldsymbol{H}'$ has a component perpendicular to the anisotropy axis. If K_u is the effective anisotropy of whatever origin, the external susceptibility is $\mu_0 M_s^2/2K_u$. The magnetization attains saturation when the applied field equals the anisotropy field, $2K_u/\mu_0 M_s$.

7.2.4 Condon domains

Domains of a different type, unrelated to electron spin, are associated with Landau diamagnetism. When the magnetization exhibits strong de Haas–van Alphen oscillations where the differential susceptibility exceeds a threshold value of 1, the sample will break up into domains magnetized parallel and antiparallel to the applied field, in order to minimize the energy associated with the dipole field.

7.3 Reversal, pinning and nucleation

Hysteresis would never exist unless there was some chance of a ferromagnetic specimin getting stuck in a metastable configuration with a remanant magnetization and higher energy than the absolute minimum energy configuration, which is reached by cooling from above the Curie temperature in zero applied field. In order to make some sense of the complex problem of hysteresis, we first examine the magnetization reversal in a single-domain particle or thin-film element, which can be calculated analytically. Then we look into how domain theory allows us to formulate the elementary processes involved in coercivity in multidomain samples.

Figure 7.8

Progress towards narrowing the gap between the anisotropy field H_a and coercivity H_c (after Kronmüller and Fähnle (2003)).

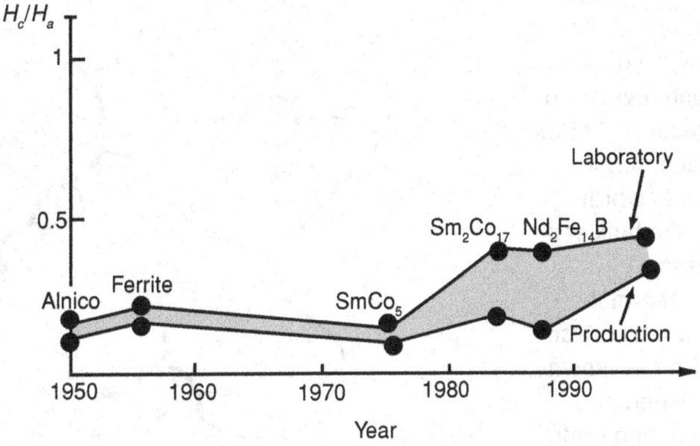

In 1947 William Fuller Brown proved rigorously that the coercivity for a homogeneous, uniformly magnetized ellipsoid obeys the inequality

$$H_c \geq (2K_1/\mu_0 M_s) - \mathcal{N} M_s, \tag{7.19}$$

a surprising result known in micromagnetics as **Brown's theorem**. The first term is the anisotropy field, and the second is the demagnetizing field. In reality, coercivity in bulk material is *never* this large. It is generally a long struggle, following the introduction of a new hard magnetic material, to achieve a coercivity that exceeds 20–30% of the anisotropy field (Fig. 7.8). This apparent contradiction between theory and practice is known as **Brown's paradox**. The explanation here, as in other cases – for example Earnshaw's theorem – where there is an apparent conflict between theory and experiment, is that the assumptions of the theory are not met in practice. All real materials are *inhomogeneous*, and magnetization reversal is initiated in a small nucleation volume around a defect.

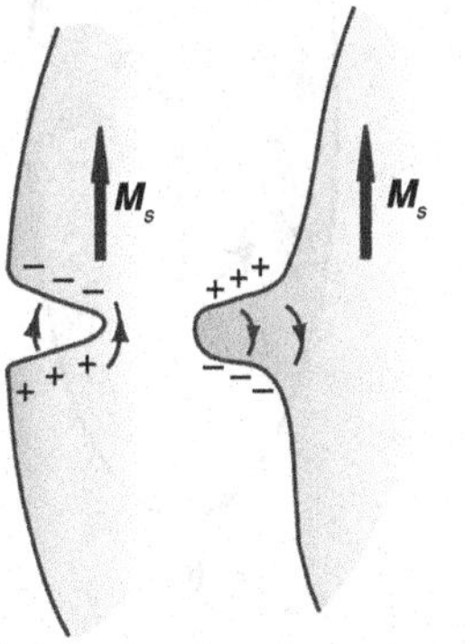

Local stray fields near a surface pit or bump. The region prone to reversal is shaded.

Surface asperities are sources of strong local demagnetizing fields. These surface defects often act as nucleation centres because, in the second quadrant of the hysteresis loop, the reverse magnetic field H is enhanced in their vicinty. Once a small nucleus of volume $V \approx \delta_w^3$ has formed, the wall may propagate outwards, growing from the nucleation volume. Otherwise the new wall may become pinned at some other defect, as illustrated in Fig. 7.9.

Very small magnetic particles must be single-domain; they do not benefit energetically from wall formation if they are below a certain critical size. Consider a spherical particle of radius R with cubic anisotropy, which forms two 90° domain walls in order to almost eliminate the stray field. The cost of creating the two walls, $2\pi R^2 \sqrt{AK_{1c}}$, must be offset by the gain in demagnetizing energy of the sphere, $-\frac{1}{2}\mu_0 \mathcal{N} V M_s^2$. The demagnetizing factor $\mathcal{N} = 1/3$ gives the following expression for the maximum single-domain size.

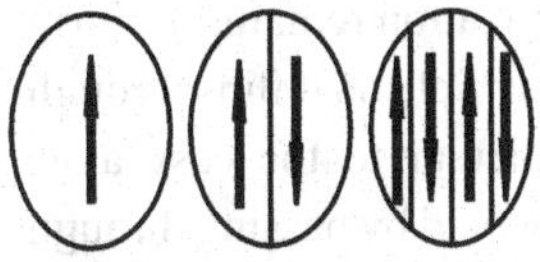

A magnetic particle reduces its energy by forming domains.

$$R_{sd} \approx 9\sqrt{AK_{1c}}/\mu_0 M_s^2. \tag{7.20}$$

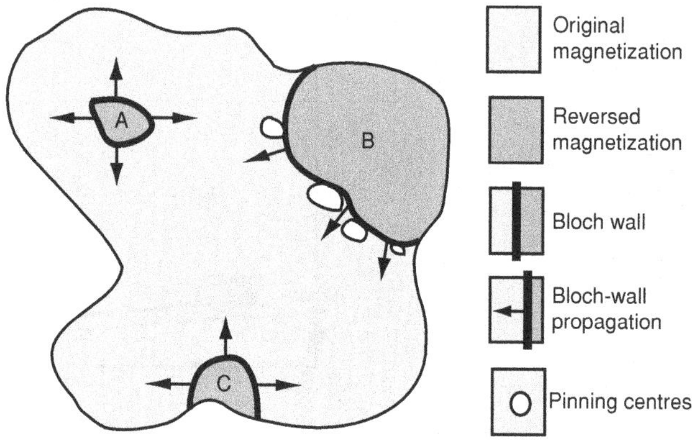

Figure 7.9

Processes involved in magnetization reversal in the second quadrant of the hysteresis loop. A is a reverse domain which nucleates in the bulk at a defect, or from a spontaneous thermal fluctuation. B is a reverse domain which has grown to the point where it is trapped by pinning centres and C is a reverse domain which nucleates at a surface asperity.

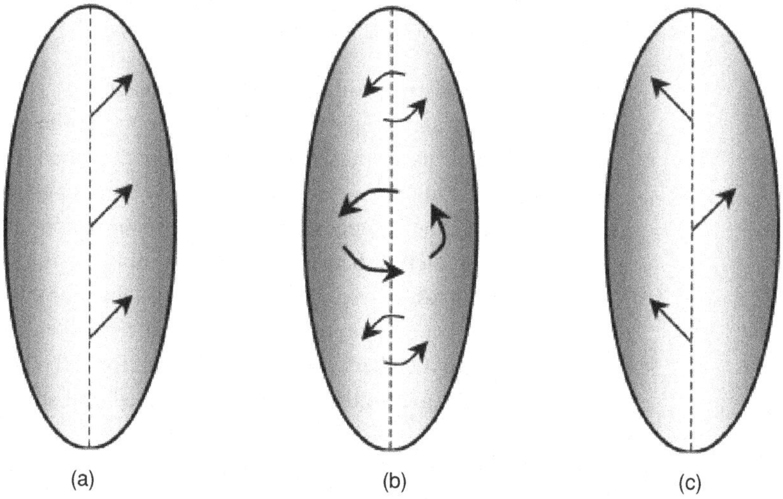

Figure 7.10

Demagnetization processes in a homogeneous ellipsoid: (a) coherent rotation, (b) curling, (c) buckling. Coherent rotation is the reversal mode for the smallest, single-domain particles.

Two or three different modes of reversal can occur in a single-domain particle. They are investigated theoretically for infinite cylinders and homogeneous ellipsoids of revolution by looking for the first deviation from a uniformly magnetized state when a field is applied along an easy axis in a direction opposite to the magnetization. They are illustrated in Fig. 7.10.

First is the coherent rotation mode where the magnetization remains uniform everywhere, and it rotates in unison, increasing the stray field as it flips through a configuration where the magnetization is perpendicular to the easy axis. Second is the curling mode, which avoids creating stray field by passing through a vortex state where the magnetization lies everywhere parallel to the surface (Fig. 7.10(b)). This costs exchange energy. The vortex state is the lowest energy state of soft magnetic particles which are larger than the coherence radius. In long prolate ellipsoids, a third reveral mode known as buckling may occur, which is a combination of the other two that creates less stray field than coherent rotation.

The simplest model of magnetization reversal, the Stoner-Wohlfarth model, assumes the *coherent* mode of magnetization reveral. The magnetization is supposed to remain uniform as its orientation changes with time during reversal. However, even if a particle is smaller than R_{sd}, magnetization reversal does not have to be coherent. Other possible reversal modes, curling and buckling, are illustrated in Fig. 7.10. Curling is the one which arises in spherical particles which are larger than the coherence radius.

7.3.1 Stoner-Wohlfarth model

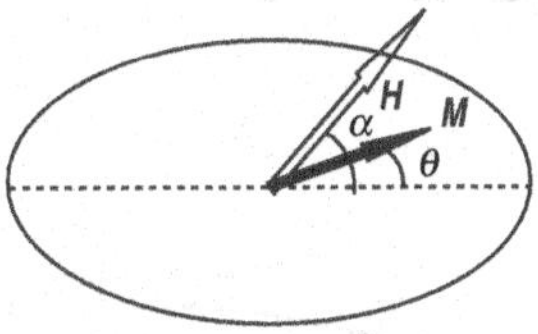

A Stoner-Wohlfarth particle.

Despite the restrictive and often unrealistic condition of coherent magnetization reversal, the Stoner–Wohlfarth model is a beacon of light on the complex landscape of hysteresis in real materials. It is the simplest analytical model which exhibits hysteresis. Imagine a Stoner–Wohlfarth particle, a uniformly magnetized ellipsoid with uniaxial anisotropy of shape or magnetocrystalline origin in a field applied at an angle α to the anisotropy axis. The energy density is

$$E_{tot} = K_u \sin^2\theta - \mu_0 M H \cos(\alpha - \theta). \tag{7.21}$$

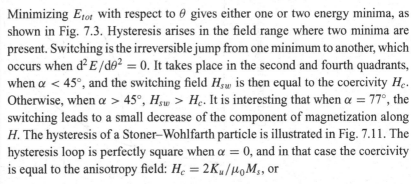

Peter Wohlfarth 1924–1988.

Minimizing E_{tot} with respect to θ gives either one or two energy minima, as shown in Fig. 7.3. Hysteresis arises in the field range where two minima are present. Switching is the irreversible jump from one minimum to another, which occurs when $d^2E/d\theta^2 = 0$. It takes place in the second and fourth quadrants, when $\alpha < 45°$, and the switching field H_{sw} is then equal to the coercivity H_c. Otherwise, when $\alpha > 45°$, $H_{sw} > H_c$. It is interesting that when $\alpha = 77°$, the switching leads to a small decrease of the component of magnetization along H. The hysteresis of a Stoner–Wohlfarth particle is illustrated in Fig. 7.11. The hysteresis loop is perfectly square when $\alpha = 0$, and in that case the coercivity is equal to the anisotropy field: $H_c = 2K_u/\mu_0 M_s$, or

$$H_c = (2K_1/\mu_0 M_s) + [(1 - 3\mathcal{N})/2]M_s. \tag{7.22}$$

Here, K_u is the sum of the magnetocrystalline anisotropy K_1 and the shape anisotropy $\frac{1}{4}\mu_0 M_s^2(1 - 3\mathcal{N})$ (5.63), which are assumed to have the same axis. Equation (7.22) is consistent with Brown's theorem (7.19) since the demagnetizing factor $\mathcal{N}$ lies between 0 and 1.

An array of noninteracting particles with a random distribution of anisotropy axes is a crude model for a real polycrystalline magnet. The hysteresis loop is plotted in Fig. 7.12 in terms of the reduced variables, $m = M/M_s$ and

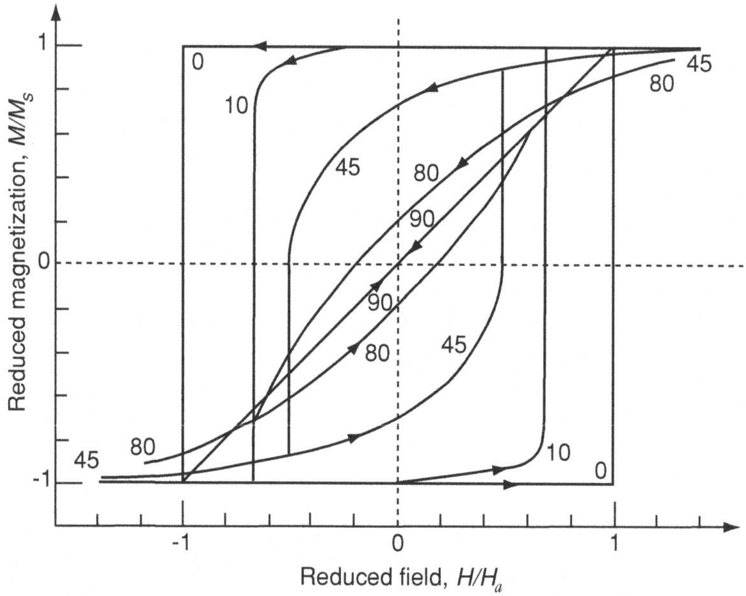

Figure 7.11

Magnetization curves for the Stoner–Wohlfarth model for various angles α between the field direction and the easy axis. Note the square loop when H is applied along the easy axis, and the lack of hysteresis in the perpendicular direction. H is given in units of the anisotropy field H_a.

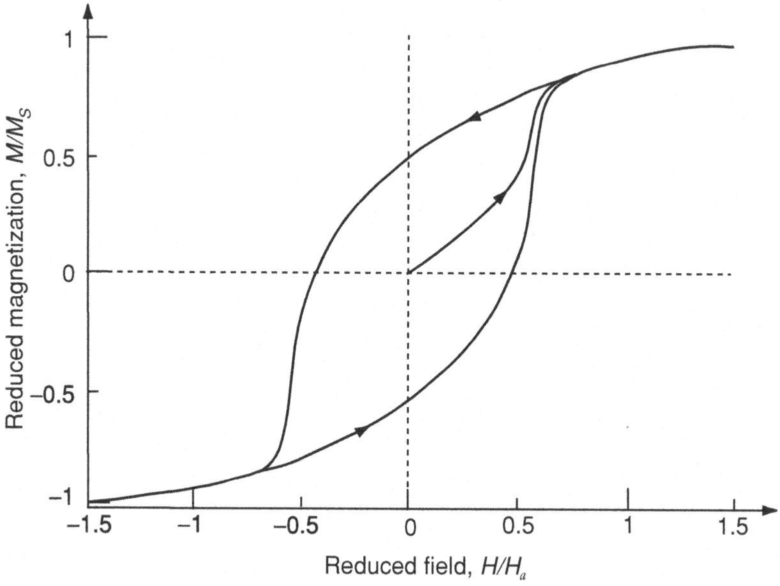

Figure 7.12

Hysteresis loop for a randomly oriented array of Stoner–Wohlfarth particles.

$h = H/H_a$, where H_a is the anisotropy field $2K_u/\mu_0 M_s$. The remanence for the array of particles is $m_r = \frac{1}{2}$; and the coercivity is $h_c = 0.482$. The remanent coercivity h_{rc}, defined as the reverse field needed to reduce the remanence to zero is 0.524.

If the anisotropy directions are distributed at random within a plane, which may be the case for some particulate recording media, the remanence is $m_r = (2/\pi) = 0.637$, $h_c = 0.508$ and $h_{rc} = 0.548$.

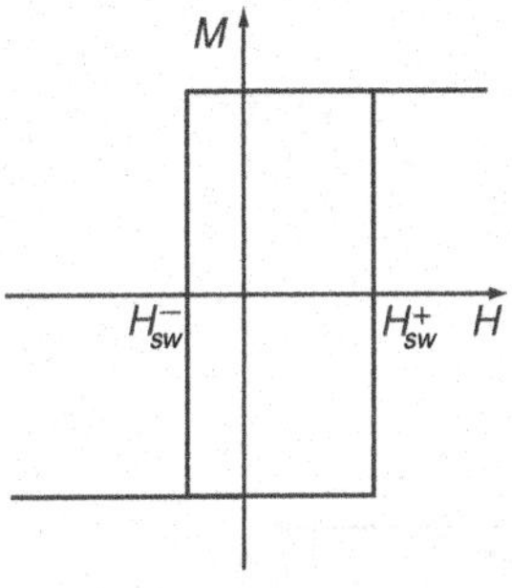

An elementary 'hysteron' used in Preisach modelling of hysteresis.

Wohlfarth also pointed out a simple relation between two remanence curves for the system of noninteracting particles. The remanence on the initial magnetization curve M_r, and is obtained by applying a field H to the virgin state, and reducing it to zero. The remanence M_{rd} is obtained in a reverse field after saturating the magnetization. They are related by

$$2M_{ri}(H) = M_r - M_{rd}(H). \tag{7.23}$$

Deviations from a linear plot of M_{ri} versus M_{rd}, known as a Henkel plot, are evidence for interparticle interactions.

A popular way of modelling hysteresis curves is the Preisach model, where a real hysteresis loop is decomposed as a superposition of elementary rectangular loops known as hysterons with different switching fields. These can be regarded as the loops of a distribution of Stoner–Wohlfarth particles with differing anisotropy. Interactions among the particles mean that the positive and negative switching fields H_{sw}^+ and H_{sw}^- may be different.

7.3.2 Reversal in thin films and small elements

The magnetization of soft ferromagnetic films usually lies in the plane of the film, in order to minimize the demagnetizing field. In-plane demagnetizing factors are small (Fig. 2.8). A weak, in-plane uniaxial anisotropy K_u may be induced deliberately to control the reversal process by magnetic annealing or off-axis deposition. The thin-film element is equivalent to a Stoner–Wohlfarth particle, for which the coherent rotation of the magnetization is confined to a plane.

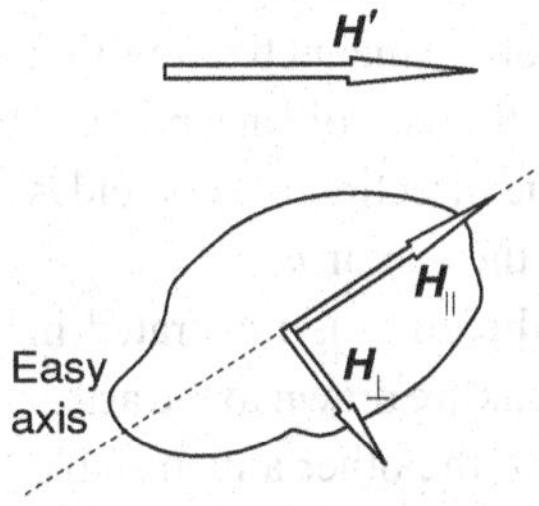

A magnetic field applied to a small particle or thin-film element with easy-axis anisotropy. The field is resolved into two components.

Applying the theory of the previous section, a square loop with $H_c = 2K_u/\mu_0 M_s$ is obtained when the field is applied along the easy axis. When it is applied in the transverse, in-plane hard direction, there is no hysteresis, but saturation occurs in the same field – the anisotropy field. The applied field can be resolved into components $H\cos\alpha$ and $H\sin\alpha$ along the easy and hard directions, and these components are again normalized by the anisotropy field to give the reduced variables $h_\parallel$ and $h_\perp$. Equation (7.21) becomes

$$E_{tot} = K_u[\sin^2\theta - 2h_\parallel\cos\theta - 2h_\perp\sin\theta]. \tag{7.24}$$

The equilibrium angles θ are determined by the condition $\mathrm{d}E_{tot}/\mathrm{d}\theta = 0$, giving

$$\frac{h_\perp}{\sin\theta} - \frac{h_\parallel}{\cos\theta} = 1,$$

Switching occurs when the energy minimum becomes unstable, $\mathrm{d}^2E_{tot}/\mathrm{d}\theta^2 = 0$. Solving these two equations gives the parametric equations for the switching field h_{sw}: $h_{\perp sw} = \sin^3\theta$, $h_{\parallel sw} = -\cos^3\theta$. Eliminating θ, the switching field is given by the Stoner–Wohlfarth asteroid drawn in Fig. 7.13:

$$h_{\parallel sw}^{2/3} + h_{\perp sw}^{2/3} = 1. \tag{7.25}$$

Figure 7.13

Stoner–Wohlfarth asteroid. The asteroid indicates where switching occurs. Inside the asteroid, the equilibrium orientation in a normalized field h is given by the tangent construction. θ is either β or β'.

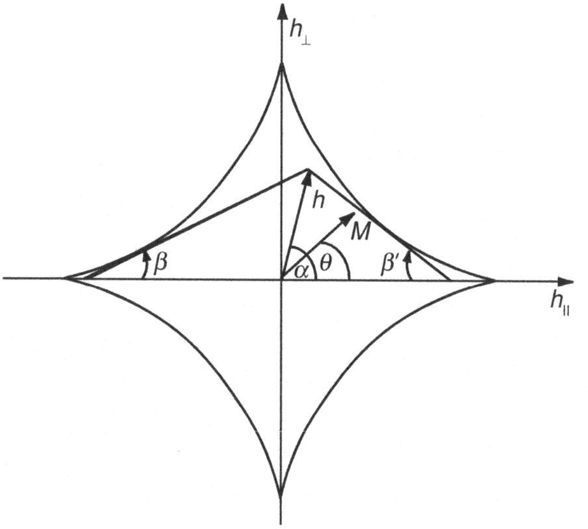

Outside the asteroid, there is a single energy minimum; the direction of magnetization responds to the applied field by continuous rotation. At any point inside the asteroid, the energy (7.23) has a double minimum, one stable, the other metastable. On crossing the asteroid by changing the magnitude or direction of the applied field, there may be a 'catastrophe' where the magnetization of the particle jumps to a new minimum with a different direction. In other words, the asteroid is the locus of points where a bifurcation of the free energy surface occurs. Magnetization switching never takes place within the asteroid, only on its surface. The remarkable properties of the asteroid emerge when we address the question of how to determine the direction of magnetization in a given field $\boldsymbol{h}$. The slope $\mathrm{d}h_{\perp sw}/\mathrm{d}h_{\parallel sw}$ of the tangent to the asteroid at a point θ_0 deduced from the parametric equations is $\tan\theta_0$, so the direction $\mathbf{e}_M$ coincides with the *tangent* to the asteroid drawn from the tip of the vector $\boldsymbol{h}$.

The construction to determine the magnetization direction is illustrated in Fig. 7.13. From a field point in the control plane, tangents are drawn to the asteroid. One represents the stable energy minimum $\theta = \beta$, the other a metastable minimum or a maximum $\theta = \beta'$. Hysteresis can be traced out in an oscillating field applied along some direction, or in a rotating field. Loops as a function of the variables $h_\parallel$ and $h_\perp$ are drawn in Fig. 7.14.

Irreversible, catastrophic discontinuities in magnetization are known as **Barkhausen jumps**. In the present model, a Barkhausen jump takes place whenever the field crosses the asteroid, approaching it from inside.

Coherent rotation of the magnetization is an *assumption* of the Stoner–Wohlfarth theory. It must be emphasized that other reversal modes may be possible, which have less coercivity. One example for a thin film is shown in Fig. 7.15. The film is initially magnetized along an out-of-plane hard direction, and the field is reduced. The two in-plane easy directions have the same energy,

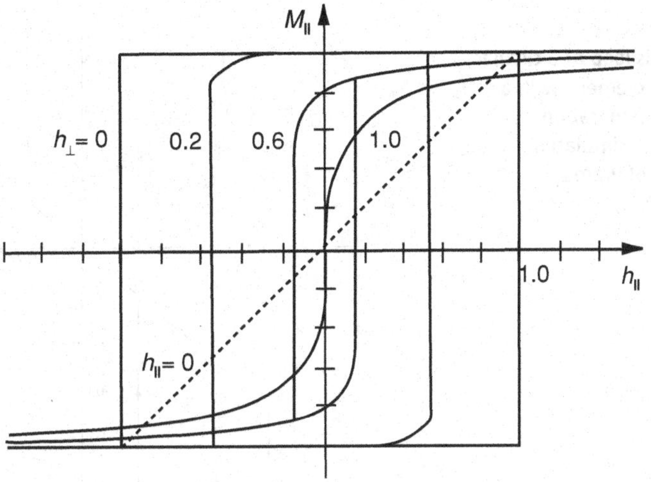

Figure 7.14

Switching curves for thin films as a function of the field applied in-plane. The hard-axis curve ($h_\parallel = 0$) is shown dashed.

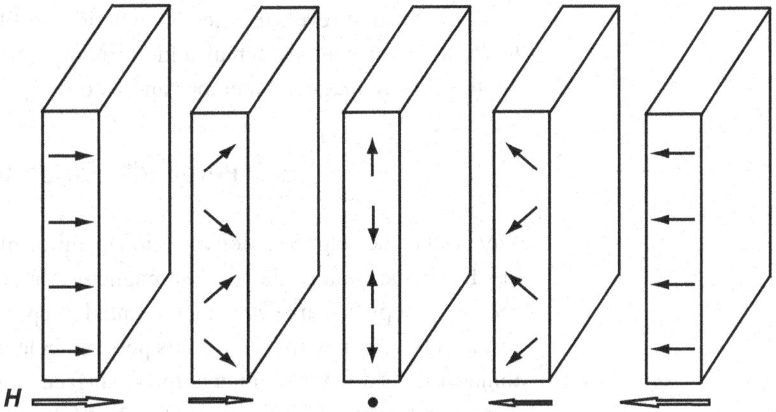

Figure 7.15

Magnetization reversal in a thin film when the field is applied in the perpendicular hard direction.

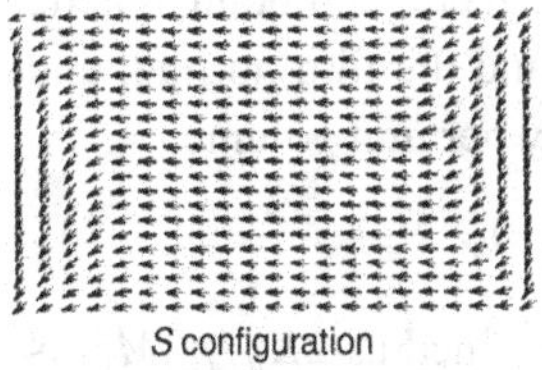

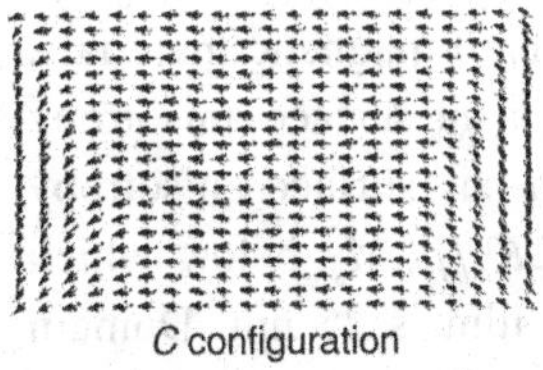

The *C* and *S* states which may appear during reversal of a thin film element in an in-plane field.

and initially a periodically modulated structure appears, which settles into a structure of in-plane strip domains, not a single perpendicular domain. There is no coercivity.

Whereas small thin-film ferromagnetic elements tend to have a collinear structure and reverse coherently, as described by the Stoner–Wohlfarth model, larger elements with negligible anisotropy tend to adopt a vortex state. A circular dot has four possible configurations, with clockwise or anticlockwise chirality and the spins at the centre can point up or down out of the plane. Starting from one of these, an in-plane field pushes the vortex reversibly towards the edge of the dot, and the magnetization then switches irreversibly to saturation as shown in Fig. 7.16. The vortex nucleates again when the field is reduced. Double vortices and various metastable states, named C, S and W states because of the shape of the spin configuration, can arise in differently shaped thin films, and they all exert an influence on the reversal process.The C states lead to coercivity and should be avoided if complete switching is sought in low field.

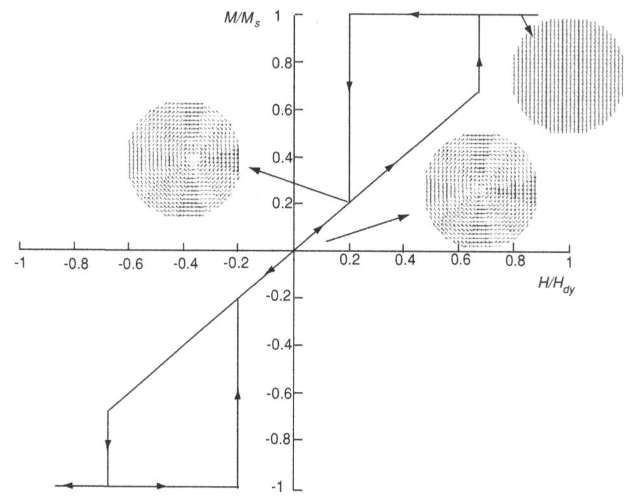

Figure 7.16

Hysteresis loop of a circular thin-film element with a vortex configuration. (Computer simulation, courtesy of Pramey Upadhyaya.)

Magnetization reversal in ferromagnetic thin-film elements is an important topic, both for a fundamental understanding of the micromagnetics and for applications in magnetic memory and recording.

7.3.3 Perpendicular anisotropy

We saw that the magnetization direction in uniformly magnetized thin films usually lies in the plane of the film for magnetostatic reasons. However, perpendicular anisotropy can arise when an oriented or epitaxial film of a hard magnetic material is, grown with its easy axis perpendicular to the film plane. In the very thinnest films a few nanometers thick, surface anisotropy can sometimes lead to perpendicular magnetization (§8.2.2). Multilayer stacks with perpendicular anisotropy can be built up of alternating thin ferromagnetic and nonmagnetic layers. If ϑ is the angle between the magnetization and the film normal, and there is some perpendicular anisotropy K_u, the energy per unit volume,

$$E_{tot} = K_u \sin^2 \vartheta + \tfrac{1}{2}\mu_0 M_s^2 \cos^2 \vartheta,$$

has a minimum at $\vartheta = 0$ when the quality factor Q, defined as $2K_u/\mu_0 M_s^2$, is greater than unity; the condition for perpendicular anisotropy is $K_u > \frac{1}{2}\mu_0 M_s^2$. This condition is only satisfied by oriented films of hard magnetic materials, and materials with a low magnetization M_s or dominant surface anisotropy $K_u = K_s/\mathrm{t}$, where t is the film thickness. If a film has surface anisotropy of 1 mJ m^{-2} and $\mu_0 M_s = 1$ T, the thickness $\mathrm{t} = 2K_s/\mu_0 M_s^2$ below which the magnetization tends to lie perpendicular to the film is 25 nm. Domain formation increases the critical thickness. The perpendicular magnetization of films with $Q < 1$ tends to break into maze domains with equal areas of $\uparrow$ and $\downarrow$ magnetization in zero applied field (Fig. 7.2(c)), in order to reduce the demagnetizing energy $-\frac{1}{2}\mu_0 M_s^2$ (§7.3.2).

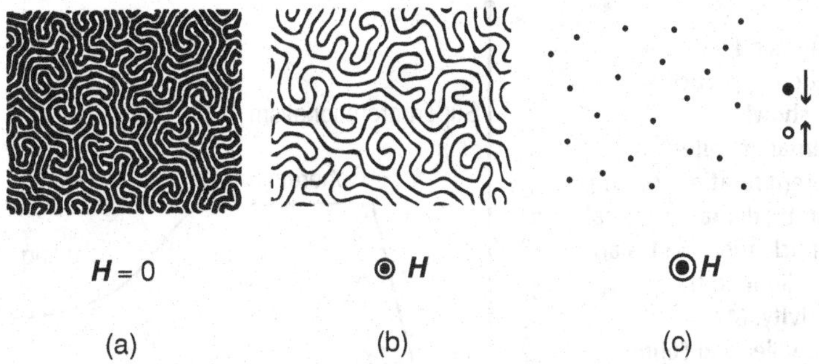

Figure 7.17

Magnetization and domain structure of a thin film with perpendicular anisotropy. Maze domains give way to bubble domains on increasing magnetic field.

A perpendicular field progressively increases the width of the ↑ domains, and narrows the ↓ domains up to a point known as strip-out, where the narrow ↑ stripes break up into short strips and small 'bubbles', which are cylindrical islands about a micrometre in diameter that extend throughout the film thickness (Fig. 7.17). Further increase of the field reduces the domain diameter, until the bubbles eventually disappear at saturation. Bubbles can be manipulated by suitable external fields and guided around magnetically defined tracks. In some circumstances, it is possible to form an hexagonal bubble lattice. Bubble domains in shift registers formed the basis of a sophisticated, but ultimately uncompetitive nonvolatile memory technology developed in the late 1970s.

Perpendicular anisotropy was important for magneto-optic recording. It is essential for modern perpendicular recording media (Chapter 14).

7.3.4 Nucleation

In order to calculate reversal properly, we need to include exchange and demagnetizing energy terms in the expression for E_{tot}. Starting from a uniformly magnetized ellipsoidal sample, the nucleation field H_n is defined as the field where the first deviation from the uniformly magnetized state appears. Coercivity is conventionally positive, whereas the nucleation field is negative. Brown actually proved that $H_n \leq -2K_1/\mu_0 M_s + \mathcal{N}M$. Since nucleation must precede reversal, $H_c \geq -H_n$ so his theorem (7.19) follows.

Nucleation of the coherent reversal mode is deduced from an eigenmode analysis of the linearized micromagnetic equation:

$$H_n = -\frac{2K_1}{\mu_0 M_s} - \frac{1}{2}(1 - 3\mathcal{N})M_s, \tag{7.26}$$

which is the Stoner–Wohlfarth expression for the coercivity (7.21) when the anisotropy energy $E_a = K_1 \sin^2\theta$. If higher-order terms are included, for example $E_a = K_1 \sin^2\theta + K_2 \sin^4\theta$, the nucleation field is unchanged, but when $0 < K_1 < 4K_2$, $H_c > -H_n$.

The other common nucleation mode for ellipsoidal particles is curling, Fig. 7.18. This mode creates no demagnetizing field, but it costs exchange energy to

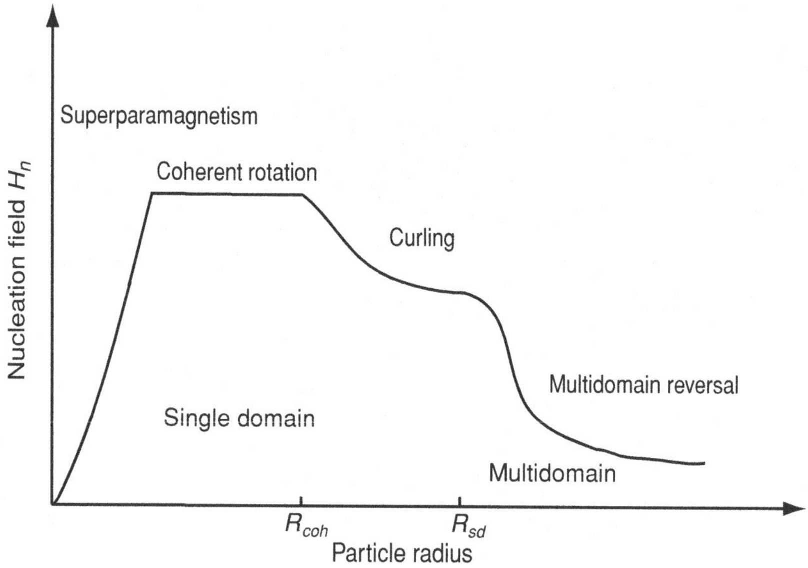

Figure 7.18
Nucleation field as a function of particle size, showing superparamagnetism, coherent rotation, curling and multidomain reversal. The nucleation field is an upper limit on the coercivity. (After Kronmüller and Fähnle (2003).)

produce the deviation from a uniformly magnetized state. The nucleation field for curling is

$$H_n = -\frac{2K_1}{\mu_0 M_s} + \mathcal{N}M - k_c M_s \left(\frac{R_0}{R}\right)^2, \tag{7.27}$$

where $R_0 = \sqrt{8\pi A/\mu_0 M_s^2}$. A typical value with $A = 10^{11}$ J m^{-1} and $M_s = 10^6$ A m^{-1} is $R_0 = 10$ nm. The factor k_c ranges from 1.08 for a long cylinder to 1.48 for a flat plate. The curling therefore controls magnetization reversal when specimen dimensions are larger than about 15 nm.

Critical micromagnetic dimensions for some ferromagnetic materials are summarized in Tables 7.1 and 8.1. In samples much larger than R_{sd} there is no reason to expect that the magnetization reversal from the saturated state will be a coherent and uniform process. Real specimens are always inhomogeneous to some extent, and they have a surface. Since it takes little energy to move domain walls, the critical step in magnetization reversal is often the creation of a small reverse domain by a spontaneous fluctuation at a weak point in the system. The minimum size of such a nucleus is of order δ_w^3. Once nucleated, the reverse domain tends to expand under the action of the reverse field, unless the domain walls are pinned by defects (Fig. 7.9).

7.3.5 The two-hemisphere model

The effect of exchange coupling in an inhomogeneous material is illustrated by a simple model.

Consider a ferromagnetic sphere, Fig. 7.19, made up of two hemispheres, α, β, which have different anisotropies, K_α, K_β. Assuming first that

Figure 7.19

A sphere made of two halves, α, β, showing (a) the uniform state, (b) coherent rotation, (c) incoherent rotation, and (d) the mode where the soft hemisphere switches before the hard one when $K_\alpha \gg K_\beta$ unless the radius of the sphere is less than the exchange length l_{ex}.

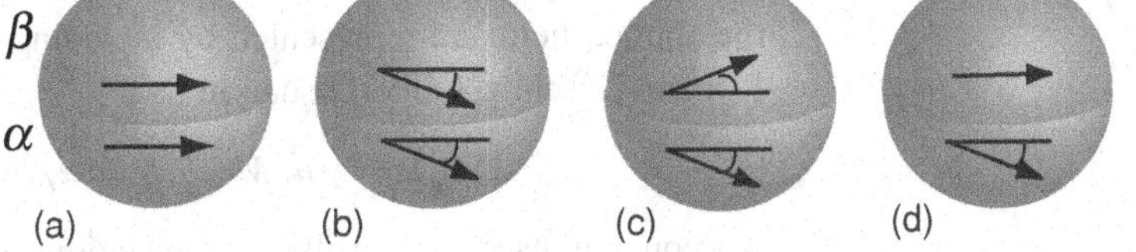

$K_\alpha = K_\beta = K$, and that the magnetic moment of each half is located at $\pm R/2$, the energy per unit volume in an external field H for small deviations, θ_α, θ_β, is

$$E = \left[A/R^2 - \tfrac{1}{24}\mu_0 M^2\right](\theta_\alpha - \theta_\beta)^2 + \tfrac{1}{4}[2K_1 - \mu_0 M H](\theta_\alpha^2 + \theta_\beta^2). \quad (7.28)$$

The first term represents the increase of exchange energy and the decrease in dipole energy associated with an incoherent reversal mode. The second term represents the anisotropy and Zeeman energies. Coherent and incoherent modes of reversal are illustrated in Fig. 7.19(b) and (c).

The nucleation field for the coherent reversal mode is $H_n = 2K/\mu_0 M_s$, whereas that for the incoherent mode is given by (7.27) with $\mathcal{N} = 1/3$, $k_c = 1$ and $R_0 = \sqrt{8A/\mu_0 M_s^2}$. The reversal is incoherent if particles are larger than the coherence radius, $R_{coh} = \sqrt{24A/\mu_0 M_s^2}$. If, on the other hand, we assume $K_\alpha = K$ and $K_\beta = 0$, independent reversal of the soft hemisphere occurs when $H \approx \frac{1}{8}M_s$ unless the sphere is very small. When R is less than the exchange length (Eq. (7.8)) the soft hemisphere can no longer reverse independently. It is stiffened by exchange coupling with the hard sphere. This exchange stiffening effect operates in soft regions of dimensions of order $4l_{ex} \approx 10$ nm (Table 7.1).

7.3.6 Switching dynamics

When a ferromagnetic sample with moment $\mathfrak{m}$ is placed in a magnetic field $\boldsymbol{H}$ which is inclined to $\mathfrak{m}$, it experiences a torque $\boldsymbol{\Gamma} = \mu_0 \mathfrak{m} \times \boldsymbol{H}$. The gyromagnetic ratio $\gamma = g\mu_B/\hbar$ is the ratio of magnetic moment to angular momentum, hence we have the gyroscopic equation

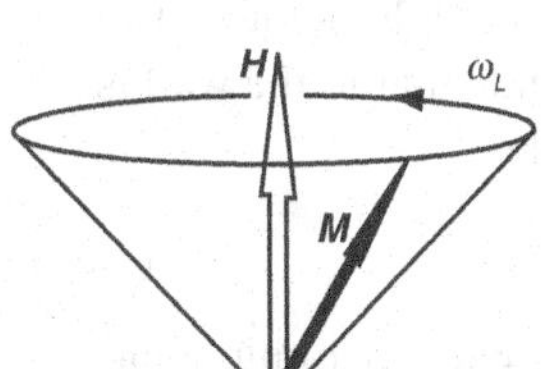

Precession of magnetization in a magnetic field.

$$\mathrm{d}\boldsymbol{M}/\mathrm{d}t = \gamma\mu_0 \boldsymbol{M} \times \boldsymbol{H},$$

which describes the precession of the magnetization around the applied field. The Larmor precession or ferromagnetic resonance frequency

$$\omega_L = \gamma\mu_0 H$$

corresponds to a frequency of 28 GHz T^{-1} when $g = 2$ and $\gamma = -e/m_e$.

When there is uniaxial anisotropy, represented by an anisotropy field $H_a = 2K_1/\mu_0 M_s$, the Larmor precession frequency becomes

$$\omega_L = \gamma\mu_0(H + H_a). \quad (7.29)$$

Losses, without which the moment would never be able to align itself with the applied field, are represented by including a damping term (§9.2.2). The Landau–Lifschitz–Gilbert equation

$$\mathrm{d}\boldsymbol{M}/\mathrm{d}t = \gamma\mu_0 \boldsymbol{M} \times \boldsymbol{H} - (\alpha/M_s)\boldsymbol{M} \times \mathrm{d}\boldsymbol{M}/\mathrm{d}t \tag{7.30}$$

is frequently used to describe magnetization dynamics. If the field is applied in the plane of a ferromagnetic film, perpendicular to the easy axis, the magnetization tries to precess out of the plane and it is then subject to a large demagnetizing field of order of a tesla, which accelerates the rotation of the magnetization towards the field (Fig. 9.7). The time taken to turn through $\pi/2$ is of order $(4 \times 28 \times 10^9)^{-1} \approx 0.01$ ns. Coherent reversal is an inherently fast process.

7.3.7 Domain-wall motion

Reversal involving domain-wall motion is far slower. An applied field exerts a pressure on the walls. If it is applied parallel to the easy z axis of a uniaxial magnet, the wall will move in the x-direction. The change in Zeeman energy per unit area of wall is $2\mu_0 MH\delta x$, hence the pressure is $2\mu_0 MH$. In small fields, the wall velocity is essentially proportional to the pressure:

$$v_w = \mu_0 \eta_w (H - H_p), \tag{7.31}$$

where η_w is the wall mobility and H_p is the small field needed to depin the wall. Mobilities range from 1 to 1000 m s^{-1} mT^{-1}, but a value of 100 m s^{-1} mT^{-1} is representative for thin films of a soft material like permalloy. A field of 0.1 mT will move the walls at about 10 m s^{-1}. The linear increase of wall velocity with field collapses at velocities beyond about 100 m s^{-1}. Wall dynamics are important in materials such as laminated electrical steel, where the walls move more slowly, oscillating at 50 or 60 Hz (§12.1.1). Eddy currents control their motion.

To obtain an expression for the mobility consider a single domain wall. If the flux through the cross section of thickness t perpendicular to the wall is Φ, the wall velocity is related to the rate of change of Φ by

$$\mathrm{d}\Phi/\mathrm{d}t = 2\mu_0 M_s \mathrm{t} v_w. \tag{7.32}$$

Since $j = \sigma E \propto (\mathrm{d}\Phi/\mathrm{d}t)/\mathrm{t}$ and p_w, the power dissipated per unit length of wall is proportional to j^2/σ, it follows that $p_w \propto \sigma(\mathrm{d}\Phi/\mathrm{d}t)^2$. A full calculation, valid for slow-moving, undistorted walls, yields

$$p_w = \sigma G \left(\frac{\mathrm{d}\Phi}{\mathrm{d}t}\right)^2, \quad \text{where} \quad G = \frac{4}{\pi^3} \sum_{n\ \mathrm{odd}} \frac{1}{n^3} = 0.1356. \tag{7.33}$$

The work done per unit length and per unit time on the wall is $H(\mathrm{d}\Phi/\mathrm{d}t)$. Hence $H = \sigma G(\mathrm{d}\Phi/\mathrm{d}t)$. Using (7.32) to write $\mathrm{d}\Phi/\mathrm{d}t$ in terms of the wall velocity v_w,

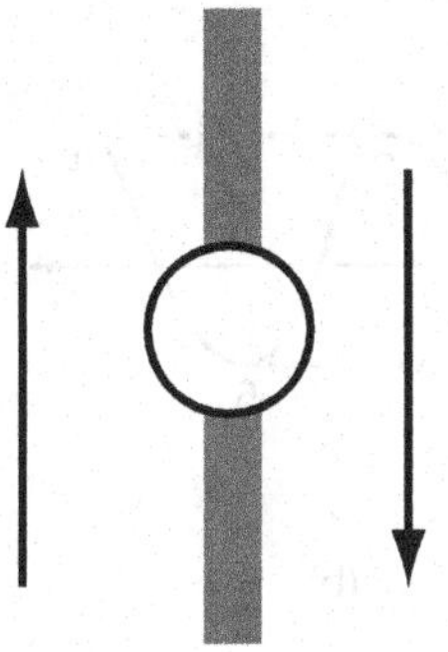
Domain wall pinned at a defect.

we obtain an expression for the wall mobility in (7.31):

$$\eta_w = \rho/2G\mu_0^2 M_s t. \qquad (7.34)$$

For example, the domain-wall mobility in electrical sheet steel, t = 350μm, $\mu_0 M_s = 2.0$, $\rho = 50 \times 10^{-8}$ Ω m is 2.1 m s^{-1} mT^{-1}. Since the mobility varies as the inverse of film thickness, velocities in thin films are much higher.

For some purposes it is useful to associate an effective mass per unit area with the domain wall

$$m_w = 2\pi/\mu_0 \gamma^2 \delta_w. \qquad (7.35)$$

A typical value for a soft magnet is 10 μg m^{-2}. Domain-wall velocities can be measured by examining the phase lag between magnetization and applied field in a picture-frame experiment, or by timing the movement of a wall down a ferromagnetic wire, detecting the change of flux in pick-up coils or the change in anomalous Hall effect as it passes.

On account of the energy per unit area associated with the domain wall, $\gamma_w = 4\sqrt{AK}$, it tends to get pinned at defects, especially planar defects, where A or K differs from the bulk values. **Strong pinning** arises when these defects have a dimension comparable to the domain wall width δ_w. Planar defects are the most effective pinning centres because the whole wall finds itself with a different energy when it encompasses the defect. The planar defect acts either as a trap or barrier to wall motion, depending on whether γ_w is lower or higher in the defect than it is in the bulk. Line and point defects are less effective pinning centres, but they do best when their diameter is comparable to δ_w, and when there is pronounced contrast in K or A between the bulk and the defect region. Contrast is high at a void, where $K = A = 0$.

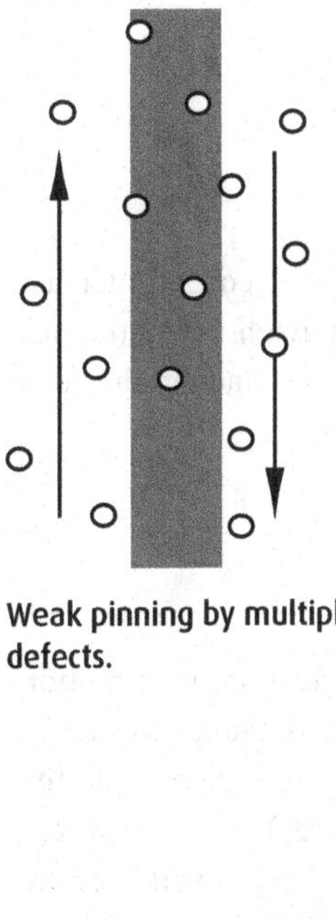
Weak pinning by multiple defects.

Weak pinning occurs when many small defects, particularly point defects, are distributed throughout the wall. The energy for weak pinning derives from statistical fluctuations in the numbers of defects contained in the wall.

Inevitably, there will always be some distribution of defects in any sample of magnetic material. Suppose the energy per unit area depends only on the wall position represented by a coordinate x, and the applied field H. Then

$$E_{tot} = f(x) - 2\mu_0 M_s H x. \qquad (7.36)$$

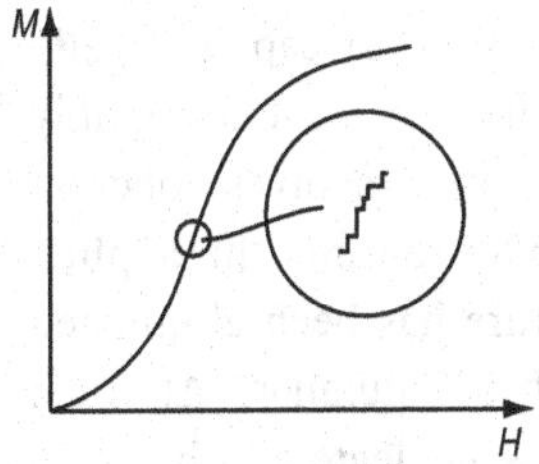

Barkhausen jumps.

It can be seen from Fig. 7.20 how a hysteresis loop results from an energy landscape with many minima. At a local energy minimum, $\mathrm{d}f(x)/\mathrm{d}x = 2\mu_0 M_s H$. The wall jumps from points with the same $\mathrm{d}f(x)/\mathrm{d}x$ on increasing field, the magnetization changing discontinuously in a **Barkhausen jump**. The hysteresis loop of a macroscopic sample consists of many discrete jumps. These can be observed directly in a sensitive measurement of the magnetization of a ferromagnetic wire. Heinrich Barkhausen heard the jumps in an experiment he carried out in 1919, when he filled a long pick-up coil with

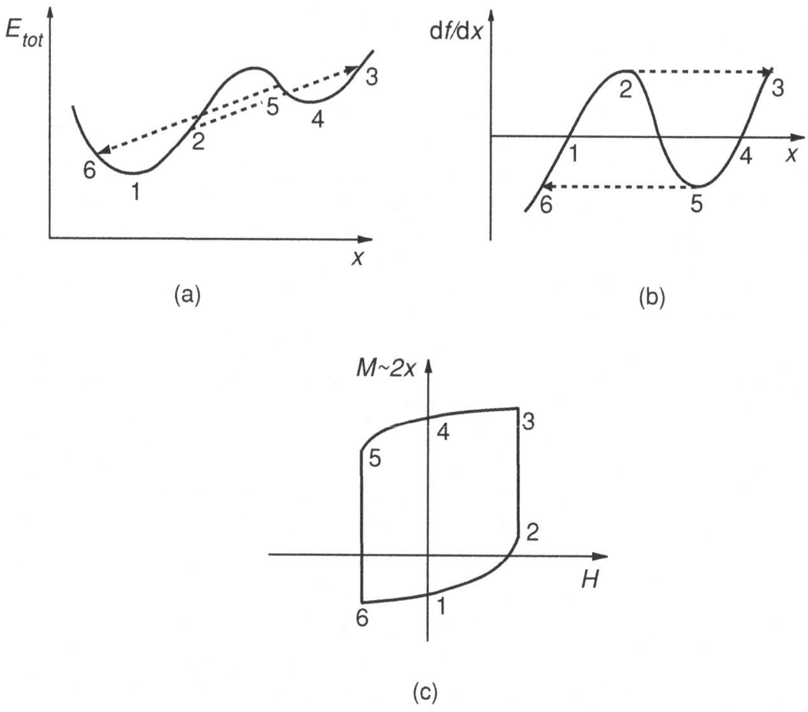

Figure 7.20
(a) Energy as a function of wall position, (b) the equilibrium condition, $\mathrm{d}f(x)/\mathrm{d}x = 2\mu_0 M_s H$ and (c) a hysteresis loop due to field cycling.

a bundle of nickel wires, and detected the flux jumps in the coil on increasing field using an amplifier connected to a loudspeaker. Each discontinuous jump of magnetization induced a small EMF in the coil, and produced a click.

7.3.8 Real hysteresis loops

Heinrich Barkhausen, 1881–1956.

Hysteresis loops in real materials exhibit features of nucleation, wall motion and coherent rotation. A schematic loop for a material with cubic anisotropy is shown in Fig. 7.21. The reversible linear segment $1 \to 2$, known as the initial magnetization curve or virgin curve where $0 \leqslant (M/M_s) \lesssim 0.1$, is where the walls are pinned, but bow out from the pinning centres in a reversible way and snap back to their original places when the field is removed. Segment $2 \to 3$ involves irreversible Barkhausen jumps as the domain walls sweep erratically through the sample, eventually eliminating all but the one most favourably oriented domain. Segment $3 \to 4$ is again reversible, as it involves coherent rotation of the magnetization of the remaining domain towards the applied field direction. This region, where the domain structure has been eliminated and $0.9 \lesssim (M/M_s) < 1.0$, is known as the approach to saturation. At some point on the reverse segment $4 \to 5$, reverse domains nucleate and begin to propagate, eventually reducing the sample to a multidomain state with no net

Figure 7.21

Initial magnetization curve and demagnetization curve in the first and second quadrants of a cubic ferromagnet.

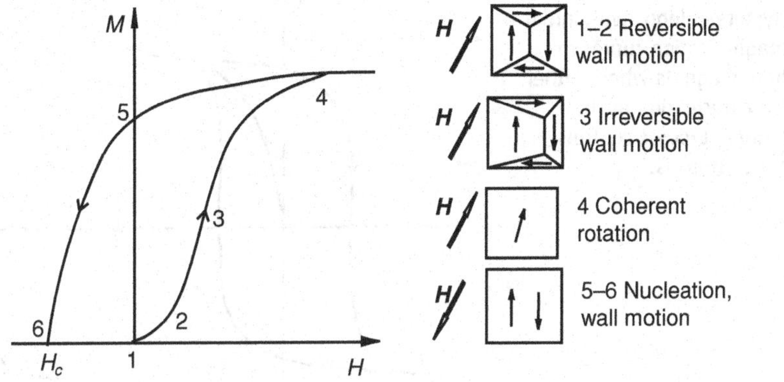

magnetization at the coercive field H_c. This multidomain state is a minimum in the energy landscape far removed from the virgin state 1, which is now forever inaccessible, unless the ferromagnet is reborn by heating above the Curie point, and cooling in zero field.

In bulk material, it is practically impossible to calculate a complete hysteresis loop precisely. An empirical approach for hard magnets is to use the Kronmüller equation, inspired by the Stoner–Wohlfarth model,

$$H_c = \alpha_K(2K_1/\mu_0 M_s) - N_{eff} M_s,$$

where α_K and N_{eff} are empirical parameters, to be determined from experiment.

Hard magnetic materials whose magnetization process is governed by nucleation or pinning are readily distinguished by their initial magnetization curves. The domain walls move freely through a nucleation-type magnet, which has a high initial susceptibility, but they are constantly being trapped in a pinning-type magnet, so the initial susceptibility is small until the depinning field is reached (Fig. 7.22). Coherent rotation, the dominant reversal mechanism in a uniaxial system, is provided by the dependence of coercivity on the angle between the applied field and the easy axis. This varies as $\cos\alpha$ in the Stoner–Wohlfarth model, whereas if nucleation is involved, it is only the component of field along the easy axis which is effective at creating reversal, hence there is a $1/\cos\alpha$ dependence.

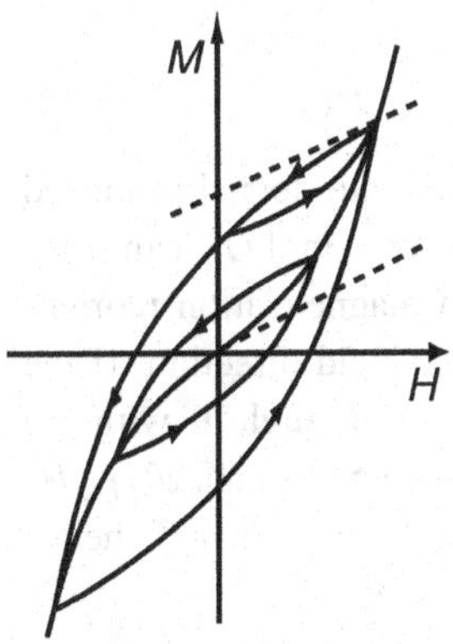

An elementary, low-field hysteresis loop of a soft material is composed of parabolic segments. The lower dashed line shows the initial susceptibility χ_i.

In *soft* magnetic materials, the hysteresis in the initial region in fields which are small compared with the saturation coercivity is described by the empirical Raleigh laws enunciated by John Strutt in 1887. These are, starting from a state $M_1(H_1)$ attained by reducing field,

$$M(H) - M_1 = \chi(H - H_1) + \nu(H - H_1)^2, \tag{7.37}$$

where $H > H_1$, and starting from a state $M_2(H_2)$ attained by increasing field,

$$M(H) - M_2 = \chi(H - H_2) - \nu(H - H_2)^2, \tag{7.38}$$

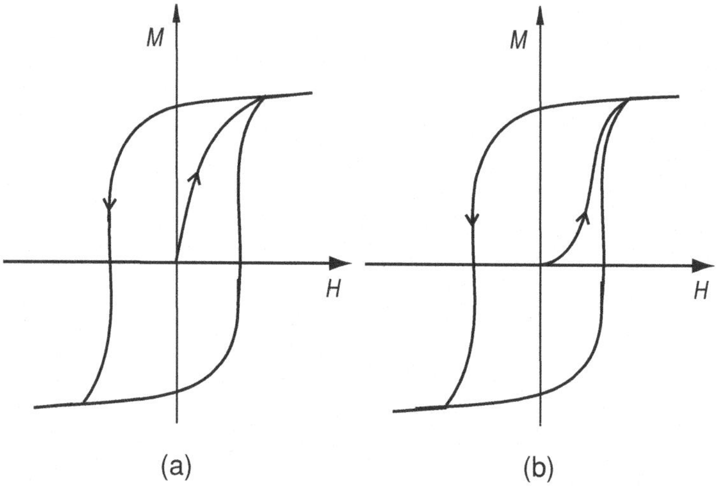

Figure 7.22

Hysteresis loops with initial magnetization curves for hard magnets where either (a) a nucleation or (b) a pinning process controls the hysteresis.

where $H < H_2$. In each case, there is the sum of a linear, reversible response proportional to H, and a nonlinear, irreversible term varying as H^2. The remanence remaining after applying a field H is $\nu H^2/2$. The elementary hysteresis loop is therefore made up of parabolic segments. Here χ_i represents the reversible initial susceptibility, and the term in ν is the irreversible response to the field. The initial magnetization curve, measured in an alternating field is

$$M = \chi_i H + \nu H^2. \tag{7.39}$$

Originally formulated to describe the magnetic properties of steel, the microscopic origin of these Raleigh laws is the deformation and pinning of domain walls by defects.

The high-field approach to saturation of the magnetization curve is represented by an empirical expression

$$M = M_s(1 - a/H - b/H^2 - \cdots) + \chi_0 H, \tag{7.40}$$

where the last term is a small high-field susceptibility due to field-induced band splitting, which is called the paraprocess. The term in $1/H$ can arise from defects, whereas the term in $1/H^2$ arises from magnetization reorientation when the anisotropy axis is misaligned with the field direction. If the misalignment angle is α and the magnetization makes an angle θ with the easy axis, minimizing the energy (7.21) yields $(\alpha - \theta) \approx 4K_1 \sin 2\theta/\mu_0 H$, but $M = M_s \cos(\alpha - \theta) \approx M_s(1 - (\alpha - \theta)^2)$ when $(\alpha - \theta)$ is small, hence $b = 16K_1 \sin^2 2\alpha/\mu_0$.

7.3.9 Time dependence

The stable state of a bulk ferromagnet or an ensemble of superparamagnetic particles is one with no net magnetization. The loop is not a static object, fixed

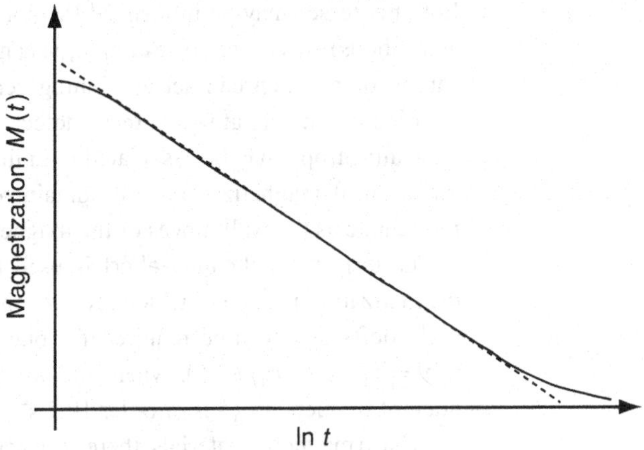

Figure 7.23

The time dependence of the magnetization of a ferromagnet. The logarithmic variation is indicated by the dotted line.

in time. The magnetic states observed around the hysteresis loop are metastable, and the loop looks different depending on whether the field is swept slowly or rapidly. The slower the scan, the lower the coercivity. Furthermore, the magnetization in a fixed field evolves continuously with time, and the variation is most pronounced in the vicinity of magnetization switching in the second and fourth quadrants, where the ferromagnet totters at the edge of metastable equilibrium. The time variation of the magnetization is approximately logarithmic, an effect known as magnetic viscosity (Fig. 7.23):

$$M(t) \approx M(0) - S_v \ln(t/\tau_0). \tag{7.41}$$

This empirical expression cannot be valid for either very short or for very long times. It follows from a sum of exponential decays with a flat distribution of energy barriers Δ. The magnetic viscosity coefficient S_v can be regarded as a magnetic fluctuation field whose magnitude is proportional to temperature.

Another way of describing the time dependence is as a sum of exponential decays, each with its own relaxation time τ, via the Fourier integral:

$$M(t) = M(0) \int_0^\infty P(\tau) \mathrm{e}^{-t/\tau} \mathrm{d}\tau. \tag{7.42}$$

The probability distribution $P(\tau)$ may be taken to be a gamma distribution,

$$P(\tau) = \frac{1}{\tau_0 \Gamma(p)} \left(\frac{\tau}{\tau_0}\right)^{p-1} \mathrm{e}^{-t/\tau_0},$$

where τ_0 and p are the parameters that specify the mean value and width of the distribution and $\Gamma(p)$ is the gamma function (mean $= p\tau$, variance $\sigma^2 = p\tau_0^2$). The advantage of this formulation is that (7.42) can be integrated analytically. The magnetization processes associated with the time dependence (7.41) are the usual suspects – domain-wall motion and magnetization rotation.

Both processes may be influenced by mobile atomic defects in the crystal (or amorphous) structure. For example, vacancies and interstitial defects such as carbon or nitrogen can act as pinning centres, and the wall motion may be retarded by the rate at which these defects can move. In alloys such as Ni–Fe, local anisotropy may be associated with the alignment of Fe–Fe pairs with the direction of magnetization in a domain or in a domain wall. The pairs can reorientate as the wall moves or the magnetization rotates.

The term magnetic after-effect is used in connection with time-dependent magnetization processes which are governed by physical diffusion of atomic-scale defects. The time required for one to jump from one site to the next is $\tau = \tau_0 \exp(-\varepsilon_a/k_B T)$, where the activation energy ε_a is 1–2 eV and the attempt frequency τ_0^{-1} is of order 10^{15} s^{-1}.

In hard magnetic materials, there is an activated process associated with magnetization reversal which is unconnected with defects, but leads to a variation of magnetization similar to (7.41). This is the nucleation of reverse domains, which requires a spontaneous thermal fluctuation leading to the reversal of a volume of order δ_w^3. This acts as a nucleus from which the magnetization reversal spreads out.

In order to avoid gradual creep of the magnetization, permanent-magnet arrays are often aged before use by annealing them at a temperature some 50–100 K above the operating temperature, in order to relax any easily excited magnetization reversal.

FURTHER READING

G. Bertotti, *Hysteresis in Magnetism*, San Diego: Academic Press (1998). A comprehensive monograph.

G. Bertotti and I. D. Mayergoryz, *The Science of Hysteresis*, San Diego: Academic Press (2006). A three-volume treatise covering all aspects of the theory and phenomenology of hysteresis, not just in magnets but in other physical and social systems as well.

A. Hubert and R. Schäfer, *Magnetic Domains*, Berlin: Springer (1998). An informative, complete and profusely illustrated work.

A. Aharoni, *Introduction to the Theory of Ferromagnetism*, Oxford: Oxford University Press (1996). Ideosyncratic and readable account of ferromagnetism in the continuum approximation.

W. F. Brown, *Micromagnetics*, New York: Wiley Interscience (1963). The classic monograph; difficult to find.

H. Kronmüller and M. Fähnle, *Micromagnetism and the Microstructure of Ferromagnetic Solids*, Cambridge: Cambridge University Press (2003). A modern treatment of the relation of micromagnetism to microstructure, illustrated with examples ranging from soft nanocomposites to rare-earth permanent magnets.

A. P. Malozemoff and J. C. Slonczewski, *Magnetic Domains in Bubble Materials*, New York: Academic Press (1979). A detailed account of the physical basis of a technology that was not scalable.

EXERCISES

7.1 Show that the first term in (7.5) is equivalent to (7.6).

7.2 What is the relation between A and $\mathcal{J}$ for a bcc material? If $T_C = 860$ K, $a_0 = 0.36$ nm and $S = 1$, what are the values of the two exchange constants?

7.3 Deduce (7.20) by equating the cost of creating a domain wall in a spherical particle with uniaxial anisotropy to the decrease of energy resulting from the creation of a two-domain state, assuming it has half the demagnetizing energy of the uniformly magnetized state.

7.4 Check that (7.20) is dimensionally correct.

7.5 Deduce an expression for H_c in the Stoner–Wohlfarth model for a spherical particle with $E_a = K_1 \sin^2\theta + K_2 \sin^4\theta$ and $4K_2 > K_1 > 0$.

7.6 (a) Show that the slope at remanence of the hysteresis loop in Fig. 7.12 is 2/3.
(b) Deduce equation (7.25) for the Stoner–Wohlfarth switching curve. Justify the geometrical construction for the angle θ.

7.7 Verify that the expressions for δ_w (7.17), γ_w (7.18) and m_w (7.35) are dimensionally correct.

7.8 Estimate the width of the maze domains that form in a thin 001 cobalt film.

7.9 Deduce expressions for the pinning field for a film of thickness t, (a) with a scratch of depth s and width s across its surface, assuming $s > \delta_w$ and (b) with a random distribution of n nonmagnetic inclusions per unit volume of radius r, assuming $r \ll \delta_w$.

7.10 How would you determine the two parameters α_K and N_{eff} in the Kronmüller-equation?

7.11 Show that the coercivity of an ensemble of Stoner–Wohlfarth particles aligned with a common easy axis varies as $1/\cos\alpha$.

7.12 Show that the magnetization of an array of ferromagnetic single-domain particles decreases logarithmically with time when the applied field is reversed. Assume a uniform distribution of barrier heights Δ^+.

7.13 Deduce an expression for the depinning field in the weak pinning limit. Assume the pinning centres are vacancies with volume a^3.

7.14 Use the Stoner–Wohlfarth asteroid to describe the behaviour of a particle in a rotating field h of magnitude shown in Fig 7.13.

8 Nanoscale magnetism

Nanoscale magnets have at least one dimension in the nanometre range. They exhibit size-specific magnetic properties such as superparamagnetism, remanence enhancement, exchange averaging of anisotropy and giant magnetoresistance when the small dimensions become comparable to a characteristic magnetic or electrical length scale. Thin films are the most versatile magnetic nanostructures, and interface effects such as spin-dependent scattering and exchange bias influence their magnetic properties. Thin-film stacks form the basis of modern magnetic sensors and memory elements.

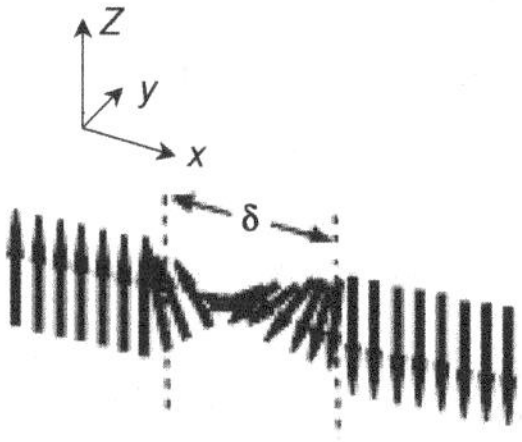

A domain wall where all spins in a *y–z* plane are parallel.

Matter behaves differently down in the nanoworld, where the length scales of interest range from about 1 nm up to about 100 nm. The atomic-scale structure of matter can usually be ignored, but the mesoscopic dimensions of the magnetic nano-objects are comparable to some characteristic length scale, below which the physical properties change. We have already encountered one important nanoscale object in bulk magnetic material – the domain wall. It is extended in two directions, but not in the third; the domain wall width δ_w is one of the characteristic lengths that concern us here.

The number of small dimensions in a nanoscale magnet may be one, two or three. Some examples of each are illustrated in Fig. 8.1. The one-small-dimension class includes magnetic thin films, which are at the heart of many modern magnetic devices. Magnetic and nonmagnetic layers can be stacked to make thin-film heterostructures, such as spin valves and tunnel junctions. The films are usually grown on a macroscopic substrate.

Two small dimensions give a nanowire. These can exist as separate acicular (needle-shaped) nano-objects or they can be embedded in a matrix, to form a nanocomposite. The wires themselves may be coated or segmented, with layers of different compositions.

Three small dimensions define a nanoparticle. Again the magnetic particles may be separate and dispersed, or embedded in a medium to form some kind of composite. The materials in the bulk heterostructure may both be magnetic, or one magnetic and the other not. By suitable engineering of the nanocomposite, unique combinations of magnetic properties or magnetic and nonmagnetic properties may be achieved that are unattainable in any homogeneous bulk material. Magnetic nanoparticles can be structured in lines or planes. Examples of the latter are patterned thin films or multilayers, and granular magnetic recording media such as Co–Pt–Cr films.

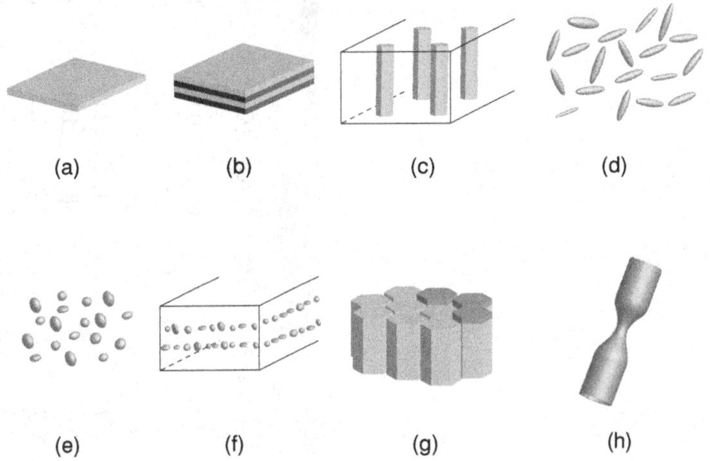

Figure 8.1

Examples of magnetic nanostructures: one small dimension: (a) thin film and (b) multilayer stack; two small dimensions: (c) nanowire array and (d) acicular particles; three small dimensions: (e) nanoparticles, (f) nanocomposite, (g) thin-film recording medium and (h) nanoconstriction.

A general feature of nanostructures is the large proportion of surface or interface atoms. The fraction of atoms on the surface is $2a/t$ for an unsupported film of thickness t and interatomic spacing a; it is $2\pi ra/\pi r^2 = 2a/r$ for a nanowire of radius r and $4\pi R^2 a/(4/3)\pi R^3 = 3a/R$ for a nanoparticle of radius R. For example, if $a = 0.25$ nm, the fraction of surface atoms in a film, wire or particle of dimension 10 nm is 5, 10 or 15%, respectively. These surface atoms and the reduced dimensions alter the magnetic properties of the material.

8.1 Characteristic length scales

It is convenient to express magnetic length scales in terms of the exchange length introduced in §7.1.1,

$$l_{ex} = \sqrt{\frac{A}{\mu_0 M_s^2}}, \tag{7.8}$$

and the dimensionless magnetic hardness parameter

$$\kappa = \sqrt{|K_1|/\mu_0 M_s^2}. \tag{8.1}$$

The exchange length reflects the balance of exchange and dipolar interactions. It is in the 2–5 nm range for most practical ferromagnetic materials. The hardness parameter κ is the dimensionless ratio of anisotropy to dipole energy. It should be greater than 1 for a permanent magnet, and much less than 1 for a good temporary magnet.

Expressions for characteristic micromagnetic lengths are gathered in Table 8.1, together with their values for a spectrum of magnetic materials where κ ranges from 0.01 to 4.3. Some of these lengths were discussed in

Table 8.1. Characteristic magnetic length scales (in nm)

Length	Expression	Fe	Co	Ni	NiFe	$Fe_{90}Ni_{10}B_{20}$	CoPt	$Nd_2Fe_{14}B$	$SmCo_5$	$Sm_2Fe_{17}N_3$	CrO_2	Fe_3O_4	$CoFe_2O_4$	$BaFe_{12}O_{17}$
κ	$\sqrt{\lvert K_1\rvert/\mu_0 M_s^2}$	0.12	0.45	0.13	0.01	0.01	2.47	1.54	4.30	2.13	0.36	0.21	0.84	1.35
l_{ex}	$\sqrt{A/\mu_0 M_s^2}$	2.4	3.4	5.1	3.4	2.5	3.5	1.9	3.6	2.5	4.4	4.9	5.2	5.8
R_{coh}	$\sqrt{24} l_{ex}$	12	17	25	17	12	17	9.7	18	12	21	24	26	28
δ_w	$\pi l_{ex}/\kappa$	64	24	125	800	900	4.5	3.9	2.6	3.7	44	73	20	14
R_{sd}	$36\kappa l_{ex}$	10	56	24	1.6	0.7	310	110	560	190	48	38	160	280
R_{eq}	$(3k_BT/4\pi BM)^{1/3}$	0.8	0.8	1.2	1.0	0.9	1.0	0.9	1.0	0.9	1.3	1.2	1.2	1.3
R_b	$(6k_BT/K_1)^{1/3}$	8	4	17	55	63	1.7	1.7	1.1	1.4	11	13	5	4

κ, hardness parameter; l_{ex}, exchange length; R_{coh}, maximum particle size for coherent rotation; δ_w, Bloch wall width; R_{sd}, maximum equilibrium single domain particle size; R_{eq}, particle radius for which $\mathfrak{m}B = k_BT$ in 1 T at 300 K; R_b, superparamagnetic blocking radius at 300 K.

the previous chapter. The quantity δ_w is the width of a 180° Bloch wall. The coherence radius R_{coh} is the maximum size of a uniformly magnetized particle, where magnetization reversal takes place by coherent rotation. In other words the direction of magnetization $\boldsymbol{e}_M$ does not depend on $\boldsymbol{r}$; $\boldsymbol{M} = (M_\theta, M_\phi)$. The maximum size of a particle that will be single-domain in equilibrium is R_{sd}. The lengths R_b and R_{eq} depend on temperature. The first is the superparamagnetic blocking radius. Particles smaller than R_b undergo spontaneous thermal fluctuations of the magnetization direction. The second gives the size of particles for which the thermal energy $k_B T$ and Zeeman energy $-\mathfrak{m}\cdot\boldsymbol{B}$ are equal in magnitude. It is not a fundamental length scale, rather an indication of when a linear response can be expected. The values of R_b and R_{eq} listed in Table 8.1 are for room temperature, 300 K and 1 T.

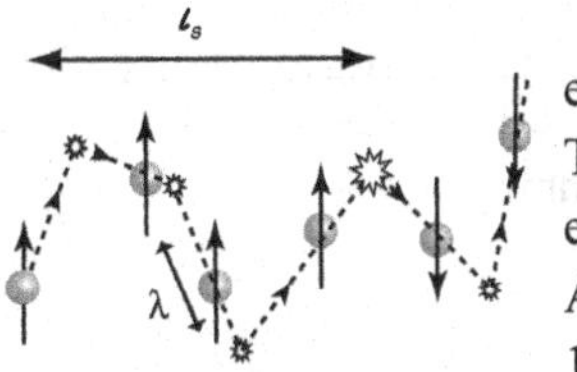

The spin diffusion length l_s is much greater than the mean free path λ, the average distance an electron travels between collisions.

Length scales for transport measurements are the mean free path of the electron λ, the inelastic scattering length λ_{el} and the spin-diffusion length l_s. These are, respectively, the mean distances travelled by an electron before it experiences a scattering event that modifies its momentum, energy or spin state. All three are spin-dependent in magnetic metals, and typically fall in the range 1–100 nm, which are the dimensions of interest here.

Another length scale relevant for magnetotransport is the cyclotron radius, $r_{cyc} = m_e v/eB$. However, this is a few micrometres in 1 T in typical metals, where the electron velocity v is $\approx 10^6$ m s^{-1}. It only approaches the nanometre scale in very high fields, or in semimetals or semiconductors, where the electron density, and hence the Fermi velocity, is low.

Finally, there are quantum length scales. The wavelength $\lambda_F = 2\pi(3\pi^2 n)^{-1/3}$ for a $3d$ metal with 0.6 s-like free electrons per atom is only 0.05 nm. Longer Fermi wavelengths in semiconductors facilitate the fabrication of quantum wells and quantum dots. Furthermore, in magnetic quantum phenomena such as Landau quantization, a magnetic length $l_B = \sqrt{\hbar/eB}$ appears. Numerically, $l_B = 26/\sqrt{B}$ nm when B is in teslas.

8.2 Thin films

The intrinsic magnetic properties – magnetization, Curie point, anisotropy, magnetostriction – may differ appreciably in thin films and bulk material. As an example, the magnetostriction of iron, shown in Fig. 8.2, changes sign at a film thickness of 20 nm, and it approaches the bulk value only in films a few tens of nanometres thick. Many of these differences result from the special environments of surface and interface atoms, and the strain induced by the substrate. The lattice parameters of a perfectly relaxed film will differ from those of the bulk. In $3d$ metal films, the separation of the surface planes tends to be a few per cent greater than in the bulk. Surface atoms are missing some of their neighbours, so their vibrational amplitude is enhanced and exchange

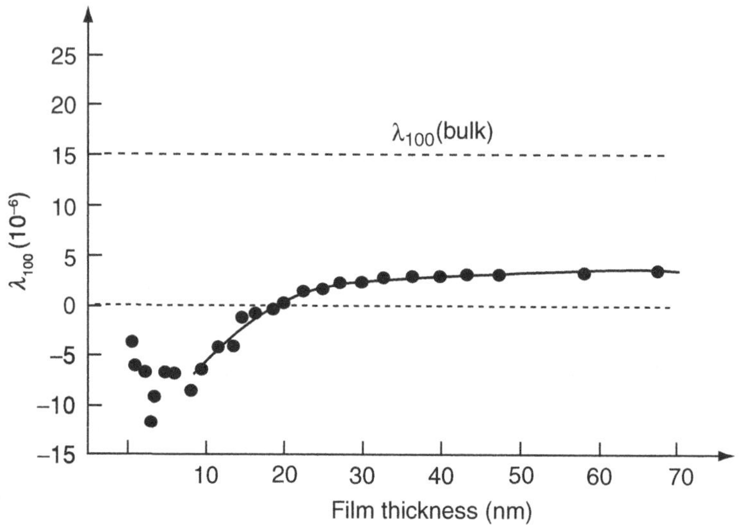

Figure 8.2

Magnetostriction of iron as a function of film thickness – an example of how physical properties change in the nanoworld. (after D. Sander, A. Enders and J. Kirschner, *J. Magn. Magn. Mater* 189, 519 (1999))

interactions are weakened. The bands are narrower at the surface, and the local density of states and the local magnetic moments may be enhanced. These effects are limited to the first one or two monolayers. Sometimes, the substrate influences the electronic structure and magnetic moment of the first atomic layers at the interface.

Clean surfaces can only survive in ultrahigh vacuum. On exposure to air, they are instantly covered by a layer of adsorbed gas, which modifies their electronic and magnetic properties. A monolayer forms in a matter of minutes in a vacuum of 10^{-5} Pa. These effects are sensitive to crystallographic orientation, and whether it is a single-crystal or a polycrystalline film. A cap layer serves to protect an underlying film from these effects.

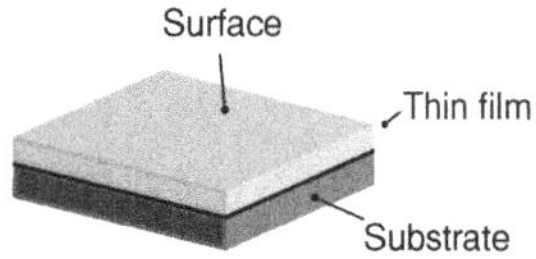

A thin film on a substrate.

Magnetic films with thicknesses ranging from a single monolayer to upwards of 100 nm may be grown on crystalline or amorphous substrates by a variety of physical or chemical methods, described in Chapter 10. An epitaxial single-crystal film is one that grows in perfect atomic register with a single-crystal substrate. An oriented film has one specific crystal axis oriented perpendicular to the substrate. The region of the film near the interface may be highly strained if there is a difference in lattice parameters between film and substrate. The compressive or expansive biaxial strain is accompanied by a strain of opposite sign in the direction normal to the substrate. Excess strain is relaxed in thicker films by atomic-scale dislocations as the film eventually adopts its equilibrium lattice parameters far from the substrate. When the lattice mismatch is too great (>4%), or if the substrate is amorphous (e.g. glass) it has less ability to dictate the structure of the film growing on it. Substrates can be chosen to influence the crystallographic or magnetic texture of the film directly, or else a thin seed layer can be deposited first to do the job.

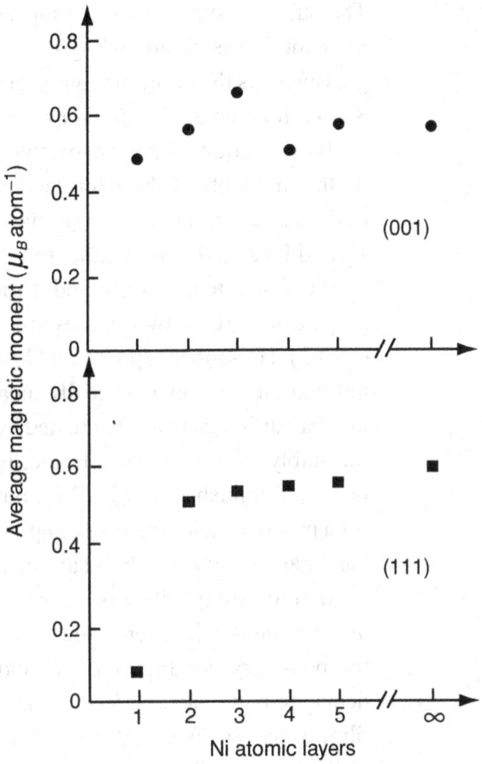

Figure 8.3

Variation of the average magnetic moment of a thin film of Ni on Cu (001) or (111). (J. Tersoff and L. M. Falicov, *Phys. Rev B* 26 6186 (1982).)

Epitaxial films may even have a different crystal structure to bulk material. For example, iron grown on copper (100) is fcc, whereas bulk iron is bcc. There is great scope to manipulate the structure and lattice parameters of solids in thin-film form, by suitable choice of substrate and preparation conditions.

8.2.1 Magnetization and Curie point

Dramatic modifications of magnetization are found in films a few monolayers thick. Vanadium and rhodium become ferromagnetic in films 1–2 monolayer thick, although they are nonmagnetic in bulk. Metals such as palladium which have an enhanced paramagnetic susceptibility, become ferromagnetic when deposited on a substrate of iron or nickel. The example of nickel on copper is illustrated in Fig. 8.3.

The magnetic properties of iron are notoriously structure-sensitive (Fig. 5.15) While the bcc form is ferromagnetic, fcc iron may be nonmagnetic, antiferromagnetic or ferromagnetic, depending on the lattice parameter. When grown epitaxially on fcc copper, the magnetic properties of an iron film depend on the substrate temperature during deposition. Ambient temperature gives ferromagnetic fcc iron films, but when the substrate is cooled, they are antiferromagnetic.

The surface layer of a bcc film has been found to have a zero-temperature moment that is about 20% greater than the bulk. The iron moment falls progressively as the coordination is increased, from 4.0 μ_B for an isolated atom to 3.3 μ_B for a chain, 3.0 μ_B for a plane to 2.25 μ_B in the bulk.

Hybridization with the d-orbitals of the substrate has an important influence on the moments of the interface layers. A monolayer of iron on bcc tungsten (100), for example, is antiferromagnetic, with a moment of 0.9 μ_B but when a second layer is deposited, the iron becomes ferromagnetic.

The Curie temperature and critical behaviour are expected to change on going from three to two dimensions. According to the Mermin–Wagner theorem (§5.4.3), Heisenberg spins should never order magnetically in two dimensions, but two dimensions here really means a monolayer. Furthermore, the perpendicular surface anisotropy created by the broken symmetry means that the spins inevitably take on an anisotropic, Ising-like character. The Curie temperature is often diminished in thin films, but it does not necessarily extrapolate to zero for a monolayer. There are examples among the rare-earths, for example, where band narrowing at the film surface actually increases T_C slightly.

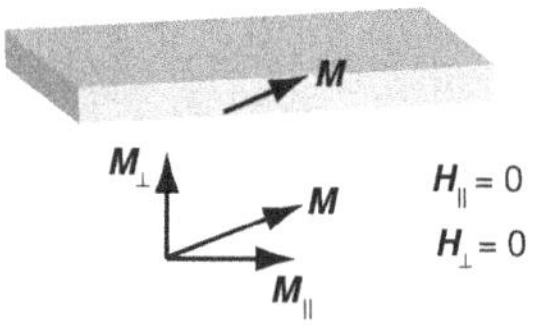

Neither component of a uniformly magnetized thin film produces a stray field.

An interesting feature of a uniformly magnetized thin film is that *it produces no stray field*, whatever its direction of magnetization. This can be seen from the boundary conditions on B and H (§2.4.2). If magnetized in-plane, the demagnetizing factors $\mathcal{N}_x$, $\mathcal{N}_y$ are zero, hence H_x and H_y are zero in the film. Since $H^{\|}$ is continuous at the interface, H is zero outside the film as well. However, if magnetized out-of-plane, $\mathcal{N}_z = 1$, $H = -\mathcal{N}_z M$. Since $B = \mu_0(H + M) = 0$ and $B^{\perp}$ is continuous, it follows that $H^{\perp} = 0$ outside the film. For any other angle, the magnetization can be resolved into parallel and perpendicular components, so H is always zero outside the film. If any useful stray field is to appear above the film, the magnetization must vary on a scale comparable to the film thickness. Magnets need to be block-shaped in order to be effective. This is as true for macroscopic permanent magnets as it is for nanoscale recording media.

8.2.2 Anisotropy and domain structure

Besides the usual magnetocrystalline anisotropy due to the single-ion (crystal field) and two-ion (dipolar) terms, there are also the '3s' contributions in a thin film – *s*hape, *s*urface and *s*train.

Shape anisotropy was introduced in §5.5.1. The principal components $(\mathcal{N}_x, \mathcal{N}_y, \mathcal{N}_z)$ of the demagnetizing factor of a uniformly magnetized thin film are (0, 0, 1), so the anisotropic contribution to the self-energy $-\frac{1}{2}\mu_0 M_{sz}^2$ in the demagnetizing field of a thin film is

$$E_d = \tfrac{1}{2}\mu_0 M_s^2 \cos^2 \vartheta, \tag{8.2}$$

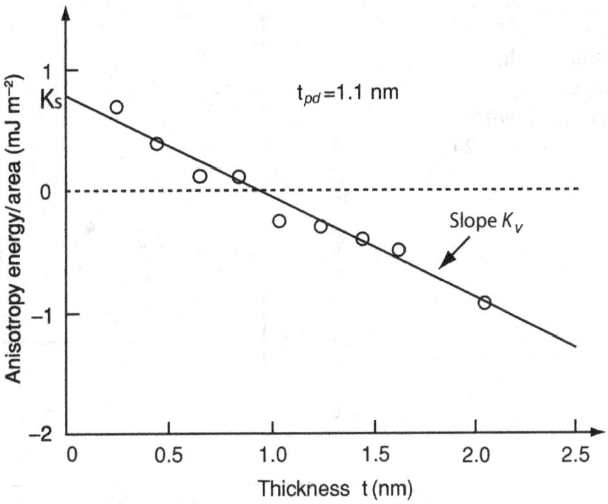

Figure 8.4

Plot to determine the surface and volume anisotropy terms for a Co–Pd multilayer (F. J. A. den Broeder, W. Horing, P. J. H. Bloemen *et al.*, *JMMM* 93, 562 (1991).)

where ϑ is the angle between the magnetization direction and the surface normal. This is equivalent, within a constant, to the leading term in the usual expression for the anisotropy energy $E_a = K_u \sin^2\theta$ (5.62), where the shape anisotropy constant $K_u = K_{sh} = -\frac{1}{2}\mu_0 M_s^2$. Values for Fe, Co and Ni are −1.85, −1.27 and −0.15 MJ m^{-3} respectively (Table 5.4). These are relatively large numbers, and since K_{sh} is negative, the demagnetizing field produces fairly strong easy-plane anisotropy in a thin film. A permanent magnet needs to have another source of anisotropy greater than this, if it is to be fabricated in any desired shape. Roughening of the film surface will lead to the appearance of a stray field, and diminish K_{sh}.

Next, there is the anisotropy in thin films and nanoparticles that comes from the surface. Surface anisotropy was first discussed by Néel in 1956. He estimated the magnitude as $K_s \approx 1$ mJ m^{-2}. It arises mainly from the single-ion mechanism – the coupling of the surface atoms to the crystal field produced by their anisotropic surroundings. Mostly it originates in the surface monolayer which has broken symmetry, but it extends to the first few monolayers which experience structural relaxation normal to the surface. The total anisotropy of a thin film is therefore the sum of terms that scale with volume and surface area. Writing $E_a = K_{eff} \sin^2\theta$, where

$$K_{eff} = K_v + K_s/\mathrm{t}$$

and t is the film thicknes, it is possible to deduce both K_v and K_s by plotting the anisotropy energy per unit area versus t for a series of films of different thickness, provided they have a common easy axis (Fig. 8.4). From the magnitude and sign of K_s, it is clear that the surface anisotropy is sufficient to impose a perpendicular magnetization direction on a film less than about a nanometre thick. The atomic density of the transition metals of about

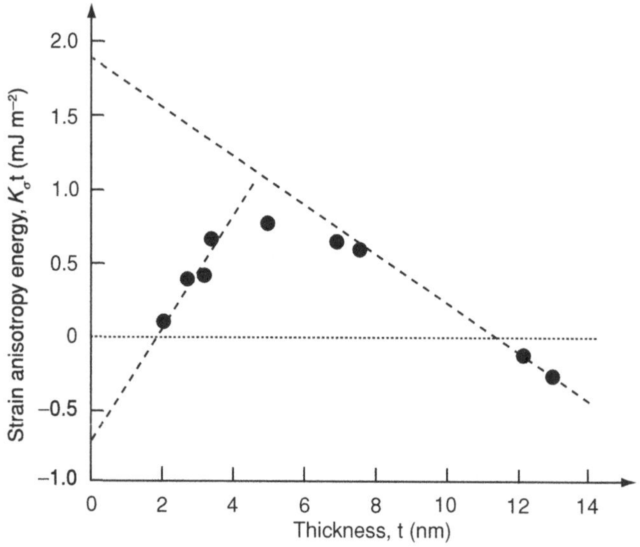

Figure 8.5

Strain anisotropy in thin films of Ni on Cu (R. Jungblut *et al.*, *Journal of Applied Physics* **75**, 6424 (1994).)

9×10^{28} m^{-3} corresponds to an atomic volume $\omega = 1.1 \times 10^{-29}$ m^3. Hence, if $K_s = 1$ mJ m^{-2} the energy per surface atom $\kappa_s = K_s \omega^{2/3} \simeq 4$ K atom^{-1}. This exceeds by an order of magnitude the dipole interaction and the bulk anisotropy energy normally found in crystals. The anisotropy of the surface monolayer corresponds to $K_{eff} = 4.5$ MJ m^{-3}. It is interesting that the bulk anisotropy of $L1_0$ compounds such as CoPt or FePd with a structure made up of alternating layers of the two constituents is of a similar magnitude (Table 11.7).

The third contribution to the anisotropy of a thin film is strain. The anisotropy energy associated with stress is $K_\sigma = (3/2)\lambda_s \sigma$ (5.68). This can have either sign, depending on the signs of λ_s and σ. For example, a 2% epitaxial compression in a material with elastic modulus 2×10^{11} N m^{-2} corresponds, according to Hooke's law, to a stress $\sigma = 4 \times 10^9$ N m^{-2}. If $\lambda_s = -20 \times 10^{-6}$, the stress anisotropy $K_\sigma = -120$ kJ m^{-3}. Epitaxial strain can extend for many monolayers, so it is possible for the stress anisotropy to outweigh the surface term in films that are about 1–10 nm thick.

Figure 8.5 illustrates the effect for a film of Ni grown on Cu. For thick films, the plot is as expected for $K_v = 15$ kJ m^{-3} and $K_s = 1.8$ mJ m^{-2}, but for films thinner than 5 nm there is a change of slope, and when $t < 2$ nm the films actually have easy-plane anisotropy. The surface contribution is overwhelmed by the contribution of the epitaxially strained layer.

Apart from surface effects on the electronic structure, the magnetization should be uniform in very thin ferromagnetic films. In thicker films with strong surface anisotropy $K_s \gg \frac{1}{2}\mu_0 M_s^2$, we would expect the surface layers to be magnetized perpendicular to the film, and the layers in the interior to be magnetized in-plane, with a progressive rotation of the magnetization $M(z)$, where

Figure 8.6

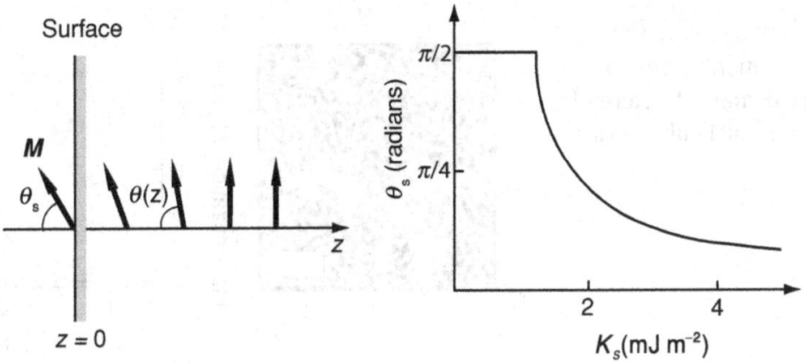

Twist of magnetization due to surface anisotropy.

$\theta = \theta(z)$, ϕ =const. As in a domain wall, the distance over which the rotation occurs is determined by the exchange stiffness A. Generally, we cannot *assume* that the response of a thin film to an applied field will be coherent rotation of uniform magnetization. $\boldsymbol{M}$ may depend on x, y.

The energy density (§7.1) of a thick film with a surface at $z = 0$ and uniaxial anisotropy K_u is

$$E = \int_{-t}^{0} \left[A \left(\frac{d\theta}{dz} \right)^2 + [K_u + K_s \delta(0)] \sin^2 \theta \right] dz,$$

where the delta function confines the surface anisotropy to the plane $z = 0$. Minimizing the integral gives the Euler equation $2A\partial^2\theta/\partial z^2 = \partial K_u \sin^2\theta/\partial\theta$, hence $\partial\theta/\partial z = \sqrt{K_u/A}\sin\theta$. The boundary condition from the integration of the delta function is $2A(d\theta/dz)_{z=0} = K_s \sin 2\theta_0$, where θ_0 is the value of θ at $z = 0$. The result is a domain wall equation

$$z = \sqrt{A/K_u} \ln\{\tan[(\theta - \theta_0)/2 + \pi/4]\}.$$

This is just the result (7.16) with a shift of the origin. The orientation of the surface layer θ_s is plotted in Fig. 8.6. When the net anisotropy is negative, or $K_u < \frac{1}{2}\mu_0 M_s^2$, the magnetization is everywhere in-plane, but when it is positive, or $K_u > \frac{1}{2}\mu_0 M_s^2$, the angle θ_0 is never quite zero, and θ only approaches $\pi/2$ asymptotically for large values of z as in a domain wall.

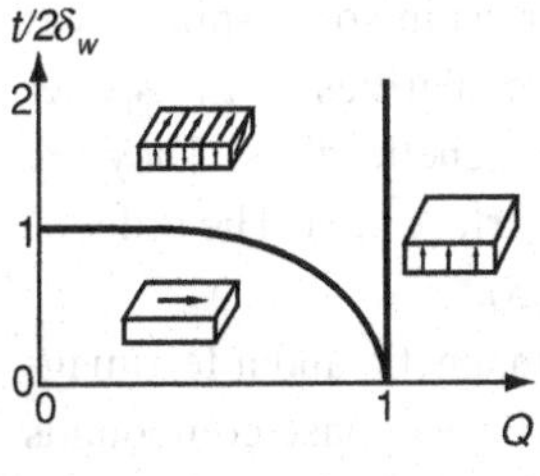

Magnetic domains in a thin film as a function of Q and film thickness t.

When considering the magnetization of thin films with a perpendicular anisotropy which may be due to epitaxial stress, surface or growth-induced texture, it is customary to use the quality factor $Q = -K_u/K_{sh}$ where K_u is the uniaxial perpendicular anisotropy, of whatever origin, and $K_{sh} = -\frac{1}{2}\mu_0 M_s^2$ is the shape anisotropy of the thin film. When $Q < 1$, the magnetization of the film lies entirely in-plane, if the film is thinner than twice the domain-wall width δ_w. Otherwise, a system of stripe domains with an alternating in and out of plane component develops (Fig. 8.7). In a polar Kerr effect image of the surface, for example, these look like maze domains, although most of the magnetization is actually in the film plane.

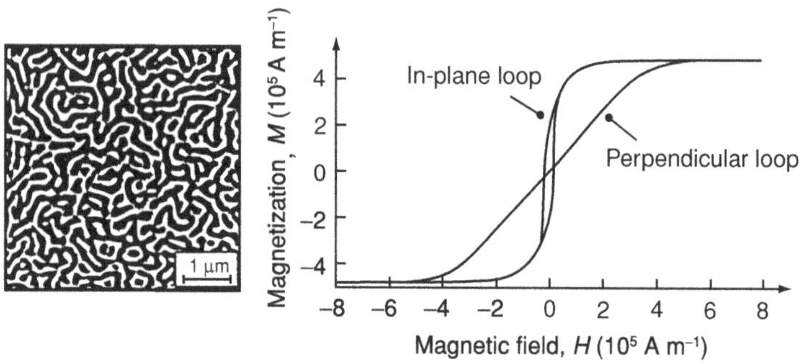

Figure 8.7

Magnetization curves and surface domain structure of a 200 nm (001) film of Ni.

Another property that can be greatly influenced by surface effects and substrate-induced strain is magnetostriction . As an example, we showed the magnetostriction of iron films in Fig. 8.2. The bulk value of λ_{100} for iron is 20×10^{-6}. Magnetostriction changes sign at a thickness of 20 nm.

8.3 Thin-film heterostructures

Magnetic multilayers are formed of alternating layers of magnetic and non-magnetic metal. When the layers are all epitaxial, the multilayer becomes a superlattice. Heterostructures can also be made of magnetic layers in direct contact one with another; there will be direct exchange coupling at the interface. Indirect exchange coupling in multilayers is mediated by spin polarization in the nonmagnetic layers, provided they are thin enough. Dipolar interactions also play a role in coupling ferromagnetic thin films which are not perfectly smooth.

8.3.1 Direct exchange coupling; exchange bias

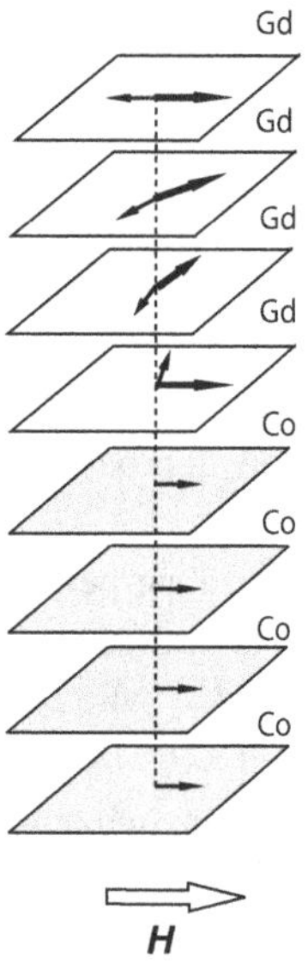

A field-controllable domain wall in an YCo_2–$GdCo_2$ bilayer.

A magnetic bilayer composed of two different ferromagnetic layers with a clean interface is expected to behave as a single ferromagnetic layer. However, if the two layers are imperfectly separated by a thin spacer, as in some spin valves and tunnel junctions, there is the possibility that irregularities in the spacer can lead to pinhole contacts between the two ferromagnetic films. They are exchange-coupled only over a small fraction of the interface area. The order of magnitude of the coupling is 0.1 mJ m^{-2} (Exercise 8.5).

More interesting are bilayers composed of a ferromagnetic and a ferrimagnetic layer, for example YCo_2 and $GdCo_2$. Both are Laves phase compounds with similar lattice parameters $a_0 = 737$ pm (see §11.3.5). They each have a ferromagnetic cobalt sublattice with a moment of 1.5 μ_B/Co, but yttrium is nonmagnetic whereas gadolinium has a moment of 7 μ_B coupled antiparallel

Figure 8.8

Shifted loop of partially oxidized cobalt particles, measured after cooling to 77 K in a 1 T field (W. H. Meiklejohn and C. P. Bean, *Phys. Rev.* 105, 904 (1956)).

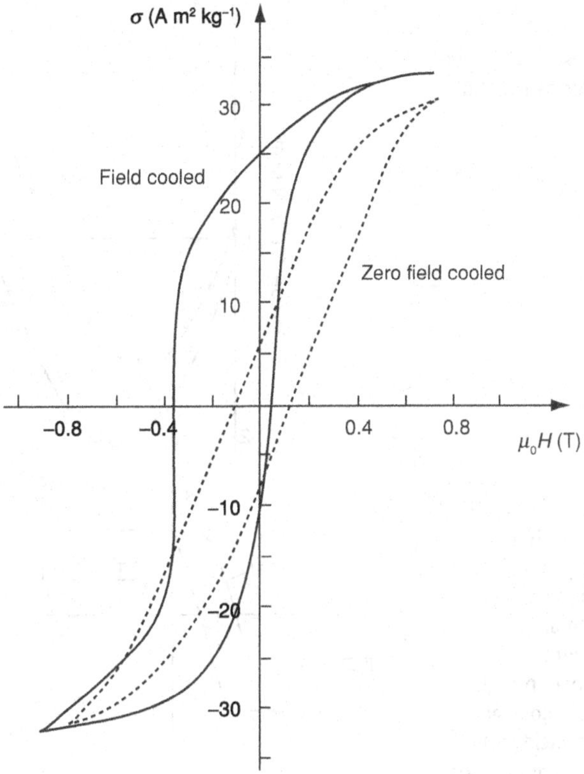

to the cobalt. The rare-earth behaves as a light transition element with three electrons in a $5d/6s$ band, so its spin moments couple antiparallel to those of heavy transition elements like Co, where the $3d^{\uparrow}$ shell is full. Applying an in-plane field to the system creates a twist in the magnetization, which resembles a Bloch wall that becomes narrower as the field increases – effectively a field-controllable domain wall.

When a ferromagnetic and an antiferromagnetic film are in contact, the coupling between them leads to an unusual unidirectional anisotropy first observed by Meiklejohn and Bean in 1956 in Co nanoparticles ($T_C = 1390$ K) coated with CoO ($T_N = 291$ K), Fig. 8.8. The CoO was cooled through T_N in the exchange field of the cobalt, which has been aligned in an external field. The effect was a shifted ferromagnetic hysteresis loop, from which the authors concluded that '*A new type of magnetic anisotropy has been discovered which is best described as an exchange anisotropy. This anisotropy is a result of an interaction between an antiferromagnetic material and a ferromagnetic material*'. A related phenomenon is rotational hysteresis, illustrated in Fig. 8.9.

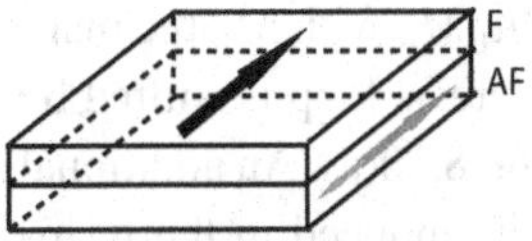

An antiferromagnetic (Af) exchange bias layer coupled to a Ferromagnetic (F) layer.

Néel later described a similar effect in a pair of coupled thin films. Exchange bias arises when the Curie temperature of the ferromagnet exceeds the Néel temperature of the antiferromagnet. Either the antiferromagnetic or the ferromagnetic layer may be on top; these are called top-pinned and bottom-pinned

Figure 8.9

Rotational hysteresis of the same particles as Fig. 8.8.

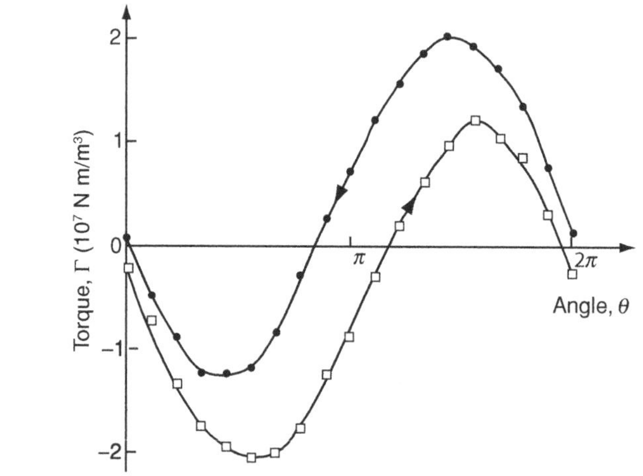

Figure 8.10

The effect of exchange bias on the hysteresis loop of a ferromagnetic layer (F) coupled to an antiferromagnetic layer (AF). The direction of the exchange field does not necessarily coincide with the antiferromagnetic axis. The loop in (a) is measured with the applied field parallel to the exchange field; the one in (b) with the applied field in a perpendicular direction.

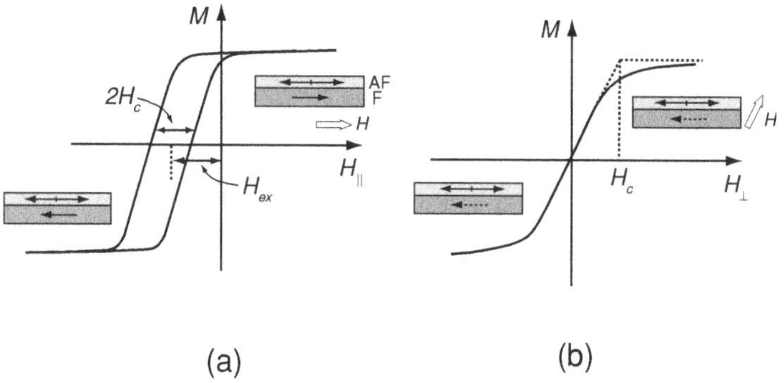

structures, respectively. The exchange bias depends on the atomic scale structure of the interface between the two layers. It is set by cooling in a field sufficient to saturate the ferromagnetic layer. An important application is in spin valves, where the magnetization direction of one ferromagnetic layer is pinned by exchange bias, while another layer remains free to flip its magnetization in a small field.

The energy of a bilayer system when the magnetization lies at an angle ϕ to the field, which was applied in plane, along the x axis is

$$E_x = -\mu_0 M_p H_x \cos\phi - K_{ex}\cos\phi. \tag{8.3}$$

It is as if the effective field acting on the ferromagnetic pinned layer of magnetization M_p is $H_{eff} = H + H_{ex}$, where $H_{ex} = K_{ex}/\mu_0 M_p$ with M_p the magnetization of the ferromagnetic pinned layer. The hysteresis loop is shifted by the unidirectional anisotropy $K_{ex}\cos\phi$, as shown in Fig. 8.10(a). An additional uniaxial anisotropy of the usual type $K_u \sin^2\phi$ may be induced in the ferromagnetic layer by the thermal treatment in the magnetic field. If a field is then applied in the transverse in-plane y-direction, the energy is

$$E_y = -\mu_0 M_p H_y \cos(\pi/2 - \phi) - K_{ex}\cos\phi + K_u \sin^2\phi.$$

Figure 8.11

The dependence of exchange bias on (a) the thickness of the antiferromagnetic layer and (b) the thickness of the ferromagnetic layer (O'Handley, 2000).

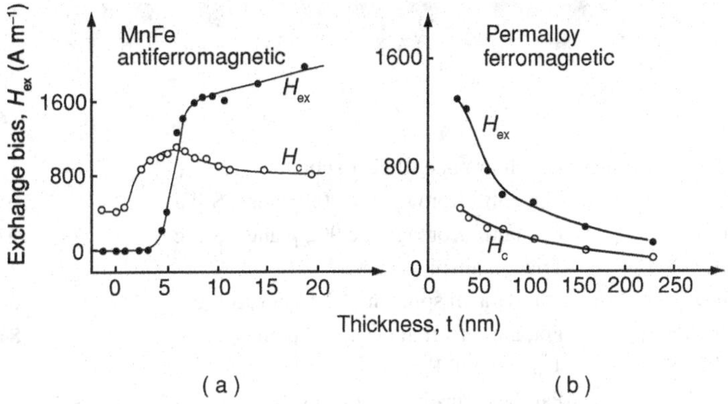

Minimizing this energy leads to the transverse magnetization curve shown in Fig. 8.10(b) where the slope at the origin $M_p d\phi/dH_y$ is $\mu_0 M_p^2/(K_{ex} + 2K_u)$. The anisotropy field H_a determined by extrapolation of the slope to saturation is $(K_{ex} + 2K_u)/\mu_0 M_p$.

The energy is often written per unit *area* of film, instead of per unit volume, as above, because exchange bias is an interfacial interaction which scales with the interface area. Replacing K_{ex} by σ_{ex}/t_p, the energy per unit surface area is

$$E_{\mathcal{A}} = -\mu_0 M_p H_x \mathrm{t}_p \cos\phi - \sigma_{ex}\cos\phi + K_u t_p \sin^2\phi, \qquad (8.4)$$

where t_p is the ferromagnetic pinned layer thickness, and the field is applied in the x-direction. The field corresponding to an anisotropy or exchange energy per unit area is $E_{\mathcal{A}}/\mu_0 M\mathrm{t}_p$.

Minimizing $E_{\mathcal{A}}$ gives

$$\sin\phi[\cos\phi + (\sigma_{ex}/2K_u\mathrm{t}_p) + (\mu_0 M_p H/2K_u)] = 0.$$

There are stable solutions at $\phi = 0$ and $\phi = \pi$, and switching occurs at $\phi = \pi/2$ if the rotation is coherent, when $H = H_{ex} = -\sigma_{ex}/\mu_0 M_p \mathrm{t}_p$. The switching is independent of K_u, but it varies as the inverse thickness of the ferromagnetic layer, as shown in Fig. 8.11 for permalloy on FeMn. Taking $M_p = 500$ kA m^{-1} for permalloy, these data give $\sigma_{ex} = 0.12$ mJ m^{-2}, which is a typical value for exchange bias (Table 8.2). The perpendicular anisotropy field obtained when the field is applied in the y-direction can be written in terms of σ_{ex} as $H_a = (\sigma_{ex} + 2K_u\mathrm{t}_p)/\mu_0 M_p \mathrm{t}_p$. It varies inversely with t_p for small thickness, but becomes independent of t_p for large thickness.

The effectiveness of the exchange bias depends on t_p, but also on the thickness and anisotropy of the antiferromagnetic layer, in the sense that there is a threshold thickness t_{af}^c which is necessary for exchange bias to become effective, as shown in Fig. 8.11(a). The threshold thickness can be used to estimate

Table 8.2. Antiferromagnetic materials for exchange bias

		T_N (K)	T_b (K)	σ_{ex} (mJ m^{-2})
FeMn	Fcc; four noncollinear sublattices; S || {111}	510	440	0.10
NiMn	Fct; antiferromagnetic 002 planes, S || a	1050[a]	≈700	0.27
PtMn	Fct; antiferromagnetic 002 planes, S || c	975	500	0.30
$RhMn_3$	Triangular spin structure	850	520	0.19
$Ir_{22}Mn_{78}$	Fct; parallel spins in 002 planes, S || c	690	540	0.19
$Pd_{52}Pt_{18}Mn_{50}$	Fct; antiferromagnetic 002 planes	870	580	0.17
$aTb_{25}Co_{75}$[b]	$T_{comp} = 340$ K	600	>520	0.33
NiO	Parallel spins in 111 planes, S ⊥<111>	525	460	0.06
αFe_2O_3	Canted antiferromagnet, S ⊥c	960	≈500	0.05

[a] Order-disorder transition
[b] Sperimagnetic; T_N is the Curie temperature.
fcc – face centred cubic; fct – face centre tetragonal

the coupling constant $\sigma_{ex} \approx t^c_{af} K_{af}$, where K_{af} is the volume anisotropy of the antiferromagnet. Typical values of $t^c_{af} = 10$ nm and $K_{af} = 20$ kJ m^{-3} give $\sigma_{ex} \approx 0.2$ mJ m^{-2}.

In seeking a nanoscale explanation of the origin of exchange bias, questions that have to be addressed are: What is the origin of σ_{ex}? Why is the blocking temperature T_b below which exchange bias is effective significantly less than T_N (Table 8.2)?

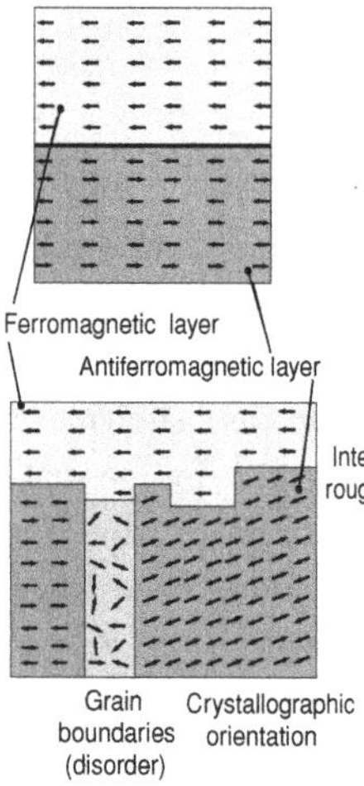

An ideal interface and a real interface.

Consider first the case of an ideal atomically flat interface between a ferromagnet and an antiferromagnet. The antiferromagnetic surface may be one with equal numbers of ↑ and ↓ spins, in which case $\sigma_{ex} = 0$ if the ferromagnetic and antiferromagnetic axes are parallel. Another possibility is a planar antiferromagnetic structure with alternating planes of ↑ and ↓ spins normal to the interface. In that case, $\sigma_{ex} = A/d$, where d is the interplanar spacing. Typical values of A and d of 2×10^{-11} J m^{-1} and 0.2 nm, respectively, give $\sigma_{ex} \approx 100$ mJ m^{-2}, three orders of magnitude too big. Only a small fraction of the atoms in the interface, perhaps one in a thousand, appear to participate effectively in the exchange coupling. The real interface is bound to be somewhat rough, defining antiferromagnetic surface regions of dimension L, containing $(L/a)^2$ atoms, where a is the interatomic spacing. The uncompensated moment of these randomly drawn regions is that of L/a atoms, so only a fraction a/L of the atoms participate in the exchange. Hence $L \approx 1000a$, or $L \approx 200$ nm. Antiferromagnetic domains are larger than this, but surface roughness on this scale is quite plausible. However, the moments of these regions themselves add randomly, giving an overall fraction $a/\mathcal{A}$ of atoms participating in the exchange, where $\mathcal{A}$ is the sample

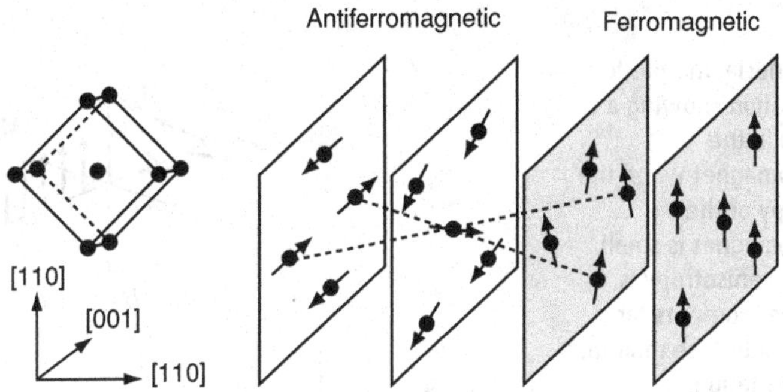

Figure 8.12

An explanation of how the ferromagnetic and antiferromagnetic axes may couple perpendicularly. There is a common [110] plane with antiferromagnetic coupling across it.

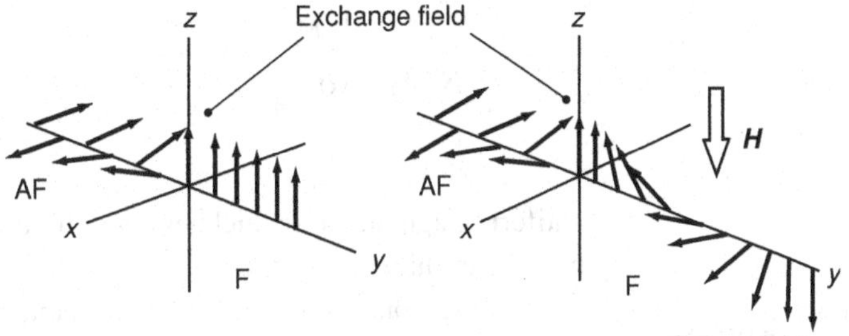

Figure 8.13

Formation of a domain wall at an interface of a soft ferromagnetic layer (F) exchange-coupled to an antiferromagnet (AF).

area. This is tiny in a macroscopic sample, and cannot account for the magnitude of σ_{ex}.

The exchange bias could arise from the regions of the antiferromagnet such as grain boundaries or other defects where the exchange is partly frustrated, and the interaction with the ferromagnet can stabilize one specific spin configuration at the defect. Another explanation is based on the idea that the susceptibility of an antiferromagnet is anisotropic, taking its greatest value when the antiferromagnetic axis lies perpendicular to an applied field (Fig. 6.2). The interfacial exchange may be represented as a molecular field H^i acting on the first plane of antiferromagnetically coupled atoms at the interface. The antiferromagnetic axis therefore tends to lie perpendicular to the ferromagnetic axis, as illustrated in Fig. 8.12. The interfacial coupling energy (Fig. 8.13) comparable to that stored in a 90° is domain wall $\frac{1}{2}\sqrt{AK_{af}}$, which is ≈ 0.3 mJ m^{-3}, the correct order of magnitude. A spin flop occurs when K_{af} is small; Fig. 8.14. The exchange field is $\frac{1}{2}\sqrt{AK_{af}}/\mu_0 M_p t_p$.

To summarize, the exchange coupling between a ferromagnetic and an antiferromagnetic film does not vanish when the structure of the interface is spin-compensated. The coupling is weakened by surface roughness. The effective surface exchange σ_{ex} is 100 times less than might be expected from an uncompensated interface, because a domain wall forms there allowing easier reversal of the ferromagnetic layer. The exchange bias vanishes below a critical

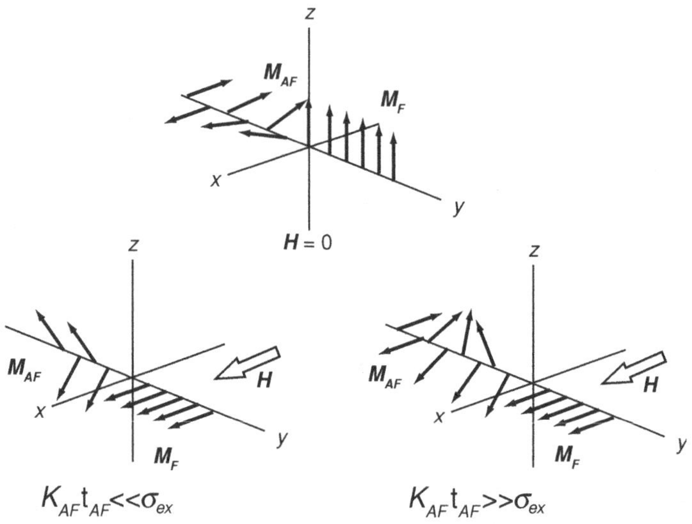

Figure 8.14

The interfacial magnetic configuration showing a spin flop in the antiferromagnet when the anisotropy of the antiferromagnet is small. When the anisotropy is large, the moments far from the interface maintain their orientation.

antiferromagnetic layer thickness. and it varies with the inverse of ferromagnetic layer thickness.

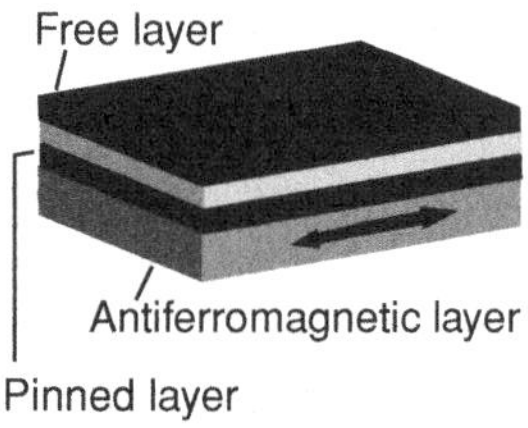

A spin valve.

Exchange bias is practically important, but still imperfectly understood. Indeed, it was the first practical application of antiferromagnetism, used in spin valves to maintain the direction of the pinned layer, while an adjacent free layer responds to a very small magnetic field. A spin valve is a sandwich of two ferromagnetic layers with a metallic or insulating spacer. The operating principle is that its resistance depends on the angle between the directions of magnetization of the free and pinned layers which is controlled by the applied field. The exchange bias is set by cooling the ferromagnetic–antiferromagnetic bilayer in a field applied during manufacture. The magnetoresistence is exploited in about a billion spin valve sensors produced each year – an illustration of the idea mentioned in Chapter 1, that practical applications of a technology need not await a perfect understanding of the physics.

8.3.2 Indirect exchange coupling

The sign of the exchange coupling in multilayers with alternating ferromagnetic and nonferromagnetic spacer layers oscillates with the thickness of the spacer, Fig. 8.15, in a way that resembles the RKKY interaction, Fig. 5.8. In fact, the true periodicity of the interaction may be masked by the fact that the spacer thickness is a discrete number of monolayers, so that a different periodicity is apparent. This is known as the aliasing effect (Fig. 8.16).

By tuning the spacer thickness, the interlayer coupling can be chosen to be ferromagnetic, antiferromagnetic or zero. For example, in an FeCo–Ru–FeCo

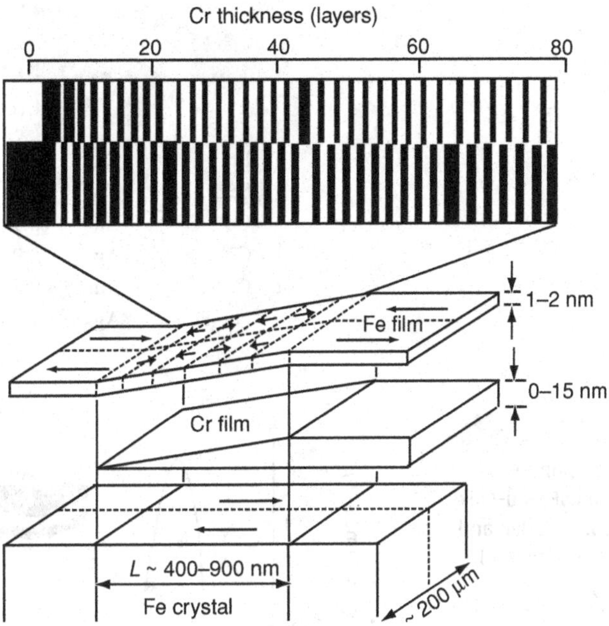

Figure 8.15

An experiment which demonstrates the oscillating spin polarization as a function of spacer thickness. The two iron layers are separated by a wedge-shaped chromium spacer. The sign of the coupling oscillates with the spacer thickness. The pattern of magnetization is shown in the iron overlayer, above two domains in the underlying iron crystal. (J. Unguris *et al.*, *Phys. Rev. Lett.* **67**, 140 (1991))

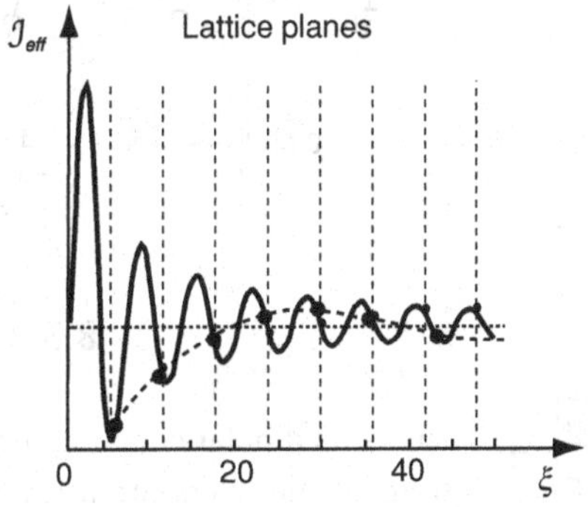

Figure 8.16

The aliasing effect.

trilayer, the coupling is zero for ruthenium thicknesses of 0.5 and 1.0 nm (Fig. 8.17). The idea of an artificial antiferromagnet (§6.1.1) can be extended to a multilayer with antiferromagnetic coupling between an even number of ferromagnetic layers of equal thickness. It exhibits many of the properties of a normal antiferromagnet including transverse susceptibility and spin flop. The ↑ and ↓ layers play the part of the ↑ and ↓ sublattices. If the ↑ and ↓ layers are of unequal thickness, the multilayer is an artificial ferrimagnet. Some values of the maximum antiferromagnetic exchange coupling are collected in Table 8.3. The optimum ruthenium thickness, for the strongest antiferromagnetic coupling of cobalt layers is ≈ 0.7 nm.

Table 8.3. Antiferromagnetic interlayer exchange coupling in multilayers

		t_s(nm)	σ_{ex}(mJ m^{-2})
Fe	Cu	1.0	−0.3
Fe	Cr	0.9	−0.6
Co	Cu	0.9	−0.4
Co	Ag	0.9	−0.2
Co	Ru	0.7	−5.0
$Ni_{80}Fe_{20}$	Ag	1.1	−0.01

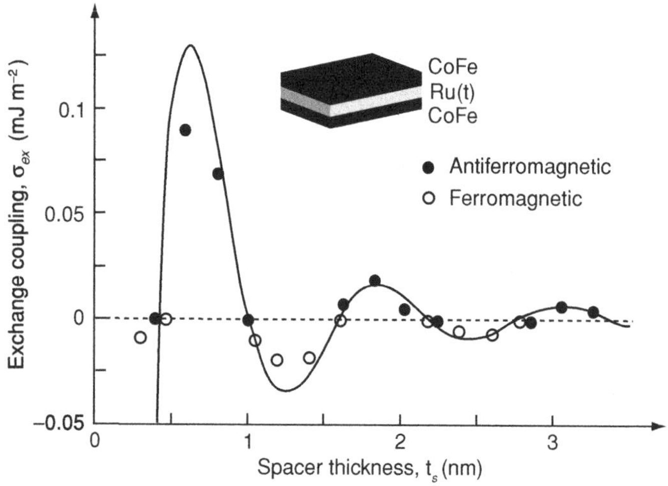

Figure 8.17 Oscillating exchange coupling in a CoFe–Ru–CoFe trilayer. (S. S. P. Parkin and D. Mauri, *Phys. Rev.* B44, 7131 (1991))

8.3.3 Dipolar coupling

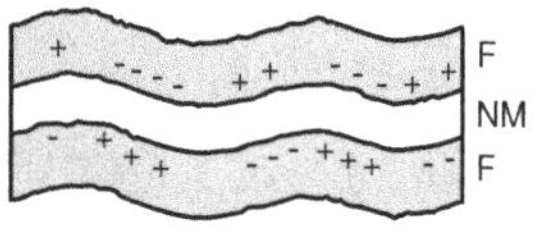

The orange-peel effect. NM is a nonmagnetic spacer layer.

There is no dipolar coupling between perfectly smooth, uniformly magnetized ferromagnetic layers because they create no stray field. However, rough surfaces do couple via dipolar fields. This is known as the orange-peel effect. When the roughness of the two layers is correlated, the coupling is ferromagnetic. The magnitude of the dipolar coupling associated with surface roughness was calculated by Kools, who found

$$\sigma_d = \frac{\pi^2}{\sqrt{2}}\frac{\delta_s^2}{l}\mu_0 M_s^2 \exp(-2\pi\sqrt{2}t_s/l), \tag{8.5}$$

where t_s is the thickness of the spacer layer, δ_s is the surface roughness of the spacer and l is the period of the roughness. Some typical values for a rough film are $t_s = 5$ nm, $\delta_s = 1$ nm $l = 20$ nm, which gives $\sigma_d = 0.03$ mJ m^{-2} for a film with $\mu_0 M_s = 1$ T. This is not insignificant compared with the exchange coupling shown in Table 8.3.

Figure 8.18

GMR in an Fe–Cr multilayer stack. The layer thicknesses are given in nanometres. (M. N Baibich *et al.*, *Phys. Rev. Letters* **61**, 2472 (1988).)

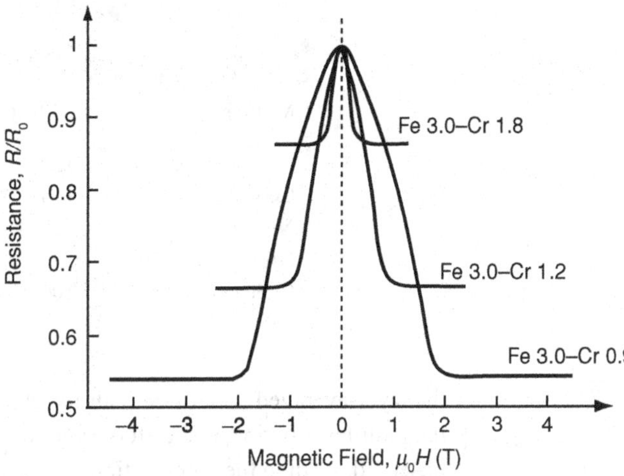

Albert Fert, 1938–.

Peter Grünberg, 1939–.

Dipolar coupling is also significant in magnetic nanopillars, where each nanostructured layer creates a stray field.

8.3.4 Giant magnetoresistance

The most important property of magnetic multilayers from a practical viewpoint is magnetoresistance. The unexpectedly large effect dubbed giant magnetoresistance (GMR) was discovered in an epitaxially grown, antiferromagnetically coupled Fe–Cr multilayer by Albert Fert and coworkers, and independently by Peter Grünberg and coworkers in 1988. Their discovery, which led to the development of the spin-valve sensor, was recognized by the award of the 2007 Nobel Prize in physics. The magnitude of the GMR effect, given by the ratio $\Delta R/R$, where ΔR is the resistance change in the field, and R is the zero-field resistance, can be some tens of per cent. Hence the effect is 'giant' by comparison with the intrinsic AMR effect discussed in §5.6.4. Some early results on an Fe–Cr multilayer are shown in Fig. 8.18. The Cr layer thickness gives antiparallel coupling of the Fe layers, so the effect of applying the magnetic field is to change from an antiparallel to a parallel alignment of adjacent layers. The magnitude of the GMR in different multilayers is summarized in Table 8.4.

The decrease of resistance on applying the field can be understood in terms of Mott's two-current model of conduction, which neglects spin-flip scattering. The $\uparrow$ and $\downarrow$ electron channels conduct in parallel, but the scattering is different for parallel and antiparallel magnetic alignment of the layers. Scattering occurs in the *bulk*, and at the *interfaces* between the magnetic and nonmagnetic layers. If we consider only bulk scattering in the ferromagnetic layers, the quantities $R_\uparrow$ and $R_\downarrow$ are the resistances of the stack for the $\uparrow$ and $\downarrow$ electrons. When

Table 8.4. Magnitude of the GMR effect in the given field at 4 K for some multilayer stacks

Multilayer	GMR (%)	$\mu_0 H$(T)
Fe–Cr	150	2.0
Co–Cu	115	1.3
NiFe–Co	25	1.5
NiFe–Ag	50	0.1
CoFe–Ag	100	0.3

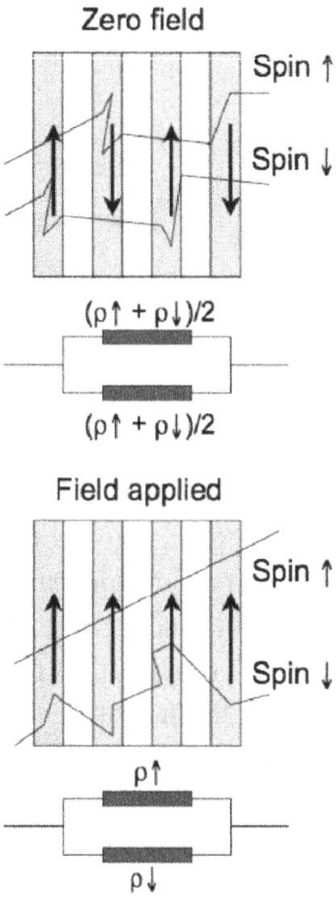

Illustration of the derivation of (8.8) and (8.9).

the layer-averaged mean free path for electrons of one spin direction exceeds the multilayer period, and differs from that of the other spin direction, GMR is observed. Adding the contributions of the two spin channels, the net resistance in the parallel state is given by

$$R_p^{-1} = R_\uparrow^{-1} + R_\downarrow^{-1}, \tag{8.6}$$

whereas, in the antiparallel state, each channel has the same resistance $(R_\uparrow + R_\downarrow)/2$, so

$$R_{ap} = (R_\uparrow + R_\downarrow)/4. \tag{8.7}$$

The magnetoresistance defined as

$$\Delta R/R = (R_{ap} - R_p)/R_{ap} \tag{8.8}$$

can be expressed in terms of the resistivity ratio $\alpha = \varrho_\downarrow/\varrho_\uparrow$ as

$$\frac{\Delta\varrho}{\varrho} = \frac{(1-\alpha)^2}{(1+\alpha)^2}. \tag{8.9}$$

According to the definition (8.8) magnetoresistance is positive when $R_{ap} > R_p$. This is the case when the stack is composed of one ferromagnetic material, and a simple nonmagnetic spacer. The value of $\alpha = \rho_\downarrow/\rho_\uparrow$ is about 5 for Co or Ni,[1] so (8.9) predicts an effect of $\sim$45%. Resistance decreases when a magnetic field is applied, unlike the classical B^2 magnetoresistance of a normal metal or semiconductor (§3.2.7).

In fact, interface scattering is usually dominant in magnetic multilayers – the interface resistance can exceed the bulk resistance by a factor 100 – but the equations can be retained by replacing the bulk values $R_\uparrow$ and $R_\downarrow$ by the interface resistances $R_{i\uparrow}$ and $R_{i\downarrow}$.

The characteristic length scale for transport with current parallel to the plane of the layers (CIP) is the mean free path λ, so the effect disappears

[1] α is related to the polarization of the conduction electrons, $P = (j^\uparrow - j^\downarrow)/(j^\uparrow + j^\downarrow) = (1-\alpha)/(1+\alpha)$. Another definition is $P = (2\beta - 1)$, where β ranges from 0 for pure ↓ current to 1 for pure ↑ current.

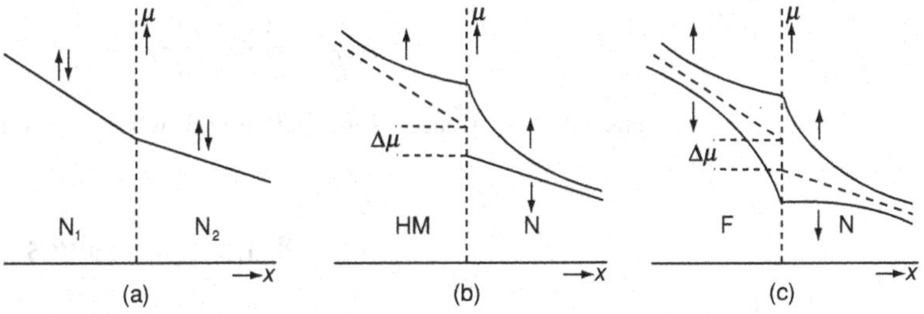

Figure 8.19

Matching the chemical potentials at the interface of: (a) two normal (N) metals, (b) a half-metal (HM) and a normal metal and (c) a ferromagnet and a normal metal. The characteristic length scale for spin accumulation in the normal metal is l_s, the spin-diffusion length.

when the nonmagnetic layer thickness is much greater than λ because the electrons do not then sample both layers. However, for transport with current flowing perpendicular to the plane (CPP), the characteristic length scale is the much-longer, spin diffusion length l_s. The build-up of spin polarization near the interface of a ferromagnet and a nonmagnetic metal whenever there is a component of current flowing across the interface is known as spin accumulation. Each of the spin channels is in dynamic equilibrium, and different chemical potentials for the $\uparrow$ and $\downarrow$ channels persist until spin-flip processes succeed in mixing electrons in the two channels.

Some different interfaces are shown in Fig. 8.19. The first is the interface of two normal metals. There, the chemical potential μ is continuous, with different slopes for the two metals, depending on their conductivity, according to (3.49). μ, which is equal to ε_F in the bulk metal at $T = 0$ K, is not actually a potential, but an energy per electron. Next, consider the interface between a half-metal ($\alpha = 0$) and a normal metal ($\alpha = \frac{1}{2}$). Suppose the current in the half-metal is composed of $\uparrow$ electrons only, as there are no $\downarrow$ states at the Fermi level. On reaching the interface, these $\uparrow$ electrons diffuse some distance into the normal metal before spin-flip scattering processes restore the equilibrium balance of equal $\uparrow$ and $\downarrow$ occupancy. The distance over which the injected spins accumulate at the interface is the spin-diffusion length. The interface between a normal ferromagnet and a nonmagnetic metal is shown in Fig. 8.19(c). Here majority spins are injected into the normal metal, as before, but there is also some change of the spin polarization on the ferromagnetic side. Far from the interface, the chemical potentials are equal, both in the ferromagnet and in the normal metal. The individual chemical potentials of the two spin channels $\mu^{\uparrow}$ and $\mu^{\downarrow}$ must be continuous at the interface if we ignore the interface resistance, but the average chemical potential $\mu(0)$ is discontinuous there. Away from the interface the slope of the chemical potential is constant, proportional to current density j; $\partial\mu/\partial x = ej/\sigma$ (3.49). The drop in μ at the interface is eV_{sa}, where V_{sa} is known as the spin accumulation voltage. It arises whenever current in separate $\uparrow$ and $\downarrow$ channels flows across an interface where there is a discontinuity in conductance for the two channels. The voltage can be calculated from $j = j^{\uparrow} + j^{\downarrow}$, writing each current in terms of the chemical potentials and integrating. The result is

$$V_{sa} = \frac{(\alpha - 1)}{2e(1+\alpha)}[\mu^{\uparrow}(0) - \mu^{\downarrow}(0)]. \tag{8.10}$$

The effect is greatest for a half-metal, when $\alpha = 0$ (or ∞).

8.3.5 Spin valves

We can think of a spin valve as a slimmed-down multilayer. Basically, it has just two ferromagnetic layers, although they may be buried in a complex thin-film stack. A general definition is any stack with free and pinned magnetic layers where there is a resistance change when the direction of magnetization of one layer is switched relative to the other.[2] The spin valve can be used as a bistable device with two states – a low-resistance state with the two ferromagnetic layers parallel, and a high-resistance state with the two ferromagnetic layers antiparallel – or else in a sensor mode where the resistance varies continuously as one layer turns relative to the other. Although (8.8) will do, a more optimistic definition of the maximum magnetoresistance is

$$MR_{max} = \frac{(R_{ap} - R_p)}{R_p} = \frac{(G_p - G_{ap})}{G_{ap}},$$

where the suffixes p and ap denote parallel and antiparallel orientations of the two ferromagnetic electrodes. Here MR_{max} is unlimited, whereas with (8.8) it cannot exceed 100%.

A pseudospin valve is a similar device without a pinning layer where two magnetic layers switch in different fields, which means they must have different coercivities. Ways of achieving different coercivities are to use different compositions or different thicknesses of the layers, or to pattern them in different shapes. Pseudospin valves have symmetric resistance and antisymmetric magnetization curves, like those shown in Fig. 8.20(a).

The exchange-biased spin valve (§8.3.1) is better for most sensor and memory applications, and it is discussed further in Chapter 14. One of the ferromagnetic layers is pinned by exchange coupling to an adjacent antiferromagnetic layer, while the other layer, the free layer, is able to rotate, with as little coercivity as possible. In suitable spin-valve structures, the coupling between the ferromagnetic layers may be reduced to almost nothing by suitable choice of spacer thickness. Switching of the relative orientation of the pinned and free layers may then occur in very small fields, giving sensitivities of about 20% mT^{-1} for current parallel to the plane of the layers, and even more for current perpendicular to the plane of the layers.

[2] A stricter definition of a spin valve is one where the spacer layer is metallic, and the resistance change is due to GMR. We refer to this as a GMR spin valve. Our definition encompasses both GMR and tunnelling magnetoresistance spin valves with exchange bias.

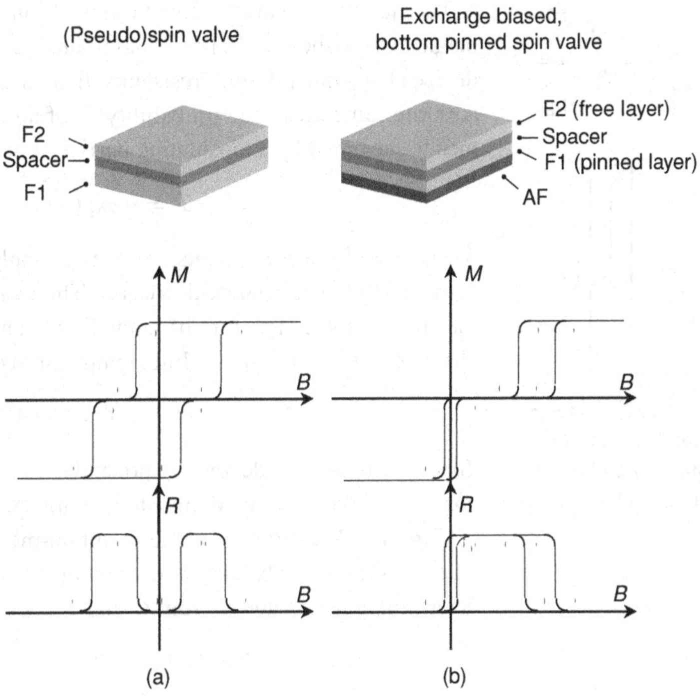

Figure 8.20

Spin valves: (a) a pseudospin valve where a difference in coercivity of the two ferromagnetic layers, F_1 and F_2, leads to a symmetric magnetoresistance curve, and (b) an exchange-biased spin valve where exchange coupling of the pinned layer F_1 to an antiferromagnetic layer, AF, leads to a shifted magnetoresistance response. The free layer, F_2, has very little coercivity, and it flips near zero field.

Any sensitivity of the free layer to fields generated by the pinned layer can be entirely eliminated by replacing it by an artificial antiferromagnet (§6.1.1), one side of which is coupled to the antiferromagnetic exchange-bias layer. The magnetization M_p for the artificial antiferromagnet (also known as a synthetic antiferromagnet) is zero.

The magnetoresistance of GMR spin valves is $\approx$10% with either definition. To achieve useful signals corresponding to resistance changes of order 100 Ω, the dimensions of a CPP GMR device should be several tens of nanometres. For CIP, the device may be extended in one direction to achieve the required resistance.

8.3.6 Magnetic tunnel junctions

Metal–insulator–metal junctions Tunnel junctions are thin-film trilayer structures in which an insulating layer, typically 1–2 nm of amorphous AlO_x or crystalline MgO, separates two metallic electrode layers. Organic spacers are also used. Ferromagnetic metallic layers are used in a spin valve structure in a planar magnetic tunnel junction (MTJ). There can also be tunnel barriers at point contacts, at the interfaces between grains in polycrystalline materials and pressed powders.

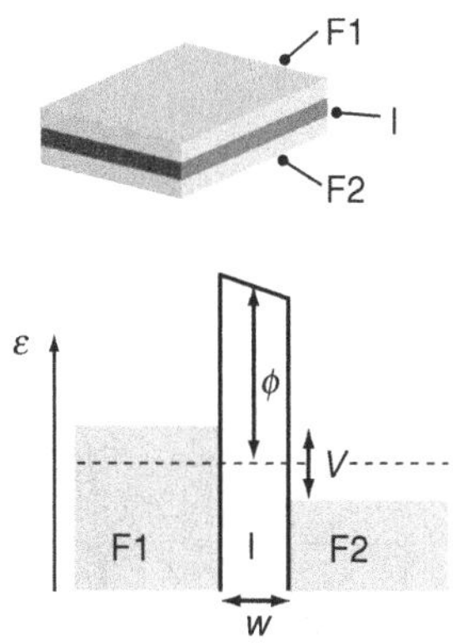

A tunnel barrier between two metals F_1 and F_2, separated by an insulator, I, with bias voltage V.

The quality of planar magnetic tunnel junctions, gauged by the tunnelling magnetoresistance (TMR) ratio, has improved spectacularly in recent years. The devices have much higher resistance than all-metal CPP multilayer structures with the same area. The probability $\mathcal{T}$ of an electron tunnelling between two metals separated by an insulating barrier of height ϕ and width w is

$$\mathcal{T} = a \exp(-bw\phi^{1/2}), \tag{8.11}$$

where a and b are constants. As bias is applied, the barrier becomes asymmetric, and the resistance decreases. The response at low bias is ohmic, but the characteristic signature of tunnelling is an additional V^3 term in the $I:V$ characteristic, which shows little temperature dependence

$$I = GV + \gamma V^3. \tag{8.12}$$

Since the resistance depends exponentially on barrier thickness, the tunnelling conductance tends to be dominated by hot spots, where the barrier is thinnest.

Simmons's treatment of the quantum-mechanical tunnelling of electrons across a symmetric barrier at low voltage leads to a formula which allows the determination of ϕ and w from G and γ:

$$G = (3e^2/2w\hbar^2)(2me\phi)^{1/2} \exp\{-(4\pi w/\hbar)(2me\phi)^{1/2}\}, \tag{8.13}$$
$$\gamma = \pi m/3\phi(ew/\hbar)^2. \tag{8.14}$$

The electrodes in magnetic tunnel junctions are ferromagnetic metals, preferably strong ferromagnets or half-metallic ferromagnets with a high degree of spin polarization. Half-metals show big effects at low temperature, but large TMR persists at room temperature only for certain Heusler alloys with high Curie points. Magnetic tunnel junctions with strong ferromagnetic cobalt or cobalt–iron-based electrodes give good results at room temperature. A TMR spin valve is produced by pinning one of the ferromagnetic layers by exchange bias with an adjacent antiferromagnetic layer, just as in a GMR spin valve.

Room-temperature TMR was first obtained with amorphous AlO_x barriers, Fig. 8.21. The quality of these devices improved steadily, but a breakthrough came in 2004 when crystalline MgO barriers were used instead where electrons of specific symmetry tunnel coherently across the insulator. When the MgO is grown epitaxially on bcc Fe–Co, the majority-spin electrons, which have Δ_1, s-like symmetry, are attenuated much less rapidly in the barrier than the minority-spin electrons, which have Δ_5, d-like symmetry, Fig. 8.22. Crystalline MgO therefore acts as a near-perfect spin filter, and huge TMR values in excess of 200% are achieved at room temperature, see Fig. 14.24.

The parallel and antiparallel configurations are usually the low- and high-resistance states of a tunnel junction. Jullière made a simple calculation of the tunnel magnetoresistance in terms of the spin polarizations P_1 and P_2 of the two ferromagnetic electrodes at the Fermi level. His result is

$$MR_{max} = \frac{2P_1P_2}{(1+P_1P_2)}. \tag{8.15}$$

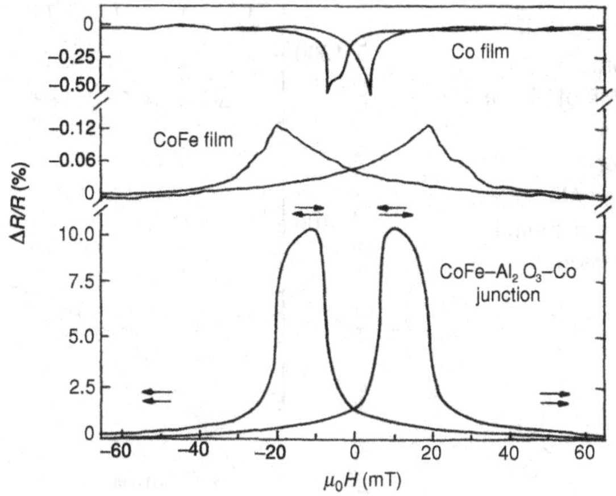

Figure 8.21

Resistance of a magnetic tunnel junction composed of a CoFe and a Co film separated by a thin layer of amorphous AlO_x. The device is a pseudo spin valve as the two layers have different coercivities, as shown by the AMR measurements in the top two panels. (J. S. Moodera, L. R. Kinder, T.M. Wong *et al., Phys. Rev. Letters*, 74, 3273 (1975).)

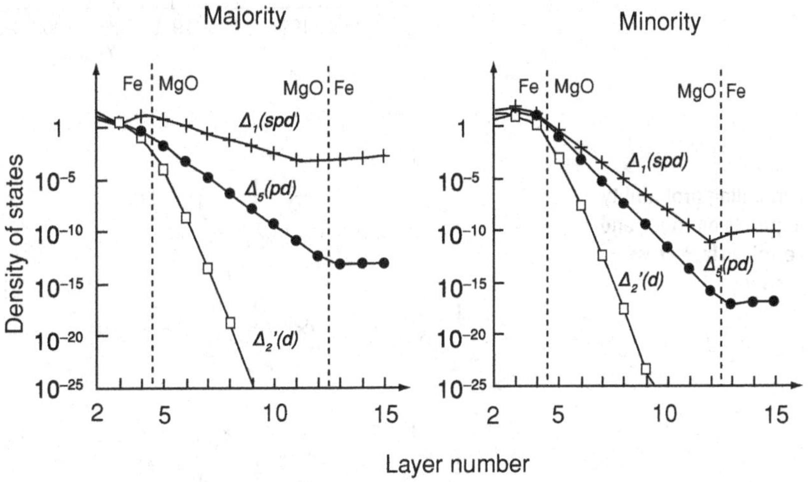

Figure 8.22

Attenuation of the density of states of the ↑ and ↓ electrons of iron as a function of thickness of an epitaxial MgO tunnel barrier Δ_1, Δ_2, Δ_5, refer to electrons in hybridized states with different symmetry. (W. H. Butler, *Phys Rev B* 63 054416 (2001).)

For identical electrodes, this reduces to

$$\frac{\Delta G}{G_{ap}} = \frac{2P^2}{(1+P^2)}. \tag{8.16}$$

The result is based on the supposition that the transmission probability of electrons across the barrier is simply proportional to the product of the initial and final densities of states of the appropriate spin. No account is taken of spin filtering due to different symmetry of the ↑ and ↓ electrons, which may be a fair assumption for polycrystalline electrodes and amorphous barriers. Hence, if G is the conductance, and $\mathcal{N}_1$ and $\mathcal{N}_2$ are the densities of states:

$$G_p \propto \mathcal{N}_{1\uparrow}\mathcal{N}_{2\uparrow} + \mathcal{N}_{1\downarrow}\mathcal{N}_{2\downarrow}, \tag{8.17}$$
$$G_{ap} \propto \mathcal{N}_{1\uparrow}\mathcal{N}_{2\downarrow} + \mathcal{N}_{1\downarrow}\mathcal{N}_{2\uparrow}. \tag{8.18}$$

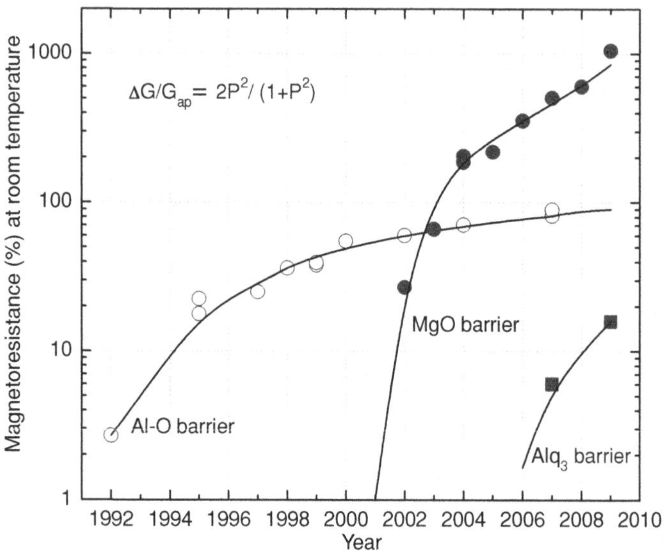

Figure 8.23

Progress in room temperature TMR of planar tunnel junctions. Characteristcs of exchange-biased AlO_x, Alq_3 and MgO magnetic tunnel junctions are shown.

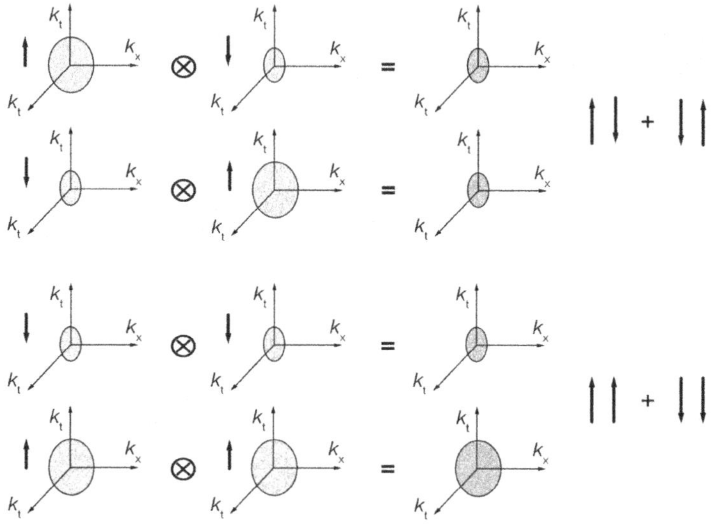

Figure 8.24

Tunnelling probability depends on the ↑ and ↓ Fermi surface cross sections.

Setting $MR_{max} = (G_p - G_{ap})/G_{ap}$ and $P = (\mathcal{N}_\uparrow - \mathcal{N}_\downarrow)/(\mathcal{N}_\uparrow + \mathcal{N}_\downarrow)$, where $\mathcal{N}_{\uparrow,\downarrow}$ are the densities of states at the Fermi level, gives the Jullière formula (8.15) (Exercise 8.3). As the bias across the tunnel barrier is increased, the tunnelling probability should reflect the changing density of states on either side of the barrier. It may even change sign at high bias.

Although widely used to derive values of P, the Jullière formula is a drastic approximation. Coherent tunnelling in moderately correlated electronic systems does not depend on the density of states at the Fermi level, but rather on

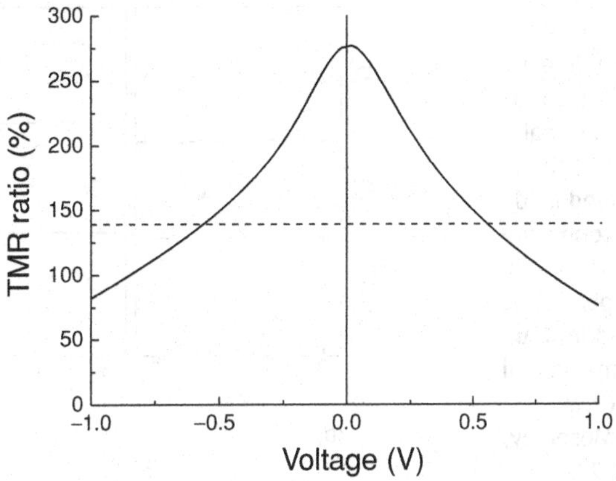

Figure 8.25

Bias dependence of the magnetroresistance of an MgO magnetic tunnel junction. The maximum voltage that can be generated in this case is 400 mV. (Data courtesy of Kaan Oguz.)

the convolution of the Fermi surfaces of the electrons on either side of the barrier, which are quite different for ↑ and ↓ electrons (Fig. 5.15). From Fig. 8.24 it can be seen that $G_p \propto \mathcal{S}_\uparrow + \mathcal{S}_\downarrow$, $G_{ap} \propto 2\mathcal{S}_\downarrow$, where $\mathcal{S}$ is the cross sectional area of the Fermi surface. Hence

$$\frac{\Delta G}{G} = \frac{\mathcal{S}_\uparrow - \mathcal{S}_\downarrow}{2\mathcal{S}_\downarrow}. \tag{8.19}$$

Besides coherent tunnelling, the density of interfacial states, metal–insulator bonding, carrier mobility and band symmetry may all play a role.

The TMR falls off with increasing bias due to excitation of magnons and phonons, which tend to randomize the spin polarization. An applied bias of 0.5 V may be enough to reduce the TMR by half, Fig. 8.25. This can be a problem, because it limits the voltage signal $\Delta V = V(\Delta R/R)$ that can be generated by the resistance change of the device.

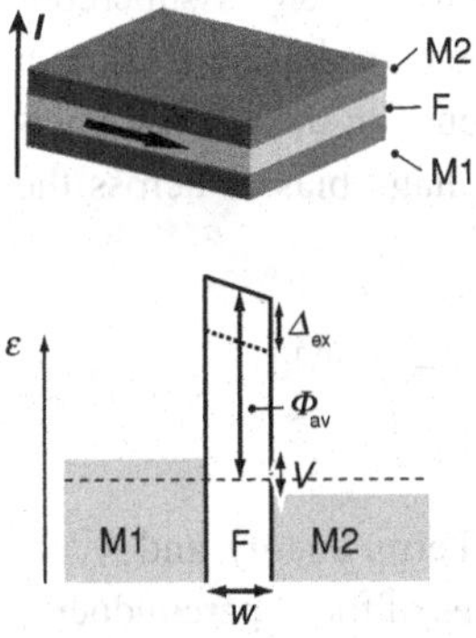

A spin filter. The ferromagnetic insulating layer, F, presents different barrier heights for ↑ and ↓ electrons from electrodes M1 and M2. In EuO, the difference is 0.54 eV. Electrons tunnelling from the unpolarized electrode, M1, emerge predominantly with ↑.

Ferromagnetic spin filter A related thin-film tunnelling structure is a spin filter, which uses a ferromagnetic insulator as the tunnel barrier sandwiched between two nonmagnetic metal electrodes. The barrier heights for ↑ and ↓ electrons differ by about 0.3 eV due to spin splitting of the unoccupied conduction band of the insulator by $s-d$ interaction. As a result of the exponential dependence of tunnelling on $\phi^{1/2}$, substantial spin polarization can be achieved. First demonstrated for the low-temperature $4f$ ferromagnet EuS as the tunnel barrier (§11.5.1), the effect has been observed at room temperature using a ferrimagnetic oxide such as $NiFe_2O_4$ or $CoFe_2O_4$. Note that the spin filter is not a spin amplifier. It does not increase the product of incident intensity and spin polarization of the incident current, but it just removes the ↓ electrons from the transmitted beam, by reflecting them at the interface.

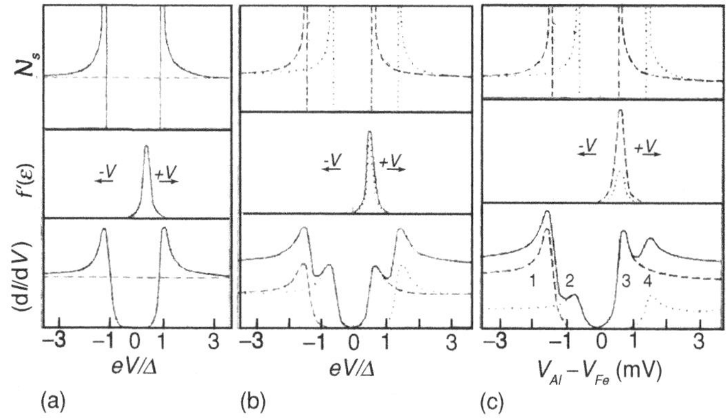

Figure 8.26

(a) Top $\mathcal{N}_s(\varepsilon)$, middle $f'(\varepsilon)$ and bottom dI/dV for a superconductor–normal metal junction, (b) the same in an applied field and (c) a superconductor–ferromagnet junction. The degree of spin polarization may be deduced from the heights of the maxima of the last curve. (P. M. Tedrow and R. Meservey, *Phys. Rev. Lett.* **26**, 192(1971))

Metal–insulator–superconductor junctions Tunnelling between a thin-film superconductor and a ferromagnet in a magnetic field was investigated in the elegant Tedrow–Meservey experiment. A layer of aluminium was used as the superconducting electrode, and it was partly oxidized to form the tunnel barrier. The ferromagnetic film was deposited on top. A magnetic field was applied during the measurement. The advantage of using a thin aluminium film was that it has a higher critical temperature, and a much higher critical field than bulk material, so it was possible to use an external field of several teslas to resolve the positions of the peaks marked 1–4 in the joint density of states.

In the case of strongly correlated metallic electrodes, such as superconductors, the tunnelling current between two metallic electrodes depends on a convolution of their densities of states, and the difference in their Fermi energies, as well as the barrier characteristics and the voltage bias V across the junction. In general

$$I(V) \approx \int_{-\infty}^{\infty} \mathcal{N}_1(\varepsilon - eV)\mathcal{N}_2(\varepsilon)[f(\varepsilon - eV) - f(\varepsilon)]\mathrm{d}\varepsilon,$$

where $f(\varepsilon)$ is the Fermi function.

In normal metals, the bias is small compared to the Fermi energy, and $\mathcal{N}_n(\varepsilon)$ may be taken as constant. The effective density of states of the superconductor $\mathcal{N}_s(\varepsilon)$ has a gap $2\Delta_s$ at the Fermi energy:

$$I(V) \approx \mathcal{N}_n(\varepsilon_F) \int_{-\infty}^{\infty} \mathcal{N}_s(\varepsilon)[f(\varepsilon - eV) - f(\varepsilon)]\mathrm{d}\varepsilon.$$

The voltage derivative $\mathrm{d}I/\mathrm{d}V$ gives the conductance:

$$G \approx \mathcal{N}_n(\varepsilon_F) \int_{-\infty}^{\infty} \mathcal{N}_s(\varepsilon)[f'(\varepsilon - eV)]\mathrm{d}\varepsilon.$$

The quantities $\mathcal{N}_s(\varepsilon)$, $f'(\varepsilon + eV)$ and their convolution are shown in Fig. 8.26. The interest of this is that the degree of spin polarization of the electrons near the Fermi level can be deduced from the amplitudes G_i of the four peaks in the dI/dV versus V curve. The spin polarization of the ferromagnet is

$$P = \frac{(G_4 - G_2) - (G_1 - G_3)}{(G_4 - G_2) + (G_1 - G_3)}. \tag{8.20}$$

Other methods of determining spin polarization are presented in §14.1.2.

8.4 Wires and needles

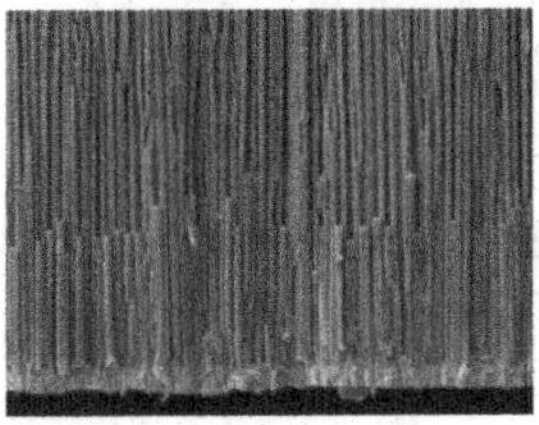

Electrodeposited Co nanowires in a porous Al_2O_3 membrane. The cathode is at the bottom, and the wires have grown to fill about 4 μm of the pores.

Magnetic nanowires can be patterned from thin films by lithographic techniques. Another way to make them is by electrodeposition into a porous template, such as an aluminium oxide membrane. In some cases, like Co–Cu, a single electrochemical bath can be used to produce segmented nanowires by toggling between two different potentials (§15.2.1).

The magnetization of ferromagnetic nanowires and acicular (needle-shaped) nanoparticles normally lies along the long axis. There is no incentive to form domains because the demagnetizing energy is already zero in the single-domain case, where the demagnetizing factor $\mathcal{N} = 0$. It is possible, however, to nucleate a reverse domain at one end of a wire, and then measure the domain-wall velocity as a function of applied field by timing the propagation of the reversal down the wire. Time-dependent changes of magnetization are detected with pick-up coils, by the magneto-optic Kerr effect or by making use of the anomalous Hall effect in wires with perpendicular magnetization. Domain-wall velocities can be high, $\approx$100 m s^{-1}, in thin-film nanowires.

Acicular magnetic particles exhibit coercivity associated with shape anisotropy provided their dimensions are of order the coherence radius R_{coh} (Table 8.1). Acicular particles of CrO_2 and γFe_2O_3 used in particulate magnetic recording media (tapes and floppy discs) had an aspect ratio of $5-10$ ($\mathcal{N} < 0.1$). Dimensions are typically $30 \times 30 \times 300$ nm. In the Stoner–Wohlfarth model, the effective anisotropy due to shape is $K_{sh} = [(1 - 3\mathcal{N})/4]\mu_0 M_s^2$ (5.63), so the limiting value for a long wire, $\mathcal{N} = 0$, gives $K_{sh} = \frac{1}{4}\mu_0 M_s^2$. The corresponding limit on the coercivity is the anisotropy field $2K_d/\mu_0 M_s$, which is $M_s/2$. This limit is never realized in practice. For example, particles of CrO_2 with $M_s = 0.5$ MA m^{-1} and $\mathcal{N} = 0.05$ have $K_{sh} = 67$ kJ m^{-3} and an anisotropy field of 213 kA m^{-1}. Typical coercivities for commercial powders are 50 kA m^{-1}. Reversal is an *incoherent* process in real acicular crystallites, proceeding by curling, or growth of a reversed nucleus.

Alnico magnets developed during the period 1930–1970 were the world's first artificial magnetic nanostructures. They are still manufactured in limited quantities as general-purpose magnets, and for special applications. Alnicos are obtained by spinodal decomposition of a quarternary cubic alloy, which leads to growth of needle-like nanoscale regions of ferromagnetic

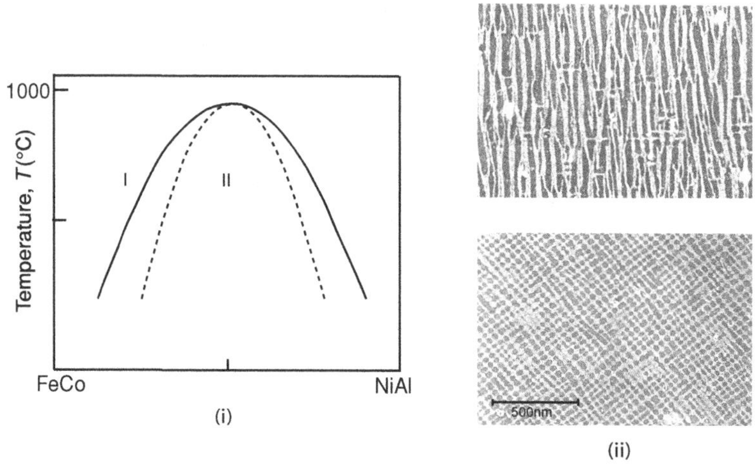

Figure 8.27

A schematic Fe–Co/Ni–Al phase diagram showing the solid-solution region (I) and the two-phase region (II). Metastability occurs between the solubility line (solid) and the spinodal line (dashed). A magnetically oriented spinodal nanostructure of alnico is illustrated, where Fe–Co needles are embedded in a Ni–Al matrix.

Fe–Co in a matrix of nonmagnetic Ni–Al – a nanoscale composite, Fig. 8.27. By conducting the heat treatment in a magnetic field, it is possible to align the long dimensions of the ferromagnetic regions in the direction of the field. The best alnicos offer quite high remanence, but limited coercivity. Generally, a permanent magnet needs a coercivity of at least half the remanence if it is to perform efficiently. A true permanent magnet is one that can be made into *any* desired shape, hence H_c must be greater than M_s. The upper limit of $M_s/2$ offered by shape anisotropy, even in an ideal nanostructure, is never enough.

The domain walls in nanowires patterned from thin films of a soft ferromagnet such as permalloy differ greatly from the Bloch walls found in bulk material. The magnetization is constrained to lie in the plane of the film by shape anisotropy and the magnetization in the domains must lie along the axis of the wire. Head-to-head (or tail-to-tail) domains form, and the walls are of two main types. One is a Bloch or Néel wall, where the magnetization at the centre lies transverse to the axis of the strip, the other is a vortex wall. They are illustrated in Fig. 8.28. Depending on dimensions, other types may form such as a wall with two vortices of opposite chirality. If the wire has width w and thickness t, the nature of the wall depends on the dimensionless ratio $\mathrm{r} = w\mathrm{t}/l_{ex}^2$; transverse walls are favoured when $\mathrm{r} \lesssim 100$.

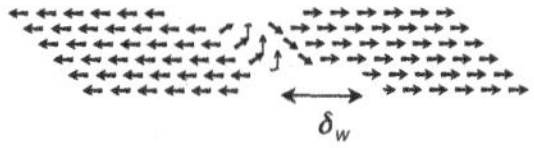

A domain wall confined by a geometric potential.

The striking feature of geometrically constrained walls is that they can be very narrow. The wall width δ_w is determined, not by the ratio of exchange to anisotropy as in (7.16), but by the width of the wire:

$$\delta_w = cw,$$

where $c \approx 1$ for the transverse wall and $c \approx 3\pi/4$ for the vortex wall. The walls can be driven by magnetic fields or by spin-polarized currents flowing in the wires. Schemes have been proposed for magnetic memory and logic based on moving the walls around permalloy tracks (Chapter 14). The walls will tend to be pinned at notches or protruberances.

Figure 8.28

Domain walls in soft magnetic strips: (a) the head-to-head transverse wall and (b) the vortex wall.

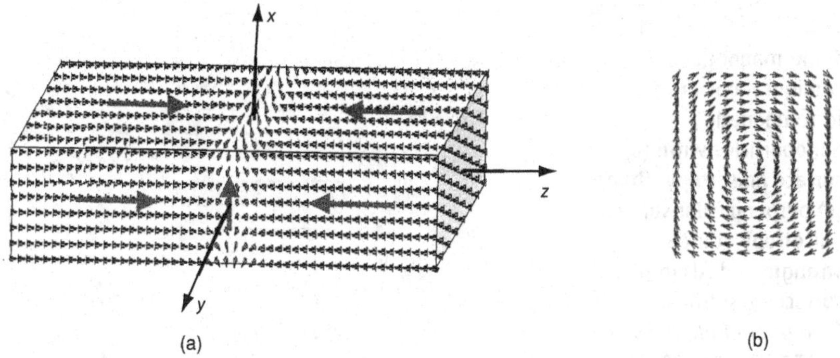

The energy per unit area of the confined wall is just due to exchange. It is

$$\gamma_w = cAw.$$

Geometrically confined domain walls can have different configurations, which may be degenerate or lie close in energy. For example, an isthmus constriction could trap a wall which is Bloch- or Néel-like, with either possible chirality. Spontaneous thermal fluctuations among these configurations can occur at ambient temperature.

8.5 Small particles

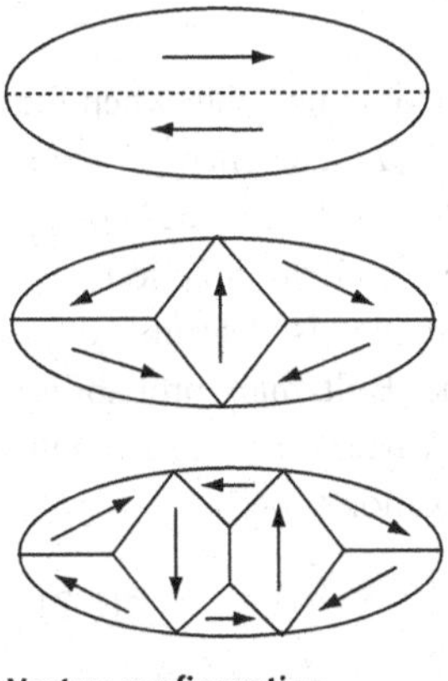

Vortex configuration of thin-film permalloy elements.

Small ferrimagnetic particles appear naturally in igneous rocks, and ferromagnetic and ferrimagnetic particles can be synthesized by a variety of chemical methods. The smallest magnetic particles exhibit superparamagnetism, behaving like paramagnetic macrospins. Larger ones adopt magnetic configurations which are governed by the balance of anisotropy, exchange and magnetic dipole interactions. Submicrometre spots of magnetic materials can be patterned from thin films. When magnetocrystalline anisotropy is negligible, the small elements tend to adopt configurations where the magnetization is oriented as far as possible parallel to the surface (§7.1.6). They can do this because the exchange length, which is the length scale over which the direction of magnetization of a ferromagnet can adapt to dipolar fields, is only 2–5 nanometres (Table 8.1). Vortex configurations are found in nanoparticles and thin-film spots of soft magnetic material of order 100 nm in size, as well as the *C* and *S* configurations (§7.3.2).

Surface anisotropy can also influence the magnetic configurations of ferromagnetic nanoparticles, when the exchange is not too strong. Some examples are shown in Fig. 8.29. These effects are unimportant in nanoparticles of 3*d* metals and alloys with Curie points above room temperature, but they may be significant for rare-earth alloys with low Curie points, or actinide ferromagnets like US, where the single-ion anisotropy is exceptionally strong.

Figure 8.29

Some magnetic configurations in ferromagnetic nanoparticles with: (a) no surface anisortopy, (b) and (c) perpendicular surface anisotropy of increasing strength and (d) in-plane surface anisotropy (L. Berger *et al.*, *Phys Rev B*. **377** 104431 (2008).)

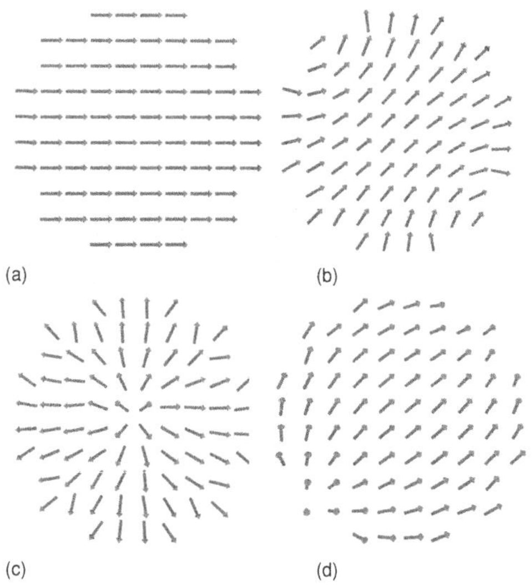

The spin configurations shown in Fig. 8.29 were determined numerically, by atomic scale Monte-Carlo simulations with simulated annealing from above the Curie point in order to find the lowest-energy state. The spin configurations have names: (b) is a throttled or flower state, (c) is a hedgehog and (d) is an artichoke.

8.5.1 Superparamagnetism

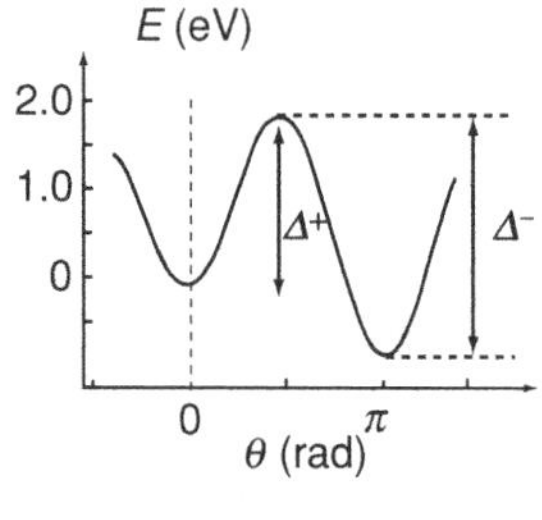

Energy barrier to magnetic reversal of a uniaxial superparamagnetic particle in an applied field.

Tiny ferromagnetic particles of radius $R \lesssim 10$ nm become unstable when the energy barrier to magnetic reversal is comparable to k_BT. The energy barrier $\Delta = K_uV$ becomes asymmetric in an applied field $\Delta \Longrightarrow \Delta \pm \mu_0 \mathrm{m}H \cos\theta_0$, where θ_0 is the angle between the moment and the applied field direction.

Néel proposed that the relaxation time for a spin flip is determined by the product of an attempt frequency τ_0^{-1} and the Boltzman probability $\exp(-\Delta/k_BT)$ that the particle has the thermal energy necessary to surmount the barrier. The inverse spin-flip frequency is the relaxation time:

$$\tau = \tau_0 \exp(\Delta/k_BT), \tag{8.21}$$

where τ_0^{-1} is of order 1 GHz, which is the ferromagnetic resonance frequency in the demagnetizing field. There is progressive, but exponentially rapid slowing down of the magnetic relaxation around some blocking temperature $T_b < T_C$.

Blocking is not a phase transition, but a continuous, albeit very rapid variation of $\tau(T)$, Fig. 8.30. A commonly used criterion for blocking

$$\Delta/k_BT = 25, \tag{8.22}$$

Table 8.5. Superparamagnetic relaxation times for cobalt particles

Radius (nm)	Temperature (K)	Relaxation time
3.5	260	332 s
3.5	300	10 s
3.5	340	0.6 s
3.5	380	76 ms
3.0	300	1.9 ms
4.0	300	223 h
5.0	300	L $\times 10^{12}$ y

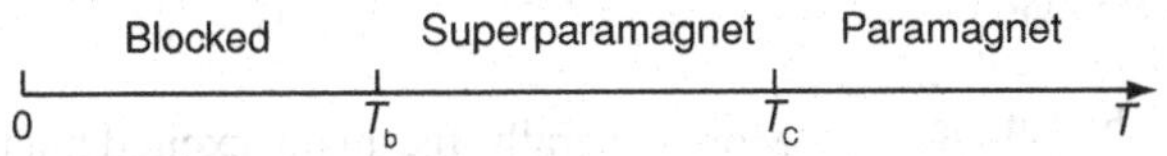

Figure 8.30

Temperature scale for the behaviour of a uniaxial ferromagnetic nanoparticle. Neither T_C nor T_b is a perfectly sharp transition and the latter depends on the time scale of the measurement which is used to decide whether or not the particles are blocked.

corresponds to $\tau \approx 100$ s which is about the time needed for a magnetic measurement. The barrier $\Delta \sim 1$ eV. The origin of Δ may be uniaxial magnetocrystalline anisotropy $K_1 V$, shape anisotropy $K_{sh} V$, or surface anisotropy $K_s \mathcal{A}$. To appreciate the power of the exponential, consider cobalt particles of different radii at different temperatures shown in Table 8.5. In cubic particles there are three or four easy axes depending on the sign of K_{1c}. The energy barrier is $K_{1c}V/4$ when K_{1c} is positive or $K_{1c}V/12$ when it is negative (see p. 170).

In the superparamagnetic region $T_b < T < T_C$, the particle behaves like a Langevin paramagnet with a giant, classical moment $\mathfrak{m}$ (4.21). The macrospin moment of a cobalt particle of radius 3.5 nm is about $3 \times 10^4 \mu_B$. The susceptibility is then

$$\chi = \mu_0 n \mathfrak{m}^2 / 3k_B T, \qquad (4.22)$$

where n is the number of particles per cubic metre. The practical test for superparamagnetism is the superposition of anhysteretic reduced magnetization curves as a function of H/T over a wide range of temperature between T_b and T_C. If the curves do not superpose, the particles are not superparamagnetic.

When an assembly of superparamagnetic particles is cooled in a field H, there is a slight bias in the direction of magnetization of the blocked particles below T_b. The particles are more likely to be trapped in the minimum where their magnetization lies along the easy axis, in a direction more or less parallel to the magnetic field, than in the opposite direction. This leads to the thermoremanent magnetization (TRM) M_{tr}:

$$M_{tr} = \chi H = \mu_0 n H \mathfrak{m}^2 / 3k_B T_b. \qquad (8.23)$$

The famous example provided by cooling of basalt in the Earth's magnetic field H_e. is discussed in §15.5.4.

An ensemble of superparamagnetic particles has no net magnetization, but the particles exhibit a time-dependent magnetic response to a change of applied

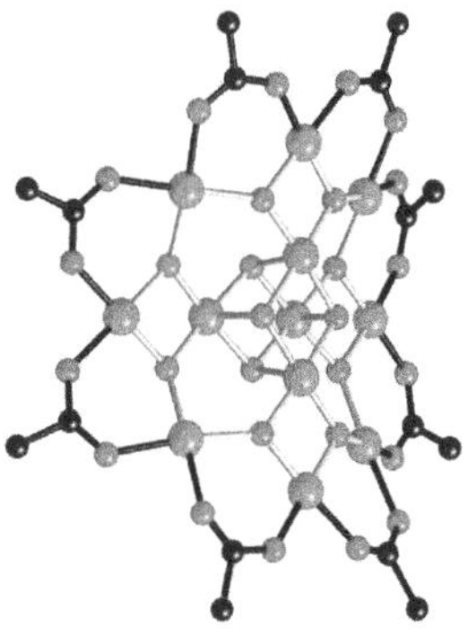

The Mn_{12} molecular magnet. The Mn ions are the large spheres, with ↑ Mn^{3+} around the outside and ↓ Mn^{4+} in the middle.

magnetic field. The time dependence is given by (7.39), which is valid for times which are neither very short, nor extremely long.

Nanoparticles, which are not quite small enough to be thermally excited over the energy barrier, can nevertheless exhibit spontaneous coherent fluctuations of their magnetic moment around the energy mimimum. These excitations take the place of long-wavelength spin waves, which cannot be excited because the particle size fixes a maximum possible wavelength. Setting $K_u V \sin^2\theta = k_B T$, it follows that the average angle of deviation of the magnetization is $\theta \approx (k_B T/K_u V)^{1/2}$. The magnetization of the particle is $M_s \cos\theta \approx M_s(1 - \theta^2/2)$. Hence there is a linear decline of the magnetization with temperature due to these collective excitations:

$$M \approx M_s\left(1 - \frac{k_B T}{2K_u V}\right). \tag{8.24}$$

More generally, thermally excited fluctuations between degenerate modes of opposite chirality are to be expected in particles larger than about 10 nm with ferromagnetic exchange, where the collinear ferromagnetic configuration is no longer the one with lowest energy. Normal spin waves cannot be excited at low energy since the maximum wavelength cannot exceed twice the size of the particle.

8.5.2 Quantum dots

The ultimate 'zero-dimensional' magnetic nanostructure is a dot so small that it contains few electrons, or maybe even a single one. The capacitance of a sphere of radius r is $C = 4\pi\epsilon_0 r$. The potential of a single electron on the sphere is $V = e/C$; for example, if $r = 14.4$ nm, $V = 100$ mV. This Coulomb barrier to adding charge to the nanodot capacitor is known as Coulomb blockade.

The electron content of the quantum box can be controlled by tunnelling electrons, one at a time, by adjusting the bias of the adjacent electrodes, which may be ferromagnetic.

The quantum dot is really an artificial atom, with a square-well potential rather than a Coulomb potential. At low temperature the unpaired spin moment of the dot can form a Kondo singlet state with electrons in nonmagnetic electrodes. Spin-polarized electron flow across it can be regulated by adjusting the potential of a gate. This is a magnetic single-electron transistor. Pairs of these magnetic quantum dots are candidate q-bits for a quantum computer.

8.5.3 Molecular clusters

Molecular magnets consist of several transition metal ions surrounded by organic and inorganic ligands. Their dimensions are a few nanometres, The best-known example is Mn_{12} acetate ([$Mn_{12}O_{12}(CH_3COO)_{16}(H_2O)_4$] which is a cluster of twelve manganese ions. Eight are Mn^{3+} ions and four are Mn^{4+} ions. The interaction between Mn^{3+} and Mn^{4+} is strongly antiferromagnetic,

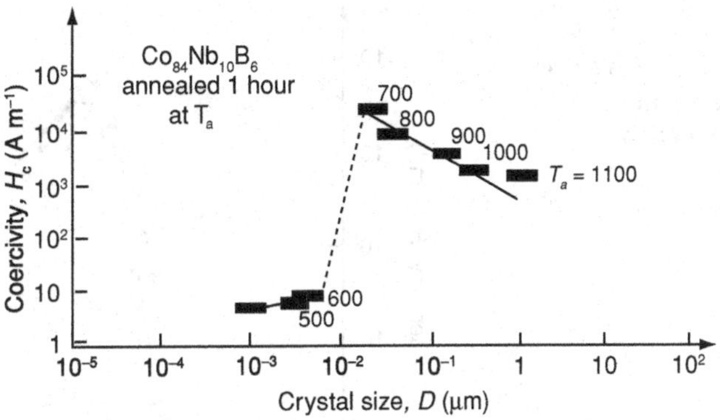

Figure 8.31

Coercivity of a partially recrystallized amorphous Co–Nb–B alloy. (G. Herzer, *IEEE Trans. Magn.* **26**, 1397(1990)).

so the net spin of the molecular cluster is $S = [8 \times 2 - 4 \times (3/2)] = 10$, giving a moment of 20 μ_B per cluster. The overall symmetry of the molecule is tetragonal, and there is strong uniaxial anisotropy due to the crystal field

$$\mathcal{H} = B_2^0\hat{\mathbf{O}}_2^0 + B_4^0\hat{\mathbf{O}}_4^0 + B_4^4\hat{\mathbf{O}}_4^4,$$

which produces an overall crystal field splitting of about 300 K. The $\pm|10\rangle$ doublet is the ground state, and if the material is magnetized at low temperature (~1 K), it remains in the $-|10\rangle$ state because relaxation to the $+|10\rangle$ state is extremely slow. In a reverse field, however, it is possible for the molecule to tunnel into an excited state, and a square, staircase hysteresis loop results. Similar magnetization dynamics are observed for other molecular magnets, and even for isolated rare earth ions with large J, the ultimate atomic nanomagnets.

8.6 Bulk nanostructures

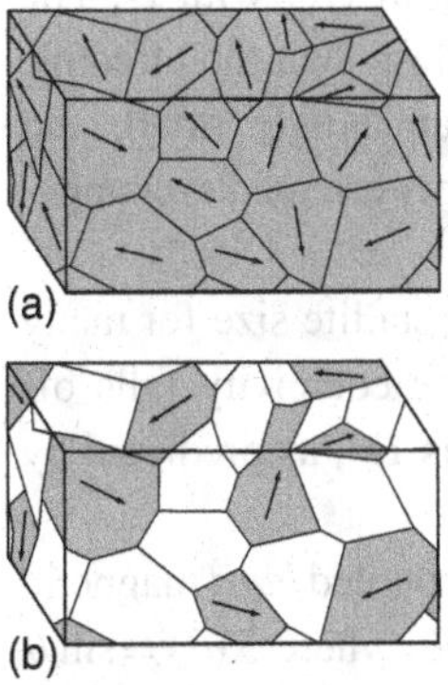

(a) single- and (b) two-phase magnetic nanostructures. The easy axis in the harder phase is marked. The crystallites are exchange coupled across the grain boundaries.

Nanostructured magnets composed of one or more ferromagnetic phases may be obtained directly from the melt by rapid quenching, or by annealing an amorphous precursor produced by melt quenching or hydrogen treatment, Fig. 8.31. These nanostructured materials may exhibit magnetic properties that are quite different to those of bulk material.

8.6.1 Single-phase nanostructures

In a single-phase nanostructure, the bulk anisotropy can be greatly reduced by exchange coupling of nanocrystallites with different anisotropy axes. Exchange averaging of anisotropy arises when:

(1) crystallites are single-domain, with a crystallite size D much less than the domain wall width δ_w;

(2) there is exchange coupling across the grain boundaries (decoupled crystallites would be superparamagnetic).

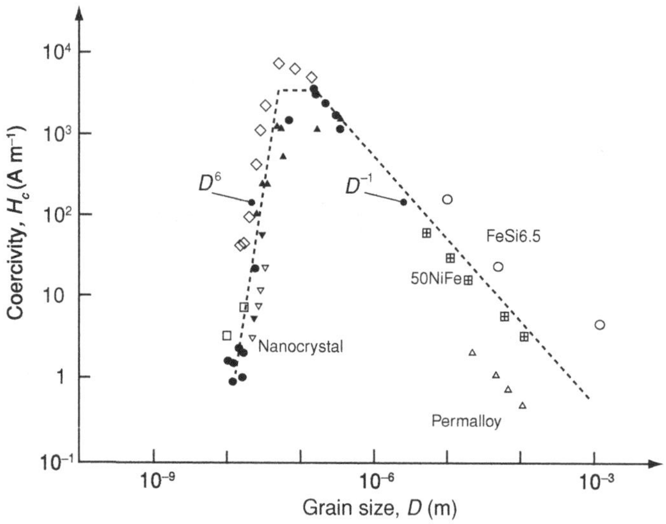

Figure 8.32

Coercivity versus grain size for a range of soft magnetic materials.

Exchange averaging is effective over the length scale δ_w. A volume δ_w^3 includes $N = (\delta_w/D)^3$ crystallites. The total anisotropy of this volume, obtained by adding N randomly oriented contributions of $K_1 D^3$ is $\sqrt{N} K_1 D^3$, hence

$$\langle K \rangle = K_1 (D/\delta_w)^{3/2}, \tag{8.25}$$

but we must use this value for $\langle K \rangle$ *consistently* in (7.16) for δ_w, $\delta_w = \pi \sqrt{A/\langle K \rangle}$. This gives $\delta_w = \pi^4 A^2 / K_1^2 D^3$. From (8.25) the effective anisotropy is

$$\langle K \rangle = K_1^4 D^6 / \pi^6 A^3. \tag{8.26}$$

The sixth-power law represents a very rapid variation of $\langle K \rangle$ with D. The coercivity is expected to scale with $\langle K \rangle$, and it will be less than the effective anisotropy field $2\langle K \rangle / M_s$. Hence H_c can be made vanishingly small, and the permeability can be very large in systems with randomly oriented exchange-coupled nanocrystals.

Figure 8.32 shows the variation of coercivity with crystallite size for many soft magnets. Above the critical single-domain size, the coercivity falls off as $1/D$, whereas in the submicrometre region it varies as D^6, as predicted by (8.26).

The remanence of a randomly oriented collection of decoupled *hard* magnetic crystallites is $\langle M_r \rangle = \int_0^{\pi/2} M_s \cos\theta \, P(\theta) d\theta \big/ \int_0^{\pi/2} P(\theta) d\theta$, where $P(\theta) = \sin\theta$ for random orientation. Hence

$$\langle M_r \rangle = \tfrac{1}{2} M_s. \tag{8.27}$$

Exchange coupling of hard nanocrystallites leads to remanence enhancement where M_r is greater than $M_s/2$. For example, optimally quenched $Nd_{14}Fe_{80}B_6$

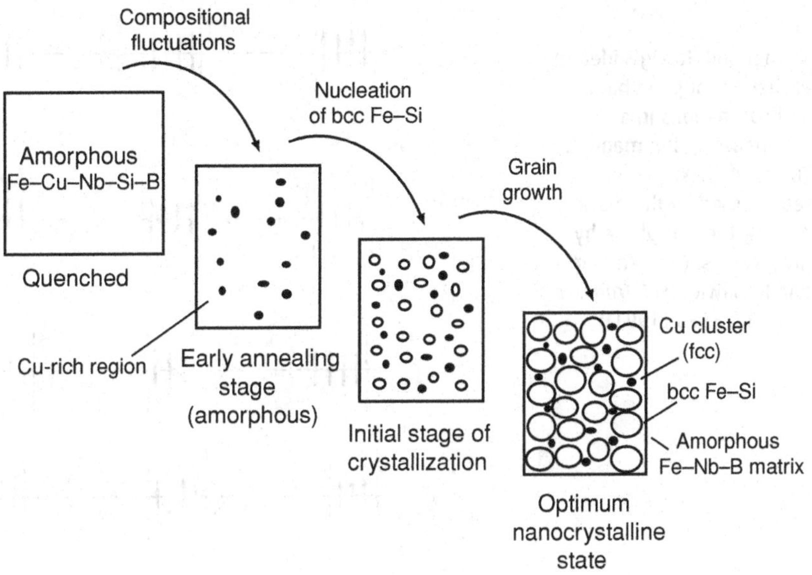

Figure 8.33

Recrystallization of amorphous Fe–Cu–Nb–Si–B to obtain a two-phase crystalline–amorphous soft nanocomposite. R. C. O'Handley *et al.*, *J. Appl. Phys.* 57, 3563 (1985)).

(Magnequench), which consists of exchange-coupled crystallites with $D \approx$ 50 nm shows a slightly enhanced remanence $\mu_0 M_r = 0.85$ T for a material with $\mu_0 M_s = 1.61$ T.

8.6.2 Two-phase nanostructures

Two-phase amorphous–crystalline structures may be obtained by partial recrystallization of an amorphous precursor, Fig. 8.33. If υ_c is the volume fraction of the crystalline phase which has anisotropy K_1 and the amorphous phase is assumed to have no anisotropy, then it follows from the argument leading to (8.26) that

$$\langle K \rangle = \upsilon_c^2 K_1^4 D^6 / \pi^6 A^3.$$

The great interest of these two-phase soft nanostructures is that the two phases may have opposite sign of magnetostriction, λ_s. The composition is chosen so that volume fractions of the two phases make $\langle \lambda_s \rangle = 0$ as required for a good soft material. The crystalline fraction may have a larger magnetization than the amorphous one, so zero magnetostriction can be combined with high magnetization. An example is 'Finemet', $Fe_{73.5}Cu_1Nb_3Si_{15.5}B_7$.

Exchange-coupled hard–soft nanocomposites are another possibility. Here δ_w is too small to average the effective anisotropy to zero (Table 8.1), but by exchange coupling to a soft phase with higher magnetization than the hard phase, it is possible to augment the remanence, achieving greater isotropic remanence than would be possible from the hard phase alone ($\mu_0 M_s$ for

Figure 8.34

Kneller and Hawig's idea of exchange coupling hard and soft regions in a two-phase spring magnet. The hard phase is represented by the black arrows, the soft phase by grey arrows. (E. F. Kneller and R. Hawig, *IEEE Trans. Mag.* **27**, 3588 (1991))

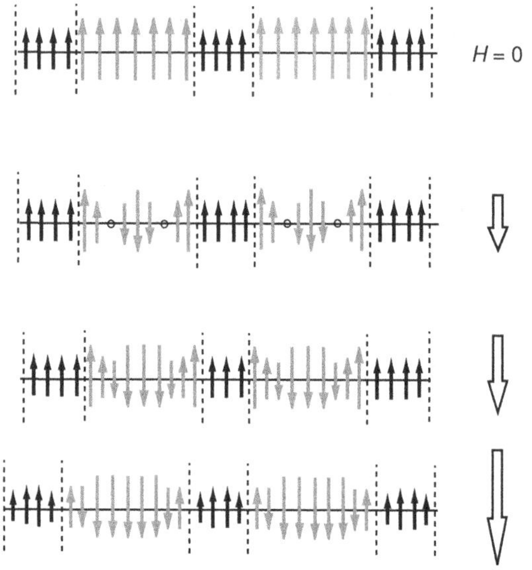

Figure 8.35

Hysteresis behaviour of a spring magnet
(a) optimized two-phase nanostructure,
(b) overcoupled nanostructure, (c) hard phase alone and
(d) constricted loop due to two independent phases
(after E. F. Kneller and R. Hawig, *IEEE, Trans. Mag.* **27**, 3588 (1991).)

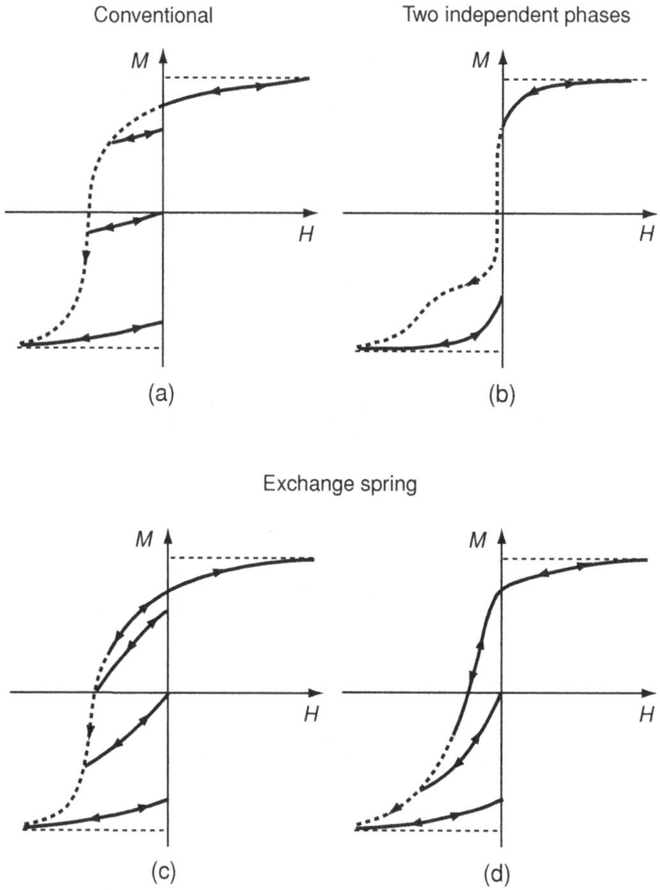

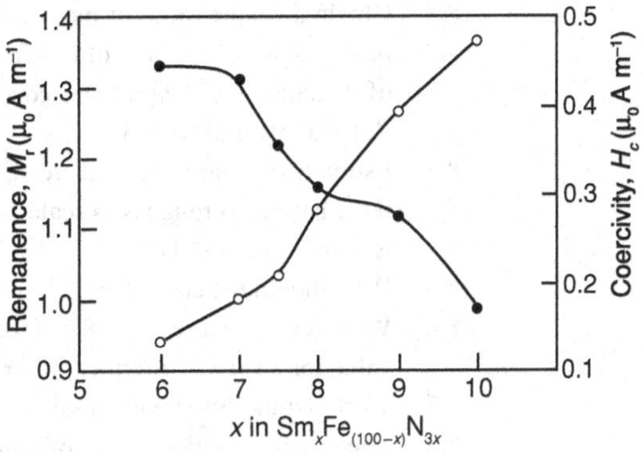

Figure 8.36

Coercivity and remanence as a function of composition in a two-phase nanocomposite of $Sm_2Fe_{17}N_3$/Fe. The crystallite size is about 20 nm.

$Nd_2Fe_{14}B$, Fe and $Fe_{70}Co_{30}$ is 1.61, 2.15 and 2.45 T, respectively). In two-phase hard–soft nanostructures, exchange stiffening leads to 'spring magnet' behaviour, illustrated in Fig 8.34. The hysteresis is illustrated in Fig. 8.35.

In any such two-phase nanocomposite there is an inevitable trade-off between remanence and coercivity. This is illustrated for the hard–soft $Sm_2Fe_{17}N_3$–Fe nanocomposite in Fig. 8.36.

FURTHER READING

E. Steinbeiss, Thin film deposition techniques, in *Spin Electronics*, M. Ziese and M. J. Thornton (editors), Berlin: Springer (2001).

S. Maekawa and T. Shinjo (editors), *Spin-dependent Transport in Magnetic Nanostructures*, Boca Raton CRC Press (2002). Experiment and theory of GMR and TMR junctions.

D. Sellmyer and R. Skomski (editors), *Advanced Magnetic Nanostructures*, New York: Springer (2006).

R. C. O'Handley, *Modern Magnetic Materials*, New York: Wiley (2000).

T. Shinjo (editor), *Nanomagnetism and Spintronics*, Oxford: Elsevier (2009).

D. Galteschi, R. Sessoli and J. Villain, *Molecular Nanomagnets*, Oxford: Oxford University Press (2006).

EXERCISES

8.1 Estimate the magnetization of an antiferromagnetic thin film 10 nm thick, which grows with its c axis perpendicular to a single-crystal substrate. Assume the antiferromagnetic structure consists of a series of oppositely aligned ferromagnetic c-planes.

8.2 By considering a bilayer of 100 nm YCo_2 and 100 nm of $GdCo_2$ estimate the field that must be applied to create a domain wall 10 nm wide at the interface. Take A as 10^{-11} J m^{-1}.

8.3 Obtain an expression for the coupling energy J_s in mJ m^{-2} which results from pinholes in the structure of two ferromagnetic layers separated by a spacer layer of thickness t if the pinholes cover a fraction f of the interface area. Evaluate J_s for a material with $A = 2 \times 10^{-11}$ J m^{-1}, if $\mathrm{t} = 2$ nm and $f = 1\%$.

8.4 Estimate the magnetic field created at a distance z from a ferromagnetic surface if the surface is rough on a scale λ, with relief of depth δ_λ when the ferromagnet is magnetized in-plane.

8.5 Why should the curve for $\mu^{\downarrow}$ have zero slope at the interface in Fig. 8.19(b)?

8.6 Work out the expression (8.10) for the spin accumulation voltage. Calculate its value for a Co–Cu interface, where $j = 10^{10}$ A m^{-2}.

8.7 Work out the Jullière expression for TMR (8.15).

8.8 The resistance ratio R_{ap}/R_p of a tunnel juction with identical electrodes is 400%. Use the Jullière formula to deduce the spin polarization P of the ferromagnet. What is the magnetoresistance? What will be the magnetoresistance if one of the electrodes is replaced by cobalt? (P(Co) = 45%).

8.9 Estimate the thermoremanent magnetization of a basalt containing 1 vol% of magnetite (Fe_3O_4) in the form of particles 50 nm in diameter, for which the blocking temperature is 600 K. Take M_s for the magnetite as 400 kA m^{-1}.

8.10 Calculate the domain-wall width and the effective anisotropy constant $\langle K \rangle$ in an assembly of exchange-coupled grains with $D = 20$ nm having $K_1 = 10^4$, $A = 10$ pJ m^{-1} and $M_s = 1$ MA m^{-1}. Give an upper limit to the expected coercivity.

8.11 What is the spin-wave gap in Kelvin for a cobalt particle of radius 10 nm? By how much is the magnetization reduced by coherent fluctuations of the magnetization at a temperature equivalent to the spin-wave gap?

9 Magnetic resonance

Resonance arises when the energy levels of a quantized system of electronic or nuclear moments are Zeeman split by a uniform magnetic field and the system absorbs energy from an oscillating magnetic field at sharply defined frequencies, which correspond to transitions between the levels. Classically, resonance occurs when a transverse AC field is applied at the Larmor frequency. Resonance methods are valuable for investigating the structure and magnetic properties of solids, and they are used for imaging and other applications. The resonant moment may be an isolated ionic spin or free radical, as in electron paramagnetic resonance (EPR), or a nuclear spin as in nuclear magnetic resonance (NMR). Otherwise it can be the ordered magnetization as in ferromagnetic resonance (FMR). Resonant effects are also associated with spin waves, and domain walls. The related techniques of Mössbauer spectroscopy and muon spin resonance provide further information on hyperfine interactions in solids.

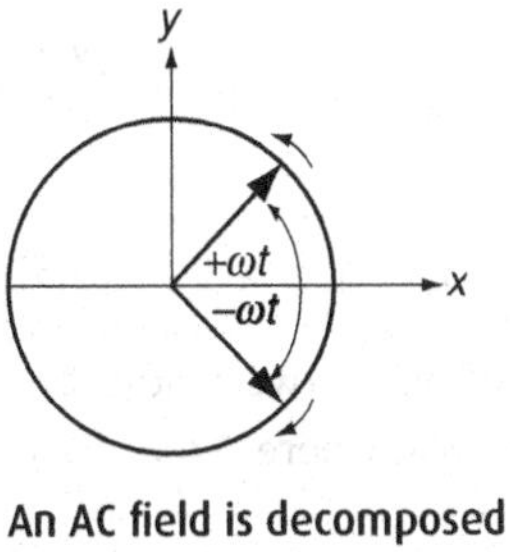

An AC field is decomposed into two counter-rotating fields.

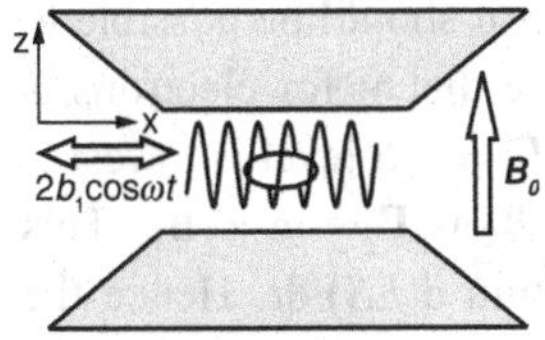

A typical magnetic resonance experiment.

A magnetic system placed in a uniform magnetic field $\boldsymbol{B}_0$ may absorb electromagnetic radiation at a precisely defined frequency $\nu_0 = \omega_0/2\pi$ which falls in the radio-frequency or microwave range. The phenomenon is related to the Larmor precession of the magnetic moment, introduced in §3.2.2. In order to observe the resonance, an experimental geometry with *crossed* magnetic fields is needed. The steady uniform field defines the z-direction, while a high-frequency AC field $b_x = 2b_1 \cos \omega t$ is applied in the perpendicular plane. It is helpful to think of b_x as the sum of two counter-rotating fields $2b_1 \cos \omega t = b_1(\mathrm{e}^{\mathrm{i}\omega t} + \mathrm{e}^{-\mathrm{i}\omega t})$. Resonance occurs when the precession is synchronized with the clockwise or anticlockwise component. No resonance occurs when $\boldsymbol{b}_1$ is parallel to $\boldsymbol{B}_0$.

There is a vast literature on magnetic resonance. It formed the basis of the fifth age of magnetism, which flowed from understanding the quantum mechanics of angular momentum, and the development of microwaves for radar in the Second World War. The resonant system is an ensemble of free radicals or ions with unpaired electron spins in electron paramagnetic resonance (EPR) – also known as electron spin resonance (ESR). The entire coupled magnetic moment may resonate in ferromagnetic resonance (FMR), or else it can be the sublattice moments which precess in antiferromagnetic resonance (AFMR). The nuclei carry tiny moments that resonate at relatively low frequencies in nuclear magnetic resonance (NMR). Other resonances are related to spin waves, domain walls and conduction electrons. In magnetically ordered material, it may be possible to observe the resonance without recourse to an external field B_0, making use of the internal demagnetizing or hyperfine fields.

These are all remarkable physical phenomena in their own right, but from our viewpoint magnetic resonance is interesting for the insight it provides into the magnetism of solids, and for applications such as high-frequency switching of magnetization and magnetic resonance imaging (MRI).

The resonant magnetic systems are usually small quantum objects – ions or electrons or nuclei with unpaired spin – so it is natural to adopt a picture of resonant transitions between quantized, Zeeman split energy levels. Nevertheless, the classical picture of excitation at the natural Larmor precession frequency, which is needed for macroscopic magnets, provides invaluable insights for the quantum systems too.

Think of the simplest case of an ion with magnetic moment $\mathfrak{m}$ which is associated with an electronic angular momentum $\hbar \boldsymbol{S}$. The constant of proportionality is the gyromagnetic ratio γ:

$$\mathfrak{m} = \gamma \hbar \boldsymbol{S}, \tag{9.1}$$

where γ has units of $s^{-1}\,T^{-1}$ (hertz per tesla) and $\boldsymbol{S}$ is dimensionless. Both $\mathfrak{m}$ and $\boldsymbol{S}$ are vector operators. The equation, which reads like a classical vector equation, really means that all of the corresponding matrix elements of $\mathfrak{m}$ and $\boldsymbol{S}$ are proportional. The Zeeman interaction $\mathfrak{m} \cdot \boldsymbol{B}_0$ in the steady field $\boldsymbol{B}_0$ applied along Oz is represented by the Hamiltonian

$$\mathcal{H}_Z = -\mathfrak{m} \cdot \boldsymbol{B}_0 = -\gamma \hbar B_0 S_z. \tag{9.2}$$

Eigenvalues are a set of equally spaced energy levels at

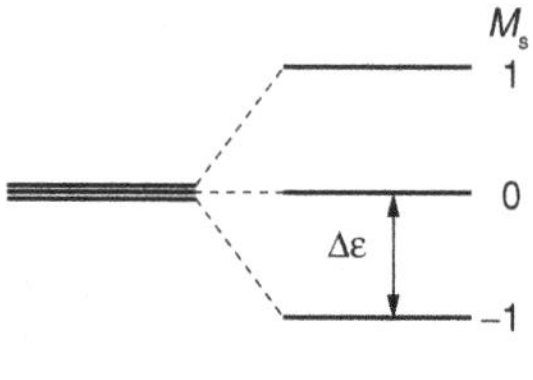

Zeeman split energy levels for an electronic system with $S = 1$.

$$\varepsilon_i = -\gamma \hbar B_0 M_s; \qquad M_s = S, S-1, \ldots, -S. \tag{9.3}$$

The level spacing is $\Delta\varepsilon = \gamma \hbar B_0$. Magnetic dipole transitions between adjacent levels can be expected for radiation of angular frequency ω_0, where $\Delta\varepsilon = \hbar\omega_0$. Hence the resonance condition

$$\omega_0 = \gamma B_0 \tag{9.4}$$

does not depend on Planck's constant, which suggests it should be possible to deduce the same result by a classical argument. Note that γ for electrons is negative on account of their negative charge so the $M_s = -S$ level is lowest.

The torque on the magnetic moment $\mathfrak{m}$ in a field $\boldsymbol{B}_0$ is $\boldsymbol{\Gamma} = \mathfrak{m} \times \boldsymbol{B}_0$. This is equated to the rate of change of angular momentum $\mathrm{d}(\hbar \boldsymbol{S})/\mathrm{d}t$. Hence the equation of motion is[1]

$$\frac{\mathrm{d}\mathfrak{m}}{\mathrm{d}t} = \gamma \mathfrak{m} \times \boldsymbol{B}_0. \tag{9.5}$$

[1] This and similar equations appear with a negative sign, if the convention is adopted that e, rather than $-e$, is the charge on the electron.

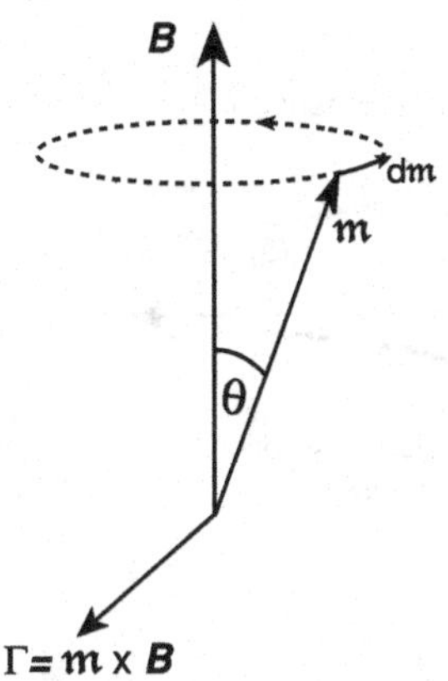

Precession of a magnetic moment in an applied field.

Now dm, the change of m in a short time interval dt, is a vector perpendicular to both m and $\boldsymbol{B}_0$, hence the moment precesses around the field, at angular frequency

$$\omega_0 = \gamma B_0.$$

This is the classical Larmor precession and resonance occurs when the field $\boldsymbol{b}_1$ turns at the Larmor frequency.

The requirement that $\boldsymbol{b}_1$ be applied perpendicular to $\boldsymbol{B}_0$ for resonant absorption also follows from quantum mechanics. The Zeeman Hamiltonian (9.2) in matrix notation is diagonal with eigenstates $| M_S\rangle$. Adding an extra field in the z-direction merely changes the eigenvalues, but does not induce any transition between the states, because the off-diagonal matrix elements which mix different states are all zero. However, if $\boldsymbol{b}_1$ is applied in the x-direction, the Hamiltonian becomes

$$\mathcal{H} = -\gamma\hbar(B_0\boldsymbol{S}_z + b_1\boldsymbol{S}_x). \tag{9.6}$$

The matrix representing $\boldsymbol{S}_x$ (§3.1.4) has non-zero off-diagonal elements $[n, n \pm 1]$. It can be expressed in terms of the ladder operators S^+ and S^- so it mixes states with $\Delta M_s = \pm 1$. At resonance, the AC magnetic field provokes transitions between the states which differ by $\Delta M_s = \pm 1$. This is known as the dipole selection rule.

9.1 Electron paramagnetic resonance

The Larmor precession frequency for electron spin is $f_L = \omega_L/2\pi = (ge/4\pi m_e)B$. Since $g = 2.0023$, the value of γ for free electrons, $-(ge/2m_e)$, is 176.1×10^9 s^{-1} T^{-1} and f_L is 28.02 GHz T^{-1}. Resonance occurs in the microwave range for fields produced by laboratory electromagnets. X-band ($\sim$9 GHz) microwaves with wavelength $\Lambda = c/\nu = 33$ mm are commonly employed, so the resonance is at about 300 mT. Sometimes Q-band ($\sim$40 GHz) radiation is used and the resonance field is correspondingly bigger. The sample is placed in a resonant cavity at the end of a waveguide, in a steady field. The cavity, operates in a TM_{100} mode and delivers the requisite transverse magnetic field $\boldsymbol{b}_1$.

Zeeman splitting of the energy levels for an isolated electron is $\gamma\hbar B_0 = g\mu_B B_0$, an energy that is small compared with k_BT when $B_0 = 300$ mT. ($\mu_B/k_B = 0.673$ K T^{-1}) so the equilibrium population difference between the $M_S = \pm\frac{1}{2}$ sublevels is tiny. The spin polarization $(N_\uparrow - N_\downarrow)/(N_\uparrow + N_\downarrow)$ is

$$P = (1 - \mathrm{e}^{-g\mu_B B/k_BT})/(1 + \mathrm{e}^{-g\mu_B B/k_BT}).$$

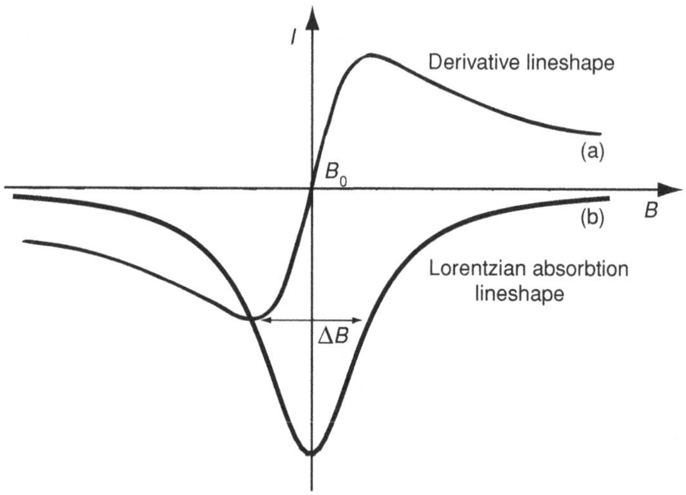

Figure 9.1

(a) An EPR trace showing the derivative of the microwave absorption obtained when sweeping the DC field at a constant rate. (b) The absorption line obtained by integration.

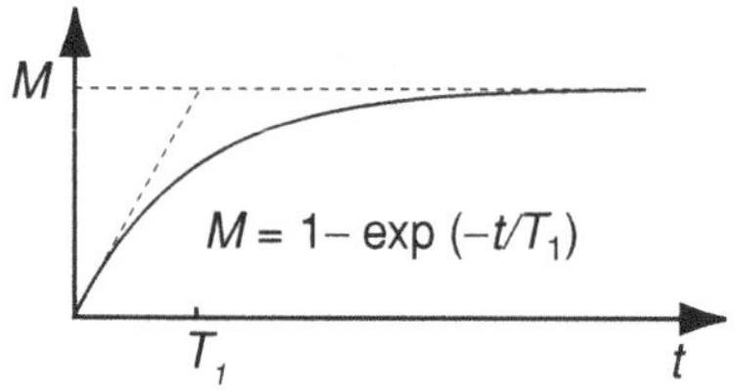

Spin-lattice relaxation of the magnetization towards its equilibrium value.

Conventionally the $\uparrow$ states are always those with their *moments* parallel to the applied field; they are those with $M_S = -\frac{1}{2}$ in this case.[2] The value of $P \approx g\mu_B B_0/2k_B T$ in 300 mT at room temperature is only 7×10^{-4}, so a sensitive detection method is needed to observe the resonance. It is often more convenient to sweep the magnetic field rather than the microwave frequency. Sensitivity is increased by using field modulation coils and detecting the absorbed power at the modulation frequency with a lock-in amplifier. The measured trace is the derivative of the absorption as a function of field (Fig. 9.1). The absorption line is the integral of this signal.

Measured parameters in EPR are the intensity of the resonance, its position B_0 which is normally expressed as an effective g-factor $g_{eff} = \hbar\omega_0/\mu_B B_0$, where ω_0 is the resonance frequency, and the linewidth ΔB (full width at half maximum). The cavity resonance is very sharp, so the linewidth is determined by the sample.

Absorption of radiation is a dynamic process, which tends to equalize the Boltzmann populations of the levels. This tendency is counter-balanced by the desire of the spin system to regain its thermal equilibrium. The temperature T of the system is defined by the crystal lattice, so the exchange of energy between the spins and the lattice which is involved in thermalization is known as spin-lattice relaxation. The linewidth ΔB is inversely proportional to the spin-lattice relaxation time T_1. If T_1 is very short, the line becomes too broad to observe, whereas if T_1 is very long, the line is sharp, but its intensity becomes vanishingly small because the populations of the $\uparrow$ and $\downarrow$ states remain equal; there is no dissipation of energy. The order of magnitude of T_1 is provided by the

[2] The $\uparrow$ and $\downarrow$ states are referred to as 'spin up' and 'spin down' or more correctly as 'majority spin' and 'minority spin'. The meaning is that the moments of the $\uparrow$ electrons are aligned with the applied field, but their spin angular momentum is in the opposite direction, because of the negative charge of the electron.

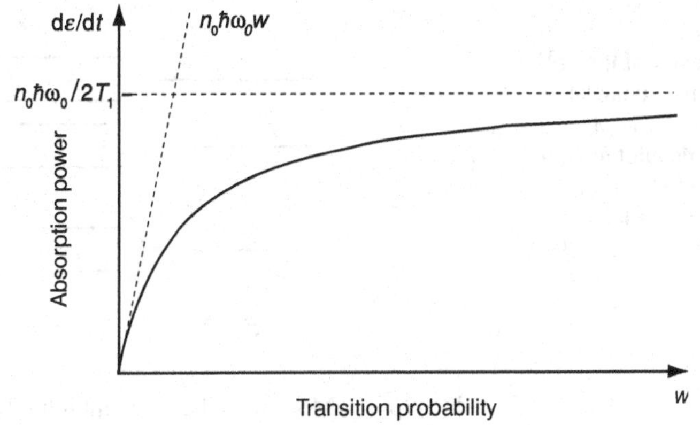

Figure 9.2

The rate of absorption of electromagnetic energy in a continuous-wave magnetic resonance experiment. The quantity w is proportional to the microwave power.

uncertainty relation $\Delta\varepsilon\Delta t \approx \hbar$, so if $\Delta B = 1$ mT, $\Delta\varepsilon = g\mu_B\Delta B \approx 2 \times 10^{-26}$ J, $T_1 \approx 5 \times 10^{-9}$ s.

The probability w of transitions between the $\pm\frac{1}{2}$ levels stimulated by the microwave field is a quantity which is proportional to microwave power and identical for transitions in either sense. The rates of change of populations are

$$\frac{dN_\uparrow}{dt} = w(N_\downarrow - N_\uparrow) \quad \text{and} \quad \frac{dN_\downarrow}{dt} = w(N_\uparrow - N_\downarrow). \tag{9.7}$$

Subtracting these equations, and setting $N = N_\uparrow - N_\downarrow$, we find $dN/dt = -2wN$, which gives $N(t) = N(0)e^{-2wt}$. The populations tend to equalize at long times. The energy ε of the system is $N_\downarrow\hbar\omega_0$, so $d\varepsilon/dt = -\hbar\omega_0 wN(t)$. The rate of change of energy tends to zero at long times.

However, when we switch off the microwave power, the populations can be expected to relax to thermal equilibrium with longitudinal time constant T_1, so that $N(t) = N_0(1 - e^{-t/T_1})$, where N_0 is the equilibrium population difference. Taking relaxation into account, the rate of change of population imbalance becomes

$$\frac{dN(t)}{dt} = -2wN(t) + \frac{N_0 - N(t)}{T_1}. \tag{9.8}$$

There is a similar equation for the magnetization, since $M = N\mu_B/V$. In equilibrium, $dN(t)/dt = 0$, so $N(t) = N_0/(1 + 2wT_1)$. The rate of absorption of electromagnetic energy $N(t)\hbar\omega w$ is then

$$\frac{d\varepsilon}{dt} = \frac{N_0\hbar\omega_0 w}{1 + 2wT_1}, \tag{9.9}$$

which is plotted in Fig. 9.2. At low power, the rate of absorption is proportional to w, but at high power it saturates at a value proportional to $1/T_1$.

Spin-orbit interaction is the mechanism by which the spin system couples to the lattice phonon bath. Good EPR spectra are obtained with ions where the orbital moment is quenched or absent. The latter are S-state ions with half-filled shells, such as free radicals ($^2S_{1/2}$), Mn^{2+} or Fe^{3+} ($^6S_{5/2}$) and Eu^{2+} or

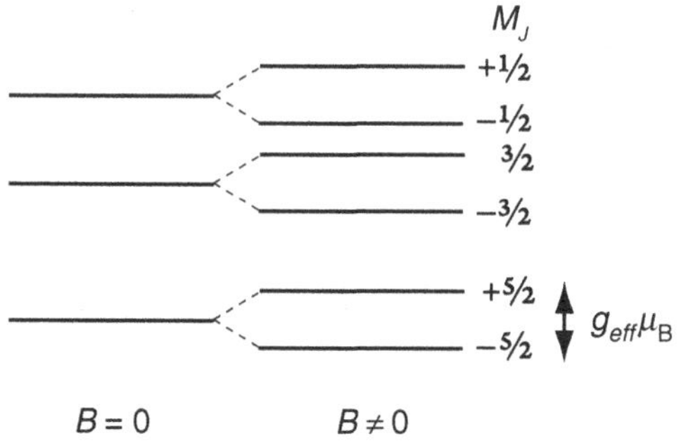

Figure 9.3

Energy levels of the Ce^{3+} ion split by a uniaxial crystal field. The $M_J = \pm\frac{5}{2}$ Kramers doublet ground state looks like a $\pm\frac{1}{2}$ doublet with a large effective g-factor in EPR.

Gd^{3+} ($^8S_{7/2}$). Moreover, the resonant ions should be dilute in the crystal lattice to minimize dipolar and exchange interactions between them, which broaden the resonance linewidth and lead to dephasing of the spins.

The outer electrons of an ion interact strongly with the surrounding ions – the crystal-field interaction was discussed in §4.4. A crystal field of second order may include a term A_2^2 which mixes states where M_S (or, more generally, M_J) differs by 2. The fourth- or sixth-order crystal field may mix states where M_S differs by up to 4 or 6. These interactions are effective when $J > \frac{1}{2}$, $J > \frac{3}{2}$ and $J > \frac{5}{2}$, respectively. Although it is the ground state that is involved in EPR, the effect of the crystal field is to create a zero-field splitting of the energy levels which modifies the effective g-factor of the lowest energy level, and makes it anisotropic with respect to the crystal axes. The example of Ce^{3+}, a $4f^1$ Kramers ion with $J = \frac{5}{2}$, is shown in Fig. 9.3.

It is common practice in EPR to replace the Hamiltonian of the system by an effective spin Hamiltonian which describes how the ground-state energy level splits in a magnetic field. An effective spin S is chosen, so that the magnetic degeneracy is $2S + 1$. Terms in the spin Hamiltonian reflect the crystal symmetry of the resonant ion. Examples of terms to add to the Zeeman term in order to build the spin Hamiltonian are:

$\mathsf{D}S_z^2$ for uniaxial symmetry;
$\mathsf{E}(S_x^2 - S_y^2)$ for an orthorhombic distortion;
$\mathsf{D}_c(S_x^4 + S_y^4 + S_z^4)$ for cubic symmetry.

Consider, for instance, the case of an ion with $S = 1$ in a site having uniaxial symmetry, with the field B_0 applied along the crystal axis. The spin Hamiltonian is

$$\mathcal{H}_{spin} = \mathsf{D}S_z^2 - g_{eff}\mu_B B_0 S_z. \tag{9.10}$$

The effect of the crystal field is to create fine structure in the EPR spectrum as shown in Fig. 9.4.

There is another interaction, of order 0.1 K at most, which modifies the splitting of the electronic ground state. This is the hyperfine interaction with the nucleus. The nucleus may possess quantized angular momentum $\hbar I$ when

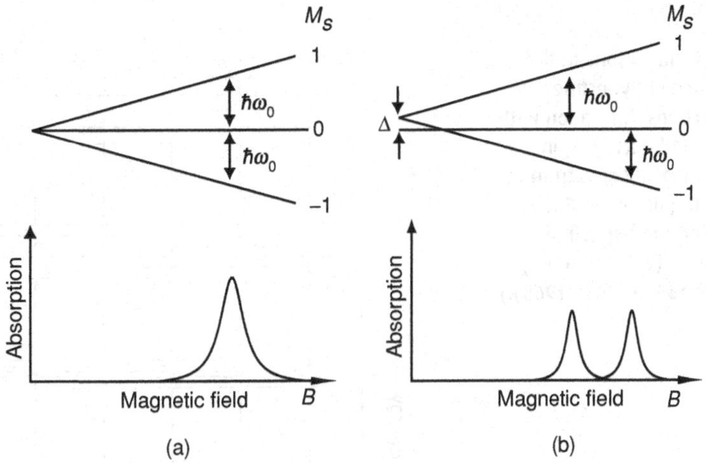

Figure 9.4

Energy levels and EPR absorption for an ion with $S = 1$: (a) without and (b) with a crystal field interaction $D > 0$.

the nuclear spin quantum number $I \neq 0$. The corresponding magnetic moment $\mathfrak{m}_n = g_n \mu_N I$ is about a thousand times smaller than the electronic moment; g_n is the nuclear g-factor, a number of order 1, and μ_N is the nuclear magneton:

$$\mu_N = e\hbar/2m_p = 5.0508 \times 10^{-27} \text{ A m}^2, \tag{9.11}$$

where m_p is the proton mass. A magnetic field separates the $2I + 1$ degenerate nuclear energy levels with $M_I = I, I - 1, \ldots, -I$.

The unpaired electrons of a magnetic ion create a magnetic field at the nucleus, known as the hyperfine field. This ranges up to about 50 T in 3d ions, and it can be ten times larger for some rare-earths because of the 4f orbital contribution. These are huge magnetic fields, albeit in a very small volume. Hyperfine interactions are therefore of order 10^{-1}–10^{-3} K. They dominate the specific heat below 1 K and they give rise to hyperfine structure in EPR, NMR and Mössbauer spectra. The interactions are represented by the term $\mathsf{A}\boldsymbol{I} \cdot \boldsymbol{S}$ in the spin Hamiltonian, with the hyperfine constant A having units of energy. Each degenerate M_S level splits into $(2I + 1)$ sublevels with energy $g\mu_B B_0 M_S + \mathsf{A}M_I M_S$. Microwave transitions only occur between levels obeying the dipole selection rule $\Delta M_S = \pm 1$ or $\Delta M_I = 0$, since the frequencies required to induce transitions between the nuclear levels lie in the radiofrequency range – MHz rather than GHz. The resonances therefore occur at

$$\begin{aligned} \hbar\omega &= [g\mu_B B_0(M_S + 1) + AM_I(M_S + 1)] - [g\mu_B B_0 M_S + AM_I M_S] \\ &= g\mu_B B_0 + AM_I. \end{aligned} \tag{9.12}$$

Each EPR line therefore splits into $2I + 1$ hyperfine lines, as shown in Fig. 9.5.

EPR is normally applied to magnetic ions in insulators. However, it is possible to obtain a signal known as conduction electron spin resonance (CESR) from the free electrons in metals and semiconductors, provided the relaxation time is not too short.

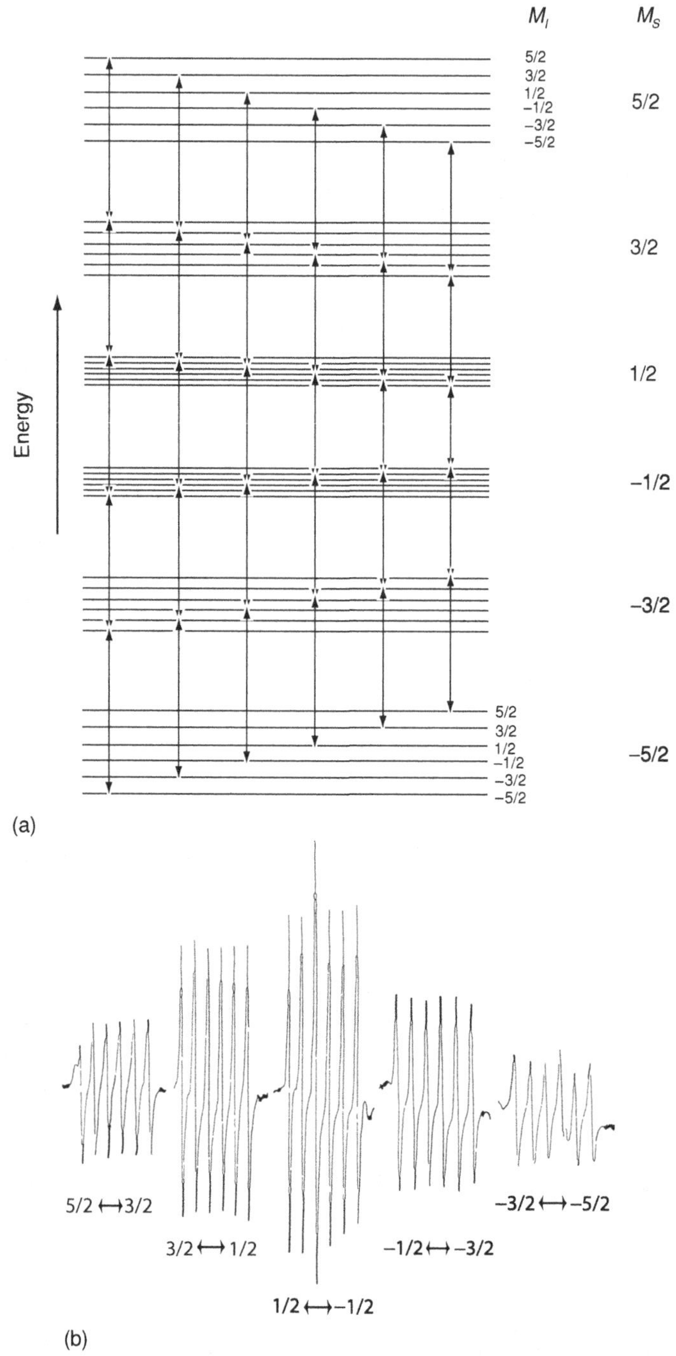

Figure 9.5

(a) EPR transitions in the presence of hyperfine interactions, for an ion with $S = \frac{5}{2}$ and nuclear spin $I = \frac{5}{2}$. (b) EPR spectrum of Mn^{2+} impurities in Ga_2O_3, showing the hyperfine structure. (V. J. Folen, *Phys. Rev. B* **139**, A1961 (1965).)

9.2 Ferromagnetic resonance

Resonance is also observed in the microwave frequency range when a *ferromagnet* is subject to a uniform field $\boldsymbol{B}_0$ and a transverse high-frequency field $\boldsymbol{b}_1$. The system will be treated classically, like a giant spin, or macrospin. The magnetization of the sample is first assumed to remain uniform.

In the absence of damping the equation of motion is

$$\mathrm{d}\boldsymbol{M}/\mathrm{d}t = \gamma(\boldsymbol{M} \times \boldsymbol{B}_0'). \tag{9.13}$$

The magnetization precesses around the z axis at the Larmor frequency $f_L = \omega_0/2\pi$, where $\omega_0 = \gamma B_0$. Since the magnetization of the ferromagnet is largely due to the spin moments of the electrons, $\gamma \approx -(e/m_e)$, resonant frequencies for FMR are similar to those for EPR. The same apparatus is used for both.

If the precession is to be detected by resonant absorption of microwaves, high-frequency radiation has to be able to penetrate the specimen. This poses no difficulty for insulators such as ferrimagnetic oxides, but the skin depth of metallic iron at 10 GHz, for example, is only of order a micron so thin films or fine powder samples have to be used for metallic materials (12.2).

Furthermore, in ferromagnets we have to distinguish clearly the external field $\boldsymbol{H}' = \boldsymbol{B}'/\mu_0$ and the internal field $\boldsymbol{H} = \boldsymbol{H}' + \boldsymbol{H}_d$ present inside the ferromagnet. The demagnetizing field $\boldsymbol{H}_d = -\mathcal{N}\boldsymbol{M}$, where the demagnetizing tensor is assumed to be diagonal:

$$\mathcal{N} = \begin{bmatrix} \mathcal{N}_x & 0 & 0 \\ 0 & \mathcal{N}_y & 0 \\ 0 & 0 & \mathcal{N}_z \end{bmatrix}.$$

Provided $b_1 \ll B_0$, the magnetization is $\boldsymbol{M} \approx M_s\mathbf{e}_z + \boldsymbol{m}(t)$, where $\boldsymbol{m}(t) = \boldsymbol{m}_0\mathrm{e}^{\mathrm{i}\omega t}$ is the small in-plane component. The demagnetizing field is therefore

$$\boldsymbol{H}_d = -\mu_0[\mathcal{N}_x m_x\mathbf{e}_x + \mathcal{N}_y m_y\mathbf{e}_y + \mathcal{N}_z(m_z + M_s)\mathbf{e}_z]. \tag{9.14}$$

The oscillating components of magnetization $m = m_0\mathrm{e}^{\mathrm{i}\omega t}$ in the xy-plane are

$$\frac{\mathrm{d}m_x}{\mathrm{d}t} = \mu_0\gamma(m_y H_z - M H_y) = \mu_0\gamma[H_0' + (\mathcal{N}_y - \mathcal{N}_z)M]m_y, \tag{9.15}$$

$$\frac{\mathrm{d}m_y}{\mathrm{d}t} = \mu_0\gamma(-m_x H_z + M H_x) = -\mu_0\gamma[H_0' + (\mathcal{N}_x - \mathcal{N}_z)M]m_x, \tag{9.16}$$

where $M_z = M$. The external field $B_0' = \mu_0 H_0'$ is applied in the z-direction. Solutions of these equations exist when

$$\begin{vmatrix} \mathrm{i}\omega & \gamma\mu_0[H_0' + (\mathcal{N}_y - \mathcal{N}_z)M] \\ -\mu_0\gamma[H_0' + (\mathcal{N}_x - \mathcal{N}_z)M] & \mathrm{i}\omega \end{vmatrix} = 0,$$

which leads to the Kittel equation for the resonance frequency:

$$\omega_0^2 = \mu_0^2\gamma^2[H_0' + (\mathcal{N}_x - \mathcal{N}_z)M][H_0' + (\mathcal{N}_y - \mathcal{N}_z)M]. \tag{9.17}$$

Table 9.1. g-factors for metallic ferromagnets

Fe	2.08
Co	2.17
Ni	2.18
Gd	1.95

Special cases are:

- a sphere, $\mathcal{N}_x = \mathcal{N}_y = \mathcal{N}_z = \frac{1}{3}$, $\omega_0 = \gamma\mu_0 H_0'$;
- a thin film with H_0' perpendicular to the plane $\mathcal{N}_x = \mathcal{N}_y = 0$, $\mathcal{N}_z = 1$; $\omega_0 = \gamma\mu_0(H_0' - M)$;
- a thin film with H_0' in plane $\mathcal{N}_y = \mathcal{N}_z = 0$, $\mathcal{N}_x = 1$; $\omega_0 = \gamma\mu_0[H_0'(H_0' + M)]^{\frac{1}{2}}$.

Magnetocrystalline anisotropy also influences the ferromagnetic resonance frequency, so to the demagnetizing field in the above expressions may be added the anisotropy field, $H_a = 2K_1/\mu_0 M_s$. A sphere, for example, with the z axis as the easy anisotropy axis and $\mathcal{N}_x = \mathcal{N}_y = \mathcal{N}_z = \frac{1}{3}$ has resonance frequency $\omega_0 = \gamma\mu_0(H_0 + 2K_1/\mu_0 M_s)$. It is possible to observe ferromagnetic resonance in *zero* external field for a single-domain particle, or a crystal of high-anisotropy material magnetized along Oz.

For a spherical sample with cubic anisotropy (§5.5.2), when H_0' is applied along [100], (9.17) applies with $K_1 = K_{1c}$. When H_0' is applied along [111]

$$\omega_0 = \gamma\mu_0(H_0' - 4K_{1c}/3\mu_0 M_s - 4K_{2c}/9\mu_0 M_s), \tag{9.18}$$

whereas if H_0 is applied along [110]

$$\omega_0 = \gamma\mu_0[(H_0' - 2K_{1c}/\mu_0 M_s)(H_0' + K_{1c}/\mu_0 M_s + K_{2c}/2\mu_0 M_s)]^{\frac{1}{2}}. \tag{9.19}$$

In the case of a thin film with an easy axis perpendicular to the plane of the film, the expression for the resonance frequency is

$$\omega_0 = \gamma\mu_0(H_0' + 2K_1/\mu_0 M_s - M_s). \tag{9.20}$$

Ferromagnetic resonance can therefore provide a measurement of M_s and K_i as well as γ. An advantage of the method is that the magnetization, or magnetic moment per cubic metre, is determined with no need to know the sample volume. The gyromagnetic ratio is related to the g-factor; $\gamma = -g\mu_B/\hbar$. Values of g for the metallic ferromagnets are given in Table 9.1. The ratio of orbital moment to spin moment is $\frac{1}{2}(g - 2)$.

The instantaneous field in the sample is uniform in an FMR experiment provided the wavelength of the microwaves is much greater than the sample size. At 10 GHz, $\lambda = 3$ cm, so the condition is satisfied for millimetre-size samples. However, the giant-spin assumption of uniform magnetization throughout the sample is not generally valid. Nonlinear magnetostatic modes may be excited.

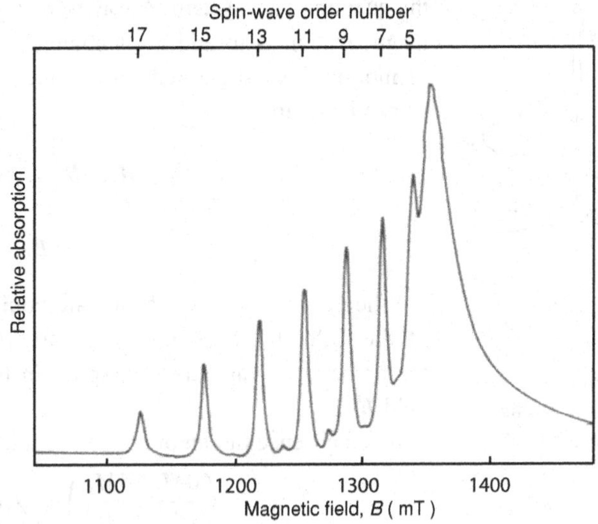

Figure 9.6

Spin-wave resonance spectrum of permalloy (R. Weber and P. Tannenwald, *IEEE Trans. Maps.* 4, 28 (1968)).

An example is the standing spin waves excited in thin ferromagnetic films when the steady field B_0 is applied normal to the film surface, Fig. 9.6. The in-plane radio-frequency field can excite modes with an odd number of half-wavelengths. Those with an even number do not couple with the field. Ferromagnetic spin waves follow the dispersion relation (5.56) $\hbar\omega_q = D_{sw}q^2$, where $q = n\pi/\mathrm{t}$; n is an integer and t is the film thickness. Equation (9.17) becomes

$$\omega_q = \gamma\mu_0(H_0' - M) + D_{sw}(n\pi/\mathrm{t})^2/\hbar.$$

The spin-wave stiffness can be determined in this way.

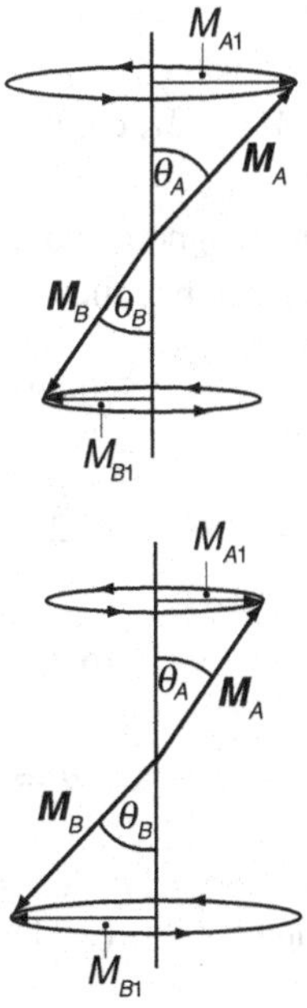

Precession mode for antiferromagnetic resonance.

9.2.1 Antiferromagnetic resonance

An antiferromagnet is composed of two equal and opposite sublattices, each of which is subject to an anisotropy field and an exchange field. The exchange field on the 'A' sublattice, for example, is $H_{exA} = -n_{AB}M_B$ (6.1), where n_{AB} is the molecular field coefficient. Solving the equations of motion for m_x, m_y for the two sublattices leads to the solution

$$\omega_0 = \gamma\mu_0[H_a(H_a + 2H_{ex})]^{\frac{1}{2}}. \qquad (9.21)$$

Molecular field coefficients may take values of up to 100 T in antiferromagnets, so the resonance frequencies are very high, often in the 10^2–10^3 GHz range.

9.2.2 Damping

Free precession of the magnetization in the internal field at frequency $\omega_0 = \gamma\mu_0 H$ cannot go on for ever. Eventually, the magnetization must align with the field. In EPR and NMR, this process involves spin-lattice relaxation of

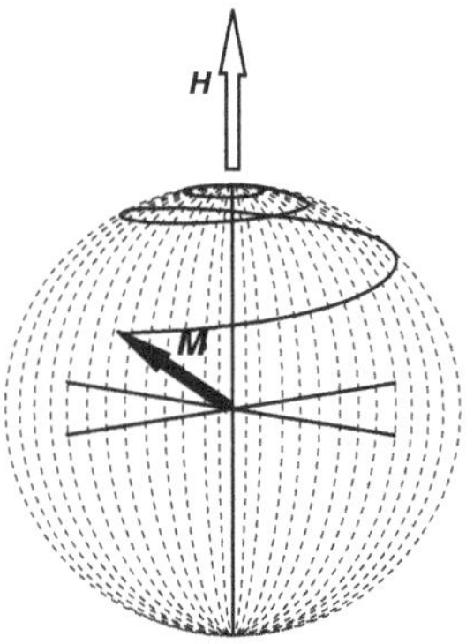

Precession of the magnetization in a field showing the effect of damping.

the quantum spin system. A way to represent the process for the macroscopic magnetization is to add a phenomenological damping term to the equation of motion. Two suggested forms, due to Landau and Lifschitz and Gilbert respectively, are:

$$\frac{\mathrm{d}\boldsymbol{M}}{\mathrm{d}t} = \gamma \boldsymbol{M} \times \boldsymbol{B}_0 - \gamma \frac{\lambda}{M} \boldsymbol{M} \times \boldsymbol{M} \times \boldsymbol{B}_0; \tag{9.22}$$

$$\frac{\mathrm{d}\boldsymbol{M}}{\mathrm{d}t} = \gamma \boldsymbol{M} \times \boldsymbol{B}_0 - \frac{\alpha}{M} \boldsymbol{M} \times \frac{\mathrm{d}\boldsymbol{M}}{\mathrm{d}t}. \tag{9.23}$$

When $\alpha \ll 1$, the two forms are equivalent, with $\lambda = \alpha$. A typical value of the Gilbert damping is $\alpha = 0.01$. The effect of damping is to make the precessing magnetization spiral in towards the direction of the applied field $\boldsymbol{B}_0$.

In the absence of damping, the equation of motion in component form is

$$\left(\frac{\mathrm{d}M_x}{\mathrm{d}t}, \frac{\mathrm{d}M_y}{\mathrm{d}t}\right) = \gamma \mu_0 H (M_y, -M_x). \tag{9.24}$$

Differentiating with respect to time,

$$\left(\frac{\mathrm{d}^2 M_x}{\mathrm{d}t^2}, \frac{\mathrm{d}^2 M_y}{\mathrm{d}t^2}\right) = \gamma \mu_0 H \left(\frac{\mathrm{d}M_y}{\mathrm{d}t}, -\frac{\mathrm{d}M_x}{\mathrm{d}t}\right) = \gamma^2 \mu_0^2 H^2 (-M_x, M_y). \tag{9.25}$$

Hence

$$\frac{\mathrm{d}^2 M_x}{\mathrm{d}t^2} = -\omega_0^2 M_x; \quad \frac{\mathrm{d}^2 M_y}{\mathrm{d}t^2} = \omega_0^2 M_y; \quad \frac{\mathrm{d}^2 M_z}{\mathrm{d}t^2} = 0. \tag{9.26}$$

The solution is a uniform precession

$$M_x = M_s \sin\theta \exp \mathrm{i}\omega_0 t; \quad M_y = M_s \sin\theta \exp(\mathrm{i}\omega_0 t + \pi/2); \quad M_z = M_s \cos\theta. \tag{9.27}$$

where the real parts represent the physical components of the magnetization. When Gilbert damping is taken into account, the equations of motion reduce to

$$\frac{\mathrm{d}M_x}{\mathrm{d}t} = \omega_0' \left[M_y + \alpha \frac{M_y M_z}{M_s}\right], \tag{9.28}$$

$$\frac{\mathrm{d}M_y}{\mathrm{d}t} = \omega_0' \left[-M_x + \alpha \frac{M_x M_z}{M_s}\right], \tag{9.29}$$

$$\frac{\mathrm{d}M_z}{\mathrm{d}t} = \omega_0' \left[-M_s + \alpha \frac{M_z^2}{M_s}\right], \tag{9.30}$$

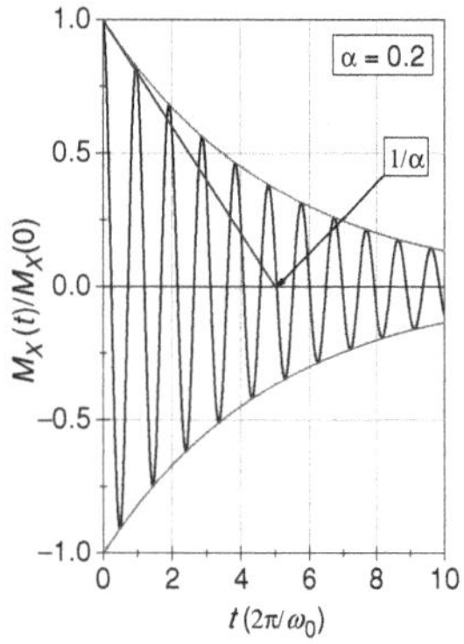

Gilbert damping of a resonant oscillation.

where $\omega_0' = \omega_0/(1+\alpha^2)$. These are spiralling solutions with $\omega \neq \omega_0$ and $\theta = \theta(t)$. Differentiating the expression (9.27) for M_x,

$$\frac{\mathrm{d}M_x}{\mathrm{d}t} = \mathrm{i}\omega M_s \sin\theta \exp \mathrm{i}\omega t + M_s \frac{\mathrm{d}\theta}{\mathrm{d}t} \cos\theta \exp \mathrm{i}\omega t \tag{9.31}$$

$$\frac{\mathrm{d}M_x}{\mathrm{d}t} = \omega M_y + \frac{M_y M_z}{\sin\theta M_s} \frac{\mathrm{d}\theta}{\mathrm{d}t}. \tag{9.32}$$

Hence $\omega = \omega_0'$ and $\mathrm{d}\theta/\mathrm{d}t = \omega_0' \alpha \sin\theta$.

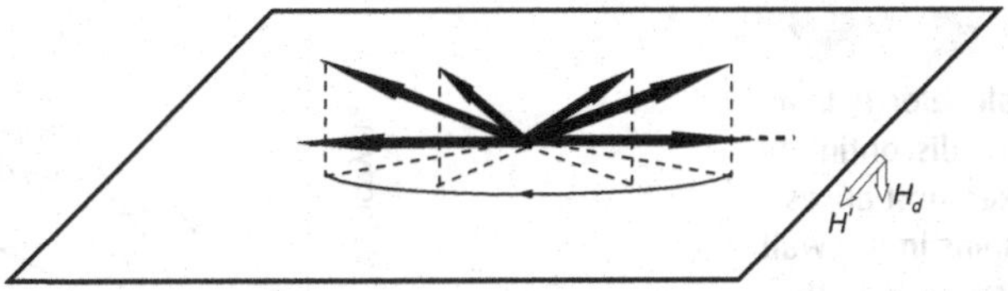

Figure 9.7

Switching of the magnetization of a thin-film element. The switching makes use of the precession of the magnetization in the demagnetizing field when the magnetization acquires an out-of-plane component as it begins to process around H'.

When $\alpha \ll 1$, the motion is lightly damped, and many precessions take place before alignment is achieved. When $\alpha \gg 1$, the motion is overdamped. Critical damping is when $\alpha = 1$. Then if a reverse field is applied, switching takes a time $t \approx 2/\gamma\mu_0 H$. Switching is rapid. For example, a moment with $\theta = 20°$ in a reverse field takes four precessions to reach $\theta = 170°$. If $\mu_0 H = 10$ mT, $t = 1$ ns when $g = 2$ ($\gamma = 1.76 \times 10^{11}$ s^{-1} T^{-1}). For switching the films it is possible to make use of the demagnetizing field, Fig. 9.7. Rapid switching of the magnetization of ferromagnetic thin film elements is important for spin electronics (§14.4).

9.2.3 Domain wall dynamics

Consider a 180° Bloch wall separating two domains with magnetization up or down along an anisotropy axis taken as the z axis. The magnetization turns in the yz-plane, making an angle θ with Oz, as shown in Fig. 7.6. From (7.15)

$$\frac{d^2\theta}{dx^2} = \frac{\pi^2}{\delta_w^2}\sin\theta\cos\theta, \tag{9.33}$$

where $\delta_w = \pi\sqrt{A/K_1}$ is the domain-wall width. The solutions are of the form $d\theta/dx = \pi\sin\theta/\delta_w$. Suppose now that a field $\boldsymbol{H}$ is applied along the positive z axis. This tends to drive the wall along Ox, exerting a pressure $2\mu_0 H M_s$ on it. If the field is applied for a time t, the wall acquires a velocity v_w, which we now calculate.

A consequence of $\boldsymbol{H}$ is to exert a torque on the spins in the wall. They precess around Oz with angular velocity $\omega_z = \mu_0\gamma H$. After a short time, they all make a small angle ϕ with the z-plane. The magnetization of the wall acquires a component along) O_x, $M_x \approx M_s\phi\sin\theta$. The sheet of magnetization in turn creates a demagnetizing field $H_x = -M_x$, and it is this field that produces the torque needed to move the wall. The magnetization in the wall precesses around Ox with angular velocity $\omega_x = d\theta/dt = \mu_0\gamma H_x = -\mu_0\gamma M_s\phi\sin\theta$. The effect is to move the entire wall along Ox with velocity v_w. The angle ϕ accumulated in the time needed to damp the oscillation remains constant after $\boldsymbol{H}$ is switched off. In the moving wall, θ is a function of $x' = x - v_w t$. Hence

$$\frac{d\theta}{dx'} = -\frac{1}{v_w}\frac{d\theta}{dt} = \frac{1}{v_w}\mu_0\gamma M_s\phi\sin\theta. \tag{9.34}$$

Figure 9.8

Domain-wall velocity in a thin film. The discontinuity at the Walker limit arises when the spins in the wall are driven faster than the Larmor precession frequency.

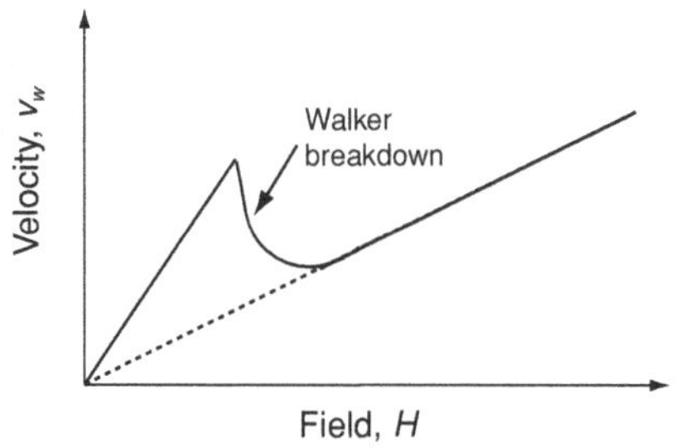

Since $d\theta/dx = \pi \sin\theta/\delta_w$ and $\phi = \mu_0\gamma H t_0$, it follows that $v_w = 2\mu_0^2\gamma^2 H M_s \delta_w/w_0\alpha$. The ratio of impulse per unit area to velocity is an effective domain-wall mass per unit area (§7.3.7):

$$m_w = \frac{2\pi}{\mu_0\gamma^2\delta_w} \tag{9.35}$$

This is known as the Döring mass. From data in Table 7.1, values of m_w for Fe and $Nd_2Fe_{14}B$ are 4×10^{-9} kg m^{-2} and 4×10^{-8} kg m^{-2}, respectively. Domain-wall dynamics is determined by this mass.

At low fields, the wall velocity v_w is proportional to the driving field beyond the depinning field H_p, with a constant of proportionality η_w known as the domain-wall mobility (7.31). In thin films, the mobility is limited by the Gilbert damping parameter α, so that $\eta_w = \mu_0\gamma\delta_w/\alpha$ at high driving fields. The velocity of the wall collapses as the precession frequency of the spins approaches the ferromagnetic resonance frequency $\gamma B_0/2\pi$. The spins in the moving wall precess at a frequency v_w/δ_w; Walker breakdown occurs when this frequency hits the ferromagnetic resonance frequency, Fig. 9.8.

9.3 Nuclear magnetic resonance

The nucleus is the collection of protons and neutrons at the heart of every atom, which constitutes 99.98% of the mass of solids, liquids and gases. The magnetic properties of these and other elementary particles are listed in Table 9.2.

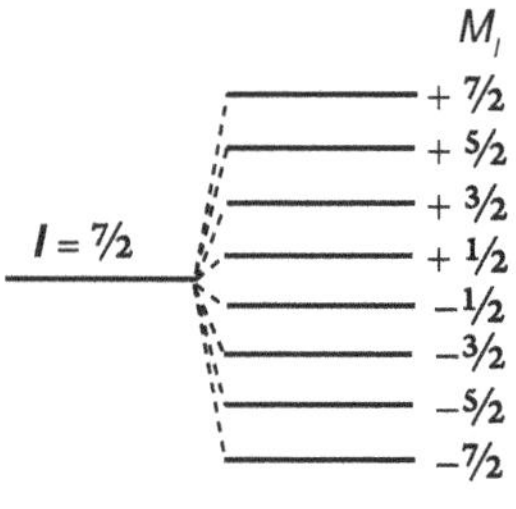

Splitting of nuclear energy levels of ^{59}Co ($I = \frac{7}{2}$) in the hyperfine field.

While the proton and the neutron possess the same spin angular momentum as the electron, $\hbar/2$, their magnetic moments are far smaller because of their much greater mass. The magnetic moments associated with the spin angular momentum can be written as $\mathfrak{m}_n = g_n\mu_N I$, where the nuclear g-factors are $g_n = 5.586$ and $g_p = -3.826$ for the proton and neutron, respectively. The nuclear magneton μ_N is $e\hbar/2m_p$, 5.051×10^{-27} A m^2. The interacting protons and neutrons in the nucleus form an entity with angular momentum $I\hbar$, having $2I + 1$ degenerate states labelled with a magnetic quantum number M_I, analogous to the many-electron states of an atom. Whereas the excited states of the

Table 9.2. Magnetism of some elementary particles

		Charge	m/m_e	$\tau_{1/2}$	$I(\hbar)$	m	f_L(Hz T^{-1})
Proton	p	e	1836	Stable	$\frac{1}{2}$	2.793 μ_N	42.58 × 10^6
Neutron	n		1836	10.3 m	$\frac{1}{2}$	−1.913 μ_N	29.17 × 10^6
Electron	e	$-e$	1	Stable	$\frac{1}{2}$	−1.001 μ_B	27.99 × 10^9
Positron	e$^+$	e	1	Stable[a]	$\frac{1}{2}$	1.001 μ_B	27.99 × 10^9
Muon	μ^+	e	206.7	2.2 µs	$\frac{1}{2}$	0.00484 μ_B	135.5 × 10^6
Muon	μ^-	$-e$	206.7	2.2 µs	$\frac{1}{2}$	−0.00484 μ_B	135.5 × 10^6
Photon	ϕ			Stable	1		

[a] Positrons combine with electrons in condensed matter to produce two γ photons, each of energy 0.511 MeV.

many-electron atom lie 1–100 eV above the atomic ground state, the excited states of the many-nucleon nucleus lie 10 keV–10 MeV above the nuclear ground state. We rarely need to consider any but the nuclear ground state. Unlike electronic moments, which are usually negative (oppositely directed to the angular momentum on account of the negative electronic charge) the nuclear moments are usually positive.

On applying an external magnetic field, B_0 in the Oz-direction, the $2I + 1$ magnetic levels are Zeeman split which leads to the establishment of the Boltzmann populations. A small net magnetization is induced in the direction of the field, given by Curie's law. In a field of 1 T, the energy splitting for the proton $g_p\mu_N B$ is 2.8×10^{-26} J or 2 mK. The difference in population of the two levels is therefore less than 1 part in 10^5 at room temperature.

As for EPR, resonant transitions between the energy levels require an AC field to be applied in the xy-plane, perpendicular to the uniform field. The frequencies required for NMR are in the radio-frequency range (Table 9.3) rather than the microwave range, so the samples can be excited in a resonant coil with a few turns, rather than a waveguide.

The nucleus is surrounded by the electron shells of its atom, and by the other atoms in the sample, which have the effect of creating or modifying the electric and magnetic fields acting at the nucleus. NMR is very widely used in organic chemistry as a fingerprint spectroscopy for nonconducting organic compounds in the liquid state. The diamagnetic susceptibility of the inner electron shells tends to shield the nucleus slightly from the applied field, leading to a chemical shift of the resonance to slightly higher frequency ω_0. Chemical shifts are measured in parts per million, but the resonance linewidth may be a hundred times less, so a rich fund of molecule- and bond-specific chemical information is available. Modern high-resolution spectrometers use superconducting magnets which deliver B_0 in the range 12–20 T, and operate in the 500–800 MHz range.

In metals, the paramagnetic susceptibility of the conduction electrons shifts the resonance in the opposite direction. This is the Knight shift, discussed below. It is an effect of order 1%.

Table 9.3. Some nuclei of interest for magnetic resonance

Nucleus	I^{parity}	(%)	m (μ_N)	f_L(MHz T^{-1})	Q (10^{-28} m^2)
^{1}H	1/2$^+$	99.9885	2.793	42.58	
^{2}H	1$^+$	0.0115	0.857	6.54	0.0029
^{13}C	1/2$^+$	1.07	0.702	10.71	
^{14}N	1$^+$	99.632	0.404	3.08	0.0204
^{17}O	5/2$^-$	0.038	−1.893	5.77	−0.0256
^{19}F	1/2$^+$	100	2.627	40.05	
^{23}Na	3/2$^+$	100	2.217	11.26	0.104
^{27}Al	5/2$^+$	100	3.641	11.09	0.14
^{29}Si	1/2$^-$	4.6832	−0.555	8.46	
^{31}P	1/2$^+$	100	1.132	17.24	
^{33}S	3/2$^+$	0.76	0.643	3.27	−0.0678
^{53}Cr	3/2$^-$	9.501	−0.474	2.41	−0.150
^{55}Mn	5/2$^+$	100	3.468	10.54	0.330
^{57}Fe	1/2$^+$	2.19	0.091	1.38	
^{59}Co	7/2$^+$	100	4.616	10.10	0.420
^{61}Ni	3/2$^-$	1.14	−0.750	3.81	0.162
^{63}Cu	3/2$^+$	69.17	2.226	11.29	−0.220
^{87}Rb	3/2w^+	27.835	2.750	13.93	0.134
^{89}Y	1/2$^-$	100	−0.137	2.09	
^{105}Pd	5/2$^-$	22.33	−0.639	1.95	0.660
^{143}Nd	7/2$^-$	12.81	−1.063	2.32	−0.630
^{147}Sm	7/2$^-$	15.0	−0.813	1.76	−0.259
^{157}Gd	3/2$^-$	15.65	−0.339	2.03	1.350
^{159}Tb	3/2$^+$	100	2.008	9.66	1.432
^{163}Dy	5/2$^+$	24.9	0.676	1.95	2.648

9.3.1 Hyperfine interactions

The atomic nucleus is a point probe of electric and magnetic fields at the very heart of the atom. Atoms in different crystallographic sites may be distinguished by their hyperfine interactions, which result from coupling of the electric and magnetic moments of the nucleus with these fields. Nuclei with $I \neq 0$ have a magnetic moment $g_n\mu_N I$, and the Zeeman splitting of the $(2I + 1)$ magnetic levels denoted by the nuclear magnetic quantum number $M_I = I, I-1, \ldots, -I$ results from the action of the hyperfine field B_{hf} at the nucleus.

The complete Hamiltonian is

$$\mathcal{H}_{hf} = -g_n\mu_N \boldsymbol{I}\cdot\boldsymbol{B}_{hf} - eQV_{zz}\{[3I_z^2 - I(I+1)] + \eta(I_x^2 - I_y^2)\}/[4I(2I-1)], \tag{9.36}$$

where the first term is the magnetic hyperfine interaction and the second term represents the interaction of the electric quadrupole moment of the nucleus with the electric field gradient V_{zz}. Higher-order moments of the nuclear charge are

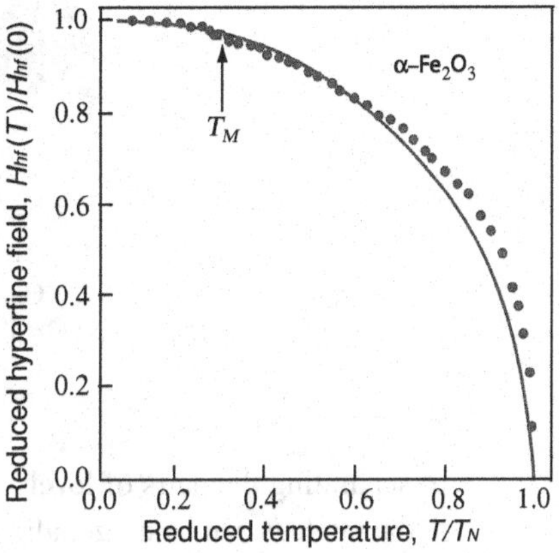

Figure 9.9

Temperature dependence of the hyperfine field of an antiferromagnet. The data are for α-Fe_2O_3 which exhibits a spin reorientation transition at T_M. (F. van der Woude *et al., Phys. Stat. Sol.* 17, 417 (1966).)

negligible. There are several contributions to the hyperfine field B_{hf}; one is the Fermi contact interaction of the nucleus with the unpaired electron density at its site. Unpaired electrons are largely in 3d and 4f shells which have no electron density at the nucleus, but they polarize the 1s, 2s and 3s core shells which do have some charge density there. The core polarization contribution is largest for 3d elements; it is about -11 T μ_B^{-1} in iron and cobalt, and 4 T μ_B^{-1} in the rare-earths. A further contribution comes from the spin-polarized 4s or 6s conduction electrons. For non-S state ions there are also orbital and dipolar contributions, $B_{orb} = -2\mu_0\mu_B\langle r^{-3}\rangle\langle \boldsymbol{L}\rangle/h$ and $B_{dip} = -2\mu_0\mu_B\langle r^{-3}\rangle\langle \boldsymbol{S}\rangle\langle 3\cos^2\theta - 1\rangle/h$, produced by the unquenched orbital angular momentum and the non-spherical atomic spin distribution, respectively. In non-S-state rare-earths, these contributions reach values of several hundred teslas. At lattice sites which do not have cubic symmetry, there may also be a dipolar contribution from the moments of the atoms on the rest of the lattice, which is of order 1 T.

Normally, *no* magnetic hyperfine splitting is observed in the paramagnetic state. The reason is that the fluctuations of the atomic moment in a paramagnet are much faster than the nuclear Larmor precession frequency in the hyperfine field $f_L \approx 10^9$ Hz. Dilute paramagnetic salts of non-S-state rare-earth ions may be an exception. There the Larmor precession frequency is several gigahertz on account of the orbital and dipolar contributions, and if the crystal field stabilizes a $\pm M_J$ ground state with $M_J > 3$ the fluctuation time can be slow at low temperature because the $M_J \to -M_J$ transitions are suppressed on account of the large change of orbital angular momentum $\Delta M_J = 2J$ involved.

The magnetic hyperfine field in magnetically ordered material faithfully follows the ordered moment, and it falls to zero at the Curie or Néel temperature. Accurate values of the critical exponent β are obtained in this way, which is especially valuable for antiferromagnets, Fig. 9.9.

The second term in (9.36) represents the electrostatic coupling of the nuclear quadrupole moment Q with the electric field gradient at the nucleus $V_{ij} = \mathrm{d}^2V/\mathrm{d}x_i\mathrm{d}x_j$. Any nucleus with $I \geq \frac{3}{2}$ has a quadrupole moment, and the electric quadrupole interaction, represented by the second term in (9.36), has the effect

Table 9.4. Some values of the Knight shift

	$\mathcal{K}$ (%)
^{22}Na	0.11
^{23}Al	0.16
^{53}Cr	0.69
^{63}Cu	0.24
^{105}Pd	−3.00

of separating the pairs of levels with different $|M_I|$. The electric field gradient at the nucleus can be diagonalized by a suitable choice of axes; only two of the three components (V_{xx}, V_{yy}, V_{zz}) are independent. Conventionally, the biggest is labelled V_{zz} and the asymmetry parameter is defined as $\eta = (V_{xx} - V_{yy})/V_{zz}$. These quantities in S-state ions are related to the second-order crystal field acting on the electronic shell: $V_{zz} \approx A_2^0$, $\eta \approx A_2^2$. The inner electron shells greatly amplify the electric field gradient produced by the lattice $(V_{zz})_{latt}$, and they shield the atomic charge $(V_{zz})_{val}$ contribution.

$$V_{zz} = (1 - \gamma_\infty)(V_{zz})_{latt} + (1 - R)(V_{zz})_{val}, \qquad (9.37)$$

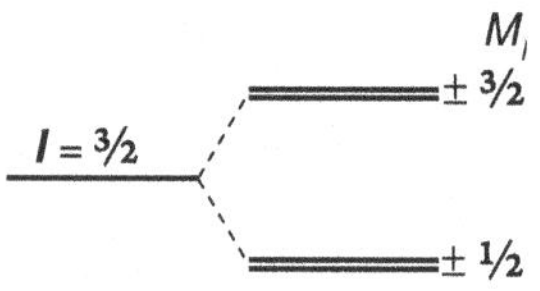

Quadrupole splitting of the energy levels of a nucleus with $I = \frac{3}{2}$.

modifying the field gradient at the nucleus. The factors $\gamma_\infty$and R are the Sternheimer antishielding and shielding factors, respectively. For example, values for ^{57}Fe are $\gamma_\infty = -9.14$ and $R = 0.32$.

The absorption of radio-frequency radiation by nuclei such as ^{14}N which are subject to an electric field gradient is nuclear quadrupole resonance. As for zero field splitting in epr, no applied magnetic field is required. The order of magnitude of the magnetic and electric hyperfine interactions is 10^{-6} eV (10 mK). Some nuclei of interest for nuclear quadrupole resonance are included in Table 9.3.

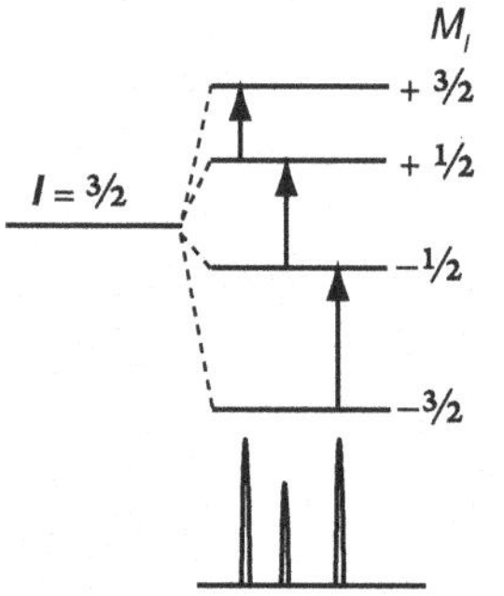

A nucleus with $I = \frac{3}{2}$ subject to magnetic and electric hyperfine interactions, and the corresponding hyperfine spectrum.

The Knight shift of a paramagnetic metal is related to the magnetically induced hyperfine field, which may be expressed in terms of the hyperfine coupling constant A and the susceptibility. The energy of the nucleus in an applied field B_0 is

$$\varepsilon = (-\gamma_n \hbar B_0 + \mathsf{A}\langle s_z \rangle) M_I. \qquad (9.38)$$

The first term is the Zeeman interaction of the nucleus with the applied field, and the second is the interaction with the electrons in the spin-polarized conduction band. The nuclear gyromagnetic ratio is $\gamma_n = g_n e/2m_p$. Now $M = ng\mu_B\langle s_z\rangle = \chi_P B_0/\mu_0$, where n_c is the conduction electron density, g is the electronic g-factor and χ_P is the Pauli susceptibility. Hence $\varepsilon = -\gamma_n \hbar B_0(1 + \mathcal{K})M_I$, where

$$\mathcal{K} = -\mathsf{A}\chi_P / n_c g \mu_B \gamma_n \hbar \qquad (9.39)$$

is the Knight shift. Some values of $\mathcal{K}$ are given in Table 9.4.

9.3.2 Relaxation

Consider the magnetization M^n of a system of nuclei: $M^n = n\langle \mathfrak{m}_n \rangle$, where n is the number of nuclei per unit volume. When the system is perturbed, it tends to return to its equilibrium state where M_0^n is along Oz with a characteristic relaxation time in a similar way to the magnetization of a system of electrons discussed in §9.1. However, the nuclei are weakly coupled to the lattice, so the longitudinal, spin-lattice relaxation time T_1 is much longer than it was for electrons: $M_z^n(t) = M_z^n(0) + [M_0^n - M_z^n(0)][1 - \exp(-t/T_1)]$. Thus

$$\frac{dM_z^n(t)}{dt} = \frac{M_0^n - M_z^n(t)}{T_1}. \qquad (9.40)$$

The torque acting on a nuclear moment $\mathfrak{m}_n$ is $\mathfrak{m}_n \times \boldsymbol{B}$, which is equal to the rate of change of angular momentum $d(\hbar I)/dt$. Hence $d\mathfrak{m}_n/dt = \gamma_n \mathfrak{m}_n \times \boldsymbol{B}$. The nuclear magnetization for an ensemble of nuclei M^n is $\langle \mathfrak{m}_n \rangle$. The z component of the equation of motion in the absence of irradiation is

$$\frac{dM_z^n}{dt} = \gamma_n(\boldsymbol{M}^n \times \boldsymbol{B})_z + \frac{M_0^n - M_z^n}{T_1}. \qquad (9.41)$$

In other words, the nuclear magnetization precesses around Oz, while relaxing towards the equilibrium value M_0^n. The values of T_1 for protons range from milliseconds, up to about 1 s in pure water.

Spin-lattice relaxation in metals mostly involves the conduction electrons. The inverse relaxation time $1/T_1$ is proportional to temperature, and to the Knight shift, a result known as the Korringa relation:

$$\frac{1}{T_1} = \mathcal{K}\left(\frac{\gamma_n}{\gamma}\right)^2 \frac{4\pi k_B T}{\hbar}. \qquad (9.42)$$

The equations of motion for the x and y components are different from (9.41). There is no nonzero equilibrium value for these components, and they decay with a relaxation time known as the transverse or spin–spin relaxation time T_2. Hence

$$\frac{dM_x^n}{dt} = \gamma_n(\boldsymbol{M}^n \times \boldsymbol{B})_x - \frac{M_x^n}{T_2},$$
$$\frac{dM_y^n}{dt} = \gamma_n(\boldsymbol{M}^n \times \boldsymbol{B})_y - \frac{M_y^n}{T_2}. \qquad (9.43)$$

T_2 is a measure of the time for which the moments contributing to $\boldsymbol{M}_x^n$ and $\boldsymbol{M}_y^n$ precess in phase with each other. It is the spin dephasing time, due to the fact that nuclear moments in different parts of the sample all experience slightly different magnetic fields, and therefore precess at different rates. If we consider that the local field fluctuations are due to the dipole fields of nearby nuclei, $H_{dip} \approx \mathfrak{m}_n/4\pi r^3$. Taking $r \approx 0.2$ nm and $\mathfrak{m}_n = \mu_N$ gives $\mu_0 H_{dip} \approx 60\,\mu$T. The dephasing time T_2 is the time taken to precess through a radian in the

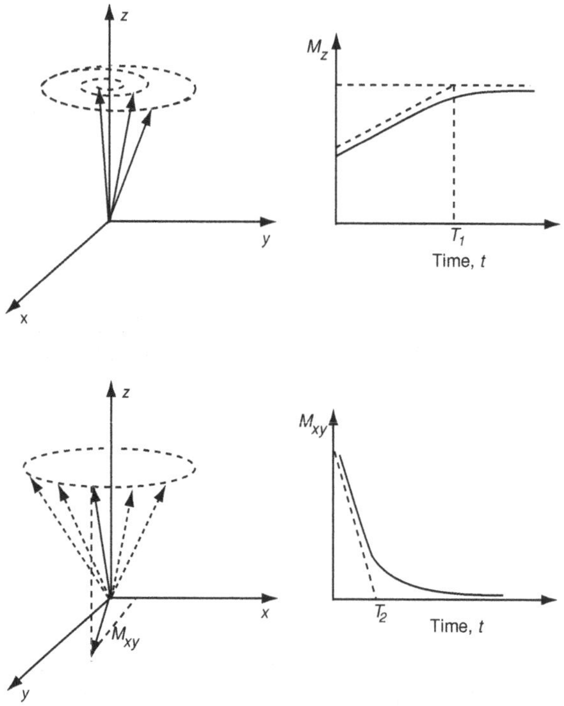

Figure 9.10

Illustration of the basic relaxation processes in NMR. The longitudinal relaxation time T_1 for alignment of the moments with the field direction is longer than the dephasing time T_2 of the in-plane magnetization due to slightly different precession frequencies of the different moments.

random field, at an angular frequency $\omega = \gamma_n \mu_0 H_{dip}$. Hence $T_2 \approx \hbar/\mu_0 \mu_N H_{dip} \approx 30\ \mu s$.

Inhomogeneity of the applied field B_0 may also contribute to T_2. The combined time constant T_2^* is given by

$$\frac{1}{T_2^*} = \frac{1}{T_2} + \frac{1}{T_2^{inho}}. \tag{9.44}$$

Unlike the relaxation of the longitudinal component M_z^n, there is no exchange of energy with the surroundings associated with the transverse components M_x^n and M_y^n. Equations (9.41) and (9.43) are known as the Bloch equations. They are phenomenological relations, first proposed by Felix Bloch in 1946. The longitudinal and transverse relaxation effects are illustrated in Fig. 9.10.

Adding in the AC field, of which we consider only the clockwise-rotating component $\boldsymbol{b}_1(t) = b_1(\mathbf{e}_x \cos \omega t + \mathbf{e}_y \sin \omega t)$ which can excite the resonance, Bloch's equations become

$$\frac{dM_x^n}{dt} = \gamma_n \left(M_y^n B_0 - M_z^n b_1 \sin \omega t \right) - \frac{M_x^n}{T_2}, \tag{9.45}$$

$$\frac{dM_y^n}{dt} = \gamma_n \left(M_z^n b_1 \cos \omega t - M_x^n B_0 \right) - \frac{M_y^n}{T_2},$$

$$\frac{dM_z^n}{dt} = \gamma_n b_1 \left(M_x^n \sin \omega t - M_y^n \cos \omega t \right) + \frac{M_0^n - M_z^n}{T_1}.$$

Rotating frame It is helpful to think of resonance experiments in a set of axes x', y', z' that rotate at an angular velocity $\boldsymbol{\Omega}$ relative to the laboratory frame x, y, z. The time derivatives of the unit vectors $\mathbf{e}_i$, $i = x, y, z$ in the rotating frame are $\mathrm{d}'\mathbf{e}_i/\mathrm{d}t = \boldsymbol{\Omega} \times \mathbf{e}_i$. The time derivative of a general vector $\boldsymbol{A}$ in the rotating frame is related to its derivative in the stationary frame by

$$\frac{\mathrm{d}'\boldsymbol{A}}{\mathrm{d}t} = \frac{\mathrm{d}\boldsymbol{A}}{\mathrm{d}t} + \boldsymbol{\Omega} \times \boldsymbol{A}.$$

We are interested in the magnetization $\boldsymbol{M}$, subject to a uniform field $\boldsymbol{B}_0$ along Oz. The z axis is common to the laboratory and rotating frames. Thus

$$\frac{\mathrm{d}'\boldsymbol{M}}{\mathrm{d}t} = \gamma_n \boldsymbol{M} \times \boldsymbol{B}_0 + \boldsymbol{\Omega} \times \boldsymbol{M} = \gamma_n \boldsymbol{M} \times \left(\boldsymbol{B}_0 - \frac{\boldsymbol{\Omega}}{\gamma_n}\right).$$

It is as if the magnetization in the rotating field were subject to an effective magnetic field $\boldsymbol{B}_0' = (\boldsymbol{B}_0 - \boldsymbol{\Omega}/\gamma_n)$. When $\Omega = \gamma_n B_0$ is the Larmor precession frequency ω_0, there is no effective field and the magnetization appears to be stationary in the rotating frame.

The in-plane rotating field $b_1(t)$ becomes a static field b_1 directed along Ox in a frame rotating with angular velocity ω. Rewriting the Bloch equations in the rotating frame with $z' = z$, we find

$$\begin{aligned}
\frac{\mathrm{d}M^n_{x'}}{\mathrm{d}t} &= \gamma_n M^n_y B_0' - \frac{M^n_{x'}}{T_2},\\
\frac{\mathrm{d}M^n_{y'}}{\mathrm{d}t} &= \gamma_n\left(M^n_{z'} b_1 - M^n_{x'} B_0'\right) - \frac{M^n_{y'}}{T_2},\\
\frac{\mathrm{d}M^n_{z'}}{\mathrm{d}t} &= -\gamma_n M^n_{y'} b_1 + \frac{M^n_0 - M^n_{z'}}{T_1},
\end{aligned} \tag{9.46}$$

where $B_0' = (\omega_0 - \omega)/\gamma_n$. These equations can be solved to give expressions for $M_{x'}$, $M_{y'}$, $M_{z'}$ in the steady state, when the time derivatives are zero. Furthermore, provided we stay in the limit of low excitation field b_1, far from saturation ($\gamma b_1 \ll T_1, T_2$), the resonance is independent of T_1:

$$\begin{aligned}
M^n_{x'} &= \frac{\gamma_n b_1'(\omega_0 - \omega)T_2^2}{1 + (\omega_0 - \omega)^2 T_2^2} M^n,\\
M^n_{y'} &= \frac{\gamma_n b_1 T_2}{1 + (\omega_0 - \omega)^2 T_2^2} M^n,\\
M^n_{z'} &= M^n,
\end{aligned} \tag{9.47}$$

where $M^n = \chi^n B_0/\mu_0$. The in-phase component of the magnetization $M_{x'}$ and the out-of-phase component $M_{y'}$ are plotted in Fig. 9.11.

Written in terms of susceptibility, the magnetization of the system subject to an in-plane oscillating field of amplitude $2b_1$ in the laboratory frame is

$$M_x = 2b_1(\chi' \cos \omega t + \chi'' \sin \omega t).$$

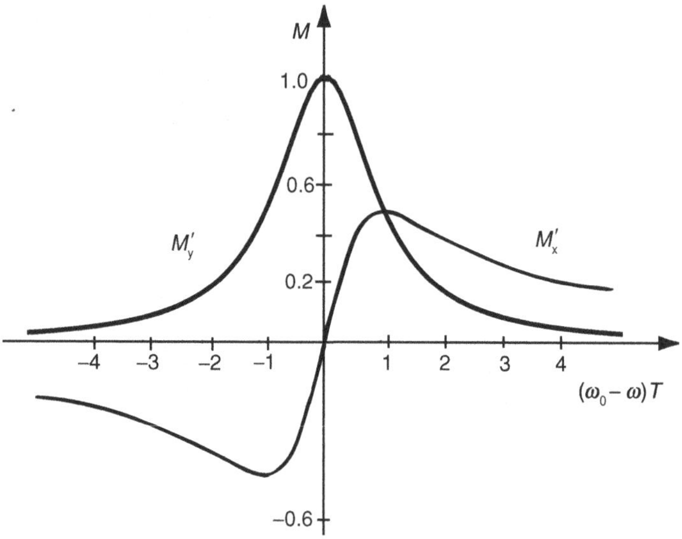

Figure 9.11

The Lorentzian lineshape of the resonance at the Larmor precession frequency ω_0. The in-phase component $M_{x'}$ is dispersive, while the out-of-phase component $M_{y'}$ is absorptive.

The components of the magnetization in the rotating frame $M_{x'}$ and $M_{y'}$ are then proportional to χ' and χ'', the components of the complex susceptibility $\chi = \chi' - i\chi''$. The power absorbed per unit volume of the system is $\mathcal{P} = \omega\chi'' b_1^2/\mu_0$.

The longitudinal relaxation time in the rotating frame is $T_{1\rho}$. It reflects molecular motion in macromolecules.

Equations (9.41) and (9.43) are also applicable to EPR and FMR where they are known as the Bloch–Bloembergen equations. T_1 and T_2 are then the spin-lattice and spin–spin relaxation times for the electronic system The resonance linewidth is given by $\Delta B = 2/(\gamma T_2)$: it is also related to the Gilbert damping parameter α by $\Delta B = 2\alpha\omega_0/\gamma$.

9.3.3 Pulsed NMR

Most modern NMR spectrometers apply the radio-frequency field in precisely timed bursts, rather than continuously. $b_1(t)$ is stationary in the rotating-axis frame at resonance, so it is possible to precess the magnetization of the sample M^n around Ox' through any desired angle in the $y'z'$-plane by choosing the length of the pulse. A π pulse, for example, reverses the magnetization, whereas a $\pi/2$ pulse causes precession through a quarter turn, and brings the magnetization into the xy-plane. The pulse lengths are generally shorter than the relaxation times T_1 and T_2.

A single $\pi/2$ pulse is especially useful, as it flips the magnetization from the z axis to the xy-plane, where it continues to precess at the Larmor frequency; Larmor precession does not depend on the angle between the magnetization and the field $\boldsymbol{B}_0$. As the magnetization precesses, it induces a radio-frequency signal

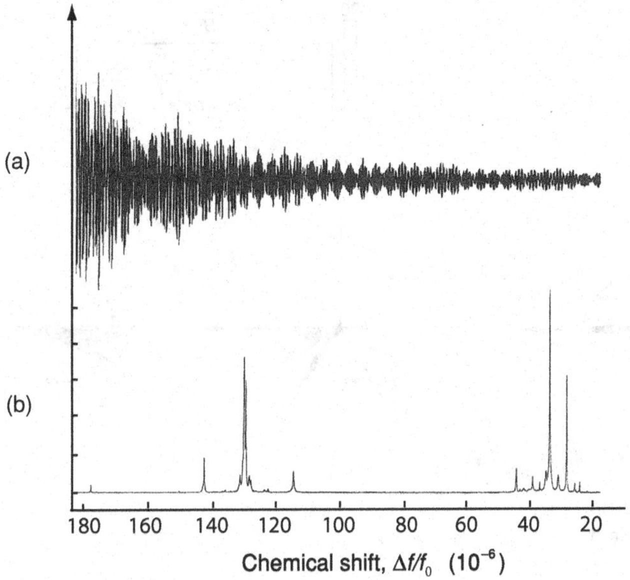

Figure 9.12

(a) A free induction decay for a sample with protons present in several different environments. (b) The Fourier transform shows the spectrum of component frequencies in the free induction decay. The horizontal axis shows the chemical shift of the resonance frequency in parts per million. (Data courtesy of V. J. McBrierty.)

in the coil that was used to create the pulse, which is damped by the progressive dephasing of the contributions of individual nuclei which see slightly different magnetic fields. This signal is known as the free induction decay, Fig. 9.12. It allows T_2 to be measured directly. A fast Fourier transform of the free induction decay gives the frequency spectrum, and the chemical shifts of the constituent nuclei can be identified.

Spin echo The benefits of high resolution might seem to depend on our ability to build a magnet capable of producing a perfectly uniform magnetic field B_0 over the whole volume of the sample. Luckily, that is not true, thanks to an ingeneous pulse sequence invented by Erwin Hahn in 1950. The spin-echo method uses two pulses. First a $\pi/2$ pulse switches the magnetization into the plane along Oy and a free induction decay is measured which reflects both the inherent field fluctuations in the sample, as well as any inhomogeneities in field produced by the magnet. Then after a time τ a π pulse is applied, which flips the spins around Ox and reverses their order as they continue to precess, as shown in Fig. 9.13. The spins which were precessing faster now find themselves lagging behind their slower counterparts, but they catch up the delay, and all spins find themselves coming together along the y-direction after a time 2τ, producing the 'spin echo'.

T_1 can be measured by first applying a π pulse to flip the magnetization, and then using a $\pi/2$ pulse after a variable time delay τ to determine the magnitude of the magnetization measured in the free induction decay $M(\tau)$. A curious feature of the nonequilibrium population distributions that can be achieved with spin systems is spin temperature. This is the fictitious temperature T^* at which the population distribution measured at some instant would be in

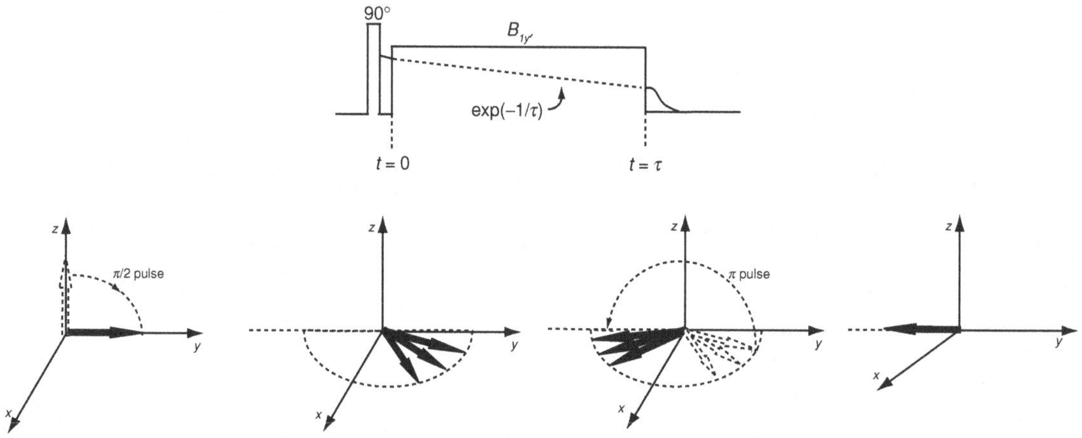

Figure 9.13

The spin-echo measurement.

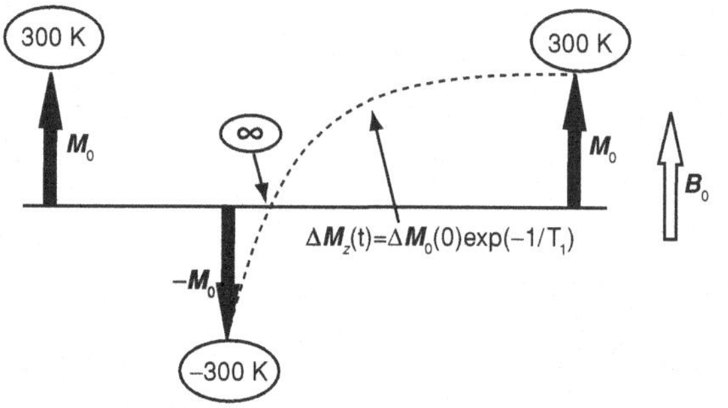

Figure 9.14

Recovery of the magnetization in the direction of the uniform field B_0 following a π pulse at room temperature. Some spin temperatures are circled.

equilibrium. The population of the M_I sublevels is described by a Boltzmann distribution with the temperature equal to T^* rather than T. The spins can equilibrate with each other at a temperature quite different from the lattice temperature when $T_2 \ll T_1$. After inverting the magnetiztion with a π pulse at 300 K, for example, the spin temperature flips to -300 K. It then becomes increasingly negative, diverging to $-\infty$ as the moment crosses the xy-plane, and it eventually reaches 300 K again when $t \gg T_1$, as shown on Fig. 9.14.

Spin locking of the moments in the rotating frame is achieved with a $\pi/2$ pulse along Ox', followed by a $\pi/2$ phase shift of the radio-frequency field, so that it lies along Oy'. This aligns M_0 and b_1 in the rotating frame, allowing measurement of $T_{1\rho}$. The tendency for the magnetization to dephase is suppressed, and the large population imbalance in relation to the field produces a low effective spin temperature $T^* = T(b_1/b_0)$. For example, if $B_0 = 1$ T and $b_1 = 1$ mT, the spin temperature is 300 mK.

Many other pulse sequences devised by practitioners of NMR are described in specialist texts.

Finally, we return to the concept of motional narrowing. When some component of the hyperfine field acting on a nucleus is fluctuating because of temperature or diffusive motion, motional averaging affects the spectrum provided the fluctuation frequency is comparable to or greater than the spectral width arising from the interactions. For magnetic hyperfine interactions, the fluctuation frequency is the Larmor precession frequency. We have seen how paramagnetic fluctuations of the magnetization are averaged out when their frequency exceeds the Larmor frequency in the hyperfine field. Similarly, line broadening associated with dipole fields created by neighbouring nuclei in liquids, for example, averages out when the diffusion frequency exceeds the frequency associated with the dipole field. This permits the observation of very narrow resonance lines in liquids, with relative linewidths of 10^{-8}. A similar effect is achieved in solids by the technique of magic angle spinning, where anisotropic interactions such as dipole coupling and quadrupole coupling which vary as $(3\cos^2\theta - 1)$ are averaged out by spinning the sample around an axis inclined at an angle of $\cos^{-1}(1/\sqrt{3}) = 54.7°$ with Oz.

9.4 Other methods

9.4.1 Mössbauer effect

Mössbauer spectroscopy is based on the 1958 discovery by Rudolf Mössbauer that it is possible for a nucleus of ^{191}Ir to decay without recoil from its first excited state to its ground state by γ-ray emission, whenever the nucleus belongs to an atom bound in a solid. Momentum has to be conserved *on average*, but lattice momentum must be taken up by the creation of phonons. In quantum mechanics, zero-phonon, one-phonon, two-phonon, ..., events all have finite probabilities. The recoilless fraction f_M is the fraction of decays where γ-rays are emitted without recoil in a zero-phonon process:

$$f_M = e^{(-\varepsilon_\gamma \langle x^2 \rangle / \hbar^2)}, \tag{9.48}$$

where $\langle x^2 \rangle$ is the mean-square thermal displacement of the nucleus and ε_γ is the energy of the γ-ray. A similar *Debye–Waller factor* governs the intensity of elastic X-ray and neutron scattering in solids. The Mössbauer effect is simply due to these zero-phonon events having a finite probability.

Mössbauer spectroscopy is based on the γ-rays emitted or absorbed in zero-phonon processes in transitions between the nuclear excited state and the ground state. Like optical spectroscopy, it requires a source, an absorber and a means of modulating the emitted photon energy. The source is a solid containing a radioactive isotope which decays with a half-life $\tau_{1/2}$ to populate a low-lying nuclear excited state, which in turn decays rapidly by emitting an unsplit gamma line. The conditions are stringent. The excited state must be at no more

Table 9.5. Some suitable isotopes for Mössbauer spectroscopy

Isotope	Abundance (%)	source	$\tau_{\frac{1}{2}}$	$t_{\frac{1}{2}\gamma}$ (keV)	$t_{\frac{1}{2}}$ (ns)	g_n	I^{parity}	Q	g_e	I_e^{parity}	Q_e
^{57}Fe	2.12	^{57}Co (ec)	270d	14.4	98.0	0.0906	$1/2^-$	–	−0.155	$3/2^-$	0.16
^{61}Ni	1.2	^{61}Co (β^-)	99m	67.4	5.3	−0.750	$3/2^-$	0.162	0.47	$5/2^-$	−0.30
^{119}Sn	8.6	^{119m}Sn (IT)	293d	23.9	18.0	3.359	$1/2^+$	–	2.35	$3/2^+$	−0.13
^{149}Sm	13.8	^{149}Eu (ec)	90d	22.5	7.3	−0.672	$7/2^-$	0.075	−0.620	$5/2^-$	0.40
^{151}Eu	47.8	^{151}Sm (β^-)	90y	21.6	9.6	3.465	$5/2^+$	0.903	2.587	$7/2^+$	1.51
^{155}Gd	14.8	^{155}Eu (β^-)	4.8y	86.5	6.5	−0.259	$3/2^-$	1.270	−0.515	$5/2^+$	0.16
^{161}Dy	18.9	^{161}Tb (β^-)	6.9d	25.7	29.1	−0.481	$5/2^-$	2.507	0.59	$5/2^+$	1.36

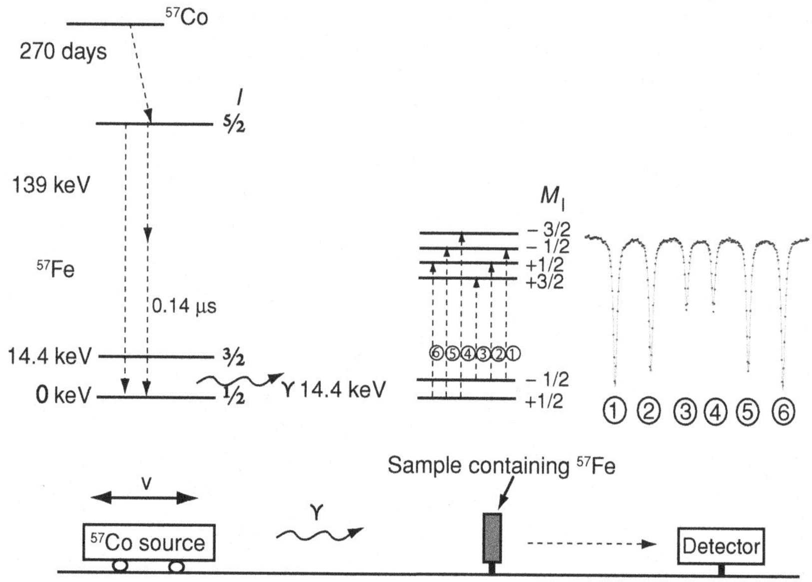

Figure 9.15

Nuclear decay scheme of ^{57}Co which undergoes (n,γ) decay, populating the 14.1 keV excited state of ^{57}Fe. These γ-rays are resonantly absorbed in an absorber containing ^{57}Fe. Hyperfine structure in the absorption spectrum is revealed by modulating the energy of the source by moving it with constant acceleration with an electromagnetic transducer. A six-level hyperfine spectrum of a ferromagnetic αFe absorber is illustrated.

than about 30 keV, or else the recoilless fraction at room temperature becomes vanishingly small. Nature must provide a convenient radioactive precursor, if we are to avoid recourse to a synchrotron source. Some useful nuclei, and their radioactive precursors are listed in Table 9.5. Luckily for magnetism, the best example is an iron isotope, ^{57}Fe. Details are featured in Fig. 9.15. The source is ^{57}Co, which has a half-life $\tau_{1/2}$ of 270 days, and the energy ε_γ of the transition from the excited state to the ground state is 14.4 keV. Mössbauer spectroscopy is usually measured in transmission; a single-line source is energy-modulated by moving it with velocity v ($\approx$ cm s^{-1}) so that it undergoes a Doppler shift $\Delta\varepsilon = \varepsilon_\gamma v/c$. The absorption of γ-rays is measured as a function of velocity. The absorption linewidth determined by the lifetime $t_{1/2}$ of the nuclear excited state is 0.19 mm s^{-1} for the 98 ns I =3/2 excited state of ^{57}Fe.

Both magnetic hyperfine splitting and electric quadrupole interactions are observed, for ^{57}Fe on account of the quadrupole moment of the excited state. The corresponding energy splittings are shown in Fig. 9.15. In addition, there is another interaction, unmeasurable in NMR, known as the isomer shift. It arises because the nucleus is a slightly different size in its excited state, and there is a shift of the resonance line due to the Coulomb interaction, depending on the difference in electron density at the nucleus in the source and absorber. Different absorbers have different isomer shifts δ_{IS} relative to a source, and it is possible to infer the charge state of the absorber ion, for example, Fe^{2+} or Fe^{3+}. Further details are provided in §10.2.3.

9.4.2 Muon spin rotation

Muons are unstable spin-$\frac{1}{2}$ particles with charge $\pm e$ which have a mass $m_\mu = 206.7\ m_e$. Their half-life is $\tau_\mu = 2.2\ \mu s$. The positive muon is quite useful as a probe of magnetic solids because it occupies an interstitial site where it experiences the local magnetic field. Negative muons are like heavy electrons and they bind closely to atomic nuclei, which is useful for cold fusion. Muon beams are produced in accelerators, where high-energy protons collide with a target producing pions, which decay into muons 26 ns later:

$$\pi^+ \longrightarrow \mu^+ + \nu_\mu.$$

Remarkably, the muons produced are fully spin polarized; this is because the pion has no spin and the muon neutrino ν_μ has its spin antiparallel to its momentum and so the muon also has its spin antiparallel to its momentum.

The muons are created with energy in the MeV range, but they are rapidly thermalized on entering a solid specimen without loss of spin polarization. The final resting place of the muon in the sample is an interstitial site far from the track of radiation damage produced in the early stages of thermalization. After a time t there is a probability proportional to $1 - e^{-t/\tau_\mu}$ that the muon will have decayed into a positron and two neutrinos:

$$\mu^+ \longrightarrow e^+ + \nu_e + \bar{\nu}_\mu.$$

The positron emerges in a direction related to the spin direction of the parent muon. If a magnetic field is applied in the perpendicular direction, the muon precesses around this field at its Larmor frequency of 135 MHz T^{-1} before decaying to emit the positron.

There might only be a single muon in the sample at a given instant, but by averaging over many events, curves showing the intensity oscillations of the positron flux in fixed detectors can be recorded and the precession frequency of the muon is determined. The muon precesses equally well around the internal magnetic field present at its interstitial site and the main use of muon spin rotation (μSR) in solids is to study these local fields, which are in the range

10^{-4}–1 T. The spin depolarization of the muon can be used to probe spin dynamics both above and below the Curie temperature of a ferromagnet.

FURTHER READING

A. Guimares, *Magnetism and Magnetic Resonance in Solids*, New York: Wiley-Interscience (1998). An introduction to magnetic resonance in solids.

C. P. Schlichter, *Principles of Magnetic Resonance* (third edition), Berlin: Springer (1990).

A. Abragam, *Principles of Nuclear Magnetism*, Oxford: Oxford University Press (1961). The definitive text.

A. Abragam and B. Bleany, *Electron Paramagnetic Resonance of Transition Ions*, Oxford: Oxford University Press (1970).

N. N. Greenwood and T. C. Gibb, *Mössbauer Spectroscopy*, London: Chapman and Hall (1971). An account of the principles, followed by data on many materials for all the main resonances.

EXERCISES

9.1 Sketch the EPR spectrum of an Fe^{3+} ion ($S = \frac{5}{2}$) in a site with crystal field parameter $\mathsf{D} = -0.05$ K. Assume a microwave frequency of 9.0 GHz.

9.2 Why does the entire, instantaneous hyperfine field split the electronic energy levels in an EPR experiment, while only its thermal average is effective for splitting the nuclear energy levels in an NMR experiment?

9.3 Estimate the Walker breakdown field for a single-crystal film of $Nd_2Fe_{14}B$, which has its c axis perpendicular to the plane of the film. Assume a damping factor $\alpha = 0.1$.

9.4 Sketch the Mössbauer spectrum for ^{155}Gd at 4K. The hyperfine field is 35 T.

9.5 Discuss the possibility of measuring the gravitational acceleration of photons using the Mössbauer effect.

10 Experimental methods

Why think? Why not try the experiment?
John Hunter 1728–1793

Central to most magnetic measurements is the generation and detection of magnetic fields. Atomic-scale magnetic structure is best probed by neutron diffraction, while other atomic-scale element-specific information is provided by spectroscopic methods. Domain-scale magnetization measurements are made by magneto-optic methods or magnetic force microscopy, whereas macroscopic measurements of magnetization are made in open or closed circuit by a variety of methods. Spin-wave and other excitations are best explored by inelastic neutron scattering. Numerical methods of investigation are of growing importance for understanding the static and dynamic behaviour of real magnetic materials and magnetic systems.

Magnetism is an experimental science. Experiments serve to inspire and refine physical theory, besides providing all the quantitative information on which the applications depend. The traditional image of apparatus on a laboratory bench does not tell the whole story; some magnetic measurements are now conducted at national or multinational institutes built around large-scale facilities for generating high magnetic fields, neutron beams or intense streams of synchrotron radiation. Computers have evolved in the opposite sense, from central facilities to benchtop instruments for data acquisition, display and modelling. Numerical computations and simulations may be regarded as an experimental tool for investigating a model reality, where complex magnetic behaviour at the atomic, micromagnetic or system level can be investigated with the aid of a computer workstation.

10.1 Materials growth

Materials are the foundation of experimental magnetism; practical applications depend on the nature and form of the material, whether it is a few micrograms of permalloy in a thin-film sensor or a ton of $Nd_2Fe_{14}B$ in a large permanent magnet for resonance imaging. Although naturally occuring magnetic materials are always interesting, the great majority of specimens examined in the laboratory are synthetic, and they are often products of a complex fabrication process. It is appropriate to provide a brief account of how magnetic materials are made.

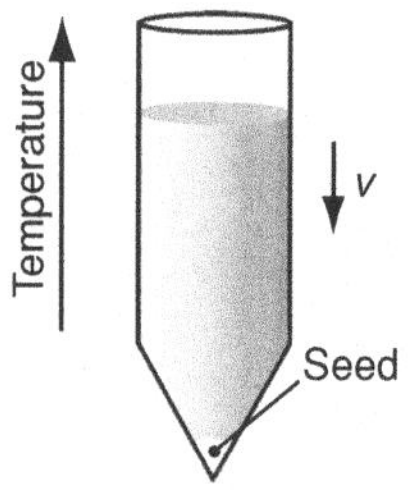

Bridgeman method: the crucible is lowered slowly through a thermal gradient spanning the melting point. A nucleus forms at the cool pointed tip of the crucible and grows to consume the melt.

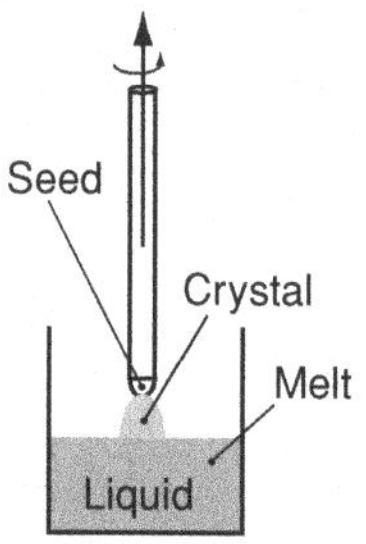

The Czochralski method: a seed is dipped into a slightly supercooled liquid, and slowly rotated while it is extracted to pull a crystal from the melt.

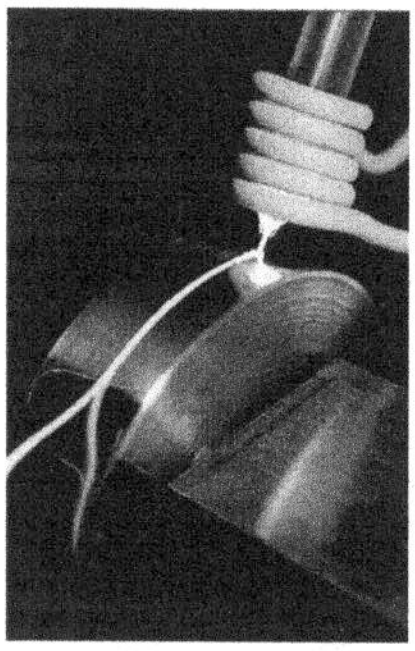

Melt spinning: a metallic glass is produced by quenching the melt on a rapidly rotating copper wheel.

10.1.1 Bulk material

Rather different approaches are taken for metallic and nonmetallic materials. Metallic alloys and intermetallic compounds are usually melted first in an arc furnace, where a DC argon arc is struck between a tungsten or graphite electrode and the sample, which lies in a hollow in a water-cooled conducting hearth, or in a radio-frequency induction furnace, where the sample is placed in a few-turns, water-cooled coil and heated by the eddy currents producd by some kilowatts of 150 kHz power. Subsequent heat treatments to produce atomically ordered or phase-segregated microstructures are usually carried out in a resistance furnace, where the temperature and atmosphere can be precisely controlled. When single-crystal samples are required, they may be grown from a seed crystallite in the melt, using the Bridgeman or Czochralzki methods. Crystallization from the melt is a two-step process: initially one or more tiny crystalline nuclei form as a result of a random fluctuation of the atomic positions in the supercooled liquid, then the nuclei grow at a rate dependent on the degree of undercooling and quickly consume the melt.

Amorphous metals demand a different approach. A multicomponent melt is rapidly quenched on a spinning copper wheel, for example, which leaves little time for the nuclei to grow. The surface velocity is of order 50 m s^{-1}. Melt spinning works well at a deep eutectic in the compositional phase diagram, where the melt can be quenched almost instantaneously to a temperature below the glass transition, which is the point where long-range diffusive motion of the atoms is frozen out. An amorphous metal produced in this way is known as a metallic glass. Mechanical alloying of constituent elements in a high-energy ball mill is an alternative means of producing a highly disordered bulk metal.

Nonmetals, especially oxides, are freqently prepared by ceramic methods. Mixtures of powders of appropriate precursors with the correct cation ratio, for example Fe_2O_3 and CoO to make $CoFe_2O_4$, are repeatedly ground and sintered to achieve dense, uniform material by solid-state diffusion. Precursors like carbonates or acetates which have a low decomposition temperature can be used to produce finely divided oxide as a first stage, or else solid solutions can be formed directly as precipitates (gels) from ionic solution.

Ceramics are usually refractory, with high melting points. Crystals may be grown from the melt in an image furnace where infrared radiation is focussed onto a small section of a sintered polycrystalline rod by two parabolic mirrors and the molten zone moves along the rod. Other crystal growth methods include chemical vapour transport, and the flux method, where a mixture of oxides is melted with a flux such as PbF_2 which has no solid solubility in the required crystal. On slow cooling, oxide crystals nucleate and grow throughout the melt. They are extracted by dissolving the flux.

Single crystals are indispensable for complete characterization of the anisotropic magnetic properties. Certain techniques like neutron diffraction or measurements of elastic constants require quite large crystals,

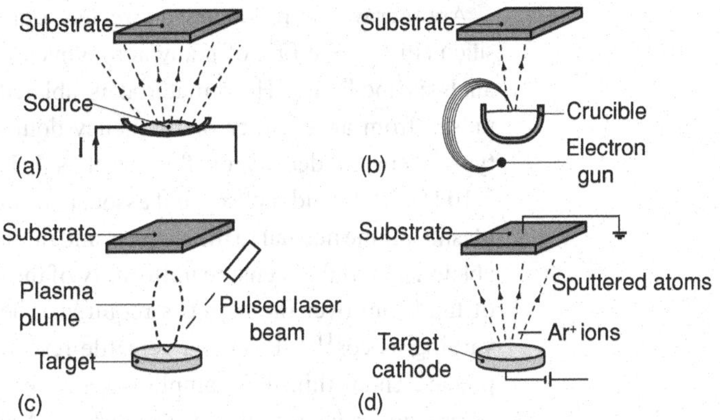

Figure 10.1

Methods for preparing thin films by a vapour or plasma condensing on a substrate: (a) thermal evaporation, (b) e-beam evaporation, (c) pulsed-laser deposition, (d) sputtering.

$(1\ \text{mm})^3$–$(10\ \text{mm})^3$. Crystal growth is something of an art; its practitioners are star supporting actors in the author lists of numerous publications.

10.1.2 Thin films

Magnetic thin films are commonly prepared by a physical vapour deposition method, where the source of material is separated by a distance d_{ss} from the substrate on which it is deposited, which is often heated in the range 400–1000°C to facilitate the growth process. The deposition chamber is evacuated to a pressure P. At low pressure, the atomic species from the source arrive at the substrate without collision, but at higher pressure they are thermalized by collision with gas atoms in the chamber. The mean free path of the atomic species in thermal equilibrium is determined by the pressure. The kinetic theory of gases leads to a handy numerical relation $\lambda \simeq 6/P$ at room temperature, where λ is the mean free path of the atom in millimetres, and P is the pressure in pascals.[1]

Some preparation methods are summarized in Fig. 10.1. The simplest is thermal evaporation from a source of molten metal in a resistively heated boat. The method is restricted to materials with moderately low melting points to avoid contamination by the boat, usually made of graphite, molybdenum or tungsten. Electron-beam evaporation overcomes this problem by using a focussed beam of energetic electrons, typically 10 kV, 10 mA, to melt a small pool in a block of source material which is held in a water-cooled crucible. Contamination is avoided because the molten pool is effectively contained in a solid crucible of the same material. For alloys and multilayers, multiple sources with individual power supplies are required. The source–substrate distance is usually at least 300 mm.

[1] 1 pascal (abbreviation Pa) is 1 N m^{-2}. In vacuum systems, the unit mbar is often used (1 mbar = 100 Pa). An older unit based on the density of Hg is the torr (1 torr = 136 Pa = 1.36 m bar).

Another way of producing a directed source of material is pulsed-laser deposition (PLD – the first of many acronyms attached to thin-film production and analysis methods). Here material is ablated from the target by nanosecond pulses from an excimer or frequency-doubled Nd-YAG[2] ultraviolet laser. A typical energy density on the target is 1 J cm^{-2}, and the repetition rate is $\sim$10 Hz. The rapid melting and associated shock wave creates a highly directed plasma plume normal to the surface, and the substrate is positioned to collect the ablated material. To ensure uniformity of the deposit, some rotation or rastering of the beam over the target is required. The deposition is highly directional, varying as $\cos^{11}\theta$. PLD is a versatile research method, equally well suited to prepare small thin-film samples of a range of different metals or insulators. Deposition rates are of order 1 nm s^{-1}. A drawback of PLD is the tendency for micrometre-size droplets ejected from the target to litter the growing film. These can be controlled by working at an energy density close to the ablation threshold, by using fully dense targets and by setting the substrate parallel to the plasma plume in an off-axis deposition geometry. A pulsed electron source is an alternative to the pulsed laser.

High-quality epitaxial films need to be grown very slowly, monolayer by monolayer. Deposition rates are less than 1 nm s^{-1}. This demands ultrahigh vacuum (UHV), in the range 10^{-7}–10^{-9} Pa, to avoid contamination of the growing film by residual gas in the chamber. The time taken for a monolayer to accumulate can be estimated from the kinetic theory of gases. The density of molecules is P/k_BT, where P is their partial pressure. The root-mean-square velocity is given by $\langle v\rangle = (3k_BT/\mathrm{m})^{1/2}$, where m is mass of the molecule. Those reaching the substrate of area $\mathcal{A}$ within a time δt are contained in a volume $\langle v\rangle\mathcal{A}\delta t$. Of these, $\frac{1}{6}$ are moving in the right direction, towards the substrate, so the time taken for a monolayer of a species with lattice parameter a to form is

$$\delta t = (12k_BT\mathrm{m})^{1/2}/Pa^2. \tag{10.1}$$

Oxygen, for example, has $a \simeq 0.2$ nm, and the time taken to collect a monolayer at 10^{-5} Pa is about a minute, hence the need for UHV to avoid any trace of contamination.

The well-controlled growth of epitaxial films in UHV, $P < 10^{-8}$ Pa, is known as molecular-beam epitaxy (MBE). The pressure of the evaporating species is 10^{-6}–10^{-4} Pa, so the mean free path of the atoms emitted from the evaporation source, given by the $(6/P)$ relation, is much greater than the size of the chamber; they travel without being scattered to the substrate where they are rapidly thermalized at the substrate temperature. High temperature promotes the mobility of the adatoms on the surface and favours the layer-by-layer growth needed for epitaxy (Franck–van der Merwe growth). At lower substrate temperatures,

[2] YAG is yttrium aluminium garnet, the nonmagnetic analogue of YIG, yttrium iron garnet discussed in §11.6.6.

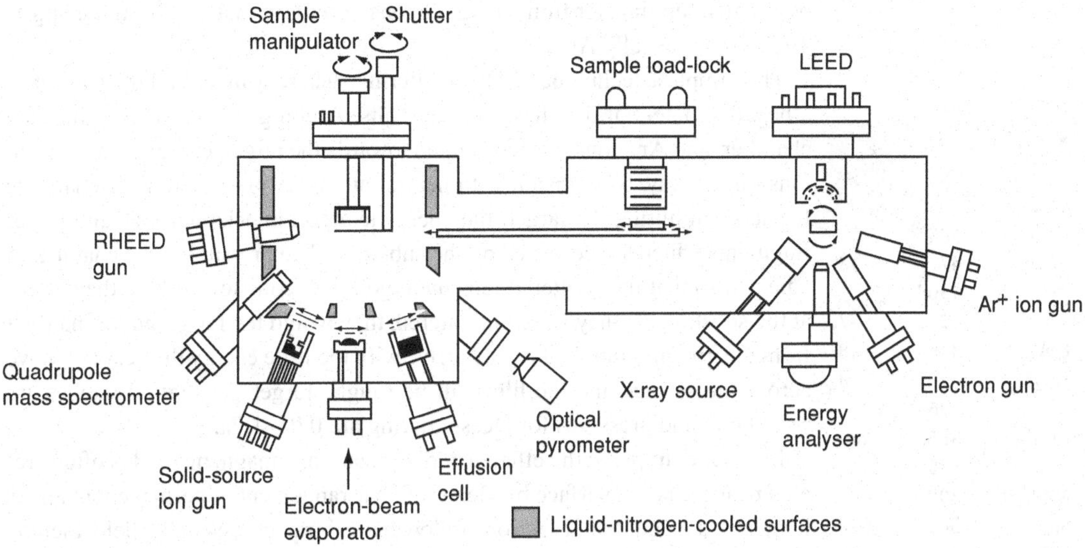

Figure 10.2

A typical laboratory MBE system.

an island-like growth mode (Volmer–Weber growth) is commonly found. An intermediate mode starts with a continuous monolayer, followed by island-like growth (Stransky–Krastanov growth).

MBE is the method of choice for growing the highest-quality, defect-free semiconductor or metal films. Thermal or electron-beam evaporation sources may be used, or else special effusion cells which create a molecular beam by breaking down gaseous precursors – trimethyl gallium and arsine are used for GaAs, for example. MBE is a tool for depositing magnetic materials found in research laboratories, rather than an industrial production method. A typical research system (see Fig. 10.2) is equipped to monitor the growth and to analyse the films without removing them from the vacuum. Techniques include reflection high-energy electron diffraction (RHEED) to monitor the growth mode and lattice parameter of the growing film, low-energy electron diffraction (LEED) to show the crystallographic structure of the growth planes and Auger electron spectroscopy (AES) to provide chemical analysis of the top nanometre or two of the film. Facilities for point-probe analysis by techniques such as atomic-force microscopy (AFM) or scanning tunelling microscopy (STM) and high-energy electron spectroscopy for chemical analysis (ESCA) may be built into the vacuum chamber. Other forms of analysis such as transmission-electron microscopy (TEM) for atomic imaging of cross sections, wide-angle X-ray scattering (XRD) for precise structural information and transport or magnetic measurements generally have to be conducted after the film has been removed from the vacuum chamber. It is often protected by a thin capping layer.

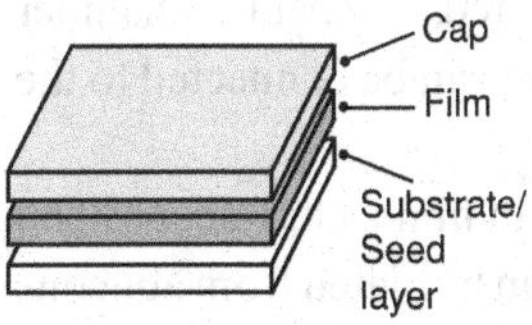

A magnetic thin film stack.

The most widely used method for making magnetic thin films is sputtering. Unlike evaporation methods, sputtering involves nonthermal transport based on momentum transfer from energetic ions to remove material from the target. The ions are usually Ar^+.

Substrate lattice parameters (a,c) (pm) and susceptibility (10^{-6})		
MgO	420	−11
Al_2O_3	477, 1304	−18
$SrTiO_3$	391	−7
$LaAlO_3$	379	−18
TiO_2	459, 296	4
$NdGaO_3$	386, 385	2100
$MgAl_2O_4$	808	−15
Si	543	−4
SiO_2	490, 539	−18
YSZ	513	−8

The simplest technique is DC sputtering, where a metallic target is set at a negative potential of a few hundred volts. Argon gas is introduced into the chamber, and Ar^+ ions are accelerated towards the target, creating more argon ions on the way, by collision with neutral atoms. A plasma glow discharge is formed. On hitting the target, the energetic argon knocks out ions and larger fragments which are collected on the substrate. The pressure in the chamber is adjusted so that the sputtered ions undergo a few collisions before they arrive at the substrate. If they were unscattered, they would tend to resputter material from the growing film, but if they arrive with too little energy they cannot move across the surface and the film will be rough. Target–substrate distances are ~ 100 mm and pressures for DC sputtering are 0.05–1 Pa.

In order to improve the efficiency of ionization, a magnetic field is often created near the target surface by means of an arrangement of permanent magnets known as a magnetron. Electrons follow helical trajectories in the field, thereby increasing their probability of collision with the argon. Enhanced growth rates of order 10 nm s^{-1} are thereby achieved. If the target is ferromagnetic it must be thin enough to be easily saturated by the flux available from the magnetron, so that a stray field will still be present near its surface. Reactive sputtering is carried out by mixing a gas such as O_2 or N_2 with the Ar in order to produce thin films of oxides or nitrides from metal targets. More control over the sputtering process is achieved by ion-beam deposition, where an ionized argon beam is generated separately, and focussed onto the target.

To make oxide or other insulating films directly, the method of radio-frequency sputtering is employed. The power supplies commonly operate at 13.56 MHz. For part of the cycle, Ar ions bombard the target; for the rest of the cycle, electrons neutralize the build-up of positive charge. Electrons also ionize the argon to create the plasma. An argon pressure of 0.02 Pa is sufficient to maintain a radio-frequency discharge.

Sputtering is a well-controlled method of thin-film growth. By fixing the gas flow rate, power and substrate bias it is possible to reproduce the growth conditions. The method is as well suited for industrial production as it is for laboratory research. Large targets or planetary motion of the substrate are required to achieve uniform deposits over wafers 150–300 mm in diameter. A typical system for making metallic thin-film stacks has six different metal targets, automatic wafer handling and the possibility of in-situ sputter cleaning of the substrates (Fig. 10.3). If oxide layers are required, a separate chamber designed for radio-frequency or reactive DC sputtering can be connected to the metals chamber.

Of the chemical methods of producing magnetic thin films, electrodeposition is widely used. It is discused in §15.2. Metallic films can be plated from aqueous solutions of metal ions which are not too electronegative. Relatively thick films of permalloy ($Ni_{78}Fe_{22}$) can be prepared for magnetic shields, but monolayers

Figure 10.3

A six-target sputtering tool with automatic wafer loading.

can also be grown, or removed by reversing the voltage. A current density of 1 μA mm^{-2} deposits a monolayer in about 5 s; 1 mA mm^{-2} gives a deposition rate of about 50 nm s^{-1}.

Other chemical methods use vapours of organometallic precursors, which are introduced into a chamber where they are decomposed thermally, or with ultraviolet light. These methods are variants of chemical vapour deposition (CVD).

Once the magnetic thin films or thin-film stacks have been produced, they normally have to be patterned into small device structures. Lithographic techniques have been borrowed from the semiconductor industry. Optical lithography, with ultraviolet light is good for producing structures larger than about 0.5 µm in the laboratory, although much smaller structures can be produced by this method in industry. Electron-beam lithography, often in a scanning electron microscope, is used to prepare submicrometre structures down to about 30 nm. Both techniques involve transferring the pattern to a layer of polymer resist and using ion-milling or lift-off in a solvent to produce the structure. It is important to make uniform magnetic structures with smooth edges to ensure controllable switching and avoid unwanted nucleation or pinning centres. When special one-off structures are required, a focussed ion beam, usually of Ga^+ ions, can be used to mill structures down to about 10 nm in size. The centre of gravity of magnetic research is shifting from bulk materials to microscopic thin-film devices. Methods of preparing them are becoming more widespread in universities and research centres world-wide.

Nanoparticles are usually produced by wet chemical methods. Particle size can be controlled from a few nanometres up to many micrometres, spanning the superparamagnetic, single-domain and multidomain regions. Monodisperse few-atom clusters are best created in UHV, with mass selection by a mass spectrometer.

Special processes have been optimized over the years to create specific magnetic self-assembled nanostructures. Good examples are alnico magnets, Sm_2Co_{17} magnets and thin-film recording media.

10.2 Magnetic fields

10.2.1 Generation

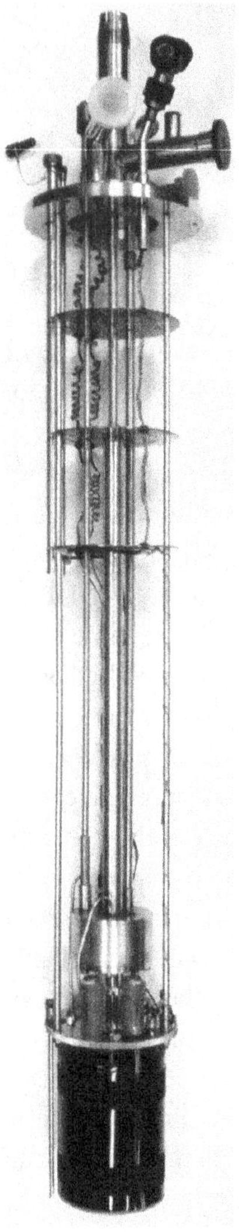

6 MA m^{-1} superconducting solenoid with variable-temperature insert in the bore for immersion in a liquid helium bath.

All methods of generating static or low-frequency magnetic fields are based on either electric currents or permanent magnets. Soft iron can be used to concentrate or direct the flux. Some schemes for producing uniform magnetic fields were introduced in Chapter 2. Here we discuss the practicalities. Magnetic fields used in measurements on ferromagnetic materials must overcome the demagnetizing field and comfortably exceed the coercivity, which can be 1 MA m^{-1} or more in a rare-earth magnet. The field needed to magnetize a hard material to saturation is typically three times the coercivity. Larger fields, of order 10 MA m^{-1}, may be required to study high-field magnetization processes along the hard axes and to determine anisotropy constants from magnetization curves.

The principle of generating magnetic fields from electric currents is the Biot–Savart law (2.5), which gives the field due to a current element $I\,dl$. For a single-turn coil of radius a carrying current I, the field on the axis at a distance z from the centre is therefore $H = a^2I/2(a^2 + z^2)^{3/2}$. When $z \gg a$, the coil is equivalent to a dipole of moment $\mathfrak{m} = \pi a^2 I$. Integration gives the field on the axis of short solenoid:

$$H = nI(\cos\theta_1 - \cos\theta_2)/2, \tag{10.2}$$

where θ_1 and θ_2 are the angles subtended at each end and n is the number of turns per unit length. For a long solenoid, $\theta_1 = \pi$, $\theta_2 = -\pi$, and we recover (2.20) $H = nI$.

The units of H, A m^{-1}, carry a sense of what is needed to generate the field. The free-space flux density B_0 corresponding to 1 MA m^{-1} is $10^6\mu_0 =$ 1.26 T. Air-cored resistive solenoids can produce continuous fields of up to 100 kA m^{-1} without cooling. Much larger fields are available from superconducting solenoids cooled with liquid helium or by a closed-cycle refrigerator. The solenoids are wound with multifilamentary NbTi or Nb_3Ge wire and the maximum fields of 10–15 MA m^{-1} are limited by the critical current of the type II superconductor. Higher continuous fields are available from Bitter magnets made from perforated copper pancake segments which constitute a large helical coil with $n \approx 10^3$ m^{-1}, cooled by a continuous flow of water at high pressure. These magnets exist in only a few special institutes such as the

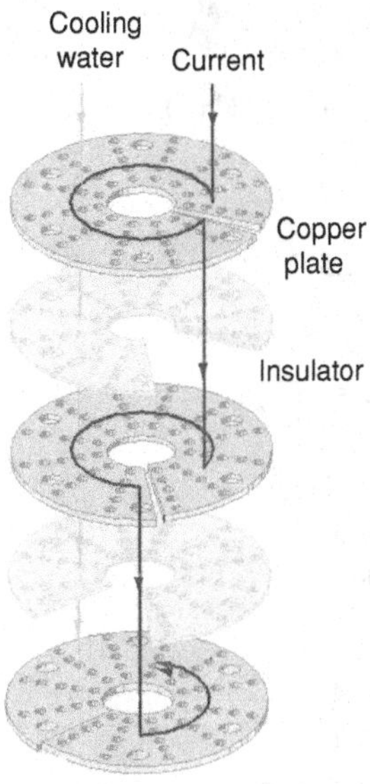

A segment of the coil for a water-cooled Bitter magnet.

high-magnet-field laboratories in Talahassee and Grenoble. Typically they dissipate 15 MW to generate a field of 20 MA m^{-1} using a current of 20 kA. Hybrid magnets composed of a Bitter coil inset in the bore of a large superconducting coil hold the record for continuous fields, which is 36 MA m^{-1} (45 T).

Still higher fields can be achieved provided their duration is limited to a fraction of a second. The coil is then energized by discharging a multimegajoule capacitor bank. The chief limitation is the yield strength of the wire under the pressure of the confined magnetic field. Reusable coils can generate fields of up to 70 MAm^{-1} lasting for tens or hundreds of milliseconds, but to achieve fields in excess of 100 MA m^{-1} it is necessary to sacrifice a coil with every shot, which lasts a matter of microseconds. The highest fields, of order 1 GA m^{-1}, are generated by momentary flux compression with high explosives. They are not much use for physical measurements!

Another approach to creating ultrashort-pulse fields is to use a bunch of high-energy electrons from a particle accelerator.

High-frequency magnetic fields are associated with electromagnetic radiation, as we have seen in Chapter 9. The root-mean-square H-field in ambient daylight is only 10 A m^{-1}, but intense, ultrashort laser pulses can deliver fields as high as 1 MA m^{-1}.

Returning to the laboratory or industrial scale, pulsed fields of 3–5 MA m^{-1} lasting tens of milliseconds are able to magnetize rare-earth permanent magnets. Short pulse magnets are also useful for determining anisotropy fields by the singular-point method where an anomaly in the time derivative of the magnetization marks the anisotropy field. But the workhorse for most magnetization measurements remains the venerable electromagnet. Flux generated by large water-cooled coils is confined by a yoke of soft iron and concentrated in an airgap using tapered iron or cobalt–iron pole pieces. The technology reached its apogee in 1928 with a 120 ton electromagnet at Bellevue, near Paris, which produced fields in excess of 4 MA m^{-1}.

Competition for the electromagnet comes from compact permanent-magnet flux sources, Fig. 13.14, which achieve comparable field levels without any need for a power supply or cooling water. It is relevant to remark that the magnetization of $Nd_2Fe_{14}B$, for example, is 1.3 MA m^{-1}, which is the value of the equivalent surface current. Permanent magnets are always the better solution in small spaces.

An electromagnet with tapered pole pieces capable of generating 1.8 MA m^{-1}.

A summary of the different magnetic flux sources is provided in Table 10.1, together with the airgap flux density $B_0 = \mu_0 H$ in teslas. Two reasons for quoting fields from now on in teslas rather than MA m^{-1} are firstly that magnetic circuits are conveniently discussed in terms of flux which is conserved, and secondly that the field H is always multiplied by μ_0 when it interacts with matter.

Costs are of order €25k for an electromagnet or permanent magnet flux source, €50k for a superconducting solenoid, €200k for a pulsed-field installation and >€1M for a Bitter magnet with a 15 MW power supply.

Table 10.2. Production of high magnetic fields

Method	Duration	Maximum field (T)
Air-core solenoid	Steady	0.2
Permanent magnet	Steady	0.1–2
Electromagnet	Steady	0.5–2.5
Superconducting solenoid	Steady	2–23
Bitter magnet	Steady	15–35
Hybrid magnet	Steady	40–45
Discharge coil	100 ms	25–80
Discharge coil	10 μs	50–100
Expendible coil	1 μs	>100
Implosive flux compression	< 1 μs	1000

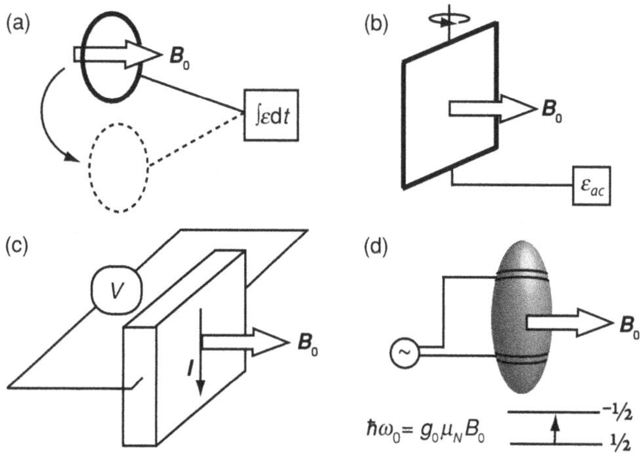

Figure 10.4

Some methods of measuring magnetic fields: (a) search coil, (b) rotating coil fluxmeter, (c) Hall probe and (d) nuclear magnetic resonance probe.

10.2.2 Measurement

Next we must consider how magnetic field is measured. Field-measuring instruments are known as 'gaussmeters' (10^4 gauss = 1 tesla). The magnitude and direction of a uniform, steady field can be determined absolutely using a search coil or a rotating-coil fluxmeter (Fig. 10.4). The principle is that a transient or alternating emf $\mathcal{E}$ is induced according to Faraday's law, which follows from (2.48)

$$\mathcal{E} = -N\mathrm{d}\Phi/\mathrm{d}t, \tag{10.3}$$

where N is the number of turns on the coil of area $\mathcal{A}$. The flux density in air $B_0 = \Phi/\mathcal{A}$ is deduced by integrating the emf, $B_0 = (1/N\mathcal{A})\int \mathcal{E}\mathrm{d}t$ as the search coil is removed from the uniform field to a region where the field is zero. When measuring a pulsed field, there is no need to move the coil. Likewise, a coil rotating about an axis perpendicular to $\boldsymbol{B}_0$ generates an alternating emf $\mathcal{E} = -N\mathcal{A}\omega B_0 \sin\omega t$. These measurements are absolute, but inconvenient.

In practice, a Hall effect or magnetoresistance sensor is generally used to measure magnetic field. The active area of the semiconducting Hall probe is of order 1 mm^2, and it generates a voltage $V_H = B_0 I/n_c e\mathrm{t}$, where I is the sensing current, t is the thickness of the semiconductor and n_c is the carrier density. The probe needs to be calibrated, and possibly corrected for temperature fluctuations, but the Hall voltage is linear in magnetic field. Accuracy is about 1%. Magnetoresistance sensors, discussed in §14.3, can be based, for example, on spin valves with crossed anisotropy. Like fluxgate sensors, which depend on the asymmetric saturation of soft magnetic wires or thin-films elements (§12.4), they are good for fields below 100 A m^{-1} but they have the advantage of a high bandwidth.

Much greater accuracy and precision is possible by measuring the NMR frequency of rubidium vapour, for example, which is $\nu = 13.93$ MHz T^{-1} for the ^{77}Rb nucleus, or even water, where the resonance frequency is 42.58 MHz T^{-1} for the ^{1}H nucleus, better known as the proton.

10.2.3 Magnetic shielding

Shielding is necessary to create a magnetic-field-free environment for sensitive instruments such as high-resolution electron microscopes. It is also needed for measurements of very weak magnetic fields, such as those emanating from the brain. Two strategies to eliminate static and low-frequency fields are passive shielding and active shielding. Passive shielding involves surrounding the shielded volume by one or more boxes of high permeability material such as permalloy, which diverts the flux (12.2). Superconducting shields, which allow no *change* of flux in the enclosed volume, may be used for small spaces. Active shielding uses a sensor to detect one component of the field, which is then compensated by a current flowing in a pair of Helmholtz coils. Three orthogonal pairs can cancel the magnetic field vector completely.

At high frequency ($\geq$ 100 kHz) a simpler approach is to surround the shielded volume by a continuous wire mesh known as a Faraday cage. Induced currents in the conducting circuits tend to cancel the changes to which they are due.

10.3 Atomic-scale magnetism

10.3.1 Diffraction

The scaffold for magnetism in solids is the atomic-scale structure of the crystal. The electronic structure of the atoms, together with the crystalline (or amorphous) structure of the solid determine the atomic moments, the exchange and

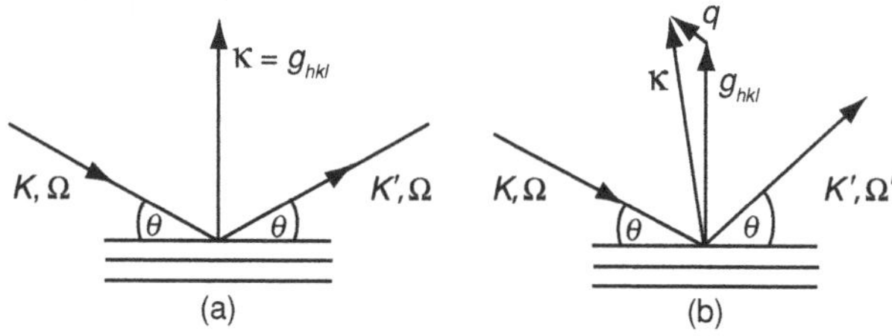

Figure 10.5

Scattering of a beam of radiation from a crystal: (a) elastic scattering in the Bragg geometry where $K' - K = \kappa = g_{hkl}$; (b) inelastic scattering in the vicinity of a Bragg peak, where $\kappa = g_{hkl} + q$.

dipolar interactions and the crystal fields, which are the ingredients of collective magnetic order (6.14).

The intrinsic magnetic properties are probed by diffraction and spectroscopy, which are respectively the elastic and inelastic scattering by the solid of a beam of particles or electromagnetic radiation. If the *wavevector* and *energy* of the incident and scattered beams are $(\boldsymbol{K}, \hbar\Omega)$ and $(\boldsymbol{K}', \hbar\Omega')$, complete information about the solid is contained in the differential scattering cross section:

$$\sigma_{diff} = \frac{d^2\sigma(\kappa\omega T)}{d\kappa\, d\omega}, \tag{10.4}$$

where σ is the total scattering cross section and $\kappa = \boldsymbol{K}' - \boldsymbol{K}$, $\omega = \Omega' - \Omega$ (Fig. 10.5). Commonly used particle beams are neutrons and electrons. Polarization is another relevant variable, especially with electromagnetic radiation, such as X-rays. Sometimes it is possible to detect the wavevector $\boldsymbol{q}$ and energy $\hbar\omega_q$ for the excited species directly by measuring the ejected electron, in photoelectron spectroscopy for example. The spins of the incident and scattered or excited particles can be analysed. Different methods probe different aspects of the generalized susceptibility. Together, they provide a rich description of the crystal and magnetic structure, the electronic structure and the distribution of spin and orbital moments, as well as the excitation spectra from which, for example, elastic constants, interatomic exchange and crystal field parameters can be determined.

Diffraction methods are used to study both crystal structures and magnetic structures of solids. A beam of radiation is needed whose wavelength is comparable to the interatomic spacing (≈ 0.2 nm). The radiation is scattered by the atomic electrons or nuclei, and in the case of neutrons, by the magnetic moments of the electrons. Interference of the scattered waves gives rise to a number of diffracted beams in precisely defined directions relative to the crystal axes. The directions of these Bragg reflections are determined by the lattice parameters, according to Bragg's famous law

$$2d_{hkl} \sin\theta = n\Lambda, \tag{10.5}$$

where d_{hkl} is the spacing of a set of reflecting planes whose Miller indices are (h, k, l). Both incident and diffracted beams make an angle θ with the hkl planes. The scattering is elastic so that $|\boldsymbol{K}| = |\boldsymbol{K}'|$ and the scattering vector κ is perpendicular to the reflecting planes. The wavelength of the radiation Λ

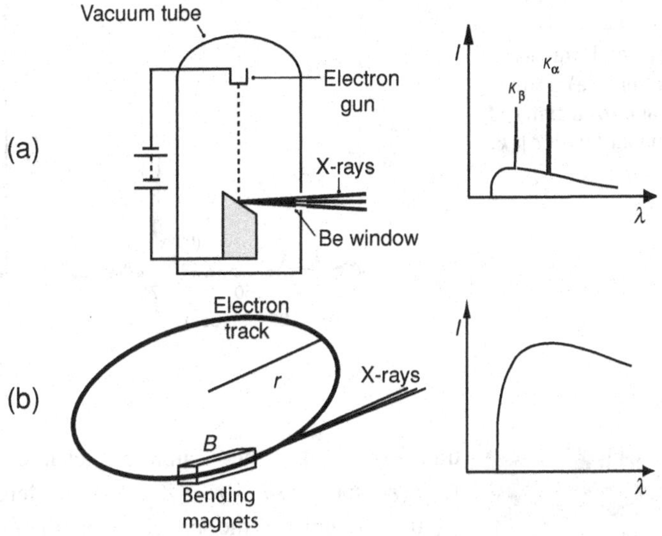

Figure 10.6

Sources of X-rays: (a) a sealed X-ray tube, (b) a synchrotron source. The intensity from the synchrotron is more than four orders of magnitude greater than the peak intensity from the X-ray tube.

Powder diffraction pattern of bcc αFe

h	k	l	d(pm)	I/I_{max}
1	1	0	202.7	100
2	0	0	143.3	20
2	1	1	117.0	30
2	2	0	101.3	10
3	1	0	90.6	12
2	2	2	82.8	6

Powder diffraction pattern of fcc γNi

h	k	l	d(pm)	I/I_{max}
1	1	1	203.4	100
2	0	0	176.2	42
2	2	0	124.6	21
3	1	1	106.2	20
2	2	2	101.7	7
4	0	0	88.1	4
3	3	1	80.8	14
4	2	0	78.8	15

is $2\pi/K$ and the integer n is known as the order of the reflection. The Bragg condition is equivalent to the requirement that the scattering vector $\boldsymbol{\kappa}$ be a reciprocal lattice vector $\boldsymbol{g}_{hkl}$, where $|\boldsymbol{g}_{hkl}| = 2\pi/d_{hkl}$.

Intensities of the Bragg reflections depend on the strength of the atomic scattering and the arrangement of atoms within the unit cell. They are proportional to the square of the complex structure factor:

$$F_{hkl} = \sum_i f_i e^{i\boldsymbol{\kappa}\cdot\mathbf{r}_i} = \sum_i f_i e^{[-2\pi i(hx_i+ky_i+lz_i)]} \tag{10.6}$$

where the sum is over the i atoms at positions $\boldsymbol{r}_i = x_i\boldsymbol{a} + y_i\boldsymbol{b} + z_i\boldsymbol{c}$ in the unit cell. f_i is the atomic scattering factor, which has dimensions of length and generally depends on $\boldsymbol{\kappa}$ or θ.

X-ray diffraction (XRD) is the standard method of crystal structure analysis. The energy $\varepsilon = \hbar\Omega = hc/\Lambda$ of electromagnetic radiation associated with a wavelength of 200 pm is 6.20 keV, which is close to the K absorption edge for Cr. The K edge corresponds to the energy needed to ionize an atom by creating a 1s hole, the L edges correspond to 2s or 2p holes etc. In a laboratory X-ray set, a target of a suitable metal in a vacuum tube is bombarded with energetic electrons. Characteristic X-ray radiation is emitted as electrons from outer shells de-excite to fill holes created in the inner shells (Fig. 10.6(a)). The flux from a sealed tube is typically 10^{16} photons s^{-1}. The K edge for the commonly used copper target, for example, is at 8.98 keV, and monochromatic K_α radiation with $\varepsilon = 8.04$ keV and $\Lambda = 154.2$ pm is produced when holes in the 1s shell are filled by 2p electrons. K_β radiation is produced when the 1s holes are filled by 3p electrons.

X-rays are scattered by the charge of the atomic electrons, and the appropriate atomic scattering function in (10.6) is proportional to the Fourier

Figure 10.7

X-ray powder diffraction patterns from: (a) a $SmCo_5$ powder and (b) a sintered $SmCo_5$ magnet with $c \parallel \kappa$.

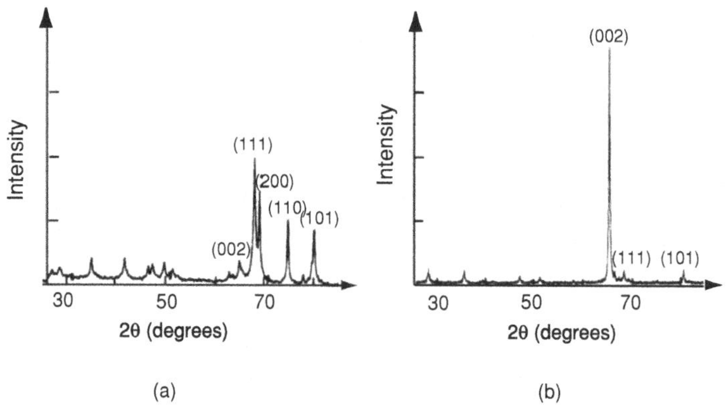

Powder diffraction pattern of hcp εCo

h	k	l	d(pm)	I/I_{max}
1	0	0	217.0	27
0	0	2	203.5	28
1	0	1	191.5	100
1	0	2	148.4	11
1	1	0	125.3	10
1	0	3	115.0	10
2	0	0	108.5	1
1	1	2	106.7	9
2	0	1	104.8	6
0	0	4	101.7	1
2	0	2	95.8	1
1	0	4	92.1	1
2	0	3	84.7	3
2	1	0	82.0	1
2	1	1	80.4	5

transform of the atomic charge distribution $\rho(\boldsymbol{r})$; $f(\boldsymbol{\kappa}) = r_e \int \rho(\boldsymbol{r}) \exp(-i\boldsymbol{\kappa} \cdot \boldsymbol{r}) d^3r = Zr_e f_X(\boldsymbol{\kappa})$. Here Z is the atomic number, $f_X(0) = 1$, and r_e is the scattering length of an electron, known as the classical electron radius, $\mu_0 e^2/4\pi m_e$. Its value is 2.818 fm.

Samples for investigation by X-ray or neutron diffraction are often in powder form. Randomly oriented crystalline powder scatters an incident monochromatic beam in a series of cones centred on the incident beam axis; each cone corresponds to a Bragg reflection from a set of randomly oriented crystallites. A point detector or linear multidetector is then used to measure the diffracted intensity as a function of 2θ in the scattering plane. A typical diffraction pattern for $SmCo_5$ powder is shown in Fig. 10.7(a). Particle alignment restricts the diffracted beams to certain directions in the cones. For example, if the c axis is well aligned parallel to $\boldsymbol{\kappa}$ as in the sintered $SmCo_5$ magnet of Fig 10.7(b), then the Bragg reflections from the $\{001\}$ family of planes dominate the powder pattern.

Far more intense fluxes of X-ray and UV radiation are available at synchrotron sources, where a beam of electrons or positrons is accelerated to a velocity close to that of light, and then constrained by a magnetic field $B \lesssim 1$ T to travel around a storage ring which may be tens or hundreds of metres in diameter. The electron energy is typically 5 GeV, or $\gamma m_e c^2$ with $\gamma \approx 10^4$. As they race around their track, the electrons emit a narrow beam of white radiation of width $1/\gamma$ radians which is linearly polarized (> 90%) in the plane of the orbit. The cutoff wavelength is $4\pi r/3\gamma^3 = 0.00714/B\gamma^2 m_e$, where r is the radius of the electron orbit. Elliptically polarized radiation can be obtained by placing a slit to collect radiation emitted just above or just below the plane of their orbit. Otherwise, circularly polarized radiation may be obtained by means of a wiggler or undulator insertion device of the type described in §13.3.2. The energy of the X-rays is selected using X-ray mirrors and single-crystal monochromators. The huge photon fluxes, $\approx 10^{17}$ photons s^{-1} in a tightly collimated beam, with a 0.1% energy bandwidth and high degree of polarization, make synchrotron radiation very suitable for both absorption and photoelectron spectroscopy over an extremely wide range of wavelength, from far infra-red to hard

Table 10.2. Some nuclear (b) and magnetic (p) neutron scattering lengths in fm (10^{-15} m) and absorption cross sections (σ_a) in barns (10^{-28} m^2). The rare earth magnetic scattering lengths are for 3+ cations. Ti–Fe are for spin-only 3+ ions and Co–Cu are for spin-only 2+ ions.

	b	$p(0)$	σ_a		b	$p(0)$	σ_a		b	σ_a
Ti	−3.4	2.7	6.1	Y	7.8	1.3		B	5.3	767
V	−0.4	5.4	5.1	Pr	4.6	8.6	11.5	C	6.6	0.004
Cr	3.6	8.1	3.1	Nd	7.2	8.8	51.2	N	9.3	1.9
Mn	−3.7	10.8	13.3	Sm	0.0	1.9	5670	O	5.8	0.0002
Fe	9.5	13.5	2.6	Gd	9.5	18.9	29400	Al	3.4	0.2
Co	2.5	8.1	37.2	Tb	7.4	24.3	23	Si	4.1	0.2
Ni	10.3	5.4	4.5	Dy	16.9	27.0	940	Sr	7.0	1.3
Cu	7.7	2.7	3.8	Ho	8.0	27.0	65	Ba	5.1	1.2

X-rays. Chemical and orbital selectivity is achieved by tuning the radiation to the appropriate atomic absorption edge.

Magnetic scattering of X-rays is typically smaller than charge scattering by a factor of 10^6, making it impracticable to observe magnetic structures using sealed X-ray tubes. However, in the vicinity of an absorption edge the magnetic effect may amount to 1% of the charge scattering, which allows tunable synchrotron sources to be used for magnetic structure determination. Magnetic X-ray diffraction is most appropriate for rare-earths like Sm and Gd where neutron diffraction is hampered by their enormous neutron capture cross sections (Table 10.2), or for micrometre-size single crystals.

Neutron diffraction is the standard method for magnetic structure analysis. Beams of neutrons are produced in specially optimized nuclear reactors or in spallation sources where pulses of GeV protons from a linear accelerator produce bursts of neutrons as they impinge on a heavy-metal target. There are only a handful of these wonderful facilities in the world but they have contributed enormously to the knowledge of magnetism.

The neutron is an uncharged particle with spin, which carries a magnetic moment of $-1.91\ \mu_N$. The neutron energy for a de Broglie wavelength Λ of 0.20 nm is $h^2/2m_n\Lambda^2 = 0.0204$ eV, which is comparable to k_BT at ambient temperature. The neutrons from a reactor are thermalized in a moderator and a narrow slice is selected from the Maxwellian energy distribution by Bragg reflection from a single-crystal monochromator. Typical reactor fluxes are 10^{19} m^{-2} s^{-1}, which have to be collimated and are then reduced by 2–3 orders of magnitude by the monochromator. Monochromatic neutron beams from a high-flux reactor are roughly 1000 times less intense than monochromatic X-ray beams from a laboratory X-ray generator.

This, and the relatively weak scattering of neutrons in solids mean that large samples, of order 1 cm^3, are needed for neutron diffraction.

The scattering of thermal neutrons by an atomic nucleus is isotropic because nuclear interactions are very short-range; for an incident neutron plane wave

$e^{i\boldsymbol{K}\cdot x}$, the scattered spherical wavelet ψ is $-(b/r)e^{i\boldsymbol{K}'\cdot r}$, where b is the scattering length of the nucleus. In fact b is different for every isotope, so the values quoted for an element in Table 10.2 are isotopic averages. The scattering cross section, σ_s, is $4\pi b^2$. Unlike X-ray scattering, where the scattering length increases with $\boldsymbol{Z}$, the number of atomic electrons, b, varies erratically across the periodic table and can even change sign, which makes it possible often to distinguish elements which are close in atomic number.

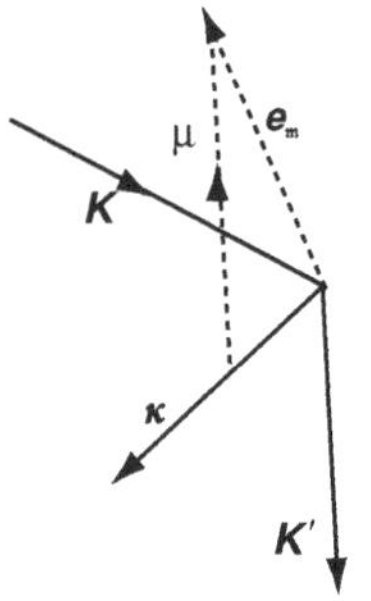

The magnetic interaction vector $\mu = e_m - \kappa(\kappa \cdot e_m)/\kappa^2$.

The neutron is also scattered by the unpaired spin density of the atomic electrons. A magnetic scattering length p is defined as $(1.91r_e)Sf_S$ for a spin-only moment. The quantity in brackets is 5.38 fm. When both spin and orbital moments are present, $p = (1.91r_e)(Sf_S + \frac{1}{2}Lf_L)$, where f_S, f_L are given by $\{J(J+1) \pm [S(S+1) - L(L+1)]\}/[2(J+1)]$. The form factors f_S and f_L are normalized to unity at $\theta = 0$. A magnetic interaction vector $\boldsymbol{\mu}$ is defined as $\mathbf{e}_m - \boldsymbol{\kappa}(\boldsymbol{\kappa} \cdot \mathbf{e}_m)/\kappa^2$, where $\mathbf{e}_m$ is a unit vector in the direction of the magnetic moment. For unpolarized neutrons, the intensities of the magnetic and nuclear scattering add, so

$$|F_{hkl}|^2 = \left|\sum_i b_i e^{(-i\boldsymbol{\kappa}\cdot\boldsymbol{r}_i)}\right|^2 + \left|\sum_i p_i \boldsymbol{\mu}_i e^{(-i\boldsymbol{\kappa}\cdot\boldsymbol{r}_i)}\right|^2, \quad (10.7)$$

whereas if the neutron beam is polarized with magnetic moment in a direction $\boldsymbol{\Lambda}$, the intensity is

$$|F_{hkl}|^2 = \left|\sum_i \{b_i + (\boldsymbol{\Lambda} \cdot \boldsymbol{\mu}_i)p_i\} e^{(-i\boldsymbol{\kappa}\cdot\boldsymbol{r}_i)}\right|^2. \quad (10.8)$$

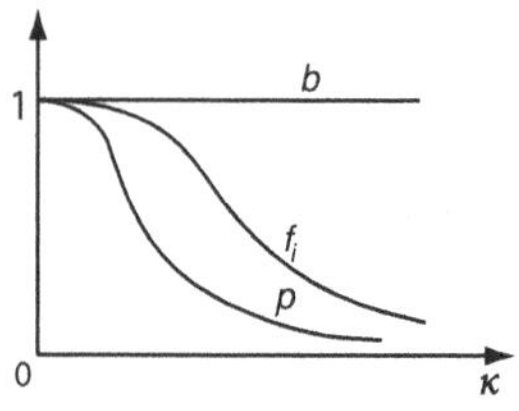

Comparison of the normalized nuclear (b) and magnetic (p) scattering factors for neutrons and the atomic scattering factor for X-rays (f_i).

Since magnetic scattering depends on the orientation of the moments relative to the scattering vector, the complete magnetic structure (magnitudes and directions of the moments in a unit cell) can in principle be determined from the positions and intensities of the magnetic Bragg reflections. Note that there is no magnetic intensity when $\mathbf{e}_m \parallel \boldsymbol{\kappa}$.

A typical neutron powder diffraction pattern is shown in Fig. 10.8, together with the least-squares fit, the Rietveld profile based on the unit cell parameters. With large unit cells containing N atoms there can be as many as $6(N+1)$ structural and magnetic parameters to refine. Powder data may be inadequate for structures more complex than that of $Nd_2Fe_{14}B$, for example, because of the limited number of Bragg reflections, and one then has to resort to single crystals and polarized neutrons.

10.3.2 Spectroscopy

Many spectroscopic techniques are available to probe the energy levels and excitations of magnetic solids. Inelastic neutron scattering, where the energy and $\boldsymbol{K}$-vector of an inelastically scattered beam are analysed in a triple-axis

Figure 10.8

Neutron powder diffraction pattern of CrO_2. Magnetic reflections are shaded.

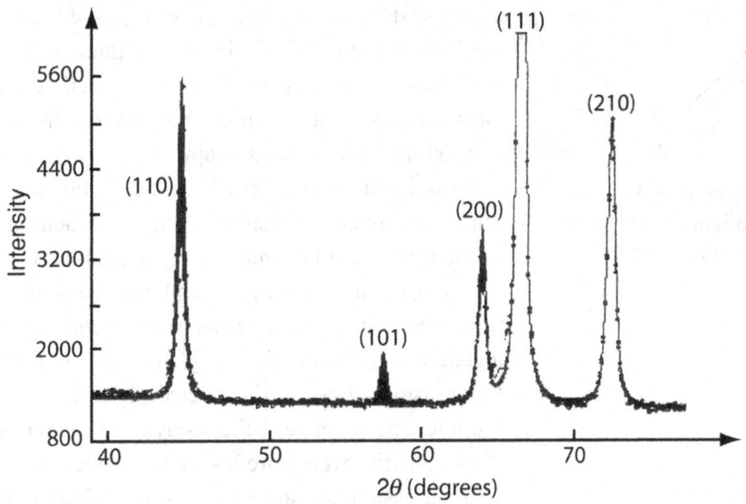

Figure 10.9

A triple-axis neutron spectrometer. The angles θ_M, θ_S and θ_A refer to the orientations of the monochromator, the source and the analyser, respectively.

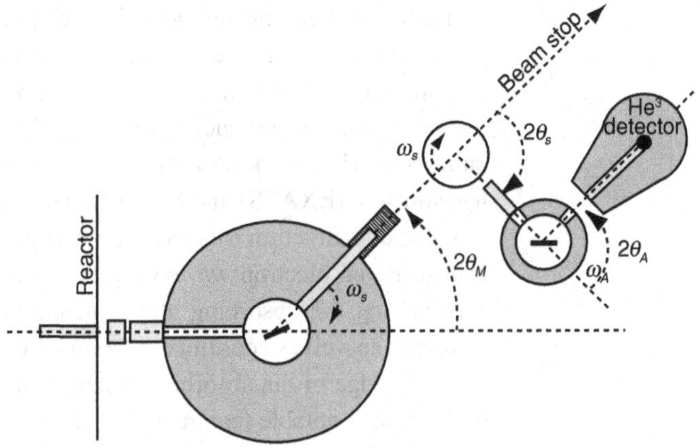

spectrometer (Fig. 10.9), is the most general method because the changes both in energy and momentum of the neutron due to thermal excitations are appreciable. The spin-wave dispersion relations $\omega(\boldsymbol{q})$ are measured in this way by scanning the instrument near a Bragg peak to collect neutrons at constant energy or constant momentum transfer. The exchange parameters $\mathcal{J}_{ij}$ in the Heisenberg Hamiltonian for different pairs of interacting neighbours are deduced by fitting these dispersion curves. Dispersionless excitations such as the crystal field excitations of the rare-earth atoms can also be investigated by inelastic neutron scattering. This is useful for metals, where optical absorption spectroscopy may be impracticable.

Another useful technique for probing the excitations in magnetic solids is Brillouin light scattering. An optical photon, which has a wavevector $K \approx 0$ (the wavelength of light is very much greater than the lattice spacing) may excite or absorb a magnon or phonon near the centre of the Brillouin zone, in

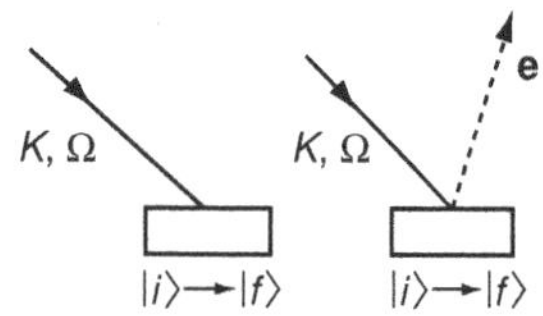

Absorption and photoemission processes for a single photon.

an ineleastic process. The range of possible $\boldsymbol{K}$-vectors accessible by inelastic light scattering is extended in two-magnon processes, where a pair of magnons with wavevectors $\pm K$ are excited. The method is good for measuring spin-wave gaps, and dispersion relations near the origin, $\boldsymbol{K} = 0$.

X-ray absorption spectroscopy is a powerful tool for investigating the magnetic and structural properties of magnetic materials. With the construction of purpose-built electron storage rings, monochromatic X-rays with a wide range of energies have become readily available. The techniques are element- and orbital-specific as the incident photon beam can be tuned to a desired absorption edge, and the core levels of different elements do not usually overlap. In a generic experiment the sample is irradiated with monochromatic X-rays and the transmitted beam, the ejected photoelectrons or the photons generated by the interaction between the X-rays and matter are analysed. When the photoelectrons arise from core levels, the process is the inverse of X-ray generation. Photoemission is a surface-sensitive technique because the electrons can only emerge unscathed from a surface layer ≈ 1 nm thick. Depending on the absorption edge studied and the detection scheme chosen, X-ray spectroscopy can provide information on quantities such as local structure, magnetic properties and the shape of the valence and conduction bands.

Two common techniques belonging to the X-ray absorption family of particular interest when working with magnetic materials are extended X-ray absorption fine structure (EXAFS) and X-ray magnetic circular dichroism (XMCD). EXAFS is a type of diffraction pattern produced by the interference of the outgoing and backscattered electron waves. It gives information on the local environment surrounding the absorbing atom, including nearest-neighbour positions and distances as well as coordination number. EXAFS spectra are usually recorded at the K-edge of the absorbing element. When compared to laboratory X-ray diffraction, a notable feature of EXAFS is that there is no need for the sample to display long-range crystalline order.

XMCD refers to the difference in the absorptive part of the refractive index of a solid for left- and right-circularly polarized light passing through a sample magnetized parallel to $\boldsymbol{K}$. Thanks to spin-orbit coupling, the incident photons transfer their angular momentum to both the orbital and spin momentum of the photoelectron as it is transferred to an unoccupied state just above the Fermi level. Of particular interest in magnetism are $2p$ ($L_{2,3}$-edges) and $3d$ ($M_{4,5}$-edges). Core level spectroscopies probe unoccupied states in the $3d$- and $4f$-orbitals. The advantage of these methods is that, with the use of the magneto-optical sum rules, it is one of the few ways to independently determine the spin and orbital moment of the unpaired electrons in an atom. For example, in $BaFe_{12}O_{19}$ it is possible not only to measure the average moments on iron, but also to detect any moment on barium or oxygen. It is also to some extent possible to distinguish between the different iron sites due to their different symmetries. The XMCD spectrum of iron is shown in Fig. 10.10.

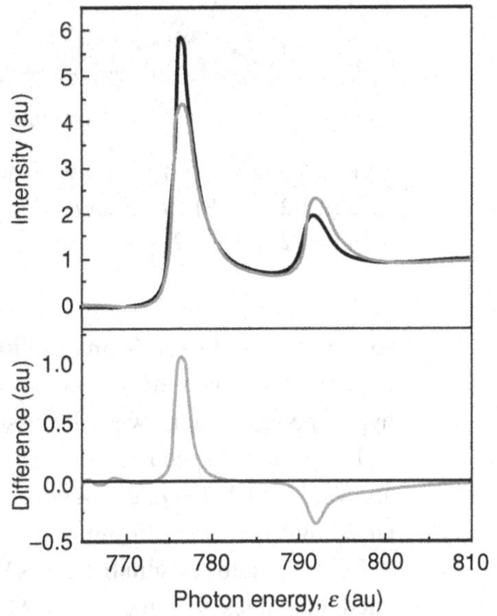

Figure 10.10
The absorption of left- or right-circularly polarized radiation in the vicinity of the L-absorption edge of iron. The difference is the XMCD signal, from which the spin and orbital moments can be delivered. The black line is μ_+ and the grey line is μ_-.

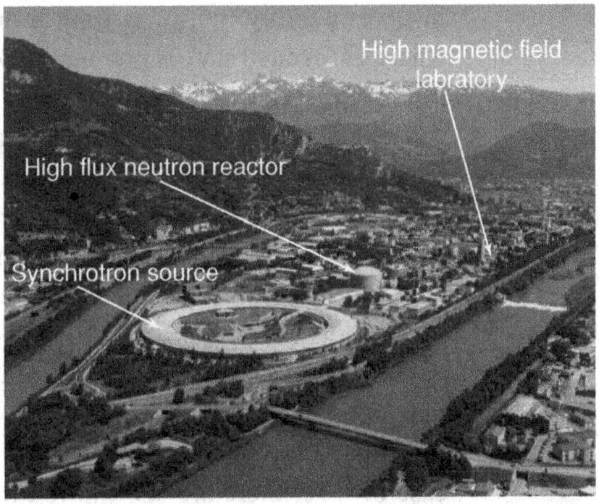

Figure 10.11
An aerial view of Grenoble, which includes the three principal large-scale facilities for magnetic research.

New methods of X-ray spectroscopy exploit the ability to analyse the polarization of the transmitted or generated photons or the spin of the ejected photoelectrons. Angular-resolved ultraviolet photoemission can be used to image a map of the spin-polarized density of states in the Brillouin zone. X-ray spectroscopy is best conducted at a synchrotron source. Large-scale facilities play an important role in magnetism research. A unique cluster of them is found in Genoble, Fig.10.11.

Hyperfine interactions The atomic nucleus is a point probe of magnetic and electric fields at the very heart of the atom. Atoms in different crystallographic sites may be distinguished by their hyperfine interactions. The degeneracy of the $(2I + 1)$ levels with $M_I = I, I - 1, \ldots, -I$, associated with the nuclear

Table 10.3. Relative intensities of Mössbauer absorption lines for ^{57}Fe. Here θ is the angle between K_γ and B_{hf}

Lines	Relative intensity	Powder average	$B_{hf} \| K_\gamma$	$B_{hf} \perp K_\gamma$
1,6 ($\pm 3/2 \longrightarrow \pm 1/2$)	$3(1+\cos^2\theta)$	3	3	3
2,5 ($\pm 1/2 \longrightarrow \pm 1/2$)	$2\sin^2\theta$	2	0	4
3,4 ($\pm 1/2 \longrightarrow \pm 1/2$)	$1+\cos^2\theta$	1	1	1

spin $I\hbar$ is raised with Zeeman splitting in the hyperfine field $\boldsymbol{B}_{hf}$. The electric field gradient raises the degeneracy of levels with different values of M_I^2. These hyperfine interactions were discussed in Chapter 9.

The principal spectroscopic techniques for measuring hyperfine interactions are *NMR* and *Mössbauer spectroscopy*. Heat capacity below 1 K is also influenced by hyperfine splitting.

Data on isotopes suitable for NMR and Mössbauer spectrosocpy were collated in Tables 9.3 and 9.5. NMR involves resonant absorption of radiofrequency radiation by the nucleus in its ground state. Mössbauer spectroscopy involves a γ-ray transition from a nuclear excited state to the ground state, where the excited state is populated by a radioactive precursor. The best-known example, ^{57}Fe, was featured in Fig. 9.15. The source is ^{57}Co, which has a convenient half-life $\tau_{\frac{1}{2}}$ of 270 days, and the energy ε_γ of the transition from the excited state to the ground state, 14.4 keV, is low enough to ensure a large recoilless fraction (9.48). Mössbauer spectra are usually recorded by modulating the energy of a single-line gamma source by moving it relative to the absorber, which is the sample of interest. The Doppler shift of the energy of the emitted line is $\Delta\varepsilon = \varepsilon_\gamma v/c$.

Only $\Delta M_I = 0, \pm 1$ transitions are allowed by the selection rule for dipole radiation, and the relative intensities depend on Clebsch–Gordan coefficients determined by the angle between $\boldsymbol{K}_\gamma$ and the nuclear quantization axis. In the case of ^{57}Fe, there are six allowed transitions between Zeeman-split nuclear levels (Fig. 9.15) and the $\Delta M_I = 0$ transitions are lines 2 and 5. The relative intensities are listed in Table 10.3. For example, when $\boldsymbol{K}_\gamma \| \boldsymbol{B}_{hf}$ the intensity of lines 2 and 5 is zero. Mössbauer spectroscopy is therefore quite useful for determining the magnetization *direction* in single-crystals or magnetically textured samples of iron compounds. Different crystallographic sites may be distinguished by their hyperfine spectra. The example of $BaFe_{12}O_{19}$ is shown in Fig. 10.12.

A variant, useful for thin films, is conversion electron Mössbauer spectroscopy. Nuclei in thin films which absorb the γ-ray may subsequently de-excite by re-emitting the 14.4 kV γ-ray or else an electron produced by a nuclear internal conversion process and an X-ray. The conversion electrons have an average escape depth of about 50 nm. Depth selection is possible with energy analysis.

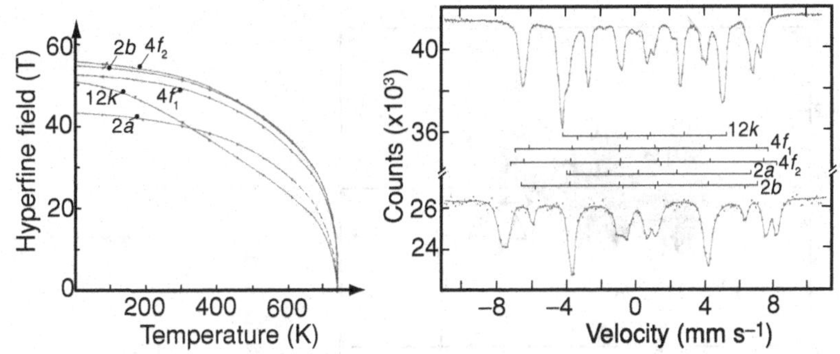

Figure 10.12

Mössbauer spectrum of $BaFe_{12}O_{19}$. Each site gives rise to a six-line hyperfine pattern. An applied magnetic field allows the ferrimagnetic sublattices to be resolved. (J. M. D. Coey *et al.*, *Rev. Sci. Instrum.* **43**, 54 (1972).)

10.3.3 Electronic structure

The electron dispersion relations $\varepsilon(\boldsymbol{k})$ of spin-polarized electrons are a complete description of the electronic structure of a solid. Photoemission spectroscopy gives partial information about the dispersion relations and the density of states $\mathcal{D}(\varepsilon)$. The degree of spin polarization of electrons near the Fermi level may be inferred from spin-polarized photoemission, or from tunnelling or point contact experiments involving a ferromagnet and a superconductor, discussed in Chapter 8. The results are surprising insofar as electrons from iron, cobalt and nickel all exhibit a positive spin polarization of about 40%, whereas the measured value for the half-metallic ferromagnet CrO_2 is >95%. The strong ferromagnets Co and Ni might be expected to show a negative polarization on account of the $3d \downarrow$ band at ε_F. Electrons emitted from the $3d$ metals may have predominant $4s$ character.

However, the main technique for exploring the electronic structure of ferromagnets is computational. Electronic structure calculations were discussed in §5.3.7 and some further examples are shown in Fig. 10.13. The local spin density approximation (LSDA) has proved to be rather reliable for calculating the zero-temperature spin-polarized density of states for $3d$ metals and $4f$–$3d$ intermetallic compounds. By introducing spin-orbit coupling it is possible to take into account the orbital moments and estimate the $3d$ band anisotropy, although the anisotropic energy is only 10^{-6} of the band energy. The LSDA method can also be used to calculate hyperfine interactions and $4f$ crystal-field coefficients. Moments of different atoms in the structure are determined from the local densities of states.

10.4 Domain-scale measurements

Techniques for visualizing domains and domain walls depend on sensing the stray field $\boldsymbol{H}$ outside the magnetic material, or the magnetization $\boldsymbol{M}$ or flux

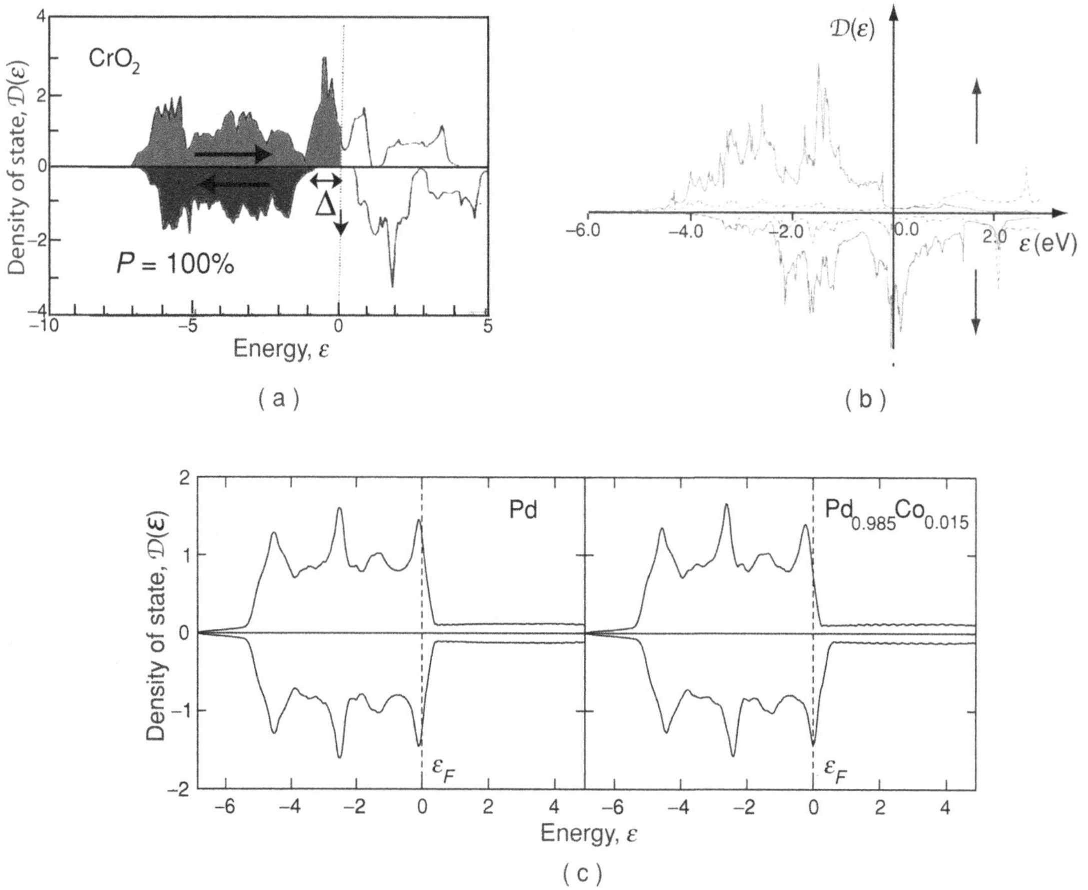

Figure 10.13

Examples of electronic structure calculations: (a) a half-metal CrO_2, (b) a ferromagnetic intermetallic compound $SmCo_5$; the solid lines are the 3*d* spin density averaged over the two Co sites and the dashed line is the 5*d* spin density of Sm – and (c) a dilute alloy PdCo; pure Pd is not magnetic, but doping with 1.5% Co introduces a giant moment of 18 μ_B per cobalt atom. (Data courtesy of Stefano Sanvito.)

density ***B*** within it. The specimen usually has to be prepared as a polished surface, foil or thin film, which begs the question whether the domains observed at the surface are actually representative of the bulk. Bulk domains can be sensed using special methods such as neutron tomography or small-angle neutron scattering. Nevertheless much useful information regarding the micromagnetic exchange and anisotropy parameters A and K_1 can be obtained from domain studies, and insight into coercivity mechanisms and magnetization reversal is achieved by simultaneous observation of the domain structure and microstructure. For thin films with in-plane magnetization, there is little distinction between surface and bulk. The domain structure is the one observed on the surface.

10.4.1 Stray-field methods

The first method for visualizing domains was developed in the 1930s by Francis Bitter. A magnetic colloid, normally a drop of oil- or water-based ferrofluid

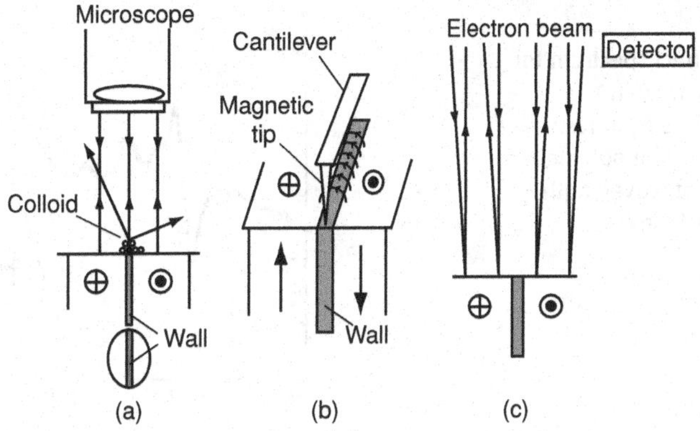

Figure 10.14

Stray-field methods of domain observation: (a) Bitter method, (b) MFM and (c) SEM with type 1 contrast.

is spread over the polished surface of a specimen, and the tiny ferromagnetic particles are drawn to the regions of maximum field gradient, thereby decorating the domain walls (Fig. 10.14).

The modern method of measuring stray fields is a scanning probe technique based in the same principle – magnetic force microscopy (MFM). A single ferromagnetic particle is mounted on a tiny silicon cantilever, or the tip of the cantilever is coated with a ferromagnetic film and it is rastered across the surface of the sample. The force derivative registered by the deflection of the cantilever or the change of its mechanical resonance frequency gives an image of the stray field gradient at the surface. The resolution achievable with MFM is about 20 nm. A problem arises when imaging soft magnetic materials, because the stray field produced by the tip may perturb the domain structure of the sample.

Reading magnetically recorded information from discs or tapes likewise depends on sensing the stray field distribution of the domain pattern imposed on the magnetic medium, using an inductive or magnetoresistive pick-up head. Magnetic recording is discussed in Chapter 14.

Scanning electron microscopy (SEM) is a workhorse of materials science. It involves rastering a sample surface with a finely focussed electron beam and detecting the secondary electrons emitted from the surface. SEM is used to image both microstructure and topology, and can provide chemical analysis on a local scale because the energies of the secondary electrons and especially those of the accompanying X-rays are characteristic of the chemical elements present. Sensitivity is good for elements with 3*s* and deeper electronic shells (Na and beyond). Lower-energy X-rays from light elements can be observed with a special windowless detector. SEM can be adapted to provide *magnetic* contrast by detecting the deflection of the secondary electrons in the stray field produced near the surface of a multidomain sample. Alternatively, the spin polarization of the secondary electrons can be monitored as the electron beam is rastered across the surface, a technique known as scanning electron microscopy with polarization analysis (SEMPA).

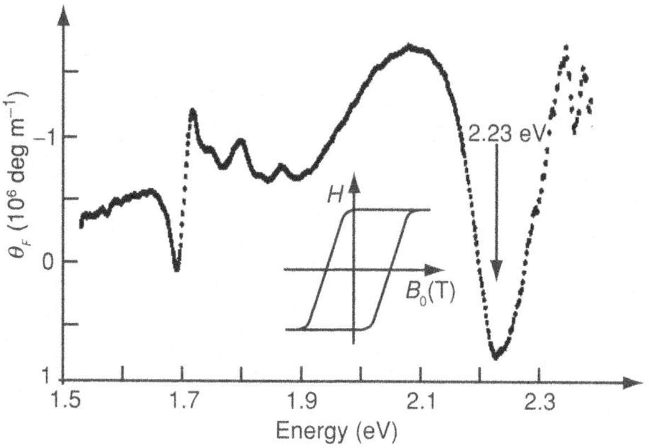

Figure 10.15

Faraday effect spectrum for $BaFe_{12}O_{19}$ at 20 K. The inset shows a hysteresis loop for the film obtained with light of wavelength 633 nm (1.86 eV).

10.4.2 Magneto-optic and electron-optic methods

The principle of the second group of domain imaging techniques is that a beam of radiation passing through a solid is influenced in some way by the ferromagnetic order of the solid. In the case of electromagnetic radiation, the magneto-optic effects depend on spin-orbit coupling, which scales with the magnetization $\boldsymbol{M}(\boldsymbol{r})$. A steady magnetic field has no influence on electromagnetic waves in free space, but light in a medium does interact with the magnetization of electrons in matter via the spin-orbit interaction. The effects are weak, but tend to be more pronounced in heavy atoms (see Table 3.4). In general, the magneto-optic effects are derived from the complex dielectric tensor ϵ_{ij} (5.86), defined by the relation $D_i = \epsilon_{ij} E_j$. If an electron beam is used in transmission electron microscopy (TEM) the interaction depends on the Lorentz force and the relevant quantity is the flux density in the material $\boldsymbol{B}(\boldsymbol{r})$.

Magneto-optic effects, which were discussed in §5.6.5, may be observed in transmission or reflection. When plane-polarized light passes through a transparent ferromagnetic medium with its wavevector $\boldsymbol{K}$ parallel to the direction of magnetization, the plane of polarization is observed to rotate by an amount proportional to the path length in the magnetic medium. This is the Faraday effect, and its discovery was the first hint of a link between light and magnetism. Faraday rotation is useful for determining the spin polarization of the conduction electrons in semiconductors and for observing domains in transparent ferrimagnets. The Faraday spectrum for a film of $BaFe_{12}O_{19}$ is shown in Fig. 10.15. The effect is greatest in the vicinity of transitions involving the Fe^{3+} magnetic ions, and it is enhanced near the absorption edge. Typical values of the Verdet constant $\mathcal{V} = k_V \mu_0 M_s$ are 10^5 radians m^{-1}. A useful figure of merit is the product $\mathcal{V} l_a$, where l_a is the absorption length, the thickness of material which will reduce the intensity of a transmitted beam by a factor e.

The Kerr effect can be used similarly to examine domains at the surface of polished ferromagnetic metals. The polar Kerr effect is the analogue of the

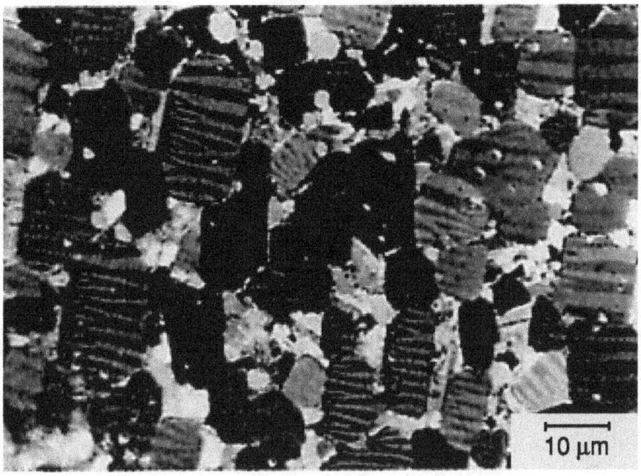

Figure 10.16

Image of the polished surface of a Nd–Fe–B sintered magnet in the Kerr microscope. The magnet is in the virgin state, and the oriented $Nd_2Fe_{14}B$ crystallites are unmagnetized multidomains. The domain contrast is due to Kerr rotation observed between crossed polarizers. (Photo courtesy of H. Kronmüller.)

Faraday effect in reflection. A beam of plane-polarized light is reflected from the surface of a ferromagnetic material with the plane of polarization rotated through a small angle θ_K of order 0.1°. The Kerr rotation for a metal is similar in magnitude to the Faraday rotation on transmission through a film of the metal thin enough to be transparent. In magneto-optic recording media, the effect was optimized with dielectric coatings. Multiple reflections take place from the top and bottom surface of the film. The Kerr microscope, a metallographic microscope modified to incorporate precise polarization analysis, will give simultaneous images of the microstructure and the domain structure. Figure 10.16 shows domains in the crystallites of an unmagnetized sintered Nd–Fe–B magnet, where the crystallite size is less than $2R_{sd}$.

Both Faraday and Kerr effects can also be used to investigate magnetization processes. When the light beam is much larger than the domain size, the initial magnetization curves and hysteresis loops reflect the net magnetization. The techniques are very sensitive, and have been used to investigate films as little as a monolayer thick. However, they do not give the absolute value of M, and are difficult to calibrate. A Kerr spectrum on a two-atomic-monolayer cobalt film is shown in Fig. 10.17.

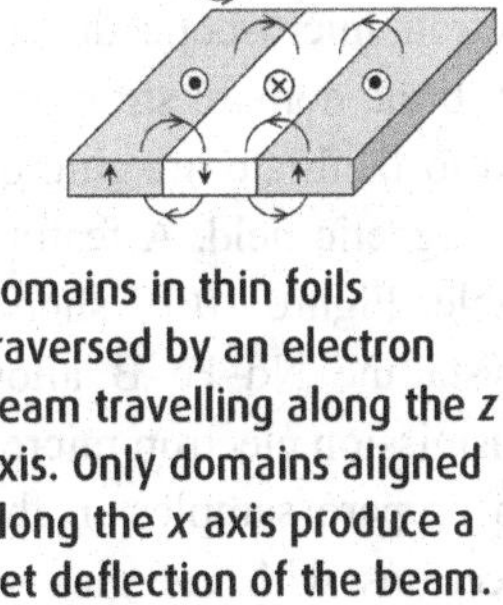

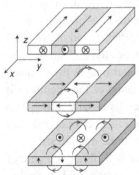

Domains in thin foils traversed by an electron beam travelling along the z axis. Only domains aligned along the x axis produce a net deflection of the beam.

In addition to the polar Kerr and Faraday effects, there exist a number of other magneto-optic imaging techniques where the intensity or phase of the polarized light is influenced by the magnetization. In the transverse Kerr effect, for example, a difference in intensity in the reflected beams polarized perpendicular and parallel to the magnetization is observed when $\boldsymbol{M}$ lies in the plane of the sample (linear magnetic dichroism). The transverse Kerr effect is useful for studying magnetization processes and domain structure in the plane of a magnetic film.

By rastering a beam of X-rays over the surface of a sample, element-specific magnetic images can be generated, which can be particularly useful for

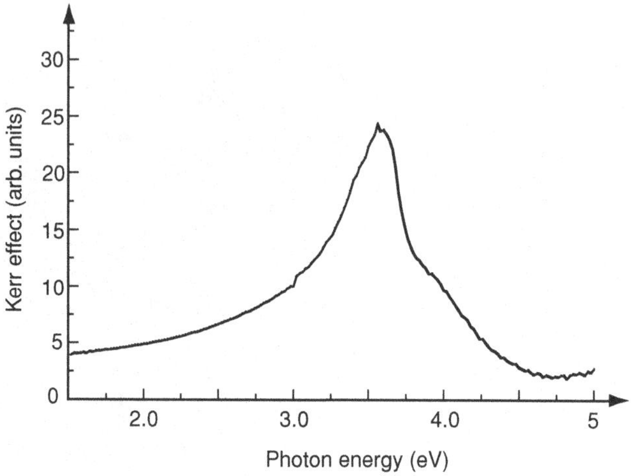

Figure 10.17

Magneto-optic Kerr spectrum from a 0.4 nm thick film of cobalt. (Data courtesy of J. F. McGilp.)

multilayer structures. Photoemission electron microscopy (PEEM) detects the spin polarization of emitted electrons and is sensitive to the projection of the magnetization in the plane of the incident X-rays. The technique is surface-sensitive, and has a resolution of order 100 nm. Magnetic circular dichroism with X-rays or ultraviolet light can also be used for element-specific magnetic imaging. The absorption coefficients, related to the imaginary parts of ϵ_{ij}, for left- and right-circularly polarized light are slightly different, which has the effect of transforming the linearly polarized incident beam into an elliptically polarized transmitted beam.

An important group of domain-observation techniques is based on TEM. The electrons experience the Lorentz force $e\boldsymbol{v} \times \boldsymbol{B}$ as they pass through a magnetized sample, and they suffer a net deflection from domains magnetized with a component of magnetization in-plane. Two methods of obtaining magnetic contrast in Lorentz microscopy are the Fresnel scheme, and the Foucault scheme, both sketched in Fig. 10.18. The former images the domain walls in a defocussed geometry as bright or dark lines. The latter images the domains themselves, with contrast that depends on the orientation of the aperture slit relative to the magnetization. A problem is that there is usually a magnetic field at the sample in a standard transmission electron microscope due to the objective lens, and this field will obviously perturb the domain structure. Instruments with special lenses have been developed to facilitate imaging of domains in specimens that are subject to almost no magnetic field. A feature of TEM is the very high spatial resolution obtainable. Figure 10.19 shows domain walls on a nanometre scale in a nanocrystalline Nd–Fe–B alloy prepared by melt spinning. A major drawback of transmission electron microscopy is the tedious preparation process require to prepare samples in the form of foils thin enough to be transparent to electrons with energies of 100–200 keV.

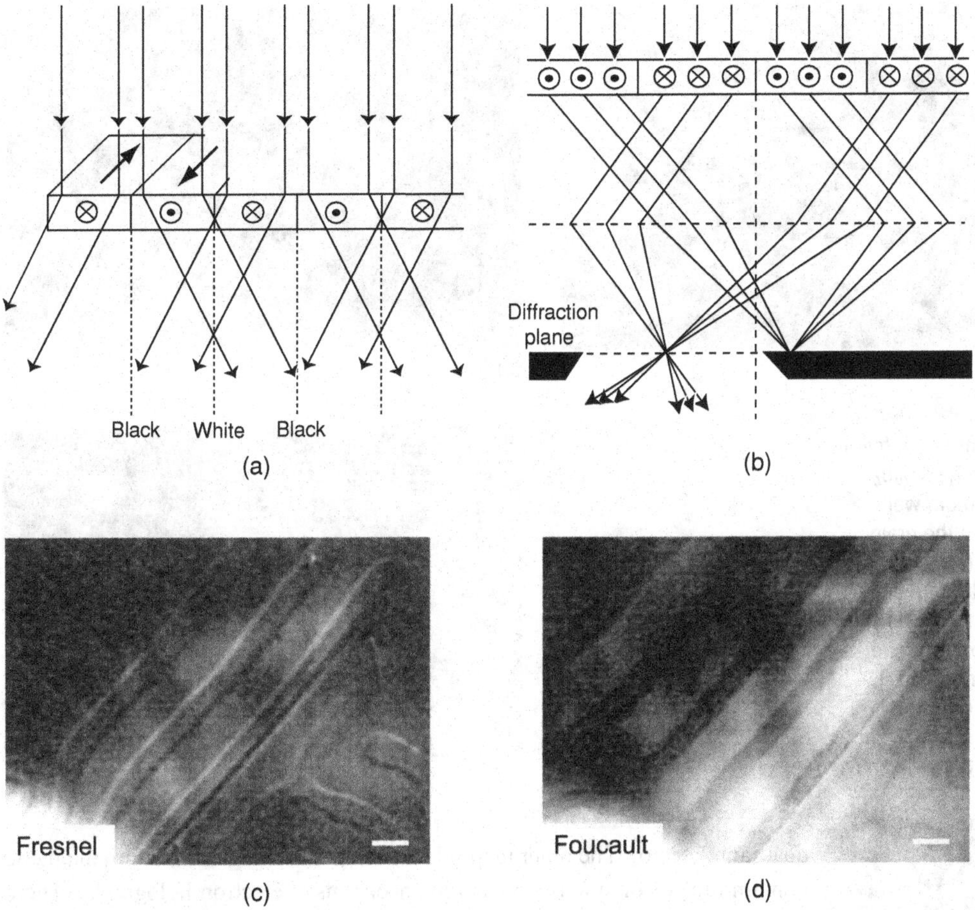

Figure 10.18

Domain imaging schemes in TEM: (a) Fresnel image and (b) Foucault image. The pictures are for a specimen of $Nd_2Fe_{14}B$. The scale bar is 100 nm. (Photo courtesy of J. Fidler.)

TEM is capable of atomic-scale resolution, and images of the real-space lattice can be obtained on suitably prepared thin specimens which is useful for examining of specimens such as cross sections of the thin film stacks used in spin electronics. Diffraction patterns and reciprocal-space images can also be obtained on very small regions (Fig. 10.19), permitting phase identification. The spots in the diffraction pattern do not appear doubled, because the deflection of the electron beam due to the Lorentz force is much less than that due to Bragg reflection. A problem with any ultrahigh resolution microscopy is to decide whether the structure visible in any particular field of view is typical of the sample, or exceptional.

Two other electron-beam methods offer more quantitative information and higher resolution. One is differential phase-contrast microscopy, where the electron beam in a scanning transmission electron microscope (STEM) is rastered across the specimen and the deflection of the beam is detected with a

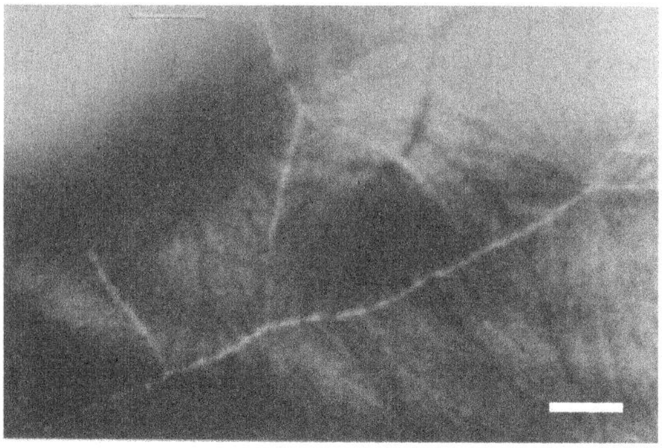
(a)

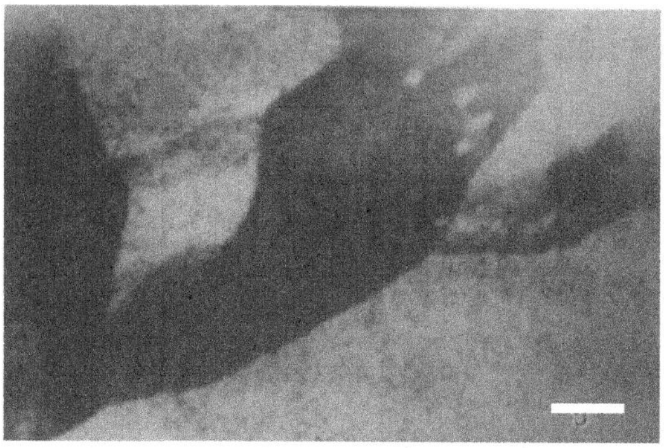
(b)

Figure 10.19

TEM image of a thin foil of melt-spun Nd_{15} $Fe_{79}B_6$ showing domain walls pinned at (a) the grain boundaries (Fresnel image) and (b) multigrain domains (Foucault image). (c) The diffraction pattern and real-space image of the atomic planes in the $Nd_2Fe_{14}B$ structure. (Photo courtesy of G. Hadjipanayis.)

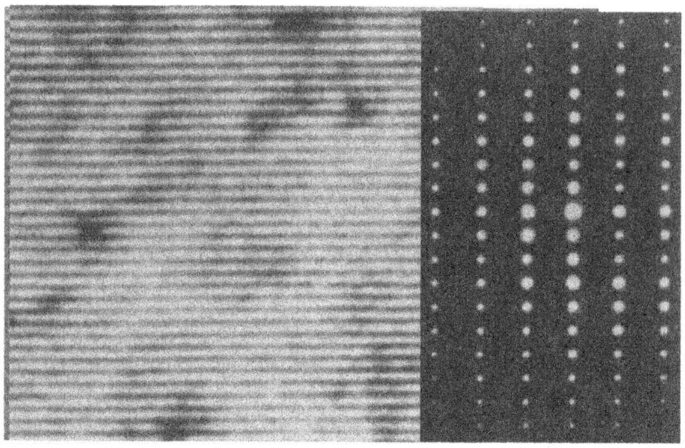
(c)

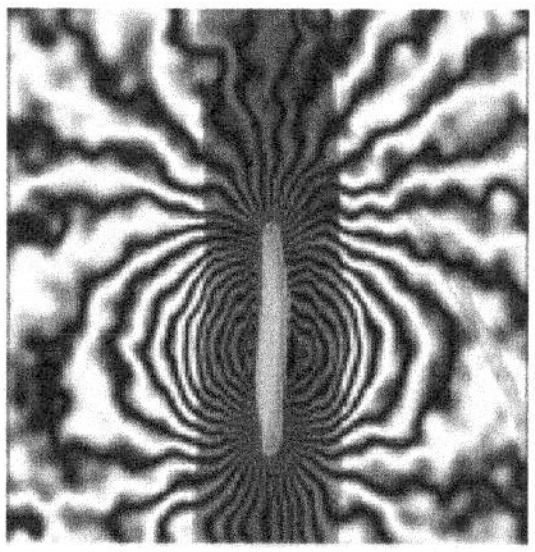

An electron hologram of a 300 nm Co platelet, showing a vortex-like magnetization distribution. A Tonumura, *Rev. Mod. Phys.* 59, 639 (1987).)

quadrant detector. The other method that can provide high-resolution magnetic domain images of suitably polarized specimens is electron holography. There is a significant phase difference ϕ between two convergent electron rays that originate at the same point, and rejoin after following different paths through a ferromagnetic specimen. This is known as the Aharonov–Bohm effect, and the phase difference is $e\Phi$, where Φ is the flux enclosed by two rays. Ferromagnetic specimens act as pure phase objects in the electron beam.

Electron holography can also be used to obtain high-resolution magnetic domain images of suitably polarized samples.

10.5 Bulk magnetization measurements

Methods for determining hysteresis loops and magnetization as a function of applied field are classified as closed-circuit or open-circuit measurements according to whether or not the sample forms part of a complete magnetic circuit.

10.5.1 Magnetization measurements: open circuit

Open-circuit measurements are easier to perform. Samples are usually small, 0.01–100 mg, but the method suffers the inconvenience that the externally applied field $\boldsymbol{H}'$ is different from the $\boldsymbol{H}$-field in the sample, because of the demagnetizing effects discussed in §2.2.4. The magnetic moment $\mathfrak{m}$ is measured and it is usually σ, the magnetic moment per unit mass, in A m^2 kg^{-1}, that is deduced. M is obtained in A m^{-1} by multiplying by the density. For thin films, it is usually the thickness rather than the mass that is known, so M is obtained directly. The units J T^{-1} kg^{-1} and J T^{-1} m^{-3} are identical to A m^{-2} kg^{-1} and A m^{-1}, respectively. Conversion factors and cgs equivalents are discussed in Appendix E and Table B. If $M(H)$ or $B(H)$ data are required as a function of the local internal field H, a fully dense sample of well-defined shape must be used so that a demagnetizing correction can be applied: $\boldsymbol{H} = \boldsymbol{H}' - \mathcal{N}\boldsymbol{M}$, where $\mathcal{N}$ is the demagnetizing factor.

The open-circuit methods for measuring magnetization as a function of applied field fall into two groups. In the first, the force on the sample is measured, whereas in the second the change of flux in a circuit is sensed as the sample is moved.

Force methods In the Faraday balance, the sample is subjected to a nonuniform horizontal field B_x which has a gradient $\mathrm{d}B_x/\mathrm{d}z$ in the (vertical) z-direction. The force on a sample of moment $\mathfrak{m}$ is given by (2.74)

$$f_z = \nabla(\mathfrak{m} \cdot \boldsymbol{B}_0) = \mathfrak{m}(\mathrm{d}B_x/\mathrm{d}z)\mathbf{e}_z, \tag{10.9}$$

where $\boldsymbol{B}_0 = \mu_0 \boldsymbol{H}'$. Hence measurement of the force on a sample freely suspended from a sensitive balance gives $\mathfrak{m}(B_0)$. When the field gradient is generated by an electromagnet with shaped pole pieces, the method is insensitive at zero field since $\mathrm{d}B_x/\mathrm{d}z = cB_x$ with the constant c determined by the shape of the pole pieces. The Faraday balance in its basic form is useless for studying permanent magnets because it cannot measure remanence, but this defect is overcome if the field gradient is produced independently by a set of gradient coils or a small permanent magnet array. The Faraday balance requires calibration with a standard sample and sensitivity of order 10^{-6} A m^2 is typical. From (10.9), this corresponds to the force in a field gradient of 1 T m^{-1} equivalent to a mass of 0.1 mg.

Thermomagnetic analysis (TMA) involves a simple magnetic balance, where the force on a sample in a field gradient produced by a permanent magnet is recorded as temperature is ramped by means of a miniature furnace; TMA is used to measure the Curie temperatures of any magnetic phases present.

The sensitivity of any physical measurement of a continuous analogue output can be enhanced if the continuous signal is converted to an alternating signal at a fixed frequency which is then sensed using a lock-in amplifier tuned to that reference. The alternating gradient force magnetometer (AGFM) is an AC

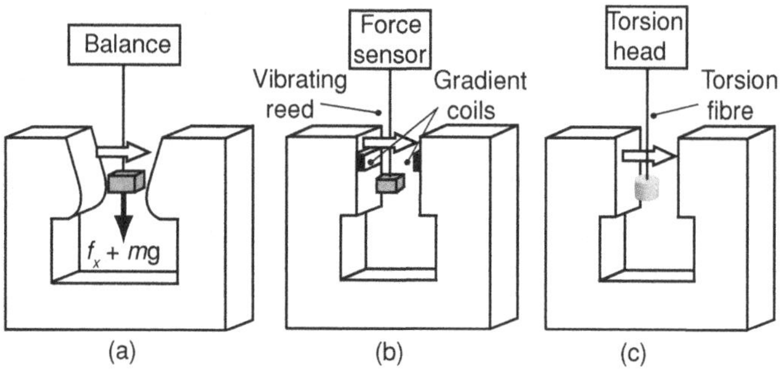

Figure 10.20

Force methods of measuring a sample in the horizontal field of an electromagnet: (a) Faraday method, (b) alternating gradient force method, (c) torque method.

version of the force magnetometer. An alternating field gradient is applied in the horizontal direction at a frequency which may be chosen to coincide with the resonant frequency of the sample support rod, a vertical vibrating reed. The applied field is uniform and also horizontal (Fig. 10.20(b)). The sensitivity of the measurement is thereby increased by some orders of magnitude to 10^{-10} A m^2, which makes it possible to measure small pieces of ferromagnetic films a few nanometres thick, deposited on a substrate.

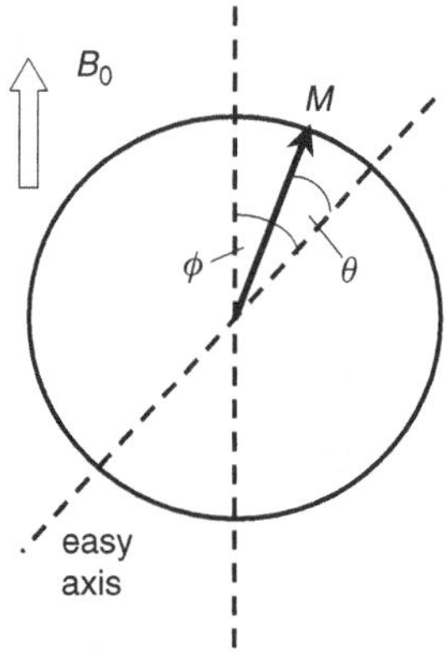

Measurement in a torque magnetometer.

In the torque magnetometer a disc-shaped, cylindrical or spherical specimen of a single-crystal or oriented magnet is suspended perpendicular to an axis of symmetry from a vertical fibre in a horizontal magnetic field (Fig. 10.20(c)). The torque $\mathbf{\Gamma}$ on the specimen is measured as the field is rotated in a horizontal plane. The instrument measures anisotropy, not magnetization. If the field is at an angle ϕ to the easy axis, as shown Fig. 10.20(c), then the energy per unit volume $E = E_a - MB_0V\cos(\phi - \theta)$, where the anisotropy energy E_a is given by the usual expression $E_a = K_1 \sin^2\theta + \cdots$, for example. In equilibrium, $\partial E/\partial\theta = 0$ and $\partial E_a/\partial\theta = -MB_0V\sin(\phi - \theta)$. This term is identified with the torque per unit volume needed to null the deflection, hence

$$\Gamma/V = -\partial E_a/\partial\theta. \tag{10.10}$$

The shape of the torque curve reflects the symmetry of the crystal. For example, $\Gamma/V = -K_1 \sin 2\theta + \cdots$ and K_1 can be deduced from the magnitude of the torque oscillations with this period. Sensitivity is typically 10^{-9} N m. For accuracy, the applied field should *exceed* the anisotropy field, and be sufficient to saturate the sample, making $(\phi - \theta)$ a small angle. This is impractical for many hard magnets, and anisotropy constants are then better deduced from the high-field magnetization curves.

Flux methods The simplest method of measuring magnetization based on the change of flux through a coil is the extraction magnetometer in which a sample positioned at the centre of a coil in the field is quickly removed to a point far from the coil (Fig. 10.21(a)). The change of flux threading the coil is obtained by integrating the induced emf, just as for the search coil. The magnetic moment of the sample is proportional to the change of flux registered on the fluxmeter. An improved pick-up coil is composed of two oppositely

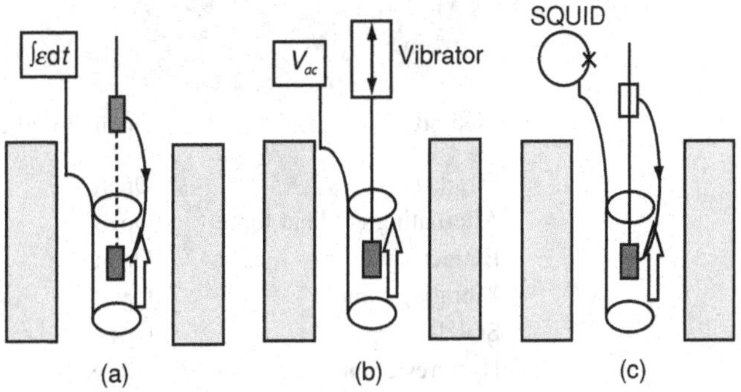

Figure 10.21

Flux methods of measuring magnetization of a sample in the vertical field of a superconducting magnet: (a) extraction, (b) VSM and (c) SQUID magnetometer.

wound segments, so changes of applied field do not register. The sensitivity of an extraction magnetometer is typically 10^{-6} A m^2.

The AC version of the extraction magnetometer is the ever-popular vibrating-sample magnetometer (VSM), also known as the Foner balance, after its inventor. Here the sample is mounted on a vertical rod and vibrated vertically about the centre of a set of pick-up coils (Fig. 10.21(b)). The coil arrangement depends on whether the applied field is vertical, as with a superconducting solenoid, or horizontal, as with an electromagnet. In either case, the upper and lower coils (pairs of coils for the horizontal applied field) are oppositely wound so that the emfs induced in them by the vibrating sample add. Two pairs of coils are used in a quadrupole configuration for the horizontal applied field to create a saddle point around which the sensitivity is independent of sample position. The vibration frequency is typically in the range 10–100 Hz and the vibration amplitude of a few tenths of a millimetre is controlled by a feedback loop. The sensitivity of a well-designed VSM is better than 10^{-8} A m^2. By using a double set of pick-up coils at right angles to each other, the moment vector in the horizontal plane can be recorded and resolved into components $\mathfrak{m}_{\|}$ and $\mathfrak{m}_{\perp}$. If the sample or the field is rotated, the quantity $\mathfrak{m}_{\perp} B_0$ can be deduced; this is equal to the torque Γ, and yields $\partial E_a/\partial\theta$ from (10.10).

A highly sensitive way of measuring the flux change through a pick-up coil is with a superconducting quantum interference device (SQUID). The flux threading a superconducting circuit is a constant, hence current flows in a pick-up coil made of superconducting wire to compensate whatever flux change occurs when the sample is extracted from it (Fig. 10.21(c)). Part of the circuit acts as a transformer to couple some flux into the active area of the SQUID. Great sensitivity is possible, 10^{-10} A m^2 or better, because the device can detect a 10^{-6} fraction of a flux quantum ($\Phi_0 = 2.1 \times 10^{-15}$ T m^2), but the measurement is time-consuming as it involves an extraction at each point and changing the field in the superconducting magnet from one measuring point to the next on the hysteresis loop, which is slow. Again, a further increase of sensitivity to 10^{-12} A m^2 is achieved by operating an AC mode.

Extreme instrumental sensitivity is important in measurements of thin films and interfaces, where the induced diamagnetic moment of the substrate may exceed that of the few micrograms of ferromagnetic film. A correction must

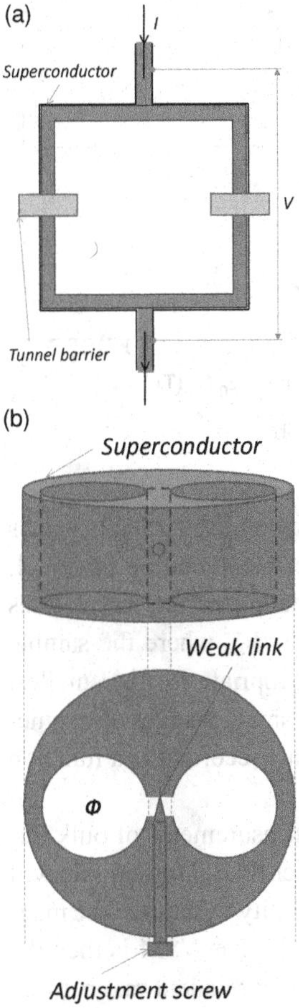

(a) A DC SQUID with two weak links (tunnel junctions).
(b) An RF SQUID with a single weak link (metallic point contact).

Table 10.4. Comparison of methods for measuring magnetization and hysteresis of magnetic materials

Method	Open/closed circuit	Typical sensitivity (A m^2)
Faraday	Open	10^{-6}
Alternating gradient force	Open	10^{-10}
Extraction	Open	10^{-6}
Vibrating sample	Open	10^{-9}
SQUID	Open	10^{-11}
Hysteresigraph	Closed	10^{-4}

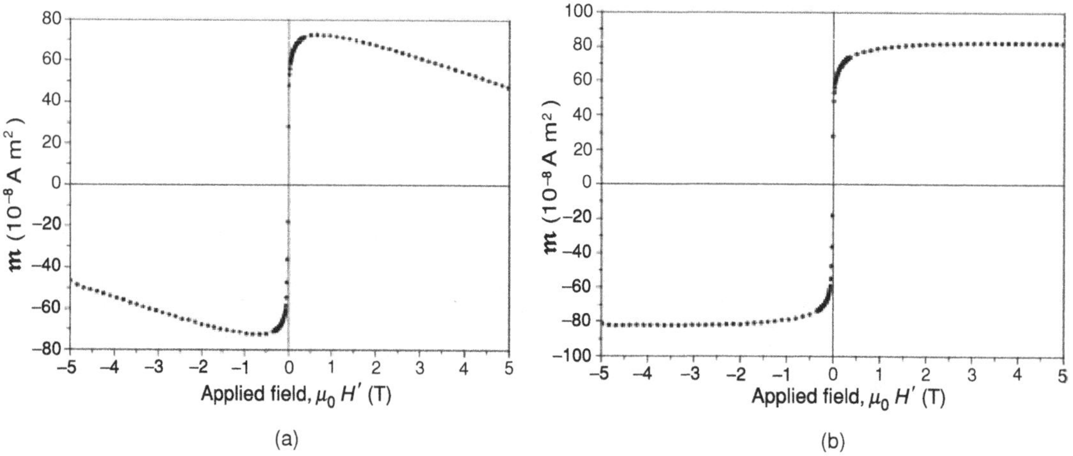

Figure 10.22

Measurement of the magnetization of a ferromagnetic thin film in a SQUID magnetometer: (a) is the signal from a thin film of Fe_3O_4 on a silicon substrate, (b) is the signal after correcting for the substrate diamagnetism. (Data courtesy of M. Venkatesan.)

be made for the susceptibility of the substrate, as shown in Fig. 10.22. Values of susceptibility for common substrate materials were included in Table 3.4. Sometimes it is necessary to measure the magnetization of tiny crystallites comparable in size to the critical single-domain size R_{sd}, where the sample mass may be less than 1 ng. Micro-SQUIDs are appropriate for the smallest specimens. Magnetic viscosity measurements, where small changes of magnetization in the second quadrant of the hysteresis loop are recorded as a function of time, require great stability and sensitivity.

High sensitivity is unnecessary in magnetization measurements of bulk ferromagnets with magnetization ≈ 100 A m^2 kg^{-1}, where a 100 mg sample will have a moment $\approx 10^{-2}$ A m^2. More important is the ability to saturate the magnetization and measure a major hysteresis loop quickly. The VSM is then the best choice. It can operate with an electromagnet or a permanent magnet flux source for measurements on most ferromagnetic materials, but a superconducting magnet may be needed for measurements on rare-earth magnets with strong anisotropy. The sensitivity of different methods for measuring magnetization is compared in Table 10.4.

An example of a magnetization measurement on an oriented spherical crystal of hexagonal YCo_5 which has easy-axis anisotropy is shown in Fig. 10.23.

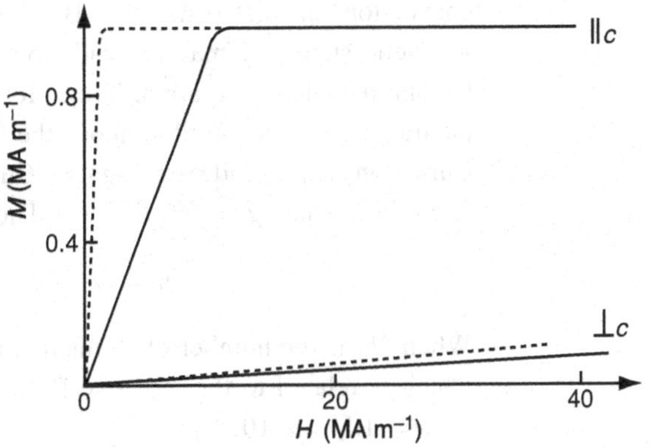

Figure 10.23

Magnetization curves of a crystal of YCo_5 measured parallel and perpendicular to the c axis. The data are plotted as a function of the external field (solid lines) or the internal field (dotted lines) after correction for the demagnetizing field.

Measurements are made for the field applied parallel and perpendicular to the easy axis. In the parallel direction, an external field $H' = M_s/3$ equal to the maximum demagnetizing field is required to achieve saturation. The curves shown by dotted lines have been corrected for the demagnetizing effect, and they are plotted as a function of $H = H' - \mathcal{N}M$, where $\mathcal{N} = 1/3$. The perpendicular magnetization curve is practically linear for this compound, and it reaches saturation when the internal field H reaches the anisotropy field $H_a = 2K_1/\mu_0 M_s$, where K_1 is the first anisotropy constant. There is no high-field slope because YCo_5, like cobalt, is a strong ferromagnet. These data may be compared with Fig. 2.11.

When the second anisotropy constant K_2 is nonnegligible, the perpendicular magnetization curve is nonlinear. Both constants can be deduced from the Sucksmith–Thompson plot of H/M versus M^2 (Exercise 5.9).

A direct measurement of the saturation field H_K can be made in a short-pulse field using the singular-point detection (SPD) technique. The saturation field H_K is equal to H_a when $K_2 = K_3 = 0$. Otherwise H_K is $2(K_1 + 2K_2 + 3K_3)/\mu_0 M_s$. The SPD method involves differentiating the signal in a pick-up coil around the sample so that $\mathrm{d}^2M/\mathrm{d}t^2$ is recorded during the pulse. The derivative $\mathrm{d}B/\mathrm{d}t$ is simultaneously recorded using a search coil so that the field H_K at which a singularity appears in the second derivative of the magnetization curve can be determined. The method can be applied to powders or single crystals.

Susceptibility Susceptibility may be deduced from the slope of the magnetization curve, generally measured in open circuit. When the sample mass is known, it is convenient to deduce σ in A m^2 kg^{-1} and the mass susceptibility $\kappa = \sigma/B_0 = M/\mathrm{d}B_0$, where d is the density. Units of the susceptibility defined in this way are J T^{-2} kg^{-1}. If the volume or density of the sample is known, the dimensionless susceptibility χ defined by $M = \chi H$ can be obtained directly. Mass and volume susceptibilities for representative materials were given in Table 3.4 and a plot for the elements was given in Fig. 3.4. Susceptibility conversions are listed in Table B. The high-field susceptibility in the ferromagnetic state χ_{hf} may be read from the slope of the magnetization curve beyond technical saturation. Values for iron-based alloys are about 10^{-3}. The paramagnetic susceptibility above the Curie point may be used to deduce the Curie constant C and paramagnetic Curie temperature θ_p if the data follow a Curie–Weiss law $\chi = C/(T - \theta_p)$. The effective local moment (4.16) is

$$\mathfrak{m}_{eff} = (3k_B C_{mol}/\mu_0 N_A)^{1/2}. \tag{10.11}$$

When N_A is the number of atoms in a mole, C_{mol} is the molar Curie constant which is related to the effective Bohr magneton number $p_{eff} = \mathfrak{m}_{eff}/\mu_B$ by $C_{mol} = 1.571 \times 10^{-6}\ p_{eff}^2$.

A common magnetic measurement is of the initial susceptibility χ_i in a small low-frequency alternating field $\approx$1–1000 A m^{-1}. A pair of precisely balanced concentric or coaxial pick-up coils is used together with a driving coil to generate the field so that no net emf is induced in the absence of a sample. High sensitivity is achieved by using a lock-in amplifier to detect the signal induced in the coil containing the sample and to determine the real and imaginary parts of the AC susceptibility $\chi = \chi' + i\chi''$ from the components of the signal in phase and in quadrature with the driving field $\boldsymbol{H}$. In aligned uniaxial magnets, χ_i is quite different when measured parallel or perpendicular to the easy axis of the crystallites. The perpendicular susceptibility $\chi'_\perp = \mu_0 M_s^2/2K_1$ is due to magnetization rotation, but the parallel susceptibility $\chi'_\parallel$ is governed by reversible domain-wall motion, and it is different in the virgin and remanent states. The lossy part of the susceptibility χ'' is dominated by irreversible wall motion. AC susceptibility is often used to determine Curie temperatures and find other magnetic phase transition temperatures.

As χ' diverges at T_C, the maximum value of the external susceptibility is limited to $1/\mathcal{N}$ by the demagnetizing field. When measuring nonellipsoidal samples with finite coils, $\mathcal{N}$ itself depends slightly on the susceptibility.

10.5.2 Magnetization measurements: closed circuit

For closed-circuit measurements, the material is usually in the form of a block or cylinder with uniform cross section and parallel faces. It is clamped between the poles of an electromagnet so that it forms part of a closed magnetic circuit. There is no need for a demagnetizing correction, as $H = H'$. The instrument in Fig. 10.24, known as a hysteresigraph or permeameter, is designed to measure $B(H)$ loops where H is the internal field in the sample. For materials which cannot be saturated in the field of the electromagnet ($\approx$2 T), the sample may be first saturated along its axis in a pulsed field, and then transferred to the hysteresigraph for measurement of the demagnetizing curve in the second quadrant. Various arrangements of coils and sensors are available to measure B, M (or $J = \mu_0 M$) and H; B can be measured by winding a coil of N turns of

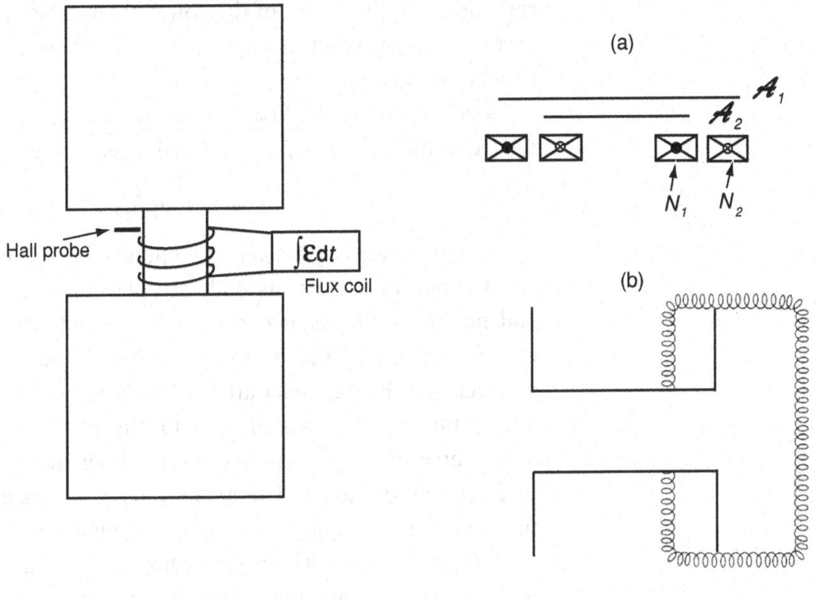

Figure 10.24

Schematic illustration of a hysteresigraph used for measuring B or M as a function of the internal field H. Insets show: (a) a compensated coil used to measure M and (b) the potential coil used to measure H.

fine wire around the sample, and bringing the leads to a fluxmeter as a tightly twisted pair. The fluxmeter is an integrating voltmeter. The cross section of the coil is practically the same as that of the magnet, $\mathcal{A}_m$, and the fluxmeter integrates $-N\mathcal{A}_m(\mathrm{d}\mathcal{E}/\mathrm{d}t)$. As shown in Fig. 10.24, H may be measured using a small Hall probe placed in the airgap close to the sample, since the parallel component $H_{\|}$ is continuous at the interface. Alternatively, a small search coil may be located in the airgap and connected to a second fluxmeter. The field in the electromagnet is swept, and B and H are recorded on a chart recorder or on a computer.

An alternative coil arrangement is used to measure the magnetization M. It consists of two concentric circular flat windings having areas $\mathcal{A}_1$ with N_1 turns and $\mathcal{A}_2$ with N_2 turns. The induced emf is proportional to $N_1[(\mathcal{A}_1 - \mathcal{A}_m)\mu_0 H + \mathcal{A}_m B] - N_2[(\mathcal{A}_2 - \mathcal{A}_m)\mu_0 H + \mathcal{A}_m B]$. If the dimensions of the two coils are chosen so that $N_1\mathcal{A}_1 = N_2\mathcal{A}_2$, the emf is proportional to $(N_1 - N_2)\mathcal{A}_m(B - \mu_0 H) = (N_1 - N_2)\mathcal{A}_m\mu_0 M$. The induced emf can therefore be integrated on the fluxmeter to give magnetization directly.

When the pole pieces of the electromagnet approach saturation, the $\boldsymbol{H}$-field is distorted in the vicinity of the sample. The measurement of H at a spot in the airgap adjacent to the sample can then give erroneous readings. The field may be deduced from the magnetic scalar potential difference $\varphi_{ab} = \int_a^b H\mathrm{d}l$ between the two ends of the magnet determined using a device known as a potential coil. This is a long flexible coil of small cross section area a, evenly wound with n turns m^{-1} of fine wire. It is connected to a fluxmeter, with one end fixed, and the change of flux is measured as the free end is moved from one point to another (Fig. 10.24(b)). Applying Ampère's law $\int H\mathrm{d}l = 0$ to a loop running down the centre of the coil with an arbitrary return path including no current-carrying conductors, $\int_a^b H\mathrm{d}l = (\varphi_a - \varphi_b)$, where a and b are points at the ends of the coil.

Since the path of the first integral runs along the centre of the coil, it can be related to the flux $\Phi = \mu_0 n \mathrm{a}\mathcal{A} \int H\mathrm{d}l$ linking the coil. Since

$$\Phi = \mu_0 n \mathcal{A}(\varphi_a - \varphi_b), \tag{10.12}$$

the coil measures φ_{ab} between two points when it is connected to a fluxmeter and brought up so that its ends are placed at the two points. The potential coil may be split into two parts and the ends embedded in the poles of an electromagnet, as shown in Fig. 10.24(b). The two ends of a potential coil connected to the fluxmeter are analogous to the two voltage probes connected to a voltmeter for normal electrical measurements. A magnet can be regarded as a source of magnetomotive force, rather like a battery which is a source of electromotive force. The analogy between magnetic and electric circuits, though not exact, is quite useful. It is summarized in §13.1.

$B(H)$ measurements on thin films, which have $\mathcal{N} \approx 0$, are carried out with a set of coils that measure the in-plane induction. The accuracy of $B(H)$ measurements, with periodic calibration of the fluxmeter, is of order 1%.

10.5.3 Magnetostriction

Magnetostriction is measured in large single crystals by means of strain gauges glued to the crystal surface, which respond to the strain in a particular direction. For thin films on a substrate of known thickness and elastic constants, the magnetorestriction coefficient can be determined from the slight flexing of the substrate when the film is magnetized, detected by long optical lever. Optical interferometry is also used.

10.6 Excitations

10.6.1 Thermal analysis

The density of states at the Fermi level may be best deduced from the electronic specific heat. At low temperatures (typically $1 < T < 10$ K) the heat capacity in J mol^{-1} K^{-1} of a nonmagnetic metal is found to vary as

$$C = \gamma_{el} T + \beta T^3, \tag{10.13}$$

where the linear term is due to electrons and the cubic term is due to phonons. The coefficient γ_{el} is related to the density of states of both spins at the Fermi

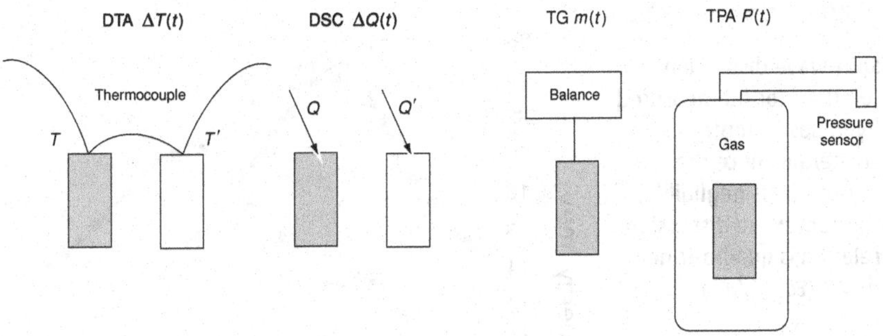

Figure 10.25

Methods of thermal analysis; the heating rate dT/dt is constant, typically 10 K minute^{-1}.

level $\mathcal{N}(\varepsilon_F)$, $\gamma_{el} = \frac{1}{3}\pi^2 k_B^2 \mathcal{N}(\varepsilon_F)$: The linear term is absent in insulators. The coefficient β is related to the characteristic temperature Θ_D in the Debye model, $\beta = 1944/\Theta_D$. Representative values for γ_{el} and Θ_D in $3d$ metals are 5 mJ mol^{-1} K^{-2} and 250 K, respectively. Further contributions to the low-temperature specific heat arise from spin-wave excitations. In general, bosons with a dispersion relation $\varepsilon = Dq^n$ give rise to a term varying as $T^{3/n}$. Since $n = 2$ for ferromagnets, a term $\alpha T^{3/2}$ must then be added to (10.5). The coefficient α is related to the exchange constant A or the exchange integrals $\mathcal{J}_i$.

Characteristic λ-anomalies are observed in heat capacity at magnetic phase transitions, and there are broad Schottky anomalies associated with crystal-field excitations.

In addition to the magnetic methods discussed in §10.4, there is a group of thermal analysis techniques which are useful for detecting magnetic phase transitions and examining gas–solid reactions. These are differential thermal analysis (DTA), differential scanning calorimetry (DSC), thermogravimetric analysis (TG) and thermopiezic analysis (TPA). In each case a uniform heating rate (often 10 K minute^{-1}) is imposed, and the temperature lag, heat flow, weight change or pressure change, respectively, in a closed volume containing the sample is detected (Fig. 10.25). Thermomagnetic analysis (TMA) is a variant of TG in which the sample is placed in a magnetic field gradient, and the apparent weight is monitored.

10.6.2 Spin waves

The method of choice for measuring spin-wave dispersion relations is inelastic neutron scattering. The entire set of $\omega(\boldsymbol{q})$ curves can be mapped out with a triple-axis spectrometer, as shown for Gd in Fig. 10.26 and Tb in Fig. 5.26. Other methods give less complete information. Inelastic light scattering or standing spin-wave resonance, for example, are restricted to the long-wavelength region of small momentum transfer.

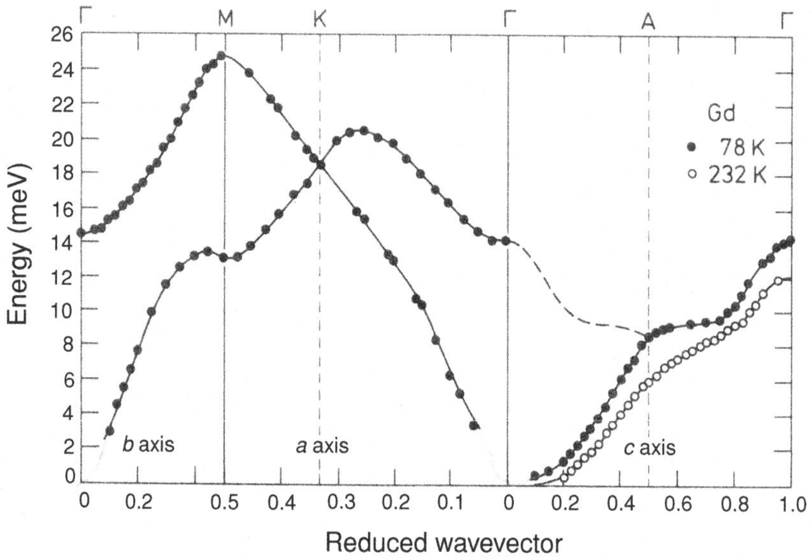

Figure 10.26

Spin-wave dispersion relations for Gd measured by inelastic neutron scattering. Since the anisotropy is negligible, the spin-wave dispersion relation is quadratic near $q = 0$ (Eq. (5.56)).

10.7 Numerical methods

10.7.1 Static fields

Numerical calculations of magnetic fields are based on solving second-order differential equations for the potential, subject to appropriate boundary conditions. When no conduction currents are present, Poisson's equation for the scalar potential (2.64) reads

$$\nabla^2 \varphi_m = \nabla \cdot \boldsymbol{M}. \tag{10.14}$$

The scalar potential is most useful for calculating the fields and forces between magnets represented by sheets of surface charge, either analytically or by numerical integration over a surface. If conduction currents are present, the vector potential must be used. In that case

$$\nabla^2 \boldsymbol{A} + \mu_0 \nabla \times \boldsymbol{M} = -\mu_0 \boldsymbol{j}, \tag{10.15}$$

which reduces to Poisson's equation for the vector potential (2.60) $\nabla^2 \boldsymbol{A} = -\mu_0 \boldsymbol{j}$ if $\boldsymbol{M}$ is uniform. The nonzero divergence at the surface is then represented by sheets of surface charge $\boldsymbol{M} \cdot \mathbf{e}_n$. The drawback in using the vector potential is that it is necessary to manipulate three vector components rather than just one for the scalar potential. However, two-dimensional problems where $\boldsymbol{B}$ is confined to the xy-plane can be expressed in terms of a single component of $\boldsymbol{A}$.

Numerical methods for solving the second-order differential equations involving the potential for real magnetic systems fall into two broad categories,

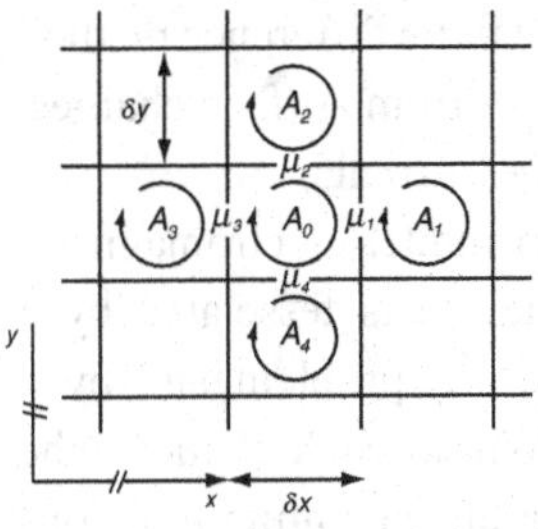

An array of points for finite-difference solution of differential equations.

finite difference and finite element. The former provide approximate solutions valid at an array of points throughout the region of interest, whereas the latter provide approximate solutions which are valid everywhere by dividing space into a mesh of three-dimensional cells. The numerical computations proceed in three stages. First is a preprocessing stage where the geometry of the device is defined and the materials properties are specified. Next comes the numerical solution of the potential equations, and finally there is a postprocessing stage, where the fields and forces of interest are extracted from the numerical results.

There are various ways of incorporating a field-dependent magnetization into the equations. An isotropic paramagnet or soft ferromagnet has a single-valued magnetization which can be written as $\boldsymbol{M} = \chi \boldsymbol{H}$ far from saturation. The magnetization of a hard magnet in its operating range can be approximated as

$$M_{\parallel} = \chi_{\parallel} H_{\parallel} + M_r,$$

$$M_{\perp} = \chi_{\perp} H_{\perp},$$

where $\chi_{\parallel} \approx 0$ and $\chi_{\perp} \approx \mu_0 M_s^2/2K$. If we set $(1 + \chi_{\parallel}) = \mu$ and $\chi_{\perp} = 0$, $\boldsymbol{B} = \mu_0(\mu \boldsymbol{H} + \boldsymbol{M}_r)$, so that (10.15) becomes

$$\nabla(1/\mu_0\mu) \times \boldsymbol{B} = \boldsymbol{j} + \nabla \times \boldsymbol{M}_r/\mu. \quad (10.16)$$

The remanence can be assimilated as an effective current $\boldsymbol{j}_r$. The second-order differential equation for $\boldsymbol{A}$ becomes

$$\nabla \times \nabla\nu \times \boldsymbol{A} = \mu_0(\boldsymbol{j} + \boldsymbol{j}_r), \quad (10.17)$$

where the reluctivity ν equals $1/\mu$. In a two-dimensional problem only the components A_z and j_z generate fields in the xy-plane, and $B_x = \partial A_z/\partial y$, $B_y = \partial A_z/\partial x$. The equation for A_z becomes

$$\frac{\partial(\nu \partial A_z/\partial x)}{\partial x} + \frac{\partial(\nu \partial A_z/\partial y)}{\partial y} = -\mu_0 j_z. \quad (10.18)$$

For a numerical solution, the partial derivatives are expressed as difference approximations. An array of points is set up in the xy-plane where values of $\boldsymbol{A}$ and ν are defined. If the separation of the points is δ, then, for example,

$$n_1(A_1 - A_0)\delta - n_3(A_0 - A_3)\delta + n_2(A_2 - A_0)\delta - n_4(A_0 - A_4)\delta = -\mu_0 j_0. \quad (10.19)$$

Hence $A_0 = [\sum \nu_i A_i + \delta^2 \mu_0 j_0]/\sum \nu_i$, and similarly for all the other cells. Taking a plausible set of A_i to begin with, the A_i values are refined by repeated scanning over all the points on the network. Then $\boldsymbol{B}$ is calculated at any point in the plane as $\nabla \times \boldsymbol{A}$. The finite difference method is slow, but there are numerical techniques to accelerate convergence.

An alternative approach involves the scalar potential φ_m. Instead of taking the curl of $\boldsymbol{B} = \mu_0(\mu \boldsymbol{H} + \boldsymbol{M}_r)$ as in (10.16), we operate with div, yielding

$$\nabla \cdot \mu \nabla \varphi + \nabla \cdot \boldsymbol{M}_r = 0. \quad (10.20)$$

This differential equation can likewise be solved by the finite difference method to give φ. If $\boldsymbol{M}_r$ is uniform there is only a contribution from surface charges, as discussed for (10.15). It is assumed that there are no currents.

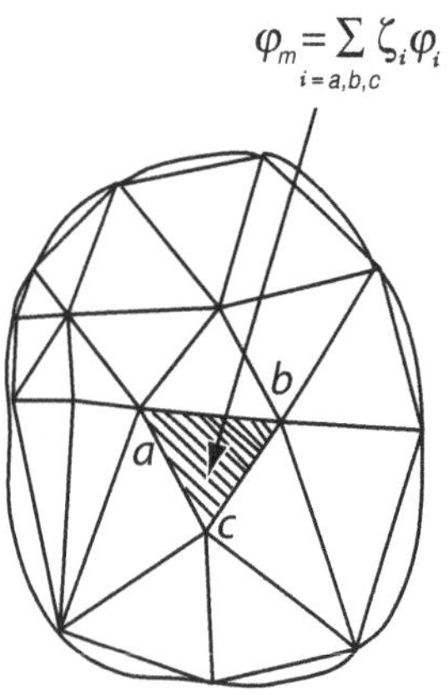

A triangular mesh used for finite-element calculations.

The finite-element method is increasingly used to model electromagnetic devices. Space occupied by and surrounding the device is tesselated by a suitably adapted triangular mesh for a two-dimensional problem and by a tetrahedral mesh for a three-dimensional problem. Within each element, the potential is described by its nodal values with appropriate weighting functions ζ_i. Taking the scalar potential in two dimensions, for example:

$$\varphi_m = \textstyle\sum_i \zeta_i \varphi_i. \tag{10.21}$$

Next, an energy functional F is defined which is an integral over Ω, the whole space of interest. The functional for the two-dimensional problem involving scalar potential is

$$F = \textstyle\int f(x, y, \varphi, \partial\varphi/\partial x, \partial\varphi/\partial y)\mathrm{d}\Omega.$$

This integral is minimized when the Euler equations are satisfied:

$$\partial f/\partial\varphi - (\mathrm{d}/\mathrm{d}x)(\partial f/\partial\varphi) = 0. \tag{10.22}$$

F is chosen so that these coincide with the differential equations which should be satisfied by the potential. For example, the functional which yields (10.22) is

$$F = -\int [\varphi\nabla \cdot \mu\nabla\varphi - 2\varphi\nabla \cdot \boldsymbol{M}_r]\mathrm{d}\Omega. \tag{10.23}$$

The energy functional is evaluated for each element in terms of the nodal potentials by substituting (10.23) in (10.22) and the functional is minimized by setting $\mathrm{d}F/\mathrm{d}\varphi = 0$ for each element, which gives a set of equations describing the entire region that provide the potential distribution. Like the finite-difference method, the procedure is iterative and it continues until the residuals are acceptably small. The required fields are obtained in the post-processing stage.

Finite-element software packages are available to model magnetic systems in two and three dimensions. Adaptive meshes are generated automatically, and the packages are a valuable design tool for the magnetic engineer.

10.7.2 Time dependence

For modelling domain structures and the dynamic response of a micromagnetic system, it is appropriate to use software which implements algorithms based on the Landau–Lifschitz–Gilbert equations (9.22) or (9.23).

A popular micromagnetics package is OOMMF, which may be downloaded free of charge at http://math.nist.gov/oommf/.

FURTHER READING

E. Steinbeiss, Thin film deposition, in *Spin Electronics*, M. Ziese and M. J. Thornton (editors), Berlin: Springer (2001).

R. Eason (editor), *Pulsed-laser Deposition of Thin Films*, New York: Wiley-Interscience (2006).

F. Fiorillo, *Measurement and Characterization of Magnetic Materials*, Amsterdam: Elsevier (2004).

G. M. Kalvius and R. S. Tebble (editors), *Experimental Magnetism* (2 vols.), Chichester: J. Wiley (1979).

T. C. Bacon, *Neutron Diffraction* (third edition), Oxford: Clarendon Press (1975). An ever-popular introduction.

N. N. Greenwood and T. C. Gibb, *Mössbauer Spectroscopy*, London: Chapman and Hall (1970). A survey for solid-state chemists.

E. S. R. Gopal, *Specific Heats at Low Temperatures*, New York: Plenum (1966). A clear elementary text.

J. Baruchel, J. L. Hodeau, M. S. Lehmann, J. R. Regnard and C. Schlenker (editors), *Neutron and Synchrotron Radiation for Condensed Matter Studies* (vols. 1 and 2), Berlin: Springer (1993). Lectures for an annual school for users of large-scale facilities.

G. Shirane, S. Shapiro and J. Tranquada, *Neutron Scattering with a Triple-Axis Spectrometer*, Cambridge: Cambridge University Press (2006). A practical guide, by experts.

P. P. Silvester and R. L. Ferrari, *Finite Elements for Electrical Engineers*, London: Cambridge University Press (1990).

J. N. Chapman, The investigation of magnetic domain structures in thin foils by electron microscopy, *Journal of Physics D* **17** 623–647 (1984).

A. Hubert and R. Schäffer, *Magnetic Domains*, Berlin: Springer (1998).

EXERCISES

10.1 Estimate the mass of material used to make a 100 nm thick film of iron (a) by thermal evaporation and (b) by e-beam evaporation.

10.2 The average intensity of solar radiation at the Earth's surface is 1.4 kW m^{-2}. Deduce the average magnetic field in the light. What energy density is required to establish a field of 1 MA m^{-2}?

10.3 Use the data in Table 5.11 to estimate the depth of penetration of light in Fe, Co and Ni.

10.4 Calculate the angular deflection of a 100 keV electron beam after it passes through 10 nm of permalloy ($\mu_0 M_s = 1.0$ T). Compare it with the magnitude of the angular deflection produced at a Bragg reflection.

10.5 Use Gauss's theorem to show that only domains magnetized along x in the marginal illustration on p. 357 produce a net deflection of an electron beam.

11 Magnetic materials

The keenest desire of matter is form

Almost all magnetically ordered materials involve $3d$ or $4f$ elements. The late $3d$ metals and their alloys, including interstitial alloys and intermetallic compounds, are frequently ferromagnetic. Large magnetocrystalline anisotropy in $3d$–$4f$ intermetallics, due mainly to the rare-earth, gives useful hard magnets. Conversely, the anisotropy can be reduced practically to zero in certain $3d$ alloys and, when magnetostriction is also vanishingly small, perfect soft ferromagnetism results. Oxides and other ionic compounds are usually insulators with localized electrons. There, antiferromagnetic superexchange coupling leads to antiferromagnetic or ferrimagnetic order. Some oxides, however, are metals, with the d-electrons forming a conduction band. Occasionally the $3d$ band is half-metallic. A few examples are included of materials showing magnetic order which involves neither $3d$ nor $4f$ electrons.

11.1 Introduction

This chapter is a catalogue of representative magnetically ordered materials. The selection is biased towards materials that are practically useful, or illustrate some interesting aspect of magnetic order. Included are the common iron-group metals and alloys, the rare-earths, intermetallic and interstitial compounds, as well as a range of oxides with ferromagnetic or antiferromagnetic interactions. The catalogue covers insulators, semiconductors, semimetals and metals. Ferromagnetic, antiferromagnetic, ferrimagnetic and noncollinear spin structures are encountred. Examples of noncrystalline metals and insulators are also included. Table 11.1 collects information on the 38 representative materials. Each is described more fully on a data sheet, where its properties and significance are indicated, and related materials are presented. In this way, it is possible to deal with about 200 magnetic materials in a fairly digestible manner.

A common format is followed on each sheet. There are two views of the crystal structure, on one of two scales to facilitate comparisons. The Strukturbericht symbols for the simpler structure-types, prototype structure, space group, lattice parameters and site occupancy all follow standard crystallographic practice. For rhombohedral structures, the corresponding hexagonal cell is shown. Enough information is given to allow the reader to draw the structure using crystallographic software such as *Crystalmaker*. Z is the number of formula

Table 11.1. Selected magnetic materials

Material	Structure		Exchange	Order	Electrons	T_c (K)	m$_0$ (μ_B/fu)	$\mu_0 M_s$ (T)	K_1 (kJ m^{-3})
Fe	Cubic	Metal	+	f	3d	1044	2.22	2.15	48
$Fe_{0.65}Co_{0.35}$	Cubic	Metal	+	f	3d	1210	2.46	2.45	18
Ni	Cubic	Metal	+	f	3d	628	0.58	0.61	−5
$Fe_{0.20}Ni_{0.80}$	Cubic	Metal	+	f	3d	843	1.02	1.04	−2
a-$Fe_{0.40}Ni_{0.40}B_{0.20}$	Amorphous	Metal	+	f	3d	535	0.90	0.82	≈0
Co	Hexagonal	Metal	+	f	3d	1360	1.71	1.81	410
CoPt	Tetragonal	Metal	+	f	3d	840	2.35	1.01	4900
MnBi	Hexagonal	Metal	+	f	3d	633	3.12	0.73	900
NiMnSb	Cubic	Metal	+	f	3d	730	3.92	0.84	13
Mn	Cubic	Metal	−	af	3d	95	0.76	–	
$IrMn_3$	Cubic	Metal	−	af	3d	960		–	
Cr	Cubic	Metal	−	af	3d	310	0.43		–
Dy	Hexagonal	Metal	+/−	he/f	4f	179/85	10.4	3.84[a]	−55000[a]
$SmCo_5$	Hexagonal	Metal	+	f	3d 4f	1020	8.15	1.07	17200
$Nd_2Fe_{14}B$	Tetragonal	Metal	+	f	3d 4f	588	37.7	1.61	4900
Y_2Co_{17}	Hexagonal	Metal	+	f	3d	1167	26.8	1.26	−340
$TbFe_2$	Cubic	Metal	+	f	3d 4f	698	6.0	1.10	−6300
$Gd_{0.25}Co_{0.75}$	Amorphous	Metal	+/−	fi	3d 4f	≈700	0.6	0.10	≈0
Fe_4N	Cubic	Metal	+	f	3d	769	8.9	1.89	−29
$Sm_2Fe_{17}N_3$	Rhombohedral	Metal	+	f	3d	749	39.0	1.54	8600

(*cont.*)

Table 11.1. *(cont).*

Material	Structure		Exchange	Order	Electrons	T_c (K)	m_0 (μ_B/fu)	$\mu_0 M_s$ (T)	K_1 (kJ m^{-3})
EuO	Cubic	Semiconductor	+	f	$4f$	69	7.00	2.36[a]	44[a]
CrO_2	Tetragonal	Half metal	+	f	$3d3d$	396	2.00	0.49	37
$SrRuO_3$	Orthorhombic	Metal	+	f	$4d$	165	1.40	0.25[a]	640[a]
$(La_{0.70}Sr_{0.30})MnO_3$	Rhombohedral	Metal/semicon.	+	f	$3d3d$	250	3.60	0.55	−2
Sr_2FeMoO_6	Orthorhombic	Metal	+	f	$3d$ $4f$	425	3.60	0.25	28
NiO	Cubic	Insulator	−	af	$3d$	525	(2.0)	–	
α-Fe_2O_3	Rhombohedral	Insulator	−	caf	$3d$	960	(5.0)	0.003	−7
γ-Fe_2O_3	Cubic	Insulator	−	fi	$3d$	985	3.0	0.54	
Fe_3O_4	Cubic	Semiconductor	−	fi	$3d$	860	4.0	0.60	−13
$Y_3Fe_5O_{12}$	Cubic	Insulator	−	fi	$3d$	560	5.0	0.18	−0.5
$BaFe_{12}O_{19}$	Hexagonal	Insulator	−	fi	$3d$	740	19.9	0.48	330
MnF_2	Tetragonal	Insulator	−	af	$3d$	68	(5.0)	–	
a-FeF_3	Amorphous	Insulator	−	sp	$3d$	29	(5.0)	–	
Fe_7S_8	Monoclinic	Semiconductor	−	fi	$3d$	598	3.16	0.19	320
$Cu_{0.99}Mn_{0.01}$	Cubic	Metal	+/−	sg	$3d$	6	0.03	–	
$(Ga_{0.92}Mn_{0.08})As$	Cubic	Semiconductor	+	f	$3d$	170	0.28	0.07[a]	2
US	Cubic	metal	+	f	$5f$	177	1.55	0.66[a]	43000[a]
O_2	Monoclininc	Insulator	−	af	$2p$	24	(0.3)	–	4600[a]
Organic ferromagnet	Orthorhombic	Insulator	+	f	$2p$	0.6	0.5	0.02	

f – ferromagnet; af – antiferromagnet; fi – ferrimagnet; caf – centred antiferromagnet; he – helimagnet; sg – spin glass; sp – speromagnet.

[a] At $T = 0$.

Table 11.2. Magnetic parameters of some useful materials

Material	T_c (K)	J_s (T)	A^b (pJ m^{-1})	K_1 (kJ m^{-3})	κ	δ_w (nm)	γ_w (mJ m^{-2})
Fe	1044	2.15	22	48	0.1	67	4.1
Co	1360	1.82	31	410	0.4	26	15
Ni	628	0.61	8	−5	0.1	140^a	0.7^a
$Ni_{0.80}Fe_{0.20}$	843	1.04	7	−2	~0	190^a	0.5^a
$SmCo_5$	1020	1.07	12	17200	4.3	2.6	57
Sm_2Co_{17}	1190	1.25	16	4200	1.8	5.7	31
CoPt	840	0.99	10	4900	2.5	4.5	28
$Nd_2Fe_{14}B$	588	1.61	8	4900	1.5	4.0	25
$Sm_2Fe_{17}N_3$	749	1.54	12	8600	2.1	3.7	41
CrO_2	396	0.49	4	25	0.4	40	1.3
Fe_3O_4	860	0.60	7	−13	0.2	73	1.2
$Y_3Fe_5O_{12}$	560	0.18	4	−50	0.3	28	1.8
$BaFe_{12}O_{19}$	740	0.48	6	330	1.3	13	5.6

[a] In very soft materials, the wall width and energy may be determined by geometric constraints.

[b] There is uncertainty in these values. A is not measured directly, and different methods of deriving it give different results.

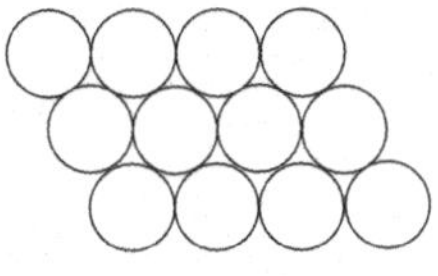

A dense-packed hexagonal layer.

Dense packed structures, hcp (top) and fcc (bottom).

units (fu) in the unit cell. The quantity d is the X-ray density; real densities are a little less. All formulae represent the atomic composition, not the weight. Intrinsic electrical and magnetic values given are for room temperature, except where stated otherwise. The moment m is in Bohr magnetons per formula. In view of its importance, the magnetization of the ferromagnets and ferrimagnets is listed in three ways: as magnetic moment per unit mass (σ), magnetic moment per unit volume ($M_s = \sigma \mathrm{d}$) and polarization ($J_s = \mu_0 M_s$). A subscript $_0$ on σ, M or J indicates a low-temperature value. Extrinsic magnetic properties such as permeability, coercivity and remanence are not included here, since they depend on the microstructure of the specimen. These matters are included in the discussion of applications in Chapters 12 and 13. Resistivities ϱ are indicative, as they depend on sample purity. For insulators, the primary bandgap ε_g is given instead. Magnetic parameters for a selection of useful materials are summarized in Table 11.2.

The atoms in metals frequently form close-packed arrays. Close packing in a plane gives a layer of filled hexagons. The most common stacking sequences of the hexagonal layers ABCABC... and ABABAB... are fcc and hcp, denoted as A1 and A3 structures, respectively. Both close-packed structures have a packing fraction $\mathfrak{f} = \pi\sqrt{2}/6 = 0.74$, and the c/a ratio for the hcp structure is $\sqrt{8/3} = 1.63$. The bcc structure, denoted as A2, is not close-packed; it has the lesser packing fraction $\mathfrak{f} = \pi\sqrt{3}/8 = 0.68$.

Table 11.3. Transition-metal alloy superstructures

Structure	Superstructure	Z	Examples
bcc	B2	2	FeCo
bcc	DO_3	16	Fe_3Al, Fe_3Si
fcc	$L1_0$	4	FePt, CoPt
fcc	$L1_2$	4	Ni_3Fe, Pt_3Co, Fe_3Pt
fcc	$L2_1$	16	Co_2MnSi
fcc	C1b	12	NiMnSb

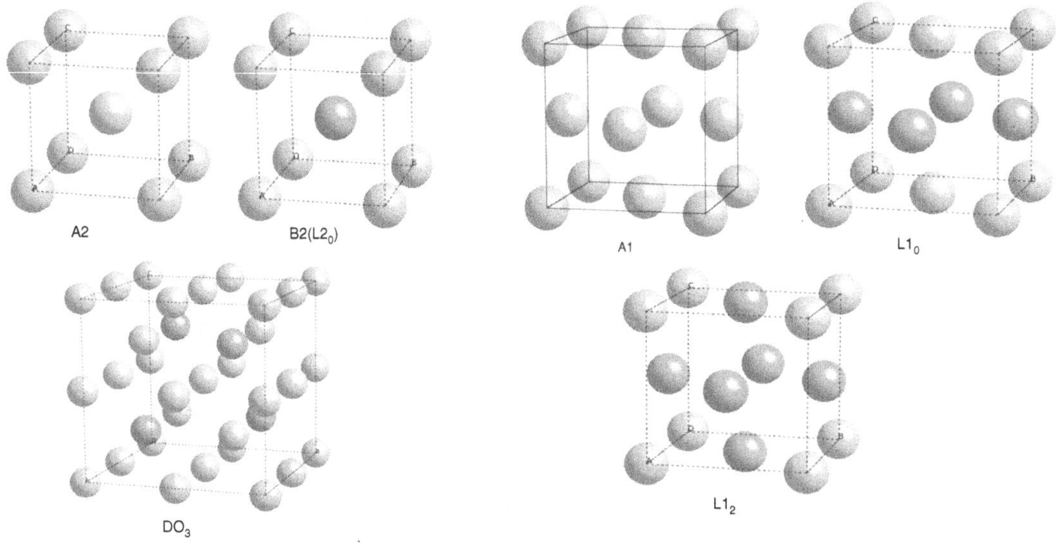

Figure 11.1

Ordered binary superstructures of the bcc and fcc crystal structures.

When metals of similar size form an alloy, either a solid solution or an ordered superstructure is possible. Some common ordered structures are listed in Table 11.3, and they are illustrated in Fig. 11.1. The ferromagnetic $3d$ metals form a series of alloys with each other and with other metals. These alloys may exist in both ordered and disordered forms. The degree of order is influenced by thermal treatment.

If two elements of quite different size and electronegativity are involved, they tend to form intermetallic compounds of well-defined composition. Atomic radii of the late $3d$ elements are about 125 pm. The atomic radii of rare-earths are around 180 pm. The $4f$ elements therefore occupy three times the volume of the $3d$ elements, and form intermetallic compounds, rather than solid solutions, with them. Intermetallics are often line compounds with a precisely defined composition, or else there is a very limited homogeneity range because they can tolerate little disorder of the constituent atoms. Some atomic radii are listed in Table 11.4, and the trends across the $3d$, $4d$, $5d$, $4f$ and $5f$ are illustrated in Fig. 11.2.

Table 11.4. Metallic radii of 3*d*, 4*f* and other elements (in pm)

Al	143	Mn	124	Zn	133	Nd	182	Dy	177	Pd	138	Pb	175
Sc	160	Fe	124	Y	181	Sm	180	Ho	177	Pt	138	Bi	155
Ti	145	Co	125	La	188	Eu	204	Er	176	Ag	160	Th	180
V	132	Ni	125	Ce	183	Gd	180	Yb	194	Au	159	U	139
Cr	125	Cu	128	Pr	183	Tb	178	Lu	173	Sn	141	Np	131

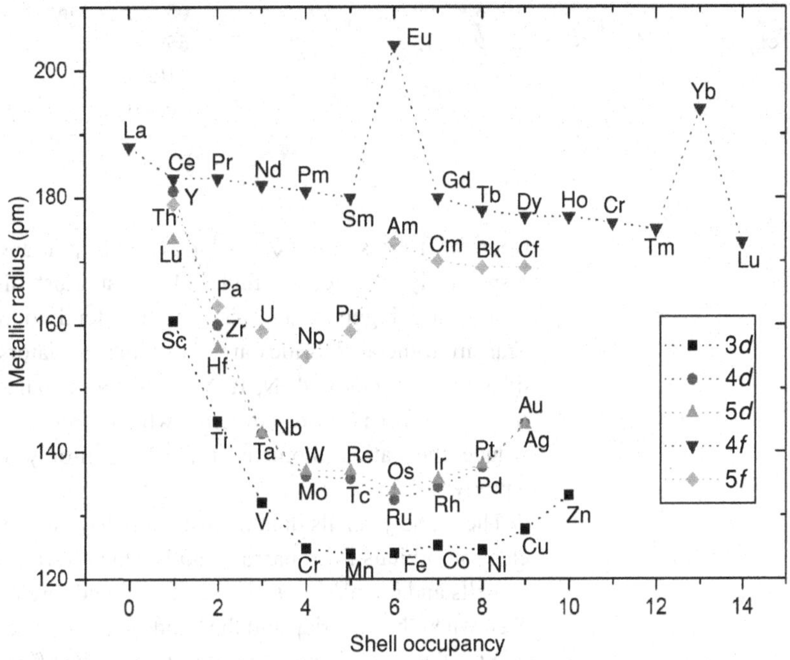

Figure 11.2
Metallic radii of all the transition-metal series.

Interstitial compounds form when a small atom with atomic radius < 100 pm, – boron, carbon or nitrogen – enters an octahedral interstitial site in a 3*d* metal or alloy. Hydrogen can also enter the structure of many rare-earth elements and their compounds. At room temperature, the hydrogen forms an interstitial lattice gas of protons (H^+) at room temperature, and produces a lattice dilation of approximately 7×10^6 pm^3 per hydrogen atom. Dilation produces striking changes of the magnetic properties of iron alloys, where the exchange is particularly sensitive to interatomic spacing.

Another important class of magnetic materials are the ionic insulators. These are often oxides, where electron transfer from the metal fills the oxygen 2*p* shell, to create an O^{2-} anion and leaves behind a positively charged metal cation with a partially filled *d* or *f* shell. Most oxide structures are based on dense-packed fcc or bcc oxygen arrays, with metallic cations occupying octahedral, and sometimes also tetrahedral interstices. Taking $r_{O^{2-}} = 140$ pm, the radii of the cations which will just fit in octahedral (six-fold) and tetrahedral

Table 11.5. Ionic radii of $3d^n$ and other ions in oxides. Low-spin state values are in brackets. The number of $3d$ electrons is in italics. The O^{2-} radius is taken as 140 pm

4-fold	n	pm	6-fold	n	pm	6-fold	n	pm	6-fold	n	pm	8-fold	pm
			Mg^{2+}	*0*	72	Al^{3+}	*0*	56					
Mg^{2+}		57	Mn^{2+}	*5*	83	Ti^{3+}	*1*	67	Ti^{4+}	0	75	Ca^{2+}	112
Zn^{2+}		74	Fe^{2+}	*6*	78(61)	V^{3+}	*2*	64	V^{4+}	1	58	Sr^{2+}	126
Al^{3+}		39	Co^{2+}	*7*	75(65)	Cr^{3+}	*3*	62	Cr^{4+}	*2*	55	Ba^{2+}	142
			Ni^{2+}	*8*	69	Mn^{3+}	*4*	65	Mn^{4+}	*3*	53		
Fe^{3+}	*5*	49	Cu^{2+}	*9*	73	Fe^{3+}	*5*	65				Y^{3+}	102
						Co^{3+}	*6*	61(55)				La^{3+}	116
						Ni^{3+}	*7*	60(56)				Gd^{3+}	105
												Lu^{3+}	98

(four-fold) sites are $(\sqrt{2}-1)r_{O^{2-}} = 58$ pm and $(\sqrt{\frac{3}{2}}-1)r_{O^{2-}} = 32$ pm, respectively. We see in Table 11.5 that most divalent and trivalent metal cations are bigger than this, so they tend to distort the oxygen lattice. Transition-metal fluorides are also ionic insulators, but the bonding in pnictides (compounds with N, P, As or Sb) and chalcogenides (compounds with S, Se and Te) is more covalent, which tends to raise the conductivity and reduce the cation moment from its spin-only value, and may destroy it entirely.

The d and f shells in ionic insulators have integral electron occupancy. The electron orbitals form narrow bands,which have a width W of about 2 eV for d shells and 0.2 eV for f shells. These bands are unable to conduct electricity, even when they overlap and the bands are not all completely full or empty. The point is that a conducting band has to include different instantaneous electronic configurations of the atoms, such as $3d^{n\pm1}$ as well as $3d^n$. For this, an electron must somewhere be transferred from one site to the neighbouring site, at an energy cost equal to U_{dd}, which is the difference in ionization energy and electron affinity of the $3d^n$ configuration in the solid. The value of U_{dd} is a few electron volts, which must be less than the total bandwidth if electron transfer is to take place. Otherwise, when

$$\frac{U_{dd}}{W} > 1$$

the material is a Mott insulator, and the electrons stay put. The d–d electron correlations turn a material which would otherwise be a metal into an insulator. A competing charge transfer process in oxides is from the filled oxygen $2p$ shell to the $3d$ shell. The electronic excitation is then $2p^6 3d^n \rightarrow 2p^5 3d^{n+1}$, and the energy cost is ε_{pd}. When $U_{dd} > U_{pd} > W$, the oxide is a charge transfer insulator. The Mott insulators tend to be found at the beginning of the $3d$ series, where the $3d$ level lies high in the $2p$(O)$-4s$(T) gap (T is a $3d$ transition

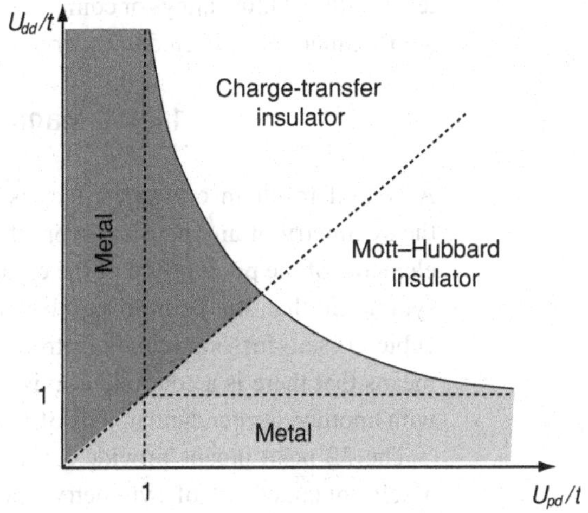

Figure 11.3

The Zaanen–Sawatzky–Allen diagram. U_{dd} and U_{pd} are the charge-transfer energies and t is the hopping integral, which determines the bandwidth (After J. Zaanen *et al.*, *Physical Review Letters*, 55, 418 (1985).).

metal), whereas the charge transfer insulators are found near the end of the series, where the $3d$ level lies near the top of the $2p(O)$ band. Figure 11.3 delineates the regions where metals or insulators are found.

Oxides are rarely perfectly stoichiometric, yet, unlike doped semiconductors, nonstoichiometric oxides remain insulating. Electrons in the $d^{n\pm1}$ configurations which form in response to oxygen excess or deficiency are immobile. They create a local distortion of the ionic lattice, known as a polaron. For polarons to hop from one site to the next, they must have the thermal energy necessary to overcome the energy barrier associated with the redistribution of the local lattice distortion.

Finally, there are a few materials whose magnetism does not fit the general picture of more-or-less localized d or f atomic moments, with interatomic Heisenberg exchange coupling. These include solid O_2, which is a molecular antiferromagnet, some organic ferromagnets, and alloys like $ZnZr_2$ whose component elements bear no atomic moment. No homogeneous liquids are known to order magnetically.

Magnetic order is a relatively low-temperature phenomenon. A histogram plot of Curie and Neel temperatures from a bibliography of magnetic materials (Fig. 1.8) shows that most order below room temperature. A Curie temperature greater than 500 K, needed for room-temperature applications, occurs in no more than about 20% of known magnetic materials. The record for a Curie temperature, $T_C = 1388$ K, is held by cobalt.

For solid solutions of alloys of magnetic and nonmagnetic atoms, (T_xN_{1-x}), the Curie temperature in mean field theory (5.26) should scale as $Z_T = Z_x$, where Z_T is the number of magnetic neighbours and Z is the coordination number. However, ferromagnetic nearest-neighbour exchange interactions do not produce long-range ferromagnetic order below the percolation threshold x_p. Weaker, longer-range interactions may lead to magnetic order at low temperature. Dilute alloys or compounds which contain less than 10% magnetic atoms cannot be expected to order magnetically at room temperature, if at all.

11.1.1 Magnetic symmetry

A central result in crystal physics is Neumann's principle, which states that the symmetry of any physical property of a solid must include the symmetry elements of the point group of the crystal. The physical property may be more symmetric than the point group, but it cannot be less. The conductivity of a cubic crystal, for example, is isotropic, with symmetry $\infty\infty m$. The notation means that there is a continuous axis of rotation ∞ lying in a mirror plane m, with another, perpendicular, axis of rotation.

The 32 point groups provide a classification of crystal symmetry. Each is a self-contained set of symmetry operations, present at every lattice point, which transform the atomic positions in the crystal lattice into an identical set of atomic positions. The point symmetry operations are the inversion centre $\bar{1}$, the mirror plane m, the rotation axes $2, 3, 4$ and 6, and the rotation-inversion axes $\bar{3}, \bar{4}$ and $\bar{6}$ ($\bar{2}$ is equivalent to m). Here the integer n indicates rotation through $2\pi/\text{n}$. The identity operation 1 is trivial and indicates no symmetry. By Neumann's principle, the point group symmetry should be reflected in the magnetic properties such as magnetocrystalline anisotropy.

Conventionally, the highest rotation axis is placed first in the point group symbol, then follows another in the perpendicular direction and/or a mirror containing the axis. If the mirror is perpendicular to the axis, it is preceded by a '/'. The seven crystal classes each have two or more point groups, as follows: triclinic 1, $\bar{1}$; monoclinic 2, m; $2/m$; orthorhombic 222, $mm2$, mmm; trigonal 3, $\bar{3}$, 32, $3m$, $\bar{3}m$; tetragonal 4, $\bar{4}$, $4/m$, 422, $4mm$, $\bar{4}2m$, $4/mmm$; hexagonal 6, $\bar{6}$, $6/m$, 622, $6mm$, $\bar{6}m2$, $6/mmm$; cubic 23, $m3$, 432, $\bar{4}3m$, $m3m$.

The eight most common point groups for inorganic crystals, namely $2/m$, $\bar{1}$, mmm, $m3m$, $4/mmm$, 222, $mm2$ and $6/mmm$, account for more than 80% of the total. Ten polar point groups, which preserve the sense of a polar vector, an arrow along the main symmetry axis, are the only ones that can support ferroelectricity. They are $1, 2, m, mm2, 3, 3m, 4, 4mm, 6$ and $6mm$.

Textured polycrystalline materials have at least one continuous ∞–fold rotation axis. Rotation by any angle θ about this axis transforms the structure into itself. There are seven such continuous Curie groups ∞, ∞m, $\infty 2$, ∞/m, ∞/mm, $\infty\infty$ and $\infty\infty m$. For example, a polycrystalline material with a uniaxial texture has cylindrical symmetry ∞/mm. The groups $\infty\infty$, ∞/m, $\infty 2$ and ∞ may be composed of either right-handed or left-handed molecules. For example, a poled polycrystalline ferroelectric belongs to ∞m.

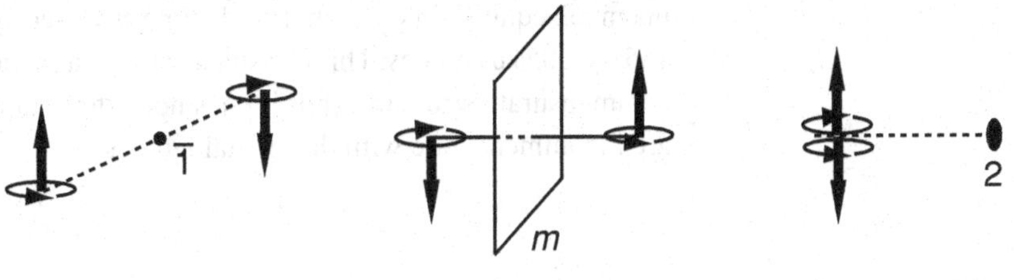

Figure 11.4

The effect of some symmetry elements on the components of a magnetic moment.

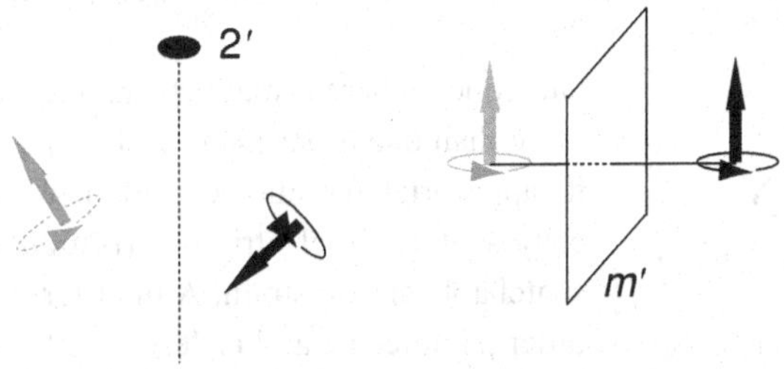

Figure 11.5

The magnetic symmetry operations 2′ and *m*′. The ′ denotes time reversal.

Crystallography was created to account for the atomic positions in crystalline solids. Magnetism adds another dimension. As it is related to *axial* vectors associated with current loops, an additional symmetry element – time reversal – appears, which is entirely absent in atomic crystals. Components of a magnetic moment subject to inversion, reflection or two-fold rotation can change sign. A two-fold rotation preserves the component of magnetization parallel to the axis, but reverses the perpendicular components, whereas a mirror reverses the components of magnetization parallel to the mirror, but preserves the perpendicular component, Fig. 11.4. Inversion preserves all the components, whereas time reversal reverses them all. Since time reversal has no influence on the atomic positions, the 32 crystallographic point groups may be considered to include all their time-reversed symmetry elements. This could be indicated by adding 1′ to the point group symbol, where a prime on any symmetry element denotes time reversal. The symmetry operations 2′ and m' in Fig. 11.5 are examples.

Any magnetically ordered crystal must be described by a magnetic point group which cannot include 1′. The 90 such point groups are listed in Appendix I. There are also 14 magnetic Curie groups. The symmetry of a polycrystalline ferromagnet in its remanent state, for example, is ∞/mm'. The magnetic structure and anisotropy are reflected in the magnetic point group. Thirty-one of them are compatible with a spontaneous ferromagnetic moment. Not one of these is cubic. Body-centered cubic (bcc) iron (crystallographic point group $m3m$) has ⟨100⟩ anisotropy and the magnetic point group is $4/mm'm'$, whereas fcc nickel (crystallographic point groups m3m) has ⟨111⟩ anisotropy, and the magnetic point group is $\bar{3}m'$.

The complete symmetry of a crystal, including the Bravais lattice and any translational symmetry elements, is given by one of the 240 space groups. The magnetic equivalents, which include time-reversed symmetry elements, are the 1651 Shubnikov groups. This classification of magnetic symmetry works well for commensurate structures, but it does not extend to magnetic structures which are incommensurate with the crystal lattices.

11.1.2 Multiferroics

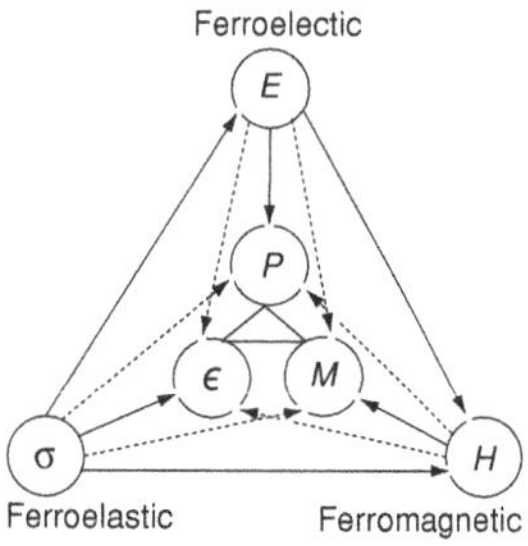

A diagram showing possible multiferroic combinations.

Our concern here is magnetism, but occasionally materials are found to exhibit more than one order parameter at the same time. Besides magnetic order, and its appropriate conjugate field, there are ferroelectrics, which exhibit electrically switchable electric polarization, and ferroelastics, which exhibit stress-switchable elastic strain. A multiferroic material exhibits at least two of these order parameters, and is described by a complicated tensor susceptibility with components of both fields. Ferromagnetism is rarely associated with ferroelectricity, and no material is known for which both magnetic and electric Curie points are above room temperature.

11.2 Iron group metals and alloys

αFe

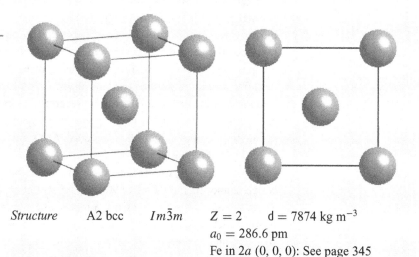

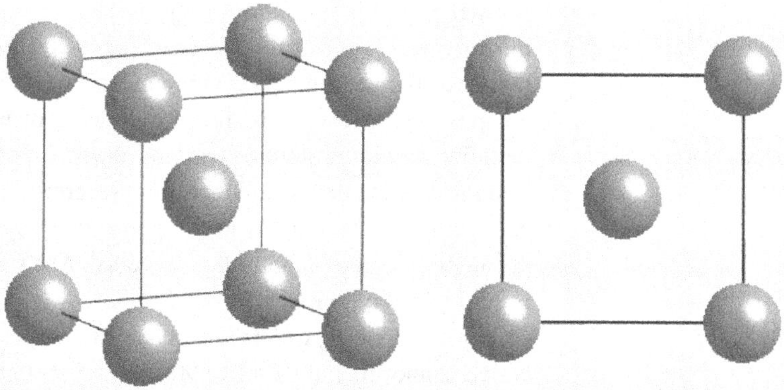

Structure A2 bcc $Im\bar{3}m$ $Z = 2$ d = 7874 kg m^{-3}
$a_0 = 286.6$ pm
Fe in $2a$ (0, 0, 0): See page 345

The bcc structure of αFe transforms to fcc γFe at 1185 K and back to bcc (δFe) between 1667 K and the melting point $T_m = 1811$ K. The αFe structure is stabilized by ferromagnetic correlations, even above the Curie point.

Electronic properties Metal $3\,d^{7.4}4s^{0.6}$ $\varrho \approx 0.05\ \mu\Omega$ m
Itinerant moments (weak ferromagnet)

Magnetic properties Ferromagnetic $T_C = 1044$ K

Atomic moment $m_0 = 2.22\ \mu_B$ atom^{-1} (spin 2.14 μ_B; orbit 0.08 μ_B)

Exchange integral $\mathcal{I} = 0.93$ eV

$\sigma = 217$ Am^2 kg^{-1}	$M_s = 1.71$ MA m^{-1}	$J_s = 2.15$ T
$\sigma_0 = 221.7(1)$ Am^2 kg^{-1}	$M_0 = 1.76$ MA m^{-1}	$J_s = 2.22$ T
$A = 21$ pJ m^{-1}	$K_{1c} = 48$ kJ m^{-3}	$K_{2c} = -10$ kJ m^{-3}
$\kappa = 0.12$	$\lambda_s = -7 \times 10^{-6}$	$\lambda_{100} = 15 \times 10^{-6}$
$\lambda_{111} = -21 \times 10^{-6}$		

Significance Iron is the most abundant element on Earth (crust, mantle and core) and it is by far the most common magnetic element in the Earth's crust, where it accounts for 2.5% of the atoms (6 wt%). One atom in 40 in the crust is iron, where it is 40 times as abundant as all the other magnetic elements combined (Fig. 1.11). The base price is about \$0.50 kg^{-1}.

Related materials The fcc form of iron, γFe, can be stabilized at room temperature by alloy additions or on an fcc substrate. Depending on lattice parameter, it can be ferromagnetic, antiferromagnetic or nonmagnetic. Electrical sheet steel is bcc $Fe_{0.938}Si_{0.062}$ (3.2 wt% Si steel). Both untextured and grain oriented, it is the mainstay of the electromagnetic industry. Global production is about 5 megatonnes per year. Sendust is a brittle ternary $Fe_{0.74}Si_{0.16}Al_{0.10}$ alloy which exhibits zero anisotropy and zero magnetostriction.

$Fe_{0.65}Co_{0.35}$ (Permendur)

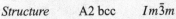

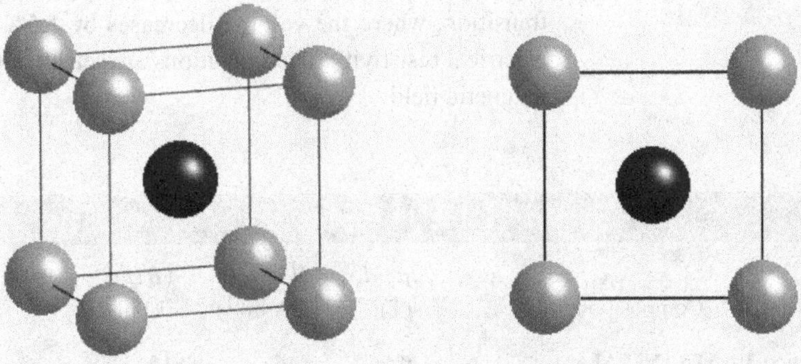

Structure A2 bcc $Im\bar{3}m$ $Z = 2$ d = 8110 kg m^{-3}

$a_0 = 285.6$ pm

Fe/Co in $2a$ (0, 0, 0)

Random bcc (α) solid solution with some tendency to B2 (α') CsCl-type of order when annealed below 920 K.

Electronic properties Metal $3d^{7.65}4s^{0.70}$ $\varrho \approx 0.08\ \mu\Omega$ m.
Itinerant moments (strong ferromagnet)

Magnetic properties Ferromagnet $T_C = 1210$ K $\langle m_0 \rangle = 2.46\,\mu_B$ atom^{-1}

Exchange integral $\mathcal{I} = 0.95$ eV

$\sigma = 240$ A m^2 kg^{-1} $M_s = 1.95$ MA m^{-1} $J_s = 2.45$ T

$K_1 \approx 20$ kJ m^{-3} $K_2 \approx -35$ kJ m^{-3} $\kappa = 0.06$

Significance Permendur has the largest room-temperature magnetization of any bulk material. It is used in place of Fe or $Fe_{0.94}Si_{0.06}$ when maximum flux concentration is needed, especially in electromagnet pole pieces and airborne electromagnetic drives. Magnetization and Curie temperature are almost constant in Fe_xCo_{1-x} for $0.65 > x > 0.50$. Ordered α'FeCo has a lower anisotropy and higher permeability, and may be used in magnetic circuits in preference to $Fe_{0.65}Co_{0.35}$. Alloy additions such as V or Cr improve the mechanical properties. Vicalloy, $Fe_{0.36}Co_{0.52}V_{0.12}$, can be drawn into wires or strips, which exhibit square hysteresis loops useful for security tags.

Related materials Alnicos are magnets with phase-separated nanostructures composed of Fe–Co needles in a nonmagnetic Al–Ni matrix. They have been used as permanent magnets since the 1930s. Heat treatment at about 900 K produces the desired spinodal structure. The Fe–Co needles are oriented by casting onto a chilled block, or by heat treatment in a magnetic field below the Curie point of Fe–Co (Fig. 8.27).

FeRh has an ordered B2 structure with $a_0 = 299$ pm. It orders ferromagnetically below 670 K, and then antiferromagnetically below 350 K at a first-order transition, where the volume decreases by 1.5% and there is a large drop in electrical resistivity. The transitions are sensitive to heat treatment and applied magnetic field.

Table 11.6. Properties of alnico alloys

	Composition	$\mu_0 M_s$ (T)	H_c (kA m^{-1})	$(BH)_{max}$ (kJ m^{-3})	
Alnico 3	$Fe_{60}Ni_{27}Al_{13}$	0.56	46	10	Original 1932 composition
Alnico 2	$Fe_{55}Co_{13}Ni_{18}\,Al_{10}Cu_4$	0.72	45	14	Isotropic
Alnico 5	$Fe_{49}Co_{24}Ni_{15}\,Al_8Cu_3Nb_1$	1.35	46	45	Cast or field-cooled from 1470 K
Alnico 8	$Fe_{31}Co_{38}Ni_{14}\,Al_7Cu_3Ti_7$	0.88	120	42	Field-annealed at 1100 K

Figure 11.6

Magnetization and permeability of Fe–Co alloys.

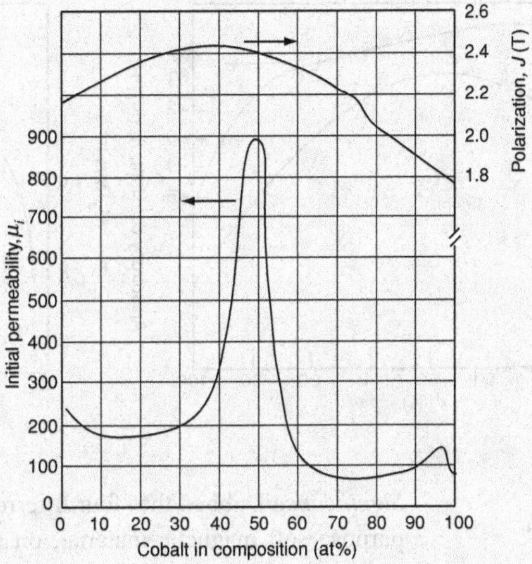

$Ni_{0.80}Fe_{0.20}$ (Permalloy)

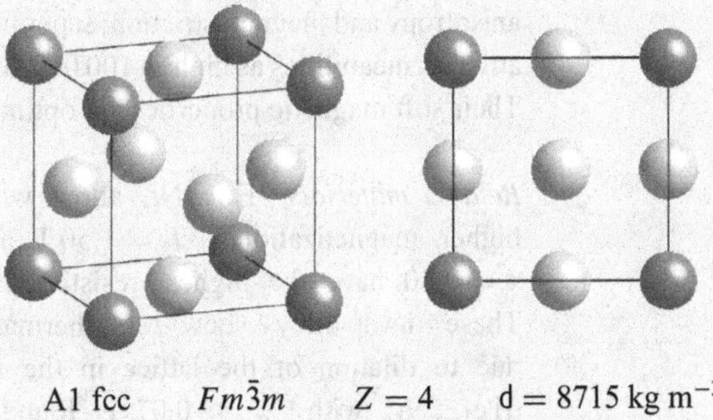

Structure A1 fcc $Fm\bar{3}m$ $Z = 4$ $d = 8715\ \mathrm{kg\ m^{-3}}$

$a_0 = 352.4$ pm

Fe/Ni in 4*a* (0, 0, 0)

A random fcc (γ) solid solution with a tendency to $L1_2$ ($\gamma'FeNi_3$-type) order depending on heat treatment.

Electronic properties Metal $3d^{9.0}4s^{0.6}$ $\varrho \approx 0.16\ \mu\Omega$ m.

Itinerant moments (strong ferromagnet)

Magnetic properties Ferromagnet $T_C = 843$ K $\langle m_0 \rangle = 1.02\ \mu_B\ \mathrm{atom^{-1}}$

Exchange integral $\mathcal{I} = 1.00$ eV

$\sigma = 95.0\ \mathrm{A\,m^2\,kg^{-1}}$ $M_s = 0.83\ \mathrm{MA\,m^{-1}}$ $J_s = 1.04$ T

$A = 10\ \mathrm{pJ\,m^{-1}}$. $K_1 \approx -1\ \mathrm{kJ\,m^{-3}}$ $\kappa \approx 0$

$\lambda_s \approx 2\ 10^{-6}$ $\alpha \approx 0.02$

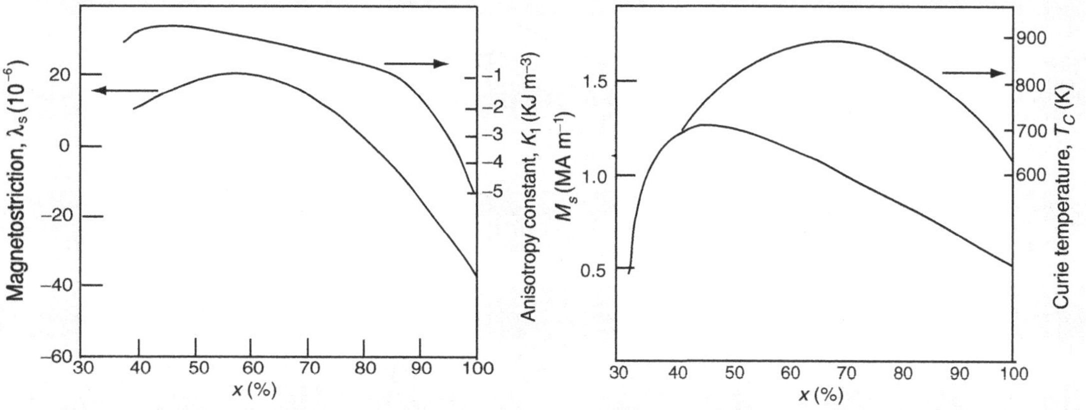

Figure 11.7

Anisotropy, magnetostriction, spontaneous magnetization and Curie temperature of $Fe_{1-x}Ni_x$ alloys.

Significance Dubbed the 'fourth ferromagnetic element', permalloy is an all-purpose soft magnetic material, on account of its near-zero anisotropy and magnetostriction. Electroplated films are used for write heads and magnetic shields. Magnetic sensors exploit the soft magnetism and anisotropic magnetoresistance (≈2%) of thin films in low fields. Their magnetic properties are improved by annealing. Small additions of Cu and Mo help cancel both the anisotropy and magnetostriction; supermalloy and mumetal, which can have relative permeabilities as high as 100 000, and are very effective magnetic shields. Their soft magnetic properties are optimized by heat treatment.

Related materials $\gamma Fe_{1-x}Ni_x$ alloys with compositions near $x = 0.50$ have higher magnetization, $J_s = 1.50$ T and $T_C = 770$ K; compositions near $x = 0.30$ have the highest resistivity $\varrho = 0.70\ \mu\Omega$ m, with $T_C \approx 670$ K. These invar alloys show zero thermal expansion near room temperature, due to dilation of the lattice in the ferromagnetic state. Phase-separated $\alpha Fe_{1-x}Ni_x$ with $x \leq 0.07$ is found together with γFe–Ni (taenite) in meteorites.

Pure nickel ($a_0 = 352.4$ pm: see page 345) is magnetically inferior to permalloy in most respects (magnetization, Curie temperature $T_C = 628$ K, low magnetostriction, cost). However, its magnetostriction $\lambda_s = -35 \times 10^{-6}$ ($\lambda_{100} = -51 \times 10^{-6}$, $\lambda_{111} = -24 \times 10^{-6}$) along with its low saturation polarization $J_s = 0.61$ T ($M_s = 488$ kA m^{-1}) is useful in magnetomechanical applications. Anisotropy is $K_1 = -5$ kJ m^{-3}, $K_2 = -2$ kJ m^{-3}. The easy direction is [111] and magnetic symmetry is $\bar{3}m'$. The specific magnetization of pure nickel, $\sigma = 54.78(15)$ A m^2 kg^{-1} at 298 K is a NIST standard for calibrating magnetometers. World nickel production is about 1.5 Mt yr^{-1}, but the metal is traded as a commodity and the price is unstable. It has fluctuated between \$5 and \$50 per kg in recent years.

a-$Fe_{0.40}Ni_{0.40}P_{0.14}B_{0.06}$ (Metglas 2826)

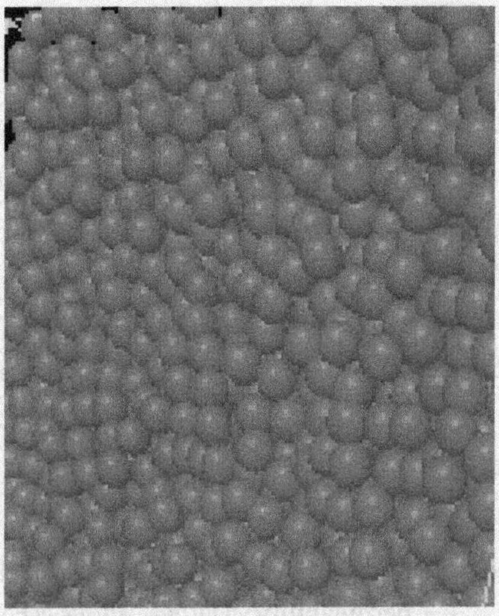

Structure Random close-packed Bernal structure d = 7720 kg m^{-3}

The packing fraction of the random close-packed structure is 0.64. Boron occupies large interstitial sites where it likes to adopt trigonal prismatic coordination.

Electronic properties Metal $\varrho \approx 1.6$ μΩ m. (maximum metallic resistivity)

Itinerant moments (strong ferromagnet)

Magnetic properties Ferromagnet $T_C = 535$ K $\langle \mathfrak{m} \rangle = 1.2\,\mu_B$ (3d atom)$^{-1}$

$\sigma = 84\,\mathrm{A\,m^2\,kg^{-1}}$ $M_s = 0.65\,\mathrm{MA\,m^{-1}}$ $J_s = 0.81$ T

$A = 8\,\mathrm{pJ\,m^{-1}}$ $K_1 \approx 0\,\mathrm{kJ\,m^{-3}}$ $\kappa \approx 0$ $\lambda_s \approx 11 \times 10^{-6}$

Significance Metglas 2826 is the amorphous analogue of $Fe_{0.50}Ni_{0.50}$. It has excellent soft magnetic properties, high resistivity and shows a large ΔE effect. Bulk anisotropy is practically zero because there is no crystal lattice. It has remarkable tensile strength > 2 GPa. Prepared by melt spinning in ribbons or sheets about 40 μm thick, uses include magnetic shielding, electromagnetic sensors, pulsed-mode power supplies, saturatable inductors and other low-frequency electromagnetic applications.

Related materials Magnetic glasses form for many combinations of transition metals and metalloids in an 80:20 ratio. Iron-rich glasses such as a-$Fe_{0.80}B_{0.20}$, a-$Fe_{0.40}Co_{0.40}B_{0.20}$ and a-$Fe_{0.81}B_{0.135}Si_{0.035}C_{0.02}$ (Metglas 2605SC) have higher polarization ($\mu_0 M_s = 1.5$–1.6 T) and higher Curie temperature (≈ 650 K). They are used in distribution transformers. These alloys exhibit isotropic linear magnetostriction, $\lambda_s = 31 \times 10^{-6}$, opposite in sign to that of crystalline iron, and can be used in sensors. Cobalt-rich metallic glasses

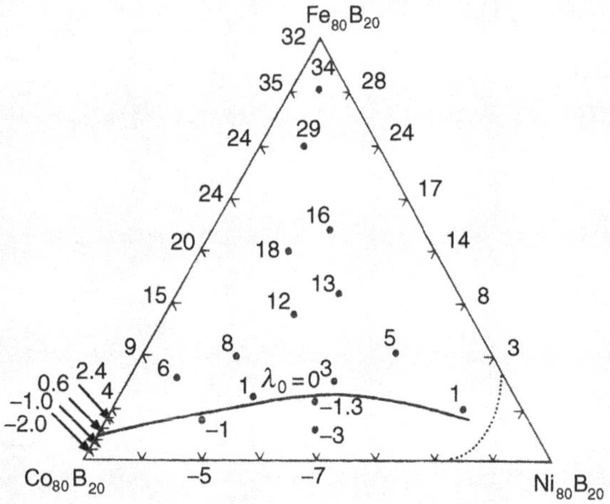

Figure 11.8

Ternary $(Fe,Co,Ni)_{0.80}B_{0.20}$ phase diagram, showing the compositions for zero anisotropy and zero magnetostriction. (R. C. O'Handley and C. P. Chou, *J. Appl. Phys.* **49**, 1659 (1978)).

(e.g. a-$Fe_{0.05}Co_{0.70}Si_{0.15}B_{0.10}$) show zero magnetostriction, and very high permeability. Finmet alloys ($Fe_{0.735}Cu_{0.01}Nb_{0.03}Si_{0.135}B_{0.09}$) are two-phase nanostructured composites composed of crystalline Fe–Si in an amorphous matrix, obtained by partial crystallization of an amorphous precursor. They show vanishingly small anisotropy and magnetostriction and very low losses, together with a polarization of 1.25 T, which is larger than permalloy. Finmet shields do not need to be annealed after shaping.

The amorphous 80:20 alloys crystallize at 700–800 K. Amorphous $Co_{0.40}Fe_{0.40}B_{0.20}$ films are used to make MgO-barrier magnetic tunnel junctions.

Co

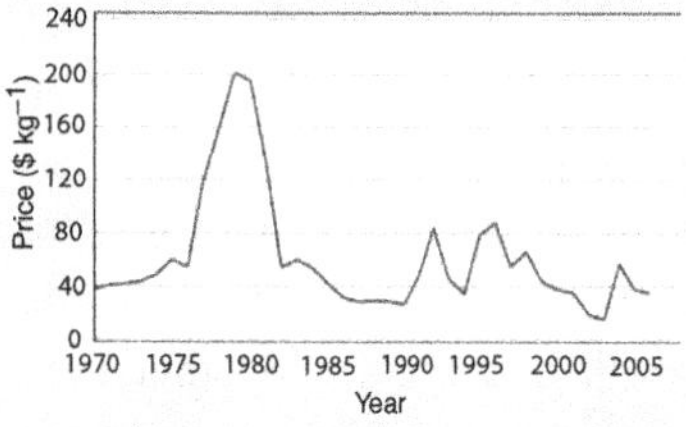

Cobalt price fluctuations ($\$\ kg^{-1}$).

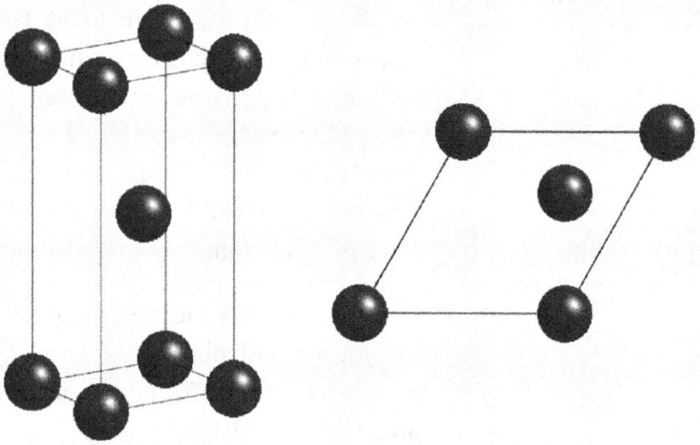

Structure A3 hcp $P6_3/mmc$ $Z = 2$ $d = 8920\ kg\ m^{-3}$

$a = 250.7$ pm $c = 407.0$ pm

Co in 2c $(\frac{1}{3}, \frac{2}{3}, \frac{1}{4})$: See page 346

The hcp structure of εCo transforms to fcc γCo at 695 K. Cobalt can be stabilized in the fcc structure in thin films.

Electronic properties Metal, $3d^{8.4}4s^{0.6}$ $\varrho \approx 0.06\ \mu\Omega$ m.
Itinerant moments (strong ferromagnet)

Magnetic properties Ferromagnet $T_C = 1360$ K
atomic moment $\mathfrak{m}_0 = 1.72\ \mu_B$ atom^{-1} (spin 1.58 μ_B; orbit 0.14 μ_B)
exchange integral $\mathcal{I} = 1.01$ eV

$\sigma = 162\,\mathrm{A\,m^2\,kg^{-1}}$	$M_s = 1.44\,\mathrm{MA\,m^{-1}}$	$J_s = 1.81\,\mathrm{T}$
$A = 31\,\mathrm{pJ\,m^{-1}}$	$K_1 = 410\,\mathrm{kJ\,m^{-3}}$	$K_2 = 140\,\mathrm{kJ\,m^{-3}}$
$\kappa = 0.45$	$\lambda_s \approx -60 \times 10^{-6}$	

Significance Fcc cobalt has the highest Curie temperature of all ferromagnets, 1388 K, but that of the hcp ε-phase is similar (1360 K). The magnetization of γCo is a little greater ($\sigma = 165$ A m^2 kg^{-1}). Cobalt is a strategic metal, unevenly distributed over the Earth, but it is roughly 100 times less abundant than iron and the price has fluctuated wildly. A total of 4000 tonnes per year, or 7% of global production, is used for magnets.

Related materials Cobalt is invaluable as an alloy addition to increase Curie temperature and anisotropy. Cobalt thin films with Cr, Pt and B additives (typical composition $Co_{0.67}Cr_{0.20}Pt_{0.11}B_{0.06}$) are used as longitudinal recording media for hard discs. These films are composed of nanoscale grains with a nonmagnetic boron-rich grain-boundary phase. Compositions with more Pt and less Cr are used for perpendicular media.

CoPt

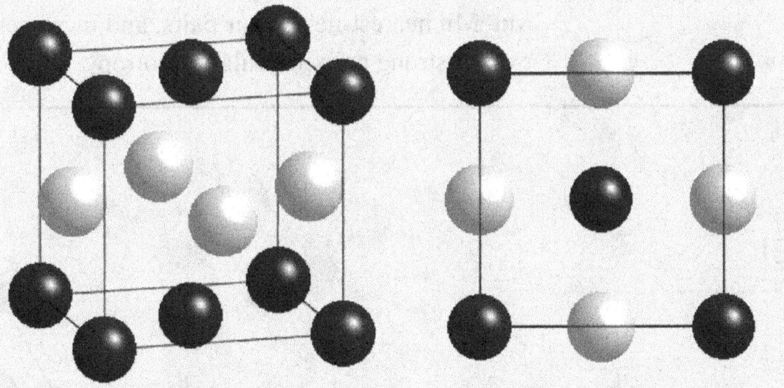

Structure $L1_0$ tetragonal (AuCu) $P4/mmm$ $Z = 2$ d $= 16040$ kg m^{-3}
$a = 377$ pm $c = 370$ pm
Co in $2a$ (0, 0, 0)
Pt in $2b$ ($\frac{1}{2}$, $\frac{1}{2}$, 0)

The face-centred tetragonal structure, produced by annealing the disordered γ-phase at $T_a = 870$ K, is always twinned, with c axes distributed along the different cube edges. Noble metal additions reduce T_a. There is a broad homogeneity range around the equiatomic composition.

Table 11.7. Ferromagnetic properties of some ordered binary AB alloys

	Order	T_C (K)	J_s (T)	K_1 (MJ m^{-3})	κ
CoPt	$L1_0$	840	1.01	4.9	2.47
FePt	$L1_0$	750	1.43	6.6	2.02
FePd	$L1_0$	749	1.38	1.8	1.10
τMnAl	$L1_0$	650	0.75	1.7	1.95
Co_3Pt	$L1_2$	1190	1.40	0.6	0.71
Ni_3Mn	$L1_2$	750	1.0	0.03	0.19

Electronic properties Metal
Itinerant moments on Co and Pt

Magnetic properties Ferromagnet $T_C = 840$ K
$\sigma = 50$ A m^2 kg^{-1} $M_s = 0.80$ MA m^{-1} $J_s = 1.01$ T
$A = 10$ pJ m^{-1} $K_1 = 4.9$ MJ m^{-3} $\kappa = 2.47$ $B_a = 6.1$ T

Significance CoPt finds wide applications as a permanent magnet in medical and military applications and in precision instruments where its exceptional corrosion resistance, ductility, machinability and good high-temperature performance justify the high cost. Energy products of up to 100 kJ m^{-3} are achieved.

Related compounds There is a large family of ferromagnetic ordered phases with the $L1_0$ or $L1_2$ structures, including the hard magnets FePd, FePt and MnAl. $L1_2$ ordering in Ni_3Mn eliminates antiferromagnetically coupled Mn–Mn nearest-neighbour pairs, and increases T_C by 500 K. Co_3Pt thin films exhibit strong perpendicular anisotropy.

MnBi

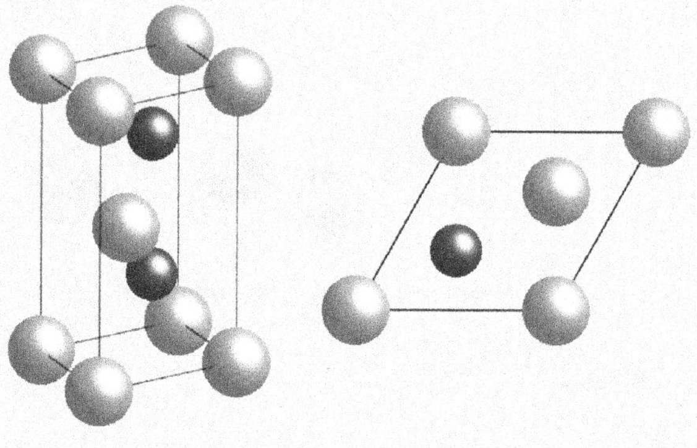

Structure $B8_1$ hexagonal (NiAs) $P6_3/mmc$ $Z = 2$ d = 9040 kg m^{-3}
$a = 428$ pm $c = 611$ pm
Mn in $2b$ $(\frac{1}{3}, \frac{2}{3}, \frac{1}{4})$
Bi in $2a$ (0, 0, 0)

Table 11.8. Ferromagnetic manganese pnictides

	T_C (K)	M_s (MA m^{-1})	
MnAs	318	0.63	Easy plane
MnSb	573	0.77	Spin reorientation at 520 K
MnBi	633	0.58	Spin reorientation at 84 K

The hexagonal NiAs structure is an ABAC... stacking of hexagonal layers of Ni (A) and As (B, C).

Electronic properties Metal with a low density of states at ε_F
Mn d electrons are strongly hybridized with Bi $6p$ states.

Magnetic properties Ferromagnet $T_C = 633$ K (first order transition)
Magnetic moment $\mathfrak{m}_0 = 3.95\ \mu_B$ fu^{-1} lies parallel to $\boldsymbol{c}$ above the spin reorientation temperature $T_{sr} = 84$ K, and perpendicular to $\boldsymbol{c}$ below T_{sr}.

$\sigma = 64$ A m^2 kg^{-1} $M_s = 0.58$ MA m^{-1} $J_s = 0.72$ T
$K_1 = 0.9$ MJ m^{-3} $K_2 = 0.3$ MJ m^{-3} $\kappa = 1.5$
$\lambda_s^c \approx 500 \times 10^{-6}$

Significance MnBi is a hard magnet with good magneto-optic properties due to the strong spin-orbit coupling of Bi: $\theta_K = 0.9°$; $\theta_F = 50°\ \mu\text{m}^{-1}$ at $\Lambda = 633$ nm. Films with perpendicular anisotropy were used to demonstrate thermomagnetic recording. Doping with Ge increases θ_K to 2.1°. Magnetostriction is very anisotropic.
Related compounds Properties of the NiAs-structure manganese pnictides are summarized in Table 11.8. NiAs itself is nonmagnetic, but NiS shows a first-order antiferromagnetic transition at $T_N = 260$ K. MnP is orthorhombic, with $T_C = 291$ K.

NiMnSb

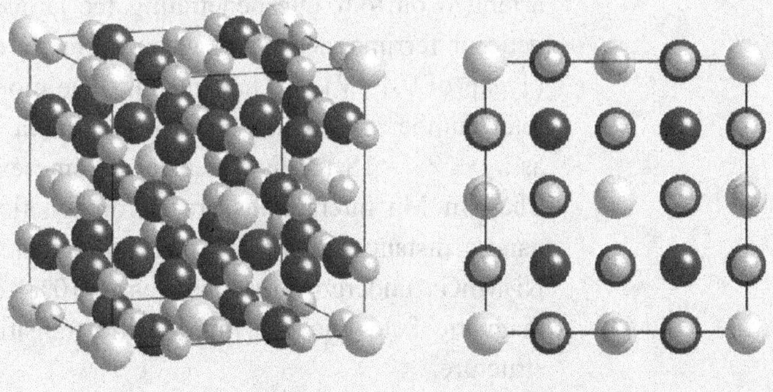

Structure $C1_b$ cubic (AgMgAs) $F\bar{4}3m$ $Z = 4$ d = 7530 kg m^{-3}
$a_0 = 592$ pm
Ni in $4a$ (0, 0, 0)
Mn in $4b$ (0, $\frac{1}{4}$, $\frac{1}{4}$)
Sb in $4c$ (0, $\frac{1}{2}$, $\frac{3}{4}$)

Table 11.9. Heusler and half-Heusler alloys

	a_0 (pm)	T_C (K)	σ_0 (A m^2 kg^{-1})	m (μ_B)
Cu_2MnIn	621	500	75	4.0
Co_2MnGa	577	694	93	4.1
Co_2MnSi^a	565	985	141	5.0
Co_2MnGe^a	574	905	116	5.1
Co_2MnSn	600	829	97	5.1
Ni_2MnGa	583	380	96	4.2
Ni_2MnSn	605	360	81	4.2
Pd_2MnSb	642	247	63	4.4
$NiMnSb^a$	592	730	93	4.0
$PtMnSb^a$	620	572	60	4.0
Mn_2VAl^a	760	730	59	2.0

[a] Half-metal

A half-Heusler alloy, with one of four interpenetrating fcc lattices vacant.

Electronic properties Half-metal, ↑ electrons at ε_F $\varrho \approx 0.6\ \mu\Omega$ m.

Magnetic properties Ferromagnet $T_C = 728$ K $m_0 \approx 4.0\,\mu_B$ fu^{-1}

$\sigma = 89\,\mathrm{A\,m^2\,kg^{-1}}$ $M_s = 0.67\,\mathrm{MA\,m^{-1}}$ $J_s = 0.84\,\mathrm{T}$ $K_1 = 13\,\mathrm{kJ\,m^{-3}}$

Significance NiMnSb and PtMnSb were the first materials to be identified as half-metals. The high spin polarization and Curie temperature of some Heusler alloys make them potentially useful for spin electronics.

Related materials There is a large family of full Heusler X_2YZ ($L2_1$ structure, $Fm\bar{3}m$) and half-Heusler XYZ alloys, in which the X,Y and Z atoms are arranged on four interpenetrating fcc lattices. Many of them are ferromagnetic or ferrimagnetic, and when perfectly ordered a few, such as Co_2MnSi (↓ gap of 0.4 eV), are half-metals. The moment per formula varies with the total number of valence electrons Z_e as $m = Z_e - 24$ for full Heuslers and as $m = Z_e - 18$ for half-Heuslers. Some examples are listed in Table 11.9. The Mn–Mn interactions are ferromagnetic, on account of large intermanganese distances ≈ 420 pm. Ferromagnetic shape-memory alloys based on Ni_2MnGa undergo large changes ($\sim$10%) of size and shape in an applied magnetic field due to a martensitic transition at 305 K from fcc to an fct structure.

αMn

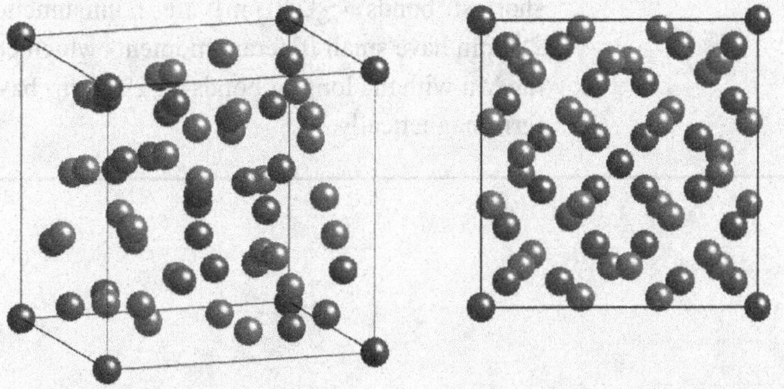

Structure A12 cubic $I\bar{4}3m$ $Z = 58$ $d = 7470$ kg m^{-3}
$a_0 = 886.5$ pm

Mn in $2a$ (0, 0, 0); $8c$ (0.317. 0.317, 0.317); $24g_1$(0.357, 0.357, 0.034); $24g_1$(0.089, 0.089, 0.282)

There is a small tetragonal distortion below the Néel point

Electronic properties Metal, $3d^{6.4}4s^{0.6}$ $\varrho \approx 1.4$ μΩ m.
Itinerant moments

Magnetic properties Noncollinear antiferromagnet $T_N = 95$ K
Atomic moment $\langle m_0 \rangle \approx 0.7\ \mu_B$ atom^{-1}
Site moments $2a$ 2.8 μ_B; $8c$ 1.8 μ_B; $24g_1$ 0.5 μ_B; $24g_2$ 0.5 μ_B

Significance Mn has the largest unit cell and the most complex structure of any element. There are partly localized moments on $2a$ and 8c sites, and delocalized moments on $24g$ sites, where the antiferromagnetic coupling is frustrated. The magnetic properties of other Mn polymorphs are given in Table 11.10.

Related materials Mn can have the largest atomic moment of any $3d$ element, but Mn additions usually decrease the moment in alloys of the ferromagnetic $3d$ elements because it couples antiparallel to them. Mn-rich alloys are antiferromagnetic on account of the direct exchange interaction in the roughly half-filled d-band. For antiferromagnetic Mn alloys, see $IrMn_3$; for more dilute ferromagnetic examples, see NiMnSb and MnBi. Generally Mn sites with the

Table 11.10. Magnetism of manganese polymorphs

			T_N (K)	
αMn	A12	Cubic	95	Noncollinear antiferromagnet
βMn	A13	Cubic		Spin liquid
γMn	A1	Cubic, stabilized with Cu	450	Antiferromagnet
γ'Mn	A5	Tetragonal, quenched	570	Antiferromagnet

shortest bonds ($\lesssim$240 pm) are nonmagnetic, those with bonds of 250–280 pm have small itinerant moments which couple antiferromagnetically and the Mn with the longest bonds ($\gtrsim$290 pm) have larger moments which couple ferromagnetically.

$IrMn_3$

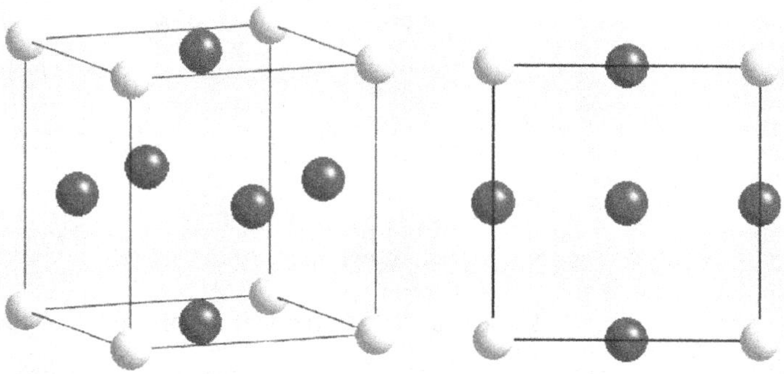

Structure $L1_2$ cubic ($AuCu_3$) $Pm\bar{3}m$ $Z = 1$ d = 11070 kg m^{-3}
$a_0 = 378$ pm,
Ir in 1*a* (0, 0, 0)
Mn in 3*b* (0, $\frac{1}{2}$, $\frac{1}{2}$)

Electronic properties Metal, $\varrho \approx 1\ \mu\Omega$ m.

Magnetic properties Antiferromagnet $T_N = 960$ K (dependent on degree of atomic order) $T_N = 730$ K in disordered ν-phase $K_{\text{eff}} = 3 \times 10^6$ J m^{-3}.

m = 2.6 μ_B Mn^{-1} No moments on Ir. The (111) planes are antiferromagnetic, with a triangular configuration and moments in-plane along $[2\bar{1}\bar{1}]$.

Significance $IrMn_3$ is frequently used for exchange bias in thin-film devices such as sensors and memory elements.

Related materials A series of Mn-based antiferromagnetic compounds have been used for exchange bias. Their Néel temperatures, blocking temperatures and interfacial exchange coupling σ with cobalt or permalloy are summarized in the Table 11.11.

Table 11.11. Antiferromagnetic Mn-based alloys

			T_N (K)	T_b (K)	σ (mJ m^{-2})
FeMn	fcc	Four noncollinear sublattices, $S \parallel [111]$	510	440	0.10
NiMn	$L1_0$	af (002) planes, $S \parallel a$	1070	700	0.27
$Pd_{0.32}Pt_{0.18}Mn_{0.50}$	$L1_0$	af (002) planes	870	580	0.17
PtMn	$L1_0$	af (002) planes, $S \parallel c$	975	500	0.30
IrMn	$L1_0$	af (002) planes $S \parallel [110]$	1145		
$RhMn_3$	$L1_2$	Triangular spin structure	855	520	0.19
$IrMn_3$	$L1_2$	fm (001) planes, $S \parallel a$	960	540	0.19

Cr

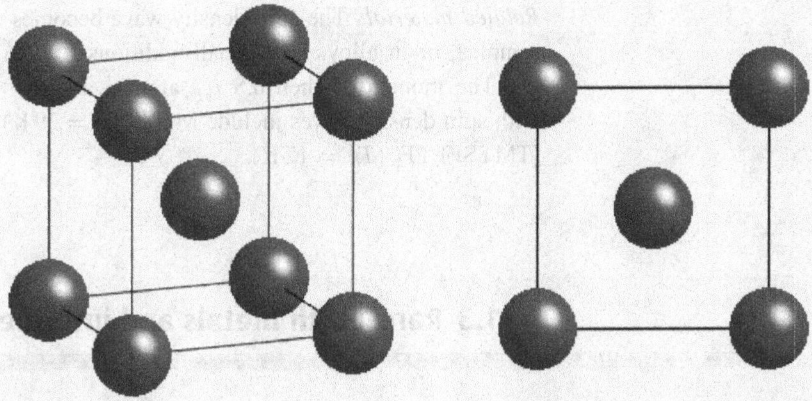

Structure A2 bcc $Im\bar{3}m$ $Z = 2$ $d = 7190\ \mathrm{kg\ m^{-3}}$

$a_0 = 288.5$ pm

Cr in $2a$ (0, 0, 0)

Electronic properties Metal, $3d^{5.4}4s^{0.6}$ $\varrho \approx 0.13\ \mu\Omega$ m.

Nested Fermi surface Itinerant moments form a spin density wave.

Magnetic properties Spin density wave antiferromagnet $T_N = 312$ K

The transition is weakly first-order $\langle m_0 \rangle \approx 0.43\ \mu_B\ \mathrm{atom^{-1}}$; 60% orbital, 40% spin. g = 1.2

Moments parallel to ⟨100⟩. Field cooling stabilizes a single antiferromagnetic domain, with a slight tetragonal distortion, $c/a > 1$

Incommensurate antiferromagnetic wavelength $\lambda = 303$ pm corresponds to a spin density wave with wavevector $Q = 0.95(2\pi/a_0)$ m$\|$ $\boldsymbol{Q}$ below T_N but m turns parallel to $\boldsymbol{Q}$ below the spin reorientation transition at 123 K

Significance Cr is the simplest example of a spin density wave antiferromagnet. The wavevector Q is determined by the nesting property of the chromium Fermi surface.

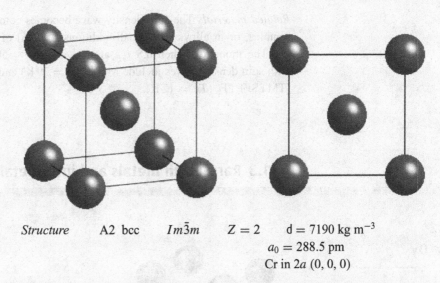

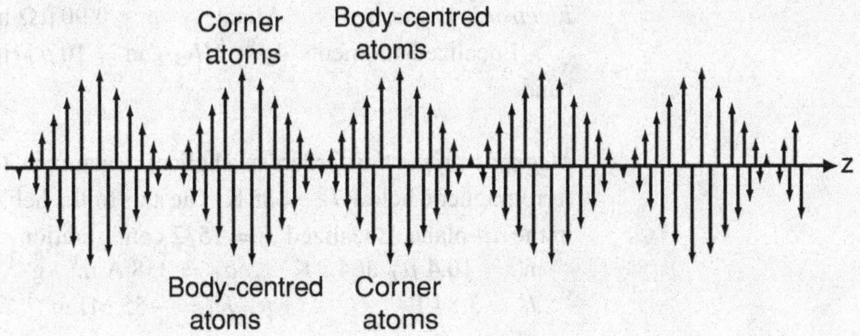

Figure 11.9

The magnetic structure of chromium, showing modulation of the magnetic structure on the two sites in the unit cell. The amplitude is 0.6 μ_B.

Related materials The spin density wave becomes commensurate in strained samples, or in alloys with small additions (≈ 2%) of Mn, Ru, Rh, Re or Os. The moment is then 0.8 μ_B atom^{-1} and $T_N \approx 500$ K. Other materials with spin density waves include MnSi ($T_N = 29$ K) and organic salts such as $(TMTSF)_2PF_6$ ($T_N = 12$ K).

11.3 Rare-earth metals and intermetallic compounds

Dy

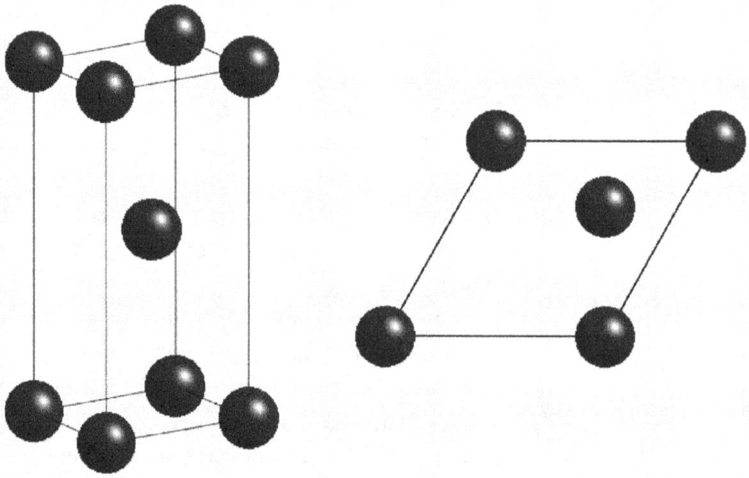

Structure A3 hcp $P6_3/mmc$ $Z = 2$ d = 8530 kg m^{-3}
$a = 358$ pm $c = 562$ pm
Dy in 2c ($\frac{1}{3}, \frac{2}{3}, \frac{1}{4}$)

Electronic properties Metal $\varrho \approx 0.90$ μΩ m
Localized moments $4f^9$; $^6H_{15/2}$ $\mathfrak{m} = 10\,\mu_B$ in $4f$ shell, 0.4 μ_B in $5d$ band.

Magnetic properties Orders in a helical structure at $T_N = 179$ K, then becomes ferromagnetic below $T_C = 85$ K. The axis of the helix is along ***c***, with moments in the ab-plane. Localized $J = 15/2$ configuration.

$\mathfrak{m}_0 = 10.4\,\mu_B$ at 4.2 K $\sigma_0 = 358$ A m^2 kg^{-1} $M_0 = 3.06$ MA m^{-1}
$J_0 = 3.84$ T $K_1 = -55$ MJ m^{-3} $K_3' = 1.5$ MJ m^{-3}

Significance Dy and Ho are the ferromagnetic elements with the largest low temperature moments and magnetization. Both show huge, hard-axis anisotropy at low temperature.

Figure 11.10

Magnetic order in the rare-earth elements. The values at the top of the figure are the first magnetic ordering temperatures.

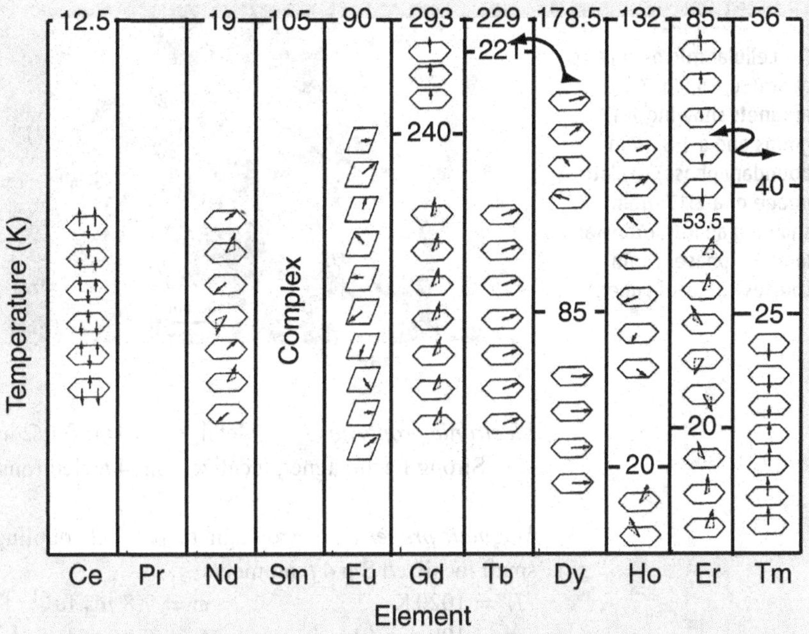

Related materials The rare-earth elements are a test-bed for localized magnetism, with properties ranging from a nonmagnetic singlet (Pr) to a nearly room-temperature ferromagnet (Gd). The magnetic ordering temperatures scale with the de Gennes factor $(g-1)^2 J(J+1)$.

$SmCo_5$

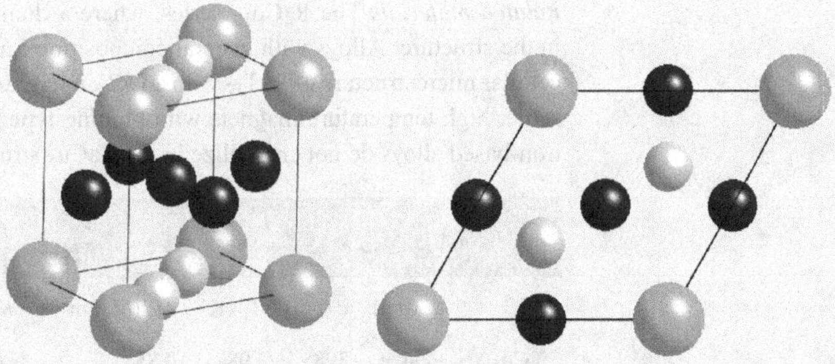

Structure D2$_d$ hexagonal ($CaCu_5$) $P6/mmm$ $Z = 1$ d = 8606 kg m^{-3}

$a = 499$ pm $c = 398$ pm

Sm in $1a$ (0, 0, 0)

Co in $2c$ ($\frac{1}{3}$, $\frac{1}{3}$, 0)

Co in $3g$ (0, $\frac{1}{2}$, $\frac{1}{2}$)

Derived from the hcp Co structure

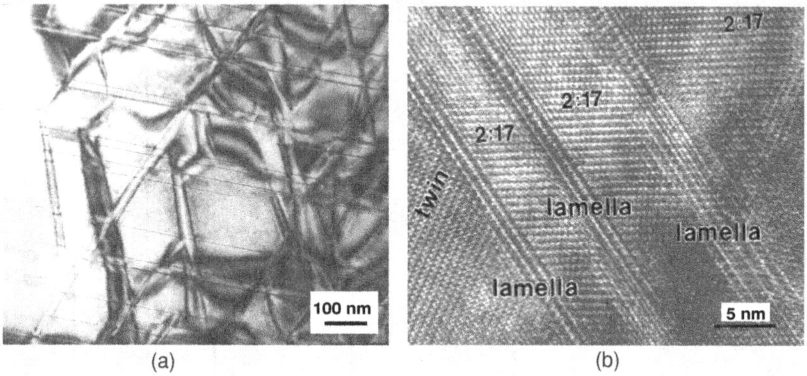

Figure 11.11

(a) Cellular microstructure of Sm(Fe, Co, Cu, Zr) $_{7-8}$ magnets showing 2:17 grains and a 1:5 grain boundary phase; (b) lattice image of a 2:17 grain showing an intragranular lamellar phase. (Data courtesy of Josef Fidler.)

Electronic properties Metal $\varrho \approx 0.8\ \mu\Omega$ m
Strong ferromagnet, localized Sm $4f$ electrons

Magnetic properties Ferromagnetic parallel coupling of delocalized Co $3d$ and small localized Sm $4f$ moments

$T_c = 1020$ K $\mathfrak{m} = 7.8\ \mu_B$ fu^{-1}
$\sigma = 100$ A m^2 kg^{-1} $M_s = 0.86$ MA m^{-1} $J_s = 1.08$ T
$K_1 = 17.2$ MJ m^{-3} $(K_1^{Sm} = 10.7$ MJ m^{-3} $K_1^{Co} = 6.5$ MJ m$^{-3})$
$\kappa = 4.3$ $B_a = 40$ T $A = 12$ pJ m^{-1} $\delta_w = 3.6$ nm

Significance The first rare-earth permanent magnet and still the one with the greatest anisotropy. About 65% comes from the Sm sublattice and 35% from Co. $SmCo_5$ magnets are used when temperature stability is critical.

Related materials The R_2Co_{17} series, where a dumbbell Co-pair replaces R in the structure. Alloys with general composition Sm(Co,Fe,Cu,Zr)$_{7-8}$ have a cellular microstructure of R_2T_{17} with an RT_5 intergranular phase. They are versatile, high-temperature magnets with pinning-type coercivity. Unfortunately iron-based alloys do not crystallize in the $CaCu_5$ structure.

Table 11.12. Cobalt-based rare-earth intermetallics with the $CaCu_5$ structure

	a (pm)	c (pm)	T_c (K)	M_s (MA m^{-1})	K_1 (MJ m^{-3})	B_a (T)	κ
YCo_5	494	398	987	0.85	6.5	15	2.7
$SmCo_5$	499	398	1020	0.86	17.2	40	4.3
$GdCo_5$	498	397	1014	0.29	4.6	32	6.6
$SmCo_4B$	509	689	470	0.67	30.2	90	7.3

$Nd_2Fe_{14}B$ (Neomax)

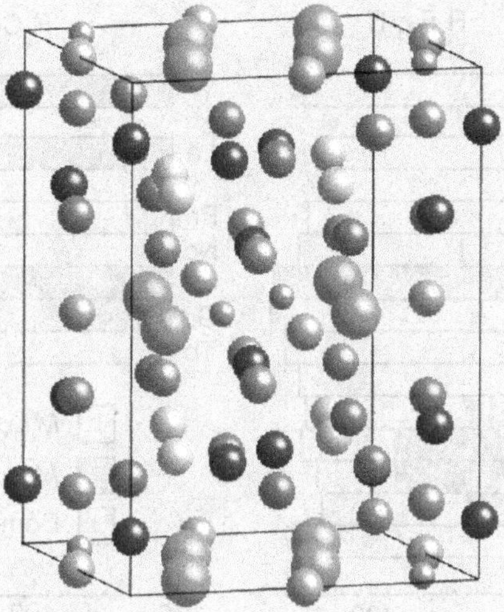

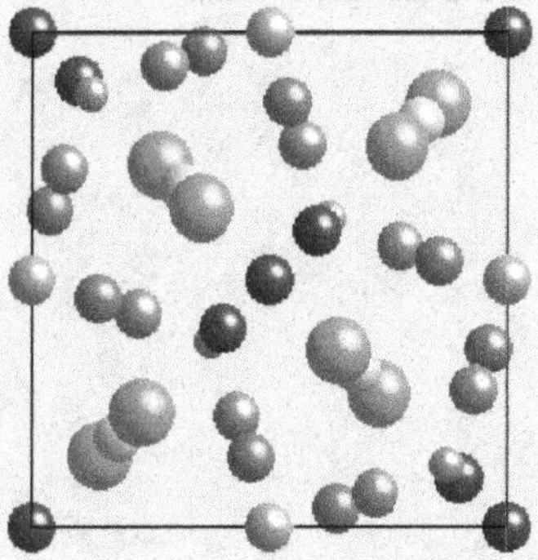

Structure Tetragonal $P42/mnm$ $Z = 4$ $d = 7760$ kg m^{-3}

$a = 879$ pm, $c = 1218$ pm

Nd in $4f$ (0.357, 0.357, 0) Nd in $4g$ (0.770, 0.230, 0)

Fe in $4c$ (0, $\frac{1}{2}$, 0) Fe in $4e$ (0, 0, 0.116)

Fe in $8j_1$ (0.098, 0.098, 0.294) Fe in $8j_2$ (0.318, 0.318, 0.255)

Fe in $16k_1$ (0.567, 0.225, 0.374) Fe in $16k_2$ (0.124, 0.124, 0)

B in $4f$ (0.124, 0.124, 0)

Double kagome layers of Fe, with Fe ($8j_2$) sandwiched between them, interleaved with planes containing Nd, B and Fe ($4c$)

Electronic properties Metal $\varrho \approx 1.0$ μΩ m.

Itinerant moments (nearly strong ferromagnet), Nd has a localized $4f^3$ core

Magnetic properties Multisublattice ferromagnet $T_C = 588$ K, $\langle m_{Fe} \rangle = 2.2\ \mu_B$ atom^{-1}, m $= 37.3\ \mu_B$ fu^{-1}

Parallel coupling of localized Nd^{3+}: moments and itinerant iron moments, Nd $4f, 4g$ 3.0 μ_B

Fe: $4c$ 1.9 μ_B; $4e$ 2.2 μ_B; $8j_1$ 2.2 μ_B; $8j_2$ 2.5 μ_B; $16k_1$ 2.1 μ_B; $8j_1$ 2.2 μ_B; $16k_1$ 2.3 μ_B

Noncollinear below $T_{st} = 135$ K, moment inclined at $\theta = 30°$ to $\boldsymbol{c}$ at $T = 0$ K.

Exchange constants $\mathcal{J}_{Fe-Fe} = 36.8$ K, $\mathcal{J}_{Nd-Fe} = 8.7$ K

$\sigma = 165$ A m^2 kg^{-1} $M_s = 1.28$ MA m^{-1} $J_s = 1.61$ T

$K_1 = 4.9$ MJ m^{-3} ($K_1^{Nd} = 3.8$ MJ m^{-3}, $K_1^{Fe} = 1.1$ MJ m^{-3}) $\kappa = 1.54$

$B_a = 7.7$ T $A = 8$ pJ m^{-1} $\delta_w = 4.0$ nm.

Significance Discovered in 1982, $Nd_2Fe_{14}B$ is the highest-performance permanent magnet; 80 000 tonnes were produced in 2010. The record energy product

Figure 11.12

Magnetic structures and spin orientations in $R_2Fe_{14}B$ and $R_2Co_{14}B$, relative to the tetragonal c axis.

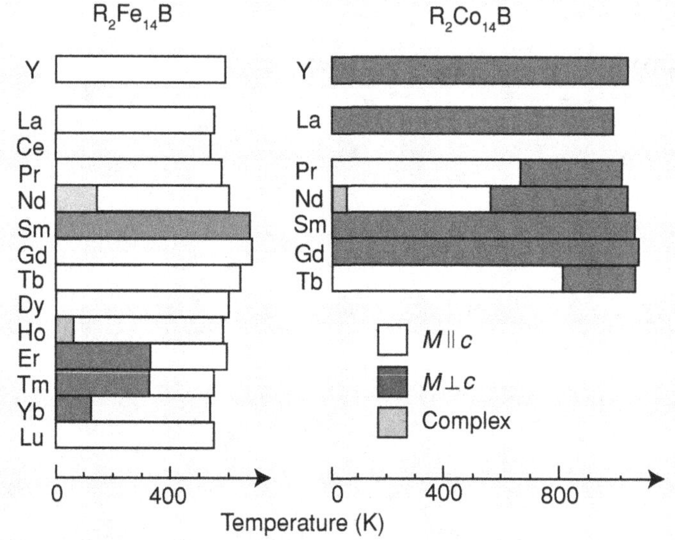

is 474 kJ m^{-3}. Main uses are in permanent magnet motors for small appliances and electric vehicles and in voice-coil actuators and spindle motors for hard disc drives. Other uses include wind generators, magnetic bearings and flux sources.

Related materials Of the entire $R_2Fe_{14}B$ isostructural series, R = La–Lu, only Pr offers comparable hard magnetic properties. Dy is added to improve coercivity especially at high operating temperatures. Isostructural cobalt compounds $R_2Co_{14}B$, R = La–Tb, show easy-plane anisotropy for the cobalt sublattices. When the rare-earth and the transition metal Fe or Co have anisotropy contributions of opposing sign, a spin reorientation occurs at a temperature where the sum is zero. Further spin reorientations may appear at low temperatures driven by higher-order terms in the rare earth anisotropy.

Y_2Co_{17}

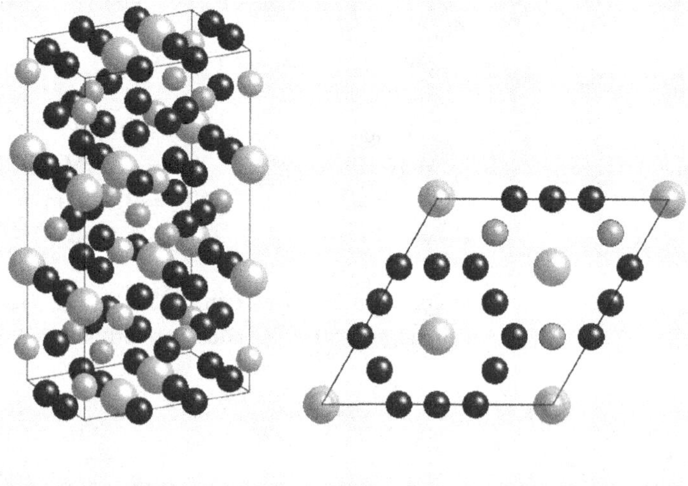

Structure Rhombohedral (Th_2Zn_{17}) $R\bar{3}m$ $Z = 3$ $d = 9003\ kg\ m^{-3}$
$a = 834$ pm $c = 1219$ pm (hexagonal cell)
Y in $6c_2$ $(0, 0, \frac{1}{3})$
Co in $6c_1$ $(0, 0, 0.097)$
Co in $9d$ $(\frac{1}{2}, 0, \frac{1}{2})$
Co in $18f$ $(\frac{1}{3}, 0, \frac{1}{3})$
Co in $18h$ $(\frac{1}{2}, \frac{1}{2}, \frac{1}{6})$

Y_2Co_{17} also crystallizes in the related Th_2Ni_{17} structure, which has a slightly different stacking sequence, giving a 50% shorter c axis and $Z = 2$; the rare-earth environment is similar

Electronic properties Metal, strong ferromagnet, $\varrho \approx 0.5\ \mu\Omega\ m$

Magnetic properties Ferromagnet $T_C = 1167\ K$ $m = 27\ \mu_B\ fu^{-1}$
$\sigma = 111\ A\ m^2\ kg^{-1}$ $M_s = 1.00\ MA\ m^{-1}$ $J_s = 1.26\ T$
$K_1 = -0.34\ MJ\ m^{-3}$

Significance Anisotropy is easy-plane, but may be tuned by substitutions. Of interest as a microwave absorber.

Related materials The compounds R_2Fe_{17} exist for all of the rare-earth series crystallizing in the rhombohedral Th_2Zn_{17} or the hexagonal Th_2Ni_{17} structure. Curie temperatures for other isostructural R–Fe series are plotted in Fig. 11.13.

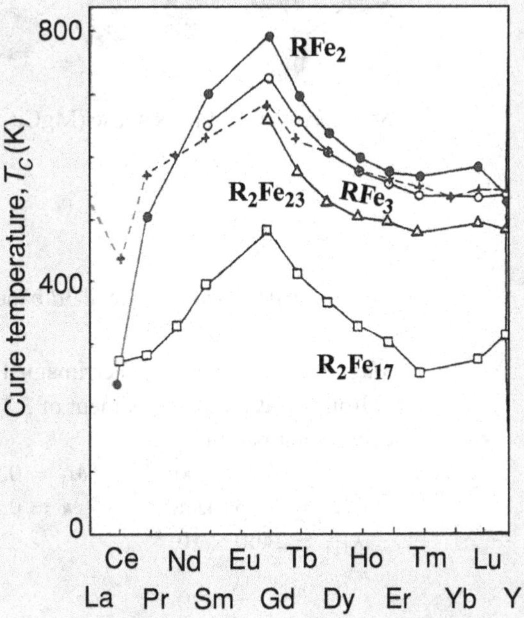

Figure 11.13
Magnetic ordering temperatures in R–Fe intermetallic compounds.

Table 11.13. Rare-earth intermetallics with the Th_2Ni_{17} or Th_2Zn_{17} structure

	a (pm)	c (pm)	T_C (K)	m (μ_B fu^{-1})	M_s (MA m^{-1})	K_1 (MJ m^{-3})	B_a (T)
Y_2Fe_{17}	848	826	327	18.6	0.48	−0.4	−1.6
Sm_2Fe_{17}	854	1243	389	22.4	0.80	−0.8	−2.0
Y_2Co_{17}	834	1219	1167	23.5	1.00	−0.3	−0.7
Sm_2Co_{17}	838	1221	1190	22.0	0.97	4.2	8.5
Gd_2Co_{17}	837	1218	1209	14.0	0.60	−0.5	−1.7

$TbFe_2$

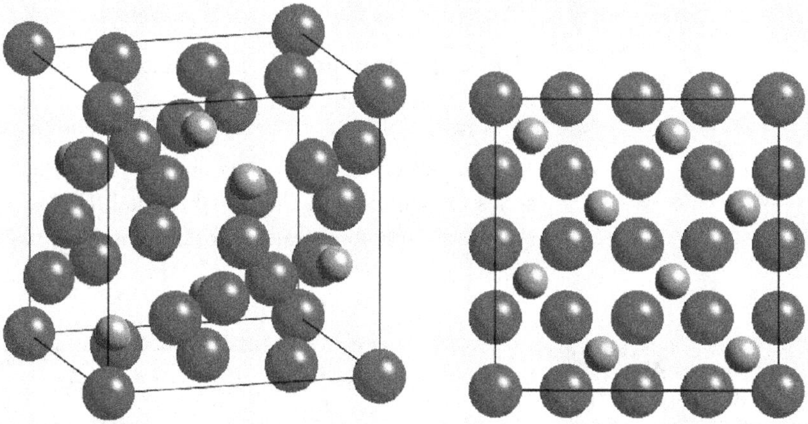

Structure C15 Laves phase ($MgCu_2$) $Fd3m$ $Z = 8$ d = 8980 kg m^{-3}

$a_0 = 737$ pm

Tb in $8a$ (0, 0, 0)

Fe in $16d$ $(\frac{5}{8}, \frac{5}{8}, \frac{5}{8})$

Electronic properties Metal $3d$ band is strong ferromagnet $\varrho = 0.6\ \mu\Omega$ m

Magnetic properties Ferrimagnet $T_C = 698$ K $m_0 = 6.0\ \mu_B$ fu^{-1}

Iron sublattice has a moment of 3.3 μ_B fu^{-1}; the localized $4f^8$ Tb moment couples antiparallel

$\sigma = 97$ A m^2 kg^{-1} $M_s = 0.88$ MA m^{-1} $J_s = 1.10$ T

$K_1 = -6.3$ MJ m^{-3} $\kappa = 0.80$ $\lambda_s = 1750 \times 10^{-6}$

$\lambda_{111} = 2400 \times 10^{-6}$

Table 11.14. Magnetization and Curie temperature of some RT_2 compounds

	Fe			Co			Ni		
	a_0 (pm)	T_C (K)	m (μ_B)	a_0 (pm)	T_C (K)	m (μ_B)	a_0 (pm)	T_C (K)	m (μ_B)
Ce	730	235	2.5	731			712		
Pr				730	49	2.8	729	15	0.9
Nd				726	105	3.5	727	16	1.8
Sm	742	688	2.7	742	227	1.3	723	21	0.2
Gd	740	796	3.6	740	404	4.8	720	79	7.1
Tb	735	698	4.5	735	238	5.7	716	40	7.7
Dy	733	635	5.8	733	146	6.9	715	27	8.8
Ho	730	608	5.5	730	87	7.7	714	20	8.8
Er	728	587	4.9	728	39	6.0	713	19	6.9
Tm	725	599	2.6	725	20	3.2	709	14	3.3

From K. H. J. Bushow, *Rep. Prog. Phys.* **40**, 1179 (1977).

Table 11.15. Magnetization and Curie temperature of some AFe_2 compounds

	a_0 (pm)	T_C (K)	m (μ_B)
UFe_2	706	158	1.1
$NpFe_2$	714	500	2.6
$PuFe_2$	719	600	2.3
$AmFe_2$	730	475	3.1

Significance $TbFe_2$ shows giant magnetostriction. When alloyed with $DyFe_2$ which has similar λ_{111} but the opposite sign of K_1 (2.1 MJ m^{-3}), it is possible to obtain highly magnetostrictive alloys $(Tb_{0.3}Dy_{0.7})Fe_2$, which are magnetically soft (Terfenol-D). Oriented bars are used as high-power transducers for underwater sonar.

Related materials The Laves phase alloys RFe_2, RCo_2 and RNi_2 exist for the whole rare-earth series. There are many possible pseudobinary solid solutions in this and other rare-earth transition-metal series of compounds. The RCo_2 compounds are on the borderline for the appearance of 3*d* magnetism; YFe_2 is a ferromagnet with $T_C = 540$ K. YCo_2 is a band metamagnet, where the magnetic moment appears at a first-order transition in an applied field. Nickel does not have a moment in RNi_2. Actinides form ferromagnetic Laves-phase compounds with iron. $ZrZn_2$ is an unusual weak itinerant ferromagnet with $T_C = 29$ K, composed of two nonmagnetic elements.

a-$Gd_{0.25}Co_{0.75}$

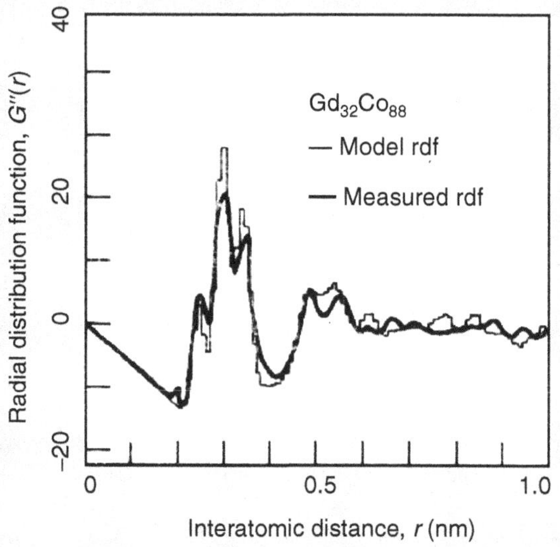

Structure Amorphous Random dense-packed structure of Gd and Co atoms d = 7800 kg m^{-3}

Electronic properties Metal $\varrho = 1.5\ \mu\Omega$ m

Magnetic properties Ferrimagnet, with antiparallel Gd and Co sublattices. $T_C \approx 700$ K, by extrapolation. The alloy crystallizes at a lower temperature.

Compensation point $T_{comp} = 320$ K

$\langle m \rangle = 0.6\,\mu_B$ fu^{-1} at $T = 0$ K

$\sigma = 10$ A m^2 kg^{-1}	$M_s = 0.08$ MA m^{-1}	$J_s = 0.10$ T
$\sigma_0 = 40$ A m^2 kg^{-1}	$M_0 = 0.31$ MA m^{-1}	$J_0 = 0.39$ T

Significance Additions of Tb allow thin films with perpendicular magnetic anisotropy to be prepared. These show a significant Kerr effect $\theta_K \approx 1°$, and can be used as magneto-optic recording media. The alloy composition x in a-$Gd_{1-x}Co_x$ can be varied continuously in the amorphous state, and it is chosen to make T_{comp} ideal for compensation-point writing.

Related materials A wide range of amorphous compositions a-$R_{100-x}Co_x$ can be produced by sputtering. The random anisotropy with non-S-state rare-earths may lead to sperimagnetic order. Likewise a-$R_{100-x}Fe_x$ alloys are asperomagnets or sperimagnets on account of the broad exchange distribution associated with the distribution of nearest-neighbour distances $P(\mathcal{J})$. The iron subnetwork becomes collinear on hydrogen absorption due to the effect of volume expansion to shift the exchange distribution to more positive values.

Figure 11.14

Compensation and Curie temperatures of crystalline and amorphous $Gd_{100-x}Co_x$.

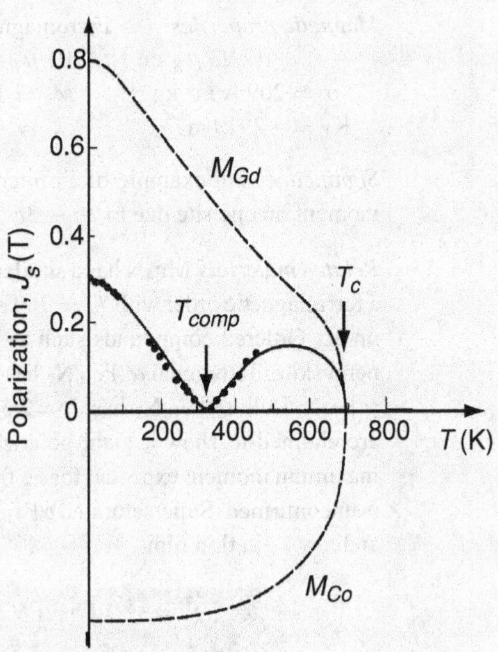

11.4 Interstitial compounds

Fe_4N

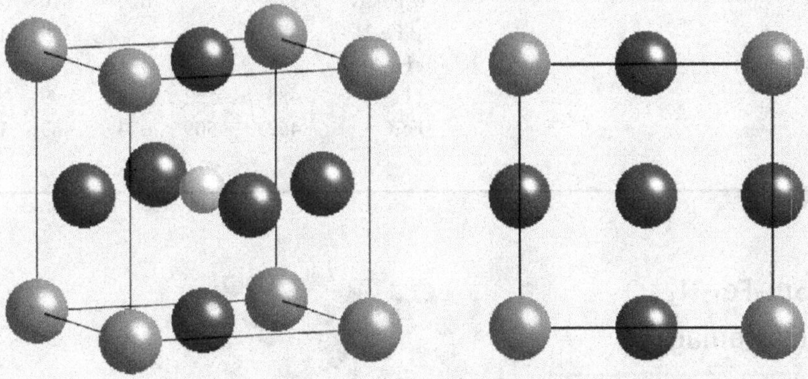

Structure E2$_1$ cubic *Pm*3*m* $Z = 1$ d = 7212 kg m^{-3}

$a_0 = 379.5$ pm
Fe in 1*a* (0, 0, 0)
Fe in 3*c* ($\frac{1}{2}$, $\frac{1}{2}$, 0)
N in 1*b* ($\frac{1}{2}$, $\frac{1}{2}$, $\frac{1}{2}$)

An fcc γFe structure stabilized by interstitial nitrogen in the body centre

Electronic properties Metal, strong ferromagnet

Magnetic properties Ferromagnet $T_C = 769$ K $\mathfrak{m}_0 = 8.88\ \mu_B\ \mathrm{fu}^{-1}$
($2.98\ \mu_B$ on $1a$, $2.01\ \mu_B$ on $3c$, $-0.13\ \mu_B$ on $1b$)
$\sigma = 209\ \mathrm{A\ m^2\ kg^{-1}}$ $M_s = 1.51\ \mathrm{MA\ m^{-1}}$ $J_s = 1.89$ T
$K_1 = -29\ \mathrm{kJ\ m^{-3}}$ $\lambda_s = -100 \times 10^{-6}$

Significance An example of an interstitial metallic structure, with a large $3d$ moment on one site due to $1a \rightarrow 3c$ charge transfer.

Related materials Mn_4N has a similar crystal structure with $a_0 = 386.5$ pm and a ferrimagnetic order with $T_c = 760$ K. Moments are $3.8\ \mu_B$ on $1a$ and $-0.9\ \mu_B$ on $3c$. Ordered compounds such as Ni_3FeN and Fe_3PtN are isomorphic with perovskite. Tetragonal $\alpha''Fe_{16}N_2$ has a structure intermediate between αFe and $\gamma'Fe_4N$. Bulk $\alpha''Fe_{16}N_2$ has $J_s = 2.3$ T with $K_1 \approx 1000\ \mathrm{kJ\ m^{-3}}$, but thin films are claimed to show a giant polarization, as high as 3.2 T. This exceeds the maximum moment expected for Fe from the Slater–Pauling curve ($2.7\ \mu_B$) and is unconfirmed. Supersaturated $\alpha Fe_{100-x}N_x$ with $x \approx 3$ has a much reduced K_1 and low λ_s in thin films.

Table 11.16. Lattice parameters and magnetic properties of iron carbide and nitrides

	a	b	c	T_C	J_s	$\langle\mathfrak{m}\rangle_0(\mu_B/\mathrm{Fe})$
$\alpha Fe_{97}N_3$	287			1010	2.2	2.20
$\alpha' Fe_{90}N_{10}$	283		312	835	2.3	2.48
$\alpha' Fe_{16}N_2$	572		629	810	2.3	2.35
$\gamma' Fe_4N$	379.5			769	1.8	2.25
ϵFe_3N	270		436	567		
ζFe_2N	483	552	443	9		0.05
Fe_3C	452	509	674	483	1.5	1.78

$Sm_2Fe_{17}N_3$ (Nitromag)

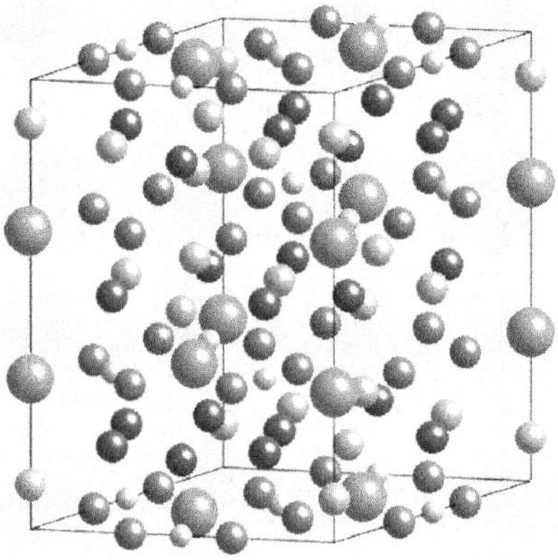

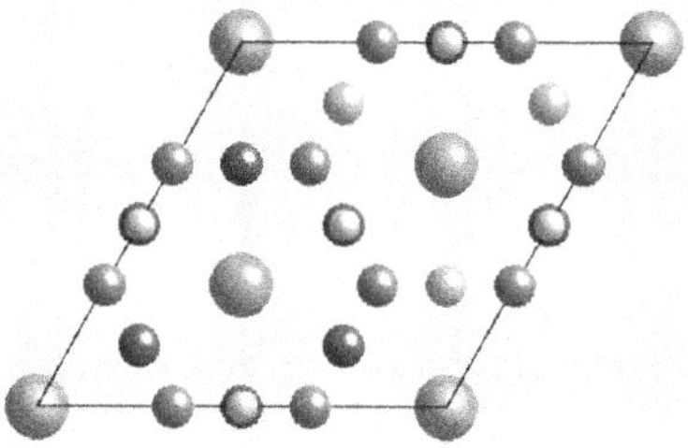

Structure	Rhombohedral	$R\bar{3}m$ $Z = 3$	d = 7680 kg m^{-3}
		$a = 873$ pm	$c = 1264$ pm
		Sm in 6*c* (0, 0, $\frac{1}{3}$)	
		Fe in 6*c* (0, 0, 0)	Fe in 18*h* ($\frac{1}{2}$, $\frac{1}{2}$, $\frac{1}{6}$)
		Fe in 9*d* ($\frac{1}{2}$, 0, $\frac{1}{2}$)	Fe in 18*f* ($\frac{1}{3}$, 0, 0)
		N in 9*e* ($\frac{1}{2}$, 0, 0)	

An interstitial compound derived from the rhombohedral Th_2Zn_{17}-type structure. Nitrogens form a triangle around the Sm.

Electronic properties Metal, strong ferromagnet

Magnetic properties Ferromagnet with parallel coupling of 3*d* moments and a small localized Sm 4*f* moment. $T_C = 749$ K m$_0 = 39\ \mu_B$ fu^{-1} (6*c* 2.8 μ_B; 9*d* 2.2 μ_B; 18*f* 2.0 μ_B; 18*h* 2.4 μ_B)

$\sigma = 160$ A m^2 kg^{-1}	$M_s = 1.23$ MA m^{-1}	$J_s = 1.54$ T
$K_1 = 8.6$ MJ m^{-3}	($K_1^{Sm} = 9.7$ MJ m^{-3}, $K_1^{Fe} = -1.1$ MJ m^{-3})	$\kappa = 2.13$
$B_a = 14$ T	$A = 12$ pJ m^{-1}	$\delta_w = 3.7$ nm

Significance Interstitial nitrogen in Sm_2Fe_{17} produces a 6% volume expansion which leads to a dramatic increase in Curie temperature of 360 K. The nitrogen also creates a large uniaxial crystal field leading to strong uniaxial anisotropy for R = Sm. Nitrogen is introduced from the gas phase into fine Sm_2Fe_{17} powder, producing material suitable for bonded magnets, but the nitride is metastable and the powder cannot be sintered.

Related materials Interstitial carbon has a similar effect. Other iron-based rare-earth intermetallics such as $ThMn_{12}$-structure compounds $RFe_{12-x}X_x$ with R = Pr, Nd and X = Si, Ti, Mo, V, $x \approx 1$, and $R_3(Fe, X)_{29}$ compounds with

Table 11.17. Some 2:17 interstitial compounds

	a (pm)	*c* (pm)	T_c (K)	M_s (MA m^{-1})	K_1 (MJ m^{-3})
Y_2Fe_{17}	848	826	327	0.48	−0.4
$Y_2Fe_{17}N_3$	865	844	694	1.17	−1.1
$Y_2Fe_{17}C_3$	866	840	660	1.00	−0.3
$Y_2Fe_{17}H_x$	852	827	475	0.75	−0.4
Sm_2Fe_{17}	854	1243	389	0.80	−0.8
$Sm_2Fe_{17}N_3$	873	1264	749	1.23	8.6
$Sm_2Fe_{17}C_3$	875	1257	668	1.14	7.4
$Sm_2Fe_{17}H_x$	861	1247	550	1.10	4.2

R = Pr, Nd, Sm and X = Ti show an increase in T_C and K_1 with interstitial nitrogen. Interstitial hydrogen in rare-earth intermetallics with H/R up to 3 forms a proton liquid. The lattice expansion leads to an increase of T_C and sometimes a change of magnetic structure.

11.5 Oxides with ferromagnetic interactions

EuO

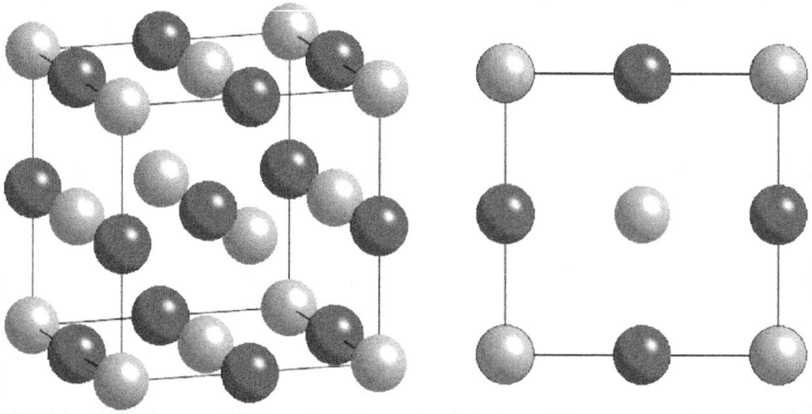

Structure B1 cubic (NaCl) $Fm\bar{3}m$ $Z = 4$ $d = 8122\ \mathrm{kg\ m^{-3}}$
$a_0 = 516\ \mathrm{pm}$
Eu in $4a$ (0, 0, 0)
O in $4b$ ($\frac{1}{2}$, 0, 0)

An fcc oxygen array with Eu^{2+} in undistorted octahedral interstices

Electronic properties Black ferromagnetic semiconductor; bandgap $\varepsilon_g =$ 1.2 eV. When nonstoichiometric, the oxide is metallic below the Curie point and it exhibits a metal–insulator transition where the resistivity increases by up to 14 orders of magnitude. Colossal magnetoresistance in the vicinity of T_C reaches $10^8\%\ \mathrm{T^{-1}}$. There is a spin splitting of the empty conduction band and a red shift of the band gap of 0.2 eV below T_C. Carriers in the paramagnetic state are magnetic polarons.

Magnetic properties Ferromagnet $T_C = 69.3$ K $\mathfrak{m}_0 = 7.0\ \mu_B$
Localized moments Eu^{2+} $4f^7$; $S = \frac{7}{2}$ Ground state $^8S - A_{1g}$
$\sigma_0 = 233\ \mathrm{A\ m^2\ kg^{-1}}$ $M_0 = 1.89\ \mathrm{MA\ m^{-1}}$ $J_0 = 2.38$ T.
$K_1 = 44\ \mathrm{kJ\ m^{-3}}$

Figure 11.15

The metal–insulator transition in $Eu_{1-\delta}O$, showing colossal magnetoresistance. (T. Penny *et al. Phys. Rev. B*, 5, 3669 (1972))

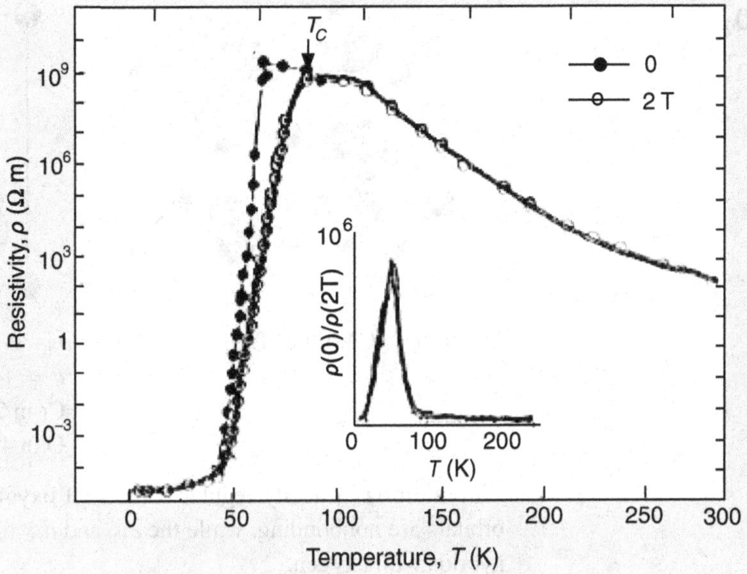

Europium forms an fcc lattice with ferromagnetic first- and second-neighbour interactions $\mathcal{J}_1$ and $\mathcal{J}_2$. The exchange involves virtual excitation of $4f$ electrons into $5d$ and $6s$ conduction band states.

Significance Stoichiometric EuO is a model ferromagnetic semiconductor with Heisenberg exchange. Electron doping increases T_C. The spin splitting of the conduction band below T_C allows EuO and EuS to be used as tunnel barrier spin filters. Cation-deficient material, or material doped with a trivalent rare-earth, is an n-type magnetic semiconductor. The low Curie temperature precludes practical applications.

Related materials GdN is a similar NaCl-structure ferromagnet with $T_C =$ 69 K; the other RN compounds have lower Curie temperatures. Europium hexaboride (EuB_6) is a ferromagnet ($a_0 = 418$ pm, $T_C = 12.5$ K), with a similar structure where the B_6^{2-} ion replaces O^{2-}. The europium chalcogenides show systematic variation of the first- and second-neighbour interactions with distance and covalency. EuTe is antiferromagnetic.

Table 11.18. Magnetic properties of europium chalcogenides

	a_0 (pm)	order	T_C	$\mathcal{J}_1$ (K)	$\mathcal{J}_2$ (K)
EuO	516	Ferro	69.3	0.60	0.12
EuS	596	Ferro	16.5	0.23	−0.11
EuSe	620	Ferri/antiferro	4.6/2.8	0.16	−0.16
EuTe	661	Antiferro	9.6	0.10	−0.21

CrO_2

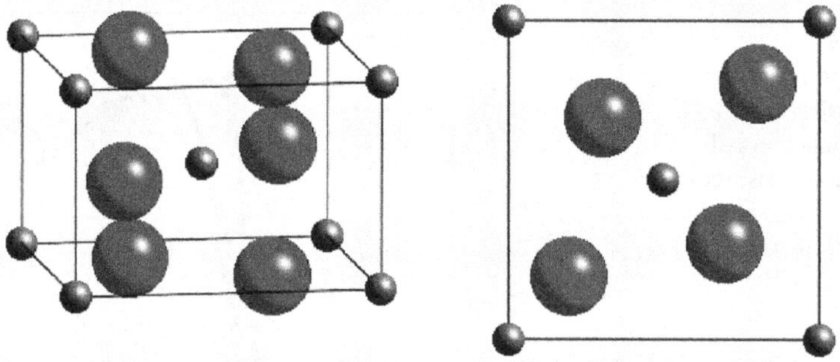

Structure C4 rutile (TiO_2) $P4_2/mnm$ $Z = 2$ $d = 4890\ \mathrm{kg\ m^{-3}}$
$a = 442$ pm $c = 292$ pm
Cr in $2a$ (0, 0, 0)
O in $4f$ (0.303, 0.303, 0)

Chromium is in nearly regular octahedral oxygen coordination, but the d_{xy}-orbitals are nonbonding, while the d_{yz} and d_{zx} form bonding and antibonding hybrids with oxygen.

Electronic properties Black half-metallic ferromagnet with a spin gap in the $\downarrow$ density of states of 0.5 eV; $\varrho \approx 2\ \mu\Omega$ m (0.03 $\mu\Omega$ m at 4.2 K). Chromium is formally Cr^{4+} $3d^2$; t_{2g}^2. One $3d$ electron is localized in the $t_{xy}^{\uparrow}$-band. The other is delocalized in a $\uparrow$ band of mixed $t_{yz}^{\uparrow}/t_{zx}^{\uparrow}$ and oxygen character.

Magnetic properties Ferromagnet $T_C = 396$ K. $m_0 = 2\,\mu_B\ \mathrm{fu}^{-1}$
$\sigma = 80\ \mathrm{A\,m^2\,kg^{-1}}$, $M_s = 0.39\ \mathrm{MA\,m^{-1}}$ $J_s = 0.49$ T
$K_1 = 25\ \mathrm{kJ\,m^{-3}}$ $\kappa = 0.36$ $A = 4\ \mathrm{pJ\,m^{-1}}$
$\mathcal{J}_{Cr-Cr} = 37.1$ K $\mathcal{J}_H = 0.9$ eV $\lambda_s = 5 \times 10^{-6}$.

Significance CrO_2 is the only binary oxide that is a ferromagnetic metal. It is the simplest half-metal. It forms acicular particles; those with a length of about 300 nm and an aspect ratio of 8:1 were used as a magnetic recording medium, especially on video tapes. The Curie temperature cannot be increased by substitution.

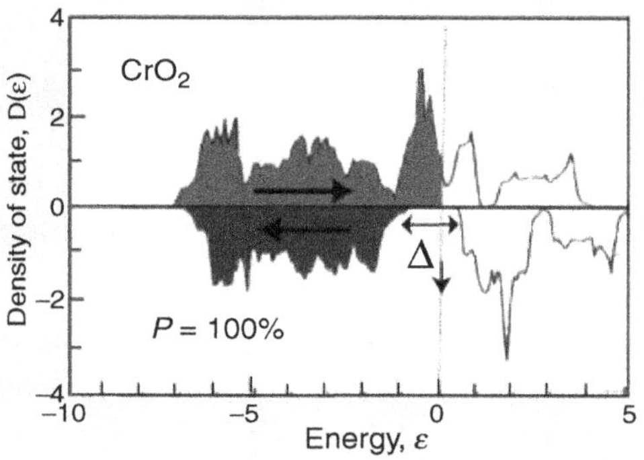

Figure 11.16

Density of states for half-metallic CrO_2, showing a spin gap in the $\downarrow$ band. *P* is the spin polarization at the Fermi level.

Related materials RuO_2 ($4d^4$) is a rutile-structure Pauli paramagnet with an unpolarized $4d$ band. VO_2 ($3d^1$) is an antiferromagnet with a metal–insulator transition at $T_N = 343$ K. TiO_2 is an insulator with a small temperature-independent paramagnetic susceptibility. There is a series of Magnelli phases Ti_nO_{2n-1} with mixed-valence titanium and sheets of oxygen vacancies. MnO_2 is an antiferromagnet with $T_N = 94$ K. SnO_2 is an n-type semiconductor.

$SrRuO_3$

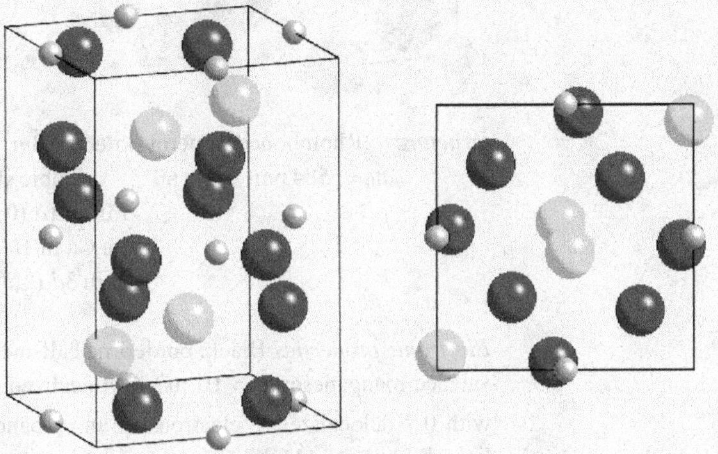

Structure Orthorhombic perovskite *Pbnm* $Z = 4$ $d = 6518$ kg m^{-3}

$a = 557.3$ pm $b = 553.8$ pm $c = 785.6$ pm

Sr in $4c$ (-0.018, 0.06, $\frac{1}{4}$)

Ru in $4b$ ($\frac{1}{2}$, 0, 0)

O in $4c$ (0.05, 0.47, $\frac{1}{4}$)

O in $8d$ (-0.29, 0.275, 0.05)

Electronic properties Black metal with spin-split Ru $4d$ t_{2g}-band of width ≈ 1 eV $\varrho \approx 4$ μΩ m.

Weak ferromagnet with low-spin Ru^{4+} $4d^4$; t_{2g}^4

Magnetic properties Ferromagnet $T_C = 165$ K $\mathfrak{m}_0 = 1.0\ \mu_B$ fu^{-1}

$\sigma_0 = 24$ A m^2 kg^{-1} $M_0 = 0.16$ MA m^{-1} $J_0 = 0.20$ T.

$K_1 = 640$ kJ m^{-3} $K_2 = -1080$ kJ m^{-3}

Significance A rare example of a ferromagnetic $4d$ metal, of no practical importance. Thin films may be grown on $SrTiO_3$.

Related compounds There is a series of Ruddlesden–Popper phases $Sr_{n+1}Ru_nO_{3n+1}$, with n corner-sharing octahedral layers, separated by an SrO layer. The $n = \infty$ end member and the $n = 3$ compound $Sr_4Ru_3O_{10}$ ($T_C = 104$ K) are ferromagnetic, $n = 1$ is a superconductor, $n = 2$ is a paramagnet with enhanced Pauli susceptibility. $CaRuO_3$ does not order magnetically.

$(La_{0.7}Sr_{0.3})MnO_3$ (LSMO)

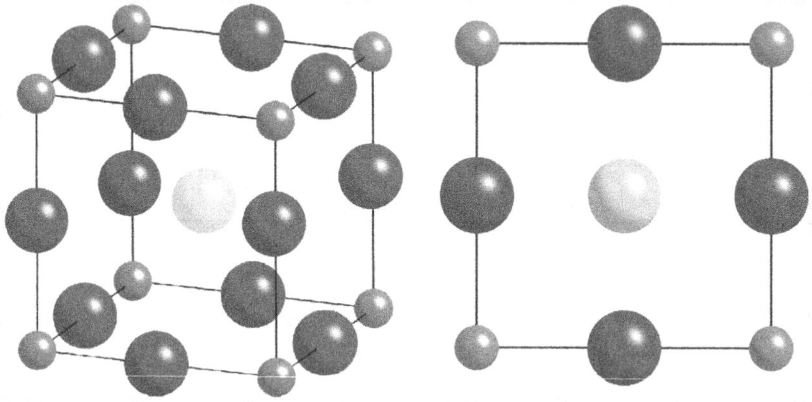

Structure Rhombohedral perovskite $R\bar{3}m$ $Z = 4$ $d = 6320\ \text{kg m}^{-3}$
$a_0 = 584$ pm $\alpha \approx 60°$ (cubic structure with $a_0 = 0.390$ pm)
Mn in $1a$ (0, 0, 0)
La,Ca in $1b$ $(\frac{1}{2}, \frac{1}{2}, \frac{1}{2})$
O in $3d$ $(\frac{1}{2}, 0, 0)$, $(0, \frac{1}{2}, 0)$, $(0, 0, \frac{1}{2})$

Electronic properties Black, borderline half-metallic ferromagnet with mixed-valence manganese; $\varrho \approx 10\ \mu\Omega$ m. Localized Mn^{4+} $3d^{3\uparrow}$; $t_{2g}^{3\uparrow}$; $S = \frac{3}{2}$ cores, with 0.7 delocalized ↑ electrons in an e_g-band. The bottom of the $t_{2g}^{\downarrow}$-band lies close to ε_F. Metal–insulator transition near T_C with colossal negative magnetoresistance. Granular samples also exhibit a low-field magnetoresistance effect. Carriers are magnetic polarons above T_C. Mn^{3+} is a Jahn–Teller ion.

Magnetic properties Ferromagnet $T_C = 370$ K $\mathfrak{m}_0 = 3.6\ \mu_B\ \text{fu}^{-1}$
$\sigma = 71\ \text{A m}^2\ \text{kg}^{-1}$ $M_s = 0.44\ \text{MA m}^{-1}$ $J_s = 0.55$ T
$K_1 = -2\ \text{kJ m}^{-3}$
$\sigma_0 = 90\ \text{A m}^2\ \text{kg}^{-1}$, $M_0 = 0.56\ \text{MA m}^{-1}$ $J_0 = 0.70$ T
The principal exchange interaction is double exchange mediated by the hopping $e_g^{\uparrow}$ electron

Significance A half-metallic oxide. LSMO has the highest ferromagnetic Curie temperature of the mixed-valence manganites. The colossal magnetoresistance effect is of limited practical importance because it is very sensitive to temperature and requires large magnetic fields. Some applications have been proposed for related materials with $T_C \approx RT$ as bolometers and position sensors.

Related materials Other half-metallic oxides include $Tl_2Mn_2O_7$, which is a half-metallic semimetal. The large family of mixed-valence manganites present very rich electronic and magnetic phase diagrams including regions of magnetic, charge and orbital order. Related ABO_3 compounds have Sr replaced by Ca (LCMO) or Ba (LBMO). Colossal magnetoresistance of LCMO was

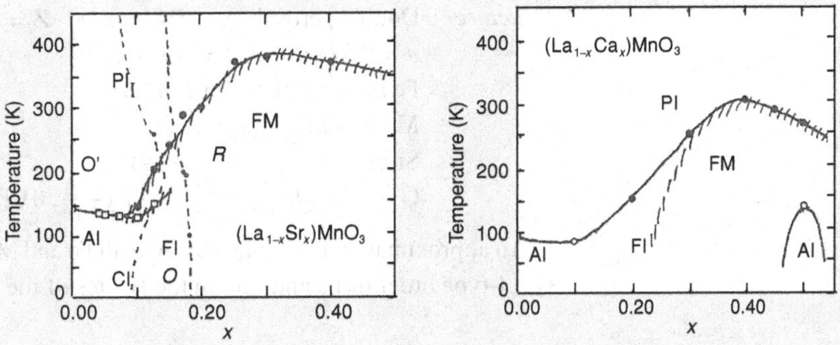

Figure 11.17

Phase diagrams of $(La_{1-x}Sr_x)MnO_3$ and $(La_{1-x}Ca_x)MnO_3$. The phases are: PI, paramagnetic insulator; AI, antiferromagnetic insulator; CI, canted antiferromagnetic insulator; FI, free magnetic insulator; and FM, paramagnetic metal. The dashed lines indicate structural phase transitions: O, O' orthorhombic, R rhombohedral.

illustrated in Fig. 5.41. Hexagonal $YMnO_3$ is both antiferromagnetic ($T_N =$ 75 K) and ferroelectric ($T_{cf} = 900$ K). Many other magnetically ordered ABO_3 perovskites exist with other 3d atoms. Those with B = Fe or Co, often order well above room temperature as, for example, the series of orthorhombic antiferromagnetic rare-earth iron perovskites $RFeO_3$. $GdFeO_3$, for example, has $T_N = 657$ K. In some cobaltites, the Co is in a low-spin state. Thanks to the numerous substitutions possible on both sites, perovskites are the largest group of magnetically ordered oxides. $LaTiO_3$ and $LaMnO_3$ are antiferromagnets with $T_N = 146$ K and 150 K, respectively. $YTiO_3$ is a ferromagnet with $T_C = 25$ K.

Perovskite structures depend on a tolerance factor related to the ionic radii of the constituents $t = (r_A + r_O)/\sqrt{2}(r_B + r_O)$, which is 1 for the ideal cubic structure. Orthorhombic distortion occurs for $0.80 < t < 0.89$ and the structure is hexagonal when $t > 1$. For intermediate values of t, it is rhombohedral.

Sr_2FeMoO_6 (SFMO)

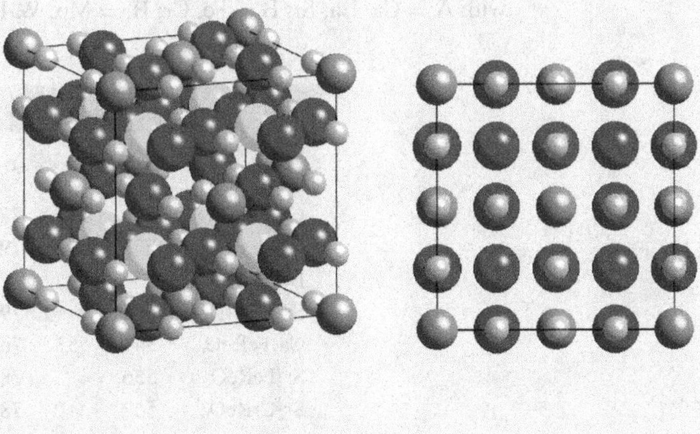

Structure Double perovskite $P4/mmm$ $Z = 2$ $d = 5714$ kg m^{-3}
$a = 557$ pm $c = 791$ pm
Fe in $4a$ (0, 0, 0) (in $Fm3m$)
Mo in $4a$ $(\frac{1}{2}, \frac{1}{2}, \frac{1}{4})$
Sr in $8c$ $(\frac{1}{4}, \frac{1}{4}, \frac{1}{4})$, $(\frac{3}{4}, \frac{3}{4}, \frac{3}{4})$
O in $24d$ $(0, \frac{1}{4}, \frac{1}{4})$, $(\frac{1}{4}, 0, \frac{1}{4})$, $(\frac{1}{4}, \frac{1}{4}, 0)$, $(0, \frac{1}{4}, \frac{3}{4})$, $(\frac{3}{4}, 0, \frac{1}{4})$, $(\frac{1}{4}, \frac{3}{4}, 0)$

An approximately fcc oxygen array with Fe and Mo in octahedral interstices; NaCl-type order of Fe and Mo on the B sites of the perovskite lattice.

Electronic properties Black ferromagnet, half-metallic when cation order is perfect $\varrho \approx 4$ μΩ m

Delocalized electron in Mo $4d^1$; t_{2g} band.

Magnetic properties Ferrimagnet $T_C = 436$ K $\mathfrak{m}_0 = 3.60\,\mu_B$ fu^{-1}
$\sigma = 35$ A m^2 kg^{-1} $M_s = 0.20$ MA m^{-1} $J_s = 0.25$ T
$K_1 = 28$ kJ m^{-3}
Localized Fe^{3+} $3d^5$; $t_{2g}^3\, e_g^2$ ion cores $S = \frac{5}{2}$
$\sigma_0 = 48$ A m^2 kg^{-1} $M_0 = 0.27$ MA m^{-1} $J_0 = 0.35$ T

The NaCl-cation superstructure means that there are no Fe–O–Fe superexchange bonds. Exchange is mediated by the delocalized ↓ 4d electron which mixes with empty Fe $3d^{\downarrow}$ states. The discrepancy between the low-temperature moment and the 4.0 μ_B fu^{-1} expected for a stoichiometric, ordered half-metal is due to Fe/Mo antisite defects.

Significance A ferromagnetic oxide with a Curie temperature significantly higher than that of any manganite.

Related compounds An extensive family of double perovskites $A_2BB'O_6$ exists with A = Ca, Ba, Sr; B = Fe, Cr; B′ = Mo, W, Re, . . .

Table 11.19. Double perovskites

	a	b	c	T_C (K)
Ca_2FeMoO_6	541	553	770	345
Sr_2FeMoO_6	558		790	436
Ba_2FeMoO_6	806			367
Sr_2FeWO_6	565		794	39†
Ca_2FeReO_6	541	553	769	539
Sr_2FeReO_6	556		787	405
Sr_2CrReO_6	552		780	635

† T_N

11.6 Oxides with antiferromagnetic interactions

NiO

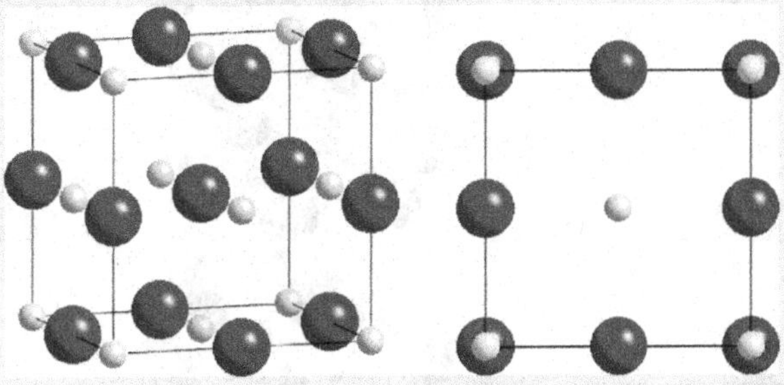

Structure B1 cubic (NaCl) $Fm\bar{3}m$ $Z = 4$ d = 6793 kg m^{-3}
$a_0 = 418$ pm
Ni in 4*a* (0, 0, 0)
O in 4*b* ($\frac{1}{2}$, 0, 0)

fcc oxygen array with Ni in undistorted octahedral interstices

Electronic properties Green insulator (black, polaron conductor when cation-deficient) $\varepsilon_g = 4.0$ eV
Localized Ni^{2+} $3d^8$; $t_{2g}^6 e_g^2$ ions $S = 1$ $^3F - A_{2g}$

Magnetic properties Type II antiferromagnet $T_N = 525$ K $m_0^A = 1.6\ \mu_B$
Ni^{-1} $\lambda_{100} = -140 \times 10^{-6}$, $\lambda_{100} = -79 \times 10^{-6}$ $V_{sw} = 38$ km s^{-1}
$\sigma_A = 56$ A m^2 kg^{-1} $M_A = 0.38$ MA m^{-1} $J_A = 0.48$ T
$K_1 = -500$ J m^{-3}

Ni^{2+} ions form an fcc lattice with partly frustrated antiferromagnetic interactions. The ⟨111⟩ directions are hard, producing a tiny rhombohedral magnetostrictive distortion below T_N. $\mathcal{J}_1 = -8$ K, but this interaction with 12 nearest-neighbour cations is frustrated; $\mathcal{J}_2 = -110$ K, and this is a strong 180° superexchange interaction with six next-nearest-neighbour cations.

Significance NiO was used for exchange bias of early spin-valve structures.

Related compounds There is a series of NaCl-structure monoxides.

Table 11.20. Antiferromagnetic monoxides

	a_0 (pm)			S	m (μ_B)	T_N (K)	θ (K)	$\mathcal{J}_1$ (K)†	$\mathcal{J}_2$ (K)†
MnO	445	R	$3d^5$	5/2	4.7	118	−610	−7.2	−3.5
FeO	431	R	$3d^6$	2	3.3	198	−570	−7.8	−8.2
CoO	426	T	$3d^7$	3/2	3.6	291	−330	−6.9	−21.2
NiO	418	R	$3d^8$	1	1.8	525	−1310	−50	−85

R – rhombohedral distortion; T – tetragonal distortion. † deduced from T_N, θ

αFe_2O_3 (hematite)

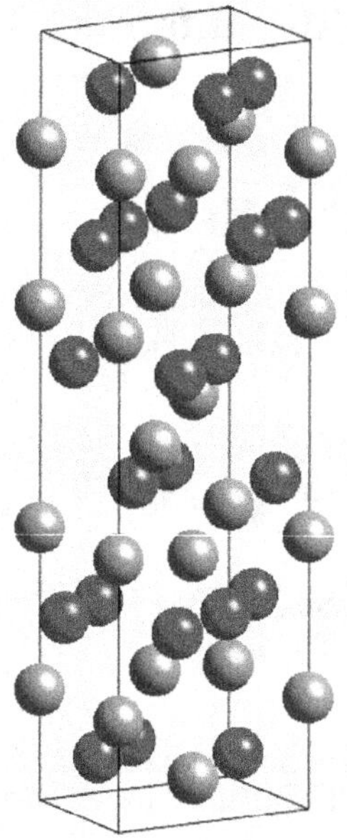

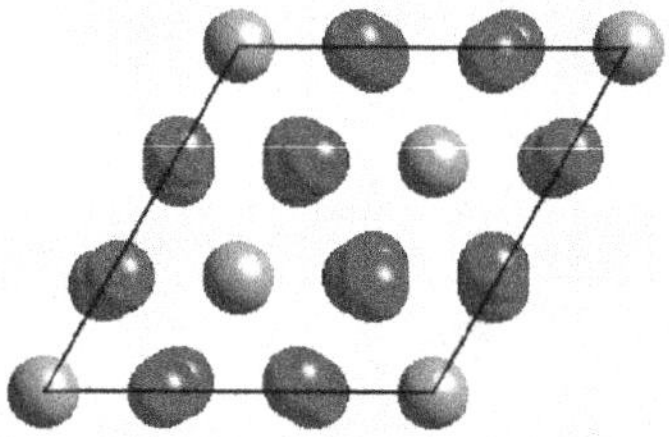

X-ray powder diffraction pattern for αFe_2O_3

h	k	l	d(pm)	I/I_{max}
0	1	2	366.0	25
2	0	0	269.0	100
2	2	0	241.0	50
0	0	6	228.5	2
1	1	3	220.1	30
2	0	2	207.0	2
0	2	4	183.8	40
1	1	6	169.0	60
2	1	1	163.4	4
0	1	8	159.6	16
2	1	4	149.4	35
3	0	0	145.2	35
2	0	8	134.9	4
1	1	9	131.0	20
2	2	0	125.8	8
0	3	6	122.6	2
2	2	3	121.3	4
1	2	8	118.9	8
0	2	10	116.2	10
1	3	4	114.1	12
2	2	6	110.2	14
0	4	2	107.6	2
2	1	10	105.5	18
1	1	12	104.2	2
4	0	4	103.8	2
2	3	2	98.9	10
2	2	9	97.2	2
3	2	4	96.0	18
0	1	14	95.8	6
1	4	0	95.1	12
4	1	3	93.1	6
0	4	8	92.0	6
1	3	10	90.8	25

Structure D5_1 corundum (Al_2O_3) $R\bar{3}c$ d $= 5260$ kg m^{-3}
$Z = 2$ (rhombohedral) $a = 252$ pm $\alpha = 55.3°$
$Z = 6$ (hexagonal) $a = 503.6$ pm, $c = 1374.9$ pm
Fe in 12c (0, 0, 0.355)
O in 18e (0.307, 0, 0.25)

The structure is an hcp oxygen array with Fe in two-thirds of the octahedral interstices

Electronic properties Red insulator $\varepsilon_g = 2.1$ eV
Localized electrons Fe^{3+} $3d^5$; $t_{2g}^3 e_g^2$ $S = \frac{5}{2}$ $^6S - A_{1g}$

Magnetic properties Canted antiferromagnet $T_N = 960$ K $m_0^A = 4.9\ \mu_B$ Fe^{-1}

Ferromagnetic (001) planes, stacked $+\ +\ --$

Exchange constants $\mathcal{J}_1 = 6.0$ K, $\mathcal{J}_2 = 1.6$ K, $\mathcal{J}_3 = -29.7$ K, $\mathcal{J}_4 = -23.2$ K, $V_{sw} = 34$ km s^{-1}.

Weak Dzyaloshinsky–Moriya interaction $\mathcal{D} \approx 0.1$ K, $\mathcal{D} \| 001$.

Spin reorientation (Morin) transition at $T_M = 260$ K; $S \| \boldsymbol{c}$ for $T < T_M$; symmetry $\bar{3}m$; $S \perp \boldsymbol{c}$ for $T > T_M$ symmetry $2/m$. K_1 changes sign at T_M.

$\sigma = 0.5$ A m^2 kg^{-1} $M_s = 2.5$ kA m^{-1} $J_s = 3$ mT
$\sigma_A = 175$ A m^2 kg^{-1} $M_A = 0.92$ MA m^{-1} $J_A = 1.16$ T
$K_1 = 9$ kJ m^{-3} $B_a = -7$ T

Significance A common rock-forming mineral, and constituent of soil. Contributes to the natural remanence of rocks. Used as iron ore, red pigment and abrasive (jewellers' rouge). Readily exhibits hysteresis on account of

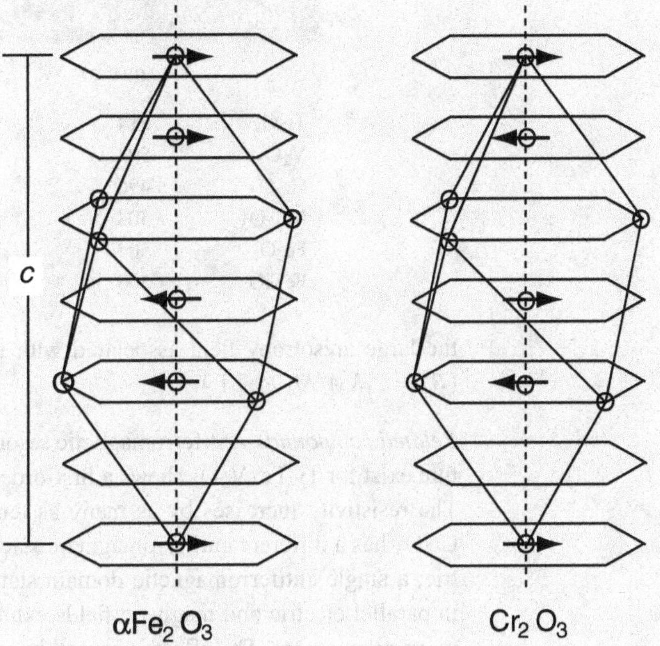

Figure 11.18

The antiferromagnetic structures of Cr_2O_3 and αFe_2O_3, above T_M. The rhombohedral cell is outlined.

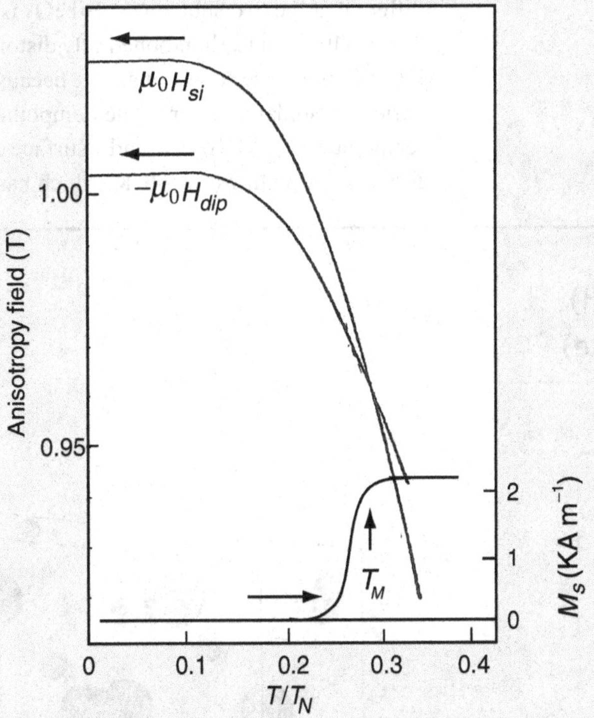

Figure 11.19

Temperature variation of the weak magnetic moment, and the two contributions to the sublattice magnetocrystalline anisotropy field in αFe_2O_3.

Table 11.21. Antiferromagnetic sesquioxides

	a (nm)	c (nm)	T_N
Ti_2O_3	514	1366	470
V_2O_3	503	1362	150
Cr_2O_3	496	1359	306
Mn_2O_3	504	1412	80
Fe_2O_3	504	1375	960
$FeTiO_3$	508	1404	58

the large anisotropy field associated with the weak ferromagnetic moment. ($B_a = 2|K_1|/M_s \approx 7$ T.)

Related compounds Antiferromagnetic sesquioxides with the corundum structure exist for Ti–Fe. V_2O_3 shows a first-order insulator–metal transition at T_N. The resistivity increases by as many as ten orders of magnitude below T_N. Cr_2O_3 has a different antiferromagnetic stacking $+ - + -$. It is magnetoelectric; a single antiferromagnetic domain state produced by cooling below T_N in parallel electric and magnetic fields exhibits a small electric-field-induced magnetic moment. The effect is caused by relative displacements of Cr^{3+} and O^{2-} in the electric field, which produces small changes of the crystal field, different at the two sublattices. $BiFeO_3$ is another canted antiferromagnet with $T_N = 640$ K, and a rhombohedrally distorted perovskite structure (see LSMO), but it is not a weak ferromagnet, because the weak moment follows a long-period cycloidal structure. The compound is also ferroelectric with transition temperature $T_{cf} = 1090$ K and a surface charge of 1 C m^{-2}. $LaFeO_3$ is an antiferromagnet with $T_N = 740$ K, which has been used for exchange bias.

αFeO(OH) (goethite)

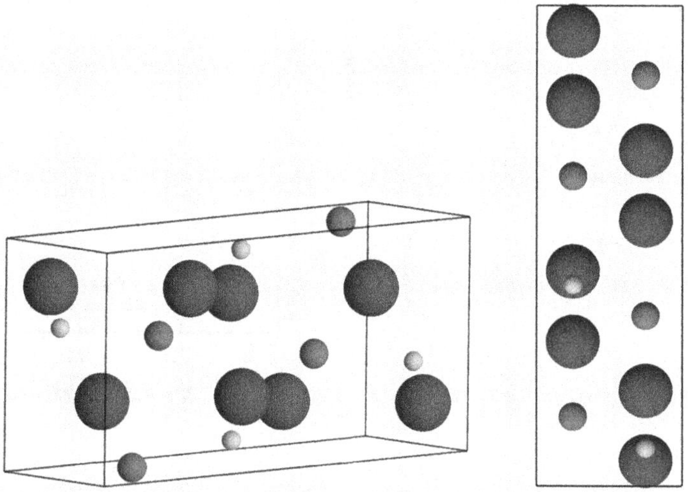

Structure Orthorhombic $Pnma$ $Z = 4$ d = 4270 kg m^{-3}
$a = 996$ pm, $b = 302$ pm, $c = 461$ pm,
Fe in $4c$ (0.145, $\frac{1}{4}$, −0.045)
O in $4c$ (−0.199, $\frac{1}{4}$, −0.288)
O in $4c$ (−0.053, $\frac{1}{4}$, −0.198)
H in $4c$ (−0.08, $\frac{1}{4}$, −0.38)

Converts to αFe_2O_3 when heated

Electronic properties Yellow-brown insulator
Localized electrons Fe^{3+} $3d^5$; $t_{2g}^3 e_g^2$ ions $S = 5/2$; $^6S - A_{1g}$

Magnetic properties Antiferromagnet $T_N = 460$ K
Ionic moment $\mathfrak{m}_0 = 4.0\ \mu_B$

The magnetic structure consists of double zig-zag iron chains, ordered antiferromagnetically with moments parallel to ***b***
$\sigma_A = 110$ A m^2 kg^{-1} $M_s = 0.47$ MA m^{-1} $J_s = 0.59$ T
$K_1 = 60$ kJ m^{-3}

Significance The main constituent of rust. Often present in tropical soils. Goethite can be superparamagnetic, even when quite well crystallized.

Related compounds There is a series of crystalline ferric hydroxides, all of which order antiferromagnetically above room temperature, except for lepidocrocate, which has a structure with well-separated planes of Fe^{3+} cations. Ferrous hydroxide has the CdI_2 structure. It is a planar antiferromagnet, where ferromagnetic c planes are coupled antiferromagnetically.

Table 11.22. Iron hydroxides

			Lattice parameters (pm)		T_c (K)
$\alpha FeO(OH)$	Goethite	$Pnma$	$a = 996$, $b = 302$, $c = 461$	af	460
$\beta FeO(OH)$	Akaganénite	$I2/m$	$a = 1053$, $c = 303$	af	295
$\gamma FeO(OH)$	Lepidocrocite	$Bbmn$	$a = 388$, $b = 1254$, $c = 307$	af	70
$\delta FeO(OH)$	Feroxyhyte	$P3ml$	$a = 293$, $c = 460$	fi	450
$Fe_5O_3(OH)_9$	Ferrihydrite	R	$a = 293$, $c = 460$	fi	450
$Fe_{1-x}(OH)_3$	Ferric gel	$P3$	Amorphous	sp	100
$Fe(OH)_2$	Amakinite	$P\bar{3}m1$	$a = 692$, $c = 1452$	af	20

af – antiferromagnet; fi – ferrimagnet; sp – speromagnet.

Fe_3O_4 (magnetite)

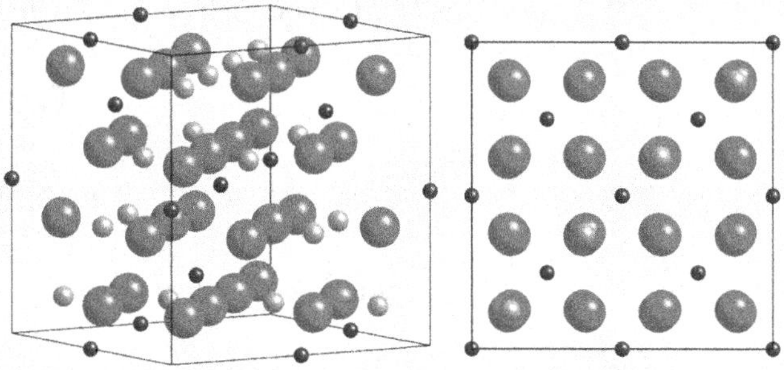

Structure $H1_1$ spinel ($MgAl_2O_4$) $Fd3m$ $Z = 8$ $d = 5195$ kg m^{-3}
$a_0 = 839.7$ pm
Fe in $8a$ $[\frac{1}{2}, \frac{1}{2}, \frac{1}{2}]$ A-site
Fe in $16d$ $\{\frac{1}{8}, \frac{1}{8}, \frac{1}{8}\}$ B-site
O in $32e$ (x, x, x), $x = 0.380$

An fcc oxygen array with Fe^{3+} in one-eighth of the tetrahedral interstices and $Fe^{2+/3+}$ in half of the octahedral interstices. $[Fe^{3+}]\{Fe^{3+}Fe^{2+}\}O_4$

Electronic properties Black, polaronic conductor $\varrho(RT) \approx 50$ μΩ m. The sixth $3d$-electron associated with Fe^{2+} on B sites is delocalized in a narrow, spin-polarized $t_{2g}^{\downarrow}$-band. Below the Verwey transition at $T_V = 119$ K, there is a charge-ordered insulating state where these electrons are localized in pairs with different but non-integral charges.

Magnetic properties Ferrimagnet $T_C = 860$ K
Localized Fe^{3+} $3d^5$; $t_{2g}^3 e_g^2$ ion cores $S = \frac{5}{2}$ ^6S on A- and B-sites
The A and B sublattices are oppositely aligned, with magnetization along $\langle 111 \rangle$. Spin-only moments for Fe^{3+} and Fe^{2+} are 5 and 4 μ_B respectively. The net moment at $T = 0$ K is $\{5 + 4\} - [5] = 4$ μ_B fu^{-1}. Exchange constants $\mathcal{J}_{AA} = -18$ K, $\mathcal{J}_{AB} = -28$ K, $\mathcal{J}_{BB} = 3$ K.

$\sigma = 92$ A m^2 kg^{-1}	$M_s = 0.48$ MA m^{-1}	$J_s = 0.60$ T
$A = 7$ pJ m^{-1}	$K_1 = -13$ kJ m^{-3}	$\kappa = 0.21$
$\lambda_s = 40 \times 10^{-6}$	$\lambda_{100} = -20 \times 10^{-6}$	$\lambda_{111} = 78 \times 10^{-6}$

Significance Very common in igneous rocks, Ti-substituted magnetite is by far the most common magnetic mineral, and the main source of rock magnetism. Outcrops of magnetite were naturally magnetized by lightning strikes. Essential constituent of lodestone, the first permanent magnet discovered by man. Now used as iron ore, pigment, toner (ink) and in ferrofluids. Biogenic magnetite is produced by bacteria, pigeons etc. A ubiquitous magnetic contaminant.

X-ray powder diffraction pattern for Fe_3O_4

h	*k*	*l*	*d*(pm)	I/I_{max}
1	1	1	485.2	8
2	2	0	296.7	30
3	1	1	253.2	100
2	2	2	242.4	8
4	0	0	209.9	20
4	2	2	171.5	10
5	1	1	161.6	30
4	4	0	148.5	40
5	3	1	141.9	2
6	2	0	132.8	4
5	3	3	128.1	10
6	2	2	126.6	4
4	4	4	121.2	2
6	4	2	112.2	4
7	3	1	109.3	12
8	0	0	105.0	6
6	6	0	99.0	2
7	5	1	97.0	6
6	6	2	96.3	4
8	4	0	93.9	4
6	6	4	89.5	2
9	3	1	88.0	6
8	4	4	85.7	8
10	2	0	82.3	4
9	5	1	81.2	6
10	2	2	80.8	4

Table 11.23. Room-temperature magnetic properties of oxide spinel ferrites

	[a]	a_0 (pm)	T_C (K)	M_s (MA/m)	K_1 (kJ/m^3)	λ_s (10^{-6})	ϱ (Ω m)
$MgFe_2O_4$	I	836	713	0.18	−3	−6	10^5
$Li_{0.5}Fe_{2.5}O_4$		829	943	0.33	−8	−8	1
$MnFe_2O_4$	I	852	575	0.50	−3	−5	10^5
Fe_3O_4	I	840	860	0.48	−13	40	10^{-1}
$CoFe_2O_4$	I	839	790	0.45	290	−110	10^5
$NiFe_2O_4$	I	834	865	0.33	−7	−25	10^2
$ZnFe_2O_4$	N	844	$T_N = 9$				1
γFe_2O_3		834	985[b]	0.43	−5	−5	~1

[a] N, normal (2+ cation on A-sites); I, inverse (2+ cation in B sites);
[b] Estimate; reverts to αFe_2O_3 above 800 K.

Related compounds Spinel itself $MgAl_2O_4$ is a nonmagnetic substrate material with $a_0 = 808$ pm. There is an important series of oxide spinel ferrites, where many cationic substitutions are possible. The magnetic properties depend on the cation distribution over A and B sites, which may be modified by thermal treatment. Chalcogenide spinels include greigite, Fe_3S_4, ($a_0 = 988$ pm, $T_C \approx 580$ K) and the *p*-type magnetic semiconductors $CuCr_2S_4$ ($a_0 = 982$ pm, $T_C = 420$ K) and $CuCr_2Se_4$ ($a_0 = 1036$ pm, $T_C = 440$ K).

γFe_2O_3 (maghemite)

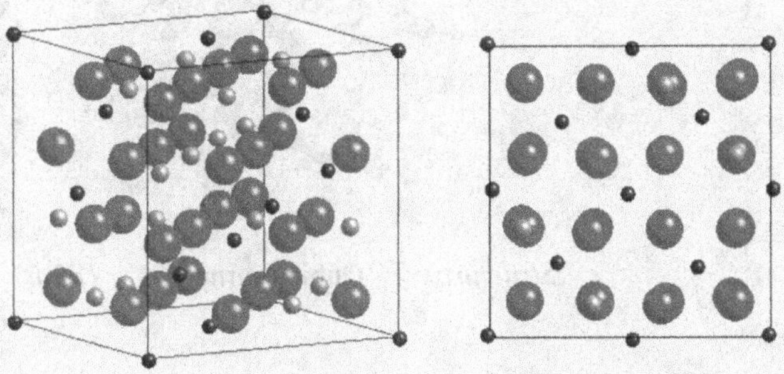

Structure Cubic defective spinel

$P4_132$ $Z = 8$ d = 4860 kg m^{-3}
$a_0 = 833.6$ pm
Fe in 8*a* [0, 0, 0] A-site
Fe in 16*d* $\{\frac{1}{8}, \frac{1}{8}, \frac{1}{8}\}$ B-site
O in 32*e* (x, x, x), $x \approx 0.25$

An fcc oxygen array with Fe^{3+} in one-eighth of the tetrahedral interstices and in one-third of the octahedral interstices. One-sixth of the B-sites of the spinel structure are vacant

$$[Fe^{3+}]\{Fe^{3+}_{5/3}\square_{1/3}\}O_4$$

The B-site vacancies may order in a $P4_1$ tetragonal cell with $c \approx 3a$ Maghemite is thermodynamically unstable, converting to hematite at about 800 K when heated in air

Electronic properties Brown insulator

Localized electrons Fe^{3+} $3d^5$; $t_{2g}^3 e_g^2$ ion $S = \frac{5}{2}$; $^6S - A_{1g}$.

Magnetic properties Ferrimagnet $T_c \approx 985$ K (estimate)

The A and B sublattices are oppositely aligned. Magnetization lies along $\langle 111 \rangle$. Symmetry $\bar{3}m'$. Net moment at $T = 0$ K is $\mathrm{m} = 5 \times (\frac{5}{3} - 1) = 3.3\,\mu_B$ fu^{-1}

$\sigma = 82$ A m^2 kg^{-1} $M = 0.40$ MA m^{-1} $J_s = 0.50$ T

$K_1 = -5$ kJ m^{-3} $\lambda_s = -9 \times 10^{-6}$

Significance Acicular γFe_2O_3 powders are widely used for flexible magnetic recording media usually with Co^{2+} surface doping to improve coercivity. Magnetic surface reconstruction (surface spin canting) reduces the magnetization of nanoparticles. Stabilized nanoparticles are used in ferrofluids, and in magnetic beads for bioassay. Maghemite is found in magnetic tropical soils.

$Y_3Fe_5O_{12}$ (YIG)

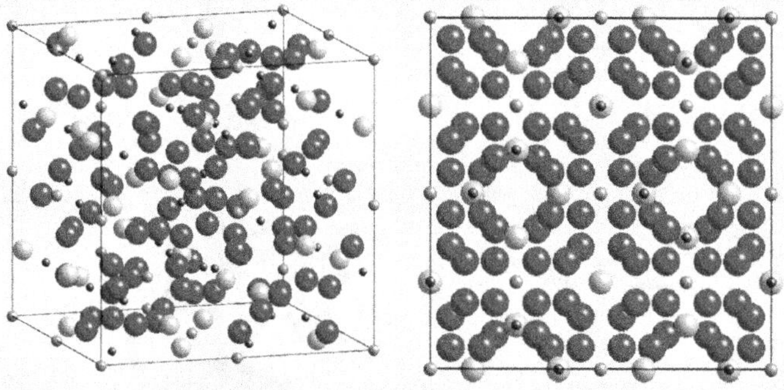

Structure Cubic (garnet) $Ia3d$ $Z = 8$ d = 5166 kg m^{-3}

$a_0 = 1238$ pm

Y in 24c $(\frac{1}{8}, 0, \frac{1}{4})$

Fe in 16a $\{0, 0, 0\}$

Fe in 24d $[\frac{3}{8}, 0, \frac{1}{4}]$

O in 96h (0.94, 0.06, 0.15)

Electronic properties Green insulator $\varepsilon_g = 2.8$ eV

Figure 11.20

Spontaneous moment of rare-earth iron garnets. (E. F. Bertaut and R. Pauthenet, *Proc. Inst. Elec. Eng. B*, 104, 261 (1957))

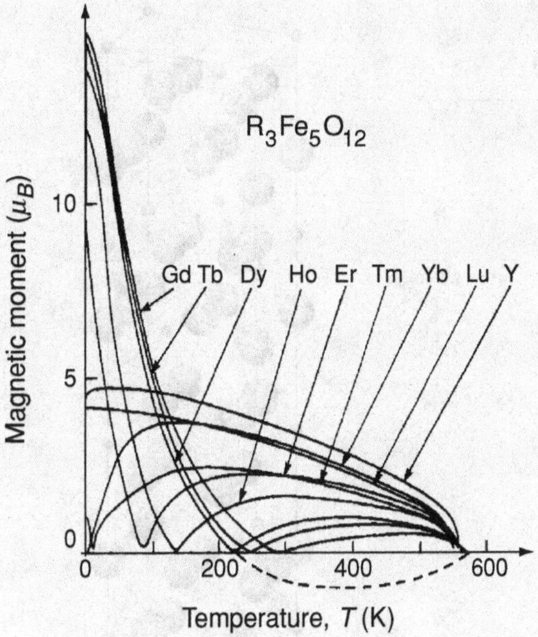

Magnetic properties Ferrimagnet $T_c = 560\,K$.

Fe^{3+} $3d^5$; $t_{2g}^3 e_g^2$ ions $S = \frac{5}{2}$; $^6S - A_{1g}$. The two ferric sublattices with tetrahedral [24*d*] and octahedral {16*a*} sites are oppositely aligned, giving a moment of 5 μ_B fu^{-1}. $Y_3[Fe_3]\{Fe_2\}O_{12}$

$\sigma = 27.6(1)$ A m^2 kg^{-1} $M_s = 0.143$ MA m^{-1} $J_s = 0.180$ T
(NIST Standard)

$A = 4$ pJ m^{-1} $K_1 = -2$ kJ m^{-3} $\lambda_{100} = 1 \times 10^{-6}$

$\lambda_{111} = -3 \times 10^{-6}$ $\kappa = 0.28$

Significance YIG has excellent high-frequency magnetic properties and a very narrow ferromagnetic resonance linewidth. It has good magneto-optic properties when doped with Bi.

Related materials The complete series of rare-earth garnets $R_3Fe_5O_{12}$ exist, with the magnetic rare-earth sublattices aligned parallel or antiparallel to the net iron moment. When the rare-earth moment exceeds 5 μ_B these materials exhibit a compensation temperature. Complex magnetic structures at low temperature are due to the anisotropy of the rare-earth sublattice. Numerous substitutions are possible on all three cation sites. Yttrium aluminium garnet (YAG) is a nonmagnetic analogue.

Natural garnets such as $Ca_3Si_3Al_2O_{12}$ are nonmagnetic. Hydrogarnet $Ca_3Al_2(OH)_{12}$ is a constituent of cement.

$BaFe_{12}O_{19}$ (M-ferrite)

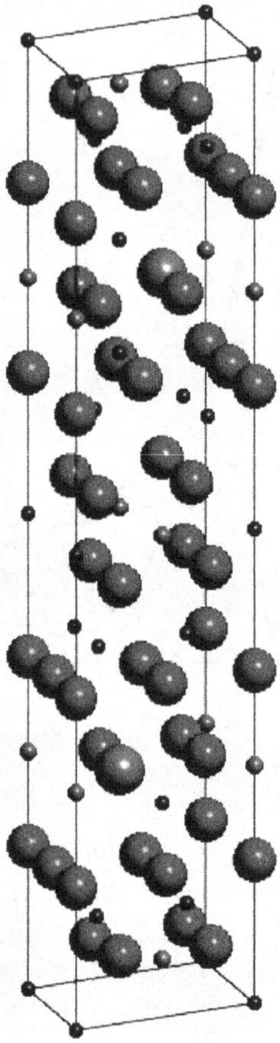

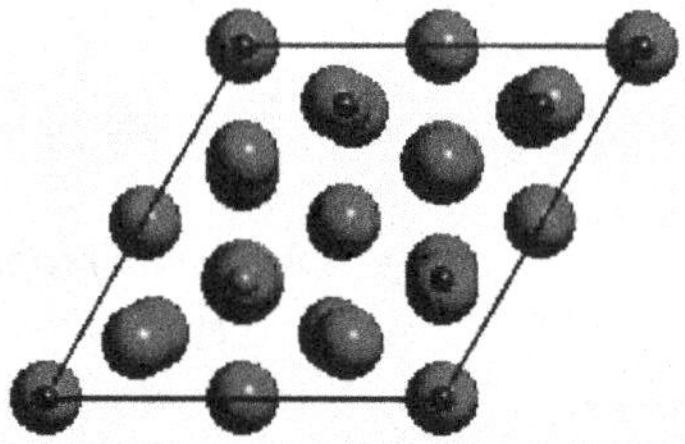

Structure Hexagonal magnetoplumbite ($PbFe_{12}O_{19}$) $P6_3/mmc$ $Z = 2$

d = 5290 kg m^{-3}

$a = 589$ pm $c = 2319$ pm

Ba in $2d$ $(\frac{1}{2}, \frac{2}{3}, \frac{3}{4})$

Fe in $2a$ $\{0, 0, 0\}$; $2b$ $\langle 0, 0, \frac{1}{4}\rangle$; $4f_1$ $[\frac{1}{3}, \frac{2}{3}, 0.028]$; $4f_2$ $\{\frac{1}{2}, \frac{2}{3}, 0.189\}$; $12k$ $\{\frac{1}{3}, \frac{1}{6}, 0.108\}$

O in $4e$ $(0, 0, 0.150)$; $4f(\frac{2}{3}, \frac{1}{3}, 0.450)$; $6h(0.186, 0.372, \frac{1}{4})$; $12k_1(\frac{1}{6}, \frac{1}{3}, 0.050)$; $2k_2(\frac{1}{2}, 0, 0.150)$

An hcp array of oxygen + barium, with Fe^{3+} in octahedral $\{12k, 4f_2$ and $2a\}$, tetrahedral $[4f_1]$ and trigonal bipyramidal $\langle 2b\rangle$ interstices

Electronic properties Brown insulator $\varepsilon_g = 1.0$ eV

Localized electrons Fe^{3+} $3d^5$; $t_{2g}^3 e_g^2$ ions $S = \frac{5}{2}$ $^6S - A_{1g}$

Magnetic properties Ferrimagnet $T_c = 740$ K

Magnetic structure is $12k^{\uparrow}$; $2a^{\uparrow}$; $2b^{\uparrow}$; $4f_1^{\downarrow}$; $4f_2^{\downarrow}$. Net moment at $T = 0$ K is $\mathfrak{m}_0 = [(6+1+1)] - (2+2)] \times 5 = 20\,\mu_B$ fu^{-1}

Table 11.24. Intrinsic magnetic properties of hexagonal ferrites at room temperature (m_0 at low temperature)

	a (pm)	c (pm)	T_C (K)	m_0 (μ_B fu^{-1})	M_s (MA m^{-1})	K_1 (MJ m^{-3})	$\mu_0 H_0$ (T)	κ
BaM	589	2320	740	19.9	0.38	0.33	1.7	1.3
SrM	589	2304	746	20.2	0.38	0.35	1.8	1.4
PbM	590	2309	725	19.6	0.33	0.25	1.5	1.4
BaW	588	3250	728	27.6	0.41	0.30	1.5	1.2
BaX	588	5570	735	47.5	0.28	0.30	1.6	1.3

Exchange constants $\mathcal{J}_{b-f_2} = -36\,\text{K}$, $\mathcal{J}_{k-f_1} = -19.6\,\text{K}$, $\mathcal{J}_{a-f_1} = -18.2\,\text{K}$, $\mathcal{J}_{f_2-k} = -4.1\,\text{K}$, $\mathcal{J}_{b-k} = -3.7\,\text{K}$

$\sigma = 72\,\text{A m}^2\,\text{kg}^{-1}$	$M_s = 0.38\,\text{MA m}^{-1}$	$J_s = 0.48\,\text{T}$
$A = 6\,\text{pJ m}^{-1}$	$K_1 = 330\,\text{kJ m}^{-3}$	$\kappa = 1.35$
$\sigma_0 = 108\,\text{A m}^2\,\text{kg}^{-1}$	$M_0 = 0.57\,\text{MA m}^{-1}$	$J_0 = 0.72\,\text{T}$
$B_a = 1.7\,\text{T}$	$K_{10} = 450\,\text{kJ m}^{-3}$	

Significance Barium and strontium M-ferrite are produced in huge quantities ($\approx$800 000 tonnes per year) as low-cost moderate-performance permanent magnets.

Related materials $SrFe_{12}O_{19}$ is very similar to $BaFe_{12}O_{19}$. A La + Co substitution for some Sr + Fe improves the magnetic properties slightly. There is a family of compounds with different stacking sequences of the basic building blocks: $BaM_2Fe_{16}O_{27}$ (W-ferrite), $Ba_2M_2Fe_{12}O_{22}$ (Y-ferrite), $Ba_2M_2Fe_{24}O_{46}$ (X-ferrite), $Ba_3M_2Fe_{24}O_{41}$ (Z-ferrite). (M represents a divalent cation.) The M, W, X and Z compounds have easy-axis anisotropy, whereas Y is easy-plane.

MnF_2

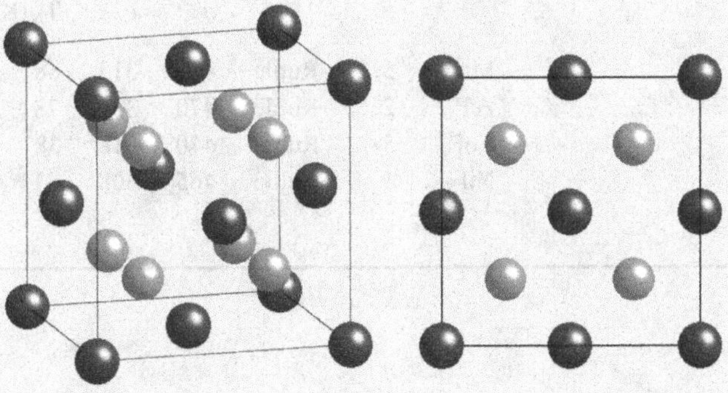

Structure C4 tetragonal, rutile (TiO_2) $P4_2/mnm$ $Z = 2$ d = 4890 kg m^{-3}
$a = 487.3$ pm $c = 313.0$ pm
Mn in $2a$ (0, 0, 0)
F in $4f$ (0.3, 0.3, 0)

Manganese is in nearly regular octahedral oxygen coordination

Electronic properties Transparent insulator $\varepsilon_g = 9$ eV

Magnetic properties Antiferromagnet $T_N = 67.5$ K Mn^{2+} $3d^5$; $t_{2g}^3 e_g^2$ ion; $S = \frac{5}{2}$ $^6A_{1g}$ $\theta = -80$ K

Identical chemical and magnetic unit cells, moments parallel to **c**. Symmetry $4'/mmm'$

Exchange constants $\mathcal{J}_1 = 0.35$ K $\mathcal{J}_2 = 1.71$ K D = 1.2 K

Significance A model antiferromagnet. Fluorine is the most electronegative element, so fluorides are ionic compounds with little covalent character in the M–F bond.

Related compounds Other antiferromagnetic difluorides and trifluorides are listed in Table 11.25. Tetragonal K_2NiF_4 is an insulator with Ni^{2+} ions ($S = 1$). It is a model two-dimensional antiferromagnet because the interlayer exchange interactions cancel. Below 97 K, the material orders in three dimensions due to uniaxial anisotropy. Rb_2CoF_4, which has stronger anisotropy, is an almost ideal two-dimensional Ising antiferromagnet. LaF_3 is a good diamagnetic matrix in which to investigate the magnetic properties of isolated rare-earth ions. The chromium trihalides are antiferromagnetic (F), metamagnetic (Cl), or ferromagnetic (Br, I) with magnetic ordering temperatures of 80, 17, 33 and 68 K, respectively.

Table 11.25. Properties of antiferromagnetic difluorides and trifluorides

	S		a	c	T_N(K)		S		T_N (K)
MnF_2	5/2	Rutile	487	313	68	CrF_3	3/2	Rhom	80
FeF_2	2	Rutile	470	331	78	MnF_3	2	Rhom	43
CoF_2	3/2	Rutile	470	318	38	FeF_3	5/2	Rhom	365
NiF_2	1	Rutile	465	308	83	CoF_3	2	Rhom	460

a-FeF_3

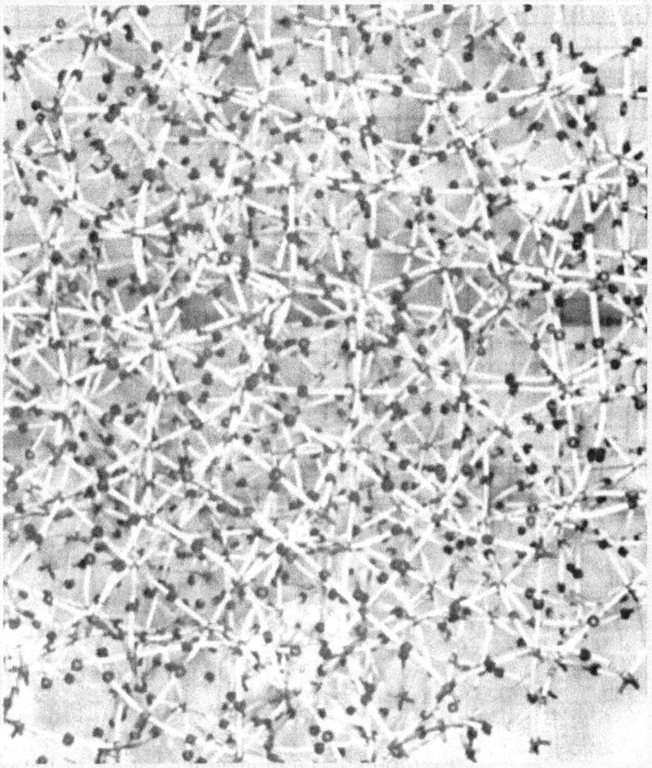

Structure Amorphous. Octahedral continuous random network with (6–2) coordination

Electronic properties Brown insulator

Localized electrons Fe^{3+} $3d^5$; $t_{2g}^3 e_g^2$ $S = \frac{5}{2}$. $^6A_{1g}$

Magnetic properties Speromagnetic $T_f = 29$ K

Spins freeze in random directions, with a tendency for antiparallel alignment of nearest neighbours. There is a difference between field-cooled and zero-field-cooled magnetization below T_f. A small remanence of ≈ 0.005 μ_B fu^{-1} is attributed to the possibility of reconfiguring groups of about 1000 spins after applying a field. The magnetic ground state is highly degenerate.

Significance A model amorphous compound with antiferromagnetic superexchange interactions. Frustration is due to the presence of three- and five-membered rings in the structure.

Related compounds Crystalline FeF_3 is an antiferromagnet with $T_N = 365$ K, $\theta_p = -610$ K. There are frustrated crystalline fluorides with pyrochlore or kagome lattices. $Fe(OH)_3 \cdot nH_2O$ is an amorphous ferric oxyhydroxide with $T_f \approx 100$ K.

Fe_7S_8 (pyrrhotite)

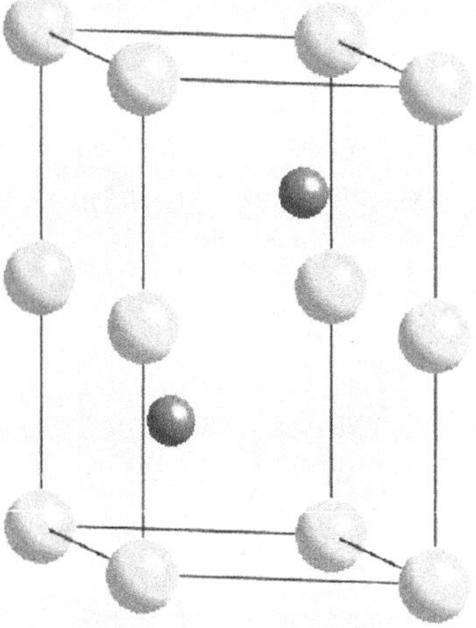

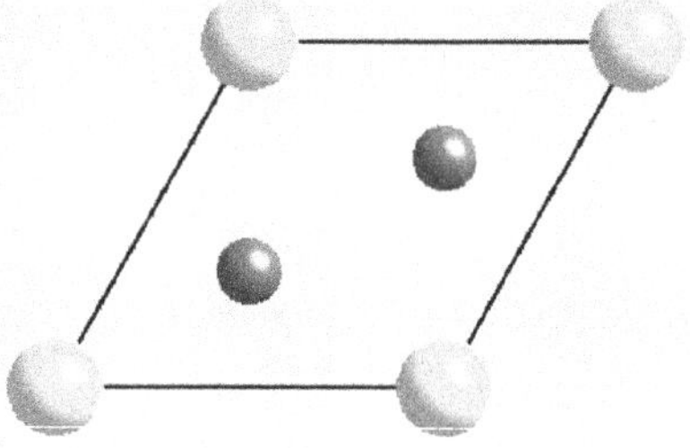

Structure Monoclinic C2/c $Z = 4$ d = 5745 kg m^{-3}

$a = 1193$ pm $b = 688$ pm $c_h = 1292$ pm $\beta = 118°$

Fe_1 $8f$ (0.126, 0.098, 0.991) Fe_2 $8f$ (0.256, 0.127, 0.246)

Fe_3 $8f$ (0.359, 0.140, 0.500) Fe_4 $4e$ (0, 0.393, 0.250)

S_1 $8f$ (0.896, 0.123, 0.876) S_2 $8f$ (0.353, 0.125, 0.124)

S_3 $8f$ (0.860, 0.125, 0.138) S_4 $8f$ (0.602, 0.124, 0.621)

Pyrrhotite has a monoclinic variant of the NiAs-type structure ($a_h = 344$ pm, $c_h = 571$ pm) with metal vacancies ordered on alternate iron planes.with an ABCD sequence in a $2\sqrt{3}a, 2a, 4c$ supercell. The vacancies are disordered at the Curie temperature, where the structure becomes hexagonal, cation-deficient NiAs-type. The phase range of $Fe_{1-x}S$ is $0.875 \leq x \leq 0.95$.

Electronic properties Metallic, itinerant ferromagnet with a low density of holes in the $3p$(S) band. Fe $3d$ electrons are hybridized with S $3p$ states. Collapses to a nonmagnetic structure under pressure $P \geq 5$ GPa, forming low-spin Fe^{II}.

Magnetic properties Ferrimagnet with ferromagnetic c-planes, coupled antiparallel to their neighbours. Moments $\perp$ $\boldsymbol{c}$. The transition at $T_c = 598$ K is weakly first order. Moments rotate out of the plane with decreasing temperature, $\theta_0 = 60°$. The moment is reduced from the $4\,\mu_B$ spin-only value for Fe^{2+} $S = 2$, by covalent admixture with sulphur to $3.16\,\mu_B$

$\sigma = 26$ A m^2 kg^{-1} $M_s = 0.15$ MA m^{-1} $J_s = 0.19$ T

$K_1 = 320$ kJ m^{-3} $\lambda_s \approx 10 \times 10^{-6}$

Significance Fe_7S_8 is a fairly common rock-forming mineral; it is exploited as an iron ore. Pyrrhotite is the most common magnetic mineral, after magnetite.

Related compounds FeS (troilite) is a rare antiferromagnetic mineral with no cation vacancies. The iron forms triangles, in a $\sqrt{3}a_h, 2c_h$ variant of the NiAs structure stable below $T_\alpha = 413$ K. Stoichiometric FeS has been found in iron meteorites. There is a family of hexagonal cation-deficient sulphides, selenides and tellurides. Fe_7Se_8 is similar to Fe_7S_8, but with a lower magnetic ordering temperature (448 K). Related hexagonal compounds are Fe_3S_4 (smythite) and Fe_4Se_4. Different vacancy superstructures have slightly different ferrimagnetic properties. The dichalcogenides FeS_2 (pyrite – fool's gold), $FeSe_2$ and $FeTe_2$ contain low-spin Fe^{2+} $3d^6$; t_{2g}^6, $S = 0$, and are diamagnetic, with a cubic, Pa$\bar{3}$ structure. Electron-rich tetragonal $FeSe_{1-x}$ is a nonmagnetic superconductor.

Table 11.26. NiAs-type iron sulphides and selenides

		$T_{c,N}$ (k)	M_s (MA m^{-1})	K_1 (MJ m^{-3})	κ
FeS	af	588	0.63	0.01	
Fe_7S_8	fi	598	0.15	0.32	
Fe_3S_4	fi	600	0.58	1.20	1.7
Fe_7Se_8	fi	448	0.07	0.25	
Fe_3Se_4	fi	314	~0		

af – antiferromagnet; fi – ferrimagnet.

Table 11.27. Pyrite-structure disulphides

	a (pm)	S			$T_{C,N}$(K)
MnS_2	610	5/2	af	sc	48
FeS_2	542	0	dia	sc	
CoS_2	553	1/2	f	metal	110
NiS_2	568	1	af	sc	50

af – antiferromagnet; dia – diamagnet; sc – semiconductor.

11.7 Miscellaneous materials

CuMn

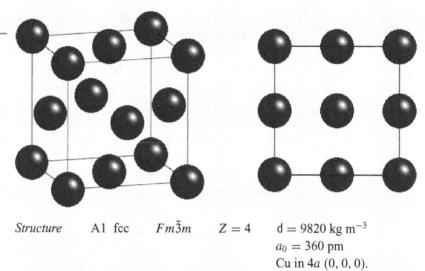

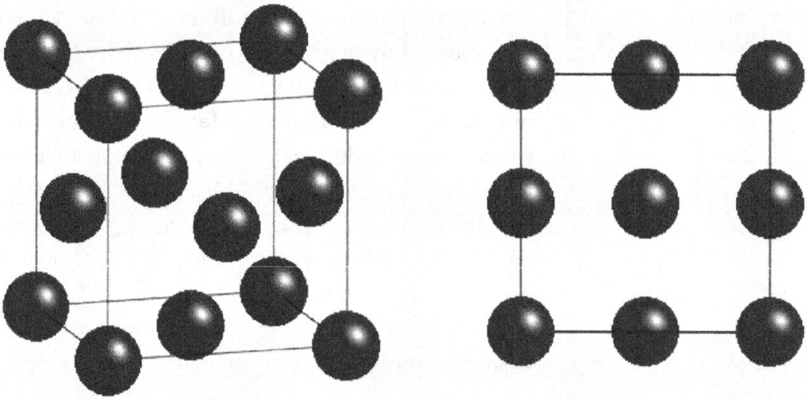

Structure A1 fcc $Fm\bar{3}m$ $Z = 4$ d = 9820 kg m^{-3}
$a_0 = 360$ pm
Cu in $4a$ (0, 0, 0).

A dilute alloy with manganese impurities dispersed in an fcc copper matrix.

Electronic properties Metal with impurity scattering due to Mn.

Magnetic properties A canonical spin glass. Spin freezing temperature and low-temperature linear magnetic specific heat scale with Mn content, $T_f \propto x, c_f \propto x$. The freezing temperature is about 2 K per % Mn. Magnetic coupling of Mn impurities is via the RKKY interaction, which leads, in the dilute limit, to a Gaussian distribution of $\mathcal{J}$ centred at $\mathcal{J} = 0$.

Significance A model system for studying the interaction of dilute magnetic impurities in an s-band metallic matrix.

Related materials **Au**Mn and **Au**Fe are examples of the numerous dilute alloy spin-glass systems. There is a tendency for the magnetic impurity atoms to cluster. Rare-earth-based spin glasses may use **Y** as a nonmagnetic crystalline host or $La_{80}Au_{20}$ as an amorphous host. $Eu_xSr_{1-x}S$ is another, insulating spin-glass system.

US

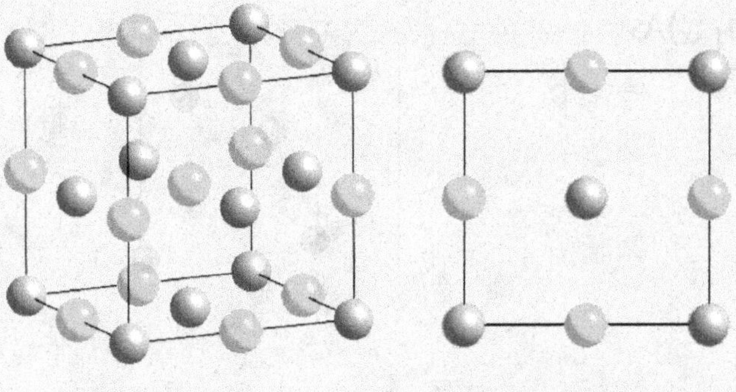

Structure Cubic B1 (NaCl) $Fm\bar{3}m$ $Z = 4$ d = 16407 kg m^{-3}
$a_0 = 548.9$ pm
U in 4*a* (0, 0, 0)
S in 4*b* ($\frac{1}{2}$, $\frac{1}{2}$, 0)

An fcc sulphur array with U in undistorted octahedral coordination

Electronic properties Dense black metal; an itinerant electron ferromagnet.

Magnetic properties Ferromagnet $T_C = 177$ K $m_0 = 1.55\,\mu_B$ fu^{-1} (orbit 3.00 μ_B; spin 1.45 μ_B)
$\sigma_0 = 32$ A m^2 kg^{-1} $M_0 = 0.53$ MA m^{-1} $J_0 = 0.66$ T
$K_{1c} = 43$ MJ m^{-3} ⟨111⟩ easy directions $\kappa = 11.1$.

Significance An example of a magnetically ordered actinide. The 5*f* moment is mainly orbital in character, and oppositely directed to the spin moment. US has the largest cubic anisotropy of any known material.

Related materials Many actinide compounds with interatomic distances $d_{AA} >$ 340 pm are found to order ferromagnetically or antiferromagnetically, usually with relatively low magnetization $M_0 < 1$ MA m^{-1} and an ordering temperature below 300 K. Some are listed in Table 11.28. Intermetallics with higher T_C form with the ferromagnetic 3*d* elements.

Oxides have the CaF_2 structure, the other compounds have the NaCl structure. In all cases $d_{AA} = a_0/\sqrt{2}$.

Table 11.28. Magnetic properties of actinide compounds

	a_0 (pm)	Order	T_C (K)		a_0 (pm)	Order	T_C (K)		a_0 (pm)	Order	T_C (K)
UN	489	af	52	UP	559	af	125	UO_2	546	af	31
NpN	490	f	87	UAs	577	af	127	NpO_2	543	af	25
PuN	490	af	19	USb	619	af	241	PuO_2	540	para	
AmN	499	para		USe	575	f	175	AmO_2	538	af	9
CmN	503	f	125	UTe	616	f	104	BkO_2	538	af	3

af – antiferromagnet; f – ferromagnet; para – paramagnet
From M. B. Brodsky, *Rep. Prog. Phys.* **41**, 1548 (1978).

$(Mn_xGa_{1-x})As$

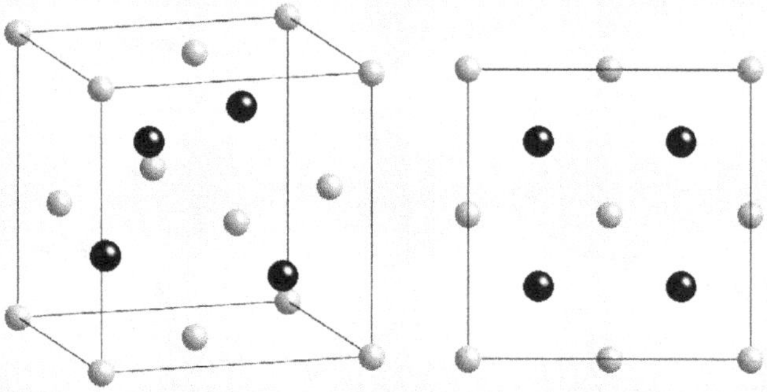

Structure Cubic ZnS-type (zinc blende) $F\bar{4}3m$ $Z = 4$ $d = 5280\,kg\,m^{-3}$
$a_0 = 565.3\,pm$
Ga, Mn in $1b$ (0.25, 0.25, 0.25)
As in $1a$ (0, 0, 0)

Electronic properties A p-type semiconductor with 2×10^{26} carriers m^{-3}, $\varrho(RT) \approx 10\ \mu\Omega$ m. Each Mn dopant introduces a hole in the As 4p band.

Magnetic properties Ferromagnet $T_C = 170\,K$ for optimal Mn-doping, $x = 0.08$, the solubility limit of Mn in GaAs.
Atomic moment $\mathfrak{m} \leq 4\ \mu_B\ Mn^{-1}$
$\sigma_0 \approx 11\,A\,m^2\,kg^{-1}$ $M_0 = 58\,kA\,m^{-1}$ $J_0 = 0.07\,T$ $K_1 = 2\,kJ\,m^{-3}$

Significance The GaAs is a semiconductor with a gap $\varepsilon_g = 1.43\,eV$. With substitutional Mn, the ferromagnetic semiconductor is hole-doped, and the ↓ polarized holes in the As $4p$ valence band mediate a ferromagnetic interaction among Mn ions. It is used with GaAs-based heterostructures (quantum wells) for spin-electronic demonstrators.

Related materials Other possible dilute magnetic semiconductors (DMS) are GaN:Mn and TiO_2:Co. However, there is often a tendency for the transition metal to form nanoscale clusters in these compounds.

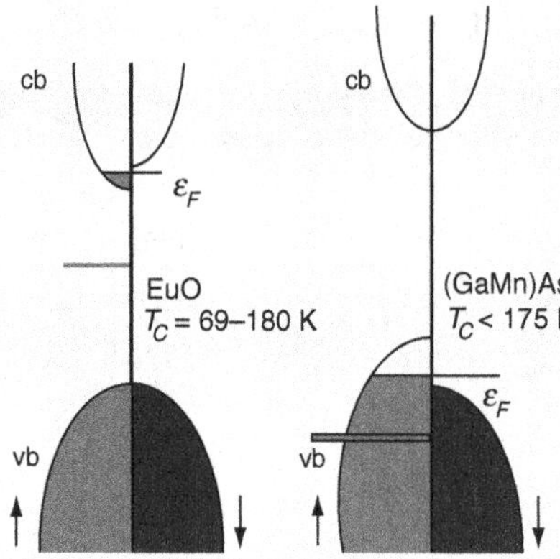

Figure 11.21

Schematic electronic structures for some magnetic semiconductors: vb – valence band; cb – conduction band.

αO_2

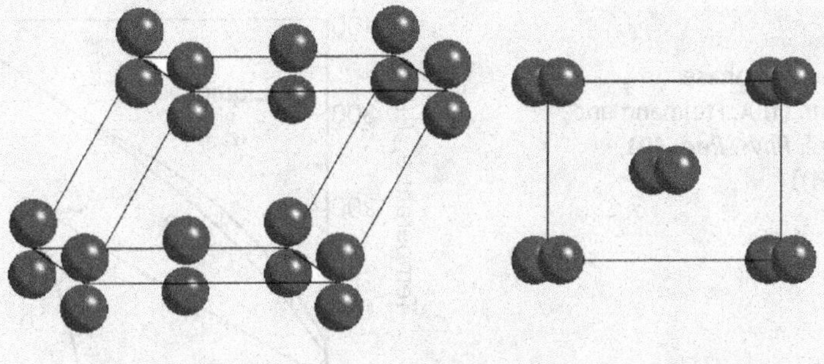

Structure Monoclininc $C2/m$ $Z = 4$ d = 1538 kg m^{-3}
$a = 537.5$ pm $b = 342.5$ pm $c = 424.2$ pm $\beta = 117.8°$
O in $4b$ (−0.055, 0, 0.133)
Close-packed (001) layers with O_2 dumbbells perpendicular to the layers.

Electronic properties Blue insulator, $\varepsilon_g = 1.0$ eV ($^3\Sigma_g^- \rightarrow^1 \Delta_g$)
Oxygen molecules O_2 have a $^3\Sigma_g^-$ triplet ground state with $S = 1, L = 0$. Holes in the $2p^5$ shells of atomic oxygen couple parallel.

Magnetic properties Antiferromagnet, $T_N = 23.9$ K, the temperature of the $\alpha \rightarrow \beta$ structural phase transition. Antiferromagnetic ab-planes with moments lying along $\boldsymbol{b}$. Weak interplane exchange gives the antiferromagnetism a quasi-two-dimensional character.

$\sigma_A = 175$ A m^2 kg^{-1}	$M_A = 0.27$ MA m^{-1}
$J_A = 0.34$ T	$\mathcal{J}_1 = -28$ K,
$\mathcal{J}_2 = -14$ K,	$\mathcal{J}_3 < -1$ K
$D \approx 6$ K	$K \approx 4.6$ MJ m^{-3}

Significance The only magnetically ordered element with no d or f electrons. Paramagnetic oxygen gas was described by Faraday, and paramagnetic liquid oxygen by Dewar.

Related materials Solid oxygen has a complex temperature–pressure phase diagram, with six different phases. β-O_2, stable from 23.9–43.8 K, is a frustrated triangular antiferromagnet exhibiting only short-range magnetic order. The γ phase, stable between 43.3 K and the melting point 54.4 K, is not magnetically ordered. The alkali metal superoxides KO_2, RbO_2 and CsO_2 are also $2p$ antiferromagnets. The compounds crystallize in the CaC_2 body-centred tetragonal structure (Ca in 0, 0, 0, C_2 in 0, $\frac{1}{2}$, $\frac{1}{2}$). Oxygen molecules are $O_2^- \lambda\, S = \frac{1}{2}$. Néel temperatures are 7, 15 and 10 K, respectively. Holes on oxygen in non-stoichiometric oxides may carry a moment with a tendency to ferromagnetic coupling.

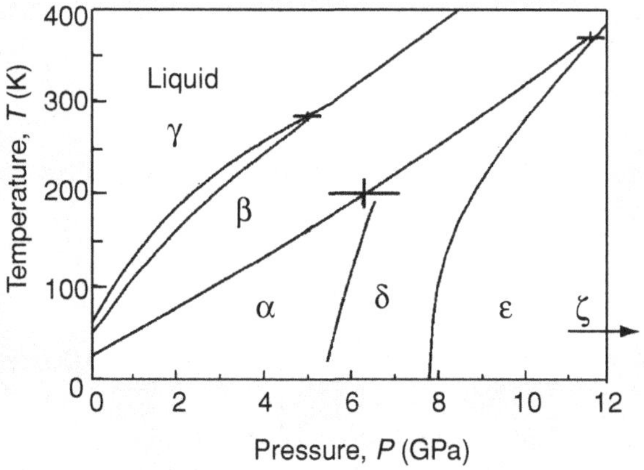

Figure 11.22

Solid oxygen phase diagram. (Y. A. Freimann and H. J. Jodl, *Phys. Rep.* **401**, 1 (2004))

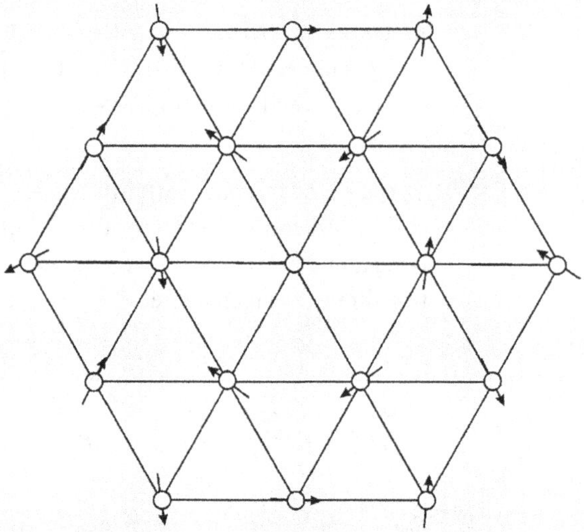

Figure 11.23

Helicoidal magnetic structure of βO_2.

Table 11.29. Phases of solid oxygen

α	$C2/m$	Blue	Antiferromagnet $m \parallel b$
β	$R\bar{3}m$	Blue	Helicoidal
γ	$Pm3n$	Blue	Paramagnetic
δ	$Fmmm$	Orange	Antiferromagnetic, ferromagnetic planes
ϵ	$A2/m$	Red	Diamagnetic
ς	$A2/m$	Metal	Superconductor $P > 96$ GPa, $T < 0.6$ K

β p-nitrophenyl nitronyl nitroxide $C_{13}H_{16}O_4N_3$ (p-NPNN)

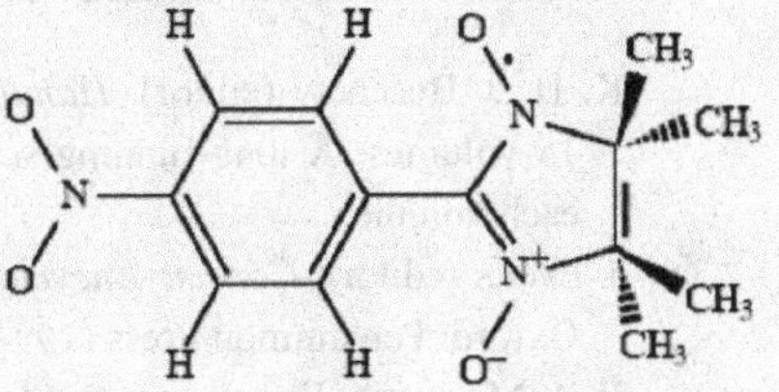

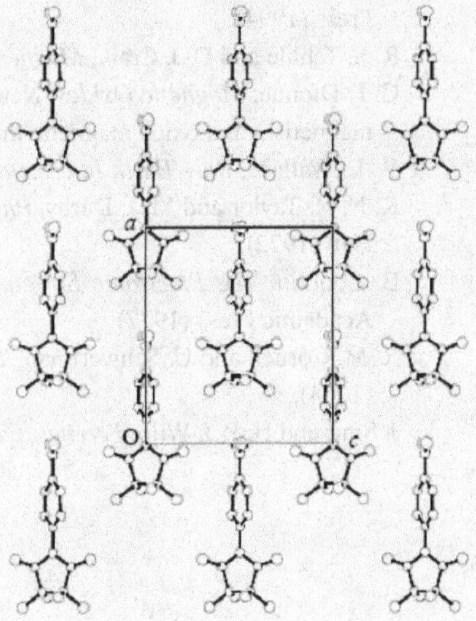

Structure Orthorhombic : $F2dd$ $Z = 8$ d $= 1416$ kg m^{-3}
$a = 1235$ pm, $b = 1935$ pm, $c = 1096$ pm

Electronic properties Insulator, $\varepsilon_g = 1.2$ eV
The free radical electron is delocalized over the ONCNO moiety.

Magnetic properties Ferromagnet $T_C = 0.6$ K $\theta_p = 1.0$ K
Atomic moment m $= 0.5\ \mu_B$ fu^{-1}
$\sigma_0 = 10$ A m^2 kg^{-1} $M_0 = 14$ kA m^{-1} $J_0 = 0.017$ T.
$\mathcal{J} = 0.6$ K

Significance The first purely organic ferromagnet. A three-dimensional $S = \frac{1}{2}$ Heisenberg ferromagnet.

Related materials The γ p-NPNN polymorph has quasi-one-dimensional ferromagnetic chains, with weak antiferromagnetic interchain coupling giving a Néel temperature of 0.65 K. The δ polymorph may be a half-metal. The highest Curie temperature in a crystalline organic material, $T_C = 1.48$ K is found in diazadamantane dinitroxide. Some π-conjugated polyethers have T_C as high as 10 K. There are numerous organometallic compounds with 3d elements which order magnetically, generally well below 100 K. The highest Curie temperature is ~370 K for vanadlium tetracyanethene, V(TCNE)$_x$.

FURTHER READING

K. H. J. Buschow (editor), *Handbook of Magnetic Materials*, Amsterdam: Elsevier, 15 volumes. A long-running series with extended reviews of specific materials in each volume.

J. Evetts (editor), *Concise Encyclopedia of Magnetic and Superconducting Materials*, Oxford: Pergammon Press (1994).

R. A. McCurrie, *Ferromagnetic Materials; Structure and Properties*, London: Academic Press (1994).

R. S. Tebble and D. J. Craik, *Magnetic Materials*, London: Wiley (1969).

G. F. Dionne, *Magnetic Oxides*, New York: Springer (2009). A traditional approach to magnetism and oxide magnetic materials.

W. E. Wallace, *Rare Earth Intermetallics*, New York: Academic Press (1973).

K. N. R. Taylor and M. I. Darby, *Physics of Rare Earth Solids*, London: Chapman and Hall (1972).

B. Coqblin, *The Electronic Structure of Rare Earth Metals and Alloys*, New York: Academic Press (1977).

R. M. Cornell and U. Schwertman, *The Iron Oxides* (second edition), Weinheim: VCH (2004).

J. Smit and H. P. J. Wijn, *Ferrites*, Eindhoven: Philips Technical Library (1959).

12 Applications of soft magnets

Don't mention the war.

Temporary magnets concentrate and guide the magnetic flux produced by circulating currents or permanent magnets. Millions of tons of electrical sheet steel are used every year in electromagnetic machinery – transformers, motors and generators. Numerous small components in magnetic circuits are made from nickel–iron alloys, which offer attractive combinations of permeability, polarization and resistivity. Insulating ferrites are particularly suitable for high-frequency applications such as power supplies, chokes and antennae, and for microwave devices.

A good soft magnetic material exhibits minimal hysteresis with low magnetostriction high polarization and the largest possible permeability. Permeability is usually referred to the internal field, because soft magnets tend to be used in a toroidal geometry where demagnetizing effects are negligible. In some range of internal field, the $\boldsymbol{B}(\boldsymbol{H})$ response is linear $\boldsymbol{B} = \mu \boldsymbol{H}$ or

$$\boldsymbol{B} = \mu_0 \mu_r \boldsymbol{H}, \tag{12.1}$$

where the relative permeability $\mu_r = \mu/\mu_0$ is a pure number. The initial permeability μ_i at the origin of the hysteresis loop is smaller than the slope B/H in slightly larger fields, where it attains its maximum value μ_{max}, as shown in Fig. 12.1. Remanence in a temporary magnet is negligible. The distinction between polarization $J = \mu_0 M$ and induction B is insignificant in applications where device design and high permeability of the soft material ensure that the $\boldsymbol{H}$-fields involved are very small.

Values of μ_{max}/μ_0 can reach a million in the softest materials. Hence $\boldsymbol{B}$ is hugely enhanced, up to a limit set by the spontaneous induction $B_s \approx J_s = \mu_0 M_s$, compared to the free-space induction $\mu_0 \boldsymbol{H}'$ that induces it. Here $\boldsymbol{H}'$ is the external applied field. Permeability and loop shape can be modified by annealing, especially in a weak external field. The distinction between susceptibility and relative permeability is insignificant when μ_r is very large; it follows from $\boldsymbol{B} = \mu_0(\boldsymbol{H} + \boldsymbol{M})$ that $\mu_r = 1 + \chi$.

Soft materials may be used for *static* or *AC* applications. The main static and low-frequency AC applications are flux guidance and concentration in magnetic circuits, including cores for transformers and inductors operating at mains frequency (50 or 60 Hz). *Forces* are exerted on current-carrying conductors in motors. Magnetic forces are also exerted between pieces of temporary magnet. Magnetostrictive transducers exert force directly. Changes

Table 12.1. Soft magnetic materials and applications

Frequency	Materials	Applications
Static <1 Hz	Soft iron, Fe–Co (permendur) Ni–Fe (permalloy)	Electromagnets, relays
Low frequency 1 Hz–1 kHz	Si steel, permalloy, finmet, magnetic glasses	Transformers, motors, generators
Audio-frequency 100 Hz–100 kHz	Permalloy foils, finmet, magnetic glasses, Fe–Si–Al powder (sendust) Mn–Zn ferrite	Inductors, transformers for switched mode power supplies, TV flyback transformers
Radio-frequency 0.1–1000 MHz	Mn–Zn ferrite, Ni–Zn ferrite	Inductors, antenna rods
Microwave >1 GHz	YIG, Li ferrite	Microwave isolators, circulators, phase shifters, filters

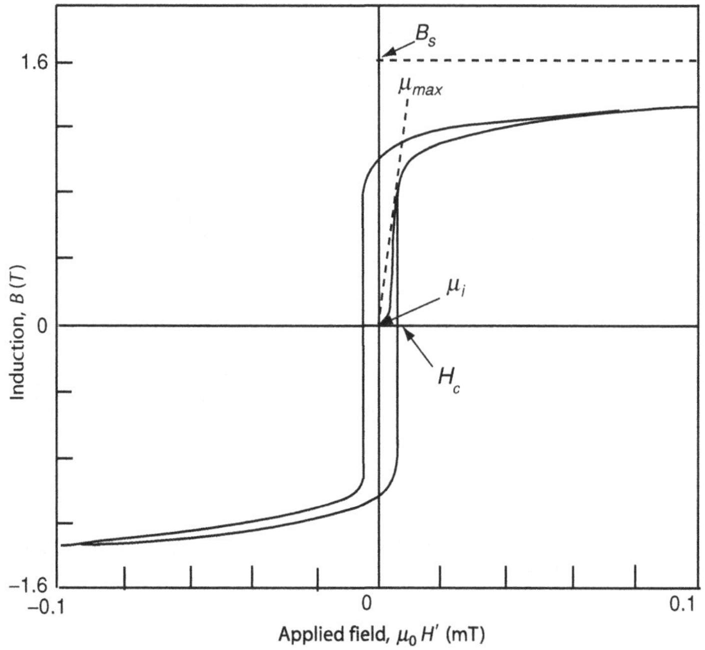

Figure 12.1

Hysteresis in a soft magnetic material. $B(H)$ and $J(H)$ are indistinguishable in small fields.

of flux produce *emf* in generators and electronic components. Metals are used up into the kilohertz range, but insulating ferrites are needed to concentrate flux and generate emfs in the radio-frequency and microwave ranges, in order to avoid eddy-current losses. Microwave applications involve the propagation of electromagnetic waves in waveguides rather than currents in electric circuits.

The higher the operating frequency, the lower the permeability and spontaneous induction of the materials used, and the smaller the fraction of saturation at which they operate. Hysteresis increases with frequency, and μ_{max} falls from about 10^4 in electrical steel to 100 or less for ferrites operating in the megahertz range. Preferred materials for the main frequency ranges are specified in Table 12.1.

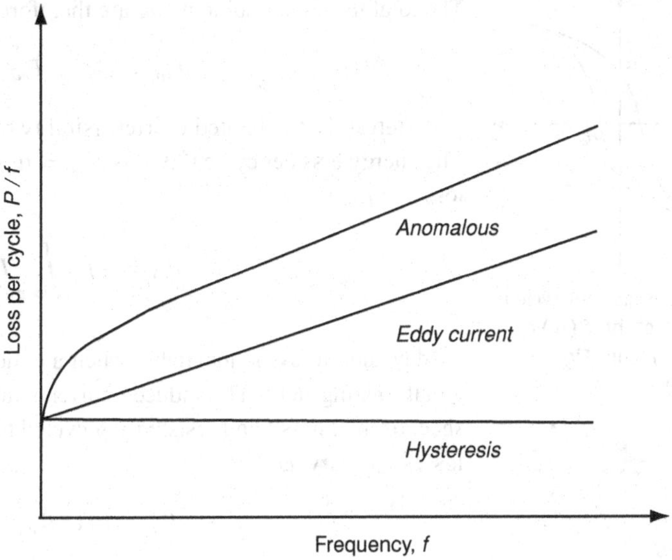

Figure 12.2

Total loss per cycle showing the three contributions.

Whenever metals are exposed to an alternating magnetic field, the induced eddy currents limit the depth of penetration of the flux. The skin depth δ_s is defined as the depth where B falls to $1/e$ of its value at the surface:

$$\delta_s = \sqrt{\frac{\varrho}{\pi \mu_r \mu_0 f}}. \tag{12.2}$$

Here ϱ is the resistivity and f is the AC frequency in hertz. For electrical steel ($\varrho = 0.5\ \mu\Omega\,\mathrm{m}$, $\mu_r = 2 \times 10^4$) the value of δ_s is 0.36 mm at 50 Hz. At 500 kHz, it is about 3.6 μm. Cores made from soft magnetic metals are often assembled from a stack of insulated laminations, where the lamination thickness is chosen to be less than δ_s, so that the applied field can penetrate each one. Insulators are untroubled by these problems of flux penetration.

12.1 Losses

12.1.1 Low-frequency losses

Energy losses are critical in any AC application. Traditionally three main sources were identified in soft metallic materials operating at low frequency, as indicated in Fig. 12.2:

- hysteresis loss P_{hy},
- eddy-current loss P_{ed},
- anomalous losses P_{an}.

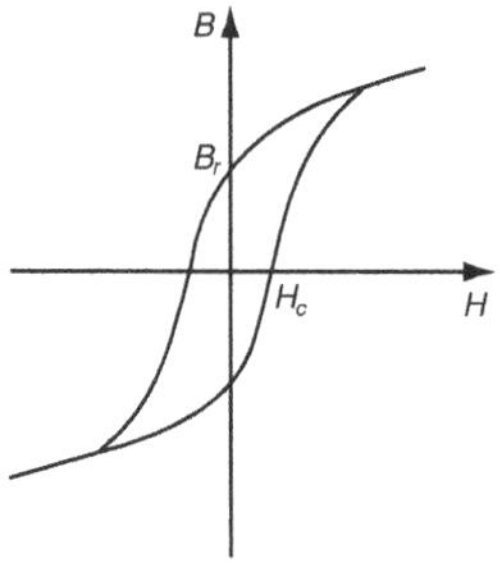

Hysteresis loss per cycle is the area of the $B(H)$ or $\mu_0 M(H')$ loop (Fig. 2.19(c)).

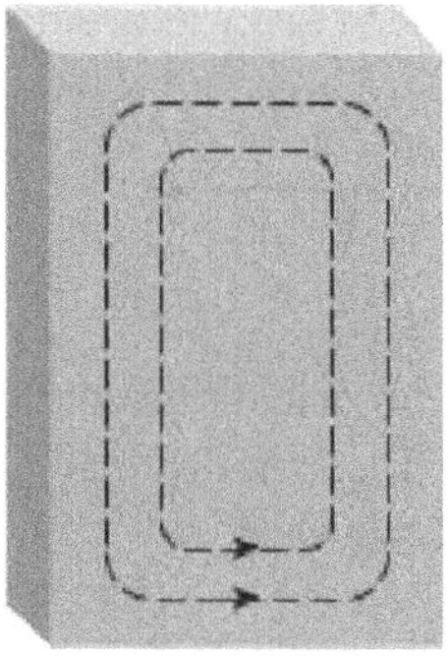

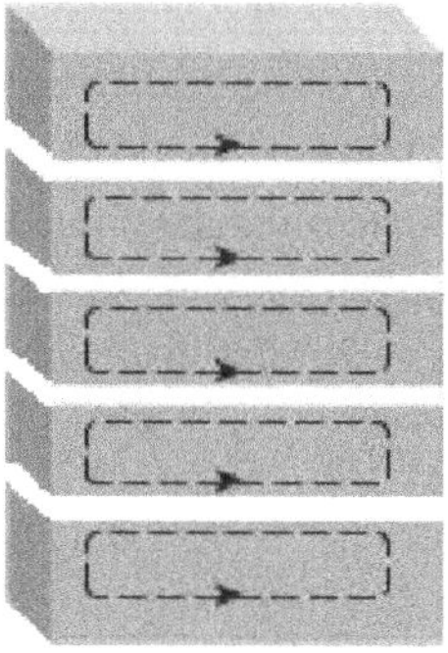

Reduction of eddy-current losses by lamination.

The total losses per cubic metre are therefore

$$P_{tot} = P_{hy} + P_{ed} + P_{an}.$$

Hysteresis loss is related to irreversibility of the static $B(H)$ or $M(H')$ loop. The energy loss per cycle (2.93) is $E_{hy} = \mu_0 \int_{loop} H' \mathrm{d}M$. At a frequency f the loss is $f E_{hy}$

$$P_{hy} = f \mu_0 \int_{loop} H' \mathrm{d}M. \tag{12.3}$$

Eddy-current loss is inevitable when a conducting ferromagnet is subject to an alternating field. The induced currents dissipate their energy as heat. In a sheet of thickness t and resistivity ϱ cycled to a maximum induction B_{max}, the losses P_{ed} vary as f^2:

$$P_{ed} = (\pi \mathrm{t} f B_{max})^2 / 6\varrho. \tag{12.4}$$

These losses can be minimized by using thin laminations or highly resistive material. For example, electrical steel $Fe_{94}Si_6$ (3 wt% Fe–Si) is usually made in laminations about 350 μm thick. It has $\varrho \approx 0.5$ μΩ m and density $\mathrm{d} = 7650$ kg m^{-3}. For $B_{max} = 1$ T, it follows from (12.4) that $W_{ed} = P_{ed}/\mathrm{d} \approx 0.1$ W kg^{-1} at 50 Hz. Lamination reduces eddy-current losses by a factor $1/n^2$, where n is the number of laminations.

Anomalous losses are whatever remains after P_{hy} and P_{ed} have been taken into account. They turn out to be comparable in magnitude to P_{ed} and they arise from extra eddy-current losses due to domain wall motion, nonuniform magnetization and sample inhomogeneity. Essentially, the anomalous losses reflect the broadening of the hysteresis loop with increasing frequency, so the separation of static hysteresis loss and anomalous loss is artificial. Think of an imaginary circuit containing a moving domain wall; the flux through the circuit is changing, so there is an emf which drives an eddy current near the moving wall.

Anomalous losses are reduced by a structure of many parallel domain walls, which decreases the distance the walls must move during the magnetization process. High-grade electrical steels are therefore laser-scribed to define a structure of narrow stripe domains. Much of the physics in that case is captured by the Pry and Bean model, Fig. 12.3. The electrical sheet is supposed to have a structure of uniformly spaced domains with separation d, which expand and contract in an AC field applied parallel to the walls. The losses in that case are

$$P_{an} = \frac{(4 f B_{\max})^2 d \mathrm{t}}{\pi \varrho} \sum_{n \text{ odd}} \frac{1}{n^3} \coth(n\pi d/\mathrm{t}), \tag{12.5}$$

which reduces to the relation (7.33) in the limit $(d/\mathrm{t}) \gg 1$. Losses for permalloy as a function of frequency are shown in Fig. 12.4.

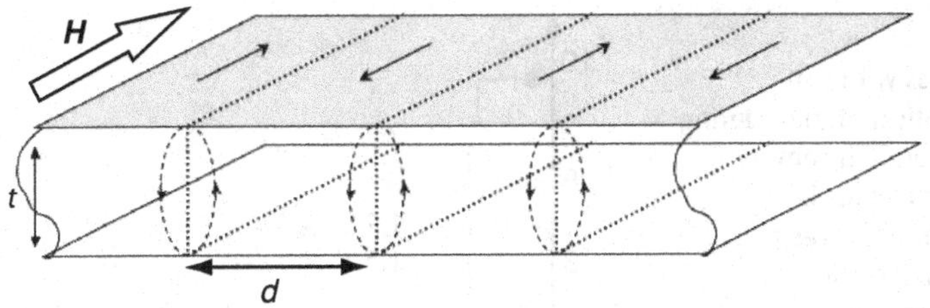

Figure 12.3

Pry and Bean model for movement of uniformly spaced domain walls. Currents are induced in the neighbourhood of the domain walls, as shown by the dashed lines.

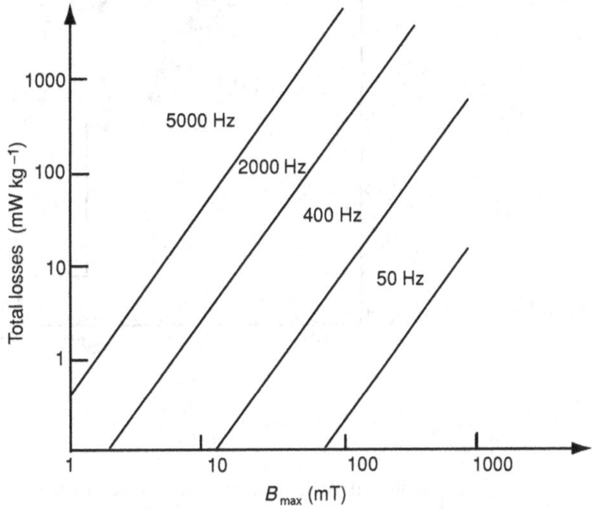

Figure 12.4

Total losses per kilogram for permalloy at different frequencies. Thickness is 350 μm.

A modern physical approach to anomalous losses, due to Bertotti, ascribes the broadening of the hysteresis loop at high sweep rates to the movement of elementary magnetic objects, identified as domain walls or groups of correlated walls. An effective field $H_{an} = P_{an}/(\mathrm{d}J/\mathrm{d}t)$ is introduced, where J is the polarization, and the anomalous losses are found to vary as $(f B_{\max})^{2/3}/\varrho^{1/2}$.

Losses in rotating fields are double those in axial fields, because the rotating field can be decomposed into two perpendicular axial components. As the polarization tends to saturation, domain walls are eliminated and the losses tend to zero.

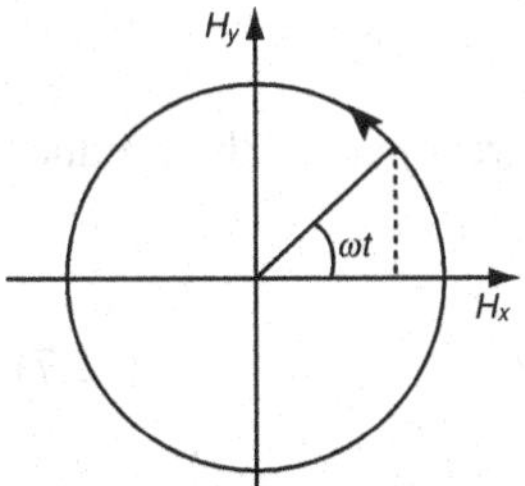

A rotating magnetic field $H = H_0 e_x \cos \omega t + H_0 e_y \sin \omega t$.

Progress in ameliorating key properties of temporary magnets operating at mains frequencies – core loss and permeability – was dramatic in the twentieth century; they improved by two and four orders of magnitude, respectively, as shown in Fig. 12.5.

12.1.2 High-frequency losses

Losses at high frequencies are best represented in terms of a complex permeability. The magnetization process involves magnetization rotation rather than

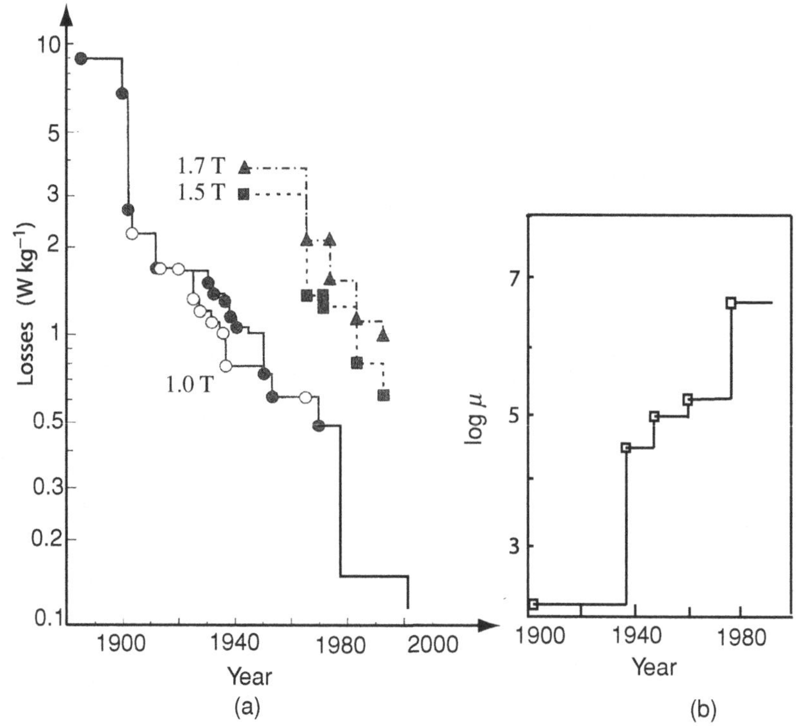

Figure 12.5

Progress with soft magnetic materials during the twentieth century: (a) total losses in transformer cores and (b) initial static permeability.

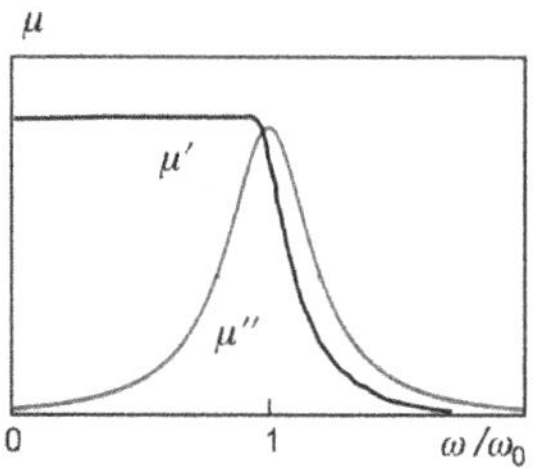

Real and imaginary parts of the permeability for an undamped resonance.

domain-wall motion, and losses in the GHz frequency range are influenced by ferromagnetic resonance. If the applied field is $h = h_0 e^{i\omega t}$, the induced flux density $b = b_0 e^{i(\omega t - \delta)}$ generally lags behind by a phase angle δ, known as the loss angle. The real parts of these expressions represent the time-dependent fields $h(t)$ and $b(t)$. The complex permeability $\mu = (b_0/h_0)e^{-i\delta}$ can be expressed as $(b_0/h_0)(\cos\delta - i\sin\delta)$, or

$$\mu = \mu' - i\mu'', \tag{12.6}$$

where $\mu' = (b_0/h_0)\cos\delta$ and $\mu'' = (b_0/h_0)\sin\delta$. The real part of the product μh is the time-dependent flux density

$$b(t) = h_0(\mu' \cos\omega t + \mu'' \sin\omega t), \tag{12.7}$$

so μ' gives the component of b that is in phase with the excitation field h, and μ'' gives the component which lags by $\pi/2$. Losses are proportional to μ'', the response in quadrature with the driving field. The magnetic quality factor Q_m is defined as $(\mu'/\mu'') = \cot\delta$ and the figure of merit is $\mu' Q_m$.

From a knowledge of $\mu'(\omega)$ and $\mu''(\omega)$, the real and imaginary parts of the frequency-dependent permeability of a linear system, it is possible to deduce the response to any small, time-dependent stimulus $h(t)$ which can be expressed as a Fourier integral

$$h(t) = \frac{2}{\pi}\int_0^\infty h(\omega)\cos\omega t\,d\omega, \tag{12.8}$$

where the Fourier components are

$$h(\omega) = \frac{2}{\pi}\int_0^\infty h(t)\cos\omega t\,dt. \tag{12.9}$$

The time-dependent response of the system is the superposition of the responses at different frequencies ω:

$$b(t) = \int_0^\infty [\mu'(\omega)\cos\omega t + \mu''(\omega)\sin\omega t]h(\omega)d\omega. \tag{12.10}$$

An equivalent expression relates $m(t)$ with $h(\omega)$ in terms of the complex susceptibility $\chi = \chi' - i\chi''$. Since $\mu_r = 1 + \chi$, it follows that

$$\mu_r' = 1 + \chi' \quad \text{and} \quad \mu_r'' = \chi''.$$

These expressions provide a complete description of the dynamic response of a linear magnetic system. Moreover, the real and imaginary parts of μ or χ are related. Knowledge of one part over the whole frequency range leads to a knowledge of the other, via the powerful Kramers–Kronig relations

$$\mu'(\omega) = \frac{2}{\pi}\int_0^\infty \frac{\mu''(\omega')\omega d\omega'}{(\omega'^2 - \omega^2)}, \quad \mu''(\omega) = \frac{-2}{\pi}\int_0^\infty \frac{\mu'(\omega')\omega d\omega'}{(\omega'^2 - \omega^2)}. \tag{12.11}$$

If the sample is anisotropic, the susceptibility and permeability are tensors related by $\hat{\mu} = I + \hat{\chi}$, which can be diagonalized by suitable choice of axes. Each component satisifes the Kramers–Kronig relations.

To calculate the energy losses, we use (2.92) for the work done on a magnetic system. The rate of energy dissipation is $P = h(t)db(t)/dt = h_0^2\cos\omega t(-\mu'\omega\sin\omega t + \mu''\omega\cos\omega t)$. Since $(1/2\pi)\int_0^{2\pi}\sin\theta\cos\theta d\theta = 0$ and $(1/2\pi)\int_0^{2\pi}\cos^2\theta d\theta = \frac{1}{2}$, the expression for the average rate of energy loss is

$$P_{av} = \tfrac{1}{2}\mu''\omega h_0^2. \tag{12.12}$$

The equivalent expression in terms of the imaginary part of the susceptibility is $\frac{1}{2}\mu_0\chi''\omega h_0^2$. The losses are necessarily positive, which explains the choice of the minus sign in the definition (12.6) of the complex permeability.

To look into the losses associated with the magnetization dynamics in more detail, we consider the damped equation of motion for coherent rotation of the magnetization introduced in Chapter 9:

$$\frac{d\boldsymbol{M}}{dt} = \gamma\mu_0\boldsymbol{M}\times\boldsymbol{H} + \frac{\alpha}{M_s}\boldsymbol{M}\times\frac{d\boldsymbol{M}}{dt}, \tag{12.13}$$

where the magnetization vector $\boldsymbol{M}$ has magnitude M_s and it precesses around the direction Oz of the static magnetic field $\boldsymbol{H}$, which incorporates the applied field, the demagnetizing field and the anisotropy field. The damping term on the right of the Landau–Lifschitz–Gilbert equation causes the magnetization to spiral inwards towards the Oz direction. Damping is essential for any measurement of static magnetization. Without damping, the spontaneous magnetization would precess perpetually around the applied field and never align with it.

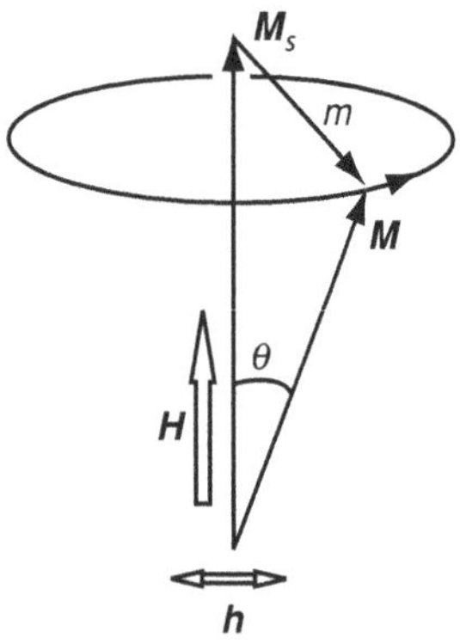

Magnetization precession.

We consider the case where a field is applied along an easy anisotropy axis. The equilibrium value of $\boldsymbol{M}$ is M_s, aligned along Oz. The instantaneous value of $\boldsymbol{M}$ is inclined at an angle θ and the deviation $\boldsymbol{m}$ is defined as $\boldsymbol{M} - \boldsymbol{M}_s$. The effective field $\boldsymbol{H}_s$, also along Oz, is the sum of the applied field $\boldsymbol{H}$ and the effective anisotropy field $\boldsymbol{H}_a \cos\theta$. The torque is $\boldsymbol{\Gamma} = \gamma\mu_0 \boldsymbol{M} \times \boldsymbol{H}_s$. Next we apply an alternating magnetic field $\boldsymbol{h} = \boldsymbol{h}_0 \cos\omega t$ in the xy-plane. In the undamped case,

$$\mathrm{d}\boldsymbol{M}/\mathrm{d}t = \gamma\mu_0 \boldsymbol{M} \times (\boldsymbol{H}_s + \boldsymbol{h}). \tag{12.14}$$

The equation is linear when the perturbation is small, that is when $h \ll H_s$ and $m \ll M_s$, Thus

$$\mathrm{d}\boldsymbol{M}/\mathrm{d}t = \gamma\mu_0 (\boldsymbol{m} \times \boldsymbol{H}_s + \boldsymbol{M}_s \times \boldsymbol{h}). \tag{12.15}$$

In component form

$$\frac{\mathrm{d}m_x}{\mathrm{d}t} - \gamma\mu_0 H_s m_y = \gamma\mu_0 M_s h_y, \tag{12.16}$$

$$\frac{\mathrm{d}m_y}{\mathrm{d}t} + \gamma\mu_0 H_s m_x = -\gamma\mu_0 M_s h_x, \tag{12.17}$$

$$\frac{\mathrm{d}m_z}{\mathrm{d}t} = 0. \tag{12.18}$$

In complex notation $\mathrm{d}/\mathrm{d}t$ is replaced by $\mathrm{i}\omega$. Setting $\omega_0 = \gamma\mu_0 H_s$ and $\omega_M = \gamma\mu_0 M_s$, the equations are written

$$\begin{bmatrix} \omega_0 & \mathrm{i}\omega & 0 \\ \mathrm{i}\omega & -\omega_0 & 0 \\ 0 & 0 & 0 \end{bmatrix} \begin{bmatrix} m_x \\ m_y \\ m_z \end{bmatrix} = \omega_M \begin{bmatrix} -h_x \\ h_y \\ h_z \end{bmatrix},$$

which can be inverted to read

$$\begin{bmatrix} m_x \\ m_y \\ m_z \end{bmatrix} = \begin{vmatrix} \kappa & -\mathrm{i}\nu & 0 \\ \mathrm{i}\nu & \kappa & 0 \\ 0 & 0 & 0 \end{vmatrix} \begin{bmatrix} h_x \\ h_y \\ h_z \end{bmatrix},$$

where

$$\kappa = \omega_0\omega_M / \left(\omega_0^2 - \omega^2\right) \tag{12.19}$$

and

$$\nu = \omega\omega_M / \left(\omega_0^2 - \omega^2\right). \tag{12.20}$$

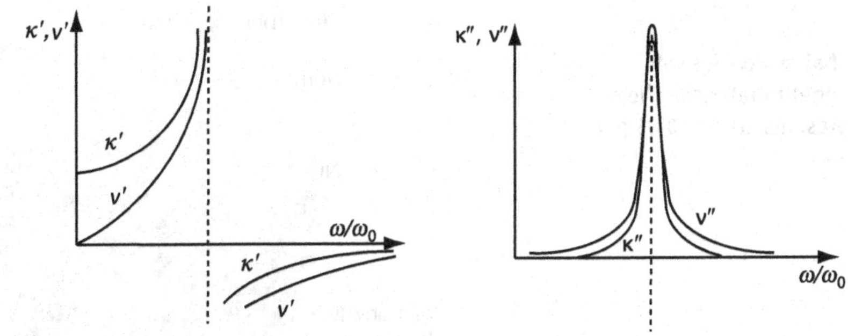

Figure 12.6

Real and imaginary parts of the susceptibilities κ and ν.

This susceptibility tensor describes the precession of the magnetization for uniaxial anisotropy in the absence of damping. Only the in-plane components need be considered, so

$$|\chi| = \begin{bmatrix} \kappa & -\mathrm{i}\nu \\ \mathrm{i}\nu & \kappa \end{bmatrix}.$$

When $\omega = 0$, $\nu = 0$, the tensor reduces to a scalar: $\chi_0 = \omega_M/\omega_0 = M_s/H_s$. As $\omega \longrightarrow \omega_0$, the ferromagnetic resonance frequency, κ, and ν diverge as shown in Fig. 12.6. The product of the static susceptibility and the resonance frequency $\chi_0\omega_0 = \omega_M$ is constant:

$$\chi_0\omega_0 = \gamma\mu_0 M_s. \tag{12.21}$$

This is Snoek's relation. The greater is the ferromagnetic resonance frequency ω_0, the lower is the static susceptibility χ_0. Unfortunately, extending the frequency response of ferrites leads to a decline in susceptibility and loss of performance.

In the case of a polycrystalline sample of volume v having crystallites of volume v_i with randomly oriented easy axes, the induced magnetization $\mathfrak{m} = \sum_i v_i \chi h = \chi_{eff} h$. The susceptibility $\chi_{eff} = \frac{2}{3}\kappa = 2\omega_M\omega_0/3(\omega_0^2 - \omega^2)$. The product of the static susceptibility and the resonance frequency in this case is $\chi_0\omega_0 = \frac{2}{3}\omega_M = \frac{2}{3}\gamma\mu_0 M_s$. This shows that it is impossible to exceed the Snoek limit in such a system, but it can be circumvented in ferrites with strong planar anisotropy. Ferrites are used as phase shifters in microwave applications above 10 GHz.

When damping is taken into consideration, a term $-(\alpha/M_s)\boldsymbol{M} \times \mathrm{d}\boldsymbol{m}/\mathrm{d}t$ is added to (12.15). The new expressions for κ and ν are

$$\kappa = \frac{\omega_M(\omega_0 + \mathrm{i}\alpha\omega)}{\omega_0^2 - (1+\alpha^2)\omega^2 + \mathrm{i}2\alpha\omega\omega_0}, \tag{12.22}$$

$$\nu = \frac{\omega_M\omega}{\omega_0^2 - (1+\alpha^2)\omega^2 - \mathrm{i}2\alpha\omega\omega_0}. \tag{12.23}$$

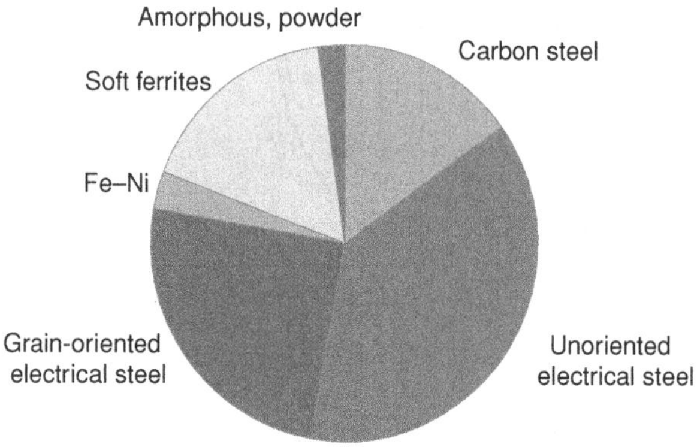

Figure 12.7

Global market for soft magnetic materials. The pie represents about 10 B$ per year.

12.2 Soft magnetic materials

The global market for soft magnetic materials is summarized in Fig. 12.7. Electrical steel for transformers and electromagnetic machinery is predominant. Produced in quantities in excess of 7 million tonnes per year, it represents about 1% of global steel production, but 95% of the tonnage and 75% of the market value of all temporary magnets. The choice of soft magnetic material is a trade-off between polarization, permeability, losses and cost. The polarization should be as large as possible for a given excitation field, and core losses must be acceptable at the operating frequency. Alloy additions such as C, Si or Al, which reduce losses and increase permeability also reduce the saturation polarization and increase cost. One alloy does not suit all needs, but there are a few widely used grades, which have been optimized over the years.

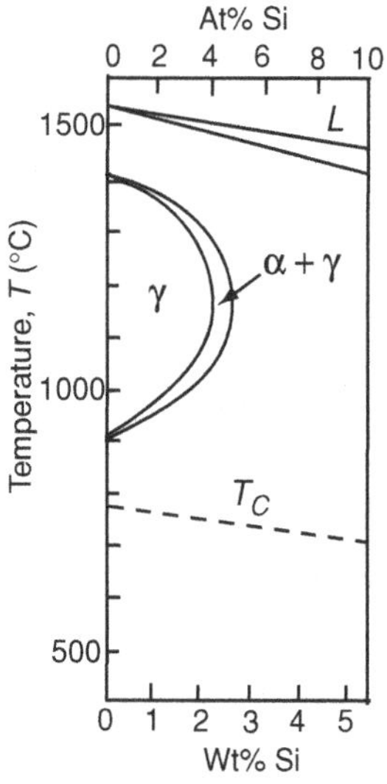

Iron-rich side of the Si-Fe phase diagram.

Low-carbon mild steel is used for cheap motors in consumer products such as washing machines, vacuum cleaners, refrigerators and fans, where losses are of little interest to the manufacturer. The customer pays for the electricity. Better electrical performance requires other alloy additions. Silicon is ideal, because 4 at% suffices to suppress the $\alpha \to \gamma$ phase transition in iron, permitting hot rolling of the sheet. Traces of carbon extend the γ phase stability region, so the silicon content is usually around 6 at% or 3 wt%. Silicon steel was invented by Robert Hadfield in 1900, who found that the 6 at% Si composition was sufficiently ductile to be rolled into thin sheets. Isotropic and grain-oriented $Fe_{94}Si_6$ in the form of sheets about 350 μm thick is produced by the square kilometre for mains-frequency electrical applications. Losses are about ten times lower than for mild steel. Silicon increases the resistivity, and reduces both the anisotropy and the magnetostriction of the iron. *Isotropic* sheet is appropriate for motors and generators, where the direction of flux changes continually during operation. In transformers, however, the axis of $\boldsymbol{B}$ is fixed, and it is beneficial to use *crystallographically oriented* sheet with an easy axis to further reduce losses. This material was developed by the felicitously named

Table 12.2. Properties of magnetic sheet steels

Material	J_s (T)	μ_r at 1.5 T	H_c ($A\,m^{-1}$)	W_{tot} ($W\,kg^{-1}$)	ϱ ($\mu\Omega$ m)
Mild steel	2.15	500	80	12	0.15
Si steel	2.12	1 000	40	2.5	0.60
Grain-oriented Si steel	2.00	20 000	5	1.2	0.50

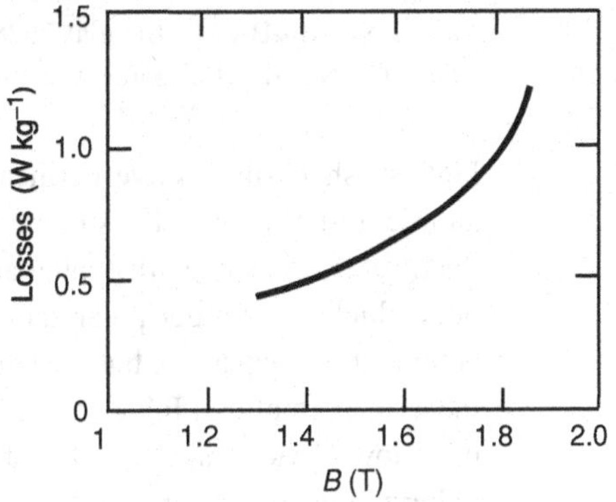

Figure 12.8

Losses as a function of operating induction for grain-oriented silicon steel.

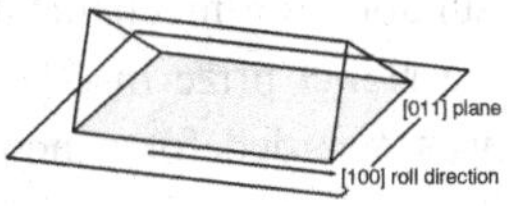

The Goss texture of grain-oriented Si–Fe.

Norman Goss in 1934. It went into mass production in 1945, and now accounts for around 20% of the volume of electrical steel produced. *Grain-oriented silicon steel* with the Goss texture is produced by an extended process of rolling and annealing which promotes secondary crystallization. The {110} planes of the secondary crystallites are parallel to the sheets and a [100] easy axis is parallel to the roll direction. The losses at 1.7 T are below 1 $W\,kg^{-1}$, but they increase as the material operates closer to saturation, Fig. 12.8.

The thickness of grain-oriented silicon steel laminations is 200–350 µm. Thinner Si–Fe sheet can be produced by planar flow casting. The thinner laminations are useful at higher frequencies, and for reducing losses for non-sinusoidal waveforms with a high harmonic content, which arise, for example, from wind farms. The properties of different magnetic steels are compared in Table 12.2.

A great range of useful magnetic properties can be achieved in the Ni–Fe alloy system. Here the economic benefit is measured not so much in thousands of tonnes as in millions of magnetic components of widely differing shapes and sizes. The outstanding feature of Ni–Fe alloys is their permeability, which is achieved by carefully controlled heat treatment. Interesting compositions, between 30% and 80% nickel, have the fcc γNi–Fe structure. A weak uniaxial anisotropy of order 100 $J\,m^{-3}$ can be induced in Ni–Fe alloys by magnetic field annealing to help control the magnetization process. Most remarkable are the permalloys with composition close to $Ni_{80}Fe_{20}$, where both anisotropy and magnetostriction fall to zero at almost the same composition. This promotes the

Table 12.3. Properties of soft materials for static or low-frequency applications

Material	Name	μ_i	μ_{max}	J_s (T)	H_c (A m^{-1}))
Fe	Soft iron	300	5 000	2.15	70
$Fe_{49}Co_{49}V_2$	V-permendur	1 000	20 000	2.40	40
$Ni_{50}Fe_{50}$	Hypernik	6 000	40 000	1.60	8
$Ni_{77}Fe_{16.5}Cu_5Cr_{1.5}$	Mumetal	20 000	100 000	0.65	4
$Ni_{80}Fe_{15}Mo_5$	Supermalloy	100 000	300 000	0.80	0.5
a-$Fe_{40}Ni_{38}Mo_4B_{18}$	Metglas 2628*SC*	50 000	400 000	0.88	0.5
$Fe_{73.5}Cu_1Nb_3Si_{13.5}B_9$	Finmet	50 000	800 000	1.25	0.5

highest, shock-insensitive permeability. A tendency towards Ni_3Fe-type $L1_2$ atomic ordering can be suppressed by Mo additions in supermalloy, and ductility is achieved with copper additions in mumetal, which is good for magnetic shielding. Larger polarization is obtained near the equiatomic $Ni_{50}Fe_{50}$ composition (hypernik), but the soft magnetic properties are slightly inferior to those of permalloy (Table 12.3). The iron-rich invar alloys, around $Ni_{36}Fe_{64}$, offer low Curie temperatures and rapid thermal variation of the spontaneous polarization, which are exploited in applications such as rice cookers and electricity meters. Their dimensional stability around room temperature due to negative volume magnetostriction makes them suitable for mechanical precision instruments. Charles Guillaume received the Nobel prize in 1920 for his discovery of invar, the only time the prize was awarded for a new magnetic material.

The Co–Fe alloys are much less versatile, and more expensive. The cost of cobalt used to fluctuate wildly, but it stabilized as more geographically diverse sources of supply became available. The great advantage of permendur, $Fe_{50}Co_{50}$, is its polarization, the highest of any bulk material at room temperature (2.45 T). Addition of 2% vanadium improves the machinability without spoiling the magnetic properties. The $Fe_{65}Co_{35}$ composition provides similar polarization with less cobalt.

Powder of iron or the brittle zero-anisotropy, zero-magnetostriction alloy *Sendust* ($Fe_{85}Si_{10}Al_5$) can be insulated and used at higher frequencies in cores. Iron powder with a particle size of a few micrometres is practically anhysteretic at all temperatures (Fig. 12.9). Relative permeability is limited to 10–100 because of the demagnetizing fields. Long telephone lines are loaded with powder-core inductors (loading coils) to balance their capacitance.

Melt-spun amorphous alloy ribbons can be produced with thickness of about 50 μm and resistivity ≈ 1.5 μΩ m. No higher resistivity can be achieved in a metal because it corresponds to a mean-free path comparable to the interatomic spacing. The compositions are close to the glass-forming ratio $M_{80}T_{20}$, where T = Fe, Co, Ni and M = B, Si. Cobalt-rich compositions exhibit zero magnetostriction, and permeability up to 10^6. Metallic glass ribbons can be wound into cores for use up to about 100 kHz, and applied, for example, in

Figure 12.9

Magnetization of diluted soft iron powder with a particle size of 6–8 μm. (Data courtesy of M. Venkatesan.)

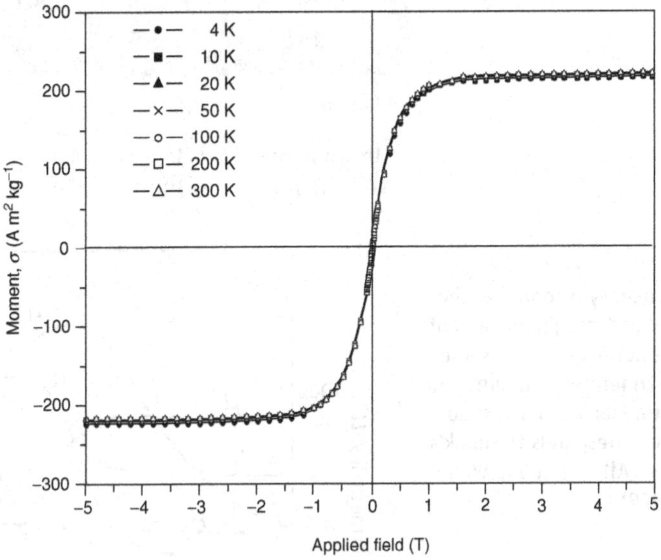

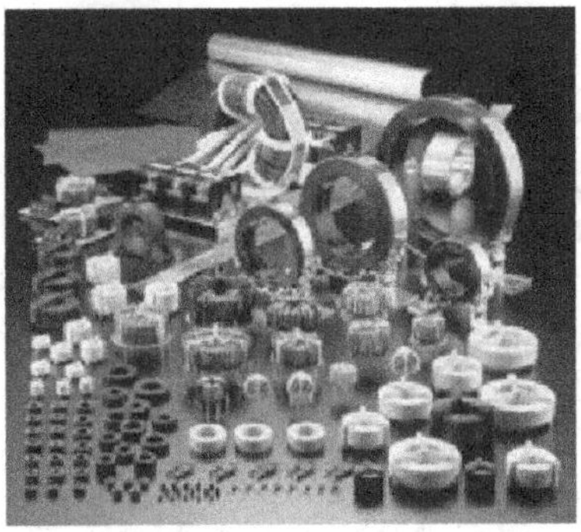

An assortment of soft-magnetic components made from finmet. (Courtesy, Hitachi metals.)

switched mode power supplies, distribution transformers, as well as in flux gate magnetometer. Partially crystallized nanocrystalline materials such as finmet (§11.2.4) offer comparable permeability, with higher polarization.

Soft ferrites have mediocre polarization and permeability compared to metallic ferromagnets – their ferrimagnetic saturation polarization is only 0.2–0.5 T – but their insulating character is a decisive advantage at high frequencies. Since eddy-current losses are not a problem, there is no need for lamination. Oxides are also much better than metals at resisting corrosion. Ceramic components have to be produced to near-net shape, and finished by grinding.

Magnetic properties of the spinel ferrites have been tailored by tuning the composition, microstructure and porosity since the 1940s when these materials first saw the light of day in the Netherlands. Mn–Zn ferrites are used up to about 1 MHz, and Ni–Zn ferrites from 1–300 MHz or more. The latter have lower polarization, but higher resistivity. Conduction is usually due to traces of Fe^{2+}, which promotes Fe^{2+}–Fe^{3+} electron hopping. Most cations make a negative contribution to the magnetostriction and anisotropy constant; $\langle 111 \rangle$ directions in the spinel lattice are usually easy, unless Co^{2+} or Fe^{2+} are present.

The frequency response of the initial permeability of a ferrite μ_i is almost flat up to a rolloff frequency f_0, associated with ferromagnetic resonance (Fig. 12.10). The losses peak at f_0. The permeability and resonance frequency vary oppositely with anisotropy, and according to Snoek's relation (12.21) their product is constant for a family of compounds with constant M_s; the higher the rolloff frequency, the lower the permeability. The figure of merit $\mu_i f_0$ is about

Table 12.4. Properties of soft materials for high-frequency applications

Material	J_s(T)	μ_i	H_c (A m^{-1})	ϱ (Ω m)	P (0.2T)
Mn–Zn ferrite	0.45	4000	20	1	200
Ni–Zn ferrite	0.30	500	300	$>10^3$	

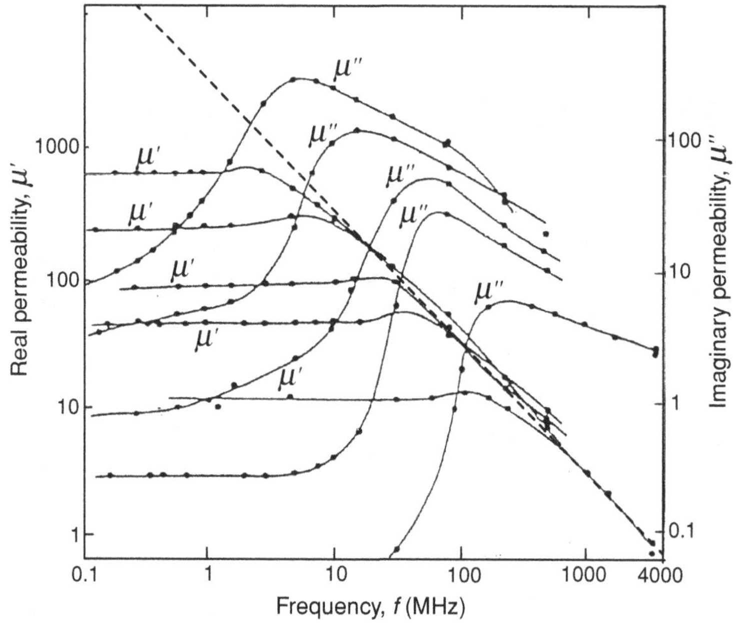

Figure 12.10

Frequency response of the real and imaginary parts of the permeability for some Ni-Zn ferrites with different permeability. The dashed line corresponds to Snoek's law. (After Smit and Wijn (1979).)

8 GHz for Ni–Zn ferrites and 4 GHz for MnZn ferrites. Hence it is not feasible for a MnZn ferrite with a permeability of 10 000 to operate above 400 kHz. Properties of high-frequency ferrites are summarized in Table 12.4.

YIG is an excellent microwave material for use in the GHz range. Highly perfect insulating crystals can be prepared which have minimal anisotropy. To measure the ferromagnetic resonance, a small sample ground and polished in the form of a sphere is placed near the end of a waveguide. The reflected signal is measured at fixed frequency as the applied field is swept through the resonance at $\omega_0 = \gamma \mu_0 H_0$, where γ is the gyromagnetic ratio $\approx$28 GHz T^{-1}. In the X-band (8–12 GHz radiation, handled in 25 mm waveguides) the linewidth of the resonance observed in an applied field of 350 kA m^{-1}, may be only 350 A m^{-1} for dense polycrystalline YIG and better than 35 A m^{-1} for a single-crystal sphere. The Q factor $\omega/\Delta\omega$ is therefore 1000 in one case and 10 000 in the other. These sharp resonances are indispensable for microwave filters and oscillators. The snag is that a magnetic field is needed to operate these components, which must be provided by an electromagnet or a permanent magnet.

12.3 Static applications

Figure 12.11

An electromagnet.

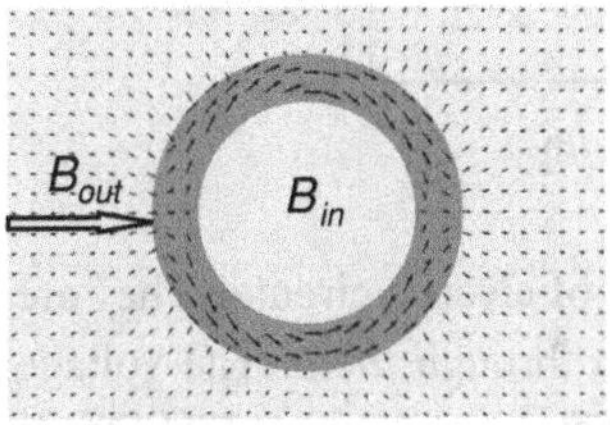

Magnetic shielding. The shielding ratio $\mathcal{R}$ is B_{out}/B_{in}.

Electromagnets consist of field coils to polarize an iron core, a yoke to guide the flux and pole pieces to concentrate flux in the airgap, Fig. 12.11. Flux guidance and concentration in electromagnets requires material with the highest polarization and very little remanence. Pure soft iron or $Fe_{65}Co_{35}$ is used. For best results, the pole pieces are tapered at an angle of 55°.

Electromagnetic relays and solenoid valves are miniature electromagnets where an iron core is magnetized and exerts a force on another temporary magnet. When the gap flux density is B_g, the force per unit area is $B_g^2/2\mu_0$ (13.13).

Passive magnetic shielding of low-frequency AC fields or weak DC fields, such as the Earth's, requires material to provide a low-reluctance flux path around the shielded volume. Reluctance is the magnetic analogue of resistance. The shielding ratio $\mathcal{R}$ is the ratio of the field outside to the field inside. Values of $\mathcal{R} \approx 100$ are achieved in low fields. The thickness of shielding material is chosen so that its polarization is unsaturated by the flux collected. It is more effective to use several thin shields rather than one thick one. DC shields are often made of permalloy or mumetal, which have negligible anisotropy and magnetostriction, and are therefore immune to shock and strain. Flexible shielding woven from cobalt-rich metallic glass tape, or finmet sheet may be used directly.

12.4 Low-frequency applications

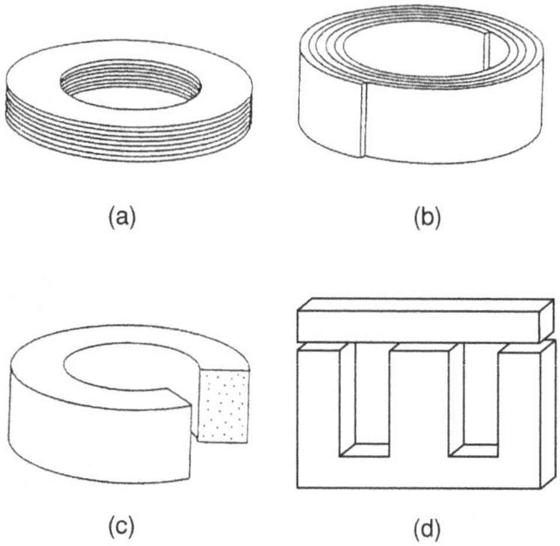

Figure 12.12

Types of cores: (a) stacked laminations, (b) tape-wound core, (c) powder core (sectioned to show internal structure) and (d) ferrite E-core.

Inductors are one of the three basic components of electric circuits. They are circuit elements that resist any change of current through them. The voltage drop across a resistanceless inductance of L henry is $V(t) = -L\partial I/\partial t$. If $I = I_0 \sin \omega t$, the resulting voltage $LI_0\omega \cos \omega t$ has a phase lag of exactly $\pi/2$. Consider the inductor as a long solenoid with cross section $\mathcal{A}$, length l and n turns per metre. The flux density in the solenoid is $\mu_0 nI$. By Faraday's law $V = -\partial\Phi/\partial t = -\mu_0 n^2 l\mathcal{A}\partial I/\partial t$; hence $L = \mu_0 n^2 l\mathcal{A}$. Filling the solenoid with soft magnetic material of permeability μ_r increases the flux density by this factor, so that

$$L = \mu_0\mu_r n^2 l\mathcal{A}. \tag{12.24}$$

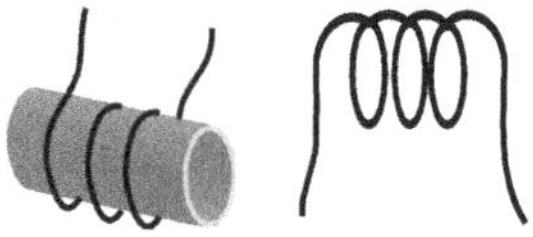

An inductor with and without a soft magnetic core.

The soft magnetic core therefore increases the electrical inertia of the inductor by orders of magnitude. Alternatively, the dimensions of a component with a given inductance can be greatly reduced by including a soft core. Some commonly used ones are shown in Fig. 12.12.

Low-frequency electrical machines include transformers, motors and generators that operate at mains frequencies of 50 or 60 Hz, or 400 Hz in the case of airborne or shipborne power. They include soft iron cores to generate and guide the flux. Eddy-current losses (12.4) are reduced by using thin laminations of material with a high resistivity. Efficiencies of well-designed transformers exceed 99%; they are probably the most efficient energy converters ever made. Core losses represent about a quarter of the total, the remainder being in the windings. The core losses in transformers nevertheless cost some 10 billion dollars per year. World-wide annual consumption of electrical energy is around 18×10^{12} kW h, which corresponds to an average rate of consumption of

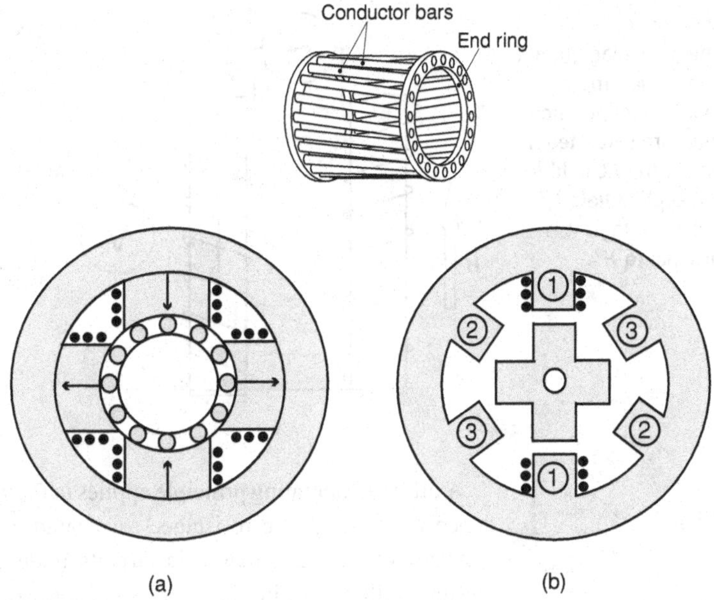

Figure 12.13

Two electric motor designs: (a) an induction motor and the squirrel cage (b) a 3/4 variable reluctance motors.

300 W per head of population. The rich billion of the Earth's people consume about ten times as much, and the poorest billion almost none at all.

Electricity is produced in power stations by large turbogenerators (~1000 MW) turning at constant frequency. The mains power is generated at a multiple of this frequency so it is unnecessary to use laminations in the rotor.

Electric motors are manufactured by the million. Whether they are excited by field coils or permanent magnets, they incorporate quantities of temporary magnets to guide the flux. Here we will describe just two designs which consume much electrical steel. The induction motor is the simplest and most rugged of all (Fig. 12.13(a)). It is manufactured in sizes ranging from 10 W to 10 MW for a myriad of domestic and industrial applications. The 'fractional horse power'[1] motors in consumer goods are usually induction motors.

The stator of this workhorse is a hollow cylindrical stack of laminations, pressed into a core, with slots to receive the field windings. The windings are energized with single-phase mains power (three-phase for industrial drives), which creates a rotating magnetic field in the centre. There is a 'squirrel cage' rotor consisting of metal bars running parallel to the axle which are short-circuited by circular end rings, and another soft-iron core is mounted in the centre. The forces on currents induced in the squirrel cage cause it to rotate with the field, but at slightly less than synchronous speed. It is an *asychronous* motor which draws maximum current at startup.

[1] 1 horse power = 746 W.

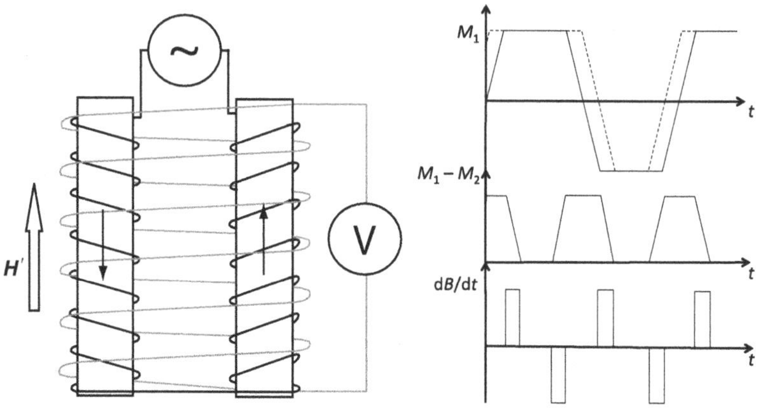

(a) (b)

Figure 12.14

(a) Schematic diagram of a fluxgate magetometer, (b) principle of operation. The cores are saturated in the sum of the AC field h and the applied field H'. The pickup voltage V is proportional to H'.

A different operating principle applies in the variable-reluctance motor. Reluctance is the magnetic resistance associated with the flux path. The analogy between electric and magnetic circuits is developed in §13.1. A 3/4 motor design is illustrated in Fig. 12.13(b). There are three pairs of stator windings, which are energized in the sequence 1–2–3 to create a rotating field. The rotor this time is just a piece of laminated soft iron with four poles in the form of a cross. Initially it is in the position shown, but when windings 2 are energized, it rotates clockwise by 60°, and so on. The motor is synchronous. It offers a high torque to inertia ratio, but it requires a precise electronic controller to power the windings.

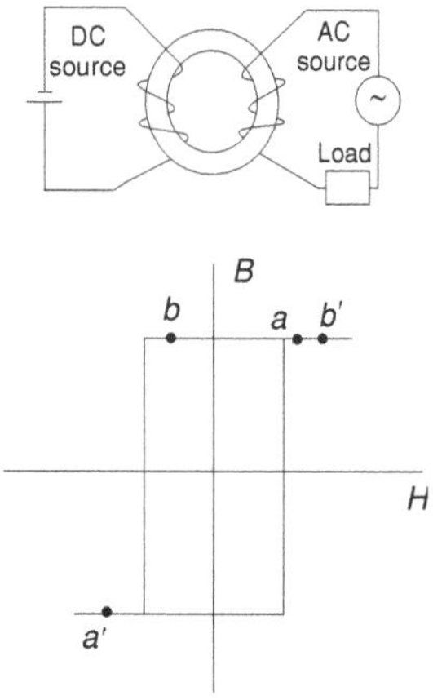

A magnetic amplifier. With no DC control current, there is a large flux swing (a–a'), but as the control current rises the flux swing ceases (b–b'') and the AC load current rises.

Magnetic amplifiers use square-loop cores which may be made from textured $Ni_{50}Fe_{50}$. When the DC control current is zero, the load current passing through the AC winding is very small because the voltage drop across the winding, proportional to $d\Phi/dt$, almost cancels the source signal. As the current from the DC source saturates the core, the change in flux in the core becomes negligible and the current in the load rises.

A related application which depends on saturable soft cores is the fluxgate magnetometer, Fig. 12.14. The magnetometer consists of two identical, parallel cores with a field winding creating an AC field sufficient to saturate the cores, h, in opposite directions. A toroidal core with a helical winding will work as well. The field, H', to be measured is parallel to the cores. It leads to unbalanced saturation of their net magnetization $M_1 - M_2$ and the changing flux is sensed in a secondary coil using a lockin detector. A response linear in applied field is obtained by flux compensation to null the signal. Fields of up to 200 A m^{-1} can be measured with an accuracy of 0.5 A m^{-1}. Typical noise figures are 100 pT Hz$^{-1/2}$.

A different family of applications of soft magnets makes use of their magnetostrictive properties, which generally depend on the direction of magnetization relative to the crystal axes. The linear saturation magnetostriction λ'_s of an isotropic polycrystalline material is an average over all directions.

Figure 12.15

A surface acoustic wave delay line.

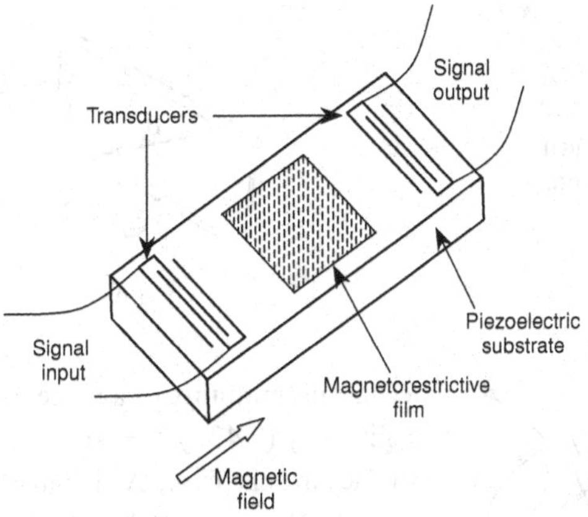

Nickel, for example, has $\lambda_s = -36 \times 10^{-6}$, so modest strains can be achieved by magnetizing nickel rods. Grain-oriented silicon steel has $\lambda_{100} = 20 \times 10^{-6}$, which is why transformers hum. Much larger effects are obtained with the mixed Tb–Dy alloy terfenol (§11.3.5), where cancellation of cubic anisotropy contributions of opposite sign for the two rare-earths means the alloy is easy to magnetize. The huge linear magnetostriction of 1500×10^{-6} makes it useful for linear actuators and ultrasonic or accoustic transducers, including sonar.

Another manifestation of magnetostriction in a soft magnetic material is the ΔE effect (§5.6.2). The dramatic softening of Young's modulus, as the domains align in a small field, can be used to tune a surface accoustic wave delay line, Fig. 12.15, of the type incorporated in military radar. The accousic wave propagates in a piezzoelectric substrate such a quartz or $LiMbO_3$, where it is excited and detected by a pair of interdigitated transducers. The beam from an array of antennae can be directed by tuning the time delays between adjacent elements of the array.

12.5 High-frequency applications

Ferrite components are extensively employed in high-frequency applications, although metallic alloys with very thin laminations offer the benefit of higher polarization at frequencies where the losses are not too severe.

Ferrite cores appear in chokes, inductors and high-frequency transformers for switched-mode power supplies. They are also used for broad-band amplifiers and pulse transformers. Losses at 100 kHz are about 50 W kg^{-1}. Another common application of ferrites is as antenna rods in AM radios, Fig. 12.16. The ferrite antenna consists of an N-turn pickup coil wound on a rod of cross-sectional area $\mathcal{A}$. The voltage induced by a radio-frequency field of amplitude b_0 and frequency ω would have amplitude $N\omega\mathcal{A}b_0$ for the coil alone. The

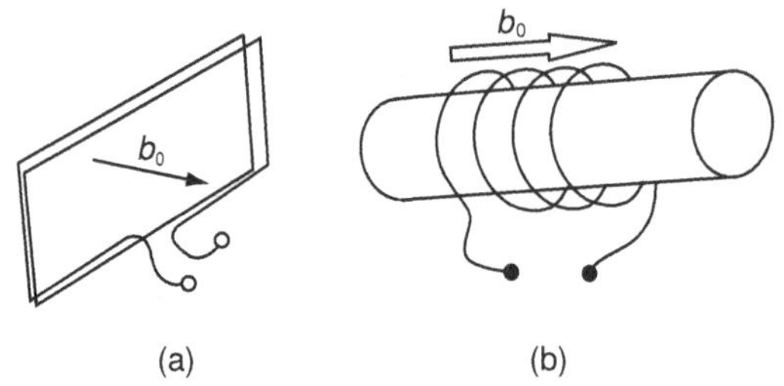

Figure 12.16

(a) A wire loop antenna, and (b) an equivalent ferrite rod with much smaller cross section.

external susceptibility of a piece of soft material with demagnetizing factor $\mathcal{N}$ was given by (2.42) $\chi'_e = \chi(1 + \chi\mathcal{N})'$.

In the present case, $\mathcal{N}$ is small and μ is very large, so that $\chi = \mu$, and $\chi\mathcal{N} \gg 1$. Hence $\mu' \sim 1/\mathcal{N}$. The amplitude of the voltage induced in the antenna is therefore $N\omega A b_0/\mathcal{N}$. The ferrite acts as a flux concentrator, so that the antenna is equivalent to a much larger bare coil of area $\mathcal{A}/\mathcal{N}$. Ni–Zn ferrite is best for this application on account of its high resistivity.

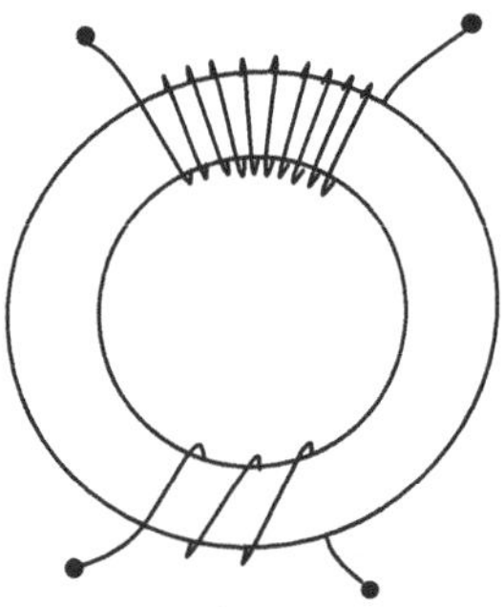

A pulse transformer.

The wide frequency range over which ferrites exhibit near constant μ' and negligible μ'' (Fig. 12.10) makes them suitable for pulse transformers. A pulse that is sharply defined in time contains Fourier components spanning a wide range of frequencies. The transformer is just a ring of soft ferrite on which both the primary and secondary are wound.

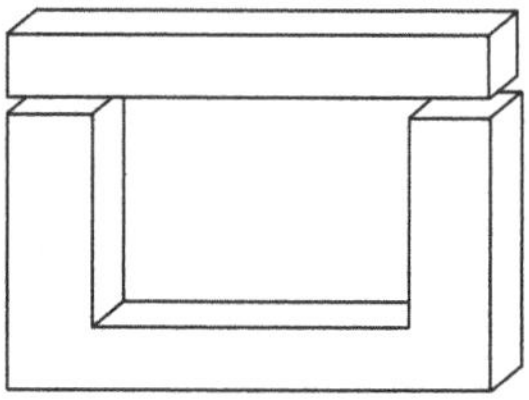

AC core with an airgap.

A single grade of ferrite can be used to produce high-frequency inductors of a standard size with different values of L by the expedient of leaving an airgap in the core. If l_m is the length of the core which has permeability μ, l_g is the length of the airgap and $\mathcal{A}$ is the cross section, the reluctance is the sum of the contributions of the ferrite and the airgap (13.5), which can be written as $(l_m/\mu_0\mu_{eff}\mathcal{A}) = (l_m/\mu_0\mu_r\mathcal{A}) + (l_g/\mu_0\mathcal{A})$. Hence

$$\frac{1}{\mu_{eff}} \approx \frac{1}{\mu_r} + \frac{l_g}{l_m} \approx \frac{l_g}{l_m}. \tag{12.25}$$

The cross section can be obtained when the permeability is degraded by the airgap, which introduces a demagnetizing field in the ferrite. The effective demagnetizing factor of the gapped core is $\mathcal{N} = l_g/l_m$.

Miniature inductors can be integrated with on-chip electronics. A planar or bilevel copper coil and a thin permalloy core can be electroplated and sputtered directly onto the chip.

Resonant filters are LC circuits which pass a narrow band of frequencies, whose width is limited by losses in the components, Fig. 12.17. The relative width of the peak gauged at the point where the value falls to $1/\sqrt{2}$ of the peak value is[2]

$$\frac{\Delta\omega}{\omega_0} = \frac{1}{Q}, \tag{12.26}$$

[2] The external susceptibility χ' in the applied field H' should not be confused with the real part of the complex susceptibility, which also features in this chapter.

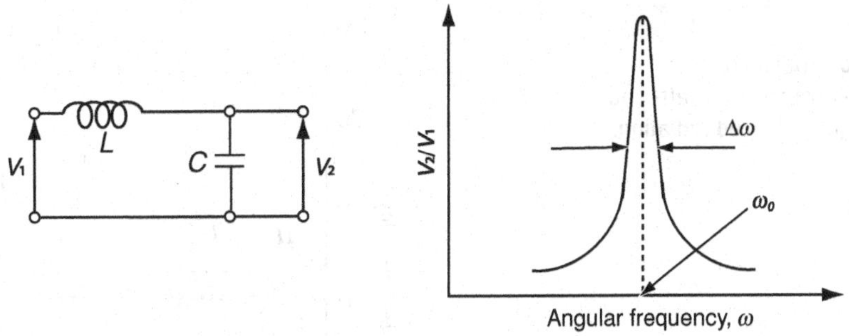

Figure 12.17

An LC filter circuit and the pass band.

where the Q factor is essentially limited by resistive losses in the inductor. Q is defined as μ'/μ''. The quality factor is maximized when the losses are equally divided between the copper windings and the magnetic material; Q is then $L\omega/2R$.

12.5.1 Microwave applications

Microwave ferrites can operate in the frequency range 300 MHz–100 GHz. At these frequencies, we need to consider the electromagnetic wave travelling in a waveguide rather than the current in a circuit. Microwave devices exploit the nonreciprocal interaction of the electromagnetic field with the ferromagnetic medium, especially in the vicinty of the ferromagnetic resonance frequency ω_0. Generally, if a uniform external field is applied along Oz and an AC field $\boldsymbol{h}'$ is applied in the xy-plane, the resulting flux density $\boldsymbol{B}$ is given by the tensor permeability

$$\begin{bmatrix} B_x \\ B_y \\ B_z \end{bmatrix} = \begin{bmatrix} \mu' & -\mathrm{i}\mu'' & 0 \\ \mathrm{i}\mu'' & \mu' & 0 \\ 0 & 0 & \mu_0 \end{bmatrix} \begin{bmatrix} H'_x \\ H'_y \\ H'_z \end{bmatrix}, \tag{12.27}$$

where $\boldsymbol{H}' = \boldsymbol{H}_z + \boldsymbol{h}$. The off-diagonal terms produce the nonreciprocal effects.

A plane-polarized wave $h = h_0 \cos \omega t$ can be decomposed into two oppositely rotating circularly polarized waves.

$$h = h_0 \cos \omega t = \frac{h_0}{2}(\mathrm{e}^{\mathrm{i}\omega t} + \mathrm{e}^{-\mathrm{i}\omega t}), \tag{12.28}$$

which generally propagate at different speeds $c/\sqrt{\epsilon\mu_+}$ and $c/\sqrt{\epsilon\mu_-}$. This magnetic circular birefringence is the microwave Faraday effect. The + and − directions are defined in relation to the polarization of the ferrite. Here μ_+ and μ_-, the effective permeabilities for right and left polarizations, are $\mu_+ = \mu'_+ - \mathrm{i}\mu''_-$ and $\mu_- = \mu'_- + \mathrm{i}\mu''_-$, Fig. 12.18. The first turns in the sense of the Larmor precession and shows a resonance at the ferromagnetic resonance frequency, the second turns in the opposite sense and does not resonate. If the plane wave is incident normal to the surface of a plaque of ferrite of

Figure 12.18

Absorption and transmission for left- and right-polarized radiation.

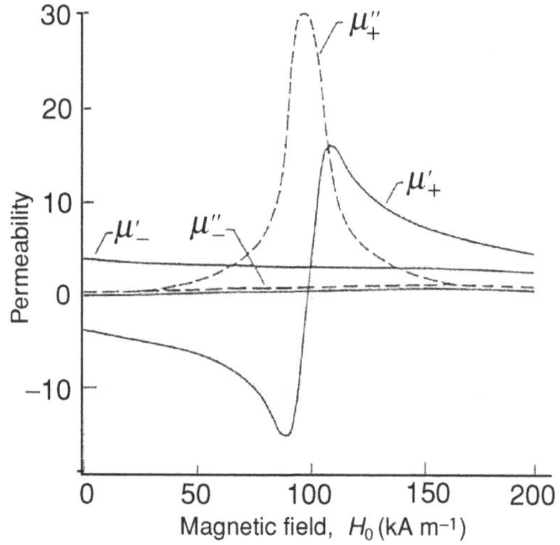

Figure 12.19

Two counter-rotating circularly polarized waves (a), which become dephased because they propagate with different velocities (b), producing the Faraday roation θ_F (c). The effect does not depend on the sense of propagation of the waves relative to the magnetization of the birefringent medium.

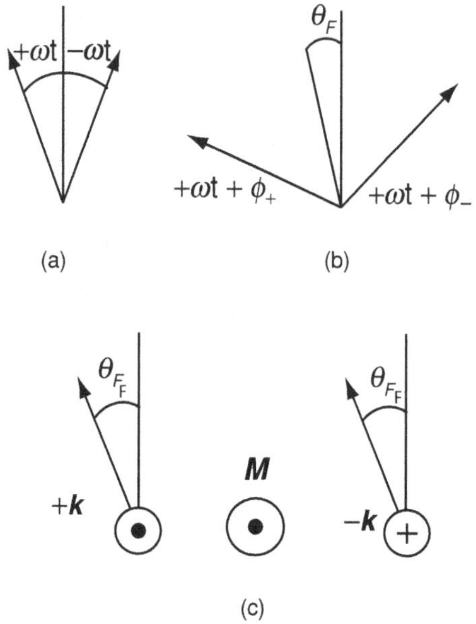

thickness t, the two counter-rotating modes become dephased by an angle $\phi_\pm = \sqrt{(\epsilon\mu_\pm)}\omega t/c$, and the plane of polarization is rotated through an angle $\theta = (\phi_+ + \phi_-)/2$ when it traverses the ferrite, as shown in Fig. 12.19. This is the microwave Faraday effect. Unlike the Faraday effect at optical frequencies, where $\mu = \mu_0$, which is of order 10^{-4} radians for $d = \Lambda$, the rotation at microwave frequencies is of order 1 radian per wavelength, so it is best to place ferrite components inside the waveguides.

The Faraday effect is nonreciprocal in the sense that the direction of rotation of the plane of polarization of the light does not depend on the direction of

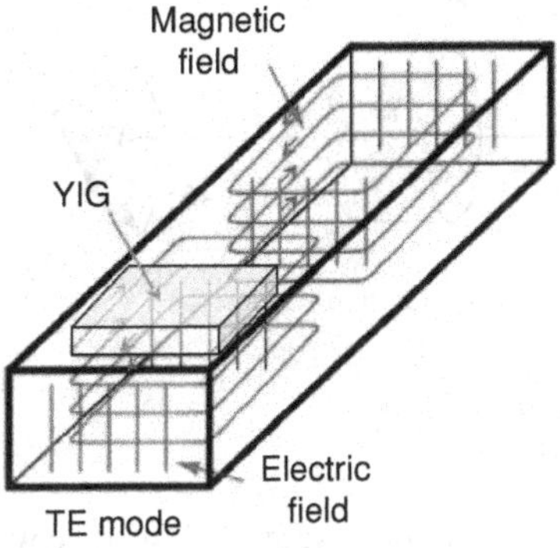

Figure 12.20

A waveguide propagating a TE_{01} mode. Filling the upper half with YIG magnetized vertically absorbs the microwaves for one direction of propagation, but not the other.

propagation of the electromagnetic waves. Reflecting the microwaves back the way they came produces double the rotation, not zero rotation.

We briefly describe devices which depend either on resonant absorption of microwaves, or on the circular birefringence far from resonance which produces the microwave Faraday effect. Microwave ferrite devices are frequently made of YIG; highly perfect polished spherical crystals may have a linewidth as low as 5 ± 1 A m^{-1} for resonance at 10 GHz. Such a sphere has $Q = f_0/\Delta f \approx 10^5$. These very sharp resonances are needed for the microwave filters and oscillators made for communications and measurement systems operating in the 1–100 GHz range.

Microwave devices exploit the difference between μ_+ and μ_- in various ways. Resonant isolators use the resonance at ω_0 to absorb signals reflected back along a waveguide, Fig. 12.20. Phase shifters exploit differences between μ_+ and μ_- above and below the resonance peak. Three- or four-port circulators transmit signals from one port to the next one, while strongly absorbing on other paths. Faraday rotators rotate the plane of polarization of the microwaves.

The isolator protects a signal source from reflections of power back down the waveguide. It operates at resonance, where there is the greatest possible difference in absorption between the + and − modes. The waveguide transmits a TE_{01} mode, where the magnetic field is parallel to the broad faces of the guide and it is circularly polarized in opposite senses in the upper and lower halves of the waveguide. The pattern of field in the guide is reversed when the wave travels in the opposite direction. If the top half of the guide is filled with YIG and a field is applied perpendicular to the broad faces so that its magnetization precesses in the plane of the $\boldsymbol{H}$-field, the wave can pass freely in the forward direction, with attenuation of about 0.3 dB, but it will be strongly alternated, by up to 40 dB in the reverse direction. The attenuation of the power in the forward sense is 7%, but in the reverse sense it is 10^4.

A circulator is a device with three or four ports which delivers the input signal at one port as the output at the next one, while the other ports remain isolated. The principle of the four-port circulator is illustrated in Fig. 12.21. The heart

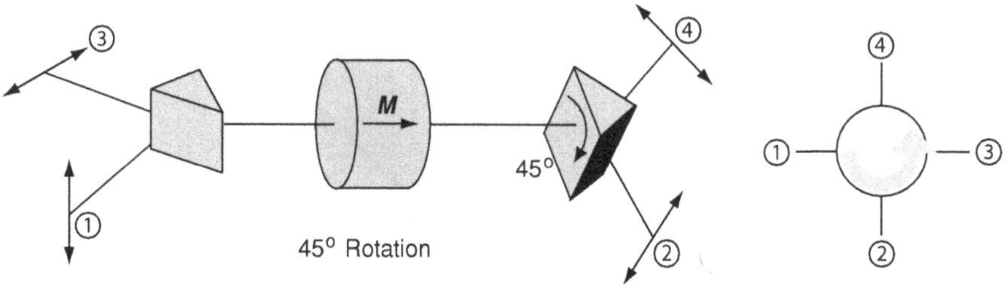

(a) (b)

In / Out	①	②	③	④
①	0	0	0	1
②	1	0	0	0
③	0	1	0	0
④	0	0	1	0

(c)

Figure 12.21

A four-port circulator: (a) illustrates the principle, (b) shows the sense of circulation and (c) is the logic table.

of the device is a YIG disc with thickness and applied field chosen to produce a Faraday rotation of precisely $\pi/4$. The device operates at a frequency where μ_+ and μ_- are sufficiently different to produce the required Faraday rotation, yet far enough from resonance to minimize losses. Two polarization splitters, the microwave equivalent of optical Wollaston prisms, are offset by $\pi/4$ as shown. Each acts to split an incident beam into two perpendicular, linearly-polarized components or conversely to combine them into a single beam. The beam splitting in the microwave device is achieved using single-mode waveguides. A linearly-polarized input at 1 becomes the output at 2. Similarly, an input at 2 becomes an output at 3, and so on. The circulator is a key component of microwave circuits, because it allows a single antenna to be used both for reception and transmission. This was essential for the development of modern telecommunications and radar equipment.

A tunable narrow-band resonant filter is made by winding two orthogonal coils on a small YIG sphere, about 1 mm in diameter. The signal from one coil will only be detected in the other at resonance, where the + mode is circularly polarized. The resonant frequency can be tuned from 1 GHz to more than 10 GHz by changing an externally-applied field.

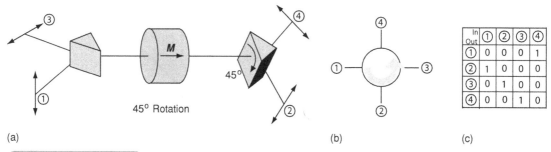

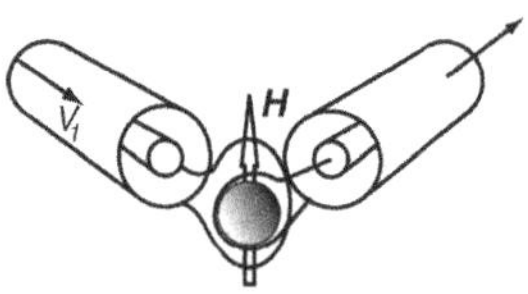

A resonant microwave filter. The device transmits signals in a narrow frequency range around the ferromagnetic resonance frequency f_0.

For higher-frequency, millimetre-wavelength microwave applications such as military phased-array radar, satellite communications or automobile anti-collision radar, the ferromagnetic resonance is not determined by an externally applied magnetic field, but by the anisotropy field of a hard magnet. For $BaFe_{12}O_{19}$, the anisotropy field $\mu_0 H_a = 1.7$ T corresponds to a resonance frequency of 36 GHz when $\mathcal{N} = 1$ (9.17). The resonant frequency can be decreased by scandium substitution or increased by aluminium substitution. These self-biased, miniature high-frequency microwave devices therefore make use of hard magnets, rather than soft magnets.

FURTHER READING

C. W. Chen, *Magnetism and Metallurgy of Soft Magnetic Materials*, New York: Dover Publications (1983). The reprint of this 1977 text provides a wealth of detailed and reliable information on almost every aspect of soft magnets.

R. M. Bozorth, *Ferromagnetism*, Piscataway: Wiley–IEEE Press (1993). The reprint of Bozorth's classic 1951 text remains a reference for metallic materials and electromagnetic applications.

P. Brissonneau, *Magnétisme et matériaux magnétiques pour l'électrotechnique*, Paris: Editions Hermès (1997). A modern text on low-frequency materials and applications.

P. Beckley, *Electrical Steels for Rotating Machines,* London: IEE (2002). A comprehensive treatment of electrical steels; including information on processing and characterization, with comparisons to other materials.

A. Goldman, *Magnetic Components for Power Electronics*, Dordrecht: Kluwer (2002). A monograph on magnetics for power electronics, covering devices, materials, components and design issues.

J. Nicolas, Microwave ferrites, in *Ferromagnetic Materials* Vol 2, E. P. Wohlfarth (editor), Amsterdam: North Holland (1980).

R. F. Soohoo, *Microwave Magnetics*, New York: Harper and Row (1985).

J. Smit and H. P. J. Wijn, *Ferrites*, New York: Wiley (1979). The standard text on ferrites and their high-frequency applications, by pioneers from Philips.

V. G. Harris, *Modern Microwave Ferrites*, IEEE Trans. on Magn. **48**, 1075–1104 (2012).

EXERCISES

12.1 Use the method of dimensions to show that $\delta \propto (\varrho/\mu f)^{1/2}$ and $P_{ed} \propto (ifB_{\max})^2/\varrho$ ((12.2) and (12.4)).

12.2 Deduce the expression for penetration depth (12.2). Estimate how thin a soft metallic film should be if it is to operate at (a) 10 kHz and (b) 1 GHz.

12.3 By considering the field at the apex of a cone of a fully saturated magnetic material produced by the surface magnetic pole density $\mathbf{J} \cdot \mathbf{e}_n$, show that the field is maximum when the half-angle of the cone is $\tan^{-1}\sqrt{2}$.

12.4 Show that the reduction of eddy-current losses in a laminated core scales as $1/N^2$, where N is the number of laminations.

12.5 Estimate (a) the average rate of consumption of electrical energy in watts, per person on Earth, (b) the number of turbogenerators on Earth and (c) the maximum possible diameter of their rotors, if the elastic limit of steel is 700 MPa.

12.6 Show that the Q-factor for an RL circuit is $L\omega/2R$.

13 Applications of hard magnets

Fail better

Samuel Beckett 1906–1989

Permanent magnets deliver magnetic flux into a region of space known as the *air gap,* with no expenditure of energy. Hard ferrite and rare-earth magnets are ideally suited to generate flux densities comparable in magnitude to their spontaneous polarization J_s. Applications are classified by the nature of the flux distribution, which may be static or time-dependent, as well as spatially uniform or nonuniform. Applications are also discussed in terms of the physical effect exploited (force, torque, induced emf, Zeeman splitting, magnetoresistance). The most important uses of permanent magnets are in electric motors, generators and actuators. Their power ranges from microwatts for wristwatch motors to hundreds of kilowatts for industrial drives. Annual production for some consumer applications runs to tens or even hundreds of millions of motors.

The flux density $\boldsymbol{B}_g$ in the airgap (equal to $\mu_0 \boldsymbol{H}_g$) is the natural field to consider in permanent magnet devices because flux is conserved in a magnetic circuit, and magnetic forces on electric charges and magnetic moments all depend on $\boldsymbol{B}$.

A static uniform field may be used to generate torque or align pre-existing magnetic moments since $\boldsymbol{\Gamma} = \mathfrak{m} \times \boldsymbol{B}_g$. Charged particles moving freely through the uniform field with velocity $\boldsymbol{v}$ are deflected by the Lorentz force $\boldsymbol{f} = q\boldsymbol{v} \times \boldsymbol{B}_g$, which causes them to move in a helix, turning with the cyclotron frequency (3.26) $f_c = qB/2\pi m$. When the charged particles are electrons confined to a conductor of length l constituting a current I flowing perpendicular to the field, the Lorentz force leads to the familiar expression $f = B_g Il$. This is the basis of operation of electromagnetic motors and actuators. Conversely, moving a conductor through the uniform field induces an emf given by Faraday's law $\mathcal{E} = -\mathrm{d}\Phi/\mathrm{d}t$, where Φ $(= \boldsymbol{B}_g \cdot \mathcal{A})$ is the flux threading the circuit of which the conductor forms a part.

Spatially nonuniform fields offer another series of useful effects. They exert a force on a magnetic moment given by the energy gradient $\boldsymbol{f} = \nabla(\mathfrak{m} \cdot \boldsymbol{B}_g)$. They also exert nonuniform forces on moving charged particles, which can be used to focus ion or electron beams or generate electromagnetic radiation from electron beams passing through the inhomogeneous field, as in a synchrotron wiggler, for example. The ability of rare-earth permanent magnets to generate a complex flux pattern with rapid spatial fluctuations ($\nabla B_g > 100$ T m^{-1}) is

Table 13.1. Classification of permanent-magnet applications

Field	Physical effect	Type	Application
Uniform	Zeeman splitting	Static	Magnetic resonance imaging
	Torque	Static	Magnetic powder alignment
	Hall effect, magnetoresistance	Static	Proximity sensors
	Force on conductor	Dynamic	Motors, actuators, loudspeakers
	Induced emf	Dynamic	Generators, microphones
Nonuniform	Forces on charged particles	Static	Beam control, radiation sources (microwave, UV, X-ray)
	Force on paramagnet	Dynamic	Mineral separation
	Force on iron	Dynamic	Holding magnets
	Force on magnet	Dynamic	Bearings, couplings, maglev
Time-varying	Various	Dynamic	Magnetometry
	Force on iron	Dynamic	Switchable clamps
	Eddy currents	Dynamic	Brakes, metal separation

unrivalled by any electromagnet. Remember that a magnet with $J \approx 1\,\mathrm{T}$ is equivalent to an Ampèrian surface current of 800 kA m^{-1}. Solenoids, whether resistive or superconducting, would have to be several centimetres in diameter to accommodate the requisite number of ampere-turns, whereas blocks of rare-earth or ferrite magnets of any size can be assembled in any desired orientation as close to each other as necessary.

Time-varying fields are produced by displacing or rotating the permanent magnets, or by moving soft iron in the magnetic circuit. A field which varies in time can be used to induce an emf in an electric circuit according to Faraday's law or to produce eddy currents in a static conductor and exert forces on those currents. If it is spatially nonuniform, it can exert a time-dependent force on a magnetic moment or a particle beam. Applications include magnetic switches and apparatus for measurements of physical properties as a function of field. The applications of permanent magnets are summarized in Table 13.1.

Other uses of magnets, in agriculture, acupuncture, pain control, rejuvenation, suppression of wax formation in oilwells or control of nucleation of limescale deposits in water pipes defy classification. Some are undoubtedly fanciful, like the persistent myth that a magnet can be beneficial or harmful depending on whether its North or South pole is presented to the patient.[1] Others are good areas for scientific investigation. A closed mind learns nothing. Figure 13.1 provides an overview of permanent magnet applications.

[1] Details of medical applications imaginatively claimed for magnets can be found in the 1811 treatise *Materia Medica Pura* by Samuel Hennemann, the father of homeopathy.

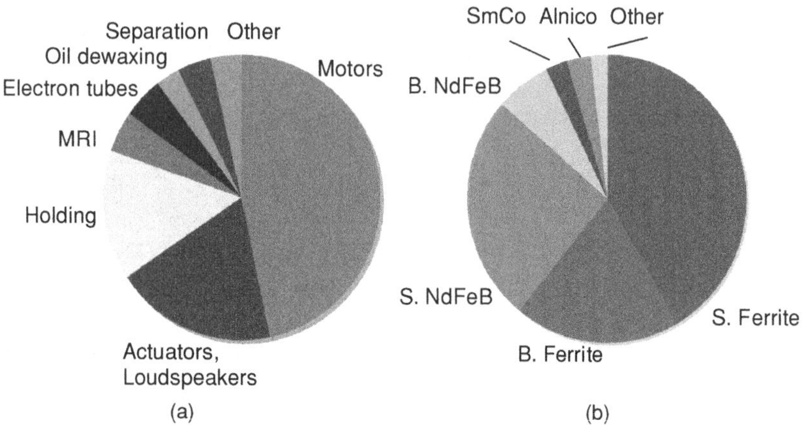

Figure 13.1
Overview of permanent magnet production by materials (a), and applications (b). The pie represents an annual market of about 6 b$.
S- sintered
B- bonded

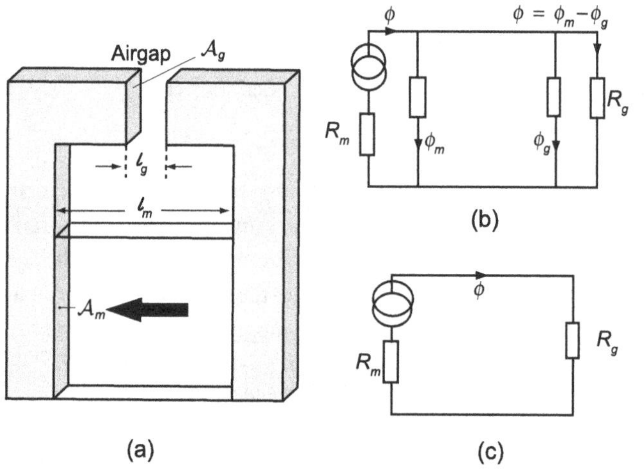

Figure 13.2
(a) A simple magnetic circuit and its electrical equivalents, (b) with and (c) without flux loss.

Permanent magnets are conventionally shown in magnetic circuits as transparent or unshaded, with a solid arrow to indicate the direction of magnetization. Soft iron is shaded.

13.1 Magnetic circuits

A magnetic circuit comprises a magnet and an airgap with soft iron to guide the flux. When flux from a permanent magnet of length l_m and cross section $\mathcal{A}_m$ is guided through soft magnetic material into an airgap of length l_g and cross section $\mathcal{A}_g$, as in the circuit of Fig. 13.2(a), $\nabla \cdot \boldsymbol{B} = 0$ gives

$$B_m \mathcal{A}_m = B_g \mathcal{A}_g, \tag{13.1}$$

provided there is no flux leakage. We make the simplifying, if unrealistic, assumptions that B_m and B_g are constants, and that the soft material in the circuit is ideal insofar as it has infinite permeability ($H = 0$). If H_m is the value of H in the magnet, Ampère's law $\oint \boldsymbol{H} \cdot \mathrm{d}\boldsymbol{l} = 0$ then gives

$$H_m l_m = -H_g l_g. \tag{13.2}$$

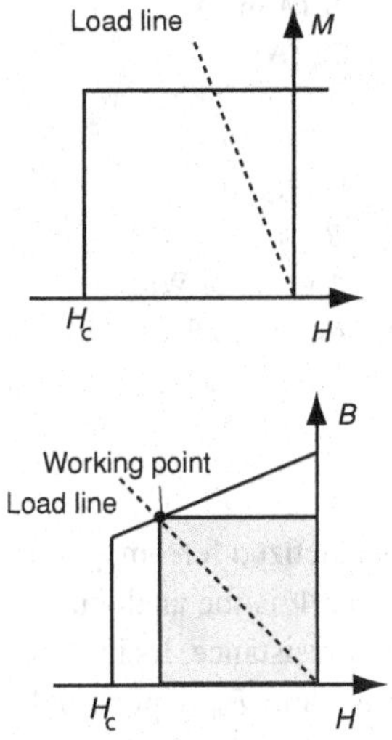

Second-quadrant $\mu_0 M(H)$ and $B(H)$ for an ideal permanent magnet. The working point and load line for maximum energy product are indicated.

Multiplying (13.1) and (13.2), and using the fact that $B_g = \mu_0 H_g$,

$$(B_m H_m) V_m = -B_g^2 V_g / \mu_0, \tag{13.3}$$

where V_m, V_g represent the volume of the magnet and the airgap, respectively. The flux density in the gap is therefore maximized when the product $B_m H_m$ is maximum; hence the emphasis on energy product as a figure of merit for permanent magnets. Dividing (13.1) by (13.2), gives the equation for the load line $B_m(H_m)$ whose negative slope is the permeance coefficient P_m:

$$\frac{B_m}{H_m} = -\mu_0 \frac{A_g l_m}{A_m l_g}. \tag{13.4}$$

The working point is where the load line intersects the B:H loop. It depends on the dimensions of the magnet and the airgap.

As materials improved, magnets grew shorter and fatter to work nearer their $(BH)_{max}$ point. This point for an isolated magnet with an ideal square magnetization loop can be calculated by maximizing $\mu_0(H_m + M)H_m$ with respect to the shape of the magnet, represented by its demagnetizing factor $\mathcal{N}$. The square $M(H)$ loop corresponds to a linear second-quadrant $B(H)$ variation with slope μ_0. Using $H_m = -\mathcal{N} M$, the result is that the optimum value of $\mathcal{N}$ is exactly $\frac{1}{2}$. Squat cylinders have replaced the bars and horseshoes of yesteryear! The working point of this ideal magnet, which has second-quadrant magnetization equal to its remanence, is at $B = B_r/2$, $H = -M_r/2$. Its permeance is μ_0, and the maximum energy product is $\mu_0 M_r^2/4$. Practical magnets have less-than-perfect loops, and their energy product always fall short of this upper limit.

Inevitably, there are flux losses in the circuit, so a factor β is introduced on the left-hand side of (13.1). Furthermore, the soft material will be less than ideal so a factor α is introduced on the left-hand side of (13.2). Typically, α is in the range 0.7–0.95, but β may be anywhere from 0.2 to 0.8. When designing a magnetic circuit, it is always good practice to place the magnets as close as possible to the airgap in order to reduce flux losses.

The design of magnetic circuits is an art, facilitated by computer simulation. A formal analogy between magnetic and electric circuits, set out in Table 13.2, is helpful. Ampère's law $\oint \boldsymbol{H} \cdot \mathrm{d}\boldsymbol{l} = 0$ in the absence of conduction currents corresponds to the electric field result $\oint \boldsymbol{E} \cdot \mathrm{d}\boldsymbol{l} = 0$ in the absence of changing magnetic fields. Corresponding continuity equations are $\nabla \cdot \boldsymbol{B} = 0$ and $\nabla \cdot \boldsymbol{j} = 0$, where $\boldsymbol{j}$ is the electric current density. Magnetic potential difference defined as $\varphi_{ab} = \int_a^b \boldsymbol{H} \cdot \mathrm{d}\boldsymbol{l}$ is measured in amperes, using the potential coil illustrated in Fig. 10.24. The permanent magnet behaves like a battery. It is the source of magnetomotive force (mmf), the magnetic analogue of emf, because it is the only segment of the circuit where the magnetic potential rises; $\boldsymbol{H}$ is oppositely directed inside the magnet and outside it. The magnetic 'battery' is energized

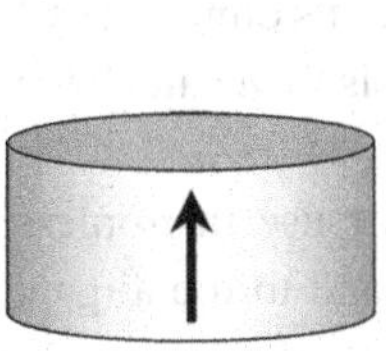
The shape of an optimized permanent magnet cylinder. The demagnetizing factor $\mathcal{N} = \frac{1}{2}$.

Table 13.2. Analogy between electric and magnetic circuits

	Electric	Magnetic
Field	$\boldsymbol{E}$ (V m^{-1})	$\boldsymbol{H}$ (A m^{-1})
Potential	φ_e (V)	φ_m (A)
Current/flux density	$\boldsymbol{j}$ (A m^{-2})	$\boldsymbol{B}$ (T or Wb m^{-2})
Potential difference	$\varphi_e = \int \boldsymbol{E} \cdot d\boldsymbol{l}$	$\varphi_m = \int \boldsymbol{H} \cdot d\boldsymbol{l}$
Continuity condition	$\nabla \cdot \boldsymbol{j} = 0$	$\nabla \cdot \boldsymbol{B} = 0$
Linear response law	$\boldsymbol{j} = \sigma \boldsymbol{E}$	$\boldsymbol{B} = \mu \boldsymbol{H}$
Current/flux	I (A)	Φ (T m^2 or Wb)
Resistance/reluctance	$R = \varphi_e / I$ (Ω)	$R_m = \varphi_m / \Phi$ (A Wb^{-1})
for a cylinder of section $\mathcal{A}$ and length l	$R = l/\mathcal{A}\sigma$	$R_m = l/\mathcal{A}\mu$
Conductance/permeance	$G = 1/R$	$P_m = 1/R_m$

as the last step in magnet manufacture, when the unmagnetized ferromagnetic component is polarized in a pulsed field.[2] Magnetic flux Φ is the analogue of current, and reluctance R_m $(= \varphi/\Phi)$ is the equivalent of resistance. Its inverse, permeance P_m is the analogue of conductance. Both R_m and P_m depend only on the physical dimensions of the circuit element. For instance, the reluctance of a short airgap is

$$R_g = \varphi_g/\Phi_g = l_g/\mu_0 A_g, \tag{13.5}$$

where φ_g is the potential drop across the gap. The magnet has an internal reluctance determined by its working point on the $B{:}H$ curve; $R_m = \varphi_m/\Phi_m$. The equivalent circuit of Fig. 13.2(b) illustrates the principle of matching the airgap reluctance to the desired working point of the magnet, which is normally the $(BH)_{\max}$ point. Figure 13.2(c) shows the equivalent circuit allowing for nonzero reluctance of the soft segments and flux losses.

Flux losses in magnetic circuits are much more severe than current losses in electrical circuits, because the relative permeability of iron $\mu_r \approx 10^3$–10^4 is much less than the conductivity of copper relative to that of air. There are many good electrical insulators, but the only true magnetic insulators are type I superconductors, for which $\mu = 0$. An approximate solution for the flux produced by a magnetic circuit may be achieved by dividing it into segments, and attributing a standard permeance to each segment according to its dimensions.[3] The permeances are added in series or parallel, and the flux is then calculated from the mmf.

Rare-earth permanent magnets are particularly suited for use in ironless circuits, where flux is confined to the magnets themselves and to the airgap.

[2] Claims of magnetic 'perpetual motion' machines appear from time to time. Some depend on drawing down the energy stored in a magnet. The energy product is reduced if the magnet does irreversible work. Such devices have an unadvertised expiry date!

[3] Tables of standard permeances are found in textbooks such as the chapter by Leupold in Coey (1996).

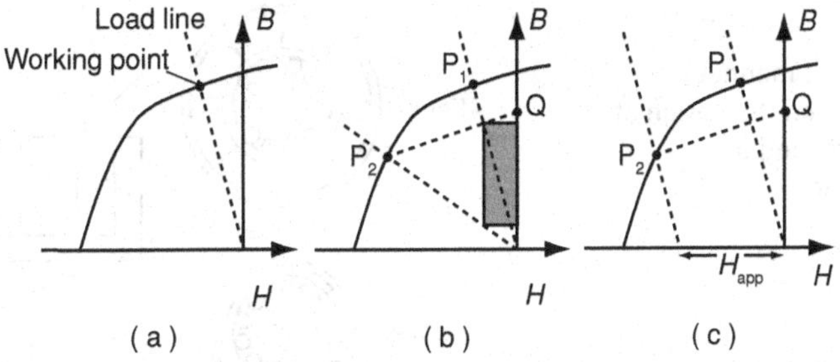

Figure 13.3

Hysteresis loops showing working points for: (a) a static application, (b) a dynamic application with mechanical recoil and (c) a dynamic application with active recoil.

Flux concentration is achieved whenever the flux density in the airgap exceeds the remanent induction of the magnet, $B_g/B_r > 1$.

13.1.1 Static and dynamic operation

Applications are classified as *static* or *dynamic* according to whether the working point in the second quadrant of the hysteresis loop is fixed or moving (Fig. 13.3). The position of the working point reflects the internal field H_m, which depends in turn on the magnet shape, airgap and any fields that may be generated by electric currents. The working point changes whenever magnets move relative to each other, when the airgap changes or when time-varying currents are present. On account of their square loops, oriented ferrite and rare-earth magnets are particularly well suited for dynamic applications that involve changing flux density. Ferrites and bonded metallic magnets also minimize eddy-current losses.

For mechanical recoil, the airgap changes from a narrow one with reluctance R_1 to a wider one with greater reluctance R_2. After several cycles, the working point follows a stable trajectory, represented by the line P_2Q in Fig. 13.3(b) whose slope is known as the recoil permeability μ_R. Recoil permeability is 2–6 μ_0 for alnicos, but it is barely greater than μ_0 for oriented ferrite and rare-earth magnets. The shaded area in Fig. 13.3(b) is a measure of the useful recoil energy in the gap, the recoil product $(BH)_u$, which will always be less than $(BH)_{max}$, but approaches this limit in materials with a broad square loop and a recoil permeability close to μ_0.

Active recoil occurs in motors and other devices where the magnets are subject to an $\boldsymbol{H}$-field during operation as a result of currents in the copper windings. The field is greatest at startup, or in the stalled condition. Active recoil involves displacement of the permeance line along the H-axis.

13.2 Permanent magnet materials

The market for hard magnetic materials is nowadays dominated by two families of materials, the hexagonal ferrites and Nd–Fe–B (Fig. 13.1(a)). Their production volumes and costs are quite different, but each holds roughly half the market.

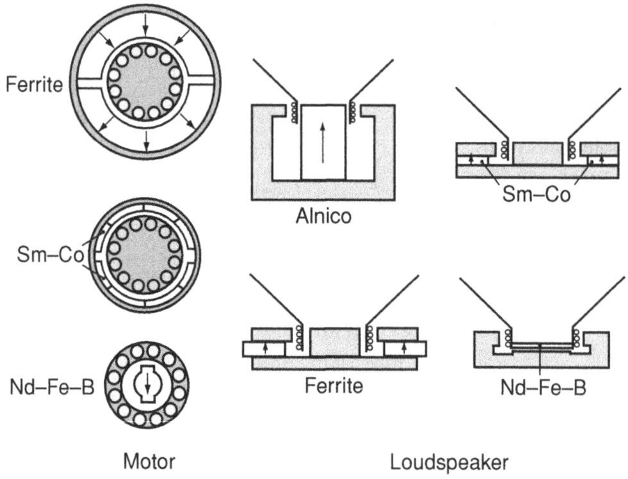

Figure 13.4

Influence of magnet properties on the design of a DC motor and a loudspeaker.

Ferrites are produced throughout the world, but Nd–Fe–B production is concentrated in China, where there are abundant rare-earth reserves. Most magnets are produced as sintered blocks or other simple shapes, but increasing quantities of polymer-bonded material, suitable for making complex shapes by injection moulding, are manufactured.

The maximum energy product doubled roughly every 12 years over the course of the twentieth century (Fig. 1.13). Consequences for permanent magnet device design have been dramatic, as illustrated in Fig. 13.4. The devices shrink as higher energy-product magnets become available, their configuration changes and the number of parts can be reduced. The advantage of magnets over coils in small structures can be appreciated by comparing a small disc-shaped magnet with a coil having the same magnetic moment. A magnetized cylinder with diameter 8 mm and height 2 mm made of a material with $M =$ 1 MA m^{-1} has $\mathfrak{m} \approx 0.1$ A m^2. The equivalent current loop, $\mathfrak{m} = I\mathcal{A}$ would require 2000 ampere-turns, an impossible demand in such a small space!

There has been no further doubling of $(BH)_{\text{max}}$ since 1996, and there seems to be little scope for further dramatic improvements of bulk material, barring the development of practical oriented hard/soft nanocomposites.

Properties of some typical magnets are shown in Table 13.3. The data are for dense oriented magnets, made from sintered uniaxial ceramic ferrite or metallic alloy powder. These magnets have a microstructure in which the c axes of the individual crystallites are aligned by applying a magnetic field during processing. All except alnico are true permanent magnets, in the sense that the magnetic hardness parameter κ, defined by (8.1), is greater than unity. The $(BH)_{\text{max}}$ values achieved fall short of the theoretical maximum $\mu_0 M_r^2/4$ because of nonideal loop shape. Two values of coercivity are listed, one is the 'intrinsic' coercivity ${}_iH_c$ measured on the $M(H)$ loop, the other, ${}_BH_c$, is the coercivity measured on the $B(H)$ loop.

Table 13.3. Properties of commercial oriented magnets

	$\mu_0 M_r$ (T)	J_s (T)	${}_iH_c$ (kA m^{-1})	${}_BH_c$ (kA m^{-1})	$(BH)_{max}$ (kJ m^{-3})	$\mu_0 M_r^2/4$ (kJ m^{-3})
$SrFe_{12}O_{19}$	0.42	0.47	275	265	34	35
Alnico 5	1.25	1.40	54	52	43	310
$SmCo_5$	0.88	0.95	1700	700	150	154
Sm_2Co_{17}[a]	1.08	1.15	1100	800	220	232
$Nd_2Fe_{14}B$	1.34	1.54	1000	900	350	359

[a] Intergrown with 1:5 phase

The world's first rare-earth permanent magnets were based on $SmCo_5$. This compound has huge uniaxial anisotropy and a high Curie temperature, but its magnetization limits the achievable energy product (§11.3.2). A family of pinning-type magnets based on Sm_2Co_{17} with iron and other additions offers better energy products. However, cobalt is expensive and supplies have been uncertain, leading to the severe price fluctuations mentioned in §11.2.5. The magnets with the best properties are now made of sintered $Nd_2Fe_{14}B$ with a high degree of crystallite orientation and the minimum amount of secondary phases. Remanence of 1.55 T and energy products of 470 kJ m^{-3} have been achieved, which approach the theoretical maxima for the $Nd_2Fe_{14}B$ phase ($\mu_0 M_s =$ 1.62 T, $\frac{1}{4}\mu_0 M_s^2 = 525\,\text{kJ}\,\text{m}^{-3}$). Premium grades containing Dy or Tb have coercivity as high as 2.0 MA m^{-1}. They resist recoil to higher temperature, at the expense of remanence, energy product and magnet cost. Both Dy and Tb are heavy rare-earths which couple with their moment antiparallel to iron. Dy is several times as expensive as Nd, but it may prove indispensable for the magnets used in motors for electric vehicles. Co additions help by increasing the Curie temperature, which is rather low for some applications, but it degrades the anisotropy.

Ferromagnetic microstructures are illustrated in Fig. 13.5. Sometimes the additional process steps needed to produce an oriented magnet cannot be justified economically. It may be cheaper to make more of an isotropic magnet with inferior magnetic properties. The c axes of the individual crystallites are random and, if interactions between crystallites are neglected, the remanence is $M_s\langle\cos\theta\rangle = \frac{1}{2}M_s$, where θ is the angle between the c axis and the direction of magnetization. Components made of ceramic or sintered metal are shaped by slicing or grinding. Polymer bonding allows for more versatility in shaping magnetic components. A coercive powder is mixed with a binder, with a fill factor $\mathfrak{f}_m$ of 60–80 vol%. The mixture is die-pressed, injection moulded, extruded or rolled into the required shape. The c axes of the crystallites in the powder may be oriented in the binder by using a magnetic field in order to augment the remanence. Since M_r is now $\mathfrak{f}_m M_s\langle\cos\theta\rangle$, where θ is the angle between the c axis and the aligning field, even a fully aligned powder

Table 13.4. Properties of sintered and bonded ferrite magnets

	M_r (kA m^{-1})	H_c (kA m^{-1})	$(BH)_{max}$ (kJ m^{-3})
Intrinsic $SrFe_{12}O_{19}$	380		45[a]
Oriented, sintered	330	270	34
Isotropic, sintered	180	310	9
Oriented, bonded[b]	240	245	16
Isotropic, bonded[c]	100	180	5

[a] Theoretical maximum.
[b] Injection moulded.
[c] Rubber bonded.

Figure 13.5

Third-quadrant $B(H)$ curves for sintered magnets.

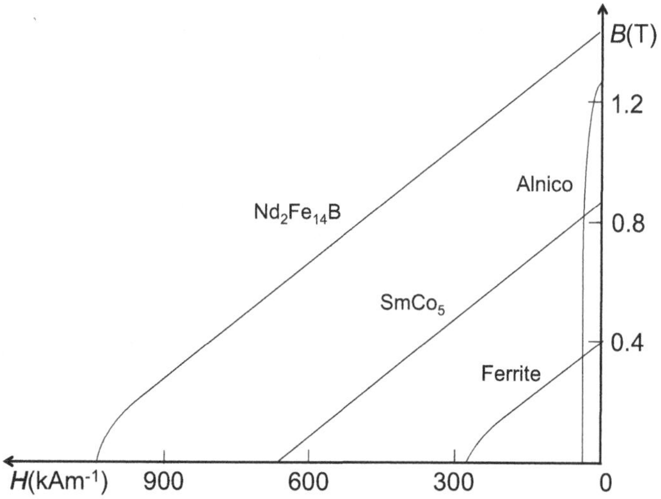

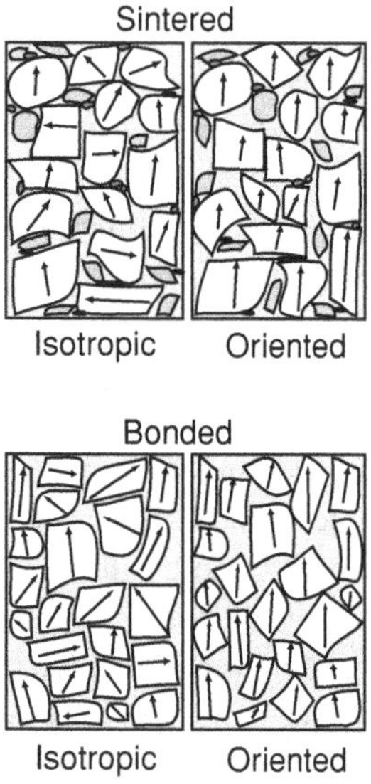

Schematic microstructure of permanent magnets.

with $\mathfrak{f}_m = 0.7$ has an energy product that is less than half of the bulk value. Table 13.4 illustrates the properties of differently processed magnets based on $SrFe_{12}O_{19}$.

The magnets in electrical machines can be subject to temperatures in excess of 100 °C during routine operation. Magnetization and coercivity naturally decline as the Curie point is approached. Temperature coefficients around ambient temperature are listed in Table 13.5. Not all the loss is recoverable on returning to ambient temperature; irreversible losses are associated with thermal cycling. Maximum service temperatures for different materials are included in the table.

Ferrites and rare-earth magnets with wide, square hysteresis loops have the property that the field of one magnet does not significantly perturb the magnetization of a neighbouring magnet. This is because the longitudinal susceptibility is zero for a square hysteresis loop, and the transverse susceptibility is M_s/H_a,

Table 13.5. Mechanical, electrical and thermal properties of permanent magnets

	d (kg m^{-3})	α (10^{-6} °C^{-1})	ϱ (μΩ m)	dM_s/dT (% °C^{-1})	dH_c/dT (% °C^{-1})	T_{max} (°C)
$SrFe_{12}O_{19}$ sintered	4300	10	10^8	−0.20	0.45	250
$SrFe_{12}O_{19}$ bonded	3600			−0.02	0.45	150
Alnico 5 cast	7200	12	0.5	−0.02	0.03	500
$SmCo_5$ sintered	8400	11	0.6	−0.04	−0.02	250
Sm_2Co_{17}[a] sintered	8400	10	0.9	−0.03	−0.20	350
$Nd_2Fe_{14}B$ sintered	7400	−2	1.5	−0.13	−0.60	160
$Nd_2Fe_{14}B$ bonded	6000		200	−0.13	−0.06	150

[a] Intergrown with 1:5 phase.
α = thermal expansion, ϱ = resistivity.

which is only of order 0.1, since the anisotropy field $H_a = 2K_1/\mu_0 M_s$ is much greater than the magnetization. Rigidity of the magnetization means that the superposition of the induction of rare-earth permanent magnets is linear and the magnetic material is effectively transparent, behaving like vacuum with permeability μ_0. Transparency and rigidity of the magnetization greatly simplify the design of magnetic circuits. There is no need to worry about the effect of one magnetic segment on another.

Magnet cost is often critical for applications. Ferrite and Nd–Fe–B magnets are both produced in large quantities – roughly 1 million tonnes of ferrite and 80 000 tonnes of Nd–Fe–B in 2010. The properties of the two are quite different, and numerous grades of Nd–Fe–B are available, featuring high coercivity or high remanence, with energy products ranging from 250 to 450 kJ m^{-3}. A very rough guide to cost is \$1 per joule of stored energy. Hence, from the data in Tables 13.3 and 13.4, the costs of oriented ferrite and Nd–Fe–B work out at about \$7 kg^{-1} and \$40 kg^{-1}, respectively. Ferrite is even cheaper than this but the cost of sintered Nd-Fe-B is about twice that estimate.

13.3 Static applications

Modern permanent magnets are ideally suited to generating magnetic fields which are comparable to, or even somewhat greater than, their remanence. These fields may be uniform or nonuniform.

13.3.1 Uniform fields

The magnetic field produced by a point dipole of moment $\mathfrak{m}$ A m^2 is nonuniform and anisotropic. It falls off as $1/r^3$. In polar coordinates with $\mathfrak{m}$

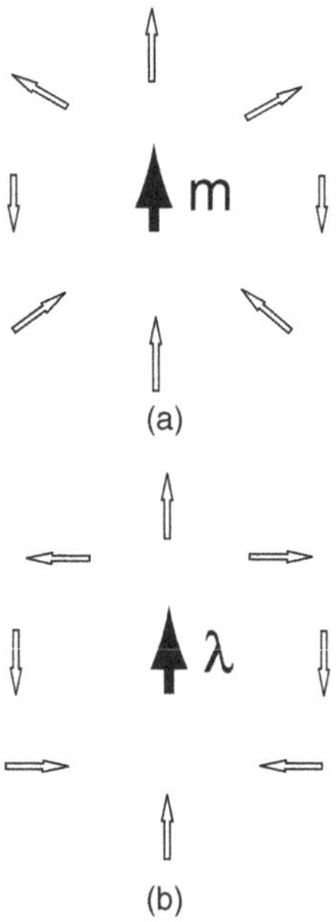

Comparison of the magnetic field produced by: (a) a point dipole m and (b) a line dipole λ.

along Oz,

$$B_r = \frac{2\mu_0 \mathfrak{m}}{4\pi r^3}\cos\theta, \qquad B_\theta = \frac{\mu_0 \mathfrak{m}}{4\pi r^3}\sin\theta, \qquad B_\phi = 0. \tag{2.10}$$

Nonetheless, a uniform field can be achieved by assembling segments of magnetic material, each magnetized in a different direction, such that their individual contributions combine to yield a field which is uniform over some region of space. The field of a magnetic dipole has the important property of scale independence. Provided the magnetization is rigid, the field generated is simply related to the remanence of the magnets via a scale-independent geometric factor $\mathcal{K}$:

$$B_g = \mathcal{K} B_r. \tag{13.6}$$

On reducing the dimensions by a scale factor ξ, the moment $\mathfrak{m}$ is reduced by a factor ξ^3, but the field which varies as r^{-3} is increased in the same proportion. Hence, the factor $\mathcal{K}$ depends on the relative dimensions, but not on the absolute size of the magnetic circuit. Scalabiliy is the secret of the success of magnetic technology.

It is not obvious how to combine dipoles to create a uniform field. One approach is to build the structure from long segments whose fields approximate those of extended dipoles with dipole moment per unit length of λ A m. The field of such a line dipole, which runs parallel to Ox and is magnetized parallel to Oz, is (2.22)

$$B_r = \frac{\mu_0 \lambda}{4\pi r^2}\cos\theta, \qquad B_\theta = \frac{\mu_0 \lambda}{4\pi r^2}\sin\theta, \qquad B_{\phi=\pi/2} = 0. \tag{13.7}$$

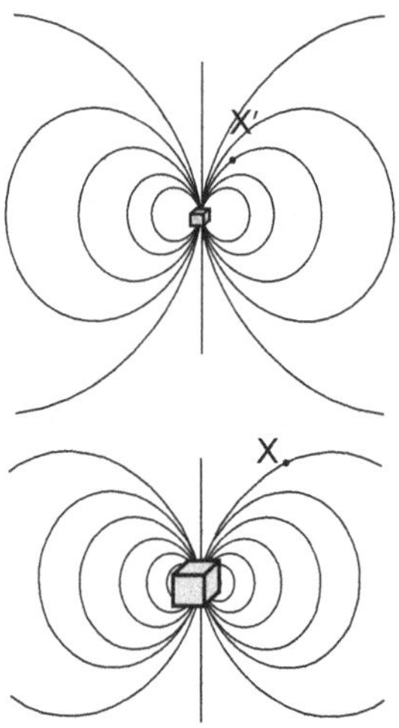

The dipole field of a block of magnetic material is scale-independent. Fields at X and X′ are the same.

The magnitude of $\boldsymbol{H}(\boldsymbol{r},\theta)$, $\sqrt{H_r^2 + H_\theta^2}$, is now independent of θ and its direction makes an angle 2θ with the orientation of the magnet as shown in the figure.

By assembling a cylindrical magnet with a hollow bore from long cylindrical segments, it is possible to generate a uniform field in a central region surrounded by magnets. Choosing the orientation of the segments appropriately, the fields all add in the centre, and they may cancel completely outside. The segments can be assembled according to many different prescriptions; some of the designs are shown in Fig. 13.6. The efficiency ε of any particular structure is defined as the ratio of the energy stored in the airgap $\frac{1}{2}\int \mu_0 H_g^2 \mathrm{d}^3 r$ to the most that can be stored in the magnets $(1/2\mu_0)\int B_r^2 \mathrm{d}^3 r$. In the case of an extended two-dimensional structure producing a uniform field in a cavity, the definition leads to

$$\varepsilon = \mathcal{K}^2 \mathcal{A}_g / \mathcal{A}_m, \tag{13.8}$$

where $\mathcal{A}_g$ is the cross sectional area of the cavity and $\mathcal{A}_m$ is the cross sectional area of the magnets. A reasonably efficient structure has $\varepsilon \approx 0.1$. The theoretical upper limit is $\frac{1}{4}$.

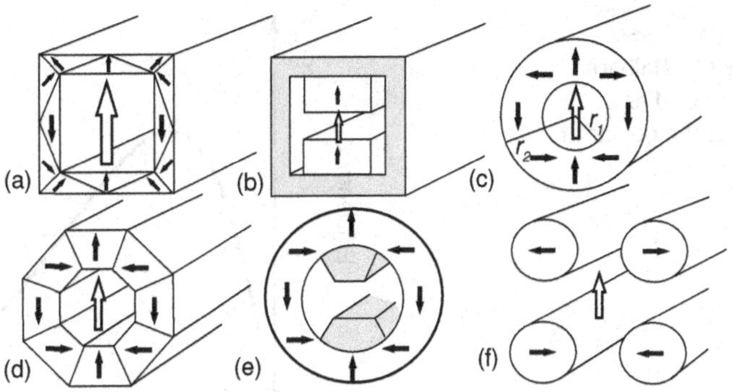

Figure 13.6

Cross sections of some permanent magnet structures discussed in the text which produce a uniform transverse field in the direction indicated by the hollow arrow. Magnets are unshaded, with an arrow to show the direction of magnetization. Soft iron is shaded.

The open square cylinder of Fig. 13.6(a) has an equipotential outer surface, and therefore produces no external field, provided the dimensions $t/r = \sqrt{2} - 1$, in which case the flux density in the air gap (the square bore) is $0.293B_r$. Multiples of this field could be achieved by nesting similar structures inside one other. This design, or a design with flat cuboid magnets and a soft iron yoke (Fig. 13.6(b)) is used to for NMR and for MRI, which combines NMR with sophisticated signal processing to yield two-dimensional tomographic images of solid objects as discussed in §15.4.5. Permanent magnet flux sources typically supply fields of order 0.3 T with homogeneity of 1 part in 10^5 for whole-body scanners. Shimming to compensate for any imperfections in the magnets is done by placing small dipoles or pieces of soft iron in appropriate places around the airgap. The fields generated by permanent magnets are lower than those of competing superconducting solenoids, but there is no need for any cryogenic installation. These systems are popular in Japan. As well as the medical applications, proton resonance is increasingly applied for quality control in the food and drug industries,

Figure 13.6(c) shows the ideal Halbach cylinder, introduced in §2.2.2, where the direction of magnetization of any segment at angular position ϑ in the cylinder is oriented at 2ϑ to the vertical axis. According to (13.7), all segments now combine to create a uniform field across the airgap in the vertical direction. Unlike the structure of Fig. 13.6(a), the radii r_1 and r_2 can take any values without creating a stray field outside the cylinder. The flux density in the airgap of this ingeneous structure is

$$B_g = B_r \ln(r_2/r_1), \tag{13.9}$$

so the geometric factor $\mathcal{K}$ is $\ln(r_2/r_1)$. Efficiency is greatest ($\varepsilon = 0.16$) when $r_2/r_1 = 2.2$. A continuously varying magnetization pattern is not easy to realize. In practice, it is convenient to assemble the device from N trapezoidal segments, as illustrated in Fig. 13.6(d) for $N = 8$. In that case, a factor $[\sin(2\pi/N)]/(2\pi/N)$ must be included on the right-hand side of (13.9). Finite

Figure 13.7

Efficiency of a Halbach cylinder plotted as a function of the geometric factor $\mathcal{K} = B_g/B_r$.

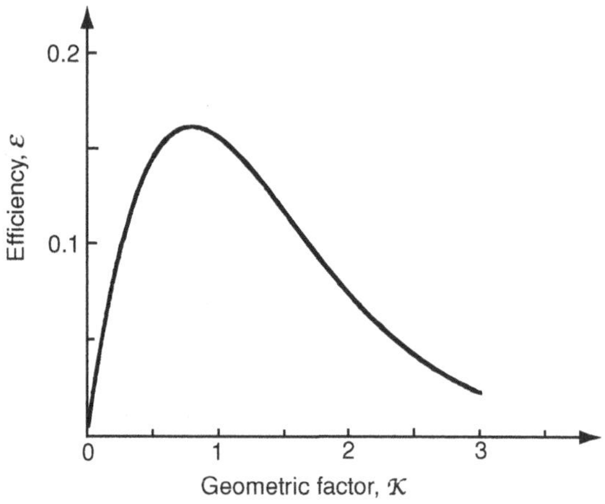

length 2 z_0 reduces the airgap flux density by a further amount

$$\left\{(z_0/2)\left[\frac{1}{z_1} - \frac{1}{z_2}\right] + \ln\left[\frac{z_0 + z_2}{z_0 + z_1}\right]\right\} B_r,$$

where $z_i = \sqrt{(z_0^2 + r_i^2)}$

$\mathcal{K} > 1$ in (13.6) corresponds to flux concentration. There is no limit in principle to the magnitude of the field that could be produced by a permanent magnet array, but in practice the coercivity and anisotropy field of the magnets sets a limit on ultimate performance. Because of the logarithmic dependence in (13.9) and the high cost of rare-earth magnets, it becomes uneconomic to use permanent magnets to generate magnetic fields that exceed about twice the remanence. The efficiency of a Halbach cylinder is plotted in Fig. 13.7.

Soft iron can be introduced into a permanent magnet circuit, either to provide a cheap return path for the flux (Fig. 13.6(b)) or to concentrate the flux in the airgap, thereby creating a larger field in a smaller volume (Fig. 13.6(e)). The extra flux density can never exceed the polarization of the soft material (e.g. 2.15 T for iron or 2.45 T for permendur), and will normally be only a fraction of that.

Large Halbach-type arrays weighing several tonnes which produce a field of order 1 T are used for magnetic annealing to set the exchange bias during manufacture of spin-valve sensors and magnetic random access memory on wafers up to 300 mm in diameter (Fig. 13.8).

A further simplification of the basic structure of Fig. 13.6(c) is the magic mangle shown in Fig. 13.6(f), which uses a few transversely magnetized rods aligned around a central bore following the 2ϑ rule. This design affords access to the field from different directions. The field at the centre for N rods which are just touching is

$$B_g = (N/2)B_r \sin^2(\pi/N). \tag{13.10}$$

When $N = 4$, $\mathcal{K} = 1$.

Figure 13.8

Large Halbach array used for magnetic annealing. (Courtesy of Magnetic Solutions Ltd.)

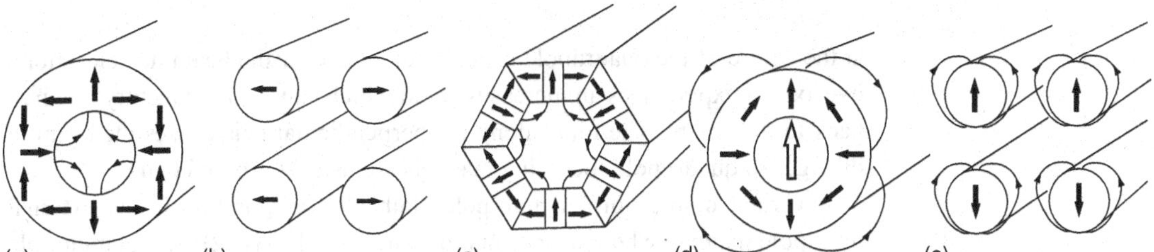

Figure 13.9

Some cylindrical magnet structures which produce nonuniform fields: (a) and (b) give a quadrupole field, (c) a hexapole field, (d) an external quadrupole and (e) a uniform field gradient.

Uniform fields can also be generated in spherical cavities using the same principles as for cylindrical cavities. A uniform field is generated in a spherical cavity when the magnitude of the polarization of a volume element at (r, ϑ, φ) in a hollow, spherical magnet is kept constant, provided its orientation varies as $(2\vartheta, \varphi)$. The expression (13.10) then acquires a factor $\frac{4}{3}$ on the right-hand side. The ultimate limit to the field that can reasonably be generated with permanent magnets is about 5 T, which has been achieved in a spherical volume of a cubic centimetre.

13.3.2 Nonuniform fields

Nonuniform fields are useful for particle beam control and focussing electrons in cathode ray tubes and other electro-optic machines. They also serve to generate microwaves and other radiation, and to exert forces in bearings, couplings, suspensions and magnetic separators. The cylindrical configurations of Fig. 13.6 may be modified to generate transverse multipole fields. Figure 13.9(a) shows an ideal quadrupole source and Fig. 13.9(b) shows a simplified vesion. Higher multipole fields are obtained by having the magnetic orientation of the segments in the ring vary as $(1 + (\nu/2))\vartheta$, where $\nu = 2$ for a dipole field, $\nu = 4$ for a quadrupole, $\nu = 6$ for a hexapole and so on. The practical hexapole shown in Fig. 13.9(c) can be used to produce ions from a plasma contained in a magnetic bottle. Halbach's original cylindrical magnets actually produced a quadrupole field for focussing beams of charged particles. The field

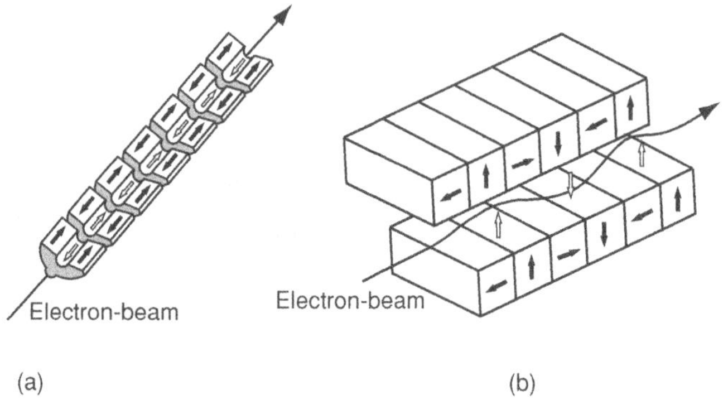

Figure 13.10

Periodic flux sources: (a) a magnet for a microwave travelling-wave tube; and (b) a wiggler magnet used to generate intense electromagnetic radiation from an electron beam.

at the centre of the quadrupole is zero, but whenever the beam deviates along one axis, it experiences an increasing field which causes its trajectory to curve back to the centre. The force along the perpendicular axis causes the beam to diverge, so quadrupole focussing magnets are used in crossed pairs.

A variant of the normal multipole Halbach configuration is the external multipole magnet, where the orientation varies as $(1 - (\nu/2))\vartheta$. The multipole field is then produced *outside* the cylinder, and it is the field inside that is zero. An external quadrupole is illustrated in Fig. 13.9(d). External multipole magnets are used as rotors of permanent magnet electric motors.

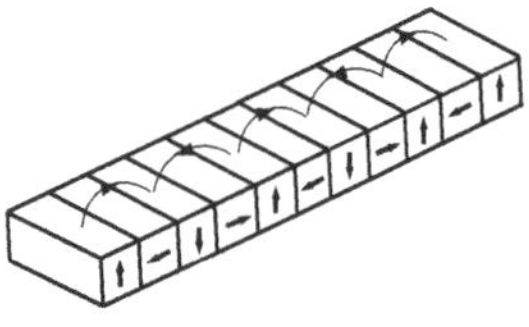

A 'fridge magnet with a one-sided flux pattern.

Unrolling a Halbach cylinder, and repeating the pattern creates a one-sided magnet, where the field turns in a direction perpendicular to the axis of the magnetized segments. There is a strong field on one side of the sheet, but none on the other. Such a configuration is used for novelty fridge magnets and other holding magnets made from bonded ferrite. The back side where the field appears depends on the sense in which the magnetization turns.

Other arrangements of permanent magnets can be devised that produce a uniform magnetic *field gradient*. Figure 13.9(e) shows an arrangement of four rods which creates a uniform field gradient in the vertical direction at the centre. Field gradients are most useful for exerting forces on other magnets.

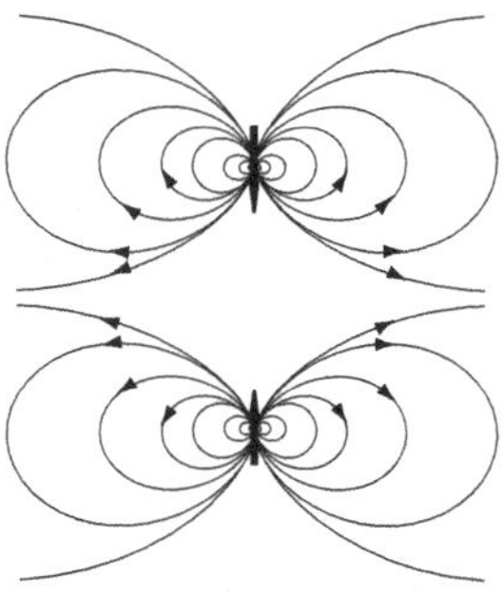

A cusp field.

Different structures have been devised to create a nonuniform field directed along the magnet axis, which may be the direction of motion of a charged particle beam. Microwave power tubes such as the travelling wave tube are designed to keep the electrons moving in a narrow beam over the length of the tube and focus them at the end while coupling energy from an external helical coil. One period of the structure in Fig. 13.10(a) generates an axial *cusp field*. Another use of cusp fields is the stabilization of molten metal flows. Axial fields of alternating direction are found in magnetic water treatment devices, for whatever reason.

Electrons in a synchrotron, moving with relativistic velocities, $v \approx c$, and energy $\gamma m_e c^2$ are guided by bending magnets around a closed track. Insertion devices or wigglers (Fig. 13.10(b)) located in straight sections of the track are permanent magnet structures which serve to generate intense beams of

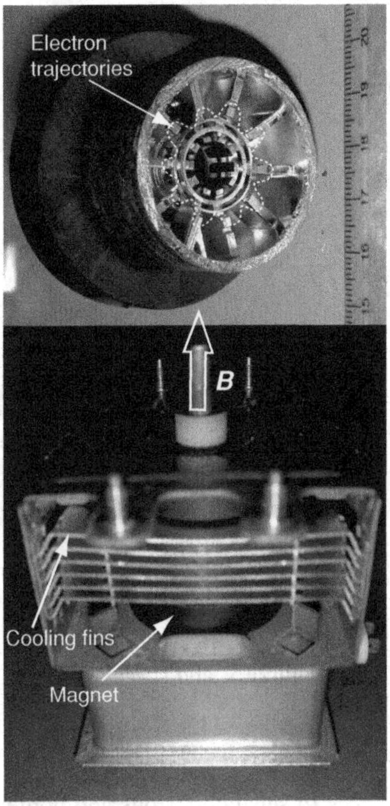

Figure 13.11

A domestic microwave magnetron. Electrons from the cathode are accelerated towards the anode in a transverse magnetic field of 90 mT produced by ferrite ring magnets. Currents circulating in the copper tines create the microwave radiation which is led to the cavity via an antenna.

polarized hard radiation (ultraviolet and X-ray) from the energetic electron beam by setting up a sinusoidal transverse field of wavelength λ. As the electrons traverse the insertion devices, they radiate at a frequency c/λ. When the frequency of oscillation in the insertion device exceeds the cyclotron frequency, the radiation becomes coherent and the wiggler is then known as an undulator.

Other devices employing permanent magnets that generate microwave radiation from electrons include klystron and magnetron sources. The magnetron in a domestic microwave oven uses ferrite magnets producing a field ≈ 0.09 T. The cyclotron frequency (3.26) of an electron in a field B is 28 GHz T^{-1}, so the frequency of radiation of an electron in this field is 2.45 GHz, corresponding to a wavelength $\lambda \approx 8$ cm which is readily absorbed by water. Water absorbs microwaves over a broad range of frequency but the 2.45 GHz band is reserved for cooking. The operation of the resonant cavity is illustrated in Fig. 13.11.

Magnetrons used in sputtering sources are permanent magnet arrays which create a magnetic field near the target, causing the electrons in the plasma to spiral, thereby increasing the ionization of the argon sputtering gas. Ion pumps for ultra-high vacuum operate similarly. Magnets increase the ionization of the residual gas used to sputter titanium, which acts as a getter.

Figure 13.12

(a) Open-gradient magnetic separation and (b) electromagnetic separation with permanent magnets.

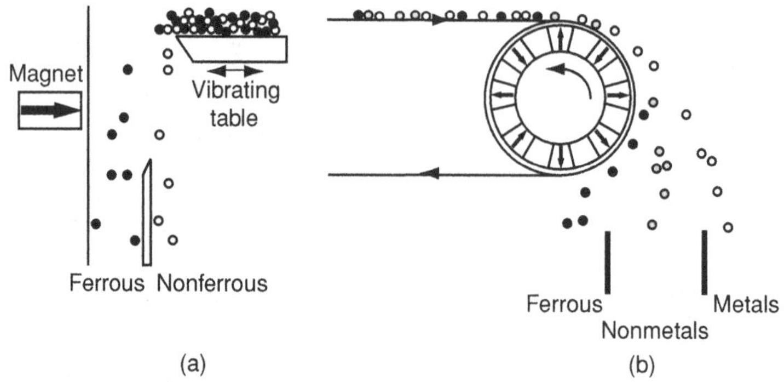

Magnetic separation The use of nonuniform magnetic fields in magnetic separation offers social benefits extending from the tiphead to the haematology laboratory. The expression for the energy of a pre-existing magnetic moment $\mathfrak{m}$ in an external field $\boldsymbol{H}'$ is $-\mu_0\mathfrak{m}\cdot\boldsymbol{H}'$, leading to the expression (2.74)

$$\boldsymbol{f} = \mu_0\nabla(\mathfrak{m}\cdot\boldsymbol{H}'). \qquad (13.11)$$

However, if the moment $\mathfrak{m} = \chi V H'$ is induced by the field in a material of volume V and susceptibility χ, the force density becomes

$$\boldsymbol{F} = \tfrac{1}{2}\mu_0\chi\nabla(H'^2). \qquad (13.12)$$

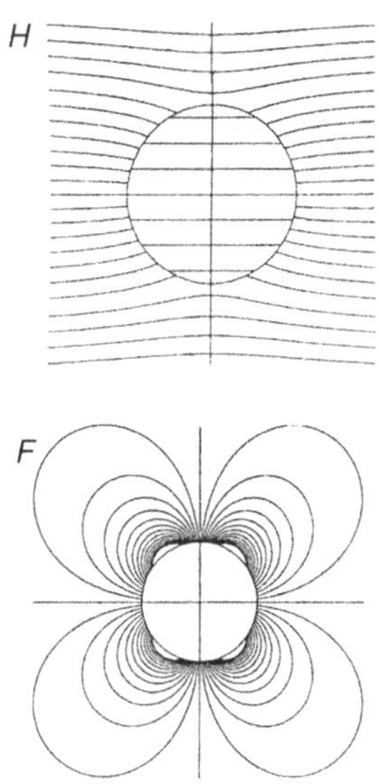

Field and force patterns around a cylindrical iron wire in a high-gradient magnetic separator.

To separate ferrous and nonferrous scrap or to select minerals from crushed ore on the basis of their magnetic susceptibility, it is appropriate to use open-gradient magnetic separation, where material tumbles through a strong magnetic field gradient, Fig. 13.12(a). Material with the greatest susceptibility will be deflected most. Since the force is proportional to particle volume (mass), the trajectory is independent of particle size and the separation is susceptibility-selective. Field gradients in open-gradient magnetic separators are about 100 T m^{-1} and separation forces on ferrous material are of order 10^9 N m^{-3}.

Permanent magnet traps are commonly used to remove ferrous scrap in industrial processes such as milling of foodgrains. A cow magnet is a specialized form of open-gradient separation, in which a magnet coated in polytetra fluoroethylene (PTFE) resides in one of the ruminant's seven stomachs, where it captures bits of barbed wire and other foreign ferrous objects.

High-gradient magnetic separation is suitable for capturing weakly paramagnetic material in suspension. A liquid containing the paramagnetic suspension passes through a tube filled with a fine soft ferromagnetic mesh or steel wool. The flux pattern of a uniform external field is distorted, creating local field gradients as high as 10^5 T m^{-1} and separation forces of upto 10^{12} N m^{-3} on paramagnetic material which remains stuck to the wires until the external field is switched off. It can then be flushed out of the system and collected. The method is applied to separate paramagnetic deoxygenated red blood cells from whole blood. A similar principle is used to separate ferrofluid particles or magnetic beads which are increasingly used in bioassays.

A different principle is employed in electromagnetic separation, where nonferrous metal such as aluminium is separated from nonmetallic material in a stream of refuse. A fast-moving conveyor belt carries rubbish over a static or rotating drum with embedded permanent magnets. The relative velocity of the magnets and the refuse may be 50 m s^{-1}. Eddy currents induced in the metal create a repulsive force and the metal is thrown off the end of the belt in a different direction to the nonmetallic waste (Fig. 13.12(b)). Deflection depends on the ratio of conductivity to density, so it is feasible to sort different metals such as aluminium or copper.

13.4 Dynamic applications with mechanical recoil

Permanent magnet structures can be adapted to produce a variable magnetic field by some change of the airgap or movement of the magnets or soft iron. The working point is displaced as the magnets move, so these devices involve mechanical recoil.

13.4.1 Variable flux sources

Figure 13.13 illustrates different types of movement that can create a variable field. Two Halbach cylinders of the sort shown in Fig. 13.6(d) with the same radius ratio $\rho = r_2/r_1$ can be nested one inside the other, Fig. 13.13(a). Rotating them in opposite senses through an angle $\pm\,\alpha$ about their common axis generates a variable field $2B_r \ln \rho \cos\alpha$. Another solution is to rotate the rods in the device of Fig. 13.6(e). By gearing a mangle with an even number of rods so that alternate rods rotate clockwise and anticlockwise though an angle α, the field varies as $B_{max} \cos\alpha$, as shown in Fig. 13.13(b). The number of rods needed for a mangle may be halved by using a sheet of soft iron to create an image. A third solution is to take a uniformly magnetized external dipole ring, which has no flux in the bore, and move one or two soft iron sheaths so that they cover part of the magnet, Fig. 13.13(c). The image in the soft iron is a normal Halbach dipole configuration with $\nu = 2$. In all cases the coercivity and anisotropy field of the material limit the maximum field available, since some magnet segments are subject to a reverse or transverse $\boldsymbol{H}$-field equal to the field in the bore.

Permanent magnet variable flux sources are compact and particularly convenient to use since they have none of the high power and cooling requirements of a comparable electromagnet. A commercial device, illustrated in Fig. 13.14, uses nested Halbach magnets made of 25 kg of $Nd_2Fe_{14}B$ to generate a 0–2.0 T field in any transverse direction in a 25 mm bore. These magnets are ideal for

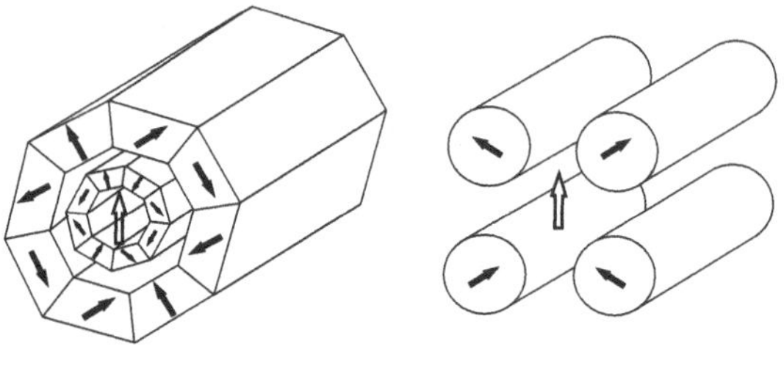

(a) (b)

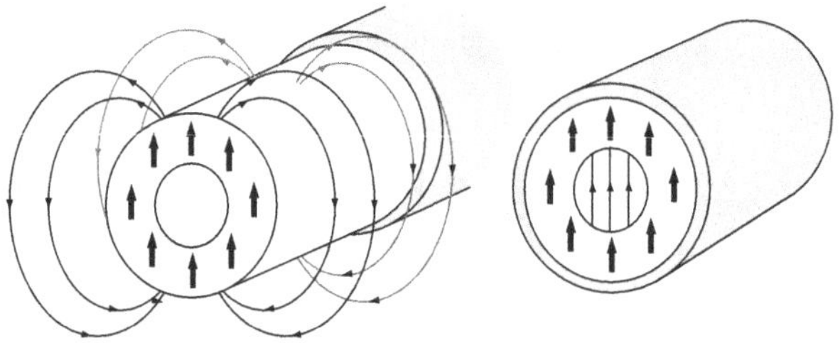

(c)

Figure 13.13

Permanent magnet variable flux sources: (a) double Halbach cylinder, (b) four-rod mangle, (c) an external Halbach cylinder with a soft iron sheath.

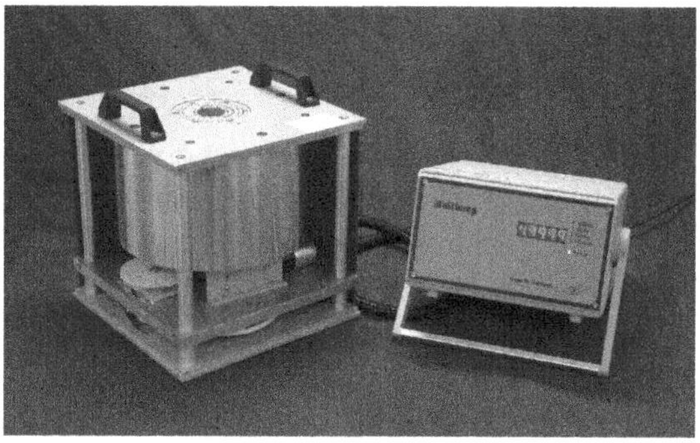

Figure 13.14

A 2 T permanent magnet variable flux source based on nested Halbach cylinders. The variable field can be in any transverse direction in the bore. The magnet shown is used for a compact vibrating-sample vector magnetometer, shown below.

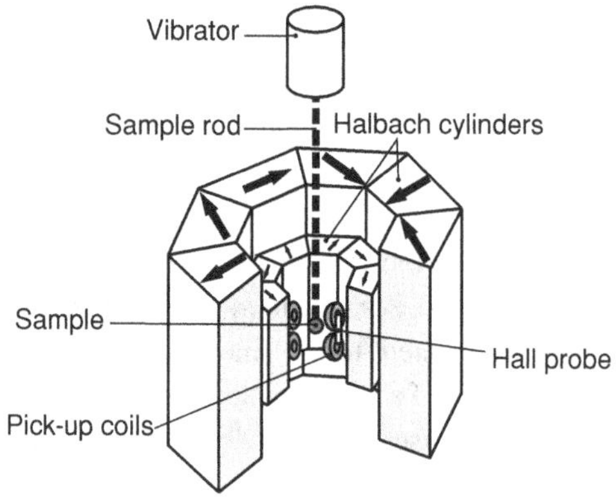

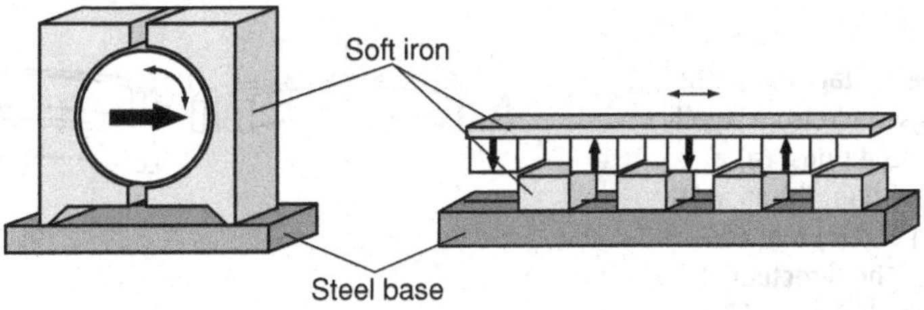

(a) (b)

Figure 13.15

Two designs for switchable magnetic clamps: (a) a rotatable magnet design; (b) a design where the magnet array is displaced laterally.

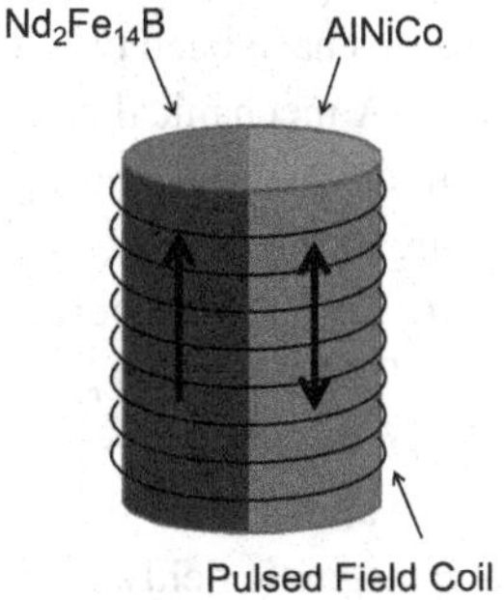

A switchable electropermanent magnet. The Alnico segment can be reversed by a field pulse.

compact instrumentation such as benchtop vibrating-sample magnetometers or magnetoresistance measurements.

The practical size limitation is imposed by (13.10). Permanent magnet variable flux sources can challenge resistive electromagnets to generate fields of up to about 2 T, but they can never compete with superconducting solenoids in the higher field range, unless regard superconductors with trapped flux as permanent magnets.

13.4.2 Switchable magnets; holding magnets

A switchable magnet is a simpler type of variable flux source. Magnetic holding devices have a piece of ferrous metal in contact with a magnet. The working point shifts from the open circuit point to the remanence point where $H = 0$ as the circuit is closed. The *maximum* force f that can be exerted at the face of a magnet may be derived by considering a toroid which is cut into two C-shaped segments that are separated by a small distance d. The energy appearing in the air gaps is $2 \times \frac{1}{2}\mu_0 H_g^2 \mathcal{A}_g d = B_g^2 \mathcal{A}_g d/\mu_0$. The work done separating the segments is $2fd$, hence the magnitude of the force per unit area is

$$\frac{f}{\mathcal{A}_g} = \frac{B_g^2}{2\mu_0}. \tag{13.13}$$

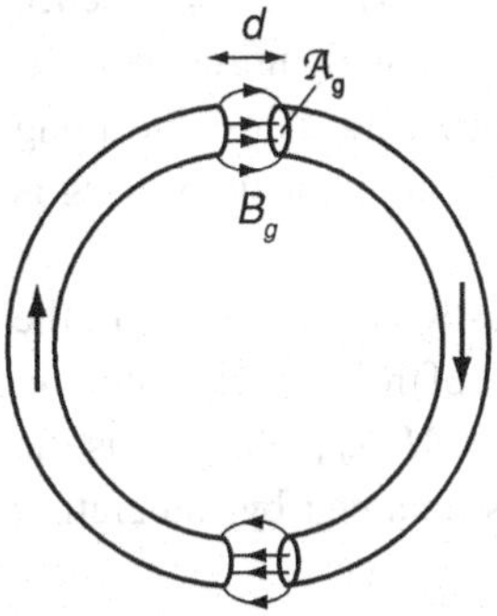

A magnetic toroid, cut and separated, creates a field in the airgap.

When the gap is narrow, $B_g = B_m$ so forces of up to 40 N cm^{-2} can be attained for $B_m = 1$ T. If large blocks of rare-earth magnet are allowed to come into contact, they are very difficult to separate!

Two switchable magnets, one used for holding components on an optical bench, the other for holding a workpiece on a machine tool are illustrated in Fig. 13.15. In other switchable designs, known as electropermanent magnets, an electromagnet is used to cancel the stray field of the permanent magnet, or switch a semihard Alnico segment.

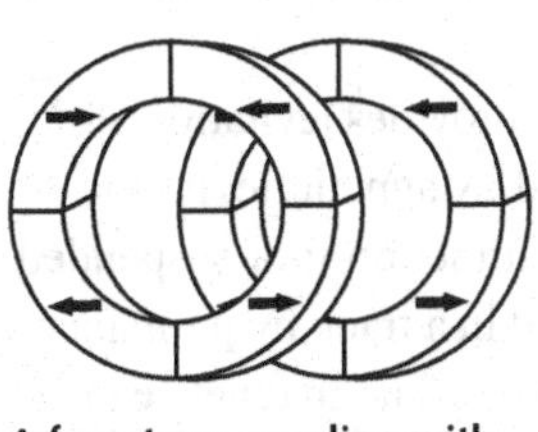

A face-type coupling with four axially magnetized segments.

13.4.3 Couplings and bearings

Permanent magnets are useful for coupling rotary or linear motion when no contact between members is allowed. Simple magnetic gears are feasible.

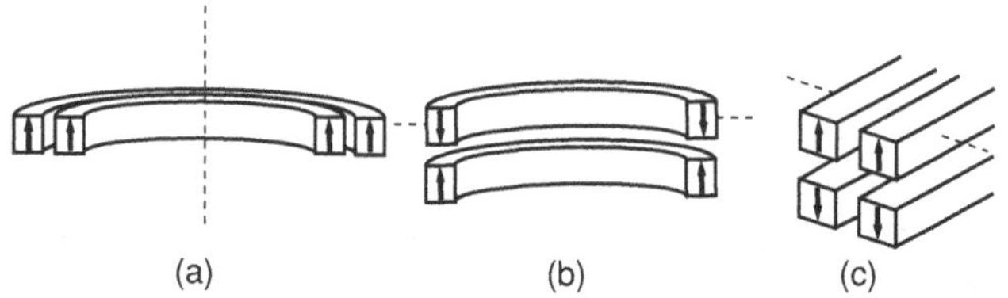

Figure 13.16

Two elementary magnetic bearings made from axially magnetized rings: (a) a radial bearing, (b) an axial bearing and (c) a linear bearing. The direction of instability is denoted by the dotted line.

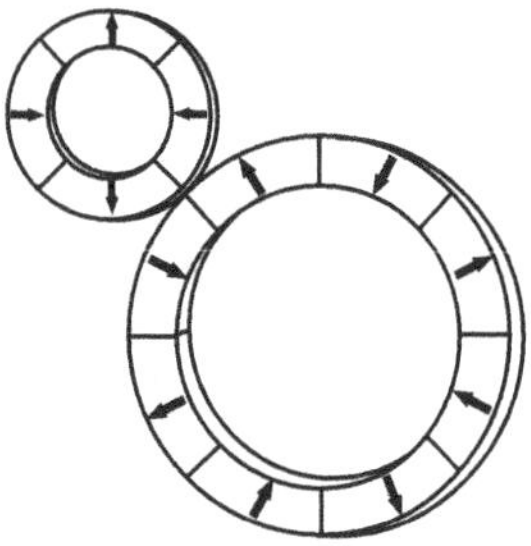

A 2:1 magnetic gear with radially magnetized segments.

Samuel Earnshaw, 1805–1888.

Magnetic bearings are well suited to high-speed rotary suspensions in turbopumps or flywheels for energy storage. Linear suspensions have been tested in prototype magnetically levitated transportation systems. A mechanical constraint or active electromagnetic support is always required when the bearing is composed only of permanent or soft magnets and an airgap, Fig. 13.16.

If the axial and radial components of the forces are $\boldsymbol{f}_z$ and $\boldsymbol{f}_r$, the corresponding diagonal stiffnesses K_z and K_r are defined as $-\mathrm{d}\boldsymbol{f}_z/\mathrm{d}z$ and $-\mathrm{d}\boldsymbol{f}_r/\mathrm{d}r$ respectively. When K_i is positive, $\boldsymbol{f}_i$ is a restoring force and the bearing is stable in that direction. In absolute stable equilibrium, $F_i = 0$ and $K_i > 0$ for all components. Sadly, this cannot be achieved just by a static magnetic field acting on a permanent magnet. It can be shown that

$$K_x + K_y + K_z = 0. \tag{13.14}$$

It is therefore impossible to achieve stability in all three directions, a result known as Earnshaw's theorem. For a cylindrical system, $2K_r + K_z = 0$, hence if K_z is positive, K_r is negative and the axial bearing is radially unstable. Conversely, if K_r is positive, K_z is negative and the radial bearing is unstable along the axis. Either a mechanical constant or an active electromagnetic bearing is needed in one direction to achieve stability.

The forces in bearings and couplings are most easily calculated using the surface charge method introduced in §2.4. Each surface of the magnet, assuming uniform magnetization $\boldsymbol{M}$, has a surface charge $\sigma_m = \boldsymbol{M} \cdot \mathbf{e}_n$, where $\mathbf{e}_n$ is the surface normal. The potential $\varphi_m(\boldsymbol{r})$ at a point $\boldsymbol{r}$ is obtained by integrating (2.56) over the surface

$$\varphi_{1m}(\boldsymbol{r}) = (1/4\pi) \int_S (\sigma_{1m}|r - r'|) \mathrm{d}^2 r'.$$

The force on an element of surface charge $\sigma_{2m}\mathrm{d}^2 r'$ on the other magnet is $-\nabla\varphi_{1m}\sigma_{2m}\,\mathrm{d}^2 r'$.

The linear magnetic bearing shown in Fig. 13.16(c) provides levitation along a track, but lateral mechanical constraint is required. Magnetic levitation of a moving vehicle (MAGLEV) may be provided by attraction to a suspended iron rail, or by repulsion from eddy currents generated in a track by permanent magnets attatched to the base of the vehicle, Fig. 13.17. The magnets can be mounted on a soft-iron plate that acts as a flux return path, and creates a configuration like a strong one-sided fridge magnet Permanent magnet levitation systems can support 50–100 times the weight of the magnets. Other levitation

Figure 13.17

A Maglev system based on eddy-current repulsion.

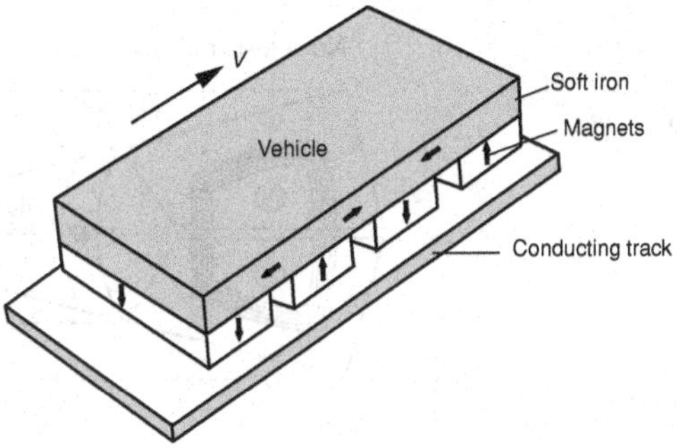

schemes are based on electromagnets or superconducting magnets. One design includes a series of embedded conducting rings in the track. Propulsion along the track is by a linear motor.

Earnshaw's prohibition of stable levitation of a permanent magnetic moment by any static field configuration applies for static magnetic fields in free space. Some ways to circumvent it are discussed in §15.3, where we return to the intriguing topic of magnetic levitation.

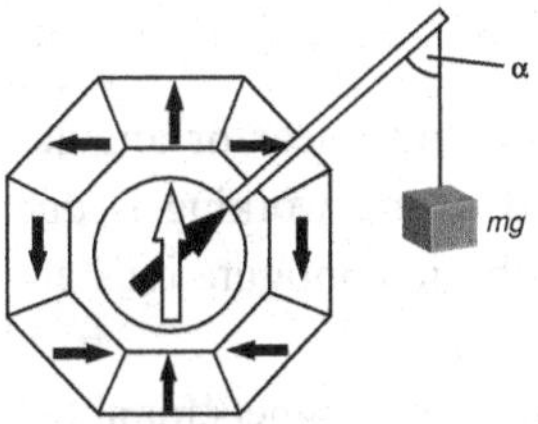

A magnetically compensated hinge.

The magnetic hinge is a bearing designed to compensate gravitational torque on an arm. The torque on an arm of length l is $\Gamma = mgl \sin\alpha$, where m is the suspended mass which may be compensated at any angular position a by the torque $\mathfrak{m}B \sin\alpha$ on a magnet of moment $\mathfrak{m}$ built into its the axle, which turns in a region of uniform field $\boldsymbol{B}$ produced by a Halbach cylinder.

13.4.4 Sensors

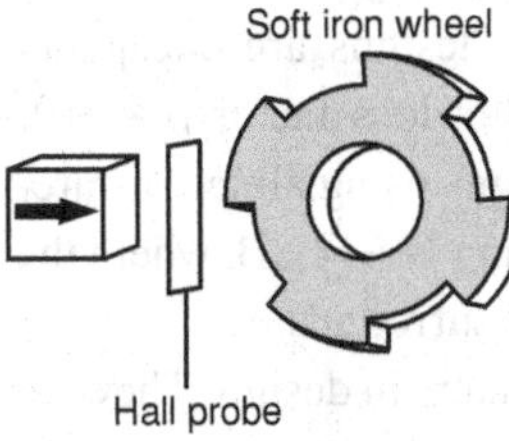

Variable-reluctance sensor based on a permanent magnet.

Magnetic sensors detect a varying field in an airgap using a Hall effect or magnetoresistance probe which delivers a voltage depending on B. Magnetic position and speed sensors used in brushless motors and automobile system controls offer reliable noncontact sensing in a hostile environment involving dirt, vibration and high temperatures. Sensors are discussed in detail in §14.3. The point to emphasize here is that a permanent magnet usually forms part of the circuit; it provides a noiseless magnetic field with no expenditure of energy. For example, a variable reluctance sensor with a magnet and a Hall probe may be used to detect the rotation of a toothed wheel.

13.5 Dynamic applications with active recoil

Motors and actuators account for much of the annually produced tonnage of permanent magnets. Designs for consumer products such as electric clocks,

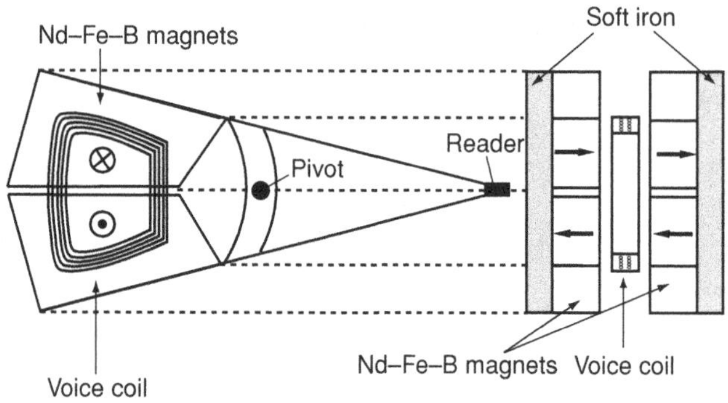

Figure 13.18

A flat voice-coil actuator for a personal-computer disc drive.

earphones, loudspeakers, cameras, kitchen appliances and hard-disc drives may be manufactured by tens or hundreds of millions every year.

13.5.1 Actuators

An actuator is a electromechanical device with a limited linear or angular displacement. Usually, the airgap is fixed, and the dynamic working is due to the $\boldsymbol{H}$-field produced by current windings. Three basic configurations are moving-coil, moving-magnet and moving-iron.

Moving-coil loudspeakers have been built with permanent magnets for about a century. Flux is directed into a radial airgap where the voice coil is suspended, attatched to a light, rigid cone. As the current passes through the coil, the force on it is proportional to B_g. The coil moves, vibrating the cone, which produces sound. Acoustic power P_v varies as B_g^2. From (13.3), this is maximized by operating at the $(BH)_{max}$ point. Large, flat ferrite ring magnets can be used with iron to concentrate the flux (Fig. 13.4). These designs are cheap but inefficient because there is much flux leakage; the flux loss factor $\beta \approx 0.4$. Stray fields are reduced in the cylindrical magnet designs using alnico or rare-earth magnets. Moving-magnet designs are feasible using $Nd_2Fe_{14}B$, where the magnet is glued to the cone and a stationary drive-coil surrounds it.

Voice-coil actuators are essentially similar to loudspeakers in design. They are used for head positioning in computer hard-disc drives and mirror positioning in laser scanners. Rapid dynamic response is ensured by the low mass of the voice-coil assembly and the low inductance of the coil in the airgap. With a cylindrical coil, radially magnetized ring magnets can be used to enhance the flux density in the airgap.

A flat-coil configuration where the coil is attatched to a lever which allows it to swing in a limited arc between two pairs of rare-earth magnets is the favoured design for the flat disc drives of portable personal computers, Fig. 13.18. Sintered, oriented Nd–Fe–B with the highest possible energy product is required

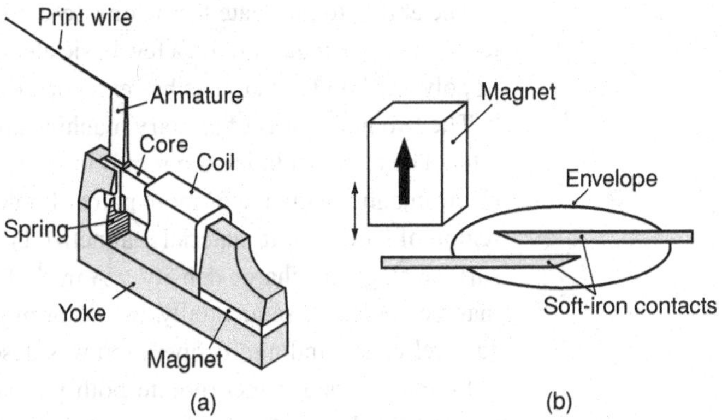

Figure 13.19

Moving-iron actuators: (a) print hammer and (b) reed switch.

($>$400 kJ m^{-3}). Access time at constant acceleration a is proportional to $a^{-1/2}$, hence to $B_g^{-1/2}$, since the force $f = B_g Il = ma$.

Moving-coil meters are rotating-coil actuators with no stringent dynamic response requirement, but very uniform airgap flux density is needed to achieve a deflection which is accurately proportional to current. The requirement for minimal temperature variation of the flux density leads to the choice of alnico magnets (Table 13.5).

Moving-magnet actuators may be linear or rotating. They offer low inertia and no flying leads. Linear reciprocating actuators with a stroke of several millimetres are used in pumps at frequencies of order 50 Hz. The mechanical system is designed to operate at its resonant frequency. Rotary actuators can be regarded as reversible electric motors with restricted travel.

Moving-iron actuators may also be linear or rotary. A design that was used for a print hammer in a dot-matrix printer is shown in Fig. 13.19(a). The hammer spring forms part of the magnetic circuit, and in the unexcited position it is held tight against the iron, with no airgap. When a current pulse passes through the solenoid, the hammer springs out. A similar principle is used in reed switches, where two flat soft-iron reeds are drawn into contact by a magnetic field (Fig. 13.19(b)). The switch can be opened by activating a solenoid to create a reverse field, or simply by moving the magnet.

13.5.2 Motors

Like actuators, motors whose operation depends on permanent magnets are produced in multiple millions. Small DC permanent magnet motors are found in domestic appliances and consumer electronics. DC servomotors power machine tools, robots and other industrial machinery. Permanent magnets can also be used to advantage in large industrial drives and for wheel drives in electric vehicles and high-speed trains.

The ability to fabricate ferrite or rare-earth magnets in any desired shape has led to many permutations on a few basic electrical machine designs. Availability of polymer-bonded magnets has extended design potential.

The two main parts of a rotary machine are the fixed stator and the revolving rotor. The classical induction motor is an AC machine in which the stator is an electromagnet with one or more pairs of poles. The term 'pole' here refers to a region of hard or soft material magnetized normal to the airgap, which bears a surface magnetic charge density σ A m^{-1}. The poles of soft-magnetic material may be energized sequentially, in two or more 'phases'. The rotor is usually a squirrel-cage winding. Its operation was described in §12.4.

DC motor designs incorporate both permanent magnets and electromagnet current windings. A permanent magnet on the stator creates a field at the current-carrying windings of the rotor. Electronic or mechanical commutation with brushes distributes current to the windings in such a way that the torque Γ on the rotor is always in the same sense. Conversely, the device will function as a generator, producing an emf $\mathcal{E}$ in the windings, if it is driven at an angular velocity ω.

The torque characteristic of the DC motor is $\Gamma(\omega)$. If the radius of the rotor winding is r, the flux density of the magnet is $\boldsymbol{B}$ and there are N conductors of length l perpendicular to $\boldsymbol{B}$, each carrying a current I, then $\Gamma = NrBIl$. The output power $\Gamma\omega$ will be equal to $\mathcal{E}I$, where the back emf $\mathcal{E}$ generated as the rotor turns is $2r\omega Bl$ per conductor. The number of conductors connected in series is $N/2$, so $\mathcal{E} = Nr\omega Bl$. If the applied voltage is $V = \mathcal{E} + IR$, where R is the resistance of the windings, the torque equation for the motor is

$$\Gamma = K(V - K\omega)/R, \tag{13.15}$$

where $K = NrBl$ is the torque constant of the motor. Torque is greatest at startup, when $\omega = 0$ and Γ is proportional to the flux density produced by the magnet. It falls to zero when $V = K\omega$.

In a DC servomotor, the torque or the angular velocity is controlled by modifying the applied voltage. Simple velocity control is based on monitoring the back emf $\mathcal{E}$, but more sophisticated control systems use a tachogenerator (a small DC generator coupled to the drive shaft) or a precise position encoder to generate voltage feedback to control the output power.

The motor design may be modified, as shown in Fig. 13.20, to eliminate the mechanical commutator which is a source of sparking and wear. In the brushless DC motor, the magnets are positioned on the rotor and the armature windings, now located on the stator, are energized in an appropriate sequence by means of power electronics. Electronically commutated motors are reliable and they are particularly suited to high-speed operation, $\omega > 100$ rad s^{-1} ($\approx$1000 rpm). Position sensors form an integral part of the device since the winding to be energized depends on the position of the rotor.

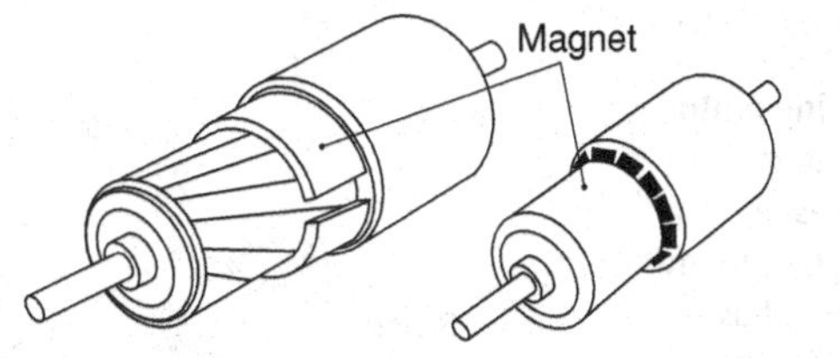

Figure 13.20

DC motor designs: (a) brush motor with magnets on the stator and (b) brushless motor with magnets on the rotor.

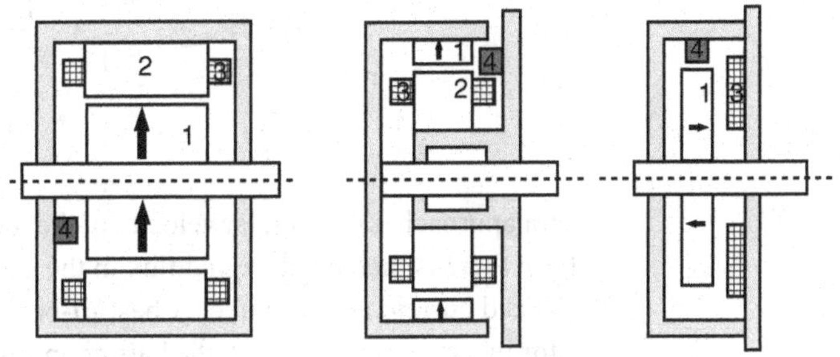

Figure 13.21

Variants of the brushless DC motor: (a) normal design, (b) cup-type (c) pancake-type.
1 – magnet; 2 – stator;
3 – stator winding;
4 – position sensor.

Variants of the design of the brushless dc motor are shown in Fig. 13.21. Flattening the rotor into a disc produces a pancake motor. The low moment of inertia means that high angular accelerations are possible, especially when Nd–Fe–B is used for the magnets. They may be embedded in the rotor so as to concentrate the airgap flux. Spindle motors in hard disc drives use a bonded Nd-Fe-B ring magnet. The cup-type rotor is another flat low-inertia design. By unrolling the armature, a linear motor is obtained.

A feature of all permanent-magnet motors is armature reaction. The currents in the armature, regardless of whether the armature windings are on the stator or on the rotor, create a field which is usually in a direction opposed to the magnetization. The working point of the magnets is shifted, as shown in Fig. 13.3(c). The effect is particularly severe at the trailing edges of the poles, and a solution is to make these sections from a grade of magnet with higher coercivity.

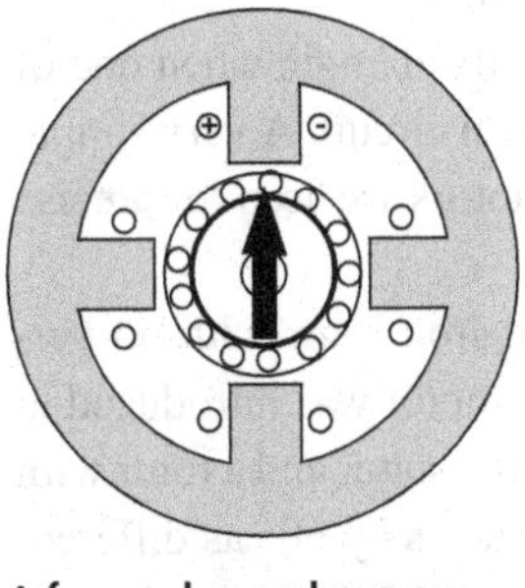

A four-pole synchronous motor with a permanent magnet rotor. A 13-bar squirrel-cage winding is built in so that the machine will operate as an induction motor for startup.

Synchronous motors run at a frequency which is a multiple of that of the AC power supply. Electronic power inverters are available which allow the frequency to be continuously variable. The basic design is that of Fig. 13.21(a), where the magnet is on the rotor, and a multiphase winding on the stator produces a field which rotates around the airgap. Torque is proportional to sin δ, where δ is the angle at any instant between the flux produced by the stator and the direction of magnetization of the rotor. A multipole rotor may be produced by embedding long magnets with different orientations.

Synchronous motors offer high efficiency and power density, but their weak point is feeble starting torque until they reach synchronous rotation speed. A classical induction motor with its squirrel-cage winding on the rotor has the opposite characteristic; the torque is maximum at startup and falls to zero as the

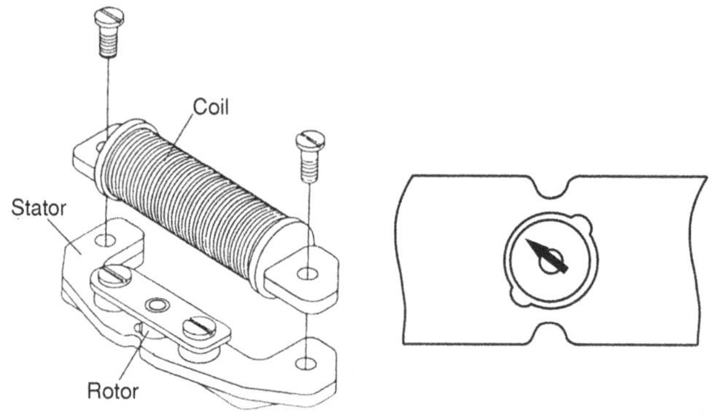

Figure 13.22

A two-pole stepping motor (Lavet motor) used in clocks and watches. In watches the magnet, made of bonded Sm_2Co_{17}, has a mass of a few milligrams.

rotor approaches the angular velocity of the rotating field produced by the stator. Then there is then no changing flux in the squirrel-cage winding and hence no induced current and no force. A best-of-both-worlds solution incorporates the rotor in a cage winding with the bars connected to conducting plates at top and bottom to help startup.

Most motor designs can be operated in an inverse mode as generators. The rotor is driven, exposing the armature windings to a time-varying magnetic field, thus generating an emf. One of the first permanent magnet devices ever to find a mass market was the bicycle dynamo. Fifty years ago, people in the developed world owned only one or two magnets. Now they own 100–200 of them, unless we count the hard disc drive on a personal computer, where there are billions more.

Large disc-type generators with many permanent-magnet poles (24–36) are employed for energy conversion in wind farms. These wind generators requires roughly 500 kg of Nd-Fe-B per MW. The alternator for the hybrid electric vehicle may also use a permanent-magnet design. Operating at 100 000 rpm, a 30 kW generator is expected to have a mass of about 10 kg.

Stepping motors are devices which rotate through a fixed angle when one of the windings is energized by a suitable electronic control circuit. A very simple two-pole motor is illustrated in Fig. 13.22. Stepping motors are used for precise position control.

It is possible to build a stepper motor using a soft-iron rotor with no permanent magnets whatsoever. This variable-reluctance motor was introduced in §12.4. The motor in Fig. 12.13(b) had salient poles on the stator and a rotor with a smaller number of protrusions or 'teeth'. The rotor turns by 60° as different pairs of windings are activated. A combination of high efficiency and small step is achieved by combining the variable-reluctance and permanent magnet stepping motors in the hybrid design, illustrated in Fig.13.23. There are six poles, with five teeth on each. The rotor has 32 teeth with a permanent magnet at the centre. As the poles are energized in sequence A,B,C, the motor turns in steps of $\frac{1}{64}$th of a revolution. A common design makes 200 steps per revolution, a 1.8° step size. With a suitable controller, it can proceed in half-steps of 0.9°.

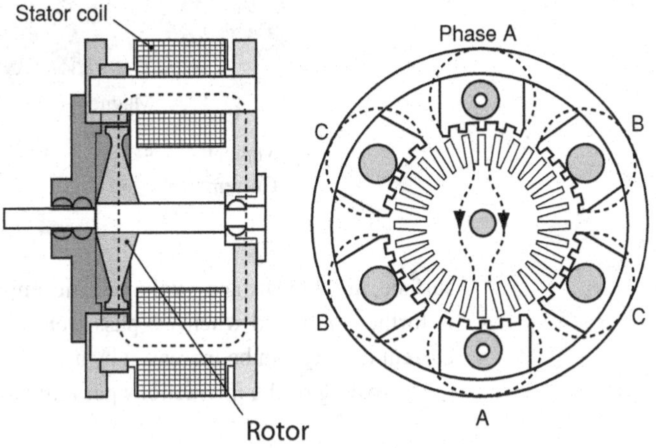

Figure 13.23

Miniature hybrid stepping motor.

13.6 Magnetic microsystems

Miniaturized mechanical drives such as microactuators and micromotors present new opportunities for permanent magnets. This is the field of magnetic MEMS. These microdevices have to exert a force on some small object such as a silicon cantilever, so it is important to understand how the forces scale as the dimensions are reduced. We have already seen that the dipole field is invariant as all the dimensions are scaled up or down by a factor ξ.

If there is a conductor in the vicinity of a magnet, the Lorentz force on it per unit volume depends on the vector product of $\boldsymbol{B}$ and current density $\boldsymbol{j}$. This, too, is scale-independent because j_c is an intrinsic property of the conductor. The properties of permanent magnet micromachines based on magnet/current forces depend only on the materials properties B_r and j_c, where j_c is the maximum current loading.

Magnet/magnet forces and current/current forces scale differently. The force on a permanent or induced moment $\mathfrak{m}$ depends on the field gradient. Consider a system of two permanent magnets. When the distances are scaled down by a factor ξ ($\xi > 1$), the field at each magnet, created by the other remains constant, but the field gradients are multiplied by ξ so the force per unit volume is *amplified*. The dynamic response improves as the system shrinks. Emfs created by electromagnetic induction depend on the rate of change of flux in a circuit, which scales as $1/\xi^2$ at constant frequency f for fields created by permanent magnets. The induced electric field scales as $1/\xi$. However, the field created by a current element given by the Biot–Savart law (Eq. (2.5)) depends on $j\delta V/r^2$. It scales as j/ξ. If the fields are created by coils, the emf and the electric fields are reduced by a further factor of ξ.

But it is really too pessimistic to assume that the current density j has to be the same in large conductors as in small ones. Heat is dissipated at the

Table 13.6. Scaling of electromagnetic interactions in microsystems

	Magnet	Current	Soft iron
Magnet	ξ	η	ξ
Current	η	$\eta_1\eta_2/\xi$	η/ξ

surface, and the surface : volume ratio improves with reducing size; the ratio is further improved when we pass from circular to planar conductors. If the current density can be increased by a factor η, which may even be greater than ξ, the scaling of the interactions per unit volume are as shown in Table 13.6.

FURTHER READING

R. Skomski and J. M. D. Coey, *Permanent Magnetism*, Bristol: IOP Publishing (1999). A monograph focussed on the physics of permanent magnetism, including chapters on experimental methods, materials and applications.

J. M. D. Coey (editor), *Rare-earth Iron Permanent Magnets*, Oxford: Clarendon Press, (1996). Physics, materials science and applications of Nd–Fe–B magnets by participants in the Concerted European Action on Magnets.

M. G. Abele, *Structures of Permanent Magnets*, New York: Wiley (1993). Theory of permanent-magnet flux sources.

P. Campbell, *Permanent Magnet Materials and their Application*, Cambridge: Cambridge University Press (1994). An accessible introduction for engineers.

R. J. Parker, *Advances in Permanent Magnetism*, New York: Wiley (1990). Engineering applications.

T. Kenjo, *Electric Motors and their Controls*, Oxford: Oxford University Press (1991). A lucid and readable account of electric motors.

O. Cugat (editor), *Microactionneurs electromagnetiques*, Paris: Lavoisier (2002). The first book on magnetic MEMS.

EXERCISES

13.1 Show that the efficiency of an extended two-dimensional magnetic structure given by (13.8) cannot exceed 1/4.

13.2 Compare the efficiencies of the cylindrical flux sources shown in Figs. 13.6(a) and 13.6(c).

13.3 Calculate the flux density at the centre of an octagonal Halbach cylinder with $r_1 = 12$ mm and $r_2 = 40$ mm. and length 80 mm, made of a grade of Nd–Fe–B having $B_r = 1.25$ T. Compare with the value for an infinite ideal Halbach cylinder (13.9). Note that a reasonable fraction of the ideal flux density is achieved in a cylinder whose length is equal to its diameter.

13.4 Design a permanent magnet assembly that will produce a flux density of 5 T in a 25 mm bore. What are the magnetic properties required of the magnet? What does your magnet weigh? Cost?

13.5 Design an undulator.

13.6 In the absence of conduction currents, $\boldsymbol{B} = -\mu_0 \nabla \varphi_m$. Show that it follows from Maxwell's equation $\nabla \cdot \boldsymbol{B} = 0$ that $\nabla \cdot \boldsymbol{f} = \nabla^2(\mathrm{m} \cdot \boldsymbol{B}) = 0$, where the force $\boldsymbol{f}$ on a magnet of moment m is $\nabla(\mathrm{m}.\boldsymbol{B})$.

13.7 Given a cylindrical permanent magnet of length 1 cm and radius r, made of a material with a polarization of 1.5 T, find the maximum value of r for which it is possible to stably levitate a tiny drop of water.

13.8 Design magnetic footware that would allow you to walk across a flat iron ceiling.

13.9 Permanent magnets arc used to limit arc discharges at high-current switches. How do they work?

14 Spin electronics and magnetic recording

Conventional electronics has ignored the spin on the electron

Spin electronics exploits the angular momentum and magnetic moment of the electron to add new functionality to electronic devices. A first generation of devices comprised magnetoresistive sensors and magnetic memory. The sensors have numerous applications, especially in digital recording. Magnetic recording uses semi-hard magnetic thin films as the recording media. Write heads are miniature thin-film electromagnets, while read heads are usually spin-valves exhibiting giant magnetoresistance (GMR) or tunnelling magnetoresistance (TMR). Magnetic random-access memory (MRAM) is based on switchable spin valve cells, similar in structure to the read head. New generations of spin electronic devices are under development in which the angular momentum of a spin-polarized current is used to exert spin transfer torque, or the flow of spin-polarized electrons is controlled via a third electrode in a transistor-like structure.

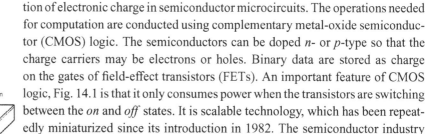

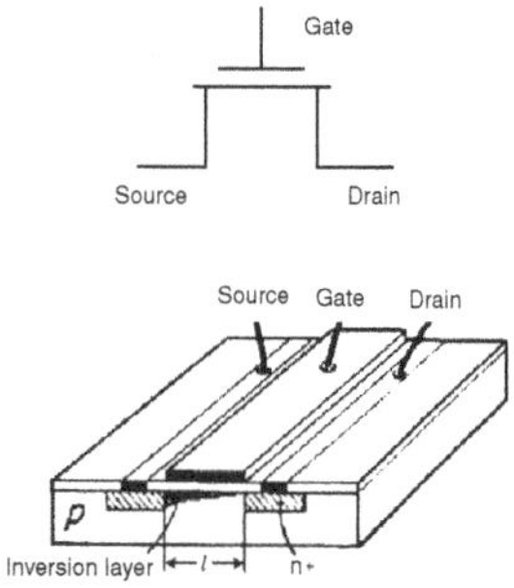

A field-effect transistor.

A hugely successful electronics technology has been built around the manipulation of electronic charge in semiconductor microcircuits. The operations needed for computation are conducted using complementary metal-oxide semiconductor (CMOS) logic. The semiconductors can be doped *n*- or *p*-type so that the charge carriers may be electrons or holes. Binary data are stored as charge on the gates of field-effect transistors (FETs). An important feature of CMOS logic, Fig. 14.1 is that it only consumes power when the transistors are switching between the *on* and *off* states. It is scalable technology, which has been repeatedly miniaturized since its introduction in 1982. The semiconductor industry follows a roadmap. The minimum feature size in silicon circuits was 45 nm in 2008, and it is projected to decrease to 22 nm in 2011. It is unclear what will take over at the end of the roadmap, when feature sizes of less than 10 nm will make CMOS unsustainable. Some form of spin-based electronics may be an option.

Semiconductor random-access memory can be static (SRAM) or dynamic (DRAM), Fig. 14.2. Both are volatile, in the sense that information stored on the gate of an FET is lost when the power is cut. SRAM requires less power and is faster, but the memory cell uses six transistors. DRAM demands periodic refreshment every few milliseconds as the charge leaks away, but it requires only one transistor per memory cell. Error rates are about one per month per gigabit of memory, but these are bound to increase as the cells become ever smaller and cosmic rays and other background radiation disturb their stored charge.

Figure 14.1

A CMOS circuit which carries out the basic logic operation NAND. It consists of four transistors, and consumes no power in the quiescent state. The output is low when both A and B are high, and if either A or B is low, the output is high.

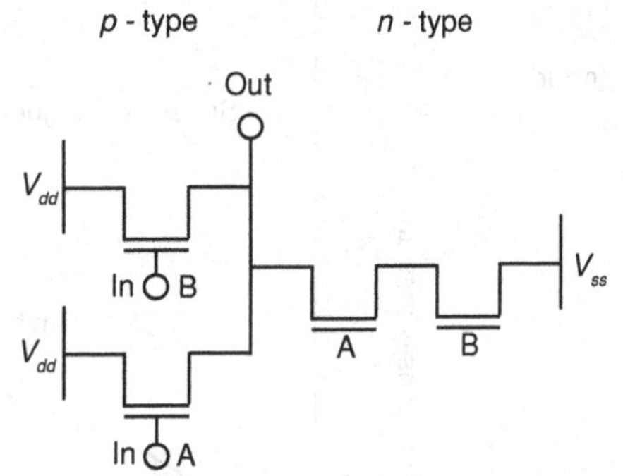

Figure 14.2

Memory cells of: (a) DRAM, consisting of a single transistor, which needs to be refreshed and (b) SRAM.

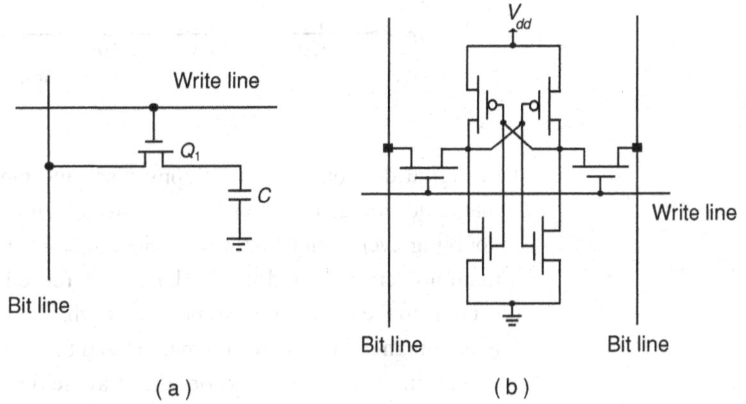

Flash is nonvolatile computer memory that can be electrically erased and reprogrammed. It is based on FET cells that have two gates instead of one. A fully insulated floating gate stores the charge. Flash is economical, but relatively slow, with limited rewritability. Solid state memory based on flash will supplement or replace magnetic hard-disc storage for some applications.

Conventional electronics works with no regard to electron spin. It treats the electron as a mobile charged particle. Charge currents surge around silicon chips in ever-smaller and faster semiconductor circuits, where the number of transistors per chip has been doubling every two years in accordance with Moore's law (Fig. 1.14). Named after the founder of INTEL, this is not a physical principle but an empirical observation based on over 30 years of industrial experience. Like any exponential growth, Moore's law must eventually come to an end. By 2020 the extrapolated dimensions of a transistor approach those of the atom! But for the present, Moore's law retains the status of a self-fulfilling prophesy. Its extrapolation sets targets for the industry road map.

Ferromagnetic information storage has been the partner of semiconductor electronics in the information revolution. Nonvolatile storage of vast quantities

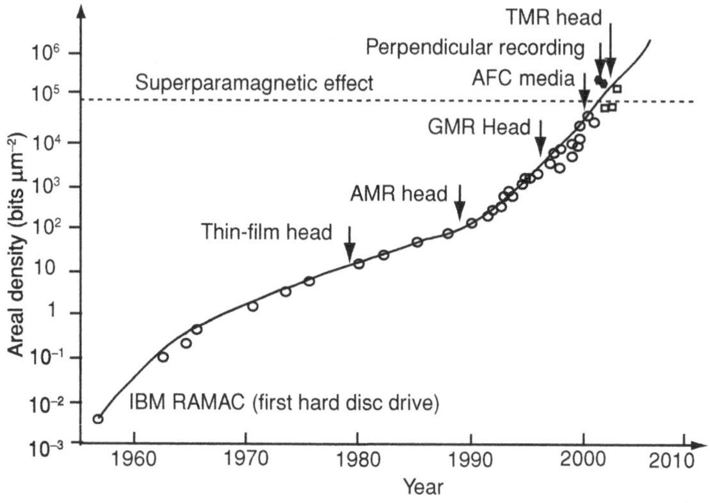

Figure 14.3

The evolution of magnetic recording on hard discs.

of digital data on hard and floppy discs and tapes has been possible because magnetic storage densities have followed their own Moore's law, with densities doubling even faster than for semiconductors (Fig. 14.3). The recorded information is erasable and the media can be reused. Magnetic recording owes its success to the nature of the dipole field, which varies as $\mathfrak{m}/r^3$, where $\mathfrak{m} \simeq Md^3$ is the magnetic moment of a magnetized bit with dimension d. As d and r are shrunk in the same proportion, the stray field remains the same. Like CMOS, magnetic recording is scalable technology. The amount of data stored on servers and hard discs around the world is astonishing, currently of order 100 exabytes (1 exabyte $= 10^{18}$ bytes) of new digital information are created and stored every year. This is the same order of magnitude as all other data existing in the world. The text of this book is a few Mbytes, and the figures account for a further 100 Mbytes. Each and every one is stored in an individually addressable magnet on a hard disc. It is sobering to realize that the numbers of magnets and transistors we manufacture in fabs exceeds the number of grains of rice and corn we grow in fields!

The relentless trend towards extreme miniaturization of electronic logic and memory has led to the consideration of schemes where the electron spin plays a key role in the operation of new devices. A highly successful first generation of spin electronics[1] was based on magnetoresistive sensors and bistable memory elements. These are two-terminal devices. Future three-terminal devices, including various types of spin transistor, may offer spin gain. The vision for spin electronics is that it will integrate magnetism with electronics at chip level, a marriage already achieved with optics in various optoelectronic devices.

[1] The term spintronics is practically synonymous with spin electronics. Magnetoelectronics is a near synonym, referring to carrying out electronic functions with magnetic elements.

14.1 Spin-polarized currents

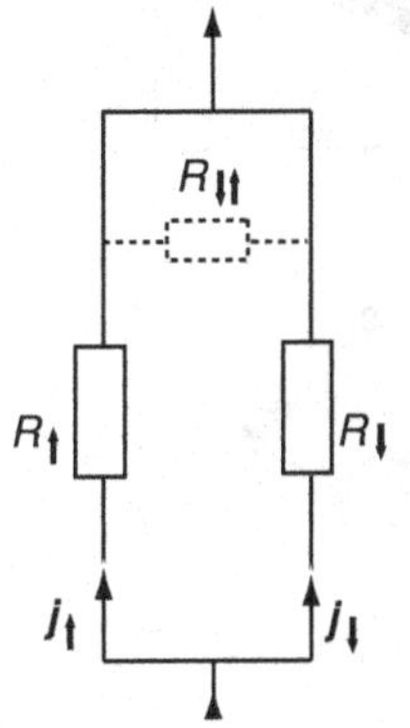

The two-current model.

Together with its charge $-e$, the electron carries quantized angular momentum $\hbar m_s$, where $m_s = \mp\frac{1}{2}$ for the $\uparrow$ or $\downarrow$ states.[2] The magnetic moment of the electron is proportional to its angular momentum, $\mathfrak{m} = \gamma\hbar m_s = \pm 1$ Bohr magneton. Spin angular momentum of the electron is the basis of both solid-state magnetism and spin electronics. An electric current is always a flow of charge; but it can also be a flow of angular momentum. The important difference is that angular momentum, unlike charge, is not conserved in an electric circuit. Electrons can be flipped from $\uparrow$ to $\downarrow$ states, or vice versa, by scattering processes which are relatively uncommon compared with the normal scattering events that modify the momentum and occasionally the energy of the electron. The comparative rarity of spin-flip scattering means that conduction can be thought of as taking place in two independent, parallel channels for $\uparrow$ and $\downarrow$ electrons – the two-current model proposed by Mott in 1936 for conduction in metals. A brief review of spin-polarized electronic conduction is appropriate, before embarking on the applications.

14.1.1 Conduction mechanisms

Conduction in metals is normally a diffusive process. The electrons are continually being scattered, and the mean free path between scattering events may be different for $\uparrow$ and $\downarrow$ electrons in ferromagnets. The conductivity for each channel is then proportional to its mean free path $\lambda_\uparrow$ or $\lambda_\downarrow$.

Consider first conduction in copper, the nonmagnetic metal having a filled $3d$ band with ten electrons per atom, and a half-filled $4s$ band with just one (Fig. 14.4). Conduction is mainly due to the $4s$ electrons which acquire a drift velocity v_d in the direction of an applied electric field, as described in §3.2.7. The current density is $j = -nev_d$, where n is the electron density. In copper, n is 8.45×10^{28} m^{-3}. Typical current densities in electronic circuits are 10^7 A m^{-2}, making v_d of order 1 mm s^{-1}.

The mean free path is the average distance travelled by an electron in the time τ between collisions, which is known as the momentum relaxation time. The mean free path is then $\lambda = v_F\tau$, where v_F is the Fermi velocity. In the free-electron model, it follows from (3.1), the de Broglie relation, and (3.37) that $v_F = (\hbar/m_e)(3\pi^2 n)^{1/3}$, which in the case of copper is 1.6×10^6 m s^{-1}. The Fermi wavelength of the electron in copper, $\lambda_F = h/m_e v_F$, is therefore 0.7 nm or about three interatomic spacings. From (3.51), the relation between

[2] The arrow points in the direction of the magnetic moment of the electron, which is opposite to the direction of its angular momentum. The $\uparrow$ electrons are the majority-spin electrons in a ferromagnet, or a paramagnet in an applied magnetic field.

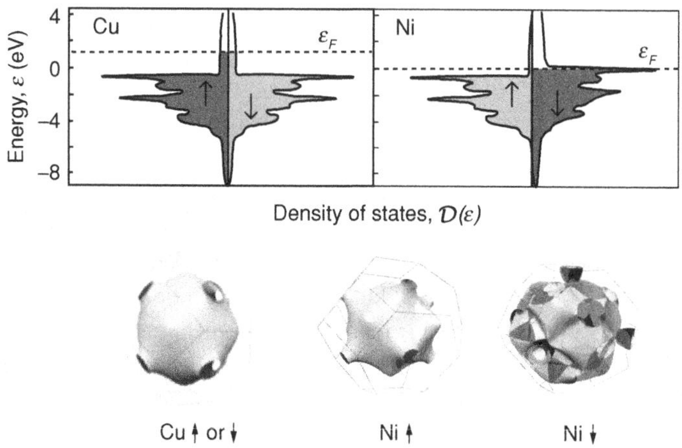

Figure 14.4

Comparison of the densities of states and Fermi surfaces of copper and nickel.

conductivity and mean free path λ is

$$\sigma = \frac{ne^2\lambda}{m_e v_F}. \tag{14.1}$$

Since the conductivity of pure copper is $\approx 10^8$ S m^{-1}, the mean free path is about 40 nm, and the momentum relaxation time is about 25 fs. A rough rule of thumb for monovalent metals is $\lambda \approx 10^{-15}\sigma$ m.

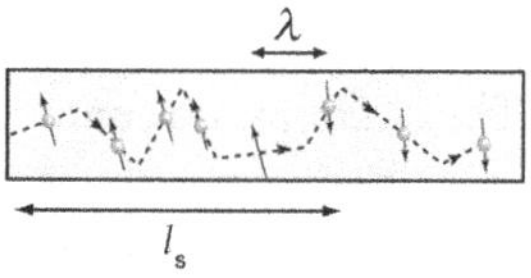

The diffusive process characteristic of electrical conduction in metals. Since spin-flip scattering is much less common than normal momentum scattering events, the spin diffusion length l_s is much longer than the mean free path λ: generally $\lambda_s \gg l_s \gg \lambda$.

An electron, like a billiard ball, will undergo several elastic collisions with momentum transfer before it experiences an inelastic collision, which involves energy loss or gain as well. It undergoes many collisions, of order $\nu = 100$ or more, before it experiences a spin-flip scattering event that involves an exchange of angular momentum with the lattice. The spin scattering time is τ_s. The distance travelled in time $t \gg \tau$ in a diffusive process is $l = \sqrt{D_e t}$, where D_e is the diffusion constant (units m^2 s^{-1}). Since electrical conduction is essentially diffusion of electrons in the direction of the applied field, the spin diffusion length is given by the three-dimensional random-walk expression:

$$l_s = \sqrt{D_e \tau_s}, \tag{14.2}$$

where the diffusion constant for electrons, which is proportional to the conductivity, is $D_e = \frac{1}{3}v_F^2\tau = \frac{1}{3}v_F\lambda$. Hence $l_s = \sqrt{\frac{1}{3}\nu\lambda^2}$. The distance travelled by the electron *along its path* between spin flip events is $\lambda_s = v_F\tau_s$. Conductivity is related to the density of states at the Fermi level $\mathcal{D}(\varepsilon_F)$ and the diffusivity D_e by the Einstein relation:

$$\sigma = \mathcal{D}(\varepsilon_F)e^2 D_e. \tag{14.3}$$

For copper, $D_e = 21 \times 10^{-3}$ m^2 s^{-1}, so the estimate $\nu = 100$ gives $l_s =$ 230 nm. The value for pure copper is larger than this, but copper in thin-film devices contains defects and diffused impurity atoms which decrease the conductivity and reduce the mean free path and spin diffusion length below

Table 14.1. Estimates of mean free paths and spin diffusion lengths at room temperature, in nanometres

	Al	Fe	Co	Ni	$Fe_{20}Ni_{80}$	Cu
$\lambda_\uparrow$	12	8	20	5	15	20
$\lambda_\downarrow$	12	8	1	1	1	20
l_s	350	50	40	10	3	200

these estimates. D_e is also reduced at finite temperatures by electron-phonon scattering.

Passing now from copper to nickel, a strong ferromagnetic metal with one less electron per atom, the $3d$ band is spin split, as shown in Fig. 14.4. The electronic configuration of nickel is $3d^{9.4}4s^{0.6}$. Conduction is still mainly by $4s$ electrons, which have a much greater mobility than their $3d$ counterparts, but the ↑ and ↓ carriers are now scattered differently. The ↑ electrons behave much as in copper since the $3d^\uparrow$ band is full and the Fermi surfaces are similar, but the ↓ electrons can be scattered into empty $3d$ states at the Fermi level. The mean free path of ↓ electrons in nickel is about five times shorter than that of their ↑ counterparts.

The spin diffusion length in nickel is only about 10 nm. Indicative values of the scattering lengths for some other metals and alloys commonly featured in thin-film devices are listed in Table 14.1.

Besides diffusion, three other modes of electron transport are encountered in solids. One is ballistic transport, where the mean free path of the electrons exceeds the dimensions of the conductor, so they traverse the conductor in a single shot, without scattering. This is what happens in small point contacts, and in highly perfect conductors with a low electron density, such as semiconductors or carbon nanotubes. There is no distinction between ballistic transport of ↑ or ↓ carriers. In diffusive transport, the electrons move ballistically between collisions.

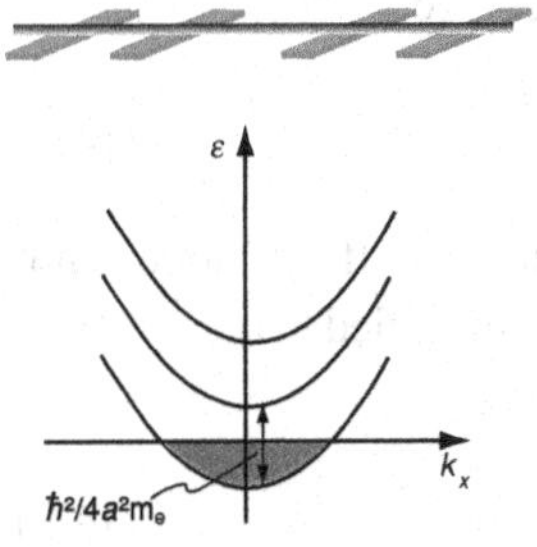

A ballistic nanowire with contact pads. The electrons moving along the wire show a free-electron dispersion $\varepsilon = \hbar^2 k_x^2/2m_e$, but the modes in the y-, z-directions are split because of quantum confinement. The current is carried by electrons in the shaded zone.

But the resistance and magnetoresistance of electrical circuits with a constriction is dominated by the constriction, the weak link in the conducting circuit. For a large contact, where the transport is diffusive, the resistance is simply $R = \varrho \mathrm{t}/\pi a^2$. Here ϱ is the resistivity of the material in the contact, a is its radius and t is the thickness. When the diffusive contact is thin, $\mathrm{t} \ll 2a$, the resistance $R_M = \varrho/2a$ is known as the Maxwell resistance. On the other hand, if the transport is ballistic, the resistance of a small contact is given by the Sharvin formula:

$$R_s = \frac{h}{2e^2}\frac{4}{a^2 k_F^2}. \tag{14.4}$$

The crossover from diffusive to ballistic transport for a thin contact occurs when $a = 0.85\lambda$. For pure copper, the crossover contact resistance is about 0.1 Ω, but it scales with the square of the resistivity, so that it can rise to a kilohm in a poorly conducting metal with $\varrho \approx 1\ \mu\Omega$ m. The ratio h/e^2 features prominently in transport theory. It has units of ohms, and magnitude 25 812 Ω. The inverse $G_0 = e^2/h$ is known as the conductance quantum.[3]

Since there is no scattering in a ballistic channel, the resistance must somehow be associated with the contacts. Considering a ballistic nanowire with radius a comparable to the Fermi wavelength, the wire behaves as an electron waveguide or two-dimensional quantum well, with only a few transverse modes separated in k_y or k_z by π/a. The wavevector along the length of the wire is unconstrained, and free-electron-like. If the potentials of the two electrodes are μ_1 and μ_2, electrode 1 injects electrons with $k_x < 0$ and electrode 2 injects electrons with $k_x > 0$. The current carried in one of the modes is $-nev$, where n is the number of electrons per unit length of wire. The electron density associated with a single k-state in a conductor of length l is $1/l$, hence the current is

$$I = -\frac{e}{l}\sum_k \frac{1}{\hbar}\frac{\partial \varepsilon_k}{\partial k} f(\varepsilon_k), \tag{14.5}$$

where $f(\varepsilon_k)$ is the probability of occupancy of a state. Replacing the sum over k-states by an integral gives

$$I = -\frac{2e}{h}\int_{\varepsilon_0}^{\infty} f(\varepsilon_k)\mathrm{d}\varepsilon_k. \tag{14.6}$$

If the number of modes ν_m is a constant over the range $\mu_1 > \vartheta > \mu_2$, then $I = (2e^2/h)\nu_m(\mu_1 - \mu_2)/e$, which leads to a conductance

$$G = \frac{2e^2}{h}\nu_m. \tag{14.7}$$

Each unpolarized mode contributes $2G_0$ to the conductance. If the mode is not a perfect transmitter, the expression for conductance is modified to the Landauer formula

$$G = \frac{e^2}{h}\sum_{i,\alpha} T_{i,\alpha}, \tag{14.8}$$

where T_i is the transmission of the ith mode for ↑ or ↓ electrons.

The third means of electron transport is tunnelling (§8.3.6). When two conductors are separated by a thin layer of insulator, or a vacuum gap where the electronic wave functions decay exponentially, there is some probability that electrons will pop out on the other side of the barrier, provided its width w is less than or of order the decay length of the electronic wave function. The probability of tunnelling through a barrier of height ϕ and width w is given by the transmission coefficient:

$$\mathcal{T} = \exp\{-2w\sqrt{2m_e e\phi}/\hbar\}. \tag{14.9}$$

[3] This is for one spin. The conductance for an unpolarized spin channel is twice as large, $2e^2/h$.

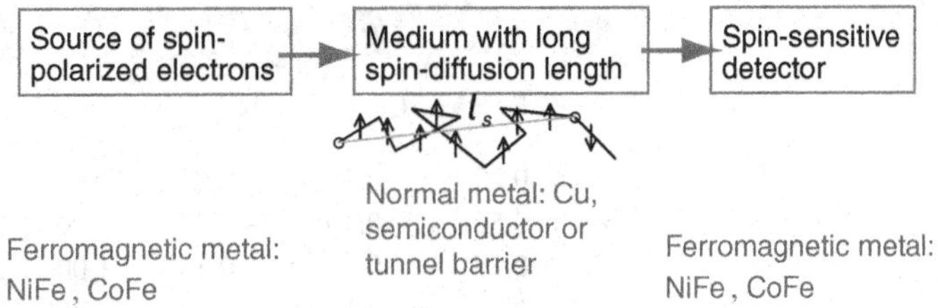

Figure 14.5

A generic spin electronic device, with a ferromagnetic spin injector and detector.

The characteristic feature of tunnelling transport is an almost temperature independent conductance $G \ll G_0$, which shows a nonlinear $I\!:\!V$ characteristic

$$I = GV + \gamma V^3, \qquad (14.10)$$

due to the deformation of the barrier in the applied electric field. Fitting this curve to the expression given by Simmonds for tunnelling through a barrier yields the effective barrier height ϕ and width w using (8.13) and (8.14).

The fourth mode of electron transport is hopping. This arises when the electrons are localized, so they are described by a confined wavepacket $\psi \sim \exp(-\alpha r)$ rather than an extended wave function $\psi \sim \exp(\boldsymbol{k} \cdot \boldsymbol{r})$. Electrons move from one site to the next by thermally assisted jumps. When the hopping to nearest-neighbour sites requires an activation energy ε_a, the conductivity is proportional to $\exp(-\varepsilon_a/k_BT)$, much like that of a semiconductor. At low temperatures, the electrons may jump further afield to find a site with almost the same energy. The conductivity is then described by Mott's variable-range hopping expression

$$\sigma = \sigma_\infty \exp -(T_0/T)^{1/4}, \qquad (14.11)$$

where $T_0 = 1.5/[k_B\alpha^3\mathcal{N}(\varepsilon_F)]$. Electron hopping is the principal transport mechanism in defective oxides and organic conductors.

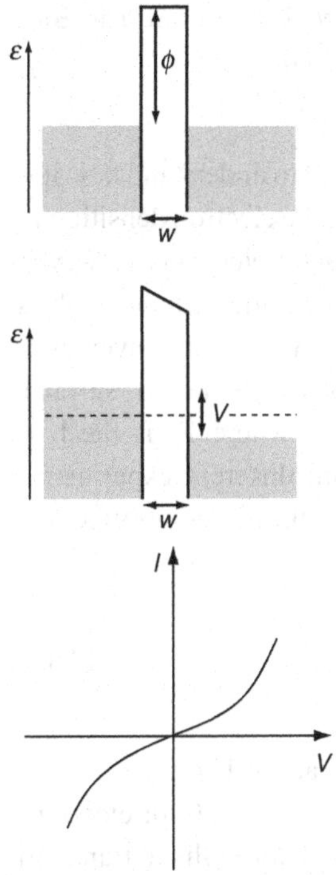

A tunnel barrier for electrons is deformed when a voltage is applied, leading to a nonlinear $I\!:\!V$ curve.

14.1.2 Spin polarization

Spin electronics depends on the creation and detection of spin polarization of mobile electrons. A simple device comprises a source of spin-polarized electrons, which are injected into a conductor of some description, where they transmit information coded in their spin polarization to a detector, Fig. 14.5. Electrical spin injection and detection are relatively straightforward for all-metal structures such as GMR spin valves, and the technique works well for the TMR spin valves with an insulating tunnel barrier, presented in Chapter 8. These spin-valve magnetoresistors are two-terminal electronic devices. Spin injection is much more problematic for semiconductors where the spin polarization of injected electrons is lost because of the impedance mismatch at the interface. Although the spin injector and detector are often ferromagnets, as in Fig. 14.5, the spin-polarized electrons may also be created or detected with polarized light.

Table 14.2. Calculated values of spin polarization

	Fe	Co	Ni	CrO_2	$La_{0.67}Ca_{0.33}MnO_3$	$Tl_2Mn_2O_7$
p^a	0.27	0.18	0.06	1.00	0.91	0.99
P_0	0.52	−0.70	−0.77	1.00	0.36	0.66
P_1	0.38	0.39	0.43	1.00	0.76	−0.05
P_2	0.36	0.11	0.04	1.00	0.92	−0.71

[a] Spin polarization calculated from 3*d* and 4*s* electron densities. (Data courtesy of Ivan Rungger, and from B. Nadgorny, I.I. Mazin, M. Osovsky *et al. Physics Review*, B**63**. 184433 (2001) and D. Singh, *Physics Review*, B**55**. 313 (1997)).

Spin polarization is the central concept here, so we need to understand how it is defined and measured. The definition (3.22) in terms of the electron densities $n_\uparrow$ and $n_\downarrow$ is equivalent to the relative spin magnetization of the electrons. However, when discussing TMR in Chapter 8, we introduced the definition which involves the densities of states at the Fermi level $\mathcal{D}_\uparrow(\varepsilon_F)$ and $\mathcal{D}_\downarrow(\varepsilon_F)$. For convenience we will drop ε_F. Transport properties involve electrons at the Fermi surface, or electrons within an energy eV_b of the Fermi level, where V_b is the bias voltage. However, the spin polarization deduced from different experiments depends on what precisely is being measured. A generalized definition at low bias is

$$P_n = \frac{v_{F\uparrow}^n \mathcal{D}_\uparrow - v_{F\downarrow}^n \mathcal{D}_\downarrow}{v_{F\uparrow}^n \mathcal{D}_\uparrow + v_{F\downarrow}^n \mathcal{D}_\downarrow}, \tag{14.12}$$

where v_F is the Fermi velocity. The density of states at the Fermi level $\mathcal{D}_{\uparrow,\downarrow}$ is weighted by the Fermi velocity raised to a power n, which is 0 for electrons ejected in a spin-polarized photoemission experiment, 1 for ballistic transport and 2 for diffusive transport or tunnelling at low bias. The sense of the velocity averaging is uniaxial for ballistic transport. Furthermore, transport measurements reflect the electron mobility, so measurements in 3*d* metals are sensitive to the polarization of 4*s* rather than 3*d* electrons. Table 14.2 lists some calculated spin polarizations for 3*d* ferromagnets, including a half-metal, CrO_2, and a semimetallic ferromagnet, $Tl_2Mn_2O_7$. The interesting point is that different definitions lead to completely different results, and they do not even necessarily agree on the sign. Only for a half-metal are they all the same $P_n = 1$.

Some methods of measuring spin polarization are indicated in Fig. 14.6. Photoemission probes the electron density at the Fermi level directly, weighted by the photoemission cross section. The electrons are ejected from the surface by ultraviolet light or soft X-rays and their spin polarization can be determined from their scattering in a heavy metal foil – a Mott detector. The energy bands can be mapped from the energy and angular dependence of the photoemitted electrons, but the method is very surface-sensitive, and the energy resolution is

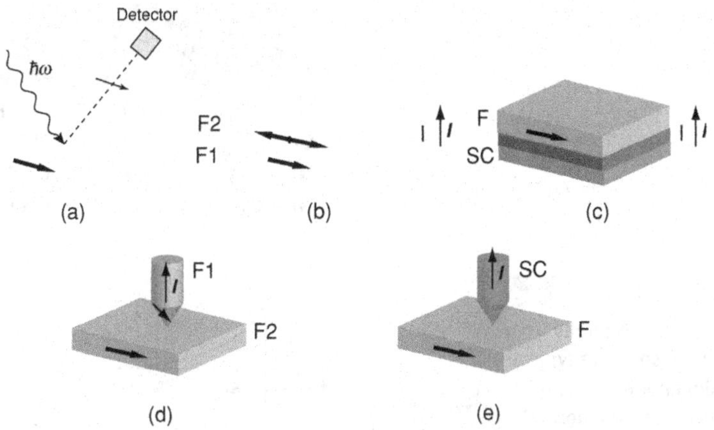

Figure 14.6

Some experimental configurations for deducing the spin polarization of a ferromagnet: (a) photoemission with polarization analysis, (b) tunnel magnetoresistance, (c) Tedrow–Meservey experiment, (d) ballistic point contact, (e) Andreev reflection. (F – ferromagnet; I – insulator; SC – superconductor.)

poor ($\gtrsim$ 100 meV). Ferromagnetic samples are generally studied in the remanent state.

The spin polarizations defined by (14.12) are intrinsic properties of the ferromagnet. However, in spin electronics we are more often interested in the spin polarization of an electric current than that of the electrons themselves. There can be a strong influence of the interfaces in a thin-film stack. Nevertheless, the Jullière model (8.15) is often used to deduce P_2, from the spin-dependent tunnelling in a junction with two ferromagnetic electrodes and an amorphous barrier. Spin filter effects depending on the symmetry of the transmitted wave functions appear with crystalline barriers. If one electrode is superconducting, spin polarization can be deduced by applying a magnetic field that is insufficient to turn the superconductor normal, as in the Tedrow–Meservey experiment (§8.3).

Ballistic point contacts between two ferromagnets can be used to estimate P_1, but it may be difficult to avoid effects of magnetostriction when the magnetic configuration changes from parallel to antiparallel. Finally, ballistic transport across a point contact between a ferromagnet and a superconductor offers a new possibility of determining the spin polarization at low temperature, from Andreev reflection.

In Andreev reflection, Fig. 14.7, an electron injected into the superconductor through the point contact with a normal metal must pair upto form a Cooper pair if the bias voltage is less than the superconducting energy gap Δ_{sc}. As a result, a hole with opposite spin is injected back into the metal. When $V < \Delta_{sc}$, the conductance of the contact is doubled compared with the value when $V > \Delta_{sc}$, on account of the current carried by the hole. For a half-metal, there are no vacant states with opposite spin near the Fermi level, so the conductance should be strictly zero when $V < \Delta_{sc}$. The polarization in the general case is

$$P_1 = \frac{1}{2}\left\{1 - \frac{G(0) - G(V > \Delta_{sc})}{G(V > \Delta_{sc})}\right\}. \tag{14.13}$$

Table 14.3. Spin polarization deduced from Andreev reflection

Fe	0.40	$Co_{50}Fe_{50}$	0.50
Co	0.40	NiMnSb	0.45
Ni	0.35	Co_2MnSi	0.55
$Ni_{80}Fe_{20}$	0.45	CrO_2	0.95

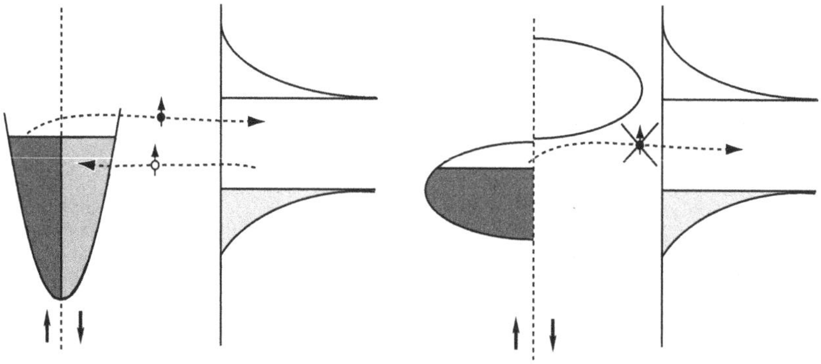

Figure 14.7

Point-contact Andreev reflection, showing typical bias dependence of the conductance for a normal metal (left) and a half-metal (right) (after R Soulen, J. M. Byers, M. S. Olovsky *et al.*, *Science*, 282, 85 (1998).)

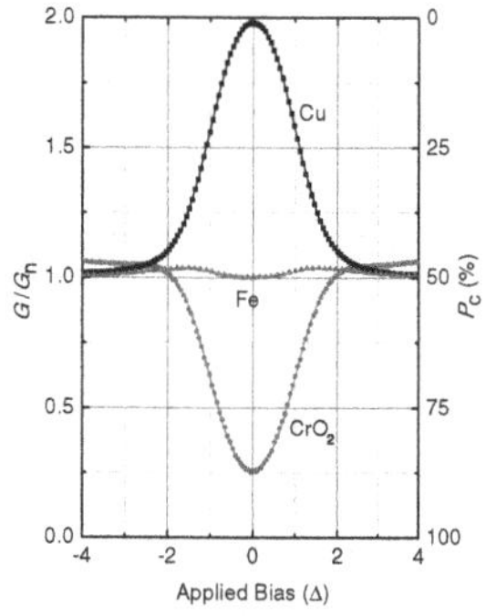

Andreev reflection for Cu, Fe and CrO_2 with a Nb point contact. The applied voltage is in units of the superconducting energy gap.

Some values of spin polarization deduced from Andreev reflection are given in Table 14.3. The method does not determine the sign.

In summary, there is no unique definition of spin polarization, except in a half-metal. Results depend on the experimental method employed, as well as on the electronic structure of the ferromagnet and the nonmagnetic materials involved, as well as their interfaces.

14.1.3 Spin injection and spin accumulation

An important concept in spin-polarized electron transport, introduced in Chapter 8, is spin accumulation. Spin populations are modified in the vicinity of a contact between a ferromagnet and a normal metal, where spin-polarized electrons diffuse into the normal metal and unpolarized electrons diffuse into the ferromagnet. The equilibrium changes in spin population are very small, $\sim k_B T/\varepsilon_F$, but the extent of the magnetic blurring on each side of the interface is the spin-diffusion length.

Significant nonequilibrium changes in spin population are set up when a current flows across the interface. Spin-polarized electrons are injected from the ferromagnet into the normal metal, and diffuse away from it. The interface, and associated changes of chemical potential are illustrated in Fig. 14.8 and in Fig. 8.19(c).

The best way to sense the small, spin-dependent voltages associated with nonequilibrium spin injection is in a nonlocal transport measurement. The

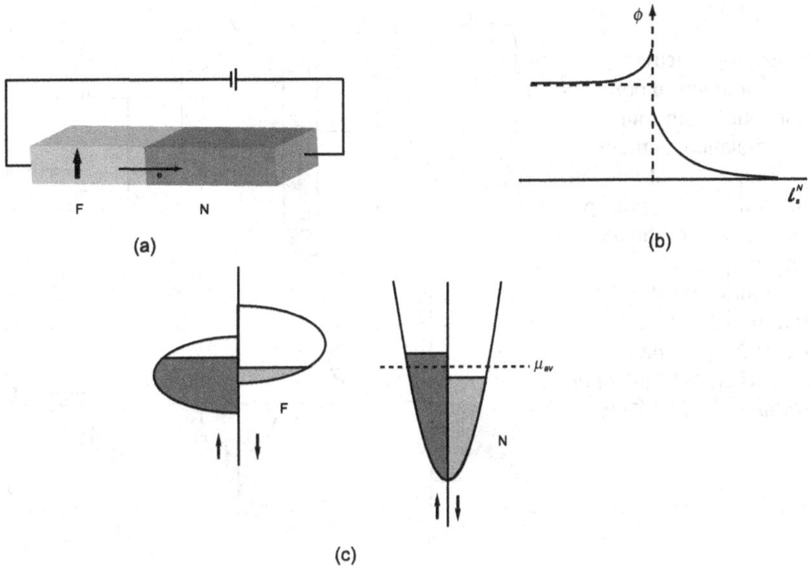

Figure 14.8

(a) Spin injection at an interface, where electrons flow from a ferromagnet (F) into a normal metal (N). (b) Spin-polarized electrons accumulate in the normal metal. (c) The nonequilibrium chemical potentials near the interface.

spin-polarized electrons are injected into the normal metal from the ferromagnet, F1, and diffuse away from it. The potential of another ferromagnetic electrode which is not in the current path, F2, is measured. The spin-polarized electrons drift away from the injector, and a spin-polarized charge cloud is set up on a scale of l_s. The potential of the sensing electrode is slightly different for the parallel and antiparallel configurations of F1 and F2, because the ↑ and ↓ electrons have different chemical potentials in the spin accumulation region, as indicated on Fig 14.9. By measuring the spin-dependent potential as a function of x, the spin diffusion length can be determined directly, although it was challenging to do this on account of the short length scale, which is of order 100 nm for a nonmagnetic metal at room temperature.

The decay of the spin accumulation voltage, $-(\mu^{\uparrow} - \mu^{\downarrow})/e$, on either side of a ferromagnet–nonmagnet (F–N) metal interface is described by the spin-diffusion equation, originally developed to describe the spreading of nuclear polarization in NMR. The diffusion equations for the two spin populations are

$$\frac{n^{\uparrow,\downarrow}(x,t)}{\tau_s} = D_e \frac{\partial^2 n^{\uparrow,\downarrow}(x,t)}{\partial x^2}, \tag{14.14}$$

where D_e is the electronic diffusion constant, and τ_s is the spin relaxation time which is related to the spin-spin relaxation time T_2 determined in a conduction electron spin resonance experiment. The surplus spin density produced by the current flowing across the interface $m(x,t) = n^{\uparrow}(x,t) - n^{\downarrow}(x,t)$ is obtained by integrating the spin-split electronic density of states $\mathcal{D}(\varepsilon)$ up to the chemical potential $\mu^{\uparrow}$ or $\mu^{\downarrow}$, and taking the difference. Hence

$$m(x,t) = \mathcal{D}(\varepsilon_F)\left[\mu^{\uparrow}(x,t) - \mu^{\downarrow}(x,t)\right].$$

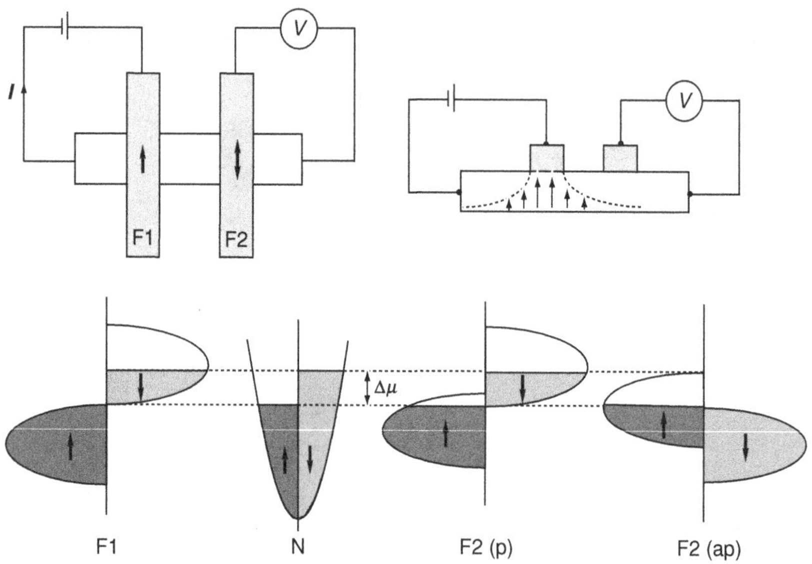

Figure 14.9

A nonlocal electrical measurement to probe the nonequilibrium spin accumulation, without passing a current through the sensing electrode. p and ap refer to the parallel and antiparallel configurations of ferromagnets F1 and F2. (After F. J. Jedema H. P. Heerscher, A. T. Flub *et al.*, *Nature* **416**, 713 (2002).)

Applying (14.14) to the difference $n^{\uparrow} - n^{\downarrow}$, the steady-state solution for the spin voltage is

$$(\mu^{\uparrow} - \mu^{\downarrow})_x = (\mu^{\uparrow} - \mu^{\downarrow})_0 \exp(-x/l_s), \tag{14.15}$$

where $l_s = \sqrt{D_e \tau_s}$, and the subscript shows where the spin voltage is evaluated.

On the ferromagnetic side of the F–N junction, the spin polarization builds up over a length scale determined by the *s–d* scattering in the ferromagnet (Fig. 8.19(c)). In pure ferromagnetic metals, l_s is about 40 nm and τ_s is of order 1 ps, but the values are smaller in alloys such as Ni–Fe or Co–Fe, where the scattering is greater.

The chemical potentials for a GMR spin valve are shown in Fig. 14.10 for the case where the nonmagnetic layer is thin enough for spin relaxation there to be neglected. On flipping the free layer from parallel to antiparallel, there is a change in spin-dependent potential and hence a change in the spin accumulation voltage V_{sa} at constant current – in other words, magnetoresistance.

The interface resistance area product RA_{int} is V_{sa}/j, which is expressed in terms of the conductivity ratio $\alpha = \sigma^{\uparrow}/\sigma^{\downarrow}$ using (8.10):

$$RA_{int} = \frac{(\alpha - 1)}{2ej(1+\alpha)}(\mu^{\uparrow} - \mu^{\downarrow})_0. \tag{14.16}$$

The GMR is

$$\frac{\Delta R}{R} = \frac{2R_{int}}{\varrho_f t_f},$$

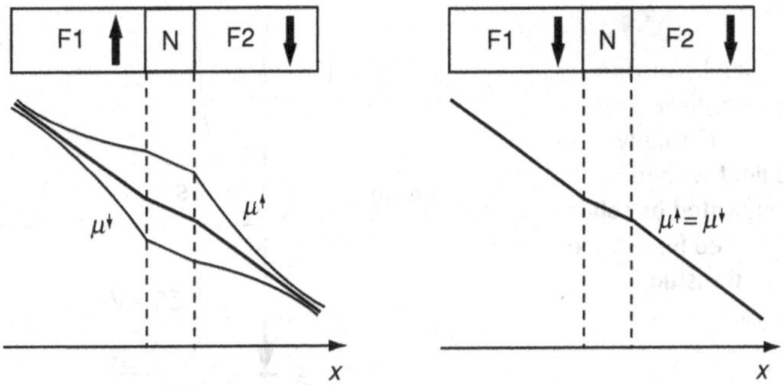

Figure 14.10

Chemical potentials for ↑ and ↓ electrons in a GMR spin valve, with parallel or antiparallel alignment of the electrodes. The spacer layer is supposed to be thin compared to the spin diffusion length. (F1, F2 – ferromagnets; N – normal metal.)

where ϱ_f is the resistivity of the ferromagnet and t_f is the total thickness of the two ferromagnetic layers. In the limit of a thin nonmagnetic spacer and thick ferromagnetic layers, the result is

$$\frac{\Delta R}{R} = \frac{(\alpha - 1)^2}{4\alpha} \frac{l_{sf}}{t_f}. \tag{14.17}$$

where l_{sf} is the spin diffusion length in the ferromagnet. The effect is maximized when the value of α is large, and the spin diffusion length in the ferromagnet is long. If $\alpha = 5$ and $l_{sf}/t_f \approx 0.25$, the magnetoresistance would be 20%. The order of magnitude of the spin accumulation voltage, even at very high current densities $j \approx 10^{12}$ A m^{-2}, does not exceed a millivolt.

In this analysis, we have assumed that the interfaces are transparent. In fact, spin-dependent scattering and reflection at the interface play a crucial role in the operation of GMR spin valves. The step in chemical potential at the interface is $\Delta\mu_i^{\uparrow\downarrow} = -eI^{\uparrow\downarrow}R_i^{\uparrow\downarrow}$, where R_i is the interface resistance. The enormous gradient of the exchange field at the interface deflects the ↑ and ↓ electrons in the adiabatic limit, as in a Stern–Gerlach experiment (Fig. 3.3). In the nonadiabatic limit of perfectly abrupt interfaces, nonadiabatic quantum reflection is important.

Spin lifetimes and diffusion lengths are much longer in semiconductors than they are in metals, but electrical spin injection and detection of spin-polarized electrons are problematic, because of the resistance mismatch between the metal and the semiconductor. Considering the two spin channel resistances adding in parallel, the metal plus contact resistance is spin-dependent, but the much larger semiconductor resistance is not. The current in each channel is almost the same, so the electrons are not significantly polarized. It is appropriate to treat scattering at interfaces in terms of resistance rather than resistivity; the scattering is integrated over the junction region where the potential changes.

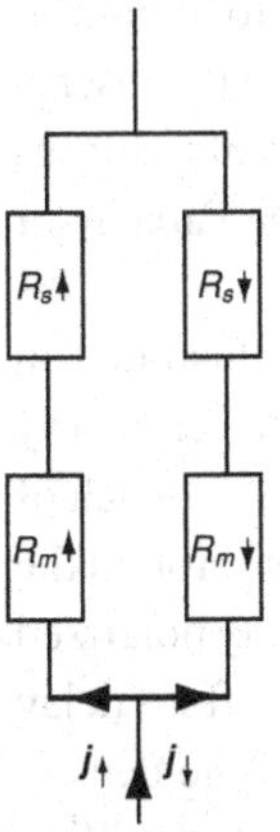

The resistance mismatch problem. Here $R_{m\uparrow} \neq R_{m\downarrow}$ but $R_{s\uparrow} = R_{s\downarrow}$ and $R_s \gg R_m$, hence $j_\uparrow \approx j_\downarrow$.

One way to overcome this problem is to introduce a highly resistive Schottky barrier or a tunnel barrier at the semiconductor–metal interface, and to inject spin-polarized hot electrons over the barrier. The barrier has a spin-dependent

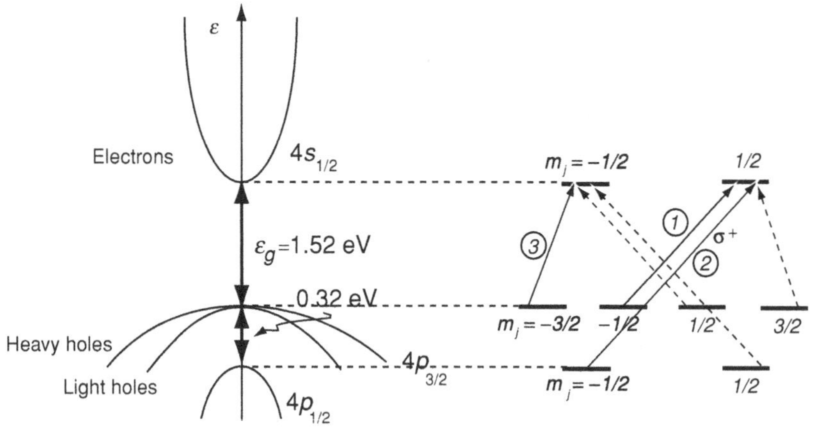

Figure 14.11

Schematic band structure of GaAs. The optical transitions for circularly polarized light with $m_j = 1$ (σ^+) are indicated by solid lines. The circled figures are the relative transition probabilities.

resistance, $R_{b\uparrow,\downarrow} > R_s$. Alternatively, optical spin injection using circularly polarized light is a solution for semiconductors with a direct bandgap, such as GaAs. The band structure is illustrated in Fig. 14.11 and the bandgap is 1.52 eV. The As $4p$ valence band is split by spin-orbit coupling into a $4p_{1/2}$ and two $4p_{3/2}$ sub-bands, which have different effective masses. When the semiconductor is pumped with circularly polarized photons, which are sufficiently energetic to excite electrons from all three valence sub-bands into the Ga $4s$ conduction band, the electrons populate the $m_j = \pm\frac{1}{2}$ states equally, giving no net spin polarization. However, if the photon energy lies between 1.52 eV and 1.84 eV, the $m_j = -\frac{1}{2}$ and $m_j = +\frac{1}{2}$ are populated in the ratio 3:1, so the net polarization of the conduction electrons p_c defined as $(n_{c\uparrow} - n_{c\downarrow})/(n_{c\uparrow} + n_{c\downarrow})$ is $p_c = (1-3)/(1+3) = -50\%$. The $4s$ electrons have a spin lifetime of order nanoseconds, whereas the $4p$ holes depolarize more rapidly, due to spin-orbit scattering, in about 100 ps. The spin-polarized carriers are detected by observing the circularly polarized luminescence emitted by recombination of electrons and holes. The spin polarization can also be detected by using the Faraday or Kerr effects.

A pump-probe experiment to determine the spin lifetime is illustrated in Fig. 14.12. The spin-polarized carriers are injected with a pulse of circularly polarized light from a laser, tuned to the semiconductor band edge. A bunch of polarized spins excited from the valence band then begin to precess around the applied field B at the Larmor frequency $\omega_L = (ge/2m_e)B$. A linearly polarized probe pulse is sent to measure their polarization by Faraday rotation after a delay τ. The g-factor in semiconductors may be quite different from 2. The density of excited electrons is of order 10^{21} m^{-3}, or fewer than one per ten million atoms, but a measurable Faraday rotation θ_F is obtained by averaging repeated pulses.The spin relaxation time is obtained from the envelope of the decay:

$$\theta_F = \exp(-t/\tau_s)\cos\left[(ge/2m_e)Bt\right]. \quad (14.18)$$

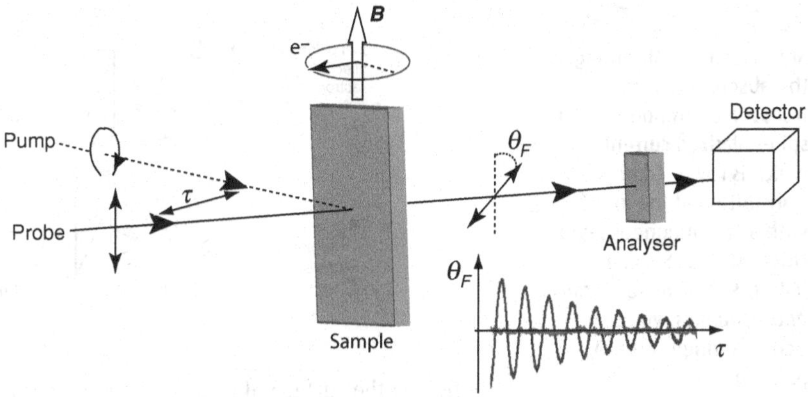

Figure 14.12
A pump-probe experiment to determine the lifetime of spin-polarized electrons in GaAs.

In order to determine the spin-diffusion length, the excited electrons drift under the influence of an electric field, and the probe beam is scanned in space along a line from the injection point. Since the spin-diffusion lengths in semiconductors are of order 100 μm, it is quite feasible to resolve the position of the diffusing packet of electrons optically. The constant D_s governing spin diffusion in semiconductors may exceed that appropriate for electron transport.

Luc Berger, 1933–.

John Slonczewski, 1928–.

14.1.4 Spin-transfer torque

Besides the flow of charge $j = -nev_d \, \mathrm{C\,m^{-2}\,s^{-1}}$, an electric current with spin polarization $0 < P_e < 1$ also carries with it a flow of angular momentum

$$j_s = n(\hbar/2)p_c v_d = j\hbar p_c/2e, \tag{14.19}$$

where the units are $\mathrm{J\,m^{-2}}$. Unlike charge, which is conserved in the total electric current, the angular momentum can be absorbed by spin-dependent scattering and other processes in a lattice, leading to a spin-transfer torque $\mathbf{\Gamma}$ which is equal to the rate of change of angular momentum in the lattice. The total angular momentum of the system electrons + lattice has to be conserved, so any loss of angular momentum of the electric current has to be balanced by a gain of angular momentum of the lattice. Even an initially unpolarized current can exert torque, if it becomes spin polarized via a spin-dependent scattering process in a ferromagnetic lattice. The spin-transfer torque is able to excite magnons and microwaves, move domain walls and reverse the magnetization of nanoscale magnetic free layers. The theory has been developed by Luc Berger and John Slonczewski, who also envisioned some of the applications. In the nanoworld, it is more effective to exert torque by spin transfer than by the magnetic fields created by currents in nearby conductors, which are known in this context as Oersted fields. Manipulation of magnetization by spin-polarized currents is one of the most exciting developments in contemporary magnetism.

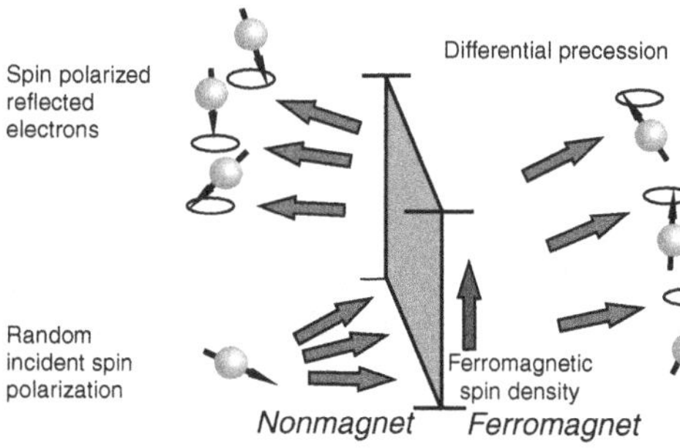

Figure 14.13

Mechanisms contributing to the absorption of the transverse component of a spin-polarized current which is reflected or transmitted at the interface with a ferromagnetic layer. (After M. D. Stiles and J. Miltat, *Spin Transfer Torque and Dynamics*, vol. 3, Berlin: Springer (2006), p. 205.)

The field at the surface of a current-carrying wire of radius r has magnitude $H = jr/2$, so the Oersted fields become increasingly ineffective, falling off as r when the dimensions are reduced. The angular momentum associated with the magnetization of the free layer in a nanopillar, which has thickness t and radius r, is $M\pi r^2\mathrm{t}/\gamma$. The rate of flow of angular momentum associated with the current is $j\hbar p_c\pi r^2/2e$. Hence the switching effect, which depends on the ratio of these quantities, varies as $jp_c/M\mathrm{t}$, *independent of* r. Spin-transfer torque is scalable, although it only becomes effective at nanoscale dimensions. Scalability is an essential requirement of any new technology which hopes to make its way into mainstream electronics.

To follow the physics underlying spin-transfer torque, consider a spin-polarized electron entering a ferromagnetic thin film which is magnetized in the z-direction (Fig. 14.13). If the electron is moving in the x-direction and is initially polarized in a direction making an angle θ with $\boldsymbol{e}_z$, it may be either reflected or transmitted at the interface. For example, at a Cu–Co interface, the majority-spin $\uparrow$ electrons are more easily transmitted than the minority-spin $\downarrow$ electrons. Reflection at the interface is a source of spin torque, especially in the nonadiabatic case. The electrons that make it into the cobalt are then subject to a huge exchange field B_{ex} in the z-direction, which is of order 10^4 T. (The exchange splitting of the 3d band is of order 1 eV.) They precess at the Larmor frequency, completing many turns as they diffuse from site to site. By the time they leave the cobalt film, the xy component of the moments is completely dephased, as the electrons have followed paths of different lengths, but their z component of magnetization is unchanged provided the thickness is less than the spin-diffusion length. Angular momentum is therefore transferred from the electron current to the cobalt at a rate $(j\hbar p_c \sin\theta)/2e$ J m^{-2}. All this has little effect on a thick cobalt layer, but it can greatly modify the magnetization of a thin one, which tends to rotate towards the direction of polarization of the incoming current. The transverse angular momentum can be absorbed in a distance which is less than the mean free path.

Another mechanism is also effective for transferring angular momentum from the electron current. The mean free paths are different for $\uparrow$ and $\downarrow$ electrons in a ferromagnet (Table 14.1), often with $\lambda_\uparrow > \lambda_\downarrow$. The minority-spin electrons

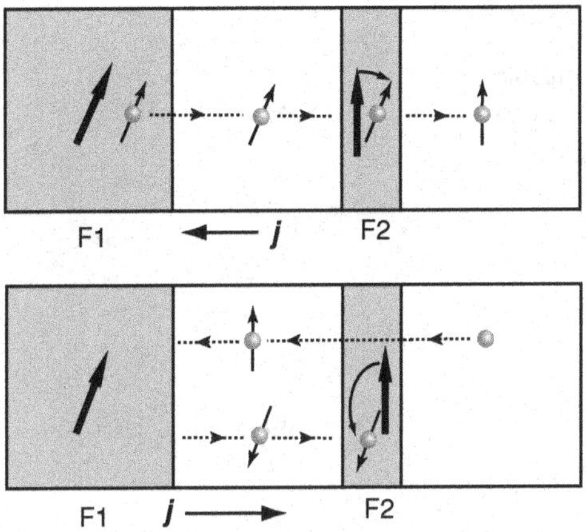

Figure 14.14

Spin-transfer torque associated with flow of angular momentum in a nanowire or nanopillar. The sense of the torque on the thin free layer, F2, depends on the current direction. The unshaded layers are nonmagnetic.

share their angular momentum with the lattice within the first nanometre or so of the interface, while their majority-spin colleagues travel much further before they are scattered. SS scattering at the interface can also contribute to the torque. Scattering must be spin-selective, but spin-flip scattering is not required. The spin-diffusion length l_s is not a relevant lengthscale.

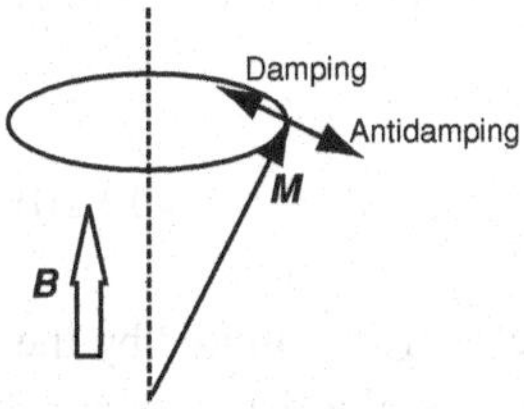

Precession of magnetization with antidamping due to spin-transfer torque.

By whatever mechanism, an unpolarized electron current becomes spin polarized after traversing about a nanometre of cobalt. The ability of ultra-thin cobalt layers to effectively impart spin polarization is widely exploited in the thin-film stacks used for spin electronics. A ballistic electron travelling at 10^6 m s^{-1} in a random direction completes a precession in a layer which is of order a nanometre thick, so the transverse component of angular momentum is very soon absorbed by the precession mechanism.

Next consider electron flow across the structure including a thick and a thin ferromagnetic layer, as shown in Fig. 14.14. Think first of electron flow in the direction from the thick layer to the thin one. The charge current is in the opposite direction, of course, because the electron has a negative charge.

The initially unpolarized electron current will be polarized in the direction of magnetization of the thick layer when it emerges. The spin polarization is not perfectly efficient; for a Co/Cu stack p_c is about 35%. The transverse component of angular momentum is later absorbed in the thin layer as we have just explained. A torque acts on F2, tending to turn the magnetization towards the orientation of the incoming spins. A parallel configuration of F1 and F2 is stabilized.

Now consider what happens when electrons flow in the opposite direction, from the thin layer to the thick one. They cross F2 acquiring spin polarization, and most of them go on to enter F1 where they ineffectually exert torque on the thick pinned layer. But some, predominantly those with spin opposite to F1, are reflected, and travel back to F1, where their transverse component of an angular momentum is absorbed, this time tending to stabilize an antiparallel configuration of F1 and F2. In summary, according to the direction of the

Figure 14.15

Precession of a macrospin with: (a) damping, (b) no net damping and (c) antidamping.

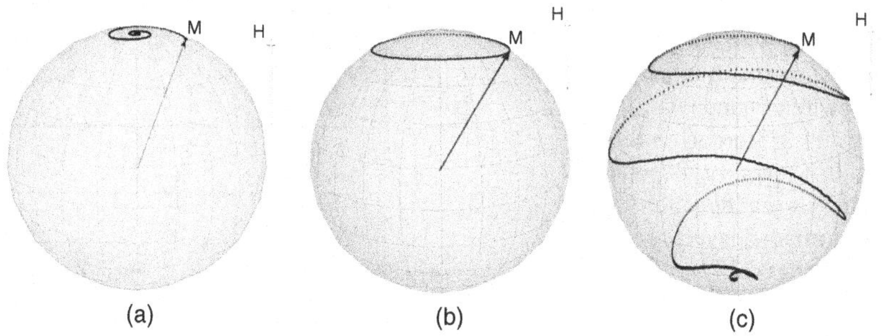

current, the spin-transfer torque tends to stabilize or destabilize the magnetic configuration of the two ferromagnetic layers.

The shortest possible time t taken to switch a free layer of thickness t can be estimated from the rate of flow of angular momentum (14.19). Equating the angular momentum delivered in time t to the angular momentum $M\mathcal{A}\hbar/2\mu_B$ associated with the magnetic moment of the free layer of cross-section a area $\mathcal{A}$ gives

$$t = \frac{M\mathrm{t}e}{j\mu_B P_e}. \tag{14.20}$$

This is independent of $\mathcal{A}$, but the switching speed is effectively limited by the maximum current density j that can be pumped through the device, which is ultimately limited to about 10^{12} A m^{-2} by problems of electromigration. In practice, the free layers must be only a few nanometres thick, which is well within the capabilities of modern thin-film technology.

To treat the dynamics of the free layer in more detail, allowing for the presence of an applied magnetic field, we make the macrospin approximation, and assume that the free layer rotates coherently like a giant spin at $T = 0$ K. The precession is described by the Landau–Lifschitz–Gilbert equation (9.23) with an extra damping-like term α' added to account for the spin-transfer torque:

$$\frac{\mathrm{dm}}{\mathrm{d}t} = \gamma\mu_0 \mathrm{m} \times \boldsymbol{H}_{eff} - (\alpha + \alpha')\mathbf{e}_m \times \frac{\mathrm{dm}}{\mathrm{d}t}. \tag{14.21}$$

The effective field $\boldsymbol{H}_{eff}$ is the sum of the applied, dipolar and anisotropy fields. According to the direction of the current, the spin-transfer torque adds to the damping torque, or subtracts from it. When the current flows in the $-\mathbf{e}_x$-direction, the spin-transfer term simply reinforces the damping. However, when the current flows in the $+\mathbf{e}_x$-direction, a dynamic instability arises when the spin-transfer term overcomes the damping. Instead of relaxing towards $\mathbf{e}_z$, the magnetization begins to spiral away from it, Fig. 14.15.

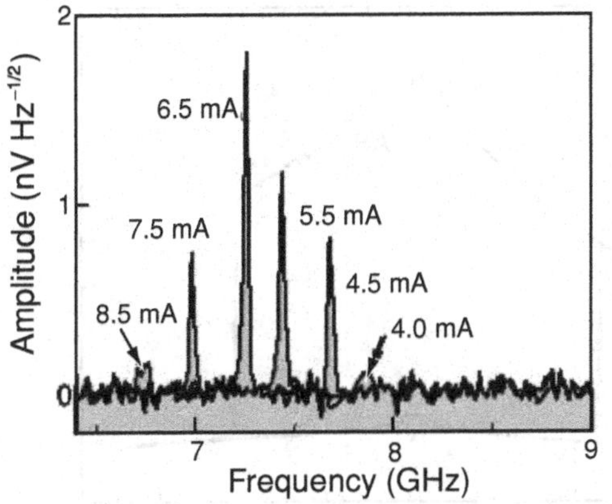

Figure 14.16

Excitation of microwaves from a GMR spin valve nanopillar by spin-transfer torque. (W. H. Rippard *et al.*, *Phys. Rev. Lett.* 92, 027201 (2004).)

The extra 'damping' term α' due to spin-transfer torque may be of either sign, according to the direction of spin-polarized current in the structure of Fig. 14.14. The macrospin dynamics when $-\alpha' > \alpha$ can take several forms. In one case, the free layer may spiral all the way to $\theta = \pi$, and be damped in the reverse direction. Switching by spin-transfer torque is particularly valuable for magnetic memory because of the scalability.

Another possibility is that the free layer may not spiral all the way to $\theta = \pi$, but precess continually in a magnetic field at some intermediate angle. The DC current thereby generates steady-state magnetic oscillations in the GHz frequency range with a frequency proportional to j (Fig. 14.16). Easily tunable on-chip microwave sources of this type offer new perspectives for magnetism and interchip communications.

The dynamics of the two-layer structure of Fig. 14.14 is surprisingly complex. The effect of temperature, which is normally to produce fluctuations in the orientation of the free layer, is enhanced by spin-transfer torque. Figure 14.17(a) shows switching of the free layer in a MTJ nanopillar structure like that of Fig. 14.14 by current. There is a current j_p, less than the switching current, where the oscillations set in. A calculated phase diagram as a function of current and applied field is shown in Fig. 14.17(b). Thermal fluctuations tend to facilitate spin-torque switching, and reduce the critical switching current density j_c.

14.1.5 Spin currents

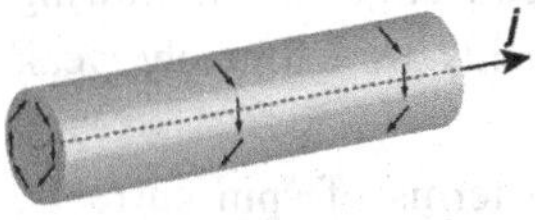

Spin Hall effect in a wire.

When an electric current passes through a conductor that contains spin-orbit scatterers, there is a tendency for spin to accumulate at the surface of the conductor. Intrinsic spin-orbit coupling may have a similar effect. This is known as the spin Hall effect. The principle is the same as for the Mott detector, which

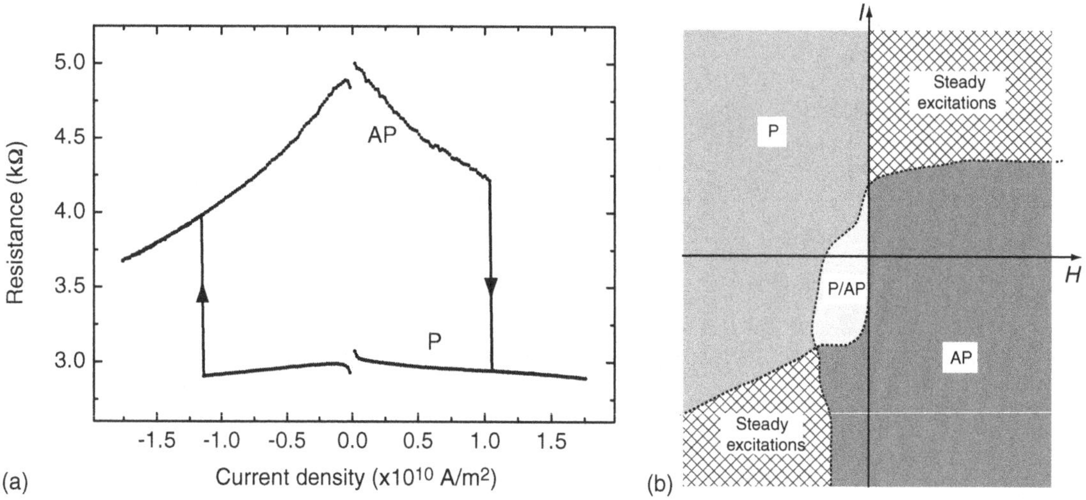

Figure 14.17

Spin-torque switching of the free layer in a nanopillar tunnel junction structure, with a single-domain free layer. P and AP refer to parallel and antiparallel configurations of the free and pinned layers. Data courtesy Kaan Oguz. (b) Calculated phaser diagram. (S. I. Kiselev *et al.*, *Nature* 425, 380 (2003).)

senses the polarization of an electron beam by passing it through a gold foil where it undergoes spin-orbit scattering, which separates the ↑ and ↓ beams. The effect is difficult to detect in a wire, because the Oersted field created by the current has the same symmetry as that due to the accumulated spin. However, the spin accumulation has been detected optically in a flat slab of GaAs, using the Kerr effect, Fig. 14.18.

The spin current in these examples, and the one that flows when the device in Fig. 14.9 is switched on, is driven by the charge current, although, unlike a normal spin-polarized electric current, it flows in a different direction. In the example of the flat slab, the transient spin current flows across the slab, and equilibrium concentrations of ↑ and ↓ electrons accumulate at opposite edges.

It is possible for spin currents to be completely divorced from charge currents, for example, when domain wall motion occurs. AC spin currents are associated with the propagation of spin waves. However, these pure spin currents can only be transient or alternating. The maximum possible transfer of angular momentum is constrained by flipping all the spins in the system. It is a challenge to devise a DC spin battery which could drive a steady flow of angular momentum, unassociated with a steady flow of charge. New concepts of spin-based computation seek to eliminate the charge current entirely, and thereby bypass the mounting energy dissipation associated with CMOS processors.

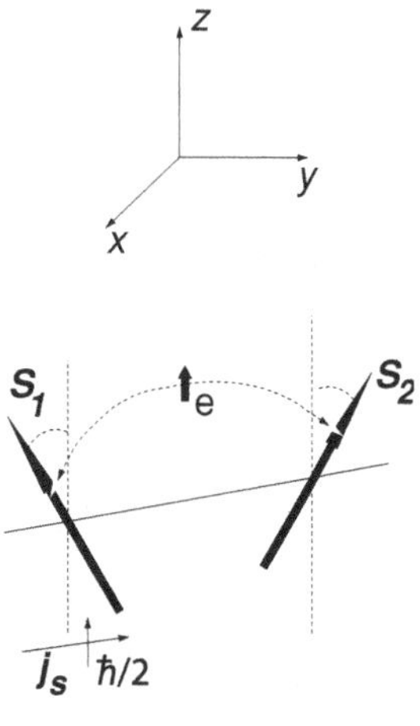

Exchange as a spin current.

A spin current, due for example to a fully polarized charge current flowing in a conductor, creates an electric field, but these fields are extremely weak (Exercise 14.6).

Exchange interactions can also be thought of in terms of spin currents. As an electron hops back and forth between two atoms, the momentum

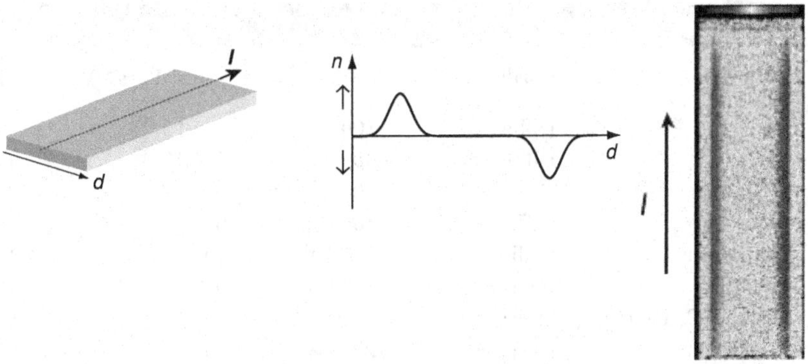

Figure 14.18

An experiment to detect the spin Hall effect in a slab of GaAs. The spin-up and spin-down densities, $n_\uparrow$ and $n_\downarrow$ are indicated by the light and dark regions. (Y. K. Kato, *Science*, 306, 1910 (2004).)

transfer is non-reciprocal, tending to align the atomic spins along a common z axis.

14.2 Materials for spin electronics

The generic spin electronic device, Fig 14.5, consists of a source of spin-polarized electrons, a transport medium and an analyser. The ferromagnetic polarizer and analyser are often Co, Fe, Co–Fe or Ni–Fe. Half-metallic Heusler alloys with high T_C such as Co_2MnSi are also used. Other half-metals of interest include oxides such as LSMO and CrO_2, but room-temperature device operation with these materials has yet to be demonstrated.

In principle, the spin transport medium can be almost anything that transmits electrons without completely destroying their polarization. In all-metal thin-film stacks, Cu or Al is often used. Heavy metals introduce spin-flip scattering, due to their strong spin-orbit interaction. Semiconductors are very effective spin transport media because there are few impurities or other electrons to scatter the spin-polarized carriers. Electrical spin injection and detection are problematic in semiconductors because of the problem of resistivity mismatch between the metal and the semiconductor. Optical injection and magneto-optic detection are feasible for semiconductors with a direct bandgap, such as III–V materials. The problem of injection and detection is less accute for organic semiconductors.

Insulators are useful as spin transport media only when they are thin enough to permit tunnelling. Ballistic conductors such as graphene, and organic conductors where spin transport is by hopping and injection is controlled by the ferromagnet–organic interface, are also potentially interesting. The organics have miserable mobility, but there is a very long spin lifetime which may find an application.

Relevant electronic properties of some representative materials are summarized in Table 14.4. There is a broad palette, with opportunities for innovation.

Table 14.4. Some materials for spin electronics

	Conduction	l_s (nm)	μ ($m^2 V^{-1}s^{-1}$)	v_F ($m s^{-1}$)	τ_s(s)
Cu	Diffusive	200	4×10^{-3}	1.6×10^6	3×10^{-12}
Co	Diffusive	40	2×10^{-3}	1.3×10^6	4×10^{-13}
Si	Diffusive	10^5–10^6	0.1	^{b}	10^{-9}–10^{-7}
GaAs	Diffusive	10^5	0.3	^{b}	10^{-9}–10^{-7}
Carbon nanotubes	Ballistic	5×10^4	10	1×10^6	3×10^{-8}
Graphene	Ballistic	1500	0.2	1×10^6	10^{-10}
Rubrene ($C_{42}H_{28}$)	diffusive	13	2×10^{-3}	^{b}	10^{-3}
Hexathiophene	Hopping	200	10^{-9}		10^{-6}
Alq_3^*	Hopping	45	10^{-14}–10^{-12}		10^{-2}–10

[a] Tris-(8-hydroxyquinoline) aluminium.
[b] Depends on the carrier concentration.

14.3 Magnetic sensors

A variety of different physical effects are exploited in magnetic sensors, which are passive devices that detect the presence of a magnetic field. They deliver an electrical voltage that is monotonic, and preferably proportional to one component of the field $\boldsymbol{B}$ or $\boldsymbol{H}$ acting at the sensor. The sensors are usually magnetoresistive structures or Hall bars, but occasionally it is advantageous to use magnetoimpedance sensors or sensors based on NMR. The latter can produce a signal proportional to the scalar magnitude of B. The great advantage of magnetic sensing is that it avoids the need for direct contact between the sensor and the object sensed. Several billion magnetic field sensors are manufactured each year, about a billion of them destined for hard-disc drives. Most of the rest are low-cost proximity sensors (§13.4.4). A car may contain as many as 100 of them, mostly integrated silicon Hall chips.

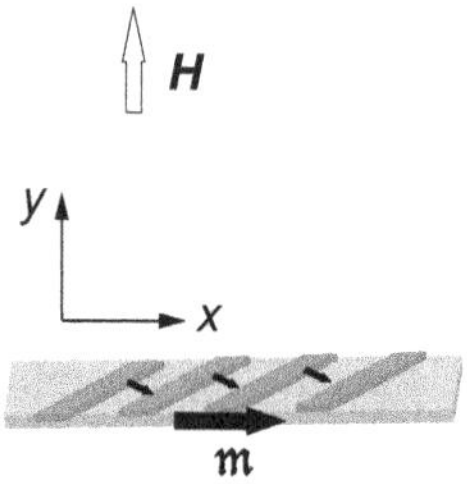

A barber pole. The dark bands are highly conducting strips which define equipotentials and the current flows as indicated by the arrows at 45° to the magnetization direction.

Characteristics of the main sensor types are summarized in Table 14.5. Principles of operation have been discussed in earlier chapters. Here we focus on aspects of thin film sensor design, and consider the critical issue of noise, which ultimately limits their sensitivity.

If a thin-film sensor, which depends for its operation on the direction of magnetization of a responsive ferromagnetic layer, is to deliver a magnetoresistive signal that is monotonic in field, the magnetization has to be able to respond continuously, rather than flip between bistable configurations ('high' and 'low' resistance states).

In the case of a single-film AMR sensor, which is usually made of permalloy, the magnetoresistive effect $\delta\varrho(\varphi) = [\varrho(\varphi) - \varrho(0)]$ is equal to $\Delta\varrho(\cos^2\varphi - 1)$, where φ is the angle between $\boldsymbol{M}$ and $\boldsymbol{j}$ (5.78). Sensitivity is greatest when

Table 14.5. Characteristics of magnetic field sensors

Sensor	Principle	Detects	Frequency Range	Field (T)	Noise/$\sqrt{\mathrm{Hz}}$	Comments
Coil	Faraday's law	$\mathrm{d}\Phi/\mathrm{d}t$	10^{-3}–10^{9}	10^{-10}–10^{2}	100 nT	Bulky, absolute
Fluxgate	Saturable ferromagnet	$\mathrm{d}\Phi/\mathrm{d}t$	0–10^{3}	10^{-10}–10^{-3}	10 pT	Discrete
Hall generator	Lorentz force	B	0–10^{5}	10^{-5}–10	100 nT	Thin film
Classical magnetoresistance	Lorentz force	B^2	0–10^{5}	10^{-2}–10	10 nT	Thin film
Anisotropic magnetoresistance (AMR)	Spin-orbit scattering	H	0-10^{7}	10^{-9}–10^{-3}	10 nT	Thin film
Giant magnetoresistance (GMR)	Spin accumulation	H	0–10^{9}	10^{-9}–10^{-3}	10 nT	Thin film
Tunnelling magnetoresistance (TMR)	Tunelling	H	0–10^{8}	10^{-9}–10^{-3}	1 nT	Thin film
Giant magnetoimpedance (GMI)	Permeability	H	0–10^{4}	10^{-9}–10^{-2}	1 nT	Wire
Magneto-optic	Faraday/Kerr effect	M	0–10^{8}	10^{-5}–10^{2}	1 pT	Bulky
SQUID (4 K)	Flux quantization	Φ	0–10^{9}	10^{-15}–10^{-2}	1 fT	Cryogenic
SQUID (77 K)	Flux quantization	Φ	0–10^{4}	10^{-15}–10^{-2}	30 fT	Cryogenic
Nuclear precession	Larmor precession	$\|B\|$	0–10^{2}	10^{-10}–10	1 nT	Bulky, very precise

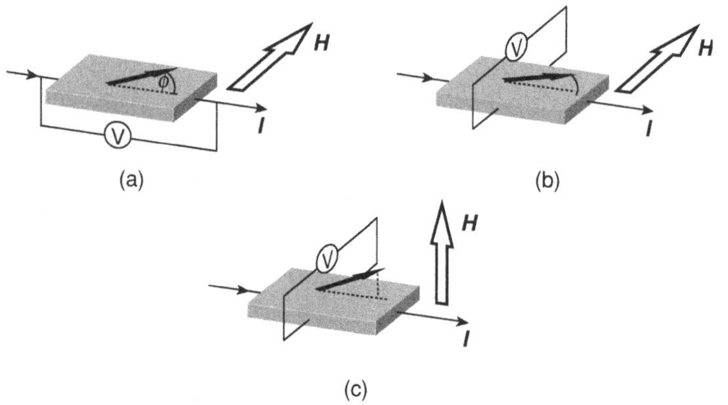

Figure 14.19

(a) AMR, (b) planar Hall and (c) anomalous Hall effect sensor configurations. The easy direction is shown by the dashed line.

$\varphi = \pi/4$. Ways of setting this angle are either to induce in-plane anisotropy by depositing or annealing the permalloy film in a magnetic field, or else to make a barber pole pattern overlaying the permalloy with stripes of a more conducting metal such as gold, which defines equipotentials across which the current flows. In either case, the sensor responds to the transverse component of the field in the plane of the film. Using the equilibrium condition that there is no net torque on the magnetization due to the internal field in the film, the demagnetizing field $-\mathcal{N}_y M \sin\delta\varphi$ balances the transverse applied field H_0'. For small rotations, $\delta\varphi = H_0/\mathcal{N}M$, and the variation of resistivity with field is

$$\delta\varrho(H) \approx \Delta\varrho H_0'/\mathcal{N}M. \tag{14.22}$$

The response is linear for small fields.

A planar Hall sensor also uses a single film of a material like permalloy. Here a transverse voltage is measured when a magnetic field is applied in-plane, perpendicular to the current (5.81). The effect is closely related to AMR, but the sensor design is a little simpler.

The anomalous Hall effect, measured in the standard Hall geometry, Fig. 14.19, can be exploited for detection of higher fields, applied perpendicular to the plane of the sensor, in which case $\mathcal{N}_z \approx 1$. Since the anomalous Hall voltage is proportional to the out-of-plane magnetization $M_\perp$ (5.79), and the demagnetizing field is $-\mathcal{N}_z M_\perp$, it follows from the condition that there is zero torque due to the internal field that $H_0' - \mathcal{N}_z M_\perp = 0$. The signal is linear, and proportional to H_0 up to saturation.

The linearization of a GMR or TMR spin-valve sensor is slightly different from the linearization of an AMR sensor. The free layer of a spin valve is normally free to rotate in the plane. The rotation in a well-designed sensor is coherent, and the free layer remains single-domain. The magnetoresistance then varies as $\sin^2\phi/2$, where ϕ is the angle between the magnetization of the free and pinned layers. Hence, the sensitivity is optimized for the crossed anisotropy configuration where $\phi = \pi/2$. This can be achieved by setting the exchange bias direction of the pinned layer perpendicular to the easy axis of the free layer. The GMR or TMR sensors are only linear for small changes of ϕ, $\Delta\phi \approx 10°$, so only a fraction of the total magnetoresistance is actually exploited in sensors such as read heads.

Giant magnetoimpedance (GMI) (§5.6.4) offers great sensitivity to low magnetic fields. In wire sensors, the circumferential anisotropy can be set by controlling the magnetostriction, or by annealing a current-carrrying wire, which creates a circumferential magnetic field. Excellent results, up to 1000%, are achieved using wires with a nonmagnetic copper core, and an electroplated permalloy sheath, about a micrometre thick. The noise sensitivity is comparable to that of good tunnel junctions. The wires are used as direction sensors in interactive computer games, for example.

A DC SQUID sensor is a superconducting ring with two weak links where flux is quantized as a multiple of $\Phi_0 = 2.1 \times 10^{-15}$ T m^2. Cooper-pair wave packets propagate on either side of the ring, accumulating a phase offset proportional to the penetrating flux, but opposite in sign. The resulting interference gives rise to a transmission probability periodic in the flux. An RF SQUID sensor contains only one weak link, often a metallic point contact. Its high-frequency inductance is again periodic in the flux linking the ring. In each case, linearization of the field sensor is achieved using a flux-locked loop.

14.3.1 Noise

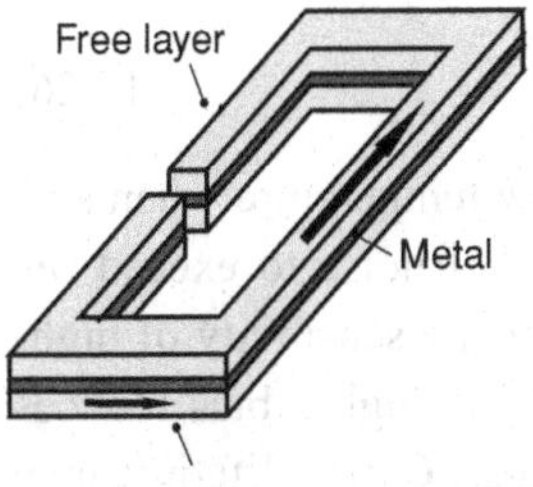

A yoke-type spin-valve sensor. The free and pinned layers are perpendicular in zero applied field.

Noise in a magnetic sensor can be electrical or magnetic in origin. It is manifest as uncontrolled random fluctuations of the voltage $V(t)$ across the device. The frequency spectrum of the fluctuations $\hat{V}(f)$ is the Fourier transform of the time series $V(t)$. The average power dissipated (per unit resistance) in fluctuations during a measuring time t_m which tends to infinity is

$$P = \lim_{t_m \to \infty} \frac{1}{t_m} \int_{-t_m/2}^{t_m/2} |V(t)|^2 \, dt = \lim_{t_m \to \infty} \int_0^{\infty} \frac{2|\hat{V}(f)|^2}{t_m} df. \tag{14.23}$$

The power spectral density of the fluctuating process is defined as

$$S_V(f) = \lim_{t_m \to \infty} \frac{2|\hat{V}(f)|^2}{t_m} (0 < f < \infty), \tag{14.24}$$

where $\hat{V} = 2\int_0^{\infty} V(t) \exp(-2\pi i f t) dt$. The voltage spectrum can be measured directly in a spectrum analyser. It is conventionally quoted for a bandwidth Δf of 1 Hz.

Electrical noise is of four main types. The first is thermal or Johnson noise, which is a characteristic of any resistor R. The spectrum is frequency-independent:

$$S_V(f) = 4k_B T R, \tag{14.25}$$

and the mean-square voltage fluctuation measured across a resistor, with no imposed current, is $\bar{V}^2 = 4k_B T \Delta f R$. For example, the root-mean-square (RMS) voltage fluctuation across a 1 MΩ resistor in a 1 kHz bandwidth at

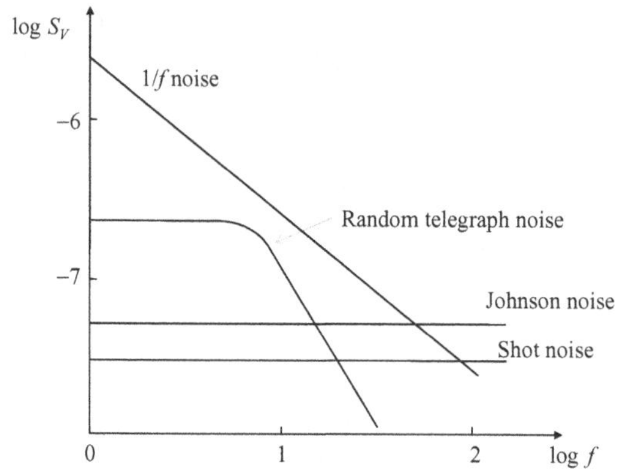

Figure 14.20
Frequency dependence of various contributions to the noise in a sensor.

room temperature is 4 μV. Hence the incentive to make electrical measurements in as narrow a bandwidth as possible at the frequency of the signal of interest, in order to improve the signal-to-noise ratio. This is the principle of the synchronous, phase-sensitive detector (lock-in amplifier), which is almost universally used for precise electrical measurements.

Shot noise is a nonequilibrium phenomenon associated with electric currents which is related to the discrete nature of electrical charge. It was first observed in vacuum tubes, where the power spectral distribution is flat in frequency $S_I(f) = 2eI$. The current noise in a bandwidth Δf is

$$I_{sn} = \sqrt{2eI\Delta f}. \tag{14.26}$$

Shot noise can be observed in tunnel junctions at low temperature when sufficient current passes through the junction for the shot noise to exceed the equilibrium Johnson noise. It may determine the ultimate sensitivity of high-frequency sensors such as read heads. Operating at too high a bias reduces the TMR (Fig. 8.25), but it also increases the shot noise. Carrier lifetime may influence shot noise at high frequency.

The signal-to-noise ratio is critical for sensor applications. Noise is frequency-dependent, Fig. 14.20. At high frequency, the noise contributions that limit performance are Johnson noise and shot noise, varying as $\sqrt{R}$ and $\sqrt{I}$, respectively. Both are known as white noise on account of their flat frequency spectrum. Another contribution called pink noise, or 1/f noise, dominates at low frequencies, where the power spectral density S_V varies as $1/f^\alpha$ with $\alpha \approx 1$. The fascination of $1/f$ noise is its ubiquity. There is an astonishing range of different natural and man-made examples, ranging from the human heartbeat ($1/f$ below 0.3 Hz) to the water level of rivers and the musical output of radio stations. The $1/f$ noise in electronic systems may be orders of magnitude greater than thermal noise at 1 Hz. In magnetic sensors, it is related to resistance fluctuations $R(t)$, which translate into voltage fluctuations when a constant current is passed. The current does not create the fluctuations, it just reveals them above the white noise. The square of the voltage fluctuations $S_V(t)$ varies as I^2, which permits the $1/f$ noise coming from a sample to be

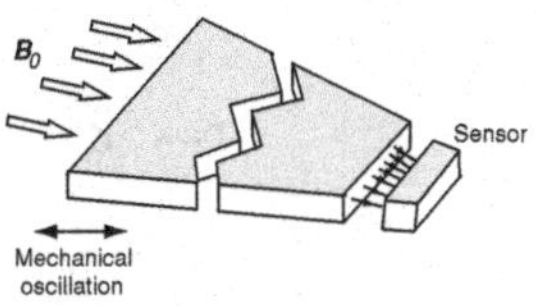

Soft magnetic flux concentrator and modulator. The signal is modulated to avoid the 1/f noise.

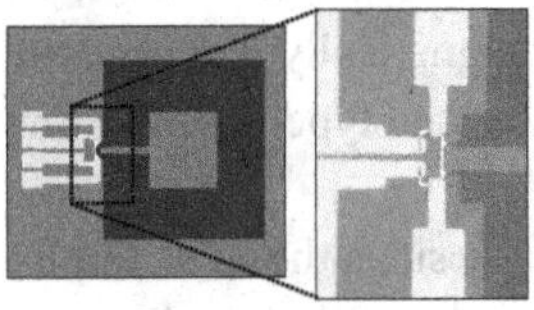

A superconducting mixed sensor. A current flows in the superconducting loop to maintain a constant flux, and the field produced by the current at the constriction is detected by a sensitive spin valve. (Courtesy of M. Pannsetier)

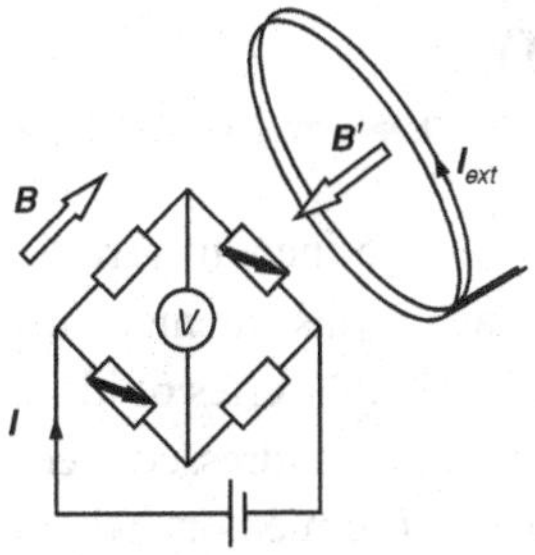

Wheatstone bridge with magnetoresistors can be used as (i) a null detector to measure small changes in B, (ii) a linear field detector, by passing a current I_{ext} in the external coil to maintain the bridge in balance.

distinguished from that arising from other sources, such as the preamplifier. In magnetoresistive sensors, there is also a magnetic contribution to $1/f$ noise that is associated with domain walls.

The $1/f$ resistance fluctuations are conventionally parameterized in terms of a phenomenological expression due to Hooge:

$$S_V(f) = \gamma_H V_a^2 / N_c f, \tag{14.27}$$

where V_a is the applied voltage, N_c is the number of charge carriers in the noisy volume ($N_c = n\Omega$, where n is the carrier density and Ω is the volume) and γ_H is a dimensionless number known as the Hooge coefficient, which allows a normalized comparison of different systems. Values of $\gamma_H \sim 10^{-3}$ are found for well-crystallized metal films, and for semiconductors. The values of γ_H for conducting magnetic sytems including GMR sensors, may be orders of magnitude greater.

The origin of the $1/f$ noise remains an open question, but the noise can be modelled in terms of an ensemble of elementary thermally activated fluctuators with a broad range of energy barriers. The task for an engineer concerned with magnetic sensors is to avoid it or reduce it as far as possible. One approach is to ensure that thin films of excellent crystalline quality are used to reduce the electrical contribution. A way to circumvent $1/f$ noise is to amplify the field and modulate it at a frequency in the kHz range, with a microcantilever for example, in order to shift the detected signal into the white noise region. beyond the range where $1/f$ noise dominates. A tapered soft-magnetic flux concentrator can be mounted onto a silicon microcantilever, for example. Another approach, suitable for very low fields, is to use a mixed sensor. Here a superconducting loop with a constriction responds to flux, which is quantized in the loop to be a multiple of $\Phi_0 = 2.068 \times 10^{-15}$ T m^2. A current I flows around the loop to maintain the flux, and it creates a field $H = I/2\pi r$ at the constriction where a sensitive spin-valve sensor is used to detect the field. The superconducting flux to field converter can be modulated thermally by heating it above its superconductivity transition temperature, at a frequency f above the $1/f$ upturn. The noise performance of a mixed sensor may be comparable to that of SQUID, and it is easier to implement.

Finally, a type of noise which is sometimes encountered in conducting magnetic thin films is random telegraph noise, where a particular two-level system is activated in a certain temperature window. This leads to a broad feature in the noise spectrum, as indicated in Fig 14.20. Some examples of noise in different magnetic structures are given in Fig. 14.21.

The magnetic noise, or electrical noise of magnetic origin, in AMR or spin-valve sensors is related to fluctuations of the magnetization, which are reflected in the magnetoresistance signal. These fluctuations can be severe when domain walls are present, and they are mitigated by ensuring that the ferromagnetic layers are single-domain, and that the rotation of the free layer is coherent. One way to achieve this is to build a permanent magnet into the sensor structure, which creates a small bias field. Otherwise, domain formation can be discouraged

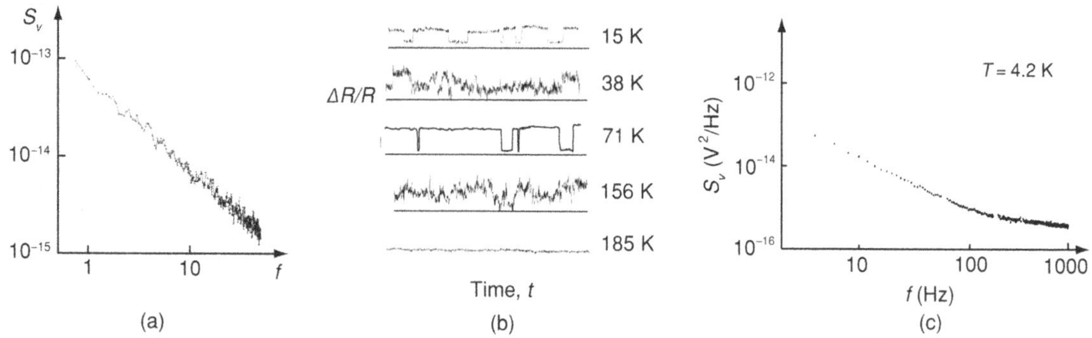

Figure 14.21

Some examples of noise in magnetic thin-film samples: (a) $1/f$ noise in a CrO_2 film, (b) random telegraph noise in a $La_{0.67}Ca_{0.33}MnO_3$ film and (c) $1/f$ and white noise in a CoFeB–MgO–CoFeB magnetic tunnel junction.

by reducing the demagnetizing field and taking advantage of shape anisotropy in a yoke-type configuration. When the sensor dimensions are very small, or the free layer is so thin that it tends to break up into magnetically independent regions, collective fluctuations of the magnetization, of the type described in §8.5, may be a significant source of noise.

The other side of the signal-to-noise coin is signal. Besides the field amplification strategies just mentioned, the sensor signal can be optimized in other ways. One is to form a Wheatstone bridge, where the four resistors are balanced in the ambient field and the voltage signal is the response to any change in the field. A linear field sensor can be constructed from such a null detector by passing a current, I_{ext}, through an external coil in such a way as to maintain the bridge in balance. The external field is then proportional to I_{ext}. Gradiometers can be constructed where two field sensitive elements in the bridge are spatially separated to detect local variations of the field against a spatially homogeneous background. With SQUIDs it is easy to wind a superconducting flux transformer to act as a planar or axial gradiometer.

14.4 Magnetic memory

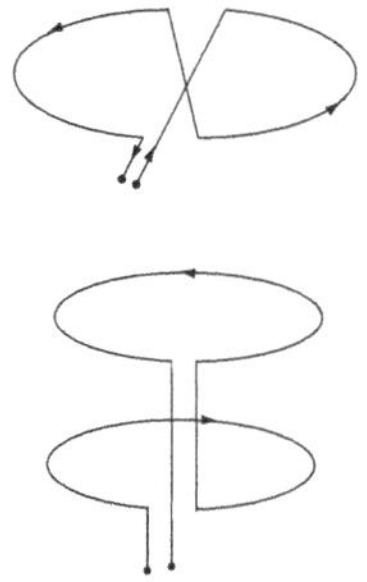

Planar and axial gradiometers for a superconducting flux transformer.

Magnetic memory is as old as digital electronics. It has the incontrovertible advantages of being nonvolatile and indefinitely rewritable. Mass memory for computers has been provided by hard discs for over 50 years. A series of ideas have been advanced for other, faster, schemes which can be addressed electrically, rather than mechanically. These included ferrite core memory which was dominant in the 1950s and 1960s, until it was superseded by semiconductor memory, permalloy plated wire memory in the 1960s, magnetic bubble memory in the 1970s and early 1980s, and magnetic random-access memory (MRAM) which has been under development since the mid 1990s. In principle, MRAM can combine the speed of SRAM with the density of DRAM and the nonvolatility of flash, with radiation hardness and reduced power consumption. The latter has always been an attraction of magnetic memory for military and

Figure 14.22

A TMR spin-valve memory cell operating on the half-select principle, with orthogonal bit and word lines (M. Johnson, (2004).)

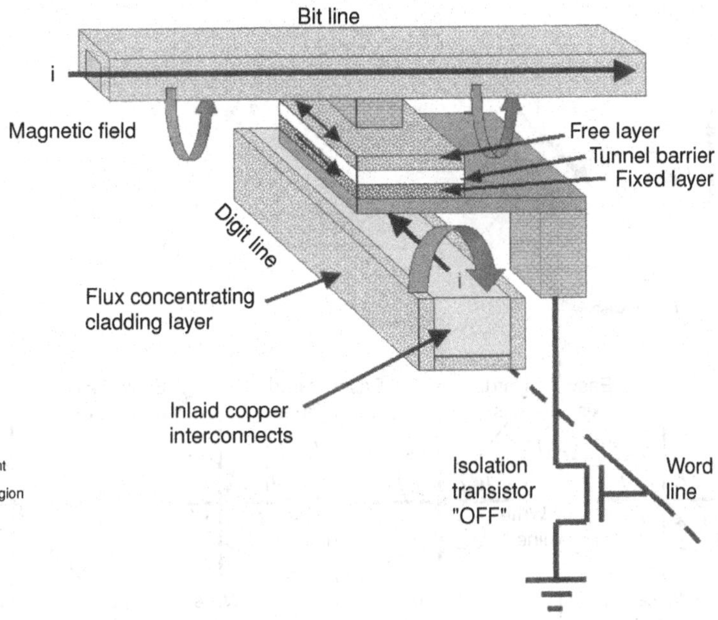

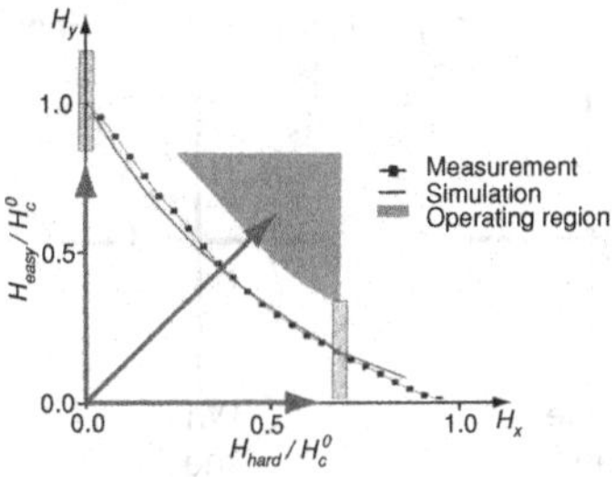

Half-select pulses, shown on the Stoner-Wohlfarth asteroid.

space applications – the computers in the *Challenger* space shuttle and the *Hubble* Space Telescope include magnetic memory. With increasing miniaturization radiation hardness could become an issue for civilian applications because of the problems associated with unshieldable background radiation.

14.4.1 Magnetic random-access memory

The tiny ferrite toroids a few hundred micrometres in diameter were the original core memory threaded on write and sense lines; they operate on the half-select principle. The same principle has been used in MRAM. Schemes under investigation include those with current-in-plane based on GMR spin valves and current-perpendicular-to-plane based on magnetic tunnel junctions. The requirement is for a bistable device where the free layer can flip very quickly between the parallel and antiparallel states, which represent the binary data, ‘0’ and ‘1’.

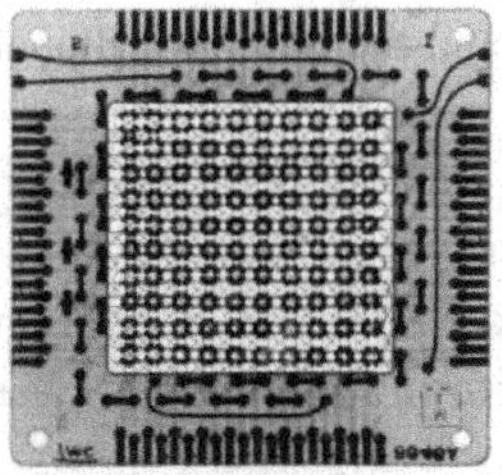

A 256-bit ferrite core memory from 1960.

Half-select switching, Fig. 14.22, uses current pulses in orthogonal bit and word lines to generate magnetic field pulses H_p. A single pulse does not disturb the state of the free layers of a line of elements, but when two pulses arrive simultaneously along the two perpendicular lines that cross at the selected element, the resultant field $\sqrt{2H_p}$ is sufficient to induce switching. This can be nicely represented on the Stoner–Wohlfarth asteroid.

The margin of error for this switching is rather small, and immense reliability is demanded of working memory devices. An improved pulse sequence known

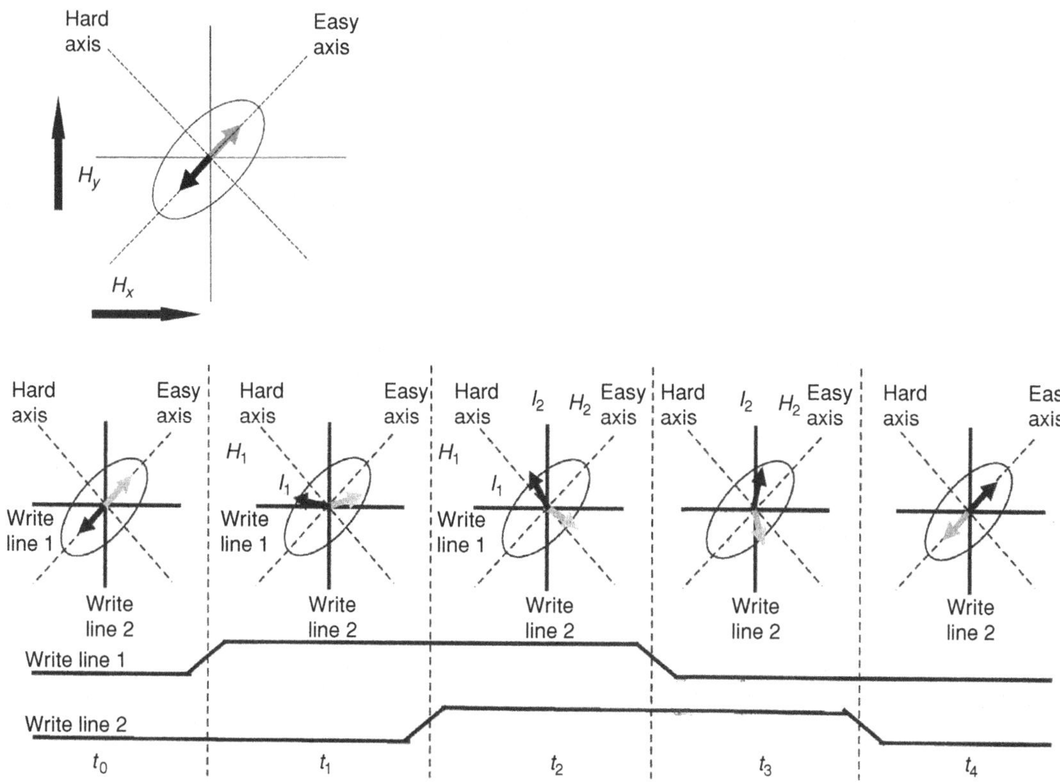

Figure 14.23

Toggle-mode sequence to enhance the reliability of half-select switching. (M. Johnson 2004)

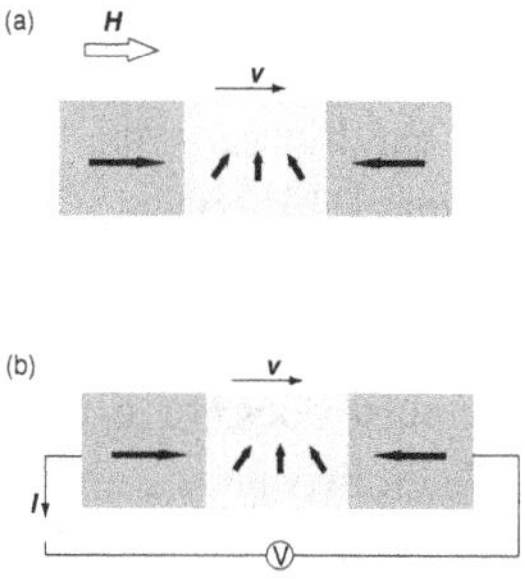

Movement of domain walls in a magnetic nanowire by (a) an applied field, and (b) spin-transfer torque.

as toggle-mode switching (Fig. 14.23) greatly enhances switching reliability; it was devised by Motorola for the first commercial MRAM chips, which appeared in 2006.

We saw in §14.1.3 that half-select switching with Oersted fields does not scale favourably as the dimensions are reduced. Possible ways around this difficulty are: (i) thermally assisted switching where the current passing through the memory cell heats it sufficiently to reduce the anisotropy, making it easier to switch and (ii) spin-transfer-torque MRAM, where the switching is accomplished by a spin-polarized current pulse surging through the spin-valve stack. TMR devices with MgO barriers offer useful voltage changes (>0.1 V) and a compact footprint with a single transistor switch, and the magnetic tunnel junction added in a back-end metallization step. The magnetoresistance characteristic of an exchange-biased TMR stack used for memory is illustrated in Fig. 14.24. These devices may be able to replace the six-transistor SRAM cell. It is a remarkable feat to be able to cover a silicon wafer 200 mm or 300 mm in diameter with a thin-film stack that includes a uniform MgO barrier layer barely 1 nm thick.

A proposed scheme for magnetic memory due to Stuart Parkin is based on the movement of a pattern of magnetized domains around a permalloy track, which constitutes a magnetic shift register. The head-to-head and tail-to-tail domain

Figure 14.24

Magnetoresistance of an exchange-biased TMR spin valve with an MgO barrier and a synthetic antiferromagnetic pinned layer. When used as a memory element, the axes of the free and pinned layers are parallel, giving an abrupt switch near zero field, as shown. When the device is used as a sensor, these axes are perpendicular, leading to a linear $R(B)$ transfer curve, as the free layer rotates coherently in the applied field. (Data courtesy of Gen Feng.)

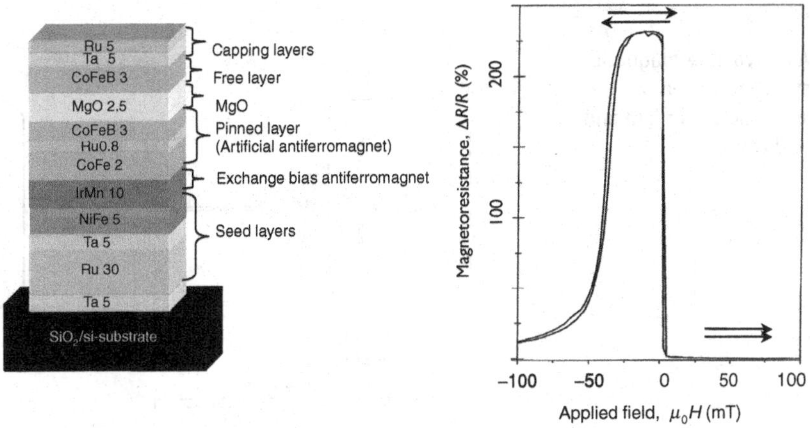

walls would move in opposite directions in an applied field, thereby annihilating the data, but the pattern of magnetization can be pushed around the track by spin-transfer torque. A current passing through a magnetized track adopts the polarization of the domain, so the walls are all pushed in the same direction. A string of bits can be moved at velocities of order 100 m s^{-1} by a train of current pulses in the magnetic racetrack memory (MRM). The scheme promises nonvolatile solid state memory, without the mechanical inconvenience of a hard-disc drive. Very high storage densities should be achievable if the registers can be built in the vertical direction. The scheme has yet to be demonstrated in practice.

14.5 Other topics

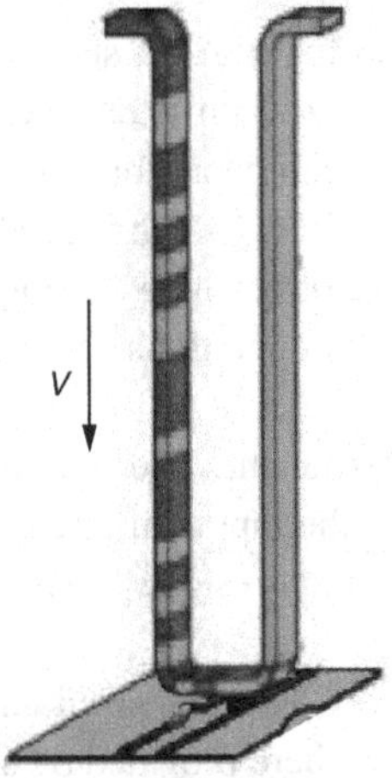

Magnetic racetrack memory. (S. S. P. Parkin *et al.*, *Science* **320**, 190 (2008).)

14.5.1 Logic

A ferromagnetic thin-film element can be pressed into service for logic as well as memory. Useful output voltages are achieved with MgO barrier magnetic tunnel junctions. An alternative implementation scheme makes use of a switchable ferromagnetic film grown on a semiconductor Hall sensor, Fig. 14.25. The stray field from a Co–Fe element acting on a two-dimensional electron gas creates a useful Hall voltage which follows the Co–Fe magnetization, reversing sign as the magnetization switches.

A generic nonvolatile magnetic switch may be applied in field-reprogrammable gate arrays. These are applications-specific CMOS logic circuits which are reprogrammable at will in order to carry out a set of logic operations required for some particular application. This avoids the need to design and produce a custom chip for each purpose. The magnetic switch delivers its output to the gate of an FET. Half-select input pulses at the terminals A and B determine the state of the device. The output of the switch, of order 0.1 V, is sufficient to pinch off the current in the channel of the FET.

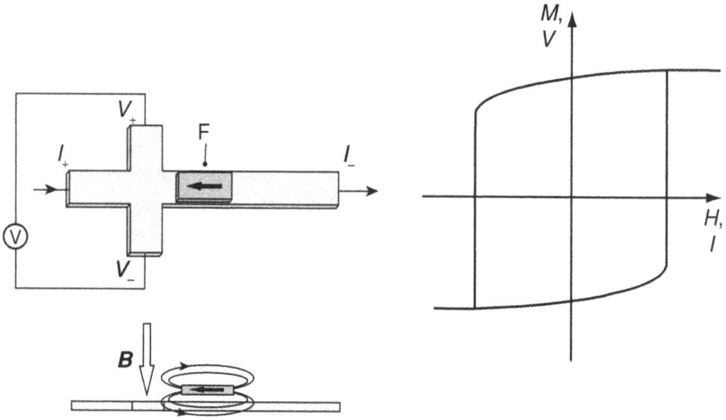

Figure 14.25

A nonvolatile magnetic switch based on a switchable thin film and a Hall cross.

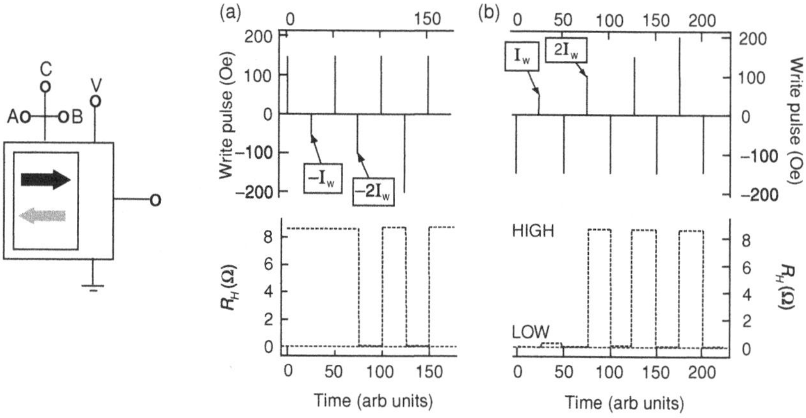

Figure 14.26

Logic operations using the third terminal C for clock pulses. The steady-state response to the input pulses at A and B demonstrate $\overline{AB}$, $\overline{A+B}$ and AB, A+B. (M. Johnson (2004)).

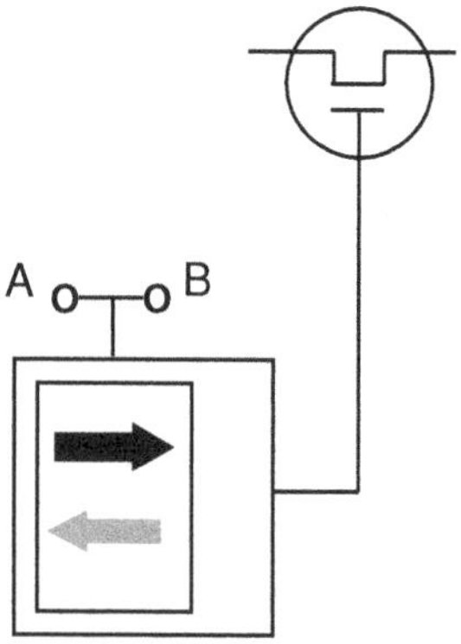

A generic nonvolatile magnetoelectric switch. The output is applied to the gate of a field-effect transistor. Switching of the magnetization is achieved by simultaneous pulses at A and B.

The magnetic switch can also be applied for Boolean logic. As it stands, the half-select switch acts as an AND gate, switching only when it receives simultaneous pulses at A and B. Reset, and the other Boolean operations OR, NAND and NOR can be conducted by adding a third terminal C, which is used for clock and reset pulses. The reset pulses can be positive or negative, setting the switch in either the high (1) or the low (0) state. All the operations can be achieved in just two clock cycles, as illustrated in Fig. 14.26.

A key requirement for any implementation of logic is fanout. The digital output from one logic gate must be sufficient to apply to the inputs of one or more contiguous gates in the next stage. Normally, fanout requires power gain, which is achieved with conventional CMOS.

An ingenious approach to magnetic logic due to Russel Coburn is based on propagating domain walls in permalloy tracks. The clock is here provided by a rotating magnetic field applied in the plane of the film, and readout is based on the Kerr effect. Annihilation of the information in the applied field is avoided by using long bits so that the head and tail of a bit are on opposite sides of a curved track, and thus they move in opposite directions in an external field.

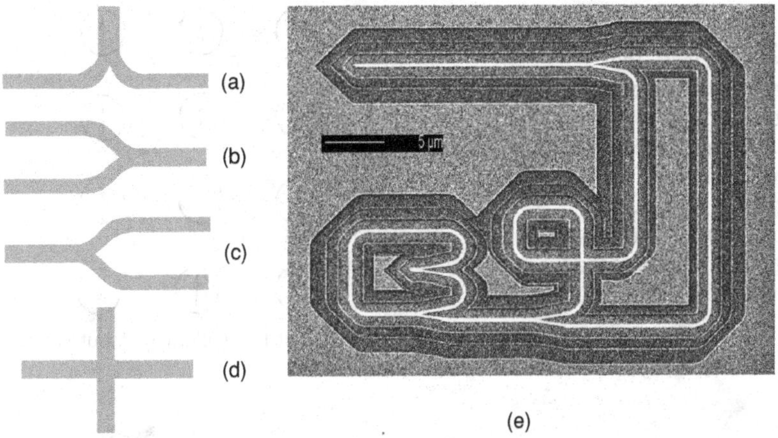

Figure 14.27

Domain wall logic in domain walls on permalloy tracks: (a) a NOT operation, (b) an AND gate, (c) fanout and (d) crossover. These operations are combined in the demonstration four element circuit (e). (Courtesy of R. Coburn.)

The rotating magnetic field moves the bits around the track. The domain walls are introduced at injection pads, and specific track shapes perform particular logic operations. Some of the structures involved are illustrated in Fig. 14.27. Generally, the four logic gates can be generated from any one of them, together with the NOT operation. The cusp acts as a NOT operator, providing a rotation of the magnetization by π, as the wall moves up to the point of the cusp, and then backs out of it. The AND gate is implemented by running two tracks together at a small angle. Taken in the opposite sense, the same structure provides fanout, creating two walls from one and taking the necessary energy from the external field. Crossover is achieved by running one track across the other; the domain walls are solitons which do not interfere with each other. Drawbacks of this elegant scheme are the large footprint of the permalloy track, and the slow clock rate available from an AC magnetic field.

Another logic scheme has been proposed which is based on three single-electron nanodots coupled by antiferromagnetic exchange. There is a weak bias field to define an easy direction. The two outer dots provide the input and the central dot is the output. The four possible inputs illustrated in Fig. 14.28 show that the device operates as a NAND gate, from which the other logic operations can be constructed.

Yet other magnetic logic schemes can be envisaged. One uses the phase of spin waves, which have the advantage that they carry pure alternating spin currents. The phase is shifted when the spin wave traverses a domain wall. By passing the spin waves around parallel arms of a narrow loop and detecting the signal at the point where they recombine, it is possible to make a logic gate based on the presence of 360° domain walls in either or both of the arms. Attractive features are that no transfer of charge is involved, and so there is no energy loss due to Joule heating although there may be energy dissipation associated with Gilbert damping. Futhermore, the spin waves could be generated on-board the chip using spin-transfer torque (Fig. 14.14).

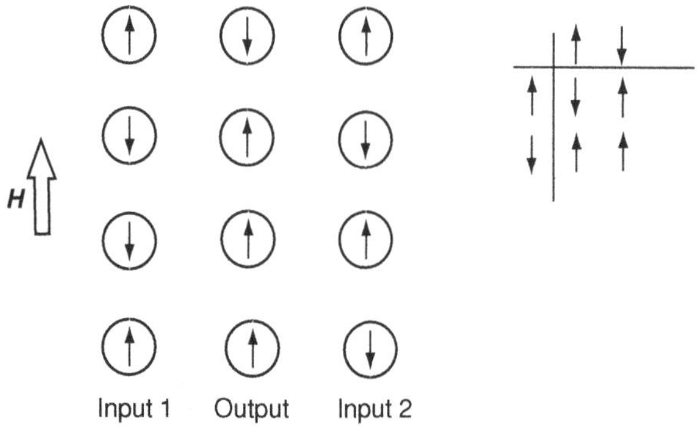

Figure 14.28

Exchange logic. The central single-electron atom is coupled antiferromagnetically to its two neighbours. The four states of the system correspond to the NAND operation shown in the truth table.

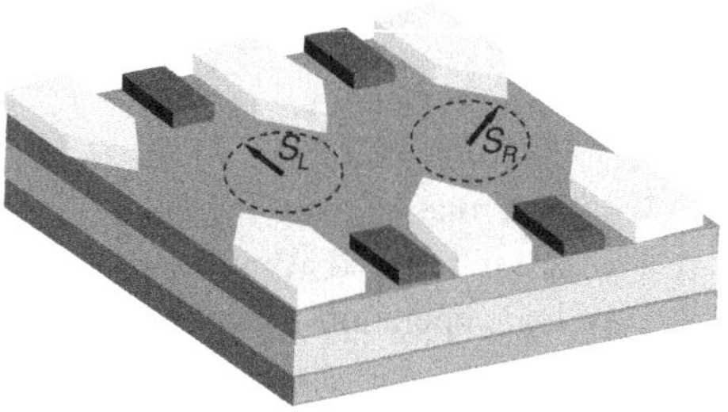

Figure 14.29

A two-q-bit structure composed of two weakly interacting single-electron spins on adjacent quantum dots.

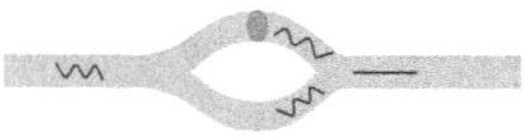

Spin-wave logic. The spin wave running through the two branches acquires a phase shift if it encounters a domain wall.

Finally, among the many candidate systems that have been considered for quantum computing, entangled q-bits based on weakly interacting spins, Fig. 14.29, meet many of the requirements. A single-electron spin on a quantum dot can be regarded as a q-bit in a state $\alpha|\uparrow\rangle + \beta|\downarrow\rangle$. The q-bits are initialized in the $|\uparrow\rangle$ state in a large magnetic field, and the individual bits are then manipulated by varying the Zeeman splitting. The two-q-bit operations, which are required for universal quantum computing are performed by tuning the weak exchange coupling between two adjacent q-bits. The coherence of the state tends to be destroyed by spin–orbit coupling and hyperfine interactions which introduce longitudinal, spin-flip T_1 relaxation processes and transverse T_2 processes which describe the decay of the superposition state.

14.5.2 Spin transistors

Spin electronics could enter a new phase if it proves possible to make a three-terminal device with spin gain, which meets the fanout requirement in logic circuitry. There have been several suggestions and demonstrations of physically interesting devices.

The first three-terminal spin electronic device was the Johnson transistor (Fig. 14.30(a)). Essentially a GMR spin valve, with the nonmagnetic metal as base electrode, the base-emitter current injects polarized spins into the base region. The floating collector samples a small voltage change depending on the

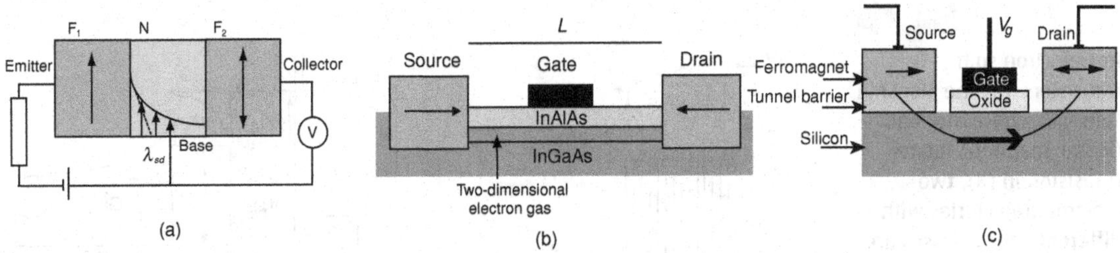

Figure 14.30

Some possible three-terminal spin devices: (a) the Johnson transistor, (b) the Datta–Das transistor, (c) the spin MOSFET.

configuration of the two ferromagnetic emitter and collector electrodes. The Johnson transistor is equivalent to the device used in the nonlocal measurement of spin accumulation.

An influential device proposal has been the Datta–Das transistor. This is an FET-like structure, with ferromagnetic source and drain. Spins are injected into the two-dimensional electron-gas channel, where transport is supposed to be ballistic. They are subject to an electric field from the gate, which looks like a magnetic field $\boldsymbol{B}^* = (\boldsymbol{v} \times \boldsymbol{E})/c^2$ to electrons moving with velocity v (3.70). This is known as the Rashba effect. The electrons undergo Larmor precession, and the source-drain current depends on the number of turns they have made by the time they arrive at the drain, which can be controlled by the gate potential. Neither the Datta–Das device, nor the related spin MOSFET, where the semiconductor channel is diffusive, made of Si or GaAs, and which promises a combination of power amplification and nonvolatile memory via the switchable ferromagnetic drain electrode, have yet been realized in practice. Nevertheless there has been significant progress towards realizing electrical spin injection and detection in these semiconductors.

A way to circumvent the resistance mismatch problem is to inject hot electrons via a Schottky or tunnel barrier, which subsequently lose energy by spin-dependent inelastic scattering in the base region of a device, Fig. 14.31. In the Monsma transistor, the base is a GMR multilayer with two Schottky barriers, at the source and drain. Another device is the magnetic tunnel transistor, where the base is a ferromagnetic electrode and there is injection from a ferromagnetic source via a tunnel barrier, and a semiconductor collector with a Schottky barrier serves as the drain. The device produces a significant magnetoconductance, defined as

$$MC = (I_{c_p} - I_{c_{ap}})/I_{c_{ap}},$$

where I_{c_p} and $I_{c_{ap}}$ are the collector current with parallel or antiparallel configurations of the two electrodes. However, in no case has gain of the spin-polarized current yet been achieved.

In summary, two grand challenges facing spin electronics are:

(i) to realize a spin amplifier – a device which can sense a spin-polarized current, and then enhance the product of its intensity and spin polarization, of either sign;

Figure 14.31

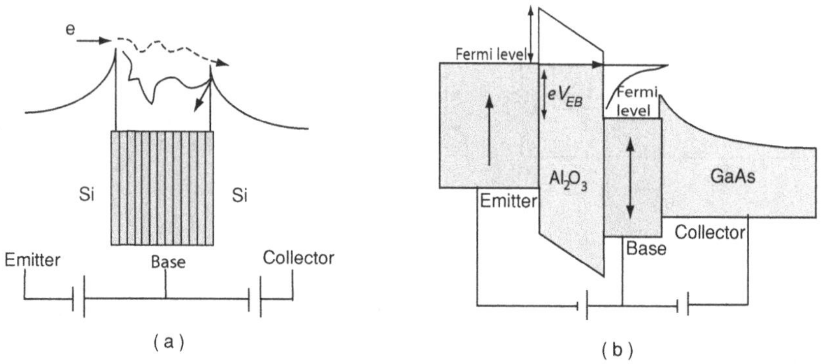

Hot-electron spin transistors: (a) the Monsma spin-valve transistor and (b) the magnetic tunnel transistor. In (a), two electron trajectories with different energy losses are illustrated.

(ii) to somehow separate the spin and charge currents so that logic operations can be performed without the ohmic dissipation of electric currents.

14.6 Magnetic recording

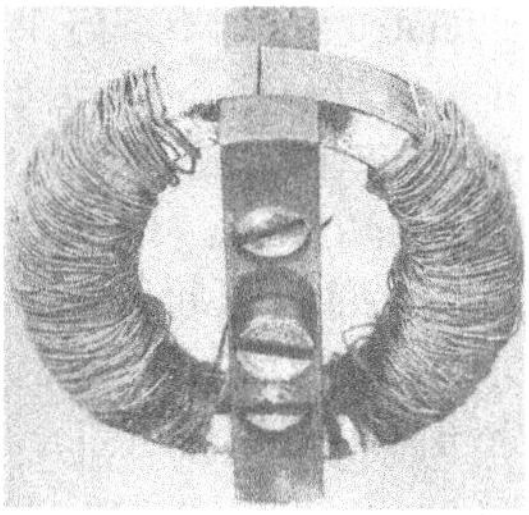

An early inductive ring head. Tape is magnetized by the stray field produced by the electromagnet near the gap. During playback, the flux changes in the ring induce an emf in the winding.

Magnetic recording has been with us for over a century. It is evolving rapidly. The first working recorders were the 'telegraphones' invented by the Danish telephone engineer Valdemar Poulsen, which date from 1898. Steel wire or a steel disc was the recording medium, and DC bias was used to enhance sound quality. The technology of analog tape recording was developed by AEG and BASF in Germany in the 1930s. Innovations incorporated in their 'magnetophon' recorders were an inductive ring recording/playback head, AC bias and cellulose acetate tape coated with a layer of fine particles of γFe_2O_3. The AC bias, which greatly enhanced recording sound quality, had been invented by Kenzo Nagai in Japan in 1938. Analog recording involves a signal of continuously varying amplitude and frequency, which must be mapped linearly onto the remanent magnetization of the medium for high fidelity. But remanence is not linear in applied field. To achieve a linear response, an AC bias is applied at an inaudible frequency of 50–100 kHz to saturate the magnetization, Fig. 14.32. The anhysteretic remanence is then proportional to the signal.

Video recording was introduced by AMPEX in the USA in 1956, and consumer versions of audio and video recording followed with the compact cassette of Philips in 1963 and the rotating head VHS system of Panasonic in 1976.

At a simpler level, records and security features can be printed on paper using ink pigmented with ferrimagnetic iron oxide particles. Special fonts giving characteristic magnetic signatures are used on cheques. Unaffected by overstamps, the records are magnetized in a horizontal field and read with an inductive read head.

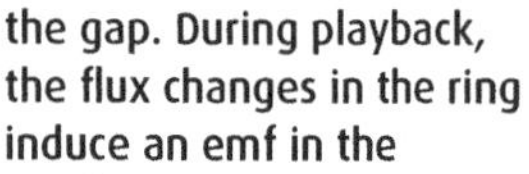

Fonts used for printing records with magnetic ink.

Digital computing demands digital recording. After some short-lived experiments with rotating drums, IBM tried a rotating disc memory, which turned

Table 14.6. Magnetic recording media

	Tape	Disc
Method	Digital, analog	Digital
Direction	In-plane	Perpendicular
Medium	Particulate, metal film	Metal film
Density	10 bits μm^{-2}	1000 bits μm^{-2}
Capacity	1–10 Tbyte	10–1000 Gbyte

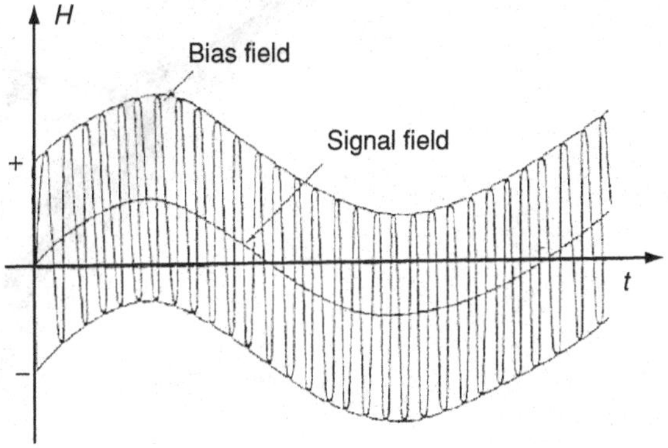

Figure 14.32
Analog audio recording with AC bias.

out to be a decisive innovation. Introduced in 1957, the original RAMAC (random access method of accounting and control) was based on a stack of 50 two-sided 24-inch platters. It weighed a tonne and boasted a storage capacity of 4.4 megabytes. RAMAC is the direct ancestor of a modern $3\frac{1}{2}$-inch hard-disc drive, which weighs a few hundred grams, and provides a terabyte of storage – an improvement in areal density and cost per byte of eight orders of magnitude!

Disc recording has been implemented in a sequence of different formats for fixed and portable storage, such as the 8-inch floppy disc, which have come and gone. This is a problem for archivists, who need a room full of obsolete hardware to access their records, unless they periodically back them all up onto ever more modern and ephemeral media. A hard copy of a book like this might be taken down from a library shelf in 2100 and thumbed by someone interested in the history of magnetism. It is doubtful that any electronically produced version will be so accessible in its original format.

A personal stereo, ca 1935. (Courtesy Orphée Cugat.)

Magnetic records are now almost all digital. Data are mainly stored on the hard discs that are used for mass memory in computers and servers – the memory banks of the Internet – as well as in consumer devices like music players and digital cameras. Tape recording retains a place for archival storage of vast quantities of digital data, as well as for low-end consumer applications. Disc and tape recording are compared in Table 14.6.

Figure 14.33

A $3\frac{1}{2}$-inch hard disc drive. (Courtesy of Seagate Technology.)

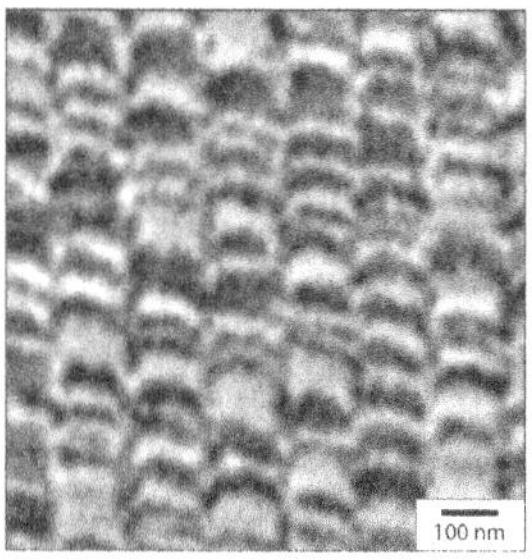

Perpendicular tracks on a hard disc imaged by high-resolution magnetic force microscopy. The width of the tracts is determined by the width of the write head. The recording density here is 300 bits μm^{-2} or 250 Gbit/square inch. (Courtesy of Nanoscan AG.)

Regardless of whether it is done on tape or disc, magnetic recording transfers information onto a moving magnetic medium. The information is coded into the current passing through a miniature electromagnet – the write head. The stray field produced by the head must be sufficient to switch the magnetization of the medium[4] by overcoming the coercivity. The recording medium is usually a thin film composed of tiny, crystalline, single-domain grains or 'particles'. Digital information is written in elongated patches of width w and length l, with $w/l \approx 5$, which are arranged in circular tracks on a disc or linear tracks on a tape. A single bit is recorded in the magnetization of N particles in the patch. For a good signal-to-noise ratio, the individual grains should be more or less magnetically decoupled. The signal-to-noise ratio in db is then of order $10 \log N$; if a value of 30 db is required, N has to be about 1000.

The head is fixed, and in contact with the lubrication layer of the medium for tape recording, but for disc recording it is mounted on a voice-coil actuator (Fig. 13.18), which scans it across the surface in a hard-disc drive, flying on an air bearing about 10 nm thick. A sophisticated feedback mechanism keeps the head on-track. The voice-coil actuator and the spindle motor driving the disc at up to 15,000 RPM are both permanent-magnet motors making use of the best grades of sintered and bonded $Nd_2Fe_{14}B$, respectively. A $3\frac{1}{2}''$ drive is illustrated in Fig. 14.33.

[4] Specialists in magnetic recording use the word media as if it were singular.

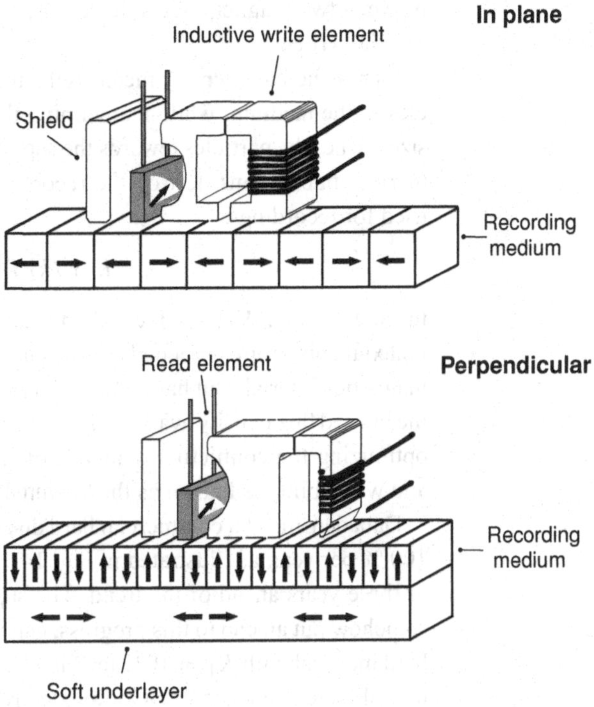

Figure 14.34

Schematic of in-plane and perpendicular magnetic recording.

Two recording modes are implemented – longitudinal and perpendicular – with the magnetic records magnetized respectively in-plane or perpendicular to the plane of the medium. Prior to 2005, when perpendicular recording was introduced on hard discs, all magnetic recording, other than magneto-optic recording, was longitudinal. Perpendicular recording permits higher areal densities of the magnetic records, but requires uniaxial anisotropy in a direction normal to the plane of the disc that is sufficiently strong for the anisotropy field to exceed the demagnetizing field:

$$2K_1/\mu_0 M_s > \mathcal{N} M_s. \tag{14.28}$$

The medium is not a uniform thin film, for which $\mathcal{N} = 1$, but it is composed of magnetically isolated columnar grains with a much smaller effective demagnetizing factor, $\mathcal{N}_{eff} \approx 0.2$.

For decades, inductive heads were used both to write and to read the data, but since the 1990s the recorded information has been read by a separate thin-film magnetoresistive sensor, which responds to the stray field created by the magnetic records. Actually it is the magnetization transitions between the recorded bits, where the stray field is greatest, that are detected. These transitions are irregular on account of the discrete nature of the particulate medium. The general scheme for magnetic recording is illustrated in Fig. 14.34. The reader and writer are merged into single thin-film stack. The reader is placed between two magnetically soft shields, one of them shared between the reader and the writer.

Increasing data density means reducing the size of the single-domain particles in the medium, while maintaining the signal-to-noise ratio. Reducing the size pushes the particles towards the superparamagnetic limit where spontaneous thermal fluctuations destroy the recorded information. The stability criterion used for recording is

$$K_1 V / k_B T > 40,$$

in other words, $K_1 V > 1$ eV. As the particle volume V decreases, stronger uniaxial anisotropy is needed to maintain stability, implying the need for harder magnetic material. But harder materials are increasingly difficult to switch with the limited flux density available from the writer. The problem of simultaneously optimizing the conflicting demands of signal-to-noise ratio, thermal stability and writeability is known as the trilemma of magnetic recording.

Data densities on commercial hard discs have been increasing at a rate of 60–100 % per year, as illustrated in Fig. 14.3. Laboratory demonstrations run two or three years ahead of the trend. The superparamagnetic limit, which should somehow put an end to this progress, can be estimated; the critical volume for a hard material with $K_1 = 10^6$ J m^{-3} is $V = 160 \times 10^{-27}$ m^3, corresponding to a particle size of about 5 nm. Optimistically, if we can take $N = 100$, there would be about 400 bits per μm^2 or 260 Gbits per square inch. The state of the art in 2012 was 2500 bits μm^{-2} (1.5 Tbits per square inch)! The superparamagnetic limit has been circumvented, for the time being, by skilful magnetic engineering. The lithographic feature sizes that now have to be implemented in recording heads are similar to those in CMOS circuits.

We will consider each component of the recording process in turn.

14.6.1 Write heads

The writer is a miniature electromagnet with an airgap. First we consider a ring head producing a horizontal field for writing on in-plane media. If the gap has width g, the horizontal and vertical components of the stray field from the ring are given by the Karlquist equations, valid when $z > 0.2g$:

$$H_x = \frac{H_g}{\pi}\left[\arctan\left(\frac{g/2 + x}{z}\right) + \arctan\left(\frac{g/2 - x}{z}\right)\right], \tag{14.29}$$

$$H_z = \frac{H_g}{2\pi}\ln\left[\frac{(g/2 - x)^2 + z^3}{(g/2 + x)^2 + z^2}\right], \tag{14.30}$$

where H_g is the field deep in the gap.

The equations can be deduced by considering two sheets of surface charge on the writer. As the medium moves past the head, there is a 'write bubble' where the horizontal field H_x exceeds the threshold for switching the magnetization

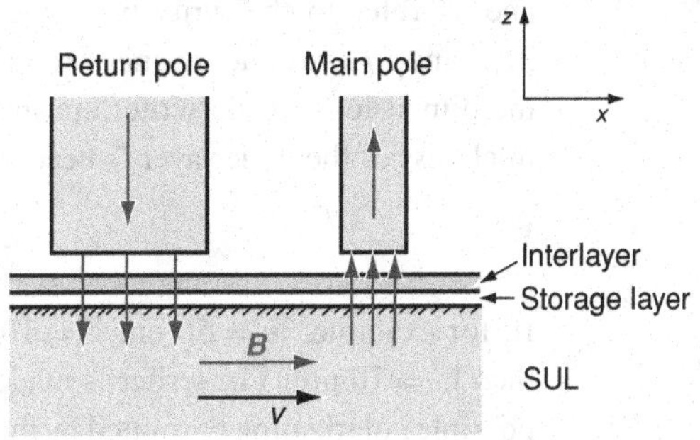

Figure 14.35

A thin-film write head for perpendicular recording.

of an in-plane record. To write effectively, a rule of thumb is that the field deep in the gap should be about three times the coercivity of the medium.

Switching the recorded bit is done with a current pulse in the inductive writer. Williams and Comstock developed a model for the recording process which predicts the width a of the recorded transition in a material with a square hysteresis loop to be

$$a = \sqrt{\frac{M_r t d}{\pi H_c}}, \tag{14.31}$$

where t is the medium thickness and d is the head–medium spacing. Hence the need to reduce these distances and increase coercivity for the highest recording densities. In hard-disc drives, the head–medium separation is about 10 nm, and the head flies on an air bearing. In tape and floppy-disc drives, the head is in direct contact with the medium.

The thin-film write head for longitudinal data recording on hard discs is a modified ring consisting of a pancake coil sandwiched between two films of a soft material such as permalloy or $Fe_{94}Ta_3N_3$, shaped to form the yoke and poles of the miniature electromagnet. The written bits are narrow bands oriented across the track, which are magnetized by the stray field at the edges of the two flat poles. The gap is less than 100 nm.

A slightly different type of head is needed to write on perpendicular media, Fig. 14.35. The flux generated by the flat coil is guided by the soft material to a single magnetic pole with its charged surface parallel to the medium in order to generate flux in the correct direction. The flux is focussed in what is effectively a narrow airgap by incorporating a soft magnetic underlayer which is physically part of the recording medium, but magnetically belongs to the head. The underlayer is separated from the medium by a thin nonmagnetic layer which breaks the exchange coupling. An image of the write pole is created in the underlayer, and the medium therefore lies at the centre of this airgap. The flux is provided with a return pole having much greater cross section than the write pole so that the flux density below it is insufficient to disturb the written information. Flux conservation requires that $B_{wp}\mathcal{A}_{wp} = B_{rp}\mathcal{A}_{rp} = B_{kl}\mathcal{A}_{kl}$, where wp, rp

and kl refer to the write pole, the return pole and the soft underlayer.[5] If M_{sw} and M_{sk} are the magnetization of the writer and the soft underlayer, and the dimensions of the writer are $w \times l$, then an expression for the minimum thickness of the underlayer t_k needed to avoid saturation is

$$t_k = \frac{M_{sw} wl}{M_{sk} 2(w + l)}. \tag{14.32}$$

If, for example, $w = 50$ nm, $l = 10$ nm, $\mu_0 M_{sw} = 2.4$ T and $\mu_0 M_{sk} = 1.0$ T, then $t_k = 10$ nm. The writer is made from the soft material having the largest possible polarization permitted by the Slater–Pauling curve, usually a nanocrystalline Fe–Co or Fe–Co–Ni alloy with $\mu_0 M_s = 2.4$ T. An estimate of the field produced at the medium is $H_w = M_{sw}\Omega/4\pi$, where Ω is the solid angle subtended by the write pole and its image at the centre of the medium. Thanks to the effect of the image pole, it is possible to achieve values of H_w approaching 1500 kA m^{-1}. Taken together, the writer and its image can be thought of as a ring head, with the medium running through the gap.

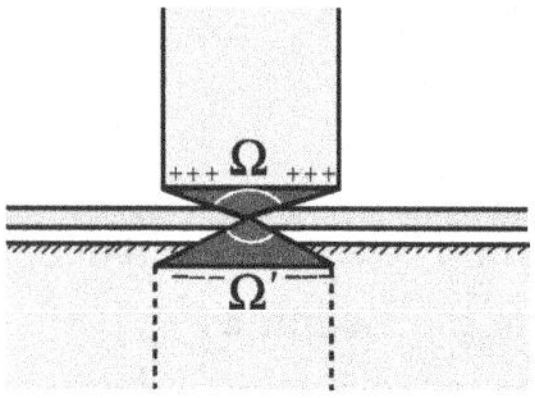

The solid angle subtended by the medium at the write pole and its image.

14.6.2 Magnetic media

Magnetic media are made of semihard materials with sufficient hysteresis to maintain a permanent record of the data, but not so much as to impede remagnetization in the field produced by the write head. The recorded information must be erasable if the medium is to be reused. Digital information is encoded in the direction of magnetization of domains located at identifiable spots on tracks in the magnetic medium. A square hysteresis loop is desirable. The emphasis in magnetic recording is therefore on controlling the nucleation of reverse domains, whereas in permanent magnetism the emphasis is on avoiding nucleation altogether.

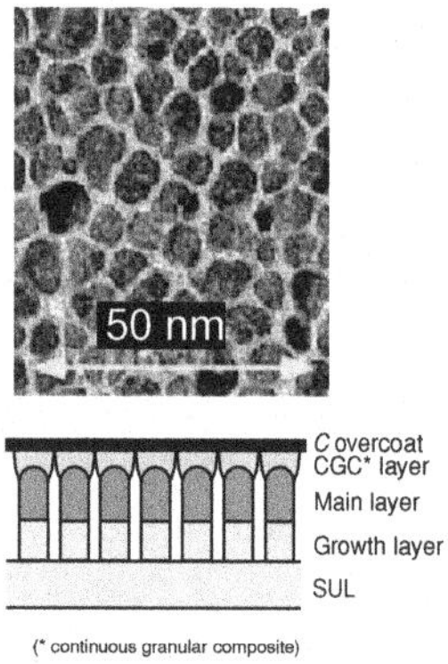

Magnetic recording medium consisting of sub-10-nm metallic grains with oxide-rich grain boundaries. The bit is recorded on a patch of grains along the track. The cross section shows the layer structure of the recording layer.

For many years, media consisted of dispersions of single-domain particles in a polymer matrix, with an easy direction of magnetization parallel to the substrate. Commonly used materials for particulate media on tapes and floppy discs were acicular (needle-shaped) γFe_2O_3, especially with Co surface-doping to increase coercivity, CrO_2 and iron metal. Acicular particles were a few hundred nanometres in length with an aspect ratio of 5:1 or 10:1 so as to provide shape anisotropy. Coercivity was about 50 kA m^{-1}.

The thin-film media on hard discs are now generally hcp cobalt-based Co–Pt alloys, with Cr, B or Ta additives to help create a regular granular nanostructure. Coercivity is about 500 kA m^{-1}. As recording densities increase, there is a need for media with higher coercivities, which can be still achieved in the Co–Pt or Fe–Pt systems. The stray field from the writer limits the usable coercivity.

[5] The soft underlayer is sometimes called the 'keeper layer', in an allusion to the obsolete practice of bridging the poles of a horseshoe magnet with a soft iron bar in order to prevent self-demagnetization.

Permanent-magnet materials like $BaFe_{12}O_{19}$ are suitable for certain magnetic records like those on credit cards or identity cards which are not intended to be erased.

In order to circumvent the superparamagnetic limit when the bit size decreases, two approaches have been adopted. One is to increase the anisotropy of the medium, while finding a way to switch it in the limited field of the writer. The other is to increase the magnetic mass of the bits.

Simplest, is to use tall slim grains which are ideal for perpendicular recording. Perpendicular media are thin-film stacks with a soft underlayer and seed layers to control grain size in the Co–Pt–Cr recording layer. The grains have a diameter of less than 10 nm, and an aspect ratio of about 3. The magnetic mass can be increased by taking two ferromagnetic layers, which are coupled antiferromagnetically via a thin ruthenium spacer. This produces the antiferromagnetically coupled bilayer (AFC) medium, which was introduced in 2001. The effective $M_r t$ is the difference for the two layers, which leads to sharper transitions (14.31), and the magnetic mass is doubled. It is also possible to exchange-couple the recording layer to an antiferromagnetic underlayer.

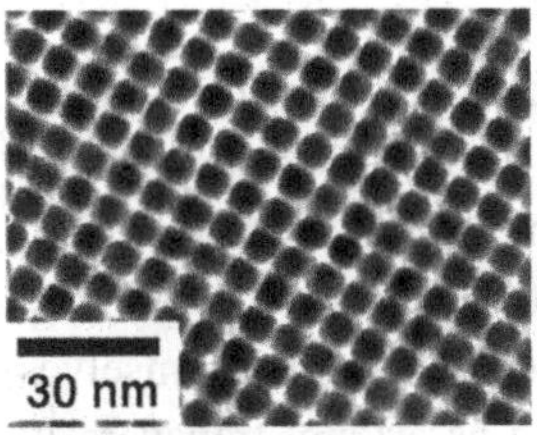

TEM micrograph of a self-assembled array of 6 nm FePt particles which could serve as a patterned medium for $N = 1$ magnetic recording. (S. H. Sun, C. B. Murray, D. Weller *et al.*, *Science* **287**, 1989 (2000).)

In order to write on hard material, one solution is to exploit the decrease of coercivity with increasing temperature. A scheme of heat-assisted magnetic recording (HAMR) has been developed where the bits are written on the medium which is locally heated by a miniature laser. Another trick is to use graded media which are soft at the surface, but have a coercivity which increases with depth. Reversal is initiated in the topmost, softer, layer, and propagates through the thickness.

Ultimate densities may demand patterned media, where single-domain particles lying at precisely determined positions along a track are individually addressed. The patterned discs are made by nanoimprint lithography from templates with a nanostructured quartz or diamond surface.

Recording densities are lower for tape recording, but the quantity of data stored in a tape casette may exceed that on a hard disc by an order of magnitude, as there is plenty of surface to write on. The cost per Gbyte is a factor of 5–10 lower. Track widths are about a micrometre. Particulate media are based on acicular particles of Co–γFe_2O_3 or iron, or else $BaFe_{12}O_{19}$ platelets.

14.6.3 Read head

In all early magnetic recording systems, the reader was the same electromagnet structure that served to write the data. The flux change in the yoke, due to changing stray field in the airgap as the medium sped by, induced an emf in the winding. This scheme, which remains in use in some simple tape recorders, was implemented in thin-film inductive heads for hard-disc recording in 1980, but it was replaced by a merged head with a separate thin film permalloy AMR read sensor in 1991.

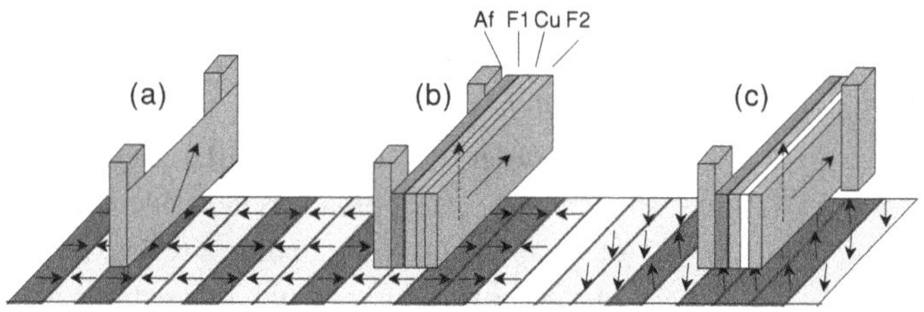

Figure 14.36

An AMR read head, a GMR spin valve read head and a TMR spin-valve read head For the first two, longitudinal magnetic records are written in the plane of the medium and the current flows in the plane of the film (CIP), whereas for the third case, the records are perpendicular to the plane of the medium, and the current flows perpendicular to the stack (CPP).

Magnetoresistive read heads are optimized linear thin-film sensors set perpendicular to the medium and the track, which respond directly to the vertical component of the stray field, rather than its time derivative as in an inductive reader. Successive generations of readers have enabled the continual miniaturization that has produced the exponential growth in storage density shown in Fig. 14.3. GMR spin valves were introduced in 1997, and TMR spin valves came with the transition to perpendicular recording in 2006. Each new generation offered an adequate signal-to-noise ratio in a smaller package.

The reader is part of the merged thin-film head, a complex multilayer structure nowadays patterned on a length scale of a few tens of nanometres. The reader must be as close to the medium as possible, no more than a bit width away in order to detect the stray field that arises at the transition from one bit to the next in longitudinal media or from the magnetized bits themselves in perpendicular media. Considering the recorded bit as a long, transversely magnetized wire, the readback signal varies as the stray field (13.7) which is proportional to the magnetic moment per unit length $\lambda = M_r l\mathrm{t}$, where l is the bit length and t is the medium thickness, and inversely proportional to r^2.

The reader is sandwiched between two relatively thick permalloy shield layers, which absorb the flux from adjacent bits. It is generally shorter than the track width. The reader itself is a complicated stack of sequences of seed layers, exchange bias layer, ferromagnetic pinned and free layers with a spacer, and cap layers on top.The pinned layer may be a synthetic antiferromagnet. The stack shown in Fig. 14.24 is a comparatively simple one! A bias field produced by segments of hard magnetic Co–Cr–Pt eliminates domain structure in the free layer, and leads to a single-domain sensor with no Barkhausen noise. The writer layers are deposited on top of this stack. A single wafer may include 20 000 heads, which have to be cut out, lapped and mounted at the tip of the slider arm of the voice-coil actuator (Fig. 13.18). Structures of different magnetoresistive readers are shown schematically in Fig. 14.36. The magnetic axes of the pinned and free layers in the spin valves are perpendicular, in zero external field.

Large current densities flow in the reader. The voltage signal in a CPP structure is

$$V = j\left(\frac{\Delta R}{R}\right)\varrho \mathrm{t}, \tag{14.33}$$

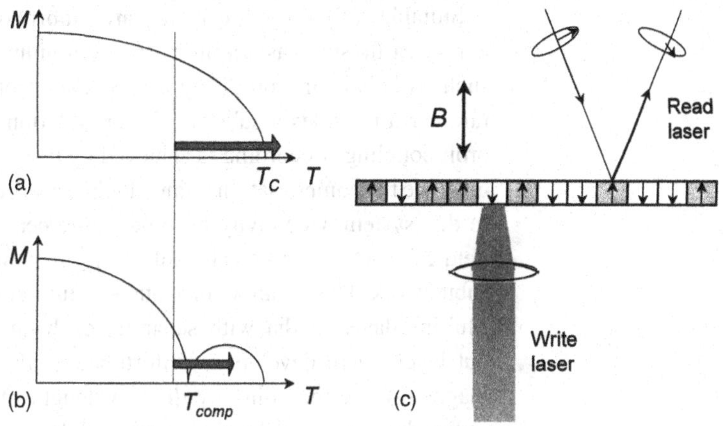

Figure 14.37

Magneto-optic recording. The principles of Curie point and compensation-point writing are illustrated in (a) and (b), respectively, and the recording method on a perpendicular medium is shown in (c).

where j is the current density, ϱ is the resistivity and t is the thickness of the TMR layer. Low resistance is required to minimize Johnson noise in any magnetoresistive sensor, which in a nanoscale tunnel junction means that the resistance-area product $R\mathcal{A} = \rho t$ must be less than 10 Ω μm^2. For this, the MgO barrier layer in Fig. 14.24 has to be less than a nanometre thick. An unwanted side effect is spin-transfer torque on the free layer.

GMR heads are still used in tape recording, which tends to lag a generation behind hard disk recording.

The hard-disc drive is a marvel of technology that sustains our data-hungry way of life. It is a triumph of sophisticated magnetics, mechanical design, nanoscale fabrication, signal processing and value engineering, which owes its persistent success to the scale invariance of the dipole field of a magnet. Scaled up, magnetic recording is like a jumbo jet flying across a landscape with its nose wheel extended to count the blades of grass a few millimetres below, missing only a few of them as it crosses the country. Scaled down, it has met our needs for nonvolatile data storage for 60 years. Everything we download from the internet is stored on hard disc drives in servers. Each year we are now recording, magnetically, more information than had been recorded in all prior human history.

14.6.4 Magneto-optic recording

A scheme which enjoyed some success at the end of the twentieth century was based on the magneto-optic readout of perpendicularly written media, Fig. 14.37. In thermomagnetic recording, the medium was heated locally by a pulsed laser to a temperature where the coercivity is small, and then cooled in a weak bias field. A thin film with perpendicular anisotropy is used in combination with readout based on the polar Kerr effect – the plane of polarization of light of a reflected semiconductor diode laser beam is rotated by an angle $\pm\theta_K$ on reflection from a $\uparrow$ or $\downarrow$ domain, where $\theta_K \approx 1°$.

Suitable magneto-optic media are amorphous $R_{1-x}T_x$ films, where R is a rare-earth such as terbium or gadolinium and T is a transition metal such as cobalt or iron. They are deposited on a transparent substrate. The rare-earth provides a substantial Kerr rotation on account of its strong spin-orbit coupling. Recording is achieved by heating either above the Curie point, or above the compensation point which is near room temperature when $x \approx 0.7$ in this system. Coercivity becomes large near T_{comp}, because the anisotropy field $2K_1/M_s$ diverges. The anisotropy is mainly provided by the rare-earth subnetwork. Use of an amorphous medium eliminates grain-boundary noise. Multiple-layer media with separate, exchange-coupled recording and read-out layers were developed. Unfortunately physical limitation is imposed on magneto-optic recording by the wavelength of light, which means that a bit cannot be much smaller than a micrometre, limiting the density to a few bits per μm^2. For this reason, magneto-optic recording is no longer competitive.

FURTHER READING

M. Ziese and M. Thornton (editors), *Spin Electronics*, Berlin: Springer (2001). A multiauthor volume which treats topics at an introductory level, with some emphasis on oxide spin electronics. The best starting point for beginners.

U. Hartmann (editor), *Magnetic Multilayers and Giant Magnetoresistance*, Berlin: Springer (1999). Readable articles focussed on magnetic multilayers and giant magnetoresistance, including one on magneto-optic recording.

M. Johnson (editor), *Magnetoelectronics*, Amsterdam: Elsevier (2004). Covers magnetoelectronics in a series of articles, including chapters on logic, tunnelling and biochips.

J. D. Stiles and J. Miltat, *Spin transfer torque and dynamics* in *Spin Dynamics in Confined Magnetic Structures* (B. Hildebrands and A. Thiaville editors), Berlin: Springer (2006), pp. 225–308. An up-to-date review.

S Maekawa (editor), *Concepts in Spin Electronics*, Oxford: Oxford University Press (2006). A monograph focussed on theoretical aspects.

S. Bandyopadhyay and M. Cahay, *Introduction to Spintronics*, CRC (2008). A comprehensive introduction with emphasis on theory.

J. Stöhr and H. C. Siegmann, *Magnetism; From Fundamentals to Nanoscale Dynamics*, Berlin: Springer (2006). A modern treatment of the subject, particularly good on spin transport, fast dynamics and electron spectroscopy. Extensive bibliography. Recommended.

L. Comstock, *Introduction to Magnetism and Magnetic Recording*, New York: Wiley-Interscience (1999). An extensive and useful introduction for engineers.

H. N. Bertram, *Theory of Magnetic Recording*, Cambridge: Cambridge Unversity Press (1994).

M. L. Plumer, J. van Eck and D. Weller (editors), *The Physics of Ultra-high Density Magnetic Recording*, Berlin: Springer (1999). A series of articles covering micromagnetic and dynamic aspects of recording, with a focus on media.

H. J. Richter, Recent advances in the recording physics of thin-film media and The transition from longitudinal to perpendicular recording, *J. Phys. D, Applied Physics* **32**, R147 (1999) and **40**, R49 (2007).

EXERCISES

14.1 Deduce the Einstein relation (14.3) in the free-electron model.

14.2 Show that the definition of spin polarization (14.12) with $n = 2$ is equivalent to the polarization of the electric current, provided the spin relaxation times $\tau_\uparrow$ and $\tau_\downarrow$ are the same.

14.3 Derive (14.22). Show that the field sensitivity of a planar Hall sensor is the same as an AMR sensor with $p \approx \pi/4$.

14.4 Estimate the dimensions below which it becomes preferable to rely on spin-transfer torque rather than Oersted fields to switch the magnetization of a ferromagnet.

14.5 Estimate the maximum thickness of a ferromagnetic free layer which can be switched in a nanosecond. Assume it is possible to pass a current of 10^{11} A m^{-2} without encountering severe problems of electromigration. Assume $P_e = 0.5$.

14.6 Give an expression for the electric field E at a perpendicular distance R from a wire carrying a flow of (fictitious) magnetic charge q_m constituting a magnetic charge current I_m. Check the dimenions of your answer. Use an analogy with the answer to Exercise 2.10 to give an expression for the electric field due to a spin current flowing in a conductor, at a perpendicular distance R from the conductor. If the spin current is due to a perfectly spin-polarized electronic charge current flowing in the conductor, estimate the maximum possible electric field that can be produced if the greatest admissable current density is $j = 10^{11}$ A m^{-2}.

14.7 Design a magnetic field sensor that makes use of the planar Hall effect. Compare its sensitivity with an AMR sensor made of the same material.

14.8 Estimate the recording density that might be achieved if $SmCo_5$ were used as the magnetic medium. What problems can you envisage in using this material?

14.9 Why is the blocking condition for magnetic recording $K_1 V/k_B T > 40$ rather than $K_1 V/k_B T > 25$ (8.22)?

14.10 Show that the polarization of the ϕ^+ photoluminesence excited by bandgap radiation in GaAs is 1/4.

15 Special topics

Disciplines grow at their boundaries. The interdisciplinary topics considered here fall into three groups. One is mainly concerned with liquids: paramagnetic liquids, ferrofluids, magnetic levitation and confinement, and magnetoelectrochemistry. The second relates to life sciences: magnetism in biology and medicine, magnetic imaging and magnetically aided diagnostics. Finally there is planetary and stellar magnetism, covering the magnetism of rocks and the Earth's magnetic field, as well as and those of other planets, the Sun and stars.

Magnetism has been a spur to human curiosity for centuries. The force field with its attractive and repulsive interactions led to dreams of levitation and perpetual motion, and hopes for cures of illness, as well as a striving for understanding. These hopes and dreams have been realized in unexpected ways. Magnetism is a mature discipline with a secure physical foundation, which allows it to engage in interdisciplinary joint ventures with other branches of science.

If perpetual motion has proved to be a pipe dream – periodically revived to peddle to gullible investors – it nevertheless finds an echo in the stationary states of quantum mechanics where the electrons occupy quantized orbits. There, they enjoy undiminished motion, at least until they exchange a quantum of energy with their environment. But they can do no work in their stationary states. Energy conservation is inviolate.

The levitated island of Laputa, from *Gulliver's Travels*. The island was supposed to contain a huge embedded lodestone.

Levitation is a more practical proposition, but again not as people imagined long ago – for example, Jonathan Swift's island of Laputa, the 'coffin of the Prophet' in Medina or the golden idol in the temple of Somnath. Static levitation *is* possible, but it is severely limited in its applications at room temperature by the feebleness of the diamagnetism of solids (Table 3.4).

Magnets are worn by golfers and tennis players; they are sold to treat ailments, drinking water, cocktails and crude oil. Seeds are reputed to germinate faster and broken bones mend better in the presence of magnetic fields. There are supposed to be malign influences as well. Paracelsus, the sixteenth-century Swiss alchemist and physician, believed the benefit of a magnet depended on which pole was presented to the patient – a belief echoed on numerous websites, and one which has caused much anguish to cancer sufferers concerned about the 'correct' definitions of the North and South poles. Yet for as long as fanciful beliefs about magnetism have flourished, they have been vigorously debunked by rational sceptics, beginning with William Gilbert himself.

Here we tread on firmer ground, looking at some interdisciplinary applications of magnetism in medcine, biology and electrochemistry, as well as phenomena in liquids. Finally, we step outside the laboratory, casting a glance on a much larger scale – planets, stars and galaxies. These special topics are a mixed bag, but many of them involve fluids, and somehow relate to magnetohydrodynamics.

15.1 Magnetic liquids

Although stable, homogeneous ferromagnetic liquids are known to exist, metallic glasses demonstrate that a crystal lattice is not a prerequisite for ferromagnetic order. It just seems that melting points always happen to exceed Curie points in metallic systems. The closest approach is the supercooled eutectic $Co_{80}Pd_{20}$, which does show signs of superparamagnetism. Interactions in nonmetallic sytems tend to be antiferromagnetic, and they are frustrated in the liquid state.

15.1.1 Paramagnetic liquids

Paramagnetic solutions of magnetic ions can be prepared with susceptibility χ ranging from effectively zero up to about 10^{-3} for multimolar solutions of Dy^{3+} or Ho^{3+}, the $4f$ ions with the greatest moments (Table 4.6). The $3d$ ions have smaller values of p^2_{eff} with the maximum for Mn^{2+} and Fe^{3+} (Table 4.7). Nitrates and chlorides are the most soluble salts. Concentrated paramagnetic solutions can be used for levitation, as discussed below. The susceptibility is sufficiently small for the demagnetizing field $-\mathcal{N}\chi H'$ to be negligible.

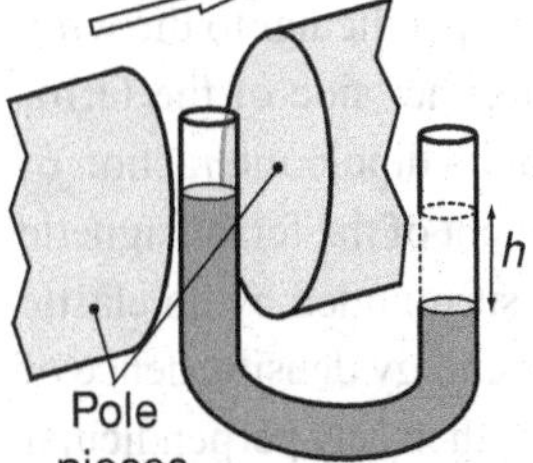

The Gouy method for measuring the susceptibility of a liquid.

A traditional method used by chemists to measure the susceptibility of liquids is the Gouy balance. A U tube of cross section a containing the liquid is placed with one arm between the poles of an electromagnet which produces a field H' in the gap, while the other arm is well away from the field. When the magnet is switched on, a paramagnetic liquid of density d is attracted into the field, and a change of level of $h/2$ is observed. A magnetic force $\boldsymbol{F}_m\delta V$, known as the field-gradient force, acts on any volume element $\delta V$of the liquid subjected to a magnetic field gradient, where the force density from (2.105) is

$$\boldsymbol{F}_m = (\chi/2\mu_0)\nabla B^2. \tag{15.1}$$

Integrating from x_1 where the applied field is zero to x_2 where B is uniform and equal to $\mu_0 H'$, since the induced magnetization is negligible, gives the force, which supports the column of liquid of height, h, and mass $h$ad. Hence

$h a d g = (\chi/2\mu_0)B^2 a$.

$$h = \frac{\chi B^2}{2\mu_0 d g}. \tag{15.2}$$

The result for a 1 molar solution of $CoCl_2$, $\chi = 60 \times 10^{-6}$, in a field of 1.5 T is $h = 5$ mm. For pure water, which is diamagnetic with $\chi = -9 \times 10^{-6}$, the level is slightly depressed as the water is repelled by the field.

The effect is quadratic in field, and therefore independent of field direction. For example, if an extended open container is placed in the horizontal bore of a superconducting solenoid, the water level is depressed in the field region relative to that outside the field. This is known as the Moses effect (although the susceptibility of the Red Sea differs little from that of pure water). For a field $B_0 = 10$ T, the depression of the water level is 37 mm.

Another curious effect of the field-gradient force is the stabilization of paramagnetic liquid tubes in water. When a drop of ink is released into a glass of water and it will disperse rapidly; this has nothing to do with atomic-scale diffusion which is a slow process at room temperature, where diffusion constants are of order 10^{-9} m^2 s^{-1}. The dispersion is due to convection imposed by the initial velocity distribution and density differences. It is possible to inhibit convection, but not diffusion, by modest magnetic field-gradient forces. The force per unit volume on a solution of concentration c mol m^{-3} and susceptibility χ_{mol} is

$$\boldsymbol{F}_m = \frac{1}{2\mu_0}[c\chi_{mol} + \chi_{water}]\nabla B^2. \tag{15.3}$$

A field of 1 T with a gradient of 10 T m^{-1} produces a force of up to about 10^4 N m^{-3} for multimolar solutions, which is similar to the magnitude of the gravitational force acting on the liquid. The diamagnetic susceptibility of water may be neglected for concentrated solutions of paramagnetic ions.

Consider now a thin iron wire stretched horizontally in a beaker of water, which is placed in a uniform vertical magnetic field. An injected paramagnetic solution forms a tube which follows the iron wire in one of two stable positions. One is above it, the other below. In a horizontal field, perpendicular to the wire, the stable positions of the injected liquid tube are at either side of the ferromagnetic wire. The behaviour is explained by the dipole–dipole interaction of the induced magnetic moment of the tube and the moment of the ferromagnetic iron wire. The paramagnetic liquid acts as if it were surrounded by an elastic membrane. The tube's surface is a contour of constant energy density, hence of constant B^2. The magnetic force, proportional to ∇B^2, then acts perpendicular to the surface.

Paramagnetic liquid tubes can be made to follow a track embedded in the base of a vessel, along which they flow almost without friction, Fig. 15.1. This is quite unlike normal pipe flow, where the flow velocity v is restricted by the radius r of the pipe according to Poiseuille's equation $v = r^2\Delta P/4\eta l$, where $\Delta P/l$ is the pressure gradient and η is the dynamic viscosity of the liquid in

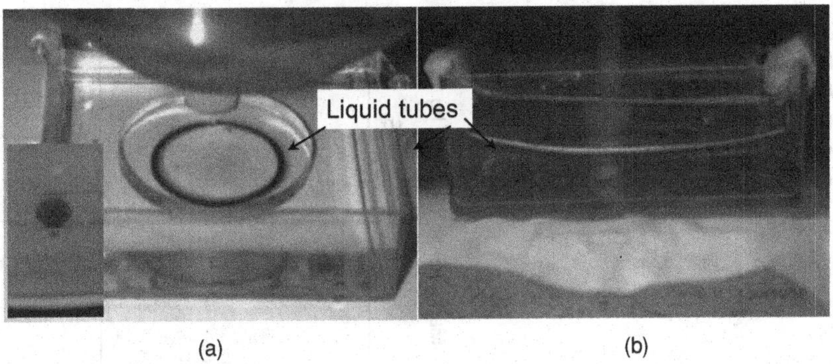

Figure 15.1

Paramagnetic liquid tubes of 1 M solutions of: (a) $ErCl_3$ suspended below an iron wire in a bath of water and (b) $CoCl_2$ following an iron track in the base of a vessel containing water. The insert shows the cross section.

N s m^{-2}. The zero-velocity boundary condition does not apply at the wall of the paramagnetic liquid tube since the liquid wall is moving. The effective radius is determined by the size of the vessel, which is many times the radius of the tube. This leads to a greatly enhanced volume flow rate, which varies as the fourth power of the radius. A consequence is the possibility of rapid mixing in magnetically confined microlitre volumes.

15.1.2 Ferrofluids and Colloids

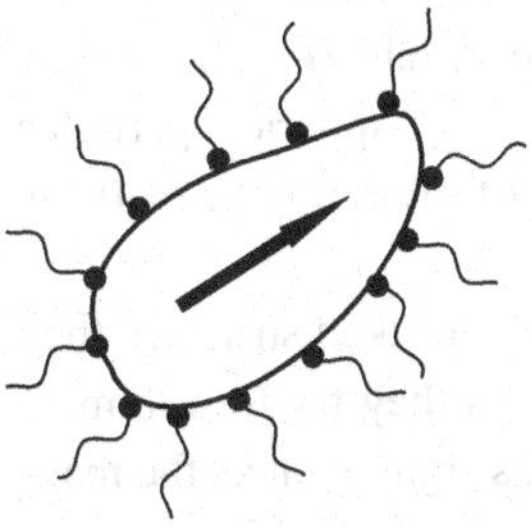

A ferromagnetic nanoparticle coated in surfactant to make it soluble in water.

A ferrofluid resembles a ferromagnetic liquid, but it is really a colloidal suspension of tiny superparamagnetic particles in oil or water. Chemical techniques were developed in the 1960s to disperse nanoparticles of magnetite or maghemite 3–15 nm in diameter in such a way that they do not agglomerate into chains under the influence of dipole–dipole interactions, when subject to an external magnetic field. In order to stabilize the colloid, it is necessary to weaken the dipole–dipole forces, which fall off as r^{-3}. Ways to keep the particles apart are to coat or embed each particle in polymer. A sheath of surfactant molecules on the surface of the oxide particles helps make them soluble in water. Otherwise charged nanoparticles can be dispersed in an ionic liquid.

Each particle has a thermal energy of order k_BT, or 4×10^{-21} J at room temperature. Besides the Néel-type superparamagnetic relaxation, described in §8.5, the particles are subject to normal Brownian motion, which helps to stabilize the colloid. A stable ferrofluid should be impervious to sedimentation under the influence of gravity, and it should neither drift in a magnetic field gradient nor agglomerate under the influence of dipole forces. These requirements impose a particle size of order 10 nm. Particles this small are single-domain.

Magnetic nanoparticles make up a fraction f of the total volume of the ferrofluid, which is at most 20%. The magnetization of magnetite is 480 kA m^{-1}, so the saturation magnetization of a commercial, magnetite-based ferrrofluid does not exceed 100 kA m^{-1}. A typical value is 50 kA m^{-1}. On account of their energy, size and separation, the particles behave like weakly

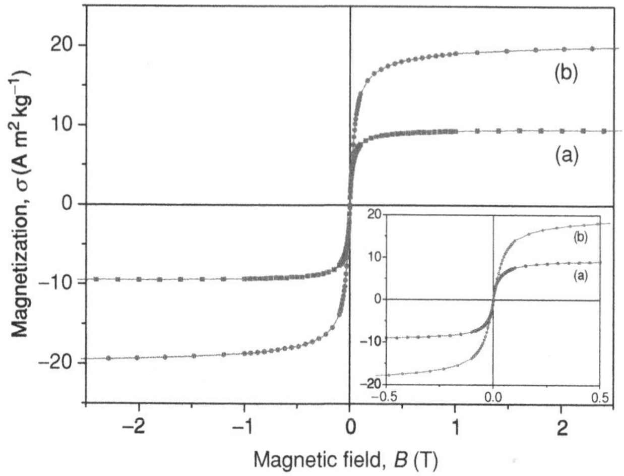

Figure 15.2

Magnetization curves for: (a) a ferrofluid and (b) magnetic microbeads containing SPIONs. (Data courtesy of Fiona Byrne.)

interacting paramagnetic macrospins, each with a moment $\mathfrak{m} \sim 10^3$–$10^5\ \mu_B$. Magnetization of the particle is given by the Langevin function (4.21) $M = M_0(\coth x - 1/x)$, where $x = \mu_0 \mathfrak{m} H / k_B T$, and the susceptibility in small fields, $x < 1$, given by the classical expression (4.22) $\chi = \mu_0 n \mathfrak{m}^2 / 3k_B T$, is in the range 5×10^{-3}–5×10^{-1}. The magnetization of the ferrofluid is $\mathfrak{f}M$. It is much larger than the susceptibility of paramagnetic liquids because of the giant moment that couples to the internal field H. Demagnetizing effects cannot now be neglected. The susceptibility in the externally applied field H' is limited by the demagnetizing field to $1/\mathcal{N}$ (§2.2.6). For well-dispersed spherical particles ($\mathcal{N} = \frac{1}{3}$), the external susceptibility is 3. Hence 90% of saturation ($x = 10$) is achieved in a field of 0.05–5 T.

A dish of ferrofluid in a vertical field adopts a peaked structure to minimize the sum of its demagnetizing and potential energies.

Ferrofluids exhibit some weird properties. One is the peaked structure they adopt in a field applied perpendicular to the surface as they try to minimize their energy in the demagnetizing field, which becomes significant as the magnetization approaches saturation. The ability of a magnetic field to control the buoyancy of objects immersed in ferrofluid is discussed in §15.3. Special ferrofluids are made by suspending nonspherical, acicular or plate-like nanoparticles in a liquid crystal.

The main application area for ferrofluids is in seals. An oil-based fluid is held in place by a suitable magnet, and if oil with a low vapour pressure is chosen, a rotary vacuum seal can be obtained. Applications include bearings for turbopumps and rotary feedthroughs for vacuum systems. Ferrofluid seals around the voice coils of loudspeakers provide damping and a path for dissipating heat. Other uses are as magnetic inks, and in magnetic levitation or separation.

A different use is made of oil-based dispersions of ferromagnetic particles in magnetorheological fluids. Here the particles are micrometre-size and multidomain, and the loading is much higher, $\mathfrak{f} \sim 70\%$. Dipolar forces become strong as the particles are magnetized. The viscosity of the fluid can therefore be increased by orders of magnitude by an applied field. These magnetorheological

Table 15.1. Properties of some commercial ferrofluids and microbeads

	Bead size (μm)	Particle size (nm)	M_s (kA m^{-1})	χ	Viscosity (Pa S)
Oil-based ferrofluid		10	36	2.2	1
Water-based ferrofluid		10	16	0.7	0.005
Magnetorheological fluid		1000	200	~0.5	100
Dynal. M280	2.8	7	13	0.32	
Myone	1.0	6	28	0.05	
Micromod	0.25	8	18	0.48	

fluids are used in mechanical clutches and suspension systems. Some typical properties are included in Table 15.1.

The superparamagnetic iron oxide nanoparticles (SPIONs) can also be dispersed in spherical polymer microbeads. These microbeads have found a variety of diagnostic and therapeutic applications (§15.4). Their magnetization curves are similar to those of ferrofluids..

15.2 Magnetoelectrochemistry

There are two distinct aspects of the interdisciplinary field where magnetism and electrochemistry meet. One is the use of electrodeposition as a means of producing magnetic films and coatings. The other is the use of magnetic fields to influence electrochemical processes.

15.2.1 Electrodeposition

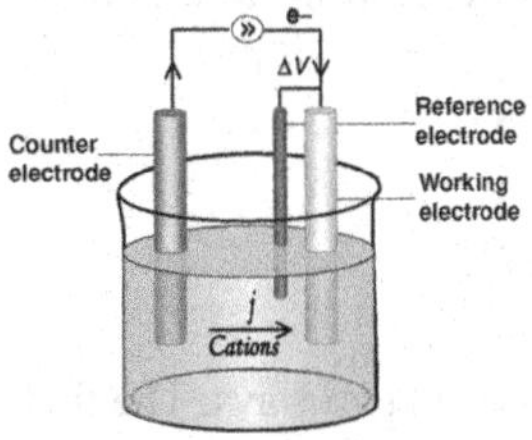

A simple electrochemical cell.

Electrodeposition is a convenient and well-established means of producing thin films of ferromagnetic metals and alloys such as Co–Fe or Ni–Fe. Permalloy is a favourite. Aqueous solutions of the metal ions, including special chemical additives to promote smoothness of the plated films, are used in an electrochemical cell which is agitated to ensure efficient plating when the current at a given potential is mass-transport limited. Conditions are chosen to ensure that any hydrogen evolved at the cathode due to the reduction of water does not spoil the quality of the deposit. Metal is deposited on the cathode, provided the applied voltage exceeds the reduction potential. For example, the reduction of nickel according to the reaction $Ni^{2+} \rightarrow Ni + 2e$ occurs at -0.25 V relative to a reference hydrogen electrode reaction. Some standard reduction potentials are listed in Table 15.2.

Metals which are not too electronegative to plate include the late transition metals from Fe to Zn, Rh, Pd, Pt as well as the noble metals Cu, Ag, Au and several others, but unfortunately not the early transition metals or the

Table 15.2. Standard reduction potentials (volts) in aqueous solution[a]

Mg^{2+}	−2.38	Mn^{2+}	−1.18	Co^{2+}	−0.28	Cu^{2+}	0.34
La^{3+}	−2.37	Zn^{2+}	−0.76	Ni^{2+}	−0.25	Pd^{2+}	0.83
Al^{3+} [b]	−1.66	Fe^{2+}	−0.44	Ag^{+}	0.22	Au^{3+}	1.42

[a] Relative to hydrogen reference electrode
[b] 0.1 M NaOH

rare-earths. The deposition rate depends on the overpotential, and it is possible to deposit alloy films such as permalloy $Ni_{78}Fe_{22}$ from a bath containing appropriate concentrations of Fe^{2+} and Ni^{2+}. The atomic composition of the bath will be quite different from the target composition of the alloy. An easy direction of magnetization of the soft magnetic film can be achieved by applying a magnetic field during deposition. The effect is similar to magnetic field annealing – a slight texture is established with Fe–Fe pairs aligned parallel to the applied field direction.

Nanowires can be obtained by plating into a porous insulating membrane which is metallized on the back, as described in Chapter 8. Microporous alumina with a dense hexagonal array of parallel pores is often used. Multilayers can be obtained from a single bath by toggling the deposition potential between two values. For example, Co–Cu layers can be deposited from a bath that is a 25 mM solution of Cu^{2+} and a 1 M solution of Co^{2+}. At 0.1 V, for example, only copper will be deposited, whereas at −0.4 V both Co and Cu will be reduced. However, the concentration of Cu in the bath is so low that the alloy is then predominantly Co.

Magnetically hard rare-earth alloys such as $SmCo_5$ cannot be plated from aqueous solutions. At the large negative voltages required, the current will consist almost entirely of protons from the water. However, phases such as CoPt can be obtained. As deposited, these alloys have a disordered fcc structure, but they only adopt the tetragonal $L1_0$ structure and develop hysteresis after annealing at about 900 K (§11.2.1).

15.2.2 Magnetic field effects

A magnetic field can influence electrochemical processes in two ways. The first is via the Lorentz force which acts on the current density $\boldsymbol{j}$ in the cell to give a body force density:

$$\boldsymbol{F}_L = \boldsymbol{j} \times \boldsymbol{B}. \tag{15.4}$$

Whenever a field is applied parallel to the electrode of an electrochemical cell, this force leads to convective stirring of the electrolyte. The transport of ions to the cathode, where they are reduced to metal, is governed by the concentration gradient ∇c, where c is the ionic concentration in moles per cubic metre. The current density $j = D\nabla c$, where D is the diffusion constant and $\nabla c = c_0/\delta$ with δ the thickness of the diffusion layer, a region a few hundred micrometres wide near the cathode where the ionic concentration falls from the average concentration c_0 in the bath to zero at the cathode surface. The stirring action of the Lorentz force reduces the thickness of the diffusion layer, and it therefore increases the mass-transport-limited current density. For typical plating current densities $j = 1$ mA mm^{-2}, the Lorentz force in 1 T is 10^3 N m^{-3}. Corrosion currents, which flow from cathodic to anodic sites some micrometres apart on the surface of a corroding electrode can be similarly influenced by magnetic fields.

The other way magnetic fields can influence the action in electrochemical cells is by means of field gradients. The force on an electrolyte containing a concentration c mol m^{-3} of ions of susceptibility χ_{mol} is given by (15.3), which follows from (2.104) when demagnetizing fields are negligible, as they always are in the solutions used in electrochemistry. The field-gradient force can be much enhanced at ferromagnetic microelectrodes, where significant forces are exerted on paramagnetic ions in solution. Values of ∇B can be as high as 10^5 T m^{-1}.

15.3 Magnetic levitation

15.3.1 Static levitation

Dreams of levitation were apparently dashed in 1842, when Samuel Earnshaw proved that levitation of a charged particle by a static configuration of electric fields was impossible. Earnshaw's theorem can be generalized to a statement that no stationary object made of charges, magnets or masses can be held in place by any fixed combination of electric, magnetic and gravitational forces. Magnets can be regarded as static distributions of magnetic charge q_m. The theorem applies whenever the energy ε of an object satisfies Laplace's equation $\nabla^2 \varepsilon = 0$, whose solutions include no isolated maxima or minima, but only saddle points.

In the magnetic case, the energy of a pre-existing dipole in free space is

$$\varepsilon_m = -\mathfrak{m} \cdot \boldsymbol{B}. \tag{2.72}$$

The force on the dipole is $\boldsymbol{f} = -\nabla \varepsilon_m$, and $\nabla \cdot \boldsymbol{f} = -\nabla^2 \varepsilon_m$. To show that this is zero, we take $\mathfrak{m}$, which has constant magnitude, outside the derivative; $\nabla \cdot \boldsymbol{f} = -(\mathfrak{m}/\mu_0)\nabla^2 \boldsymbol{B}$. Using the vector identity $\nabla^2 \boldsymbol{A} = \nabla(\nabla \cdot \boldsymbol{A}) - \nabla \times (\nabla \times \boldsymbol{A})$ and the facts that $\nabla \cdot \boldsymbol{B} = 0$ in free space and $\nabla \times \boldsymbol{B} = 0$ in the absence of electric currents, it follows that $\nabla \cdot \boldsymbol{f} = 0$. Hence the energy ε_m obeys Laplace's equation.

Figure 15.3
The Transrapid maglev train.

Imagine a small sphere around the point O where the dipole is located and use the divergence theorem.

$$\int_V \nabla \cdot \boldsymbol{f} \mathrm{d}^3 r = \int_S \boldsymbol{f} \cdot \mathrm{d}\mathcal{A} = 0. \tag{15.5}$$

The force on the dipole integrated over the surface of the sphere is zero, so if in some directions it experiences a negative, restoring, force, in others the force must be positive, tending to carry the dipole away from the position of unstable equilibrium at O.

It is a feature of the permanent magnet bearings discussed in Chapter 13 that a mechanical constraint (or active electromagnetic servo system) is generally required in one direction. The stiffness of the bearing $\boldsymbol{K}$ is defined as a vector with components $-\partial f_x/\partial x$, $-\partial f_y/\partial y$, $-\partial f_z/\partial z$, which sum to zero (13.14). An alternative scheme is employed in the 'Transrapid' maglev trains, which are capable of speeds of up to 500 km h^{-1}, namely attraction between on-board electromagnets and the base of the guideway which the cars wrap around, Fig. 15.3. The cars require a little lateral support to keep them on track.

Yet despite Earnshaw's theorem, passive levitation was *not* just a dream. His prohibition can be circumvented. What is needed is a magnetic field that is not fixed, but responds to the position of the moment $\mathfrak{m}$. This can be achieved by introducing diamagnetic material in the vicinity of the magnet; the diamagnet provides passive repulsive feedback, the field increasing as the magnet approaches it. Superconductors are the strongest diamagnets, having $\chi = -1$. The magnet creates an image in the superconductor, as shown in Fig. 2.15(b), and the force between the magnet and its image is repulsive. The repulsive force is self-regulating, increasing as the magnet approaches the superconductor, and decreasing as they move apart. Substantial masses can be levitated in this way.

A sumo wrestler standing on a magnetic plate levitated above a large disc of cuprate superconductor.

The strongest nonsuperconducting diamagnets, graphite and bismuth, have a dimensionless susceptibility that is more than 1000 times smaller than a

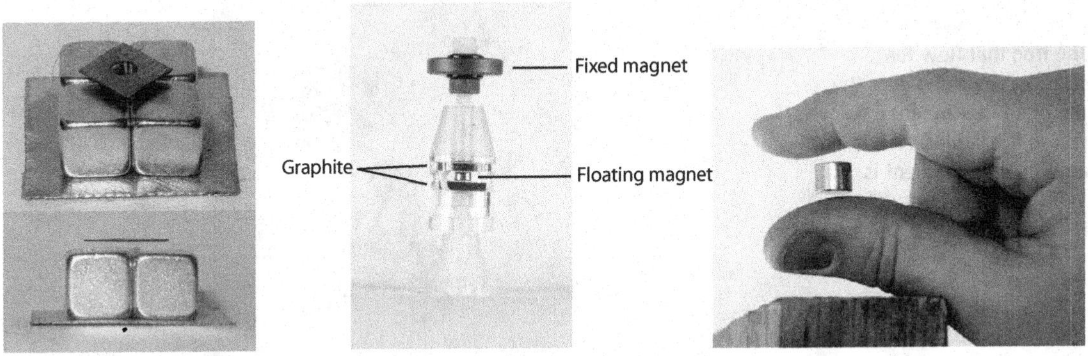

(a) (b) (c)

Figure 15.4

Stable levitation with diamagnetic material: (a) a sheet of oriented graphite floats above a permanent magnet array and (b) a small permanent magnet is suspended in a gradient field produced by the upper magnet. The equilibrium is stabilized by two graphite plates. In (c), a similar arrangement with a superconducting magnet, equilibrium is stabilized by two diamagnetic fingers. (A. K. Geim, M. D. Simon, M. I. Boemfa *et al. Nature* 400, 324 (1999).)

superconductor (Table 3.4). The weak magnetic image nevertheless enables a sheet of oriented graphite to float about a millimetre above a rare-earth permanent magnet. More generally, sheets of graphite can be used to stabilize a magnet at a position of unstable equilibrium, where K_z is negative but K_r is positive. Such passive levitation devices which operate entirely at room-temperature are illustrated in Fig. 15.4.

In principle, *any* diamagnet can be levitated by an appropriate combination of magnetic field and magnetic field gradient. The energy of a sample of volume V is $-(1/2\mu_0)V\chi B^2$, where the factor $\frac{1}{2}$ arises because the moment is induced by the field. The force in a vertical field gradient must balance the weight $-\mathrm{d}Vg$, which leads to the condition for levitation in terms of the mass susceptibility

$$B\nabla_z B = \mathrm{d}g\mu_0/\chi. \tag{15.6}$$

The levitation condition for polycrystalline graphite ($\chi_m = \chi/\mathrm{d} = -50 \times 10^{-9}$ m^3 kg^{-1}) is $B\nabla_z B = 250$ T^2 m^{-1}, whereas for water ($\chi_m = -9 \times 10^{-9}$ m^3 kg^{-1}) the requirement is 1400 T^2 m^{-1}. The former condition is satisfied close to the surface of a permanent magnet producing a field of 1 T, whereas the latter is obtainable near the end of a Bitter magnet or superconducting solenoid producing 10 T or more. It is in high-magnetic-field laboratories that it has been possible to levitate all manner of objects composed mainly of water – frogs, strawberries and spheres of salt solution, to name just a few, Fig. 15.5.

Permanent magnets can also be used to levitate watery objects, but the necessary conditions can only be achieved very close to the surface by structuring the magnets on a submillimetre scale. On another scale, it is possible to support atoms with unpaired electrons against gravity in magnetic atom traps.

We know that susceptibilities of diamagnets are relatively feeble compared with those of local-moment paramagnets. The Curie-law susceptibility of a paramagnet is $\chi = C_{mol}/T$, where $C_{mol} = 1.571 \times 10^{-6}\, p_{eff}^2$ (4.16). For example, the susceptibilities of a mole of Co^{2+} ($p_{eff} = 3.9$) or Dy^{3+} ($p_{eff} = 10.6$) at room temperature are 120×10^{-9} and 590×10^{-9} m^3 mol^{-1}, respectively. The dimensionless susceptibility of 1 M solutions of these ions is approximately

Figure 15.5

The frog that flew for science – levitated near the top of a Bitter magnet, where the field is 10 T, and the field gradient is 140 T m^{-1}. (Courtesy of L. Nelemans.)

Figure 15.6

Magnetic levitation of graphite, silicon and titanium in the fringing field of a 100 mm electromagnet. They are immersed in a 2 M $DyCl_3$ solution. The field at the centre is 1 T. (Courtesy of Peter Dunne)

the same as their mass susceptibility in $m^3\,kg^{-1}$ because 1 litre of solution has a mass of approximately 1 kg. Evidently the paramagnetism χ_{sol} of these solutions is much greater than χ_m the Pauli mass susceptibility of metals or the mass susceptibility of any diamagnetic element or compound (Table 3.4). *Immersion* in paramagnetic solution is therefore a way of levitating a wide range of materials in modest fields, and separating them if necessary (Fig. 15.6). The material in the cavity behaves as if it has susceptibility $\chi_m - \chi_{sol}$, and the condition for buoyancy now becomes

$$B\nabla_z B = -g\mu_0(\mathrm{d} - \mathrm{d}_{sol})/(\chi_m - \chi_{sol}). \qquad (15.7)$$

For example, to levitate silicon ($\chi_m = -1.8 \times 10^{-9}\,m^3\,kg^{-1}$, $\mathrm{d} = 2330$ $kg\,m^{-3}$) in air would require an enormous $B\nabla_z B$ of 6840 $T^2\,m^{-1}$, whereas the same job in a 1 M $DyCl_3$ solution requires a mere 22 $T^2\,m^{-1}$.

Ferrofluids have even larger susceptibilities than ionic solutions, of order 10×10^{-6} m^3 kg^{-1}, so they can be used to levitate and separate practically anything in a modest inhomogeneous magnetic field. Conversely, a magnet, which produces its own inhomogeneous magnetic field, will levitate spontaneously when placed in a beaker of ferrofluid (although ferrofluids are opaque so you are unable to see it).

The use of diamagnetic materials and paramagnetic liquids by no means exhausts the ways of circumventing Earnshaw's theorem. Roy Harrison discovered in 1983 that a spinning magnet in a field gradient had a small zone where it can be stably levitated. Of more practical use, however, is the levitation of a molten metal by radio-frequency eddy currents in a suitably designed cold crucible.

15.3.2 Radio-frequency levitation

High-frequency magnetic fields allow us to levitate, heat and stir conducting liquids without contact. All are handy capabilities. Together with the ability of static fields to damp motion in conducting liquid, and thanks to the action of the Lorentz force on the induced current

$$\boldsymbol{F}_{Li} = \sigma(\boldsymbol{v} \times \boldsymbol{B}) \times \boldsymbol{B}, \tag{15.8}$$

where $\boldsymbol{v}$ is the velocity of the liquid, they provide the foundations for the technology of electromagnetic processing of materials. This technology has flourished in recent decades. Magnetic damping is used to control eddies during casting of steel slabs, and convection during growth of semiconductor crystals from the melt. Heating and stirring are exploited in the induction furnace, which has changed little from the first design of Sebastian Ferranti in 1887. Interest in radio-frequency levitation is more recent, with commercial applications dating from the 1960s.

The transport equation for $\boldsymbol{B}$ in a moving conducting fluid, known as the advection–diffusion equation, is a basic equation of magnetohydrodynamics. It is obtained by combining Ohm's law, including the Lorentz force, $\boldsymbol{j} = \sigma(\boldsymbol{E} + \boldsymbol{v} \times \boldsymbol{B})$ with Faraday's law $\nabla \times \boldsymbol{E} = -\partial \boldsymbol{B}/\partial t$ to give $\partial \boldsymbol{B}/\partial t = -\nabla \times (\boldsymbol{j}/\sigma - \boldsymbol{v} \times \boldsymbol{B})$ and then using Ampère's law $\nabla \times \boldsymbol{B} = \mu \boldsymbol{j}$ to write $\boldsymbol{j}$ in terms of $\boldsymbol{B}$. Since $\nabla \cdot \boldsymbol{B} = 0$, $\nabla \times (\nabla \times \boldsymbol{B}) = -\nabla^2 \boldsymbol{B}$, the result is finally

$$\frac{\partial \boldsymbol{B}}{\partial t} = \nabla \times (\boldsymbol{v} \times \boldsymbol{B}) + \eta \nabla^2 \boldsymbol{B}, \tag{15.9}$$

where the magnetic diffusivity $\eta = 1/(\sigma\mu)$ appears as a diffusion constant, with units of m^2 s^{-1}. The terms in Maxwell's equations involving charge density and displacement current are negligible. Solutions of (15.9) are governed by the characteristic dimension l of the system and the value of a dimensionless quantity

$$\mathcal{R}_m = vl/\eta, \tag{15.10}$$

known as the magnetic Reynolds number. In liquid metals, where B is practically uninfluenced by v, we are in the limit $\mathcal{R}_m \ll 1$, (15.9) then reduces to a simple diffusion equation:

$$\frac{\partial \boldsymbol{B}}{\partial t} = \eta \nabla^2 \boldsymbol{B}. \tag{15.11}$$

The solution for an oscillating magnetic field $b_z(0) = b_0 \sin \omega t$ excited parallel to a metal surface by external currents is

$$b_z(x) = b_0 \exp(-x/\delta_s) \cos(\omega t - x/\delta_s),$$

where the penetration depth $\delta_s = \sqrt{(2\lambda/\omega)}$ is just the skin depth (12.2), since $\mu = \mu_0 \mu_r$, $\omega = 2\pi f$ and $\varrho = 1/\sigma$. The penetrating magnetic field excites eddy currents within the skin depth of the conductor, and by Lenz's law the force between the exciting and induced currents is repulsive. These are the forces exploited for radio-frequency levitation.

The induced current is given by Ampère's law $\nabla \times \boldsymbol{B} = \mu \boldsymbol{j}$, which simplifies to $j_y = -(1/\mu)\partial B_z/\partial x$ for a field applied in the z-direction. Thus

$$j_y = (b_0/\mu\delta_s) \exp(-x/\delta_s)[\cos(\omega t - x/\delta_s) - \sin(\omega t - x/\delta_s)].$$

The Lorentz force in the skin depth is $\boldsymbol{j} \times \boldsymbol{B} = j_y B_z \mathbf{e}_x$. Averaging over time, and using the fact that $\langle \cos^2 \omega t \rangle = \frac{1}{2}$, the force density is given by $\boldsymbol{F}_m(x) = (b_0^2/2\mu\delta_s) \exp(-2x/\delta_s)$. Integrating over the thickness of the metal finally gives the expression for the force per unit area, known as the magnetic pressure

$$\boldsymbol{P}_m = \frac{b_0^2}{4\mu} \boldsymbol{e}_x. \tag{15.12}$$

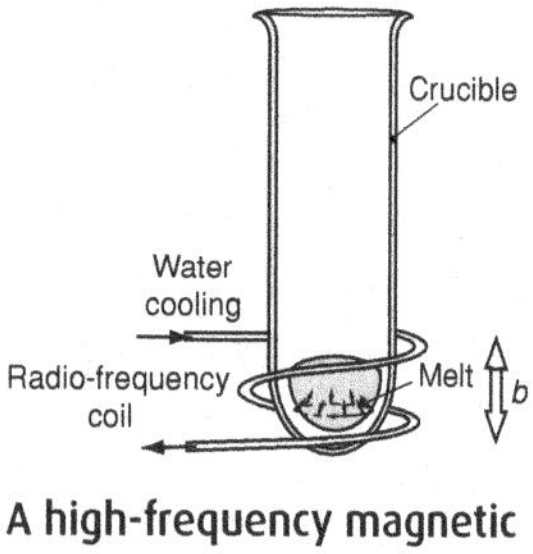

A high-frequency magnetic field penetrates the skin depth of a metal, and the Lorentz force on the induced currents creates a force normal to the metal surface, which may be used for levitation in a radio-frequency induction furnace.

The heating effect of the electric current density is $\int_0^\infty (\boldsymbol{j}^2/\sigma) \mathrm{d}x = (b_0^2/4\mu)\omega\delta_s$ W m^{-3}.

In a radio-frequency levitation furnace, Fig. 15.7, the radio-frequency coil is wound into the form of a basket in which a molten metal drop is supported by magnetic pressure normal to its surface. The surface shape is determined by the balance of gravitational, magnetic and surface-tension forces. There is a limit to the size of the drop that can be levitated because there is no magnetic pressure at the centre of the base. There the pressure due to the height of the drop has to be balanced by the surface tension alone. For this reason, large levitated drops tend to drip out along the vertical axis. If material is continuously fed from above, the effect can be exploited to create a liquid metal jet, where the flow rate can be controlled by varying the radio-frequency power.

On quite another scale, magnetic fields generated by superconducting magnets are used to confine conducting plasmas at temperatures of order 10^8 K in

Figure 15.7

A radio-frequency induction furnace. (Courtesy of Ambrell Co.)

toroidal tokamak reactors such as ITER, which are the vanguard of advanced fusion research.

15.4 Magnetism in biology and medicine

15.4.1 Magnetotaxis

Life on Earth has evolved in the presence of a weak magnetic field of 10–100 A m^{-1}. Inevitably, some creatures besides ourselves have learned to take advantage of this field. The clearest examples are magnetotactic bacteria, unicellular organisms which manufacture particles of ferromagnetic iron oxides (magnetite or maghemitite) or sulphides (greigite). The particles grow in chains in the microbe's body. Each one is of a size which would make it superparamagnetic if it stood alone, but the anisotropic dipole–dipole interaction in a chain of particles stabilizes the magnetization direction along the chain. Thus, every bacterium has a built-in compass needle, which is handed on to the next generation with the polarization direction intact, by cellular division. New particles added to the ends of the half-chain grow in the stray field there, and become magnetized parallel to their neighbours. In the language of rock magnetism, they acquire chemical remanent magnetization.

The value of magnetotaxis for bacteria is that it allows them to orient with the Earth's magnetic field lines; they swim along them at about 100 μm s^{-1}, upwards or downwards according to the polarity of their built-in magnet. Polarity is of vital importance. Anaerobic bacteria swim downwards along the field lines, reaching mud where they flourish. Those with magnets of opposite

Figure 15.8

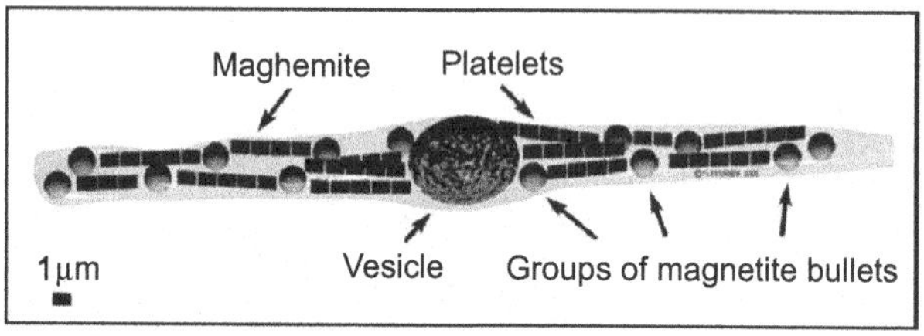

The magnetic sense organ in pigeons. It consists of three different structures, each containing magnetic material. (G. Fleissner *et al.*, *Ornithology*, 148, 5663 (2007).)

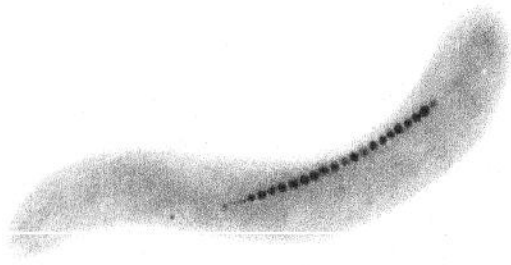

A magnetotactic bacterium. The scale bar is 1 μm.

polarity swim upwards to the surface where they perish in the toxic oxygen-rich environment, a beautiful example of natural selection. Magnetotactic bacteria can be extracted from mud and made to dance in circles in a petri dish with a rotating magnetic field.

As many as 50 other creatures, including pigeons, salmon, bats, bees and deer are thought to have a magnetic sense. The magnetotaxis is more subtle, and it may operate in conjunction with other senses to find direction. In pigeons, for example, it was thought that the force between two magnetic particles attached to nerve endings is sensed, and the direction of the axis of the magnet pair relative to the field is distinguished by its attractive or repulsive force. There is no distinction between North/South or East/West but a maximum effect is perceived when the pigeon turns through 90°. The pigeon's magnetic sense organ has been identified (Fig. 15.8). It is obviously complex, but it is not yet known exactly how it works.

Other than force or torque exerted on ferrimagnetic particles, there are two other mechanisms that might be involved in magnetotaxis. One is induced emf in large moving conducting circuits, which may operate in sharks. The other is a possible effect on radical pair reactions, where a weak magnetic field may influence the singlet–triplet interconversion rate.

Epidemiologic studies of human exposure to low-frequency fields from power lines or domestic wiring have proved inconclusive. Large static magnetic fields seem to be innocuous, with exposure provoking little response at the level of the organism.[1] Pigeons soon recover their magnetic sense, even after a spell in the bore of a 15 T magnet. Studies of the influence of high-frequency fields from mobile phones have provided no clear evidence of harm.

15.4.2 Cellular biology

Many reports can be found in the literature on effects of static or low-frequency magnetic fields on cellular processes, but very few have been reproducibly established.

[1] There is an EU directive that prescribes limits for human exposure to magnetic fields (2004/40/EC).

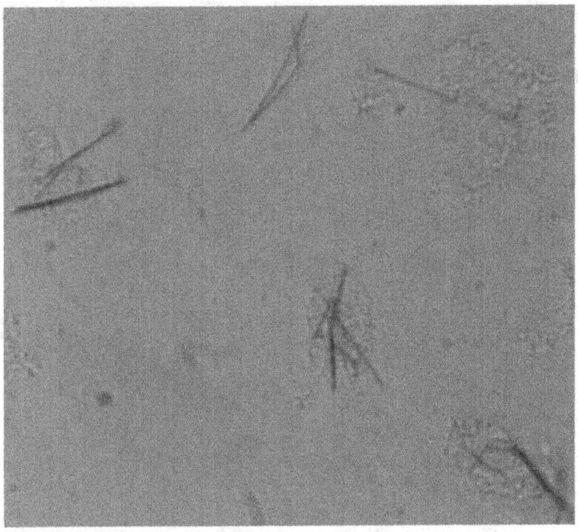

Figure 15.9

Differentiated THP1 macrophages, which have internalized nickel wires 20 μm long and 200 nm in diameter (Courtesy of A. Prina-Mello).

However, magnetic methods are beginning to make a contribution to the study of cells, and subcellular structures such as proteins and biomolecules. Generally, these studies make use of magnetic micro- or nanoparticles whose sizes are compatible with the sizes of the biological structures they are used to manipulate (10–100 μm for cells, 10–100 nm for proteins). The microparticles normally incorporate many superparamagnetic nanoparticles in a biocompatible polymer microbead, with a fill fraction $0.1 < \mathfrak{f} < 0.8$ (Table 15.1). The surface of the bead can be functionalized for a specific biochemical reaction, with an antibody for example. A single coated magnetic nanoparticle may be used to manipulate protein or similar structures. The response of the magnetic label is usually linear in the gradient of applied fields, which are of order 10–100 kA m^{-1} for miniature electromagnets, and may be larger if permanent magnets are used. Magnetic nanowires, Fig. 15.9, have the advantage that information can be recorded along the length of the label if it is segmented and the segments are permanently magnetized in opposite directions. Such magnetic barcodes can be used to label cells or proteins, and they can be detected via the characteristic patterns of stray field they produce, by using magnetoresistive sensors in a microfluidic channel, for example.

When a magnetic microbead or nanowire is attached to a cell or biomolecule, the object can be manipulated by field gradient force

$$\boldsymbol{f} \approx \nabla(\mathfrak{m} \cdot \boldsymbol{B}), \tag{15.13}$$

where $\mathfrak{m}$ is the induced moment of the bead. The studies are interesting because mechanical stress and morphology can regulate cellular functions, and it is important to be able to measure mechanical properties on the appropriate scale. Controlled stresses and forces can be applied via the magnetic labels in biological micromanipulators known as magnetic tweezers. One arrangement has three or four miniature electromagnets arranged in a circle, and the field is varied by the electric currents in microcoils, Fig. 15.10. The force needed to manoeuvre a bead through the cytoplasm of a living cell is of

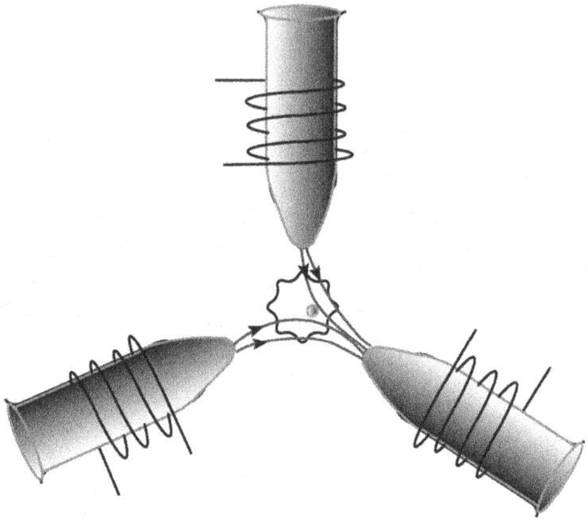

Figure 15.10

Magnetic tweezers. The currents in the coils are used to create a magnetic field that exerts force on a microbead in the cell.

order 1–10 piconewtons. For a particle of diameter 200 nm and magnetization 100 kA m^{-1}, a field gradient of order 10^4 T m^{-1} is needed to deliver the required force. Using lithographically patterned Co–Fe poles, these field gradients can be achieved over dimensions of order 10 μm.

Larger field gradients are available with small permanent magnets. The mechanical properties of individual biomolecules such as coiled DNA can be determined by attaching a magnetic bead to one end, and applying force with permanent magnet tweezers.

15.4.3 Labelling and assay

Magnetic immunoassay is a method of detecting very low concentrations of biomolecules in solution. There are two established assay procedures, both of which use magnetic microparticles. The more sensitive, two-step method uses two antibodies which attach to the molecule to be analysed. One antibody – specific antibody is attached onto the magnetic particle. The other is labelled by an optically active tag. The functionalized particles are incubated with the analyte where binding with the species in the anlyte occurs. The unbound molecules are then washed away, and in the second step the labelled antibody is incubated with the magnetic beads. The analyte is now sandwiched between two antibodies; the unreacted labels are washed away and the number of labels is measured by fluorescence or chemiluminescence. The signal increases monotonically with the quantity of analyte present, as shown in Fig. 15.11(a).

In the second, single-step, method, the solution is spiked with a known quantity of analyte which is tagged with the optically active label. The functionalized magnetic beads are incubated with the solution, and the tagged and untagged analyte molecules compete for the available antibody sites. The magnetic microparticles are then immobilized and the unbound ones are washed

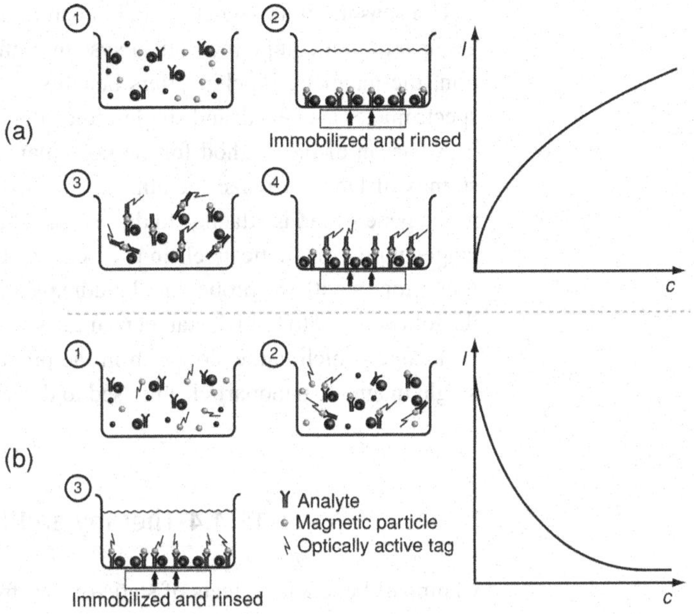

Figure 15.11
Magnetic immunoassay. (a) In the two-step method, magnetic microbeads with attached antibodies which capture the analyte, are immobilized and rinsed. Then optical markers with the same antibody are captured by the analyte, rinsed again, and the optical signal is measured. (b) In the single-step method, a known quantity of optically labelled analyte is added which competes with the analyte in solution for the antibodies on the microbeads. The beads are immobilized, rinsed and the optical signal is measured.

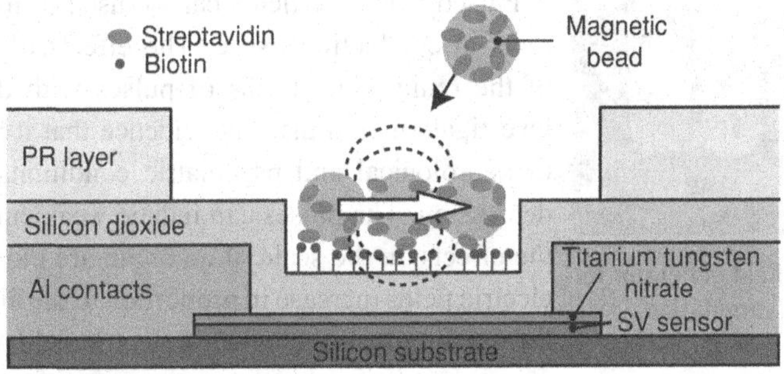

Figure 15.12
A magnetic biochip with a spin-valve sensor whose surface is functionalized with antibodies. The analyte and magnetic microparticles are captured in an incubation phase and after rinsing, a vertical magnetic field magnetizes the microbeads, creating a stray field, which can be detected by the spin valve.

out. The measured optical signal now decreases monotonically with analyte concentration (Fig. 15.11(b)).

An all-magnetic version uses a linear thin-film magnetic sensor (§14.3) rather than an optical method to detect the microbeads. The surface of the sensor is functionalized with a specific antibody, and the analyte is added, together with the functionalized magnetic beads. They are incubated together so that beads are immobilized on the sensor surface, rinsed, and the stray field of the beads is detected by the sensor, Fig. 15.12. A spin valve, AMR ring or planar Hall effect device is used, and the beads are magnetized in a perpendicular field, to which the sensor is insensitive. A linear response over several orders of magnitude, extending down to single-bead detection may be achieved. By using multiple sensor arrays, each of which is functionalized with a different antibody, it is possible to conduct many assays in parallel, with single-molecule resolution.

The sensors in a magnetic biochip can be built into microfluidic channels. Compared to a competing optical system with multiwavelength optical detection, the magnetic biochip offers benefits of cost and sensitivity. No optical spectrometer is needed, and single-molecule detection is achievable.

A variant of the method for biomolecular recognition uses complementary strands of DNA to make the attachment of the magnetic particle. A specific probe gene strand is attached to the sensor. The complementary target strand is tagged with a magnetic label, and passed over the surface of the detector, where it hybridizes with the probe. Field gradients produced by current-carrying conductors are used to hold the target near the sensor and to speed up the hybridization stage, which is the slow step in the procedure. Small beads containing a single magnetic nanoparticle are used to detect smaller biomolecules.

15.4.4 Therapy and treatment

Claims of beneficial effects of static or low-frequency magnetic fields in pain relief and treatment of inflammation are controversial. It is thought that broken bones may set more quickly if they are exposed to a magnetic field. Unfortunately, no plausible explanation of such effects has yet been advanced.

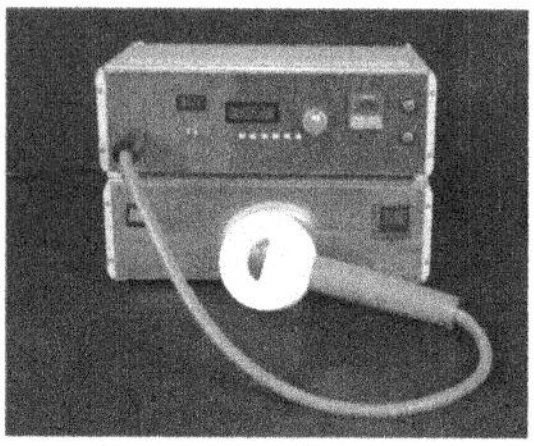

A pulsed field generator for magnetic stimulation.

Pulsed magnetic fields can be used to induce electric fields and drive currents in conducting tissue. The effects of transcranial magnetic stimulation of the brain using trains of pulses with $dB/dt \approx 10^3 - 10^6\,\mathrm{T\,s^{-1}}$ are under investigation, and there is evidence that it may be beneficial in the treatment of neurological and psychiatric conditions such as Parkinson's disease and depression. The pulses can induce weak emfs (nV–μV) at a cellular level, but the effects on the scale of an organ are more significant, because the induced electric fields increase in proportion to the dimensions (see §13.6). From (10.3), the magnitude of the induced electric field around a circuit is

$$E \approx -l\frac{dB}{dt}, \qquad (15.14)$$

where l is the dimension of the circuit. Electric fields in excess of 200 V m^{-1}, needed to induce firing of neurons, are induced in circuits some millimetres or centimetres in size. Gradient coils can be used to help confine the induced electric stimulus to some region of the brain, or some other part of the body.

Better established is the use of magnetic microparticles for hypothermia. Here the magnetic material in the bead should be conducting or exhibit hysteresis, so that it can be heated by eddy-current or hysteresis loss (§12.1) in an externally applied high-frequency field. If the beads accumulate in an area of interest, such as a tumour, heat can be applied locally to raise the temperature to about 45°C, thereby destroying the tumour.

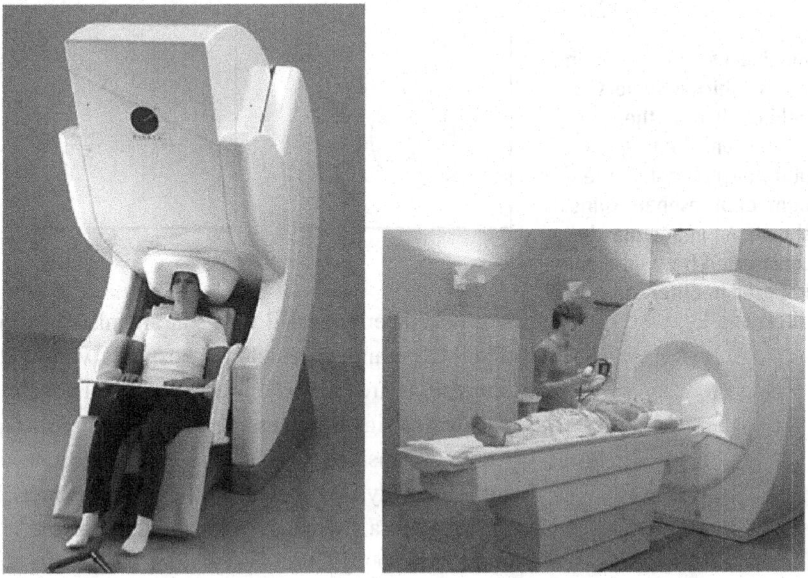

(a) (b)

Figure 15.13

Equipment for (a) magnetoencephalograpy (MEG) and (b) whole-body magnetic resonance imaging (MRI). (Courtesy of Elekta AB and Booth Radiology PLC.)

Magnetic nanoparticles may also be useful for localized drug delivery. The drug is attached to the particle surface, and they are then manipulated to the site of interest using magnetic field gradients.

15.4.5 Medical applications

Prosaic magnet applications include the use of forces between permanent magnets to secure dentures or protheses and to aid feeble muscles in the eyelids or bladder. Catheters can be guided with permanent magnets and artificial heart valves are activated by them.

Magnetic imaging is the main application area of magnetism in medicine. The magnetic field patterns produced by the electric currents flowing in muscles in the heart are detected in magnetocardiography, whereas those due to neurons in the brain are detected in magnetoencephalography (MEG). The currents are respectively of order μA and nA and the fields a few centimetres from the organs are therefore only of order of 10^{-6} or 10^{-9} Am^{-1}. The electrical activity can be measured with extended sensor arrays in magnetically shielded rooms. For example, helmets with arrays of 128 SQUID detectors have been developed for magnetoencephalograpy, Fig. 15.13(a). The main use of this technique is to pinpoint areas of the brain where abnormal electrical activity builds up during epileptic fits, as an aid in surgical treatment of the severe disorder.

Magnetic resonance imaging (MRI) is undoubtedly the most important and best-established clinical application of magnetism in medicine. Three Nobel

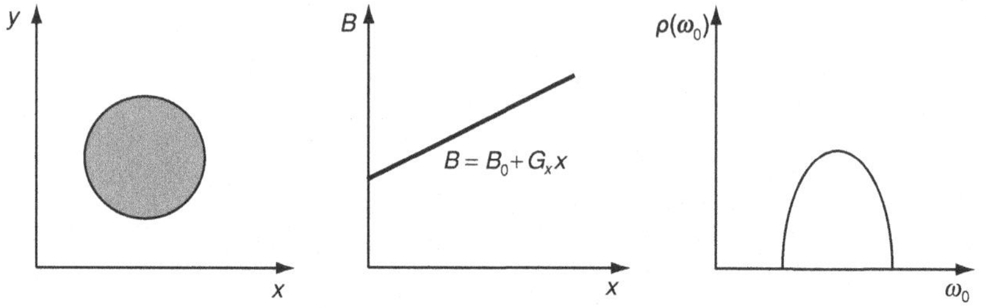

Figure 15.14

An object viewed along the z axis, which is subject to a field gradient in the x-direction. The frequency plot $\rho(\omega_0)$ recording the number of resonant spins is shown. An image may be constructed from gradients imposed in different directions.

prizes have been awarded for it, in chemistry and medicine. There are about 20 000 MRI scanners installed in hospitals worldwide, and roughly 100 million examinations are carried out every year. The technique combines NMR with sophisticated signal processing to yield two-dimensional tomographic images or three-dimensional views of a solid object, usually a part of the human body. The analysis is based on the ^{1}H resonance. Radio-frequency radiation is absorbed at a frequency corresponding to the nuclear Zeeman splitting of protons in the body. Fortunately, some 63% of the atoms in the human body are hydrogen, mainly in fat and water.

The instrument consists of a magnet, shimmed to produce a highly uniform magnetic field, a set of three-dimensional gradient coils and radio-frequency coils to excite and detect the resonace. The magnets are usually superconducting solenoids, cooled with liquid helium or a cryocooler, which generate fields of 1.0–1.5 T. Higher fields of 3 T or more provide improved resolution and contrast. Permanent magnet systems are simpler, but bulky and limited to fields of about 0.4 T.

A simplified account of the imaging principles follows. The static field in the bore of the magnet B_0 is in the z-direction, along the length of the body. The gradient coils provide linear variations of B_0 in the x-, y- and z-directions, which make the resonant frequency position-dependent. For example, if $\omega(x) = \gamma(B_0 + G_x x)$, the free-induction decay (Fig. 9.12) will contain components from each slice in the yz-plane, and the number of nuclear spins in the slice is deduced from the frequency spectrum. Initially, images were constructed from slices taken with gradients established at different directions in the xy-plane, normal to the length of the body.

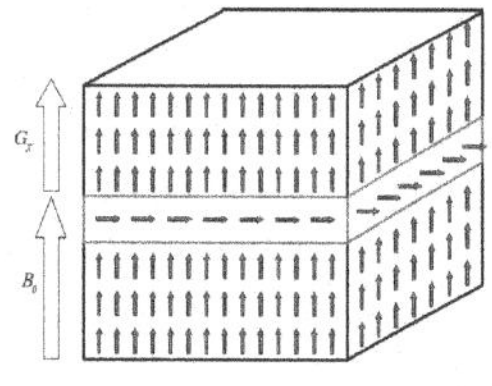

A slice of spins prepared by the 90° radio-frequency pulse and the G_x gradient pulse.

Currently MRI depends on sophisticated Fourier transform imaging techniques, introduced in the 1970s. Use is made of a slice-selection gradient and a phase-encoding gradient as well as the frequency-encoding gradient. The simplest sequence is illustrated in Fig. 15.14. It consists of a 90° radio-frequency pulse which coincides with the slice-selection gradient pulse. This flips all the resonant spins in the xz-plane by 90°. The phase-encoding gradient pulse follows, which has its gradient in the y- (or z)-direction. The final, frequency-encoding, pulse is applied in the remaining direction z (or y). The phase-encoding pulse imparts a slightly different Larmor precession frequency to the spins across the slice, and therefore endows them with a phase which

(a) (b) (c)

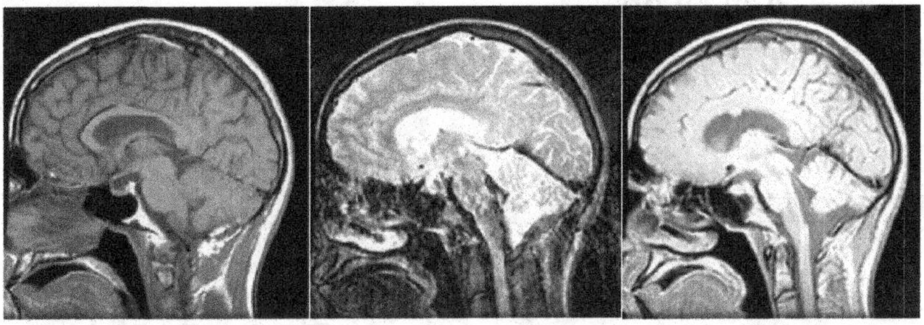

Figure 15.15

MRI images of the head based on: (a) T_1 contrast, (b) T_2 contrast and (c) spin density. (Courtesy of Y. I. Wang)

depends on its intensity or duration. Finally, the phase-encoding gradient is switched off, and the frequency-encoding gradient is applied along z, while the free-induction decay signal is recorded. The sequence is repeated 128 times with different magnitudes of the phase-encoding gradient to collect the data necessary for an image. The image is constituted by fast Fourier transforming in the x-direction to extract the frequency-domain information, and then in the y-direction to extract spatial information.

Many different pulse sequences have been devised to yield images which map T_1, T_2, T_2^* or the local spin density n. As many as ten different sequences may be used in a single MRI examination. The technique is particularly useful for imaging soft tissue and interfaces such as bone–tissue. It is useful for diagnosing soft-tissue injuries such as slipped discs and sports injuries and it is also good at identifying tumours. Some MRI images are shown in Fig. 15.15. Specific regions can be highlighted by means of magnetopharmaceutical contrast agents containing iron, manganese or gadolinium ions, which reduce the relaxation time in their vicinity. Magnetic resonance imaging is also feasible with ^{31}P and ^{13}C (Table 9.3), but these isotopes are much less abundant in human tissue than ^{1}H.

Functional magnetic resonance imaging (fMRI) is a technique that provides real-time images of the response of areas of the brain to external stimuli. It contributes to knowledge of basic neuroscience and helps in the clinical diagnosis of conditions such as stroke.

When nerve cells are active, they consume oxygen. Unlike muscles, nerves have no reserves, so the oxygen must be provided by the haemoglobin from red blood cells in nearby capillaries. There is a flow of oxygen to the active area, which peaks 2–5 seconds after the stimulus, and is known as the haemodynamic response. This leads to spatial and temporal variations in the oxygenated and deoxygenated forms of haemoglobin, which influence the proton resonance.

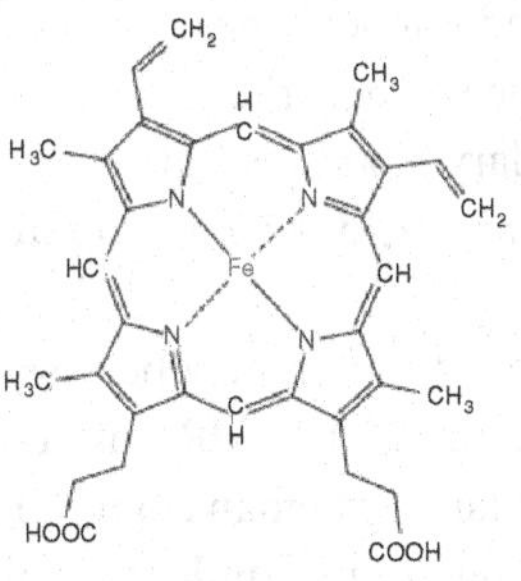

The heme group.

Haemoglobin is a large protein, with a molecular weight of about 68 000 which incorporates an iron ion in each of four subunits. The iron in the heme group is coordinated by four nitrogens at the centre of a porphyrin ring. Another protein oxygen is strongly bound at the site below the ring, and the remaining site of the coordination octahedron is occupied by the transported oxygen

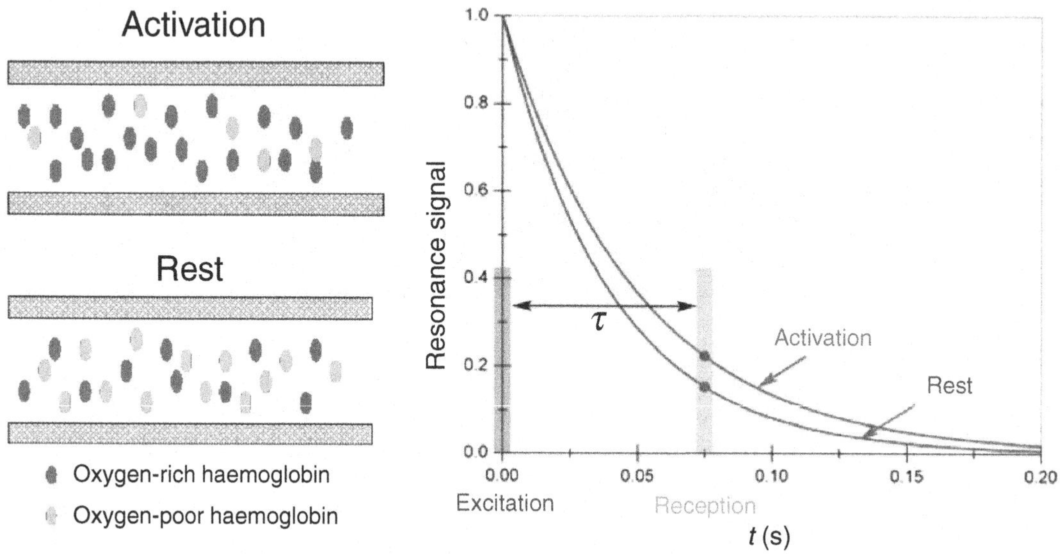

Figure 15.16

Principle of the BOLD technique, showing the blood vessels at rest and after activation, with the corresponding difference in the T_2 relaxation for the two states.

molecule in oxyhaemoglobin, or a weakly bound water molecule in deoxyhaemoglobin. Haemoglobin makes up 97% of the dry weight of red blood cells, and 0.3% of that weight is iron.

Haemoglobin is of interest magnetically, because the reduced form has Fe^{2+} in a high-spin state, $3d^6$; $t_{2g}^4 e_g^2$ with $S = 2$. There is just enough paramagnetic ferrous iron to overcome the diamagnetism of the rest of the protein. However, in the oxidized state, the iron is low-spin Fe^{III}, $3d^5$; t_{2g}^5 with a single unpaired spin $S = \frac{1}{2}$. The oxygen molecule is bound as O_2^-, which means that it too has an unpaired spin. The two form a covalent bond, resulting in no net moment. Hence haemoglobin is diamagnetic when oxygenated, but paramagnetic when deoxygenated.

The difference in magnetic properties is the basis for blood oxygen-level dependent (BOLD) imaging, introduced in 1992, Fig. 15.16. The contrast with the surroundings depends on the balance of changes of cerebral blood flow and the oxygen level of the blood itself. Imaging is based on T_2 or T_2^* contrast, and shows moderate spatial resolution of about 3 mm and temporal resolution of a few seconds. The strength of the signal varies as the square of the magnetic field, and more of the signal arises from smaller, capillary blood vessels in large fields. Hence there is a push to use high frequencies and fields of 7 T or more to improve these images.

Dramatic as images of a brain lighting up as it goes to work may be, they offer only an indirect representation of neural activity, Fig. 15.17. fMRI may be regarded as a first step towards the distant goal of explaining human cognition in terms of physical mechanisms. A much faster response with similar spatial resolution is available with magnetoencephalography, which has the advantage

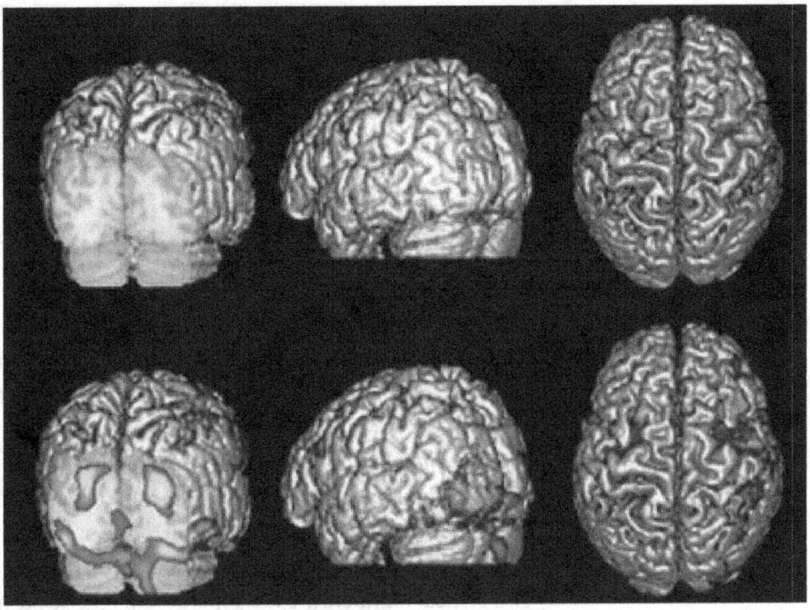

Figure 15.17

Activation of the brain of a sighted subject (top) and a blind subject (bottom) during tasks similar to reading braille. The fMRI response is superposed on surface-rendered high-resolution MRI images of the two brains. (Courtesy of N. Sadato.)

of responding directly to the neural currents. However, there are far fewer MEG systems installed worldwide than MRI systems.

15.5 Planetary and cosmic magnetism

Magnetic fields spanning 15 orders of magnitude are associated with stars and planets. These fields provide clues regarding the internal structure and history of the bodies that produce them. We know more about the magnetism of our own planet than any other, yet the Earth's magnetic field remains imperfectly understood, and we cannot predict its future course.

15.5.1 Rock magnetism

Iron represents about 5% of the mass of the continental crust, and it is present in most rocks (Fig. 1.11). Other magnetic elements are far scarcer, Table 15.3. Taking account of the fact that the atomic weight of iron is 2.2 times the crustal average we have the 40:40 rule: iron represents about $\frac{1}{40}$th of the atoms of the crust, and it is 40 times as abundant as all the other magnetic atoms together. Rocks are normally diamagnetic when they contain less than about 0.2 wt% iron, and paramagnetic otherwise, due to the presence of Fe^{2+} and Fe^{3+} ions in the crystal lattices of the mineral phases at concentrations well below the percolation threshold.

Table 15.3. Average abundances of magnetic elements in rocks (ppm)

Fe	50 000	Nd	30
Mn	950	Co	25
Cr	100	Sm	6
Ni	70	U	2

Nevertheless, rocks are heterogeneous mineral assemblages. Different silicate and oxide phases segregate at different stages in an igneous cooling process. These phases are frequently solid solutions, such as the fosterite–fayalite series of olivines $(Fe_xMg_{1-x})_2SiO_4$. Collective magnetic order can arise in iron-rich phases, where percolation paths connect the cations via common oxygen ions, forming superexchange bonds. Iron-rich oxides, which often include titanium, are found in basalts. The common phases are titanomagnetite and titanohematite. Crystallite sizes range from a few nanometres to several micrometres. Two critical numbers are the superparamagnetic particle size $2R_b$ and the single domain size $2R_{sd}$, which are around 25 and 80 nm respectively (Table 8.1). These magnetic grains may be present in quantities of order 10–100 ppm, and they impart a spontaneous magnetization of order 10 A m^{-1} to the rock.

Other iron-rich oxide phases which are magnetically ordered at room temperature are maghemite and some antiferromagnetic hydroxides, which occur in soils and sedimentary rocks. They are unstable on heating. Iron sulphides are formed by hydrothermal processes in the presence of H_2S or by the action of sulphate-reducing bacteria at the surface. Many economically valuable metal ores (Fe, Co, Ni, Cu, ...) are sulphides, and pyrite, FeS_2, is most common of all. It forms golden crystals but the iron is low-spin Fe^{II} $3d^6; t_{2g}^6$ with $S = 0$, so pyrite exhibits only a weak, temperature-independent paramagnetism.

Iron-rich oxide and sulphide minerals which are magnetic at room temperature are listed in Table 15.4. Native iron, iron–nickel $Fe_{1-x}Ni_x$ with $x \approx 0.07$ and FeS are very rare on Earth, but they are found in iron meteorites. Other iron-rich silicate and carbonate phases listed in Table 15.5 order antiferromagnetically at temperatures near or below 100 K. Iron has hardly any solubility in feldspars, the framework silicates which are the most common rock-forming minerals.

Applications of rock magnetism are based on measurements of the natural remanence, which can be acquired in several ways. Most important is the thermoremanent magnetization, acquired when igneous or metamorphic rocks cool through the superparamagnetic blocking temperature of the titanomagnetite grains. Rocks often have complex thermal and geological histories, and components of the remanence with differing stability can be acquired at different times in their past. Natural remanent magnetization can range from 10^{-5} to 10^2 A m^{-1}. Weaker remanence is acquired when grains sediment in

Table 15.4. Natural minerals, magnetically-ordered at room temperature

Mineral	Ideal formula	T_C, T_N (K)	σ_s (A m^2 kg^{-1})
Iron	Fe	1038	218
Magnetite	Fe_3O_4	853	92
Titanomagnetite	$Fe_{2.4}Ti_{0.6}O_4$	520	25
Jacobsite	$MnFe_2O_4$	570	77
Trevorite	$NiFe_2O_4$	860	51
Magnesoferrite	$MgFe_2O_4$	810	21
Maghemite	γFe_2O_3	950	84
Hematite	αFe_2O_3	980	0.5
Titanohematite	$\alpha Fe_{1.4}Ti_{0.6}O_3$	380	20
Goethite	$\alpha FeO(OH)$	400	<1
Lepidocrocite	$\gamma FeO(OH)$	470	<1
Ferroxyhite	$\delta FeO(OH)$	560	≈10
Troilite	FeS	578	<1
Pyrrhotite	Fe_7S_8	598	17
Greigite	Fe_3S_4	600	31

Table 15.5. Natural minerals, antiferromagnetically-ordered below room temperature

Mineral	Ideal formula	Structure type	T_N (K)
Fayalite	Fe_2SiO_4	Olivine; isolated SiO_4 tetrahedra	65
Andradite	$Ca_3Fe_2^{3+}(SiO_4)_3$	Garnet; isolated SiO_4 tetrahedra	12
Almandine	$Fe_3^{2+}Al_2(SiO_4)_3$	Garnet; isolated SiO_4 tetrahedra	4
Ilvaite	$CaFe_2^{2+}Fe^{3+}O(Si_4O_7)(OH)$	Pairs of SiO_4 tetrahedra	118
Ferrosilite	$FeSiO_3$	Orthoyroxene; silica chains	37
Hedenbergite	$CaFeSi_2O_6$	Pyroxene; silica chains	38
Grunerite	$Fe_7Si_8O_{22}(OH)_2$	Amphibole; double silica chains	45
Greenalite	$Fe_3Si_2O_5(OH)_4$	1:1 layer sheet silicate	17
Minnesotaite	$Fe_3Si_4O_{10}(OH)_2$	2:1 layer sheet silicate	28
Ulvospinel	$TiFe_2O_4$	Spinel	120
Ilmenite	$FeTiO_3$	Ordered corundum	40
Siderite	$FeCO_3$	Calcite	38
Rhodochrosite	$MnCO_3$	Calcite	34

the Earth's field, or when they grow to exceed the critical blocking size as a result of some low-temperature chemical process. Since we are concerned with geological time scales, 10–100 Ma, the criterion for superparamagnetic blocking, $KV/k_BT > 60$, is larger than that used for magnetic recording (>40) or laboratory measurements (>25).

The natural remanence of rocks is measured in a SQUID, or in a spinner magnetometer, where standard drill-core samples of diameter 25 mm and length

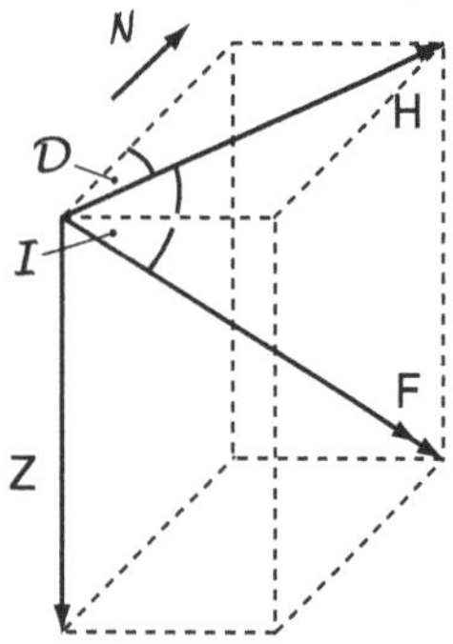

Horizontal and vertical components H, Z of the Earth's field, which has magnitude F, and direction defined by the declination (variation) D and inclination (dip) I.

Carl Friedrich Gauss, 1775–1855.

The magnetical observatory established at Trinity College, Dublin, in 1835.

20 mm are spun in a pickup coil which is shielded from any external field. The sensitivity of the spinner is 10^{-4}–10^{-5} A m^{-1}. Elaborate thermal treatments may be employed to eliminate remanence acquired after the original cooling of the rock. In this way, the direction and magnitude of the magnetic field experienced by the rock at the time when it cooled can be determined. Provided no folding has occured, the colatitude of the rock at that time is derived from (2.11) as $\theta = \text{arccot}(\frac{1}{2}\tan\mathcal{I})$. Radioactive dating of the rock by the K–Ar, Rb–Sr or Pb isotope methods then allows us to add a date stamp, and begin to tell the history not only of the Earth's magnetic field, but of the surface of the Earth itself.

15.5.2 The Earth's field

The Earth's magnetic field is dynamic. It is a vector, specified by any three of its five components, which fluctuates on time scales ranging from seconds to millions of years. The short-term fluctuations are due to currents in the upper atmosphere. They average to zero (<1 nT) over the course of a year, but on a given day a similar time sequence can be recorded at widely separated sites. This was first noticed in 1825 when daily records of magnetic observations in Kazan and Paris, cities some 4000 km apart, were compared. Gauss was then inspired to establish the *Magnetische Verein*, a world-wide association of 50 magnetic observatories coordinated from Göttingen which made meticulous measurements of the magnitude and direction of the Earth's field following a common schedule. The hope was that if enough high-quality data could be collected, an understanding of the origins of the Earth's field would follow.

An early success of the network was the proof of an *internal* source for the main field, which ranges today from 30 to 60 μT in magnitude at different spots on the globe; the average H-field is about 40 Am^{-1}. The internal origin had been postulated by Gilbert in 1600, based on his experiments with lodestone terellas. Gauss developed spherical harmonic analysis to reduce the mass of data pouring out of the magnetic observatories. Assuming that no electric currents are present at the Earth's surface, the field there can be derived from a magnetic scalar potential φ_m which satisfies Laplace's equation $\nabla^2\varphi_m = 0$. Solutions are

$$\varphi_m = \sum_{l=1}^{\infty}\sum_{m=0}^{l}\left[A_l^m r^l + B_l^m r^{-(l+1)}\right]Y_l^m(\theta,\phi), \tag{15.15}$$

where Y_l^m are the spherical harmonics, which are related to the Legendre polynomials $P_l^m(\cos\theta)$.

Here θ and ϕ are the colatitude and longitude. The A_l^m coefficients describe the contribution of sources outside the terrestrial sphere of radius $r = a =$ 6371 km, whereas the B_l^m coefficients give contribution of internal origin. It

Table 15.6. Spherical harmonic coefficients of the Earth's magnetic field (1985) in nanoteslas

		Order (n)			
Coefficient	Degree (m)	1	2	3	4
	4				169
g_n^m	3			835	−426
	2		1691	1244	363
	1	−1903	2045	−2208	780
g_n^o	0	−29877	−2073	1300	937
	1	5497	−2191	−312	233
	2		−309	284	−250
h_n^m	3			−296	68
	4				−298

turns out[2] that the As are negligible in comparison to the Bs. The potential due to internal sources is

$$\varphi_{mi} = \frac{a}{\mu_0}\sum_{l=1}^{\infty}\sum_{m-0}^{l}\left(\frac{a}{r}\right)^{l+1} P_l^m(\cos\theta)\left\{g_l^m \cos\phi + h_l^m \sin\phi\right\}, \qquad (15.16)$$

where φ is in amperes and g_l^m and h_l^m are in teslas. Some values for the leading harmonics are given in Table 15.6. About 90% of Earth's field is accounted for by a dipole of magnitude $(4a^3/\mu_0)\sqrt{g_1^{02}+g_1^{12}+h_1^{12}}$; m $=7.9\times 10^{22}$ A m^2 inclined at an angle $\theta = \tan^{-1}\{g_1^0/\sqrt{g_1^{12}+h_i^{12}}\} = 15°$ to the Earth's axis, with $\phi = \tan^{-1}\left(h_1^1/g_1^1\right)$. The first eight harmonics, or thereabouts, reflect the field produced in the Earth's core, whereas the higher terms reflect the contribution of magnetized rocks in the first 30 km of the Earth's crust, where temperatures do not exceed the Curie point of the ferromagnetic phases in minerals.

Some features of the secular variation (time dependence) of the Earth's field are quite rapid. The magnetic North pole, where the angle of dip is 90°, is moving North at an alarming rate of about 40 km y^{-1}, the magnitude of the field is decreasing by 0.1% y^{-1} and the point of zero variation on the equator has drifted westwards from the coast of Gabon to the coast of Ecuador in less than 400 years. The 'discovery' of the magnetic North pole in 1831 was an ephemeral achievement.

Edward Sabine, 1788–1883.

The other important outcome of the nineteenth-century *Magnetic Crusade*, as the Magnetische Verein became known after the addition of sites in the far-flung British Empire, was the realization by Edward Sabine in 1852 that the intensity of the short-term fluctuations followed the 11-year sunspot cycle. The short-term magnetic activity of the Earth has its origin on the Sun! Nowadays, in the

[2] A similar analysis is used in magnetoencephalography, where a is the radius of the skull.

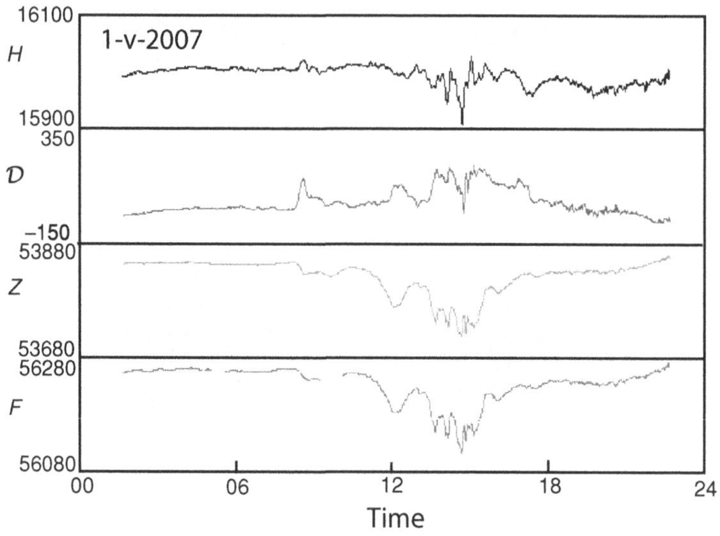

Figure 15.18

A record of fluctuations of the Earth's magnetic field, taken at Sitka in Alaska on 1 May 2007. A magnetic observatory has existed on this site since 1842. Units of H, Z and F are nanoteslas, the units of $\mathcal{D}$ are degrees. (Courtesy of USGS Geomagnetism program.)

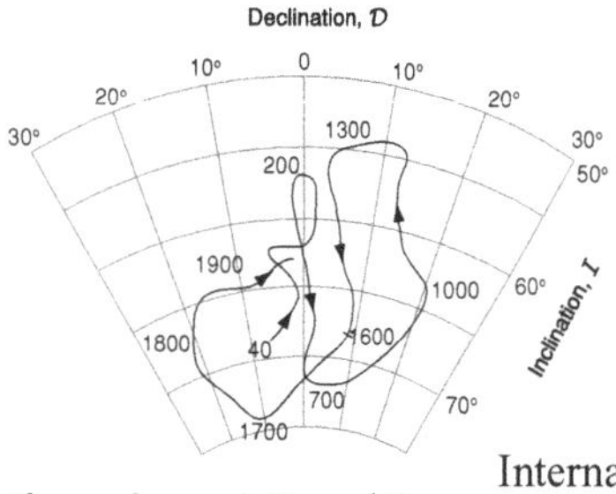

The scalar variation of the Earth's field deduced by combining observations in Paris (>1600) with measurements of the remanence of baked clay (<1600).

International Real-time Magnetic Observatory Network (INTERMAGNET), a fully automated successor stations to those established nearly 200 years ago, monitor the Earth's field with fluxgates and proton magnetometers, Fig. 15.18. This serves mainly to keep an eye on the solar weather, which influences our own. Very occasionally, magnetic storms are so severe that the emfs they induce can disable our electrical distribution networks.

The *Magnetische Verein* was one of the very earliest examples of international collaboration to solve a scientific problem; it can be regarded as a forerunner of CERN, ITER and the European Framework Programmes. Unfortunately, those involved were not to know that the origin of the Earth's field could not be inferred from a massive collection of data on its chaotic behaviour. Most of their painstaking observations were futile.

15.5.3 The Earth's dynamo

It is generally admitted that the magnetic fields of the Earth and other heavenly bodies are produced by motion of a conducting fluid core, although the details are mired in controversy. In the Earth's case, Fig. 15.19, the liquid core has inner and outer radii of 1220 and 3485 km, respectively, and is made of molten iron with minor amounts of nickel and other elements which have no strong affinity for oxygen. Less electronegative elements form oxides, which end up in the mantle.

The problem is a difficult one in magnetohydrodynamics, for which there may be no deterministic solution. The motion of the core is chaotic. We are unable to predict the secular variation. Velocities in the liquid core are thought

Figure 15.19

Internal structure of the Earth. Radii of the main structures, mean densities and temperatures at the centre and surface are given.

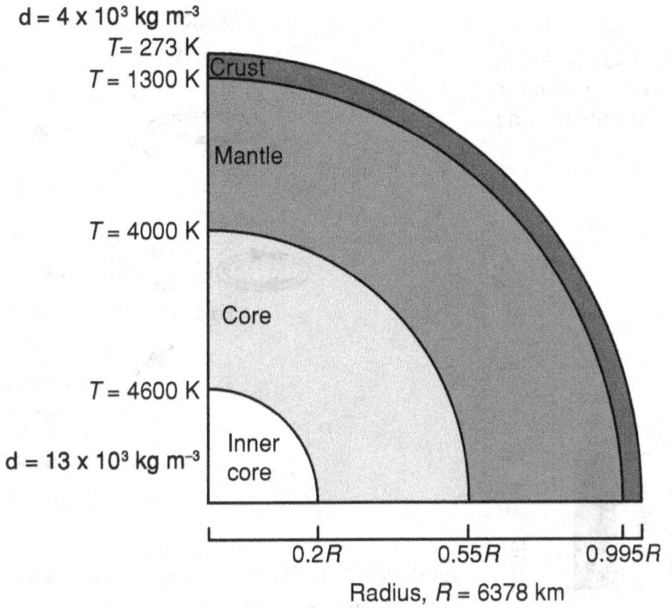

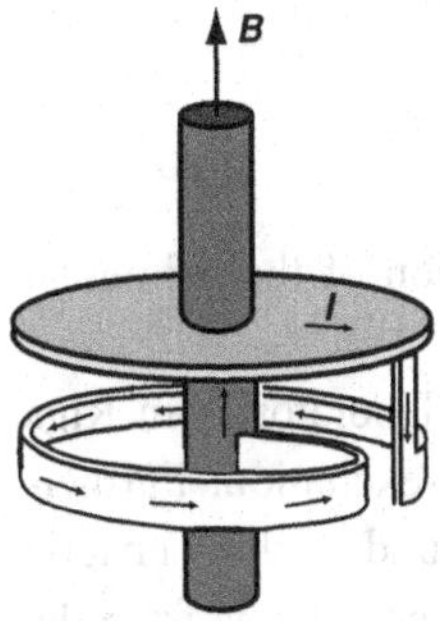

A mechanical model of a self-exciting dynamo.

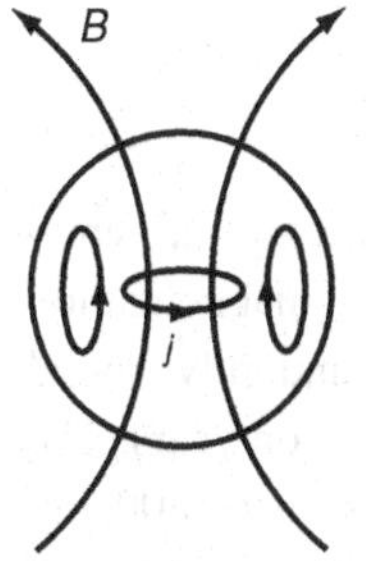

Poloidal field

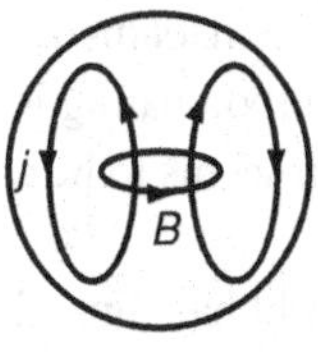

Azimuthal field

Azimuthal currents create poloidal fields, and vice versa.

to be of order $0.2\,\mathrm{mm\,s^{-1}}$ and the magnetic diffusivity $\eta = 1/\sigma\mu$ is about $2\,\mathrm{m^2\,s^{-1}}$, so the magnetic Reynolds number is given by $\mathcal{R}_m = vl/\eta \approx 100$. We are in an advection regime, where the magnetic field is carried along by the fluid medium. However, electric currents including those associated with any primordial magnetic field that was trapped when the Earth condensed from the solar nebula would decay on a timescale of 15 000 years due to resistive losses, so the currents that generate the field must somehow be sustained by dynamo action. The flows in the liquid core are driven by internal heating due to mass segregation, latent heat of crystallization and residual radioactivity of ^{40}K and U and their isotopes. Altogether about 4×10^{13} W is dissipated in these processes.

An important step towards understanding the origin of the Earth's magnetic field was the concept of a self-sustaining fluid dynamo suggested by Joseph Larmor in 1919. A mechanical model of a self-exciting dynamo is a spinning conducting disc connected to a single-turn coil. If there is some small field to begin with, an emf is generated between the axle and the rim of the spinning disc, which drives current around the circuit, thereby building up the field. The basic idea is that the fluid motion somehow stretches and twists the flux lines, thereby intensifying the magnetic field, Fig. 15.20. Just how this applies to the Earth is a matter for debate, but some constraints have been established. One is that the dynamo cannot be axially symmetric. Another is that the field in the core is quite different from the poloidal, dipolar field observed at the surface. It is thought to have a pronounced azimuthal character, generated by differential rotation between the liquid core and the mantle. The azimuthal field is generated by poloidal currents in the liquid core. In one model of the

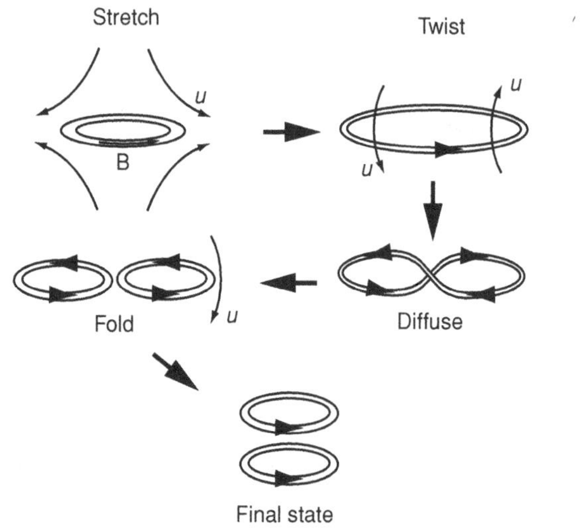

Figure 15.20

Magnetic field is intensified in a fluid core by a process of stretching and twisting flux lines. *u* is the fluid velocity.

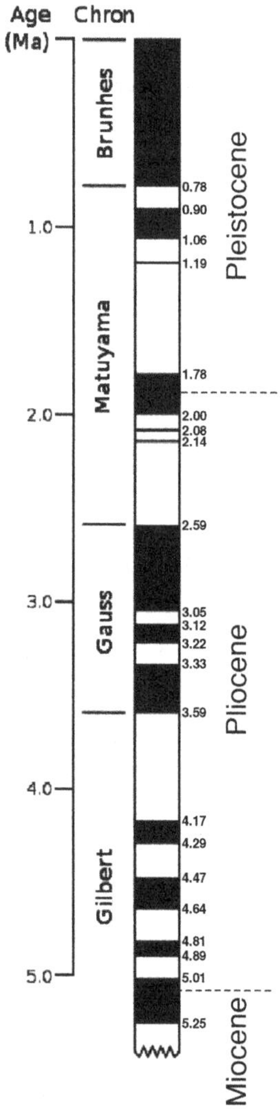

The sequence of reversals of the Earth's field over the past 5 million years.

geodynamo, turbulence leads to small-scale reorganization of the azimuthal field which creates the dipolar field.

Large-scale liquid-metal dynamos have been built in laboratories in Riga and Karlsruhe with sufficiently high $\mathcal{R}_m$ to exhibit self-excitation. Furthermore, coupled mechanical dynamo models have been found to show chaotic fluctuations and spontaneous random field reversals, which we know from the paeleomagnetic record to be salient features of the Earth's magnetic field. *Homo sapiens* has yet to experience one.

15.5.4 Paeleomagnetism

The ancient record of the Earth's field has been inferred from the natural remanence of dated basalts. A first observation is that the direction of magnetization of young rocks indicates a magnetic pole direction that scatters randomly around the geographic axis (Fig. 15.21). 'Young' for rocks is a few million years. The secular variation of the Earth's field has been measured in Paris since 1600, but the record can be extended back to Roman times by measuring the thermoremanence of baked clay in the hearths of pottery kilns, which retain a record of the field direction on the last day they were used – a nice example of archaeometry, the use of quantitative physical methods in archaeology. Radiocarbon dating is used for these comparatively recent events. The secular variation seems to be a random wander around the Earth's axis of rotation.

The most remarkable fact is that the polarity of the Earth's field has flipped randomly over geological time. Half the points in Fig. 15.21 are normal (present-day) polarity and half are reversed. The last reversal took place 700 000 years ago, and the polarity of the Earth's field is subject to more-or-less random changes on timescales of $10^5–10^6$ years. A sequence of lavas shows a characteristic pattern of normal and reversed polarity, providing the record of

Figure 15.21

Position of the Earth's magnetic pole deduced from measurements of recently formed ingeneous rocks. Half of the points have the present polarity, while the other half are reversed. On average the magnetic field is that of a

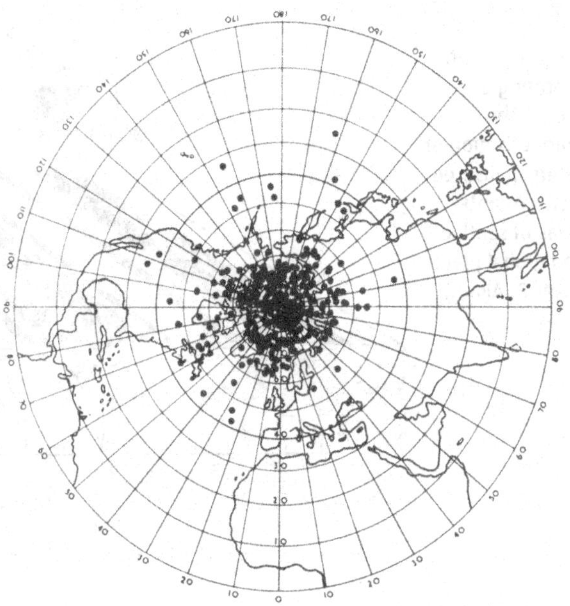

Apparent polar wander paths which are used to reconstruct the past positions of plates on the globe. Data from rocks in Europe (open circles) and North America (solid circles) can be made to coincide by closing up the Atlantic ocean.

recent reversals. Altogether some 400 reversals have been recorded in the rock records. The dipole field at a reversal changes sign over a period of a few thousand years, during which time the nondipole, higher-order harmonics are thought to be dominant. We are unable to predict these reversals. As with stock markets, past record is no guide to future performance.

The particular significance of measurements of thermoremanent magnetization of rocks is that they have established the theory of global plate tectonics. The ocean floor behaves like a giant tape recorder as new crust spreads out from the mid-ocean ridge at a rate of a few centimetres per year, Fig. 15.22. The Earth's crust is formed of plates with oceanic and continental segments. New oceanic crust forms at the mid-ocean ridges, and the plates jostle for place on the surface of the Earth, colliding where one rides over another in a subduction zone, which are regions of volcanism and earthquake activity. The random sequence of reversals provides a unique time signature. Furthermore, the magnetic colatitude θ of the ancient rocks can be collated to construct apparent polar wander paths. We know the pole does not wander far, but the plates do. These peregrinations over the face of the Earth can be reconstructed over periods of hundreds of millions of years. For example, 250 Ma ago there was a supercontinent, named Pangea, which split up into two fragments, Gondwana and Laurasia, which then broke into the continents we know today. This process has been repeated several times in the Earth's 4.5 Ga history. We know from the remanence of ancient, dated rocks that the field has been in existence for 3.5 Ga. The scientific history of the Earth is as strange as any creation myth.

Figure 15.22

Schematic representation of plates separating at a mid-ocean ridge. The pattern of magnetization of sea-floor basalts measured across the North Atlantic led to the ideas of seafloor spreading and global plate tectonics. (After D. Allan.)

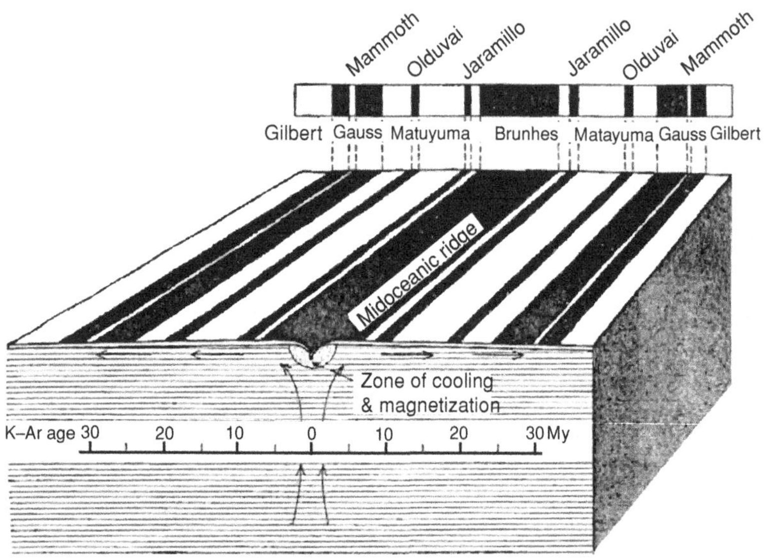

Figure 15.23

Magnetic moments of planets and moons in the solar system, plotted against their angular momentum. (After P Rochette.)

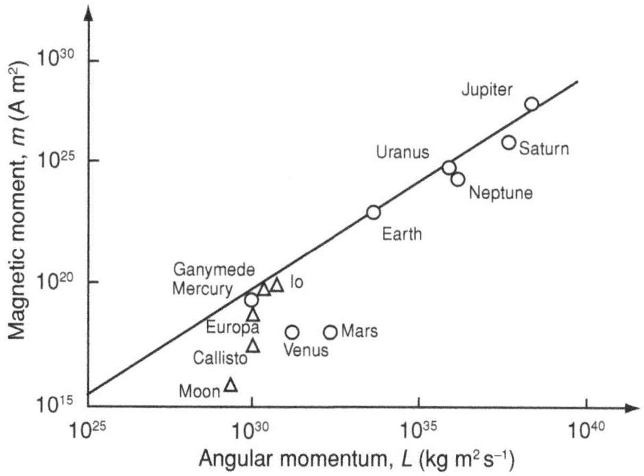

15.5.5 Planetary magnetism

The planets and moons of our solar system have been investigated with the help of magnetometers on board spacecraft. Measurements show that their magnetic moments are mostly proportional to their angular momenta, a relation known as Busse's law (Fig. 15.23). The moments are almost always directed along the axis of rotation which suggests a dynamo effect in a conducting core. The core in the gas giants is likely to be a high-pressure metallic form of hydrogen rather than molten Fe–Ni, as in Earth. The magnetic field at the surface of Jupiter is ten times as strong as that at the surface of Earth, and its magnetic moment is roughly 20 000 times as large. The weak fields of the Moon and Mars are probably related to the absence of any liquid core.

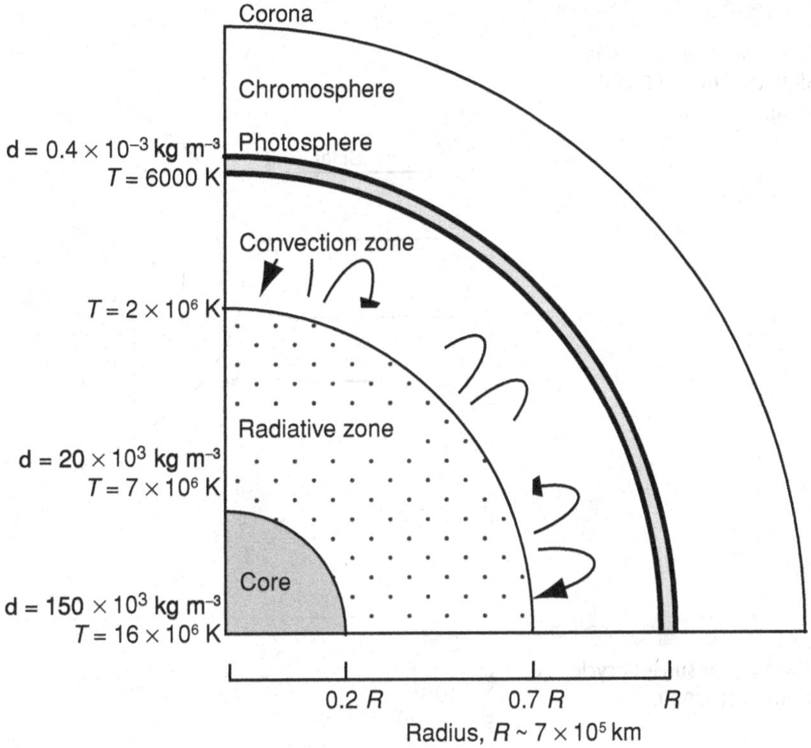

Figure 15.24

The internal structure of the Sun. Radii of the main structures, mean densities and temperatures at the centre and surface are given.

15.5.6 Solar and stellar magnetism

The ambient magnetic field in interplanetary space is 1–10 nT, whereas in interstellar space it is estimated to be 0.1 nT. The average field at the surface of the Sun is 100 μT, comparable to that at the surface of the Earth, but it can rise to over 100 mT in sunspots.

The surface of the Sun, Fig. 15.24, is less dense than air on Earth but it is seething with activity. It has a granular appearance, due to a network of convection cells, each about 10^3 km in diameter. The cells are bright at the centre, where hot plasma rises to the surface, and dark at the edges, where cooler plasma descends. The average surface temperature is 5800 K. Solar flares are spectacular displays leaping 10^5 km out from the surface. They are accompanied by ejection of bursts of energetic particles, boosting the solar wind which streams outwards from the Sun, Fig. 15.25, taking about two days to reach Earth. Particle fluxes here are 10^{12}–10^{13} m^{-2} s^{-1}. Thankfully, these energetic charged particles are deflected by the Earth's magnetic field high above the surface, and life on Earth would be impossible without the field. Some particles find their way towards the Earth at high latitudes where they dissipate their energy in collisions in the ionized upper atmosphere. The green glow of the Aurora Borealis is a manifestation of ionization of atmospheric oxygen. Sudden showers of particles provoke short-term fluctuations of the Earth's field of up

Figure 15.25

The solar wind, which is deflected by the Earth's magnetic field.

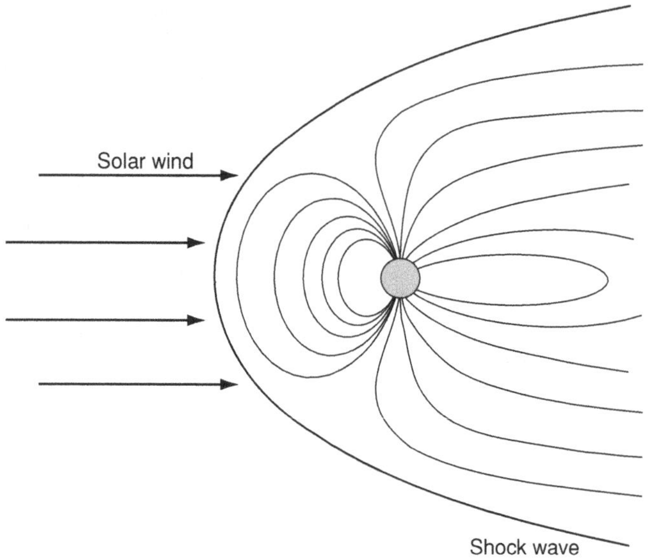

Figure 15.26

The 11-year sunspot cycle from 1760–2000.

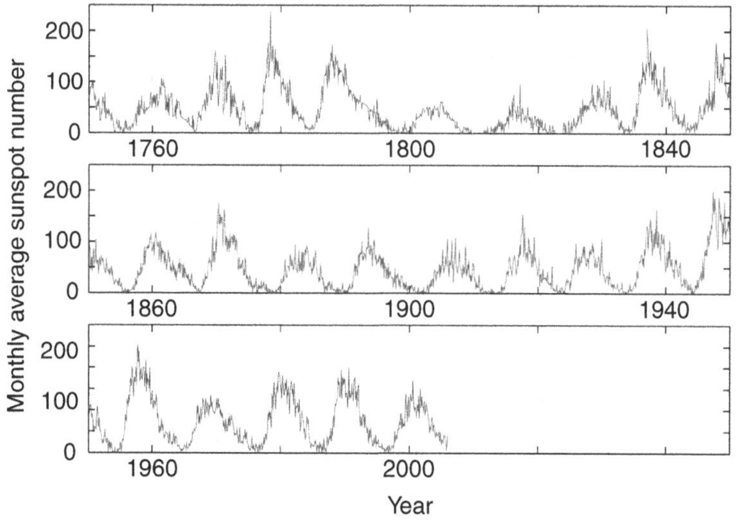

to 1 μT which can induce voltage spikes in electrical distribution networks that have led to catastrophic power failures, as well as generating intense interference at rf and microwave frequencies. The Sun's weather concerns us directly, and solar observatories provide two days warning.

Sunspots are dark patches about 10 000 km in diameter which appear in pairs near the solar equator, and last for about a week. The number of sunspots present at any one time varies from zero at the trough of the 11-year cycle to about 100 at the peak, Fig. 15.26. Sunspots were known in predynastic

China, where systematic records were kept by court astronomers from 28 BC. Galileo's observations of them, published in 1613, led to his troubles with the Roman Catholic Church. We now have some understanding of their magnetic nature. The 11-year sunspot cycle is associated with regular *reversals* of the Sun's magnetic field driven by the rotation of the sun. The average value of the Sun's field is ≈ 0.1 mT in the photosphere, but it is greater at the edge of the convection cells. Magnetic reversals occur for other stars as well.

The structure of the Sun is illustrated in Fig 15.24. The seat of thermonuclear fusion, a core of radius 200×10^3 km, is surrounded by a hydrodynamically stable radiative zone of radius 500×10^3 km, where heat diffuses by radiation into the outer, convective, zone. This is a 200×10^3 km thick shell which is in a state of ceaseless convective motion. The Sun's surface is a thin dense layer known as the photosphere, above which lies the 2500 km thick chromosphere. Finally, the chromosphere gives way to the outer layer of the solar atmosphere, known as the corona, which extends outwards to form the solar wind.

A flux tube which gas been pushed out through the surface of the Sun, forming two sunspots.

Velocities in the convective zone are about 1 km s^{-1} and the magnetic Reynolds number $\mathcal{R}_m = vl\sigma\mu$ (15.10) is huge, $\sim 10^8$. Here the characteristic length scale is 10^5 km. Magnetic field is trapped by the fluid motion, and some of the primordial field may still persist. However it is believed that there is a solar dynamo, and differential rotation of the radiative and convective zones generates an azimuthal field. The flux is amplified in flux tubes, up to about 100 mT by differential rotation. These tubes are pushed upwards towards the surface, and from time to time they burst through into the chromosphere. Sunspots are the footprint areas where the intense magnetic fields suppress convective fluid motion and the temperature falls to about 4000 K. The sunspots are often accompanied by solar flares, which may be related to collapse of the flux tubes.

Interstellar space is so vast that the dimensionless magnetic Reynolds number $\mathcal{R}_m$ can be very large indeed. There, magnetic field lines are frozen into the sparse conducting medium. The magnetic field threading a loop containing the same particles is conserved during the motion of the medium. As a cloud of interstellar matter collapses to form a star, there are three invariants, the mass $M = \frac{4}{3}\pi r^3 \mathsf{d}$, the angular momentum and the flux traversing the the cloud $\Phi = 4\pi r^2 B$. It follows that

$$B_{star}/B_{space} \approx (\mathsf{d}_{star}/\mathsf{d}_{space})^{2/3}. \qquad (15.17)$$

This equation tends to overestimate the stellar fields, which are of order 1–100 mT in many stars. The origin of the magnetic fields in the interstellar medium is an open question.

Magnetic fields of 10^8 T, a billion times that found on the Sun exist in neutron stars, which rotate at precisely defined frequencies, acting as pulsars. These extraordinary objects are composed of dense nuclear matter; they form following the rapid collapse of a star in a supernova explosion. As angular momentum is conserved when the radius of the star collapses from hundreds of thousands of kilometres to about 10 km, the rotation frequency can be as high as 1 kHz. The pulsed electromagnetic radiation we detect on Earth is produced by electrons accelerating in the rapidly changing magnetic field. The largest fields known to man, an amazing 10^{11} T, appear in a class of neutron stars known as magnetars which emit intense bursts of gamma radiation when their field is amplified by a dynamo effect during the few seconds following collapse.

Light propagating through the sparse interstellar medium undergoes Faraday rotation, where the rotation angle $\theta_F \propto \Lambda^2$, where Λ is the wavelength of light. The constant of proportionality is

$$\frac{e^2}{8\pi^2\epsilon_0 m_e^2 c^3}\int_0^d n_e B\mathrm{d}l,$$

where the product of the electron density n_e and B is integrated along the line of sight. By making some assumptions about n_e, in the vicinity of a pulsar, for example, it is possible to deduce B. Likewise, electromagnetic radiation passing through the Earth's ionosphere is subject to Faraday rotation. For example, UHF radiation (0.6 GHz, $\Lambda = 0.5$ m) completes one entire turn as it passes the Earth.

It is fitting to end with the stars. They are the source of all the iron, cobalt and nickel, and every other atomic constituent of magnetic materials, as well as all the atoms in our own bodies, except hydrogen. What felicitous miracle of organization allows one to contemplate the other, telling a story and transmitting some knowledge that will be useful for a while!

FURTHER READING

P. A. Davidson, *An Introduction to Magnetohydrodynamics*, Cambridge: Cambridge University Press (2001). A lucid and engaging introduction to MHD.

M. Yamaguchi and Y. Tanimoto, *Magneto-Science*, Tokyo: Kodansha (2006). A compendium of unusual applications of magnetism.

R. E. Rosenscweig, *Ferrohydrodynamics*, Cambridge: Cambridge University Press (1985). Everything you need to understand ferrofluids and their applications; available in a Dover reprint.

J. R. Hornak, *The Basics of MRI* http://www.cis.rit.edu/htbooks/mri. A web-book which provides an excellent introduction to a complex subject. Regularly updated.

R. B. Buxton, *An Introduction to Functional Magnetic Resonance Imaging*, Cambridge: Cambridge University Press (2002).

D. Dunlop, *Rock Magnetism*, Cambridge: Cambridge University Press (1997). An excellent monograph.

M. W. McElhinny, *Paeleomagnetism and Plate Tectonics*, London: Cambridge University Press (1973).

EXERCISES

15.1 Calculate the molarity of a solution of Cu^{2+} which has a susceptibility that is precisely zero. The diamagnetic susceptibility of water is -9.1×10^{-9} m^3 kg^{-1}.

15.2 Discuss the stability of a tube of water injected into a bath of paramagnetic solution in the configuration of Fig. 15.1(b).

15.3 A long line of n spherical single-domain magnetite particles of radius r are in contact. What is the minimum value of n necessary to form a magnetically-stable chain?

15.4 Make estimates of the minimum sizes of colloidal particles which will not aggregate: (a) under the influence of gravity in a vessel 10 cm deep, (b) under the influence of differences of magnetic field of 0.1 T and (c) as a result of dipole-dipole interactions.

15.5 Estimate the electric current flowing in the levitated frog, and explain why it does not kill the creature.

15.6 Design an axial magnetic bearing which will support a mass of 200 g.

15.7 Estimate the mass of $Nd_2Fe_{14}B$ permanent magnet that is required to levitate a 140 kg sumo wrestler above a large superconducting disc with $\chi = -1$. What is the screening current flowing in the superconductor?

15.8 What is the range of density of objects that can be floated in a water-based ferrofluid containing 20 volume % of Fe_3O_4, by applying a magnetic field.

15.9 Deduce the expression $\boldsymbol{F} = \sigma \boldsymbol{v} \times \boldsymbol{B} \times \boldsymbol{B}$ for the force per unit volume on a liquid of conductivity σ, moving with velocity $\boldsymbol{v}$ in a magnetic field $\boldsymbol{B}$. Show that this force acts to damp the motion, except when $\boldsymbol{v}$ and $\boldsymbol{B}$ are parallel.

15.10 What is the size of the largest liquid metal drop that can be supported by radio-frequency levitation?

15.11 Estimate the magnetic susceptibility of red blood cells. What is the value of ∇B^2 required for high gradient magnetic separation of red blood cells? Is this realistic? Assume the cells have radius 4.25 μm and that they move at 0.3 mm s^{-1} in a medium of viscosity 10^{-3} N s m^{-2}.

15.12 The Earth's surface is negatively charged due to lightning strikes to the extent that there is an electric field of 100 V m^{-1} at the surface. How does this influence the Earth's magnetic field?

15.13 Esimate the magnitude of the electric currents that create the Earth's magnetic field.

15.14 Estimate the energy, in electron volts, of the particles in the solar wind.

15.15 By equating the energy density of particles in the solar wind to that in the Earth's magnetic field, estimate the extent of the sheath in Fig. 15.25 as a multiple of the Earth's radius.

Appendices

Appendix A Notation

Roman symbols

a	acceleration m s^{-2}
a	interatomic spacing, lattice parameter m
a_0	cubic lattice parameter m
a	transition width m
$\boldsymbol{A}$	vector potential T m
A	exchange stiffness J m^{-1}
A_n^m	crystal-field coefficients J m^{-n}
A	hyperfine coupling constant J
$\mathcal{A}$, a	area m^2
b	scattering length (nuclear), lattice parameter m
$\boldsymbol{b}$	AC flux density T
$\boldsymbol{B}$	B-field, flux density T
B_0	free-space flux density, resonance field T
B_α	anisotropy field T
B_g	airgap flux density T
B_{hf}	hyperfine field T
B_m	flux density in a permanent magnet T
B_r	remanent flux density T
B_s	spontaneous flux density T
B_n^m	crystal-field coefficients J
$(BH)_{max}$	energy product J m^{-3}
$(BH)_u$	recoil product J m^{-3}
$\mathcal{B}_J(x)$	Brillouin function
c	concentration mol m^{-3}
c	latttice parameter m
C	Curie constant K
C'	sublattice Curie constant for an antiferromagnet K
C_A, C_B	sublattice Curie constants for a ferrimagnet K
C_m	specific heat of magnetic origin J K^{-1} m^{-3}
C_M	specific heat at constant magnetization J K^{-1} m^{-3}
C_{mol}	molar Curie constant mol K

d	distance m
d	effective spin dimensionality
d	interplanar spacing m
d_{hkl}	spacing of a set of reflecting planes of Miller indices (h, k, l) m
d	density kg m^{-3}
$\boldsymbol{D}$	electrical displacement C m^{-2}
D	dimensionality
D_e	electron diffusion constant m^2 s^{-1}
D_{sw}	spin-wave stiffness parameter J m^2
D	uniaxial crystal-field parameter J
$\mathcal{D}_{\uparrow,\downarrow}$	density of states per spin per unit volume m^{-3} J^{-1}
$\mathbf{e}$	unit vector
e_K	ellipticity
$\boldsymbol{E}$	electric field V m^{-1}
$\boldsymbol{E}'$	electric field in free space V m^{-1}
E	energy density J m^{-3}
E	enthalpy (per unit volume) J m^{-3}
E	Young's modulus Pa
E_a	anisotropy energy J m^{-3}
$E_{\mathcal{A}}$	surface energy J m^{-2}
$\mathcal{E}$	energy J
$\mathcal{E}$	electromotive force V
$\boldsymbol{f}$	force N
f	frequency Hz
f_c	cyclotron frequency Hz
f_{dip}	geometric factor
f_i	atomic scattering function m
f_L	Larmor frequency Hz
f_L	orbital form factor
f_M	recoilless fraction
f_S	spin form factor
$\boldsymbol{F}$	force density N m^{-3}
$\boldsymbol{F}_L$	Lorentz force density N m^{-3}
$\boldsymbol{F}_m$	magnetic force density N m^{-3}
F	energy functional
F	Helmholtz free energy (per unit volume) J m^{-3}
$F(\xi)$	RKKY function
F_{hkl}	structure factor m
$\mathfrak{f}$	volume fraction, packing fraction, fill fraction
$\mathfrak{f}_m$	fill factor
$\hat{\boldsymbol{g}}$	anisotropic g-tensor
$\boldsymbol{g}_{hkl}$	reciprocal lattice vector m^{-1}

g	Landè g-factor	
g_n	degeneracy of nth Landau level, nuclear g-factor	
$\mathbf{G}$	reciprocal lattice vector	m^{-1}
G	conductance	Ω^{-1}
G	de Gennes factor	
G	Gibbs free energy (per unit volume)	$J\,m^{-3}$
G_L	Landau free energy	$J\,m^{-3}$
$\mathcal{G}(r)$	radial distribution function	m^{-3}
$\mathbf{h}$	AC magnetic field	$A\,m^{-1}$
h	height	m
$\mathbf{H}$	H-field, magnetic field strength	$A\,m^{-1}$
$\mathbf{H}'$	magnetic field strength in free space	$A\,m^{-1}$
H_a	anisotropy field	$A\,m^{-1}$
H_c	coercive field	$A\,m^{-1}$
H_d	demagnetizing field, stray field	$A\,m^{-1}$
H_{dip}	dipole field	$A\,m^{-1}$
H_e	Earth's magnetic field	$A\,m^{-1}$
H_{ex}	Exchange field	$A\,m^{-1}$
H_g	gap field	$A\,m^{-1}$
H_K	saturation field	$A\,m^{-1}$
H_m	field in a permanent magnet	$A\,m^{-1}$
H_n	nucleation field	$A\,m^{-1}$
H_p	pulse field, pinning field	$A\,m^{-1}$
H_{sw}	switching field	$A\,m^{-1}$
$_BH_c$	coercivity on $B(H)$ loop	$A\,m^{-1}$
$\mathcal{H}$	Hamiltonian	
I, i	electric current	A
$\mathbf{I}$	unit tensor	
I_{c_p}	collector current (parallel arrangement)	A
$I_{c_{ap}}$	collector current (antiparallel arrangement)	A
I, M_I	nuclear spin quantum number	
$\mathcal{I}$	angle of dip	rad
$\mathcal{I}$	Stoner exchange parameter	J
$\mathbf{j}$	electric current density	$A\,m^{-2}$
j_c	conduction current density, critical switching-current density	$A\,m^{-2}$
$\mathbf{j}_m$	Amperian magnetic current density	$A\,m^{-2}$
j_s	flow of angular momentum	$J\,m^{-2}$
$\mathbf{J}$	magnetic polarization	T
$\mathbf{j} \cdot \mathbf{J}$	total angular momentum	$kg\,m^2\,s^{-1}$
$j.J$	total angular momentum quantum number	
$\mathcal{J}$	exchange constant, exchange integral	J

$\mathcal{J}_{RKKY}$	RKKY exchange J
$\mathcal{J}_{sd}, \mathcal{J}_{sf}$	exchange between localized and conduction electrons J
$\boldsymbol{k}$	electron wavevector m^{-1}
k_c	curling factor
k_F	Fermi wavevector m^{-1}
k_V	Verdet constant $T^{-1}\,m^{-1}$
$\boldsymbol{K}$	wavector of light/neutron/incident beam (of particles) m^{-1}
$\boldsymbol{K}'$	wavevector of reflected beam (of particles or radiation) m^{-1}
K	torque constant of a motor V s
K_i	stiffness constant $N\,m^{-1}$
K_s	surface anisotropy $J\,m^{-2}$
K_u, K_1, K_2, K_{eff}	anisotropy constants $J\,m^{-3}$
K_σ	stress anisotropy $J\,m^{-3}$
$\mathcal{K}$	Knight shift, geometric factor
$\hat{\boldsymbol{l}}$	orbital angular momentum operator J s
ℓ, L	orbital quantum number
$\boldsymbol{\ell}, \boldsymbol{L}$	orbital angular momentum J s
l_{ex}	exchange length m
l	length m
l_a	absorption length m
l_B	magnetic length m
l_g	airgap length m
l_m	magnet length m
l_s, l_{sf}	spin-diffusion length m
L	inductance H
$\mathcal{L}(x)$	Langevin function
m	mass kg
$m_{l,s,.j}, M_{L,S,J}$	magnetic quantum numbers
m_w	Döring mass $kg\,m^{-2}$
m^*	effective mass kg
$\mathfrak{m}$	magnetic moment $A\,m^2$
$\mathfrak{m}_0$	maximum value of $\mathfrak{m}_z$ $A\,m^2$
$\mathfrak{m}_{eff}$	effective moment $A\,m^2$
$\mathfrak{m}_n$	nuclear magnetic moment $A\,m^2$
$\boldsymbol{M}$	magnetization $A\,m^{-1}$
MC	magnetocurrent
MR	magnetoresistance
M_0	saturation magnetization (at $T = 0$) $A\,m^{-1}$
M_A, M_B	sublattice magnetizations $A\,m^{-1}$

M_L, M_S, M_J, M_I	magnetic quantum numbers
M_r	remanent magnetization, remanence $\mathrm{A\,m^{-1}}$
M_s	spontaneous magnetization $\mathrm{A\,m^{-1}}$
M_{tr}	thermoremanent magnetization $\mathrm{A\,m^{-1}}$
$\mathcal{M}$	atomic weight $\mathrm{kg\,mol^{-1}}$
n	number density of particles $\mathrm{m^{-3}}$
n	order of reflection
n	principal quantum number
n	turns per unit length (of a solenoid) $\mathrm{m^{-1}}$
n	particle density $\mathrm{m^{-3}}$
n_c	carrier electron density $\mathrm{m^{-3}}$
n_S	Stoner coefficient
$n_W, n_{AA}, n_{BB}, n_{AB}$	Weiss coefficients
N	number of particles
$\mathcal{N}$	demagnetizing factor
$\mathcal{N}_{\uparrow,\downarrow}$	density of states per atom $\mathrm{J^{-1}}$
$\hat{\boldsymbol{O}}_n^m$	Stevens operators
$\boldsymbol{p}$	electric dipole moment $\mathrm{C\,m}$
$\boldsymbol{p}$	momentum $\mathrm{kg\,m\,s^{-1}}$
$\hat{\boldsymbol{p}}$	momentum operator $\mathrm{kg\,m\,s^{-1}}$
p	magnetic scattering length m
p_c	net polarization of conducting electrons
p_{eff}	effective Bohr magneton number
$\boldsymbol{P}$	electric polarization $\mathrm{C\,m^{-2}}$
$\boldsymbol{P}_m$	magnetic pressure Pa
$P, \mathcal{P}$	power density $\mathrm{W\,m^{-3}}$
P	pressure Pa
P	probability
P	spin polarization
P_{an}	anomalous energy losses $\mathrm{J\,s^{-1}\,m^{-3}}$
P_{ed}	eddy-current energy loss $\mathrm{J\,s^{-1}\,m^{-3}}$
P_{hy}	hysteresis energy loss $\mathrm{J\,s^{-1}\,m^{-3}}$
P_m	permeance $\mathrm{Wb\,A^{-1}}$
P_v	acoustic power $\mathrm{J\,s^{-1}}$
$P_\ell^{m_\ell}(\theta)$	Legendre polynomials
$\boldsymbol{q}$	spin wave magnon wavevector $\mathrm{m^{-1}}$
q	electric charge C
q_m	magnetic charge $\mathrm{A\,m}$
$\boldsymbol{Q}$	spin density wavevector
Q	nuclear quadrupole moment $\mathrm{m^2}$
Q	heat (per unit volume) $\mathrm{J\,m^{-3}}$

Q	magneto-optic parameter	
Q	Q-factor of resonance	
Q_m	magnetic quality factor	
Q_n	multipole moment	$\mathrm{C\,m}^n$
r	distance	m
$\hat{\boldsymbol{R}}(\theta)$	rotation operator	
R	resistance	Ω
R	Sternheimer shielding factor	
R_b	superparamagnetic blocking radius	m
R_{coh}	coherence radius	m
R_h	Hall coefficient	$\mathrm{C^{-1}\,m^3}$
R_{int}	interface resistance	$\Omega\,\mathrm{m^{-2}}$
R_m	reluctance	$\mathrm{A\,Wb^{-1}}$
R_M	Maxwell resistance	Ω
R_S	Sharvin resistance	Ω
R_{sd}	single-domain radius	m
$\mathcal{R}$	shielding ratio	
$\mathcal{R}_m$	magnetic Reynold's number	
$\boldsymbol{s}, \boldsymbol{S}$	spin angular momentum	$\mathrm{kg\,m^2\,s^{-1}}$
s, S	spin quantum number	
S	entropy (per unit volume)	$\mathrm{J\,m^{-3}\,K^{-1}}$
S_v	magnetic viscosity coefficient	$\mathrm{A\,m^{-1}}$
$\mathcal{S}$	cross sectional area of a Fermi surface	$\mathrm{m^2}$
t	tolerance factor	
t	time	s
t	transfer integral	J
$t_{\frac{1}{2}}$	lifetime of an excited state	s
t_m	measuring time	s
t	thickness	m
T	temperature	K
T_1	longitudinal relaxation time	s
T_2	transverse relaxation time	s
T_2^*	combined time constant	s
T_2^{inho}	dephasing time in an inhomogeneous field	s
T^*	spin temperature	K
T_b	blocking temperature	K
T_c	ferrimagnetic Néel temperature	K
T_C	Curie temperature	K
T_{comp}	compensation temperature	K
T_f	spin freezing temperature	K
T_F	Fermi temperature	K

T_g	glass transition temperature	K
T_i	transmission of the ith mode	
T_K	Kondo temperature	K
T_M	Morin transition temperature	K
T_N	Néel temperature	K
T_{sc}	superconducting transition temperature	K
T_V	Verwey transition temperature	K
$\mathcal{T}$	kinetic energy of a single electron	J
$\mathcal{T}$	tunnelling probability	
U	Coulomb potential energy	J
U	internal energy (per unit volume)	J m^{-3}
U	Hubbard's U	J
$\mathcal{U}$	two-electron interaction	J
$\boldsymbol{v}$	velocity	m s^{-1}
v_d	electron drift velocity	m s^{-1}
v_F	Fermi velocity	m s^{-1}
v_{SW}	spin-wave velocity	m s^{-1}
v_w	domain wall velocity	m s^{-1}
$V(\boldsymbol{r})$	potential energy	J
V	voltage/potential	V
V	volume	m^3
V_b	bias voltage	V
V_H	Hall voltage	V
V_{ij}	electric field gradient	V m^{-2}
V_n^ℓ	Laguerre polynomials	
V_{sa}	spin accumulation voltage	V
$\mathcal{V}$	potential energy of a single electron	J
$\mathcal{V}$	Verdet constant	rad m^{-1}
w	width	m
w	work	J
W	bandwidth	J
W	power per unit mass	W kg^{-1}
W	work density	J m^{-3}
Y_n^m	spherical harmonics	
Z	atomic number, nunber of formula units in a unit cell, coordination number	
Z	impedance	Ω
Z_e	number of valence electrons	
$\mathcal{Z}$	partition function	
$\mathcal{Z}_m$	magnetic valence	

Greek symbols

α	Gilbert damping parameter, conductance ratio
α	thermal expansion coefficient K^{-1}
α	magnet ideality factor
β	fluxloss factor
γ	gyromagnetic ratio $C\,kg^{-1}$
γ_∞	Sternheimer antishielding factor
γ_H	Hooge coefficient
γ_n	nuclear gyromagnetic ratio $C\,kg^{-1}$
γ_w	domain wall energy $J\,m^{-2}$
Γ	torque $N\,m$
δ	diffusion layer thickness m
δ	loss angle rad
δ_B	Bloch wall width nm
$\delta_{p,q}$	Kronecker delta
δ_s	skin depth m
δ_w	domain wall width m
Δ	crystal-field splitting J
Δ_{ex}	exchange splitting J
Δ_i	impurity-level width J
Δ_{oct}	crystal-field splitting in octahedral coordination J
Δ_{sc}	superconducting energy gap J
Δ_{tet}	crystal-field splitting in tetrahedral coordination J
ϵ	strain
ϵ_{ij}	permittivity/dielectric tensor $C\,V^{-1}\,m^{-1}$
ε	efficiency
$\varepsilon, \varepsilon_M$	energy J
ε_F	Fermi energy J
ε_g	primary landgap eV
ε_{so}	spin-orbit interaction energy J
ε_Z	Zeeman energy J
η	asymmetry parameter
η	dynamic viscosity $N\,s\,m^{-2}$
η_w	wall mobility $m\,s^{-1}\,T^{-1}$
θ	colatitude rad
θ	polar angle rad
θ_F	Faraday rotation rad
θ_K	Kerr rotation rad
θ_p	paramagnetic Curie temperature K
θ	out of plane angle rad
Θ_D	Debye temperature K

κ	magnetic hardness parameter
λ	Landau damping parameter
λ	diffusion constant $m^2\ s^{-1}$
λ	dipole moment A m
λ	mean free path m
λ, Λ	spin-orbit coupling constant J
λ	wavelength (of spin waves, magnons and AC field quanta) m
λ_e	de Broglie wavelength of electron m
λ_{el}	inelastic scattering length m
λ_s	spontaneous linear magnetostriction
Λ	wavelength of light/incident particle beam m
$\boldsymbol{\mu}$	magnetic interaction vector
μ	chemical potential J per particle
μ	mobility $m^2\ V^{-1}\ s^{-1}$
μ	permeability $T\ m\ A^{-1}$
μ'	real component of permeability $T\ m\ A^{-1}$
μ''	imaginary component of permeability $T\ m\ A^{-1}$
μ_i	initial permeability $T\ m\ A^{-1}$
μ_r	relative permeability
μ_R	recoil permeability $T\ m\ A^{-1}$
ν	em radiation frequency Hz
ν	off-diagonal AC susceptibility
ν	reluctivity
ξ	correlation length m
ξ	scaling factor
ρ	(electric) charge density $C\ m^{-3}$
ρ	radius ratio
ρ_m	magnetic charge density (bulk) $A\ m^{-2}$
ρ_X	X-ray density m^{-3}
ϱ	resistivity Ω m
ϱ_{xy}	planar Hall resistivity Ω m
σ	conductivity $S\ m^{-1}$
σ	exchange bias coupling constant $J\ m^{-2}$
σ	magnetic moment per unit mass $A\ m^2\ kg^{-1}$
σ	specific moment $A\ m^2\ kg^{-1}$, $J\ T^{-1}\ kg^{-1}$
σ	stress $N\ m^{-2}$
σ	total cross section
σ_a	absorption cross section m^2
σ_d	dipolar coupling $J\ m^{-2}$
σ_{diff}	differential scattering cross section

σ_{ex}	exchange coupling J m^{-2}
σ_m	surface (magnetic) charge density A m^{-1}
σ_s	scattering cross section m^2
τ	relaxation time, period s
$\tau_{\frac{1}{2}}$	half-life s
τ_s	spin relaxation time s
ϕ	azimuthal angle rad
ϕ	colongitude rad
ϕ	potential V
ϕ_H	Hall angle rad
φ_{ab}	magnetic potential difference A
φ_c, φ_e	electric potential J C^{-1}, V
φ_m	magnetic scalar potential A
φ	field/current angle rad
Φ	magnetic flux Wb, T m^2
χ	(volume) susceptibility
χ'	external susceptibility
χ'	real component of susceptibility
χ''	imaginary component of susceptibility
χ_e	electrical susceptibility
χ_{hf}	high-field ferromagnetic susceptibility
χ_L	Landau susceptibility
χ_m	mass susceptibility m^3 kg^{-1}
χ_{mol}	molar susceptibility m^3 mol^{-1}
χ_P	Pauli susceptibility
$\varkappa$	scattering vector m^{-1}
$\varkappa$	diagonal AC susceptibility
ψ	general scalar
ψ, Ψ	wave function $m^{-3/2}$
ω	angular frequency rad s^{-1}
ω	atomic volume m^3
ω_0	resonance frequency rad s^{-1}
ω_c	cyclotron frequency rad s^{-1}
ω_q	angular frequency for phonons and magnons rad s^{-1}
ω_s	spontaneous volume magnetostriction
Ω	angular frequency of em beam/incident particle beam rad s^{-1}
Ω'	angular frequency of reflected beam of em radiation/particles rad s^{-1}
Ω	solid angle sterad
Ω	volume m^3

Appendix B Units and dimensions

B.1 SI units

We use SI throughout with the Sommerfeld convention

$$\boldsymbol{B} = \mu_0(\boldsymbol{H} + \boldsymbol{M}). \tag{B1.1}$$

Engineers prefer the Kennelly convention

$$\boldsymbol{B} = \mu_0\boldsymbol{H} + \boldsymbol{J}. \tag{B1.2}$$

Both are consistent, compatible SI units since $\boldsymbol{J} = \mu_0\boldsymbol{M}$.

The international system is based on the five basic quantities: mass (m), length (l), time (t), current (i) and temperature (θ) with corresponding units of kilogram, metre, second, ampere and kelvin. Derived units include the newton (N) $=$ kg m s^{-2}, joule (J) $=$ N m, coulomb (C) $=$ A s, volt (V) $=$ J C^{-1}, tesla (T) $=$ J A^{-1} m^{-2} $=$ V s m^{-2}, weber (Wb) $=$ V s $=$ T m^2 and hertz (Hz) $=$ s^{-1}.

Recognized multiples are in steps of $10^{\pm3}$, but a few exceptions are admited such as centimetre (cm $= 10^{-2}$ m) and angstrom (Å $= 10^{-10}$ m). Multiples of the metre are fm (10^{-15}), pm (10^{-12}), nm (10^{-9}), µm (10^{-6}), mm (10^{-3}), m (10^0) and km (10^3).

Flux density B is measured in teslas (also mT, µT). Magnetic moment is measured in A m^2 so the magnetization and the H-field are measured in A m^{-1}. From (2.73) it is seen that an equivalent unit for magnetic moment is J T^{-1}, so magnetization can also be expressed as J T^{-1} m^{-3}. The magnetic moment per unit mass, σ, in J T^{-1} kg^{-1} or A m^2 kg^{-1} is the magnetic quantity most often measured in practice in a vibrating-sample or SQUID magnetometer. The quantity μ_0 is exactly $4\pi \times 10^{-7}$ T m A^{-1} and ϵ_0 is deduced from the speed of light $c = 2.998 \times 10^8$ m s^{-1} using $c^2 = 1/(\mu_0\epsilon_0)$.

The SI system has two compelling advantages for magnetism: (i) it is possible to check the dimensions of any expression by inspection and (ii) the units are directly related to the practical units of electricity. It is the system used for undergraduate education in science and engineering world-wide. A quantitative understanding of physical phenomena requires a good grasp of the magnitudes of physical quantities. Such understanding is not fostered by confusing different unit systems. SI is the mother tongue of science. It is sensible to master your mother tongue before tackling another language. Hence the exclusive use of SI in Magnetism and Magnetic Materials.

B.2 Cgs units

Most of the primary literature on magnetism is still written using cgs units, or a muddled mixture of units where large fields are quoted in teslas and small ones in oersteds, one a unit of B, the other a unit of H! Basic cgs units are cm, g and s. The electromagnetic unit of current is equivalent to 10 A. The electromagnetic unit of potential is equivalent to 10 nV. The electromagnetic unit of magnetic dipole moment (emu) is equivalent to 10^{-3} A m^2. Derived cgs units include the erg (10^{-7} J), so that an energy density of 1 J m^{-3} is equivalent to 10 erg cm^{-3}.

The convention relating flux density and magnetization in cgs is

$$\boldsymbol{B} = \boldsymbol{H} + 4\pi \boldsymbol{M}, \tag{B1.3}$$

where the flux density or induction $\boldsymbol{B}$ is measured in gauss (G) and field $\boldsymbol{H}$ in oersteds (Oe). Magnetic moment is usually expresed as emu, and magnetization is therefore in emu cm^{-3}, although $4\pi \boldsymbol{M}$ is considerd a flux-density expression, frequently quoted in kilogauss. The magnetic constant μ_0 is numerically equal to 1 G Oe^{-1}, but its general omission from the equations makes it impossible to check their dimensions.

The most useful conversion factors between SI and cgs units in magnetism are:

1 T = 10 kG	1 G = 0.1 mT
1 kA m^{-1} = 12.57($\approx$12.5) Oe	1 Oe = 79.58 ($\approx$80) A m^{-1}
1 A m^2 = 1000 emu	1 emu = 1 mA m^2
1 MJ m^{-3} = 125.7 MG Oe	1 MG Oe = 7.96 kJ m^{-3}
1 A m^2 kg^{-1} = 1 emu g^{-1}	1 kA m^{-1} = 1 emu cm^{-3}

SI phrase	CGS translation
$B = \mu_0(H + M) = \mu_0 H + J$	$B = H + 4\pi M = H + I$
$B = 1$ T	$B = 10$ kG
$M = 1$ kA m^{-1}	$M = 1$ emu cm^{-3}
$J = 1$ T	$4\pi M = 10$ kG
$H = 1$ kA m^{-1}	$H = 4\pi$ ($\approx$12.5) Oe
$H_d = -\mathcal{N}M$ $(0 \le \mathcal{N} \le 1)$	$H_{md} = -4\pi\mathcal{N}M = -DM$ $(0 \le \mathcal{N} \le 1, 0 \le D \le 4\pi)$
$\mathrm{m} = 1$ J T^{-1}($\equiv$A m^2)	$\mathrm{m} = 1000$ emu ($\equiv$erg G^{-1})
$\sigma = 1$ J T^{-1} kg^{-1}	$\sigma = 1$ emu g^{-1}
$\chi = \partial M/\partial H$	$\chi = 4\pi\, \partial M/\partial H$
$(BH)_{max} = 1$ kJ m^{-3}	$(BH)_{max} = 40\pi$ kG Oe ($\approx$0.125 MG Oe)
$K_1 = 1$ kJ m^{-3}	$K_1 = 10^4$ erg cm^{-3}
$\varepsilon = -V\mu_0 \boldsymbol{H}\cdot\boldsymbol{M}$ J	$\varepsilon = -V\boldsymbol{H}\cdot\boldsymbol{M}$ erg
$\varphi_m = q_m/4\pi r$ A	$\chi = q_m/r$ Oe cm
$A = \mu_0 \mathrm{m} \times \boldsymbol{r}/4\pi r^3$ T m.	$A = \mathrm{m} \times \boldsymbol{r}/r^3$ G cm

B.3 Dimensions

Any quantity in the SI system has dimensions which are a combination of the dimensions of the five basic quantities, m, l, t, i and θ. In any equation relating combinations of physical properties, each of the dimensions must balance, and the dimensions of all the terms in a sum have to be identical.

B3.1 Dimensions

Mechanical							
Quantity	Symbol	Unit	m	l	t	i	θ
Area	A	m^2	0	2	0	0	0
Volume	V	m^3	0	3	0	0	0
Velocity	v	$m\ s^{-1}$	0	1	−1	0	0
Acceleration	a	$m\ s^{-2}$	0	1	−2	0	0
Density	d	$kg\ m^{-3}$	1	−3	0	0	0
Energy	ε	J	1	2	−2	0	0
Momentum	p	$kg\ m\ s^{-1}$	1	1	−1	0	0
Angular momentum	L	$kg\ m^2\ s^{-1}$	1	2	−1	0	0
Moment of inertia	I	$kg\ m^2$	1	2	0	0	0
Force	f	N	1	1	−2	0	0
Force density	F	$N\ m^{-3}$	1	−2	−2	0	0
Power	P	W	1	2	−3	0	0
Pressure	P	Pa	1	−1	−2	0	0
Stress	σ	$N\ m^{-2}$	1	−1	−2	0	0
Elastic modulus	K	$N\ m^{-2}$	1	−1	−2	0	0
Frequency	f	s^{-1}	0	0	−1	0	0
Diffusion coefficient	D	$m^2\ s^{-1}$	0	2	−1	0	0
Viscosity (dynamic)	η	$N\ s\ m^{-2}$	1	−1	−1	0	0
Viscosity	ν	$m^2\ s^{-1}$	0	2	−1	0	0
Planck's constant	$\hbar$	J s	1	2	−1	0	0

Thermal							
Quantity	Symbol	Unit	m	l	t	i	θ
Enthalpy	H	J	1	2	−2	0	0
Entropy	S	$J\ K^{-1}$	1	2	−2	0	−1
Specific heat	C	$J\ K^{-1}\ kg^{-1}$	0	2	−2	0	−1
Heat capacity	c	$J\ K^{-1}$	1	2	−2	0	−1
Thermal conductivity	κ	$W\ m^{-1}\ K^{-1}$	1	1	−3	0	−1
Sommerfeld coefficient	γ	$J\ mol^{-1}\ K^{-1}$	1	2	−2	0	−1
Boltzmann's constant	k_B	$J\ K^{-1}$	1	2	−2	0	−1

Electrical							
Quantity	Symbol	Unit	m	l	t	i	θ
Current	I	A	0	0	0	1	0
Current density	j	A m^{-2}	0	−2	0	1	0
Charge	q	C	0	0	1	1	0
Potential	V	V	1	2	−3	−1	0
Electromotive force	$\mathcal{E}$	V	1	2	−3	−1	0
Capacitance	C	F	−1	−2	4	2	0
Resistance	R	Ω	1	2	−3	−2	0
Resistivity	ϱ	Ω m	1	3	−3	−2	0
Conductivity	σ	S m^{-1}	−1	−3	3	2	0
Dipole moment	p	C m	0	1	1	1	0
Electric polarization	P	C m^{-2}	0	−2	1	1	0
Electric field	E	V m^{-1}	1	1	−3	−1	0
Electric displacement	D	C m^{-2}	0	−2	1	1	0
Electric flux	Ψ	C	0	0	1	1	0
Permittivity	ε	F m^{-1}	−1	−3	4	2	0
Thermopower	S	V K^{-1}	1	2	−3	−1	−1
Mobility	μ	m^2 V^{-1} s^{-1}	−1	0	2	1	0

Magnetic							
Quantity	Symbol	Unit	m	l	t	i	θ
Magnetic moment	m	A m^2	0	2	0	1	0
Magnetization	M	A m^{-1}	0	−1	0	1	0
Specific moment	σ	A m^2 kg^{-1}	−1	2	0	1	0
Magnetic field strength	H	A m^{-1}	0	−1	0	1	0
Magnetic flux	Φ	Wb	1	2	−2	−1	0
Magnetic flux density	B	T	1	0	−2	−1	0
Inductance	L	H	1	2	−2	−2	0
Susceptibility (M/H)	χ		0	0	0	0	0
Permeability (B/H)	μ	H m^{-1}	1	1	−2	−2	0
Magnetic polarization	J	T	1	0	−2	−1	0
Magnetomotive force	$\mathcal{F}$	A	0	0	0	1	0
Magnetic 'charge'	q_m	A m	0	1	0	1	0
Energy product	(BH)	J m^{-3}	1	−1	−2	0	0
Anisotropy energy	K	J m^{-3}	1	−1	−2	0	0
Exchange stiffness	A	J m^{-1}	1	1	−2	0	0
Hall coefficient	R_H	m^3 C^{-1}	0	3	−1	−1	0
Scalar potential	φ	A	0	0	0	1	0
Vector potential	A	T m	1	1	−2	−1	0
Permeance	P_m	T m^2 A^{-1}	1	2	−2	−2	0
Reluctance	R_m	A T^{-1} m^{-2}	−1	−2	2	2	0

B3.2 Examples

(1) Kinetic energy of a body: $\varepsilon = \frac{1}{2}mv^2$

$$[\varepsilon] = [1, 2, -2, 0, 0] \qquad [m] = [1, 0, 0, 0, 0]$$
$$[v^2] = \frac{2[0, -1, -1, 0, 0]}{[1, -2, -2, 0, 0]}$$

(2) Lorentz force on a moving charge; $\boldsymbol{f} = q\boldsymbol{v} \times \boldsymbol{B}$

$$[f] = [1, 1, -2, 0, 0] \qquad [q] = [0, 0, 1, 1, 0]$$
$$[v] = [0, 1, -1, 0, 0]$$
$$[B] = \frac{[1, 0, -2, -1, 0]}{[1, 1, -2, 0, 0]}$$

(3) Domain wall energy $\gamma_w = \sqrt{AK}$ (γ_w is an energy per unit area)

$$[\gamma_w] = [\varepsilon A^{-1}] \qquad [\sqrt{AK}] = 1/2[AK]$$
$$= [1, 2, -2, 0, 0] \qquad [\sqrt{A}] = \tfrac{1}{2}[1, 1, -2, 0, 0]$$
$$-[1, 1, -2, 0, 0] \qquad [\sqrt{K}] = \tfrac{1}{2}\frac{[1, -1, -2, 0, 0]}{[1, 0, -2, 0, 0]}$$
$$= [1, 0, -2, 0, 0]$$

(4) Magnetohydrodynamic force on a moving conductor $\boldsymbol{F} = \sigma\boldsymbol{v} \times \boldsymbol{B} \times \boldsymbol{B}$ ($\boldsymbol{F}$ is a force per unit volume)

$$[F] = [FV^{-1}] \qquad [\sigma] = [-1, -3, 3, 2, 0]$$
$$= [1, 1, -2, 0, 0] \qquad [v] = [0, 1, -1, 0, 0]$$
$$-\frac{[0, 3, 0, 0, 0]}{[1, -2, -2, 0, 0]} \qquad [B^2] = \frac{2[1, 0, -2, -1, 0]}{[1, -2, -2, 0, 0]}$$

(5) Flux density in a solid $\boldsymbol{B} = \mu_0(\boldsymbol{H} + \boldsymbol{M})$ (note that quantities added or subtracted in a bracket must have the same dimensions)

$$[B] = [1, 0, -2, -1, 0] \qquad [\mu_0] = [1, 1, -2, -2, 0]$$
$$[M], [H] = \frac{[0, -1, 0, 1, 0]}{[1, 0, -2, -1, 0]}$$

(6) Maxwell's equation $\nabla \times \boldsymbol{H} = \boldsymbol{j} + \mathrm{d}\boldsymbol{D}/\mathrm{d}t$.

$$[\nabla \times \boldsymbol{H}] = [Hr^{-1}] \qquad [j] = [0, -2, 0, 1, 0] \qquad [\mathrm{d}\boldsymbol{D}/\mathrm{d}t] = [Dt^{-1}]$$
$$= [0, -1, 0, 1, 0] \qquad = [0, -2, 1, 1, 0]$$
$$-[0, 1, 0, 0, 0] \qquad -[0, 0, 1, 0, 0]$$
$$= [0, -2, 0, 1, 0] \qquad = [0, -2, 0, 1, 0]$$

(7) Ohm's Law $V = IR$

$$= [1, 2, -3, -1, 0] \qquad [0, 0, 0, 1, 0]$$
$$+[1, 2, -3, -2, 0]$$
$$= [1, 2, -3, -1, 0]$$

(8) Faraday's Law $\mathcal{E} = -\partial\Phi/\partial t$

$$= [1, 2, -3, -1, 0] \qquad [1, 2, -2, -1, 0]$$
$$-[0, 0, 1, 0, 0]$$
$$= [1, 2, -3, -1, 0]$$

Appendix C Vector and trigonometric relations

Two products can be formed from vectors **A** and **B**:

scalar product: $\mathbf{A}\cdot\mathbf{B} = A_xB_x + A_yBy + A_zB_z = AB\cos\theta$;

vector product: $\mathbf{A}\times\mathbf{B} = \begin{vmatrix} \mathbf{e}_x & \mathbf{e}_y & \mathbf{e}_z \\ A_x & A_y & A_z \\ B_x & B_y & B_z \end{vmatrix}$

$= (A_yB_z - B_yA_z)\mathbf{e}_x - (A_xB_z - B_xA_z)\mathbf{e}_y + (A_xB_y - B_xA_y)\mathbf{e}_z$
$= AB\sin\theta\,\mathbf{e}_n,$

where the unit vector $\mathbf{e}_n$ is normal to the plane containing **A** and **B** in a direction given by the corkscrew rule, and $\mathbf{e}_x$, $\mathbf{e}_y$, $\mathbf{e}_z$ are unit vectors along the coordinate axes.

Triple products are

$$\mathbf{A}\cdot(\mathbf{B}\times\mathbf{C}) = \mathbf{B}\cdot(\mathbf{C}\times\mathbf{A}) = \mathbf{C}\cdot(\mathbf{A}\times\mathbf{B}),$$
$$\mathbf{A}\times(\mathbf{B}\times\mathbf{C}) = (\mathbf{A}\cdot\mathbf{C})\mathbf{B} - (\mathbf{A}\cdot\mathbf{B})\mathbf{C}.$$

∇ is the vector derivative $(\partial/\partial x, \partial/\partial y, \partial/\partial z)$. It acts on a scalar field ψ to produce gradψ, a vector field:

$$\nabla\psi = (\partial\psi/\partial x, \partial\psi/\partial y, \partial\psi/\partial z).$$

It acts on a vector field a vector **A** to produce div**A**, a scalar field:

$$\nabla\cdot\mathbf{A} = \partial A_x/\partial x + \partial A_y/\partial y + \partial A_z/\partial z,$$
$$\nabla\cdot\mathbf{r} = 3.$$

It acts on a vector field **A** to produce curl**A**, a vector field given by the determinant

$$\nabla\times\mathbf{A} = \begin{vmatrix} \mathbf{e}_x & \mathbf{e}_y & \mathbf{e}_z \\ \partial/\partial x & \partial/\partial y & \partial/\partial z \\ A_x & A_y & A_z \end{vmatrix} :$$

$$\nabla\times\boldsymbol{r} = 0,$$
$$\nabla\cdot\nabla\psi = \text{is a scalar: } \nabla^2\psi = \partial^2\psi/\partial x^2 + \partial^2\psi/\partial y^2 + \partial^2\psi/\partial z^2.$$

In polar coordinates,

$$\nabla\psi = \left(\frac{\partial\psi}{\partial r}, \frac{1}{r}\frac{\partial\psi}{\partial\theta}, \frac{1}{r\sin\theta}\frac{\partial\psi}{\partial\phi}\right),$$

$$\nabla\cdot\mathbf{A} = \frac{1}{r^2}\frac{\partial}{\partial r}(r^2A_x) + \frac{1}{r\sin\theta}\frac{\partial}{\partial\theta}(\sin\theta A_y) + \frac{1}{r\sin\theta}\frac{\partial}{\partial\phi}A_z,$$

$$\nabla^2\psi = \frac{1}{r^2\sin\theta}\left[\frac{\partial}{\partial r}r^2\sin\theta\frac{\partial\psi}{\partial r} + \frac{\partial}{\partial\theta}\sin\theta\frac{\partial\psi}{\partial\theta} + \frac{\partial}{\partial\phi}\frac{1}{\sin\theta}\frac{\partial\psi}{\partial\phi}\right].$$

$$\nabla\times\nabla\psi = 0,$$

$$\nabla \cdot (\nabla \times \mathbf{A}) = 0,$$
$$\nabla \times (\nabla \times \mathbf{A}) = \nabla(\nabla \cdot \mathbf{A}) - \nabla^2 \mathbf{A},$$
$$\nabla(\mathbf{A} \cdot \mathbf{B}) = (\mathbf{B} \cdot \nabla)\mathbf{A} + (\mathbf{A} \cdot \nabla)\mathbf{B} + \mathbf{B} \times (\nabla \times \mathbf{A}) + \mathbf{A} \times (\nabla \times \mathbf{B}),$$
$$\nabla \cdot (\mathbf{A} \times \mathbf{B}) = \mathbf{B} \cdot (\nabla \times \mathbf{A}) - \mathbf{A} \cdot (\nabla \times \mathbf{B}),$$
$$[\mathbf{A} \times (\nabla \times \mathbf{B})]_j = \textstyle\sum_i [\mathbf{A}_i \nabla_j \mathbf{B}_i - \mathbf{A}_i \nabla_i \mathbf{B}_j] = \sum_i \mathbf{A}_i \nabla_j \mathbf{B}_i - (\mathbf{A} \cdot \nabla)\mathbf{B}_j,$$
$$\nabla \cdot (\psi \mathbf{A}) = \mathbf{A} \cdot \nabla \psi + \psi \nabla \cdot \mathbf{A},$$
$$\nabla \times (\psi \mathbf{A}) = \nabla \psi \times \mathbf{A} + \psi \nabla \times \mathbf{A},$$
$$\int_V \nabla \cdot \mathbf{A} \mathrm{d}r^3 = \int_S \mathbf{A} \cdot \mathbf{e}_n \mathrm{d}r^2 \text{ (divergence theorem)},$$
$$\int_V \nabla \psi \mathrm{d}r^3 = \int_S \psi \mathbf{e}_n \mathrm{d}r^2,$$
$$\int_V \nabla \times \mathbf{A} \mathrm{d}r^3 = \int_S \mathbf{e}_n \times \mathbf{A} \mathrm{d}r^2,$$
$$\int_S (\nabla \times \mathbf{A}) \cdot \mathbf{e}_n dr^2 = \oint \mathbf{A} \cdot d\ell \text{ (Stokes' theorem)},$$
$$\int_S \mathbf{e}_n \times \nabla \psi \mathrm{d}r^2 = \oint \psi \mathrm{d}\ell.$$

Useful trigonometric relations are:

$$\sin^2\theta + \cos^2\theta = 1;$$
$$\sin 2\theta = 2\sin\theta\cos\theta;$$
$$\cos 2\theta = 2\cos^2\theta - 1 = \cos^2\theta - \sin^2\theta;$$
$$\tan 2\theta = 2/(\cot\theta - \tan\theta);$$
$$\sin(A + B) = \sin\theta\cos\phi + \cos\theta\sin\phi;$$
$$\cos(A + B) = \cos\theta\cos\phi - \sin\theta\sin\phi;$$
$$\mathrm{e}^{\mathrm{i}\theta} = \cos\theta + \mathrm{i}\sin\theta;$$
$$\sin\theta = (\mathrm{e}^{\mathrm{i}\theta} - \mathrm{e}^{-\mathrm{i}\theta})/2i;$$
$$\cos\theta = (\mathrm{e}^{\mathrm{i}\theta} + \mathrm{e}^{-\mathrm{i}\theta})/2;$$
$$\sinh x = (\mathrm{e}^x - \mathrm{e}^{-x})/2;$$
$$\cosh x = (\mathrm{e}^x + \mathrm{e}^{-x})/2.$$

Appendix D Demagnetizing factors for ellipsoids of revolution

α	$\mathcal{N}$	α	$\mathcal{N}$	α	$\mathcal{N}$	α	$\mathcal{N}$
0	1.000	0.20	0.749	1.40	0.249	7.00	0.035
0.01	0.985	0.25	0.703	1.50	0.232	8.00	0.029
0.02	0.968	0.30	0.661	1.60	0.219	9.00	0.024
0.03	0.953	0.40	0.588	1.70	0.207	10.0	0.020
0.04	0.940	0.50	0.526	1.80	0.194	15.0	0.010
0.05	0.925	0.60	0.476	2.00	0.173	20.0	0.0069

(cont.)

α	$\mathcal{N}$	α	$\mathcal{N}$	α	$\mathcal{N}$	α	$\mathcal{N}$
0.06	0.912	0.70	0.431	2.50	0.135	30.0	0.0034
0.07	0.899	0.80	0.394	3.00	0.109	40.0	0.0021
0.08	0.886	0.90	0.361	3.50	0.090	50.0	0.0014
0.09	0.873	1.00	0.333	4.00	0.076	70.0	0.0008
0.10	0.861	1.10	0.315	4.50	0.064	100	0.0004
0.125	0.829	1.20	0.286	5.00	0.056	200	0.0001
0.167	0.783	1.30	0.266	6.00	0.043	∞	0.0000

$\alpha = a/c$

Appendix E Field, magnetization and susceptibility

The conversion tables are printed inside the back cover. B–H conversions are valid in free space only.

Examples

An H-field of 1000 A m^{-1} is equivalent to $1000 \times 4\pi\ 10^{-3} = 12.5$ Oe.

A material of molecular weight $\mathcal{M} = 449$ with a moment of 8.6 μ_B has a specific magnetization $\sigma = 8.6\times (5585/449) = 107$ A m^2 kg^{-1}.

A magnetization $M = 1.76 \times 10^6$ A m^{-1} in a material of density 7870 kg m^{-3} is equivalent to a specific magnetization $\sigma = 1.76 \times 10^6/7870 =$ 224 A m^2 kg^{-1} or 224 emu g^{-1}.

A dimensionless SI susceptibility $\chi = 2.5 \times 10^{-6}$ in a material of density 4970 kg m^{-3} is equivalent to a dimensionless cgs susceptibility of $2.5 \times 10^{-6} \div 4\pi = 2.0 \times 10^{-7}$ and a cgs mass susceptibility $\chi_m =$ $2.5 \times 10^{-6} \times 10^3 \div (4\pi \times 4970) = 4.0 \times 10^{-8}$ emu g^{-1}.

Susceptibility of representative materials

Susceptibility	Units	H_2O	Al	$CuSO_4{\cdot}5H_2O$	$Gd_2(SO_4)_3{\cdot}8H_2O$
χ		-9.0×10^{-6}	2.1×10^{-5}	1.41×10^{-4}	2.6×10^{-3}
χ_m	m^3 kg^{-1}	-9.0×10^{-9}	7.9×10^{-9}	6.2×10^{-8}	8.7×10^{-7}
χ_{mol}	m^3 mol^{-1}	-1.62×10^{-10}	2.1×10^{-10}	1.57×10^{-8}	6.5×10^{-7}
χ_0	J T^{-2} kg^{-1}	-7.2×10^{-3}	6.3×10^{-3}	4.9×10^{-2}	6.9×10^{-1}
κ		-7.2×10^{-7}	1.70×10^{-6}	9.1×10^{-6}	2.4×10^{-4}
χ_m	emu g^{-1}	-7.2×10^{-7}	6.3×10^{-7}	4.0×10^{-6}	7.0×10^{-5}
χ_{mol}	emu mol^{-1}	-1.29×10^{-5}	1.70×10^{-5}	1.00×10^{-3}	5.2×10^{-2}

Appendix F Quantum mechanical operators

To every observable in classical mechanics, there corresponds a linear, hermitian operator in quantum mechanics. An operator $\hat{\mathbf{A}}$ is hermitian if $\hat{\mathbf{A}} = \hat{\mathbf{A}}^{\dagger}$, where $A^{\dagger}_{ij} = A^{*}_{ji}$:

position $\boldsymbol{r}$;
canonical momentum $-\mathrm{i}\hbar\nabla$;
kinetic momentum $-\mathrm{i}\hbar\nabla - q\boldsymbol{A}$;
angular momentum $-\mathrm{i}\hbar\boldsymbol{r} \times \nabla$;
angular momentum (z-component) $-\mathrm{i}\hbar\partial/\partial\phi$;
energy $(1/2m)(\mathrm{i}\hbar\nabla + q\boldsymbol{A})^2 + q\varphi_e$.

The angular momentum commutation relations are:

$$[L_1, L_2] = -[L_2, L_1]$$
$$[L_1 + L_2, L_3] = [L_1, L_3] + [L_2, L_3]$$
$$[L_1^2 + L_2] = L_1[L_1, L_2] + [L_1, L_2]L_1$$

Appendix G Reduced magnetization of ferromagnets

Reduced magnetization deduced from molecular field theory

T/T_C	$\frac{1}{2}$	1	$\frac{3}{2}$	2	$\frac{5}{2}$	$\frac{7}{2}$	∞
0	1.00000	1.00000	1.00000	1.00000	1.000000	1.00000	1.00000
0.1	1.00000	1.00000	1.00000	0.99998	0.99992	0.99964	0.96548
0.2	0.99991	0.99944	0.99833	0.99655	0.99428	0.98902	0.92817
0.3	0.99741	0.99297	0.98688	0.98019	0.97359	0.96179	0.88730
0.4	0.98562	0.97337	0.96043	0.94853	0.93815	0.92166	0.84157
0.5	0.95750	0.92657	0.01752	0.90169	0.88881	0.86006	0.78889
0.6	0.90733	0.87923	0.85599	0.83791	0.82383	0.80375	0.72588
0.7	0.82863	0.79624	0.77122	0.75262	0.73856	0.71904	0.64739
0.8	0.71041	0.67766	0.65365	0.63637	0.62358	0.60616	0.54455
0.85	0.62950	059852	0.57629	0.56051	0.54892	0.53325	0.47864
0.9	0.52543	0.49806	0.47880	0.46528	0.45543	0.44218	0.39660
0.95	0.37949	0.35871	0.34435	0.33436	0.32713	0.31747	0.28455
0.99	0.16971	0.16042	0.15400	0.14953	0.14631	0.14196	0.17198
1.0	0.00000	0.00000	0.00000	0.00000	0.00000	0.00000	0.00000

Appendix H Crystal field and anisotropy

The general expression for the anisotropy of an ion in terms of the 2^n-pole moments is:

$$\varepsilon_a = \tfrac{1}{2}Q_2A_2^0(3\cos^2\theta - 1) + \tfrac{1}{2}Q_2A_2^2\sin^2\theta\cos 2\phi + \tfrac{1}{8}Q_4A_4^0(35\cos^4\theta - 30\cos^2\theta + 3) + \tfrac{1}{8}Q_4A_4^2(7\cos^2\theta - 1)\sin^2\theta\cos 2\phi + \tfrac{1}{8}Q_4A_4^4(\sin^4\theta\cos 4\phi) + \tfrac{1}{16}Q_6A_6^0(231\cos^6\theta - 315\cos^4\theta + 105\cos^2\theta - 5) + \tfrac{1}{16}Q_6A_6^2(33\cos^4\theta - 18\cos^2\theta + 1)\sin^2\theta\cos 2\phi + \tfrac{1}{16}Q_6A_6^4(11\cos^2\theta - 1)\sin^4\theta\cos 4\phi + \tfrac{1}{16}Q_6A_6^6(\sin^6\theta\cos 4\phi).$$

The diagonal crystal-field parameters which describe the lattice environment of an ion are:

$$A_2^0 = -\frac{e^2}{16\pi\epsilon_0}\int\frac{(3\cos^2\Theta - 1)}{R^3}\rho(R)\mathrm{d}^3R,$$

$$A_4^0 = -\frac{9e^2}{1024\pi^2\epsilon_0}\int\frac{(35\cos^4\theta - 30\cos^2\theta + 3)}{R^5}\rho(R)\mathrm{d}^3R,$$

$$A_6^0 = -\frac{13e^2}{4096\pi^2\epsilon_0}\int\frac{(231\cos^6\theta - 315\cos^4\theta + 105\cos^2\theta - 5)}{R^7}\rho(R)\mathrm{d}^3R.$$

The Stevens operators are:

$$\hat{\mathbf{O}}_2^0 = [3\hat{J}_z^2 - J(J+1)],$$

$$\hat{\mathbf{O}}_2^{2c} = \tfrac{1}{2}(\hat{J}_+^2 + \hat{J}_-^2),$$

$$\hat{\mathbf{O}}_4^0 = [35\hat{J}_z^4 - 30J(J+1)\hat{J}_z^2 + 25\hat{J}_z^2 - 6J(J+1) + 3\hat{J}^2(J+1)^2],$$

$$\hat{\mathbf{O}}_4^2 = \tfrac{1}{4}\{[7\hat{J}_z^2 - J(J+1) - 5](\hat{J}_+^2 + \hat{J}_-^2) + (\hat{J}_+^2 + \hat{J}_-^2)[7\hat{J}_z^2 - J(J+1) - 5]\},$$

$$\hat{\mathbf{O}}_4^3 = \tfrac{1}{4}[\hat{J}_z(\hat{J}_+^3 + \hat{J}_-^3) + (\hat{J}_+^3 + \hat{J}_-^3)\hat{J}_z],$$

$$\hat{\mathbf{O}}_4^{4c} = \tfrac{1}{2}(\hat{J}_+^4 + \hat{J}_-^4),$$

$$\hat{\mathbf{O}}_6^0 = [231\hat{J}_z^6 - 315J(J+1)\hat{J}_z^4 + 735\hat{J}_z^4 + 105J^2(J+1)^2\hat{J}_z^2 - 525J(J+1)\hat{J}_z^2 + 294\hat{J}_z^2 - 5J^3(J+1)^3 + 40J^2(J+1)^2 - 60J(J+1)],$$

$$\hat{\mathbf{O}}_6^2 = \tfrac{1}{4}\{[33\hat{J}_z^4 - 18\hat{J}_z^2J(J+1) - 123\hat{J}_z^2 + J^2(J+1)^2 + 10J(J+1) + 102](\hat{J}_+^2 + \hat{J}_-^2) + (\hat{J}_+^2 + \hat{J}_-^2)[33\hat{J}_z^4 - 18\hat{J}_z^2 J(J+1) - 123\hat{J}_z^2 + J^2(J+1)^2 + 10J(J+1) + 102]\},$$

$$\hat{\mathbf{O}}_6^3 = \tfrac{1}{4}[(11\hat{J}_z^3 - 3\hat{J}_zJ(J+1) - 59\hat{J}_z)(\hat{J}_+^3 + \hat{J}_-^3) + (\hat{J}_+^3 + \hat{J}_-^3)(11\hat{J}_z^3 - 3\hat{J}_zJ(J+1) - 59\hat{J}_z)],$$

$$\hat{\mathbf{O}}_6^4 = \tfrac{1}{4}[(11\hat{J}_z^2 - J(J+1) - 38)(\hat{J}_+^4 + \hat{J}_-^4) + (\hat{J}_+^4 + \hat{J}_-^4)(11\hat{J}_z^2 - J(J+1) - 38)],$$

$$\hat{\mathbf{O}}_6^{6c} = \tfrac{1}{2}[(\hat{J}_+^c + \hat{J}_-^6)].$$

Some expressions for the crystal field in sites with different symmetry are:

Cubic: Fm$\underline{3}m$	$B_4^0\left[\hat{\boldsymbol{O}}_4^0 + 5\hat{\boldsymbol{O}}_4^{4c}\right] + B_6^0\left[\hat{\boldsymbol{O}}_6^0 - 21B_6^{4c}\hat{\boldsymbol{O}}_6^{4c}\right]$
Tetragonal: $4/mmm$	$B_2^0\hat{\boldsymbol{O}}_2^0 + B_4^0\hat{\boldsymbol{O}}_4^0 + B_4^{4c}\hat{\boldsymbol{O}}_4^{4c} + B_6^0\hat{\boldsymbol{O}}_6^0 + B_6^{4c}\hat{\boldsymbol{O}}_6^{4c}$
Trigonal: $3m$	$B_2^0\hat{\boldsymbol{O}}_2^0 + B_4^0\hat{\boldsymbol{O}}_4^0 + B_4^3\hat{\boldsymbol{O}}_4^3 + B_6^0\hat{\boldsymbol{O}}_6^0 + B_6^4\hat{\boldsymbol{O}}_6^4 + B_6^6\hat{\boldsymbol{O}}_6^6$
Hexagonal: $6/mmm$, $\underline{6}m2$	$B_2^0\hat{\boldsymbol{O}}_2^0 + B_4^0\hat{\boldsymbol{O}}_4^0 + B_6^0\hat{\boldsymbol{O}}_6^0 + B_6^6\hat{\boldsymbol{O}}_6^6$
Orthorhombic:	$B_2^0\hat{\boldsymbol{O}}_2^0 + B_2^{2s}\hat{\boldsymbol{O}}_2^{2s} + B_4^0\hat{\boldsymbol{O}}_4^0 + B_4^{2s}\hat{\boldsymbol{O}}_4^{2s} + B_4^{4c}\hat{\boldsymbol{O}}_4^{4c} + B_6^0\hat{\boldsymbol{O}}_6^0 + B_6^{2s}\hat{\boldsymbol{O}}_6^{2s}$
mm	$B_6^{4c}\hat{\boldsymbol{O}}_6^{4c} + B_6^{6s}\hat{\boldsymbol{O}}_6^{6s}$

Appendix I Magnetic point groups

The 31 magnetic point groups in bold type are compatible with a permanent magnetic moment, with the components specified:

triclinic ($\mathfrak{m}_x, \mathfrak{m}_y, \mathfrak{m}_z$) $\mathbf{1}$, $\bar{\mathbf{1}}$, $\bar{1}'$

monoclinic ($\mathfrak{m}_x, 0, \mathfrak{m}_z$) or ($0, \mathfrak{m}_y, 0$) $\mathbf{2}$, $\mathbf{2'}$, $\boldsymbol{m}$, $\boldsymbol{m'}$, $\mathbf{2}/\boldsymbol{m}$, $\mathbf{2'}/\boldsymbol{m'}$, $2/m'$, $2'/m$

orthorhombic 222, $\mathbf{2'2'2}$, $mm2$, $\boldsymbol{m'm'}\mathbf{2}$, $\boldsymbol{m'm}\mathbf{2'}$, mmm, $\boldsymbol{m'm'm}$, $m'm'm'$, $m'mm$

trigonal ($0, 0, \mathfrak{m}_z$) $\mathbf{3}$, $\bar{\mathbf{3}}$, $\bar{3}'$, 32, $\mathbf{32'}$, $3m$, $\mathbf{3}\boldsymbol{m'}$, $\bar{3}m$. $\bar{\mathbf{3}}\boldsymbol{m'}$, $\bar{3}'m'$, $\bar{3}'m$

tetragonal ($0, 0, \mathfrak{m}_z$) $\mathbf{4}$, $4'$, $\bar{\mathbf{4}}$, $\bar{4}'$, $\mathbf{4}/\boldsymbol{m}$, $4'/m$, $4/m'$, $4'/m'$, 422, $4'22$, $\mathbf{42'2'}$, $4mm$, $4'mm'$, $\mathbf{4}\boldsymbol{m'm'}$, $\bar{4}2m$, $\bar{4}'2m'$, $\bar{4}'2'm$, $\bar{\mathbf{4}}\mathbf{2'}\boldsymbol{m'}$, $4/mmm$, $4'/mmm'$, $\mathbf{4}/\boldsymbol{mm'm'}$, $4/m'mm$, $4'/m'mm'$

hexagonal ($0, 0, \mathfrak{m}_z$) $\mathbf{6}$, $6'$, $\bar{\mathbf{6}}$, $\bar{6}'$, $\mathbf{6}/\boldsymbol{m}$, $6'/m'$, $6/m'$, $6'/m$, 622, $6'22'$, $\mathbf{62'2'}$, $6mm$, $6'mm'$, $\mathbf{6}\boldsymbol{m'm'}$, $\bar{6}m2$, $\bar{6}m'2$, $\bar{6}m2'$, $\bar{\mathbf{6}}\boldsymbol{m'}\mathbf{2'}$, $6/mmm$, $6'/m'mm'$, $\mathbf{6}/\boldsymbol{mm'm'}$, $6/m'm'm'$, $6/m'mm$, $6'/mmm'$

cubic 23, $m3$, $m'3$, 432, $4'32$, $\bar{4}3m$, $\bar{4}'3m'$ $m3m$, $m3m'$, $m'3m'$, $m'3m$.

Formula index

Index

Table A The magnetic periodic table. Diamagnetic elements are uncoloured, paramagnets are pale grey, ferromagnets are dark grey, antiferromagnets are mid grey, and the Curie or Néel temperatures are marked. Common paramagnetic ions are indicated. Elements which bear a magnetic moment as isolated atoms are marked in bold type.

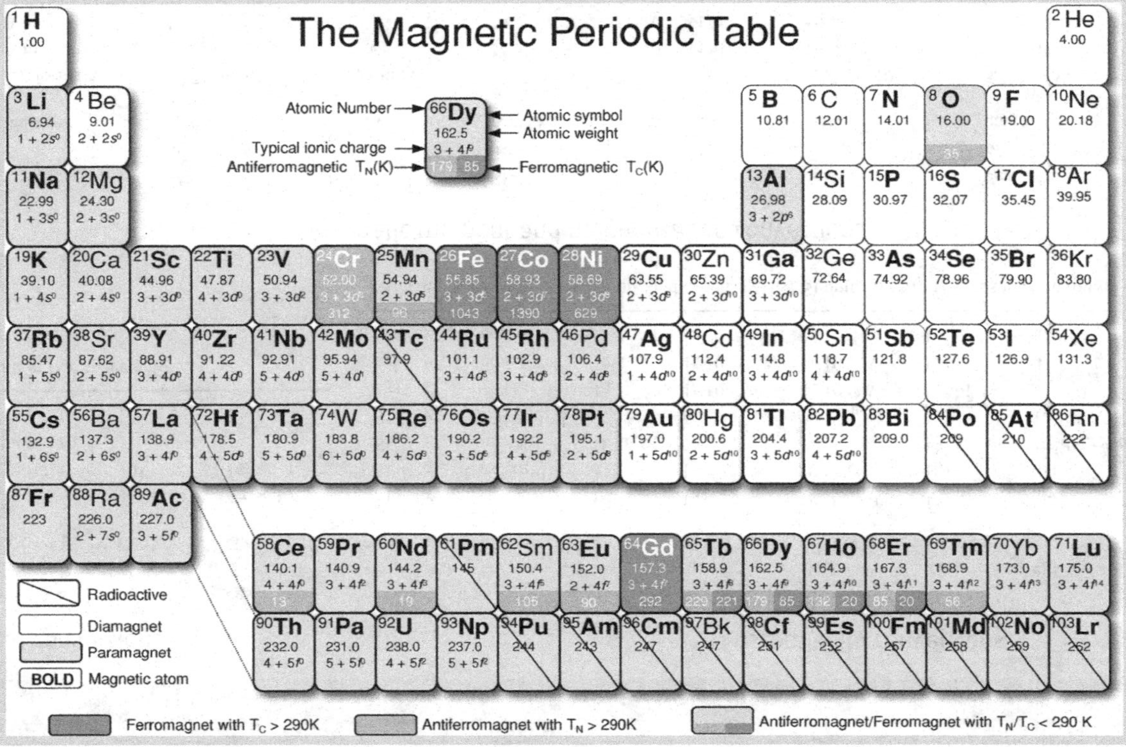

Table B SI–cgs conversion Locate the quantity you wish to convert in column A (its units are in the same row); to convert to a quantity in row B (its units are in the same column), *multiply* it by the factor in the table. Examples are given in appendix F.

Field conversions

	A ↓	B / units→	SI H A m^{-1}	SI B T	cgs H Oe	cgs B G
SI	H	A m^{-1}	1	μ_0	$4\pi \times 10^{-3}$	$4\pi \times 10^{-3}$
SI	B	T	$1/\mu_0$	1	10^4	10^4
cgs	H	Oe	$10^3/4\pi$	10^{-4}	1	1
cgs	B	G	$10^3/4\pi$	10^{-4}	1	1

B–H conversions are valid in free space only

Susceptibility conversions

	A ↓	B / units→	SI χ	SI χ_m m^3 kg^{-1}	SI χ_{mol} m^3 mol^{-1}	SI χ_0 J T^{-2} kg^{-1}	cgs κ	cgs χ_m emu g^{-1}	cgs χ_{mol} emu mol^{-1}
SI	χ		1	1/d	$10^{-3}\mathcal{M}$/d	$1/\mu_0$d	$1/4\pi$	$10^3/4\pi$d	$10^3\mathcal{M}/4\pi$d
SI	χ_m	m^3 kg^{-1}	d	1	$10^{-3}\mathcal{M}$	$1/\mu_0$	d$/4\pi$	$10^3/4\pi$	$10^3\mathcal{M}/4\pi$
SI	χ_{mol}	m^3 mol^{-1}	10^3d$/\mathcal{M}$	$10^3/\mathcal{M}$	1	$10^3/\mu_0\mathcal{M}$	10^3d$/4\pi\mathcal{M}$	$10^6/4\pi\mathcal{M}$	$10^6/4\pi$
SI	χ_0	J T^{-2} kg^{-1}	μ_0d	μ_0	$10^{-3}\mu_0\mathcal{M}$	1	10^{-7}d	10^{-4}	$10^{-4}\mathcal{M}$
cgs	κ		4π	$4\pi 10^{-3}$/d	$4\pi 10^{-6}\mathcal{M}$/d	10^4/d	1	1/d	$\mathcal{M}$/d
cgs	χ_m	emu g^{-1}	4πd	$4\pi 10^{-3}$	$4\pi 10^{-6}\mathcal{M}$	10^4	d	1	$\mathcal{M}$
cgs	χ_{mol}	emu mol^{-1}	4πd$/\mathcal{M}$	$4\pi 10^{-3}/\mathcal{M}$	$4\pi 10^{-6}$	$10^4/\mathcal{M}$	d$/\mathcal{M}$	$1/\mathcal{M}$	1

$\mathcal{M}$ is molecular weight (in g mol^{-1}), d is density (use SI units in rows 1–4, cgs units in rows 5–7)

Magnetic moment and magnetization conversions

	A ↓	B / units→	SI m A m^2	SI M A m^{-1}	SI σ A m^2 kg^{-1}	SI σ_{mol} A m^2 mol^{-1}	cgs m emu	cgs M emu cm^{-3}	cgs σ emu g^{-1}	cgs σ_{mol} emu mol^{-1}
	m	μ_B/formula	9.274×10^{-24}	5585d$/\mathcal{M}$	$5585/\mathcal{M}$	5.585	9.274×10^{-21}	5.585d$/\mathcal{M}$	$5585/\mathcal{M}$	5585
SI	m	A m^2	1	$1/V$	1/dV	$10^{-3}\mathcal{M}$/dV	10^3	$10^{-3}/V$	1/dV	$\mathcal{M}$/dV
SI	M	A m^{-1}	V	1	1/d	$10^{-3}\mathcal{M}$/d	$10^3 V$	10^{-3}	1/d	$\mathcal{M}$/d
SI	σ	A m^2 kg^{-1}	dV	d	1	$10^{-3}\mathcal{M}$	10^3dV	10^{-3}d	1	$\mathcal{M}$
SI	σ_{mol}	A m^2 mol^{-1}	10^2d$V/\mathcal{M}$	10^3d$/\mathcal{M}$	$10^3/\mathcal{M}$	1	10^6d$V/\mathcal{M}$	d$/\mathcal{M}$	$10^3/\mathcal{M}$	10^3
cgs	m	emu	10^{-3}	$10^3/V$	1/dV	$10^{-3}\mathcal{M}$/dV	1	$1/V$	1/dV	$\mathcal{M}$/dV
cgs	M	emu cm^{-3}	$10^{-3}V$	10^3	1/d	$10^{-3}\mathcal{M}$/d	V	1	1/d	$\mathcal{M}$/d
cgs	σ	emu g^{-1}	10^{-3}dV	10^3d	1	$10^{-3}\mathcal{M}$	dV	d	1	$\mathcal{M}$
cgs	σ_{mol}	emu mol^{-1}	10^{-3}d$V/\mathcal{M}$	10^3d$/\mathcal{M}$	$1/\mathcal{M}$	10^{-3}	d$V/\mathcal{M}$	d$/\mathcal{M}$	$1/\mathcal{M}$	1

$\mathcal{M}$ is molecular weight (g mol^{-1}), d is density, V is sample volume (use SI in rows 1–5, cgc in rows 6–9 for density and volume). Note that the quantity $4\pi M$ is frequently quoted in the cgs system, in units of gauss (G)

To deduce the effective Bohr magneton number $p_{eff} = m_{eff}/\mu_B$ from the molar Curie constant, the relation is $C_{mol} = 1.571\ 10^{-6}\ p^2_{eff}$ in SI, and $C_{mol} = 0.125\ p^2_{eff}$ in cgs.

Constants

a_0	Bohr radius $(4\pi\epsilon_0\hbar^2/m_e e^2)$	52.92 pm
c	velocity of light	2.998×10^8 m s^{-1}
e	elementary charge	1.6022×10^{-19} C
G_0	conductance quantum (e^2/h)	$3.874 \times 10^{-5}\ \Omega^{-1}$
h	Planck's constant	6.626×10^{-34} J s
$\hbar$	Planck's constant/2π	1.0546×10^{-34} J s
k_B	Boltzmann constant	1.3807×10^{-23} J K^{-1}
m_e	electron mass	9.109×10^{-31} kg
m_n	neutron mass	1.675×10^{-27} kg
m_p	proton mass	1.673×10^{-27} kg
m_μ	muon mass	$206.7 m_e$
N_A	Avogadro's number	6.022×10^{23} mol^{-1}
u	unified atomic mass unit	1.6605×10^{-27} kg
r_e	the electron radius	2.818 fm
R	gas constant	8.314 J mol^{-1}K^{-1}
R_0	Rydberg	$2.180 \times 10^{-18} J = 13.61$ eV
α	fine structure constant $(e^2/4\pi\epsilon_0\hbar c)$	1/137.04
ϵ_0	permittivity of free space $(1/\mu_0 c^2)$	8.854×10^{-12} C V^{-1} m^{-1}
μ_0	permeability of free space	$4\pi \times 10^{-7}$ T m A^{-1}
μ_B	Bohr magneton $(e\hbar/2m_e)$	9.274×10^{-24} A m^2
μ_N	nuclear magneton $(e\hbar/2m_p)$	5.0508×10^{-27} A m^2
Φ_0	flux quantum $(h/2e)$	2.068×10^{-15} T m^2

Unit conversions

1eV $\equiv$ 11606 K $(e/k_B) \equiv$ 8066 cm^{-1} (e/hc)

1T$\mu_B \equiv$ 0.6717 K (μ_B/k_B)

1μ_B/atom $\equiv$ 5.585 J T^{-1} mol^{-1} $(N_A\mu_B)$

1K/atom $\equiv$ 8.314 J mol^{-1} $(N_A k_B)$

Printed in the United States
By Bookmasters